Biology

Science for Life

WITH PHYSIOLOGY

FOURTH EDITION

Colleen Belk

University of Minnesota–Duluth

Virginia Borden Maier

St. John Fisher College

Boston Columbus Indianapolis New York San Francisco Upper Saddle River
Amsterdam Cape Town Dubai London Madrid Milan Munich Paris Montreal Toronto
Delhi Mexico City Sao Paulo Sydney Hong Kong Seoul Singapore Taipei Tokyo

Editor-in-Chief: Beth Wilbur
Executive Director of Development: Deborah Gale
Senior Acquisitions Editor: Star MacKenzie
Senior Development Editor: Susan Teahan
Assistant Editor: Frances Sink
Managing Editor: Michael Early
Production Project Manager: Camille Herrera
Production Management and Compositor: PreMediaGlobal
Copyeditor: Kate Brown
Proofreader: Andrea Rosenberg
Interior and Cover Designer: Hespenheide Design
Illustrators: Imagineering
Image Lead: Donna Kalal
Photo Researcher: Bill Smith Group
Manufacturing Buyer: Michael Penne
Cover Printer: Leigh Phoenix
Printer and Binder: RRD Willard
Senior Media Producer: Jonathan Ballard
Marketing Manager: Lauren Rodgers
Cover Photo Credit: sea urchins, Dimitrios/Shutterstock; wave, Eric Gevaert/Fotolia; uncompacted blastomeres, © Anatomical Travelogue/Photo Researchers, Inc.; white lotus, Adisai Chaturapitr/istockphoto; fossilized leaf, Christopher Tompkins/istockphoto; puppies, Tunart/istockphoto

Credits and acknowledgments borrowed from other sources and reproduced, with permission, in this textbook appear on p. C-1.

Library of Congress Cataloging-in-Publication Data

Belk, Colleen M.
  Biology : science for life, with physiology / Colleen Belk, Virginia Borden Maier.—4th ed.
    p. cm.
  ISBN-13: 978-0-321-76783-7
  ISBN-10: 0-321-76783-7
 1. Biology.   2. Physiology.   I. Maier, Virginia Borden.   II. Title.
  QH307.2.B43 2013b
  570.1—dc23
                                                        2011044801

1 2 3 4 5 6 7 8 9 10— DOW —15 14 13 12 11

www.pearsonhighered.com

ISBN 10: 0-321-76783-7; ISBN 13: 978-0-321-76783-7 (Student edition)
ISBN 10: 0-321-77429-9; ISBN 13: 978-0-321-77429-3 (Instructor's Resource Copy)
ISBN 10: 0-321-77433-7; ISBN 13: 978-0-321-77433-0 (A La Carte)

About the Authors

Colleen Belk and **Virginia Borden Maier** collaborated on teaching biology to nonmajors for over a decade together at the University of Minnesota–Duluth. This collaboration has continued through Virginia's move to St. John Fisher College in Rochester, New York, and has been enhanced by their differing but complementary areas of expertise. In addition to the non-majors course, Colleen Belk teaches general biology for majors, genetics, cell biology, and molecular biology courses. Virginia Borden Maier teaches general biology for majors, evolutionary biology, zoology, plant biology, ecology, and conservation biology courses.

After several somewhat painful attempts at teaching the breadth of biology to non-majors in a single semester, the two authors came to the conclusion that they needed to find a better way. They realized that their students were more engaged when they understood how biology directly affected their lives. Colleen and Virginia began to structure their lectures around stories they knew would interest students. When they began letting the story drive the science, they immediately noticed a difference in student interest, energy, and willingness to work harder at learning biology. Not only has this approach increased student understanding, but it has also increased the authors' enjoyment in teaching the course—presenting students with fascinating stories infused with biological concepts is simply a lot more fun. This approach served to invigorate their teaching. Knowing that their students are learning the biology that they will need now and in the future gives the authors a deep and abiding satisfaction.

Preface

To the Student

As you attended your classes in high school or otherwise worked to prepare yourself for college, you were probably unaware that an information explosion was taking place in the field of biology. This explosion, brought on by advances in biotechnology and communicated by faster, more powerful computers, has allowed scientists to gather data more quickly and disseminate it to colleagues in the global scientific community with the click of a mouse. Every discipline of biology has benefited from these advances, and today's scientists collectively know more than any individual could ever hope to understand.

Paradoxically, as it becomes more and more challenging to synthesize huge amounts of information from disparate disciplines within the broad field of biology, it becomes more vital that we do so. The very same technologies that led to the information boom, coupled with expanding human populations, present us with complex ethical questions. These questions include whether it is acceptable to clone humans, when human life begins and ends, who owns living organisms, what our responsibilities toward endangered species are, and many more. No amount of knowledge alone will provide satisfactory answers to these questions. Addressing these kinds of questions requires the development of a scientific literacy that surpasses the rote memorization of facts. To make decisions that are individually, socially, and ecologically responsible, you must not only understand some fundamental principles of biology but also be able to use this knowledge as a tool to help you analyze ethical and moral issues involving biology.

To help you understand biology and apply your knowledge to an ever-expanding suite of issues, we have structured each chapter of *Biology: Science for Life with Physiology* around a compelling story in which biology plays an integral role. Through the story you not only will learn the relevant biological principles but also will see how science can be used to help answer complex questions. As you learn to apply the strategies modeled by the text, you will begin developing your critical thinking skills.

By the time you finish this book, you should have a clear understanding of many important biological principles. You will also be able to critically evaluate which information is most reliable instead of simply accepting all the information you hear or read about. Even though you may not be planning to be a practicing biologist, well-developed critical thinking skills will enable you to make decisions that affect your own life, such as whether to take nutritional supplements, and decisions that affect the lives of others, such as whether to believe the DNA evidence presented to you as a juror in a criminal case.

It is our sincere hope that understanding how biology applies to important personal, social, and ecological issues will convince you to stay informed about such issues. On the job, in your community, at the doctor's office, in the voting booth, and at home watching the news or surfing the web, your knowledge of the basic biology underlying so many of the challenges that we as individuals and as a society face will enable you to make well-informed decisions for your home, your nation, and your world.

To the Instructor

By now you are probably all too aware that teaching non-majors students is very different from teaching biology majors. You know that most of these students will never take another formal biology course; therefore, your course may be the last chance for these students to see the relevance of science in their everyday lives and the last chance to appreciate how biology is woven throughout the fabric of their lives. You recognize the importance of engaging these students because you know that these students will one day be voting on issues of scientific importance, holding positions of power in the community, serving on juries, and making health care decisions for themselves and their families. You know that your students' lives will be enhanced if they have a thorough grounding in basic biological principles and scientific literacy. To help your efforts toward this end, this text is structured around several themes.

Themes Throughout *Biology: Science for Life with Physiology*

The Story Drives the Science. We have found that students are much more likely to be engaged in the learning process when the textbook and lectures capitalize on their natural curiosity. This text accomplishes this by using a story to drive the science in every chapter. Students get caught up in the story and become interested in learning the biology so they can see how the story is resolved. This approach allows us to cover the key areas of biology, including basic chemistry, the unity and diversity of life, cell structure and function, classical and molecular genetics, evolution, and ecology, in a manner that makes students want to learn. Not only do students want to learn, but also this approach allows students both to connect the science to their everyday lives and to integrate the principles and concepts for later application to other situations. The story approach will give you flexibility in teaching and will support you in developing students' critical thinking skills.

The Process of Science. This book also uses another novel approach in the way that the process of science is modeled. The first chapter is dedicated to the scientific method and hypothesis testing, and each subsequent chapter weaves the scientific method and hypothesis testing throughout the story. The development of students' critical thinking skills is thus reinforced for the duration of the course. Students will see that the application of the scientific method is often the best way to answer questions raised in the story. This practice not only allows students to develop their critical thinking skills but also, as they begin to think like scientists, helps them understand why and how scientists do what they do.

Integration of Evolution. Another aspect of *Biology: Science for Life with Physiology* that sets it apart from many other texts is the manner in which evolutionary principles are integrated throughout the text. The role of evolutionary processes is highlighted in every chapter, even when the chapter is not specifically focused on an evolutionary question. For example, when discussing infectious diseases, the evolution of antibiotic-resistant strains of bacteria is addressed. The physiology unit includes an essay on evolution in each chapter. These essays illustrate the importance of natural selection in the development of various organs and organ systems across a wide range of organisms. With evolution serving as an overarching theme, students are better able to see that all of life is connected through this process.

Pedagogical Elements

Open the book and flip through a few pages and you will see some of the most inviting, lively, and informative illustrations you have ever seen in a biology text. The illustrations are inviting because they have a warm, hand-drawn quality that is clean and uncluttered. Most important, the illustrations are informative not only because they were carefully crafted to enhance concepts in the text but also because they employ techniques like the "pointer" that help draw the students' attention to the important part of the figure (see p. 14). Likewise, tables are more than just tools for organizing information; they are illustrated to provide attractive, easy references for the student. We hope that the welcoming nature of the art and tables in this text will encourage nonmajors to explore ideas and concepts instead of being overwhelmed before they even get started.

In addition to lively illustrations of conventional biology concepts, this text also uses analogies to help students understand difficult topics. For example, the process of translation is likened to baking a cake (see p. 199). These analogies and illustrations are peppered throughout the text.

Students can reinforce and assess what they are learning in the classroom by reading the chapter, studying the figures, and answering the end-of-chapter questions. We have written these questions in every format likely to be used by an instructor during an exam so that students have practice in answering many different types of questions. We have also included "Connecting the Science" questions that would be appropriate for essay exams, class discussions, or use as topics for term papers.

In an effort to accommodate the wide range of teaching and learning styles of those using our book, each chapter includes a section entitled **A Closer Look**. Preceding each A Closer Look section is a general overview of the topic sufficient to provide nonmajors students with a basic understanding. A Closer Look then provides details that add depth to this understanding. This feature provides instructors with teaching flexibility; for example, some instructors are satisfied with a general overview of cellular respiration, including the reactants and products, and similar basics of metabolism. Others want to teach electron transport and oxidative phosphorylation. **Our format** facilitates both approaches; one can use the overview only or the overview and the Closer Look section. The Closer Look section is differentiated from the rest of the text, making it easy to include or omit from students' assigned reading.

Each chapter contains several **Stop and Stretch** questions designed to provide rest stops where students can assess whether they have understood the material they just completed reading. These questions require students to synthesize material they have read and apply their knowledge in a new situation. Many of these questions are designed to familiarize students with data analysis and help them develop skill at interpreting and understanding scientific data. Answers to these questions are provided in the appendix.

Another feature that aids in student learning is **Visualize This,** which consists of questions in figure captions that test a student's understanding of the concepts contained in the art. These questions are designed to encourage students to spend extra time with complex figures and develop a more sophisticated understanding of the concepts they present. Visualize This questions are included with three or more figures in each chapter and answers to these questions can be found in the appendix.

Each chapter ends with a feature called **Savvy Reader**. We developed this feature in response to the desire of so many of the professors teaching this course to help students learn to critically evaluate science presented in the media. We all want our students to become better, more informed consumers of science. Each Savvy Reader offers an excerpt or a summary of an article or advertisement taken from any of a variety of media sources. After reading the excerpt, the student then answers a series of questions written with the goal of helping them critically evaluate the scientific information presented in the article. Students learn to evaluate claims made in the popular press about issues ranging from the benefits of particular health care products to whether our friendships are based on similarities in genetic makeup.

Improvements in the Fourth Edition

The positive feedback garnered by previous editions assured us that presenting science alongside a story works for students and instructors alike. In the fourth edition, we have rearranged some of the content and added a new chapter and story line. Chapter 3 addresses the question of dietary supplementation while covering the essential content of nutrients and membrane transport. Chapter 4 revisits the question of body fat and health and now includes the content of cellular respiration. Chapter 5 reviews the latest information regarding the science of global warming and continues to cover the basics of photosynthesis. Story lines have been updated or replaced in other chapters compared to previous editions. Chapter 12 on Species and Races now examines the use of DNA to determine genealogy.

With the previous editions, we focused on improving flexibility for instructors via **A Closer Look** chapter subsections, helping students assess their understanding via **Stop and Stretch** and **Visualize This** questions , and providing students with opportunities to be better consumers of popular media with the **Savvy Reader**. In this edition, we have provided instructors and students with guidance in the form of **Learning Outcomes**. These outcomes describe the knowledge or skills that students should have upon completing the reading and associated activities in the text and ancillary materials. The chapter summary is organized around the Learning Outcomes, and all questions and exercises are tagged to a Learning Outcome. Exercises in the Mastering Biology platform are also tagged to these Learning Outcomes, providing students ample opportunity to master these important concepts.

Finally, users of earlier editions will notice a new style for phylogenetic trees in the text—horizontal rather than vertical, to match the standard presentation in the field of evolutionary biology.

Supplements

A group of talented and dedicated biology educators teamed up with us to build a set of resources that equip nonmajors with the tools to achieve scientific literacy that will allow them to make informed decisions about the biological issues that affect them daily. The student resources offer several ways of reviewing and reinforcing the concepts and facts covered in this textbook. We provide instructors with a test bank, Instructor Guide, and the Instructor Resource DVD which includes an updated and expanded suite of lecture presentation materials, and a valuable source of ideas for educators to enrich their instruction efforts. Available in print and media formats, the *Biology: Science for Life with Physiology* resources are easy to navigate and support a variety of learning and teaching styles.

We believe you will find that the design and format of this text and its supplements will help you meet the challenge of helping students both succeed in your course and develop science skills—for life.

As with previous editions, the overall goal of the text remains to provide a thorough overview of the essentials of biological science while trying to avoid overloading students with information. We worked closely with instructors using prior editions as well as other reviewers to pinpoint essential content to include in this edition while staying true to the book's philosophy of learning science in a story format. The development of the fourth edition has truly been a collaborative process among ourselves, the students and instructors who used prior editions, and our many thoughtful reviewers. We look forward to learning about your experience with *Biology: Science for Life with Physiology*, 4th edition.

Supplement Authors

Instructor Guide
Jill Feinstein   *Richland Community College*

Mastering Biology Quiz and Test Items
Catherine Podeszwa   *University of Minnesota–Duluth*

PowerPoint Lectures
Jill Feinstein   *Richland Community College*

"Because science, told as a story, can intrigue and

inform the non-scientific minds among us, it has

the potential to bridge the two cultures into which

civilization is split—the sciences and the humanities.

For educators, stories are an exciting way to draw

young minds into the scientific culture."

—E.O. WILSON

A Captivating Narrative

Every chapter introduces biology topics through a story, explaining important biological processes with concrete examples and applications.

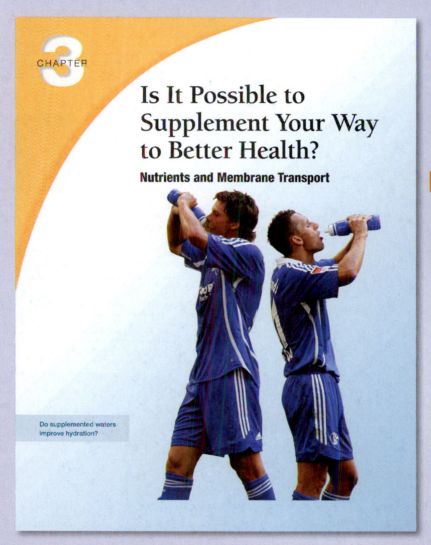

CHAPTER 3

Is It Possible to Supplement Your Way to Better Health?

Nutrients and Membrane Transport

Do supplemented waters improve hydration?

Real-life Examples

Each chapter story draws from real-life experiences, engaging the reader and making the biology more approachable. For example, Chapter 3 frames the discussion of nutrients and membrane transport around a story line on the current craze in nutritional supplements.

What Reviewers Have to Say

" A great text—right level for non-majors students with clear examples and explanation. I really like the link with the Learning Objectives, which I have not yet seen in other texts. "
—Cara Shillington, Instructor, *Eastern Michigan University*

" This is a well-written text that makes learning about biology a truly fun experience. The authors do a remarkable job of keeping the topics clear and understandable while at the same time making the concepts interesting by incorporating some very interesting stories that are both relevant and fascinating. It's not easy to write a text for students who are often resistant to concepts like evolution and natural selection; this text makes our job as instructor that much simpler. Well done. "
—Andrew Goliszek, Instructor, *North Carolina A&T University*

Updated
Table of Contents

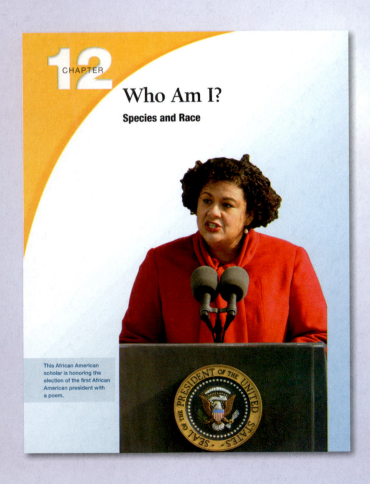

CHAPTER

12

Who Am I?

Species and Race

This African American scholar is honoring the election of the first African American president with a poem.

January 20, 2009, represented a watershed moment in American history, the inauguration of the first African American president, Barack Obama. Regardless of how you feel about his presidency, everyone can appreciate how Obama's election represented the culmination of a major transformation in American society, less than six generations removed from the legal enslavement of Africans in this country. When the African American scholar and poet Elizabeth Alexander read a verse she penned for the inauguration entitled "Praise Song for the Day," she explicitly referenced how far the nation had come toward viewing people of her race as worthy of equal rights and opportunities:

> Say it plain: that many have died for this day.
> Sing the names of the dead who brought us here,
> who laid the train tracks, raised the bridges,
> picked the cotton and the lettuce, built brick by brick the glittering edifices
> they would then keep clean and work inside of.
> Praise song for struggle, praise song for the day.
> Praise song for every hand-lettered sign,
> the figuring-it-out at kitchen tables.

African slavery was in part justified by a belief that black Africans were somehow biologically inferior to white Europeans. Even after slavery ended, the racism that helped excuse it lived on in social policies that kept blacks confined to segregated facilities and communities. Dr. Alexander's professional career has been dedicated to improving race relations and helping reduce the lingering

LEARNING OUTCOMES

LO1 Define *biological species*, and list the mechanisms by which reproductive isolation is maintained by biological species.

LO2 Describe the three steps in the process of speciation.

LO3 Explain how a "race" within a biological species can be defined using the genealogical species concept.

LO4 List the evidence that modern humans are a young species that arose in Africa.

LO5 List the genetic evidence expected when a species can be divided into unique races.

LO6 Describe how the Hardy-Weinberg theorem is used in studies of population genetics.

LO7 Summarize the evidence that indicates that human races are not deep biological divisions within the human species.

LO8 Provide examples of traits that have become common in certain human populations due to the natural selection these populations have undergone.

LO9 Define *genetic drift*, and provide examples of how it results in the evolution of a population.

LO10 Describe how human and animal behavior can cause evolution via sexual selection and assortative mating.

Less than a year after the inauguration, she learned that she was genetically more European than African.

In fact, she shares a recent ancestor with the European American comedian, Stephen Colbert.

How does this new genetic information fit into our experience of human races?

275

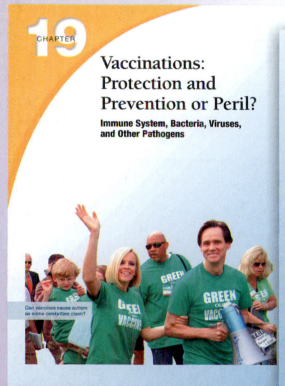

CHAPTER 19

Vaccinations: Protection and Prevention or Peril?

Immune System, Bacteria, Viruses, and Other Pathogens

Can vaccines cause autism as some celebrities claim?

LEARNING OUTCOMES

LO1 Compare the structure of bacteria to that of viruses.

LO2 Contrast bacterial replication with viral reproduction.

LO3 List the structures involved in the first line of defense against infection, and describe how they function in protecting the body against pathogens.

LO4 List the participants in the second line of defense against infection, and describe their functions.

LO5 List the two different cell types involved in the third line of defense against infection, and compare how they recognize and eliminate antigens.

LO6 Explain allergic reactions.

LO7 Explain the process immune cells undergo to be able to respond to millions of different antigens.

LO8 Describe the process used to determine whether a cell is foreign or native to the body and what happens when this process fails.

LO9 Explain the role of memory cells in helping to protect against infection.

LO10 Explain the role of helper T cells in allowing infection by the AIDS virus.

LO11 Describe the mechanism by which vaccines help confer immunity.

Should you get a flu shot? For many college students, this is one of the first health care decisions that will be made without much parental input. University health services personnel want you to, but you hear conflicting information from fellow students. The flu shot, like all vaccines, contains substances that resemble the disease-causing organisms. When you are vaccinated, your body is tricked into preparing for an attack of the disease in case one does occur.

It's not just college students who are vaccinated. Newborns and children under 6 years of age in the United States are vaccinated against 15 different diseases. The first vaccinations happen before a baby even leaves the hospital to go home and continue every few months until the child is a year and a half old. Another series of vaccinations occurs when the child is school aged, and a few additional vaccinations are given during adolescence. Virtually all parents would prefer not to subject their child to these vaccinations, which are often administered via uncomfortable injections, not to mention the inconvenience of taking children to the doctor for so many visits. But for most parents, the promise of protection from diseases makes the choice an easy one.

There are, however, a growing number of parents who are questioning whether to vaccinate their children. Fueled by Internet-obtained information and championed by celebrities who believe vaccines have hurt their children, many parents are skipping some or all vaccinations. Many college students also choose not to get the flu shot, believing the shot might make them ill or that it's not such a big deal to get the flu. To understand the risks and benefits of vaccinations, it is first necessary to learn about the causes of diseases that vaccines attempt to prevent.

Should you get a flu shot?

Why would a male be vaccinated against cervical cancer?

Are parents justified in not allowing their babies to be vaccinated?

467

Flexible
Engaging Content

Two unique features break down complex materials and give instructors flexibility in course coverage.

An Overview: DNA Fingerprinting

All of these questions were answered using **DNA fingerprinting.** This technique allows unambiguous identification of people in the same manner that traditional fingerprinting has been used in the past. To begin this process, it is necessary to isolate the DNA to be fingerprinted. Scientists can isolate DNA from blood, semen, vaginal fluids, a hair root, skin, and even (as was the case in Ekaterinburg) degraded skeletal remains.

Because often there is not much DNA present, it must first be copied to generate larger quantities for analysis. Since each person has a unique sequence of DNA, particular regions of DNA, which vary in size, are selected for copying. When the copied DNA fragments are separated and stained, a unique pattern is produced. *See the next section for A Closer Look at DNA Fingerprinting.*

A Closer Look

Heavily revised in this edition for ease of use, An Overview explains the basics of a difficult biological concept and occurs just before the related A Closer Look section, which provides the details.

A Closer Look:
DNA Fingerprinting

When very small amounts of DNA are available, as is often the case, scientists can make many copies of the DNA by first performing a DNA-amplifying reaction.

Polymerase Chain Reaction (PCR)

The **polymerase chain reaction (PCR)** is used to amplify, or produce many copies of, a particular region of DNA (**Figure 8.12**). To perform PCR, scientists place four essential components into a tube: (1) the **template,** which is the double-stranded DNA to be copied; (2) the **nucleotides** adenine (A), cytosine (C), guanine (G), and thymine (T), which are the individual building-block subunits of DNA; (3) short single-stranded pieces of DNA called **primers,** which are complementary to the ends of the region of DNA to be copied; and (4) an enzyme called *Taq* polymerase.

Taq polymerase is a DNA polymerase that uses one strand of DNA as a template for the synthesis of a daughter strand that carries complementary nucleotides (A:T base pairs are complementary, as are G:C base pairs) (Chapter 6). This enzyme was given the first part of its name (*Taq*) because it was first isolated from *Thermus aquaticus*, a bacterium that lives in hydrothermal vents and can withstand very high temperatures. The second part of the enzyme's name (polymerase) describes its synthesizing activity—it acts as a DNA polymerase.

The main difference between human DNA polymerase and *Taq* polymerase is that the *Taq* polymerase is resistant to extremely high temperatures, temperatures at which human DNA polymerase would be inactivated. The heat-resistant qualities of *Taq* polymerase thus allow PCR reactions to be run at very high temperatures. High temperatures are

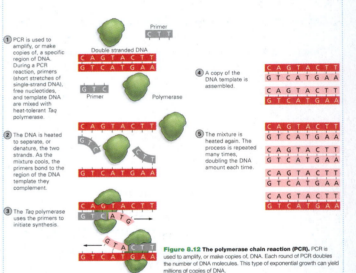

1. PCR is used to amplify, or make copies of, a specific region of DNA. During a PCR reaction, primers (short stretches of single-strand DNA), free nucleotides, and template DNA are mixed with heat-tolerant *Taq* polymerase.

2. The DNA is heated to separate, or denature, the two strands. As the mixture cools, the primers bond to the region of the DNA template they complement.

3. The *Taq* polymerase uses the primers to initiate synthesis.

4. A copy of the DNA template is assembled.

5. The mixture is heated again. The process is repeated many times, doubling the DNA amount each time.

Figure 8.12 The polymerase chain reaction (PCR). PCR is used to amplify, or make copies of, DNA. Each round of PCR doubles the number of DNA molecules. This type of exponential growth can yield millions of copies of DNA.

focus on EVOLUTION

Plant Reproduction

Land plants have a unique life cycle. Instead of one multicellular "adult" form as is found in animals, plants have two, which alternate with each other. In fact, the life cycle of plants is commonly referred to as **alternation of generations.** The two life-forms result because the product of meiosis in plants is not a sperm or egg (gamete); it is a haploid structure (containing half the typical number of chromosomes) called a **spore,** which undergoes mitosis to produce a multicellular plant called the **gametophyte** (or gamete-producing plant). The gametophyte produces sperm or eggs (or both) by mitosis. When gametes fuse, they produce a diploid embryo that divides and eventually produces spores by meiosis. This stage in the life cycle is called the **sporophyte.** The generalized life cycle is illustrated in **Figure E23.1.** Because the algal ancestors of land plants were haploid for most of their lives, but sexual reproduction requires a short period of diploidy, the alternation between two forms is a vestige of evolutionary history. In fact, the trend in land plant evolution has been that the haploid gametophyte generation has become increasingly reduced.

In mosses and other nonvascular plants, the gametophyte generation is the most prominent form, reflecting their close relationship with the green algae. Sperm and eggs are produced in structures at the tips of the plant, and sperm swim in a film of water, such as a drop of rain, to reach the eggs. The embryo develops within the female reproductive organ, eventually producing a sporophyte that emerges from the tip of the gametophyte and is dependent on it for survival. Tiny spores produced in the sporophyte are dispersed on the wind and germinate into new gametophytes if they land in moist conditions.

On land, a large haploid [...] DNA-damaging UV light, w[...] cules, is much more of a thr[...] have two copies of each ch[...] not as detrimental in a diplo[...] may be why ferns and othe[...] arose from the nonvascula[...] phyte. Spores are produce[...] dispersed by wind or water[...] tions. The small independen[...] a spore may produce both [...] a film of water to reach the e[...] out of the female reproductiv[...] comes much larger and inde[...]

The seed plants that ar[...] next step in the evolution [...] reduction in gametophyte [...] persal structure. Pine tree[...] plants, called gymnosperm[...] tophyte, both of which ar[...] for survival. Pollen, released from small cones produced by the adult sporophyte, is actually tiny, not-yet-mature male gametophyte plants. Pollen is carried by the wind (or occasionally insects) to the female reproductive structures, which are typically produced in woody cones on the sporophyte. Here, the pollen completes maturation and releases sperm to fertilize the egg. The female gametophyte, which contains the egg, is a small structure on the surface of a cone scale. The embryo begins development inside

1. Cells in the multicellular diploid plant (the sporophyte) undergo meiosis, producing a haploid spore.

Spore (1n)

2. The spore germinates into a multicellular haploid plant, the gametophyte.

3. Gametophytes produce sperm and eggs by mitosis.

4. Fertilization produces a diploid sporophyte embryo, which grows into a sporophyte.

Haploid = 1n
Diploid = 2n

Figure E23.1 Alternation of generations.

Focus on Evolution

Within the physiology chapters, Focus on Evolution features illustrate the importance of natural selection in a wide range of organisms and allow students to see that all of life is connected through this process.

Unique Savvy Reader boxes highlight the applicability of biology in everyday life. Taken from a variety of media sources, each box contains a short excerpt, relating to chapter content, followed by critical thinking questions to help students interpret and evaluate the information.

SAVVY READER

A Toolkit for Evaluating Science in the News

The following checklist can be used as a guide to evaluating science information in the news. Although any issue that raises a red flag should cause you to be cautious about the conclusions of the story, no one issue should cause you to reject the story (and its associated research) out of hand. However, in general, the fewer red flags raised by a story, the more reliable is the report of the significance of the scientific study. Use **Table 1.2** to evaluate this extract from an article by Terry Gupta in the Nashua (New Hampshire) *Telegraph*:

[Paula] Fortier, a licensed massage therapist, shared her special brew. To boiled water and tea, she adds cayenne pepper, honey and fresh squeezed lemon. Fortier's enthusiasm may be more contagious than the flu when she describes this heated, "unbeatable" combination. "Your throat opens up," she said. "You're less irritated and you sweat out some unwanted toxins." She varies the exact measures of the ingredients according to tolerance and taste. While on the topic of hot water, one of Fortier's favorite remedies for "an achy body that feels sore all over is a hot bath in a steeping tub with 2 cups of cider vinegar for 15 minutes. This draws off toxins, reduces inflammation and leaves you feeling better."

TABLE 1.2

A guide for evaluating science in the news. For each question, check the appropriate box.

Question	Preferred answer		Raises a red flag	
1. What is the basis for the story?	Hypothesis test	☐	Untested assertion No data to support claims in the article.	☐
2. What is the affiliation of the scientist?	Independent (university or government agency)	☐	Employed by an industry or advocacy group Data and conclusions could be biased.	☐
3. What is the funding source for the study?	Government or nonpartisan foundation (without bias)	☐	Industry group or other partisan source (with bias) Data and conclusions could be biased.	☐

Revised and New Savvy Reader Boxes

Savvy Reader boxes in selected chapters have been replaced with more relevant and timely media excerpts. The following are new Savvy Readers:

- Ionized Water (Chapter 2)

- Refugees from Global Warming (Chapter 5)

- What Happens When a Species Recovers (Gray Wolf's endangered species status) (Chapter 15)

SAVVY READER

What Happens When a Species Recovers

Gray Wolf's endangered species status appears close to end

A new federal bill introduced in the U.S. House of Representatives would "de-list" the Gray Wolf from Endangered Species Act protections and allow individual states to design and carry out their own wolf management and recovery plans. The legislation is co-sponsored by Rep. Jim Matheson (D) from Utah, a state that is just beginning to see increases in wolf numbers, and Rep. Mike Ross (D-Arkansas), who co-chairs the Congressional Sportsman's Caucus, made up of advocates of hunting and fishing interests.

According to Rep. Matheson, "Scientists and wolf recovery advocates agree—the gray wolf is back. Since it is no longer endangered, it should be de-listed as a species, managed as others species are—by state wildlife agencies—and time, money, and effort can be focused where it's needed."

Rep. Ross argues that officials within the regions inhabited by particular endangered species should have a greater role in management decisions related to these species. "Excessive wolf populations are having a devastating impact on elk, moose, deer, and other species, and each state has its own unique set of challenges. Both the Obama and Bush administrations have already recommended the de-listing of wolves in many states and turning their management over to state wildlife agencies because it is the right thing to do to keep our nation's sensitive ecosystem in balance."

1. What evidence does the representative from Utah provide to support his statement that "the gray wolf is back"? Is that evidence convincing?
2. Given what you've learned in this chapter, do you think it is likely that the return of the gray wolf is having a "devastating impact" on elk and moose?

Innovative
Art Program

Designed to help students visually navigate biology, this edition's photo and art program has been revised for currency and increased clarity.

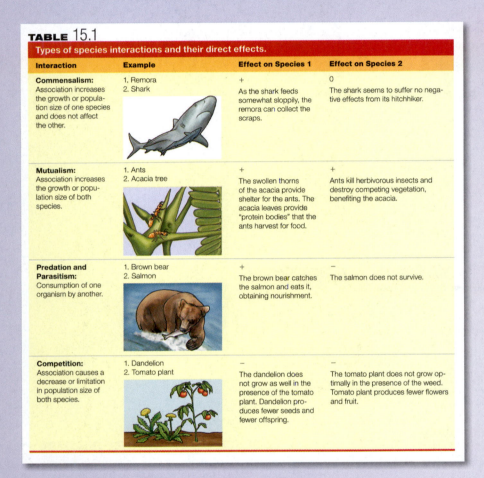

TABLE 15.1

Types of species interactions and their direct effects.

Interaction	Example	Effect on Species 1	Effect on Species 2
Commensalism: Association increases the growth or population size of one species and does not affect the other.	1. Remora 2. Shark	+ As the shark feeds somewhat sloppily, the remora can collect the scraps.	0 The shark seems to suffer no negative effects from its hitchhiker.
Mutualism: Association increases the growth or population size of both species.	1. Ants 2. Acacia tree	+ The swollen thorns of the acacia provide shelter for the ants. The acacia leaves provide "protein bodies" that the ants harvest for food.	+ Ants kill herbivorous insects and destroy competing vegetation, benefiting the acacia.
Predation and Parasitism: Consumption of one organism by another.	1. Brown bear 2. Salmon	+ The brown bear catches the salmon and eats it, obtaining nourishment.	− The salmon does not survive.
Competition: Association causes a decrease or limitation in population size of both species.	1. Dandelion 2. Tomato plant	− The dandelion does not grow as well in the presence of the tomato plant. Dandelion produces fewer seeds and fewer offspring.	− The tomato plant does not grow optimally in the presence of the weed. Tomato plant produces fewer flowers and fruit.

Illustrated Tables

Unique, comprehensive tables organize information in one place and provide easy, visual references for students.

Visual Analogies

Art throughout the book offers visual analogies to teach students by comparing familiar, everyday objects and occurrences to complicated, biological concepts and processes.

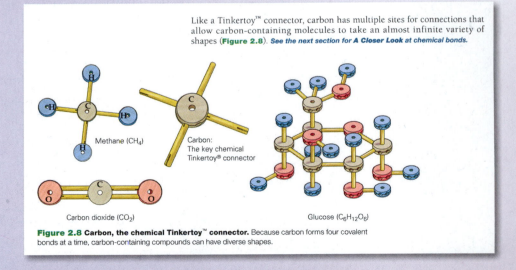

Like a Tinkertoy™ connector, carbon has multiple sites for connections that allow carbon-containing molecules to take an almost infinite variety of shapes (**Figure 2.8**). *See the next section for A Closer Look at chemical bonds.*

Methane (CH_4)

Carbon: The key chemical Tinkertoy® connector

Carbon dioxide (CO_2)

Glucose ($C_6H_{12}O_6$)

Figure 2.8 Carbon, the chemical Tinkertoy™ connector. Because carbon forms four covalent bonds at a time, carbon-containing compounds can have diverse shapes.

Dark-eyed junco

Harris's sparrow

Golden-crowned sparrow

White-crowned sparrow

White-throated sparrow

Ancestral bird

Ancestral sparrow

Black stripes on crown of head

Seven stripes on crown

White throat patch

Figure 13.21 Reconstruction of an evolutionary history. The relationships among these four species of sparrow can be illustrated by their shared physical traits compared to a more distant relative, the dark-eyed junco.
Visualize This: What traits did the ancestral bird likely have?

NEW! Phylogenetic Trees

Redesigned phylogenetic trees throughout the book offer a horizontal instead of vertical orientation, in keeping with the latest in scientific data presentation.

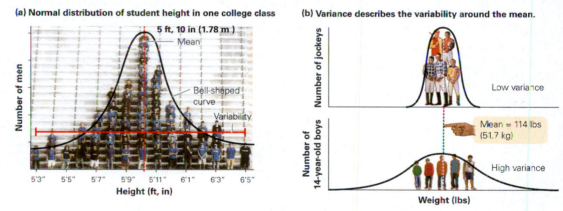

(a) Normal distribution of student height in one college class

5 ft, 10 in (1.78 m)
Mean

Bell-shaped curve

Variability

Number of men

5'3" 5'5" 5'7" 5'9" 5'11" 6'1" 6'3" 6'5"
Height (ft, in)

(b) Variance describes the variability around the mean.

Number of jockeys

Low variance

Mean = 114 lbs (51.7 kg)

High variance

Number of 14-year-old boys

Weight (lbs)

Figure 7.16 A quantitative trait. (a) This photo of men arranged by height illustrates a normal distribution. The highest point of the bell curve is also the mean height of 5 feet, 10 inches. (b) Fourteen-year-old boys and professional jockeys have the same average weight—approximately 114 pounds. However, to be a jockey, you must be within about 4 pounds of this average. Thus, the variance among jockeys in weight is much smaller than the variance among 14-year-olds.
Visualize This: Examine Figure 7.16a closely: Does an average height of 5 feet, 10 inches in this particular population imply that most men were this height? Were most men in this population close to the mean, or was there a wide range of heights?

Visualize This

Visualize This questions, accompanying select figures, encourage students to critically evaluate illustrations to further their understanding of biology.

Integrated
Learning Outcomes

New to this edition and integrated within the text and MasteringBiology,® Learning Outcomes help professors guide students' reading and allow students to assess their understanding of the text.

In Text Pedagogical Support

LEARNING OUTCOMES

LO1 Describe the relationship between genes, chromosomes, and alleles.

LO2 Explain why cells containing identical genetic information can look different from each other.

LO3 Define *segregation* and *independent assortment* and explain how these processes contribute to genetic diversity.

LO4 Distinguish between homozygous and heterozygous genotypes and describe how recessive and dominant alleles produce particular phenotypes when expressed in these genotypes.

LO5 Demonstrate how to use a Punnett square to predict the likelihood of a particular offspring genotype and phenotype from a cross of two individuals with known genotype.

LO6 Define *quantitative trait* and describe the genetic and environmental factors that cause this pattern of inheritance.

LO7 Describe how heritability is calculated and what it tells us about the genetic component of quantitative traits.

LO8 Explain why a high heritability still does not always mean that a given trait is determined mostly by the genes an individual carries.

Each Chapter Opens with Learning Outcomes

Learning Outcomes specify the knowledge, skills, or abilities a student can expect to demonstrate after reading the chapter.

Chapter Review

Mastering BIOLOGY

Go to the Study Area at www.masteringbiology.com for practice quizzes, myeBook, BioFlix™ 3-D animations, MP3Tutor sessions, videos, current events, and more.

Learning Outcomes

LO1 **Describe the relationship between genes, chromosomes, and alleles (Section 7.1).**
- Children resemble their parents in part because they inherit their parents' genes, segments of DNA that contain information about how to make proteins (pp. 148–149).

- Chromosomes contain genes. Different versions of a gene are called alleles (p. 150).
- Mutations in genes generate a variety of alleles. Each allele typically results in a slightly different protein product (p. 151).

Learning Outcomes Revisited at the End of the Chapter

Summaries for Learning Outcomes are provided at the end of every chapter to reinforce students' understanding of the main concepts covered.

Learning the Basics

1. **LO1** What types of drugs have helped reduce the death rate due to tuberculosis infection, and why have they become less effective more recently?

2. **LO4** Define *artificial selection*, and compare and contrast it with natural selection.

3. **LO8** Describe how *Mycobacterium tuberculosis* evolves when it is exposed to an antibiotic.

4. **LO2** Which of the following observations is not part of the theory of natural selection?

   A. Populations of organisms have more offspring than will survive; B. There is variation among individuals in a population; C. Modern organisms are unrelated; D. Traits can be passed on from parent to offspring; E. Some variants in a population have a higher probability of survival and reproduction than other variants do.

5. **LO2** The best definition of *evolutionary fitness* is _____.

   A. physical health; B. the ability to attract members of the opposite sex; C. the ability to adapt to the environment; D. survival and reproduction relative to other members of the population; E. overall strength

6. **LO3** An adaptation is a trait of an organism that increases _____.

   A. its fitness; B. its ability to survive and replicate; C. in frequency in a population over many generations; D. A and B are correct; E. A, B, and C are correct

shorter tusks at full adulthood than male elephants in the early 1900s. This is an example of _____.

   A. disruptive selection; B. stabilizing selection; C. directional selection; D. evolutionary regression; E. more than one of the above is correct

10. **LO8** Antibiotic resistance is becoming common among organisms that cause a variety of human diseases. All of the following strategies help reduce the risk of antibiotic resistance evolving in a susceptible bacterial population except _____.

   A. using antibiotics only when appropriate, for bacterial infections that are not clearing up naturally; B. using the drugs as directed, taking all the antibiotic over the course of days prescribed; C. using more than one antibiotic at a time for difficult-to-treat organisms; D. preventing natural selection by reducing the amount of evolution the organisms can perform; E. reducing the use of antibiotics in non-health care settings, such as agriculture

Analyzing and Applying the Basics

1. **LO3 LO6** Most domestic fruits and vegetables are a result of artificial selection from wild ancestors. Use your understanding of artificial selection to describe how domesticated strawberries must have evolved from their smaller wild relatives. What trade-offs do domesticated strawberries exhibit relative to their wild ancestors?

2. **LO3 LO5** The striped pattern on zebras' coats is considered to be an adaptation that helps reduce the

Learning Outcomes in Self-Assessment Sections

Questions in Learning the Basics and Analyzing and Applying the Basics include corresponding Learning Outcomes to encourage students to recall and formulate the most reasonable and precise answers.

Online Pedagogical Support in MasteringBiology®

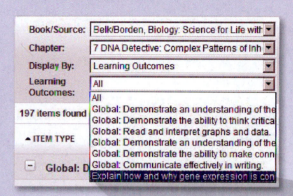

Integrated and Accessible Learning Outcomes

All MasteringBiology content, including activities and test items, are now tagged to the book specific Learning Outcomes. This unique integration of text and media enables instructors to reinforce key content in a variety of formats.

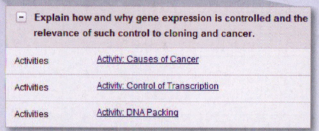

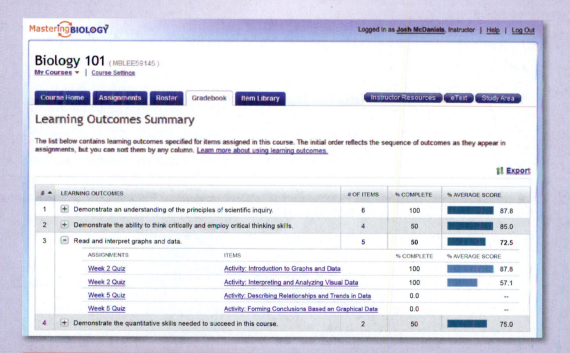

One-click Learning Outcomes Report

In MasteringBiology it's quick and easy for instructors to generate reports showing students' performance on selected Learning Outcomes. With just one click from the gradebook, instructors can see all of the Learning Outcomes associated with problems assigned in their MasteringBiology course, as well as the average score on those problems. One further click allows instructors to export this information to a Microsoft® Excel spreadsheet, which can be formatted as the instructor sees fit.

Author-created
MasteringBiology® Activities

New MasteringBiology content written by authors Colleen Belk and Virginia Borden Maier reinforce and connect the text and media.

MasteringBiology®
www.masteringbiology.com

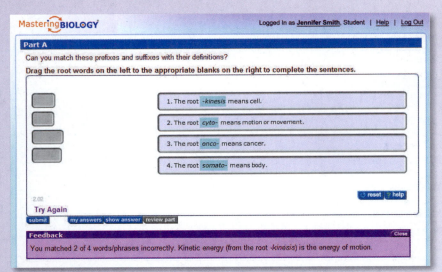

Based on the popular end of chapter section, these activities test students' understanding of key Latin and Greek roots and ask students to apply the roots in new situations.

New Video Activities support story line content and help reinforce the narrative and engaging nature of the text.

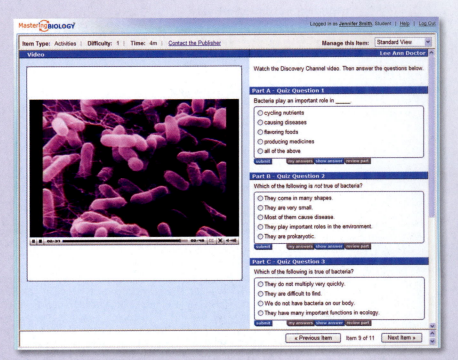

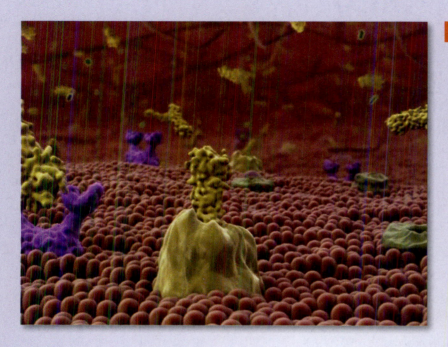

BioFlix® Animations

These animations invigorate classroom lectures and cover the most difficult biology topics, including cellular respiration, photosynthesis, and how neurons work. BioFlix include 3-D, movie-quality animations, labeled slide shows, carefully constructed student tutorials, study sheets, and quizzes.

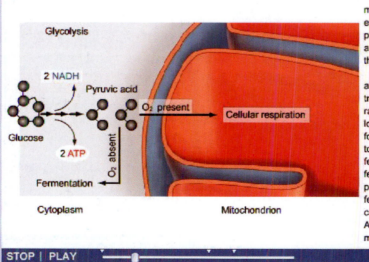

Consider the metabolism in your muscle cells. At rest or during light exercise, when oxygen is plentiful, pyruvic acid enters the Krebs cycle and continues to be metabolized through cellular respiration.

During heavy exercise, your lungs and circulatory system can't transport oxygen to your muscles rapidly enough and muscle cells run low on oxygen. The lack of oxygen forces a shift in energy metabolism toward the anaerobic pathway, fermentation. In muscle cells, fermentation results in lactate production. In some organisms, fermentation produces ethanol and carbon dioxide. However, there is no ATP production beyond the two ATP molecules generated by glycolysis.

Additional Book-specific Activities

Over 25 total book-specific activities have been added to the MasteringBiology item library.

Topics include:
- Glucose Metabolism (Chapter 4)
- The Process of Speciation (Chapter 12)
- Population Growth (Chapter 14)

Dynamic
Teaching Resources

Pearson eText

The eText gives students access to the text whenever and wherever they can access the Internet. The eText pages look exactly like the printed text and include powerful interactive and customization functions.

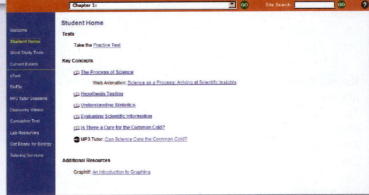

MasteringBiology® Study Area

The Study Area of MasteringBiology provides students with access to useful materials for studying independently. Study materials include flashcards, word study tools, animations, and practice quizzes.

NEW! Pre-built Assignments

Pre-built assignments make it easier for instructors to assign MasteringBiology.

Pre-built Assignment Types:
- Engaging Homework
- Activities on Tough Topics
- Build Graphing Skills
- Current Events
- Reading Quizzes

NEW! BioFlix® Player

The New Instructor's Edition DVD contains the complete BioFlix 3-D Animation Suite with a new, enhanced player. The DVD includes an "all-in-one" BioFlix lecture tool, with labels for key elements, closed captioning, drawing tool, quiz questions, and more.

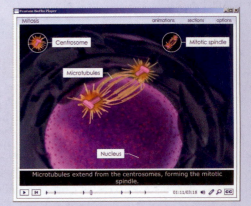

Acknowledgments

Reviewers

Each chapter of this book was thoroughly reviewed several times as it moved through the development process. Reviewers were chosen on the basis of their demonstrated talent and dedication in the classroom. Many of these reviewers are already trying various approaches to actively engage students in lectures and to raise the scientific literacy and critical thinking skills among their students. Their passion for teaching and commitment to their students were evident throughout this process. These devoted individuals scrupulously checked each chapter for scientific accuracy, readability, and coverage level.

All of these reviewers provided thoughtful, insightful feedback, which improved the text significantly. Their efforts reflect their deep commitment to teaching nonmajors and improving the scientific literacy of all students. We are very thankful for their contributions.

Karen Aguirre *Coastal Carolina University*
Marcia Anglin *Miami-Dade College*
Tania Beliz *College of San Mateo*
Merri Casem *California State University, Fullerton*
Anne Casper *Eastern Michigan University*
Dani DuCharme *Waubonsee Community College*
Steve Eisenberg *Elizabethtown Community and Technical College*
Susan Fisher *Ohio State University*
Jennifer Fritz *The University of Texas at Austin*
Anthony Gaudin *Ivy Tech Community College of Indiana—Columbus/Franklin*
Mary Rose Grant *University of Missouri, St. Louis*
Richard Jacobson *Laredo Community College*
Malcolm Jenness *New Mexico Institute of Technology*
Ron Johnston *Blinn College*
Michael L' Annunziata *Pima College*
Suzanne Long *Monroe Community College*
Cindy Malone *California State University, Northridge*
Mark Manteuffel *St. Louis Community College, Flo Valley*
Roger Martin *Brigham Young University, Salt Lake Center*
Jorge Obeso *Miami Dade College, North Campus*
Nancy Platt *Pima College*
Ashley Rhodes *Kansas State University*
Bill Rogers *Ball State University*
Michael Sawey *Texas Christian University*
Debbie Scheidemantel *Pima College*
Cara Shillington *Eastern Michigan University*
Lynnda Skidmore *Wayne County Community College*
Anna Bess Sorin *University of Memphis*
Jeff Thomas *California State University, Northridge*

Derek Weber *Raritan Valley Community College*
Jennifer Wiatrowski *Pasco-Hernando Community College*

Prior Edition Reviewers

Daryl Adams, *Minnesota State University, Mankato*
Karen Aguirre, *Clarkson University*
Susan Aronica, *Canisius College*
Mary Ashley, *University of Chicago*
James S. Backer, *Daytona Beach Community College*
Ellen Baker, *Santa Monica College*
Gail F. Baker, *LaGuardia Community College*
Neil R. Baker, *The Ohio State University*
Andrew Baldwin, *Mesa Community College*
Thomas Balgooyen, *San Jose State University*
Tamatha R. Barbeau, *Francis Marion University*
Sarah Barlow, *Middle Tennessee State University*
Andrew M. Barton, *University of Maine, Farmington*
Vernon Bauer, *Francis Marion University*
Paul Beardsley, *Idaho State University*
Donna Becker, *Northern Michigan University*
David Belt, *Penn Valley Community College*
Steve Berg, *Winona State University*
Carl T. Bergstrom, *University of Washington*
Janet Bester-Meredith, *Seattle Pacific University*
Barry Beutler, *College of Eastern Utah*
Donna H. Bivans, *Pitt Community College*
Lesley Blair, *Oregon State University*
John Blamire, *City University of New York, Brooklyn College*
Barbara Blonder, *Flagler College*
Susan Bornstein-Forst, *Marian College*
Bruno Borsari, *Winona State University*
James Botsford, *New Mexico State University*
Robert S. Boyd, *Auburn University*
Bryan Brendley, *Gannon University*
Eric Brenner, *New York University*
Peggy Brickman, *University of Georgia*
Carol Britson, *University of Mississippi*
Carole Browne, *Wake Forest University*
Neil Buckley, *State University of New York, Plattsburgh*
Stephanie Burdett, *Brigham Young University*
Warren Burggren, *University of North Texas*
Nancy Butler, *Kutztown University*
Suzanne Butler, *Miami-Dade Community College*
Wilbert Butler, *Tallahassee Community College*
David Byres, *Florida Community College at Jacksonville*
Tom Campbell, *Pierce College, Los Angeles*
Deborah Cato, *Wheaton College*
Peter Chabora, *Queens College*

Bruce Chase, *University of Nebraska, Omaha*
Thomas F. Chubb, *Villanova University*
Gregory Clark, *University of Texas, Austin*
Kimberly Cline-Brown, *University of Northern Iowa*
Mary Colavito, *Santa Monica College*
William H. Coleman, *University of Hartford*
William F. Collins III, *Stony Brook University*
Walter Conley, *State University of New York, Potsdam*
Jerry L. Cook, *Sam Houston State University*
Melanie Cook, *Tyler Junior College*
Scott Cooper, *University of Wisconsin, La Crosse*
Erica Corbett, *Southeastern Oklahoma State University*
George Cornwall, *University of Colorado*
Charles Cottingham, *Frederick Community College*
James B. Courtright, *Marquette University*
Angela Cunningham, *Baylor University*
Judy Dacus, *Cedar Valley College*
Judith D'Aleo, *Plymouth State University*
Deborah Dardis, *Southeastern Louisiana University*
Juville Dario-Becker, *Central Virginia Community College*
Garry Davies, *University of Alaska, Anchorage*
Miriam del Campo, *Miami-Dade Community College*
Judith D'Aleo, *Plymouth State University*
Edward A. DeGrauw, *Portland Community College*
Heather DeHart, *Western Kentucky University*
Miriam del Campo, *Miami-Dade Community College*
Veronique Delesalle, *Gettysburg College*
Lisa Delissio, *Salem State College*
Beth De Stasio, *Lawrence University*
Elizabeth Desy, *Southwest Minnesota State University*
Donald Deters, *Bowling Green State University*
Gregg Dieringer, *Northwest Missouri State*
Diane Dixon, *Southeastern Oklahoma State University*
Christopher Dobson, *Grand Valley State University*
Cecile Dolan, *New Hampshire Community Technical College, Manchester*
Matthew Douglas, *Grand Rapids Community College*
Lee C. Drickamer, *Northern Arizona University*
Susan Dunford, *University of Cincinnati*
Stephen Ebbs, *Southern Illinois University*
Douglas Eder, *Southern Illinois University, Edwardsville*
Patrick J. Enderle, *East Carolina University*
William Epperly, *Robert Morris College*
Ana Escandon, *Los Angeles Harbor College*
Dan Eshel, *City University of New York, Brooklyn College*
Marirose Ethington, *Genessee Community College*
Deborah Fahey, *Wheaton College*
Chris Farrell, *Trevecca Nazarene University*
Richard Firenze, *Broome Community College*
Lynn Firestone, *Brigham Young University*
Brandon L. Foster, *Wake Technical Community College*
Richard A. Fralick, *Plymouth State University*
Barbara Frank, *Idaho State University*
Stewart Frankel, *University of Hartford*
Lori Frear, *Wake Technical Community College*
David Froelich, *Austin Community College*

Suzanne Frucht, *Northwest Missouri State University*
Edward Gabriel, *Lycoming College*
Anne Galbraith, *University of Wisconsin, La Crosse*
Patrick Galliart, *North Iowa Area Community College*
Wendy Garrison, *University of Mississippi*
Alexandros Georgakilas, *East Carolina University*
Robert George, *University of North Carolina, Wilmington*
Tammy Gillespie, *Eastern Arizona College*
Sharon Gilman, *Coastal Carolina University*
Mac F. Given, *Neumann College*
Bruce Goldman, *University of Conn ticut, Storrs*
Andrew Goliszek, *North Carolina Agricultural and Technical State University*
Beatriz Gonzalez, *Sante Fe Community College*
Eugene Goodman, *University of Wisconsin, Parkside*
Lara Gossage, *Hutchinson Community College*
Tamar Goulet, *University of Mississippi*
Becky Graham, *University of West Alabama*
John Green, *Nicholls State University*
Robert S. Greene, *Niagara University*
Tony J. Greenfield, *Southwest Minnesota State University*
Bruce Griffis, *Kentucky State University*
Mark Grobner, *California State University, Stanislaus*
Michael Groesbeck, *Brigham Young University, Idaho*
Stanley Guffey, *University of Tennessee*
Mark Hammer, *Wayne State University*
Blanche Haning, *North Carolina State University*
Robert Harms, *St. Louis Community College*
Craig M. Hart, *Louisiana State University*
Patricia Hauslein, *St. Cloud State University*
Stephen Hedman, *University of Minnesota, Duluth*
Bethany Henderson-Dean, *University of Findlay*
Julie Hens, *University of Maryland University College*
Peter Heywood, *Brown University*
Julia Hinton, *McNeese State University*
Phyllis C. Hirsh, *East Los Angeles College*
Elizabeth Hodgson, *York College of Pennsylvania*
Leland Holland, *Pasco-Hernando Community College*
Jane Horlings, *Saddleback Community College*
Margaret Horton, *University of North Carolina, Greensboro*
Laurie Host, *Harford Community College*
David Howard, *University of Wisconsin, La Crosse*
Michael Hudecki, *State University of New York, Buffalo*
Michael E. S. Hudspeth, *Northern Illinois University*
Laura Huenneke, *New Mexico State University*
Pamela D. Huggins, *Fairmont State University*
Sue Hum-Musser, *Western Illinois University*
Carol Hurney, *James Madison University*
James Hutcheon, *Georgia Southern University*
Anthony Ippolito, *DePaul University*
Carl Johansson, *Fresno City College*
Thomas Jordan, *Pima Community College*
Jann Joseph, *Grand Valley State University*
Mary K. Kananen, *Penn State University, Altoona*
Arnold Karpoff, *University of Louisville*
Judy Kaufman, *Monroe Community College*

Michael Keas, *Oklahoma Baptist University*
Judith Kelly, *Henry Ford Community College*
Karen Kendall-Fite, *Columbia State Community College*
Andrew Keth, *Clarion University*
David Kirby, *American University*
Stacey Kiser, *Lane Community College*
Dennis Kitz, *Southern Illinois University, Edwardsville*
Carl Kloock, *California State, Bakersfield*
Jennifer Knapp, *Nashville State Technical Community College*
Loren Knapp, *University of South Carolina*
Michael A. Kotarski, *Niagara University*
Michelle LaBonte, *Framingham State College*
Phyllis Laine, *Xavier University*
Dale Lambert, *Tarrant County College*
Tom Langen, *Clarkson University*
Lynn Larsen, *Portland Community College*
Mark Lavery, *Oregon State University*
Brenda Leady, *University of Toledo*
Mary Lehman, *Longwood University*
Doug Levey, *University of Florida*
Lee Likins, *University of Missouri, Kansas City*
Abigail Littlefield, *Landmark College*
Andrew D. Lloyd, *Delaware State University*
Jayson Lloyd, *College of Southern Idaho*
Judy Lonsdale, *Boise State University*
Kate Lormand, *Arapahoe Community College*
Paul Lurquin, *Washington State University*
Kimberly Lyle-Ippolito, *Anderson University*
Douglas Lyng, *Indiana University/Purdue University*
Michelle Mabry, *Davis and Elkins College*
Stephen E. MacAvoy, *American University*
Molly MacLean, *University of Maine*
Charles Mallery, *University of Miami*
Ken Marr Green, *River Community College*
Kathleen Marrs, *Indiana University/Purdue University*
Matthew J. Maurer, *University of Virginia's College at Wise*
Geri Mayer, *Florida Atlantic University*
T. D. Maze, *Lander University*
Steve McCommas, *Southern Illinois University, Edwardsville*
Colleen McNamara, *Albuquerque Technical Vocational Institute*
Mary McNamara, *Albuquerque Technical Vocational Institute*
John McWilliams, *Oklahoma Baptist University*
Susan T. Meiers, *Western Illinois University*
Diane Melroy, *University of North Carolina, Wilmington*
Joseph Mendelson, *Utah State University*
Paige A. Mettler-Cherry, *Lindenwood University*
Debra Meuler, *Cardinal Stritch University*
James E. Mickle, *North Carolina State University*
Craig Milgrim, *Grossmont College*
Hugh Miller, *East Tennessee State University*
Jennifer Miskowski, *University of Wisconsin, La Crosse*
Ali Mohamed, *Virginia State University*
Stephen Molnar, *Washington University*
James Mone, *Millersville University*
Daniela Monk, *Washington State University*
David Mork, *Yakima Valley Community College*

Bertram Murray, *Rutgers University*
Ken Nadler, *Michigan State University*
John J. Natalini, *Quincy University*
Alissa A. Neill, *University of Rhode Island*
Dawn Nelson, *Community College of Southern Nevada*
Joseph Newhouse, *California University of Pennsylvania*
Jeffrey Newman, *Lycoming College*
David L.G. Noakes, *University of Guelph*
Shawn Nordell, *St. Louis University*
Tonye E. Numbere, *University of Missouri, Rolla*
Lori Nicholas, *New York University*
Erin O'Brien, *Dixie College*
Igor Oksov, *Union County College*
Kevin Padian, *University of California, Berkeley*
Arnas Palaima, *University of Mississippi*
Anthony Palombella, *Longwood University*
Marilee Benore Parsons, *University of Michigan, Dearborn*
Steven L. Peck, *Brigham Young University*
Javier Penalosa, *Buffalo State College*
Murray Paton Pendarvis, *Southeastern Louisiana University*
Rhoda Perozzi, *Virginia Commonwealth University*
John Peters, *College of Charleston*
Patricia Phelps, *Austin Community College*
Polly Phillips, *Florida International University*
Indiren Pillay, *Culver-Stockton College*
Francis J. Pitocchelli, *Saint Anselm College*
Roberta L. Pohlman, *Wright State University*
Calvin Porter, *Xavier University*
Linda Potts, *University of North Carolina, Wilmington*
Robert Pozos, *San Diego State University*
Marion Preest, *The Claremont Colleges*
Gregory Pryor, *Francis Marion University*
Rongsun Pu, *Kean University*
Narayanan Rajendran, *Kentucky State University*
Anne E. Reilly, *Florida Atlantic University*
Michael H. Renfroe, *James Madison University*
Laura Rhoads, *State University of New York, Potsdam*
Gwynne S. Rife, *University of Findlay*
Todd Rimkus, *Marymount University*
Laurel Roberts, *University of Pittsburgh*
Wilma Robertson, *Boise State University*
William E. Rogers, *Texas A&M University*
Troy Rohn, *Boise State University*
Deborah Ross, *Indiana University/Purdue University*
Christel Rowe, *Hibbing Community College*
Joanne Russell, *Manchester Community College*
Michael Rutledge, *Middle Tennessee State University*
Wendy Ryan, *Kutztown University*
Christopher Sacchi, *Kutztown University*
Kim Sadler, *Middle Tennessee State University*
Brian Sailer, *Albuquerque Technical Vocational Institute*
Jasmine Saros, *University of Wisconsin, La Crosse*
Ken Saville, *Albion College*
Louis Scala, *Kutztown University*
Daniel C. Scheirer, *Northeastern University*
Beverly Schieltz, *Wright State University*

Nancy Schmidt, *Pima Community College*
Robert Schoch, *Boston University*
Julie Schroer, *Bismarck State College*
Fayla Schwartz, *Everett Community College*
Steven Scott, *Merritt College*
Gray Scrimgeour, *University of Toronto*
Roger Seeber, *West Liberty State College*
Mary Severinghaus, *Parkland College*
Allison Shearer, *Grossmont College*
Robert Shetlar, *Georgia Southern University*
Cara Shillington, *Eastern Michigan University*
Beatrice Sirakaya, *Pennsylvania State University*
Cynthia Sirna, *Gadsden State Community College*
Thomas Sluss, *Fort Lewis College*
Brian Smith Black, *Hills State University*
Douglas Smith, *Clarion University of Pennsylvania*
Mark Smith, *Chaffey College*
Gregory Smutzer, *Temple University*
Sally Sommers, *Smith Boston University*
Anna Bess Sorin, *University of Memphis*
Bryan Spohn Florida, *Community College at Jacksonville, Kent Campus*
Carol St. Angelo, *Hofstra University*
Amanda Starnes, *Emory University*
Susan L. Steen, *Idaho State University*
Timothy Stewart, *Longwood College*
Shawn Stover, *Davis and Elkins College*
Bradley J. Swanson, *Central Michigan University*
Joyce Tamashiro, *University of Puget Sound*
Jeffrey Taylor, *Slippery Rock University*
Martha Taylor, *Cornell University*
Glena Temple, *Viterbo*
Tania Thalkar, *Clarion University of Pennsylvania*
Alice Templet, *Nicholls State University*
Jeffrey Thomas, *University of California, Los Angeles*
Janis Thompson, *Lorain County Community College*
Nina Thumser, *California University of Pennsylvania*
Alana Tibbets, *Southern Illinois University, Edwardsville*
Martin Tracey, *Florida International University*
Jeffrey Travis, *State University of New York, Albany*
Michael Troyan, Pennsylvania State University
Robert Turgeon, *Cornell University*
Michael Tveten, *Pima Community College, Northwest Campus*
James Urban, *Kansas State University*
Brandi Van Roo, *Framingham State College*
John Va ghan, *St. Petersburg Junior College*
Martin Vaughan, *Indiana State University*
Mark Venable, *Appalachian State University*
Paul Verrell, *Washington State University*
Tanya Vickers, *University of Utah*
Janet Vigna, *Grand Valley State University*
Sean Walker, *California State University, Fullerton*
Don Waller, *University of Wisconsin, Madison*
Tracy Ware, *Salem State College*
Jennifer Warner, *University of North Carolina, Charlotte*
Lisa Weasel, *Portland State University*

Carol Weaver, *Union University*
Frances Weaver, *Widener University*
Elizabeth Welnhofer, *Canisius College*
Marcia Wendeln, *Wright State University*
Shauna Weyrauch, *Ohio State University, Newark*
Wayne Whaley, *Utah Valley State College*
Howard Whiteman, *Murray State University*
Vernon Wiersema, *Houston Community College*
Gerald Wilcox, *Potomac State College*
Peter J. Wilkin, *Purdue University North Central*
Robert R. Wise, *University of Wisconsin, Oshkosh*
Michelle Withers, *Louisiana State University*
Art Woods, *University of Texas, Austin*
Elton Woodward, *Daytona Beach Community College*
Kenneth Wunch, *Sam Houston State University*
Donna Young, *University of Winnipeg*
Michelle L. Zjhra, *Georgia Southern University*
John Zook, *Ohio University*
Michelle Zurawski, *Moraine Valley Community College*

Student Focus Group Participants
California State University, Fullerton:

Danielle Bruening	Mahetzi Hernandez
Leslie Buena	Heidi McMorris
Andrés Carrillo	Daniel Minn
Victor Galvan	Sam Myers
Jessica Ginger	David Omut
Sarah Harpst	Jonathan Pistorino
Robin Keber	James W. Pura
Ryan Roberts	Samantha Ramirez
Melissa Romero	Tiffany Speed
Erin Seale	Tristan Terry
Nathan Tran	
Tracy Valentovich	
Sean Vogt	

Fullerton College:
Michael Baker

The Book Team

We are indebted to our editor Star MacKenzie. Star is a delight to work with—insightful, funny, kind, and generous with her time. Star is truly committed to producing an excellent book that meets the needs of students and instructors. We feel fortunate to have had her guidance throughout the project.

We also owe a great deal of thanks to our organized and thoughtful development editor, Susan Teahan. Susan is a creative and careful editor whose enthusiasm for the stories and interest in the underlying science added much to this revision.

This book is dedicated to our families, friends, and colleagues who have supported us over the years. Having loving families, great friends, and a supportive work environment enabled us to make this heartfelt contribution to nonmajors biology education.

COLLEEN BELK AND
VIRGINIA BORDEN MAIER

Brief Contents

Contents

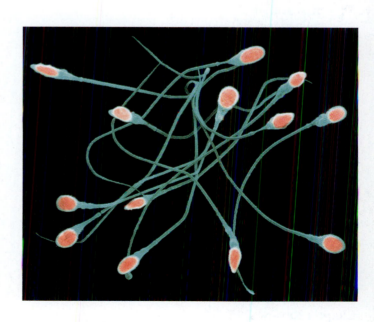

CHAPTER 24

Growing a Green Thumb 596
Plant Physiology

CHAPTER
1

Can Science Cure the Common Cold?

Introduction to the Scientific Method

Another cold! What
can I do?

We have all been there—you just recover from one bad head cold and on a morning soon after you notice that scratchy feeling in your throat that signals a new one is about to begin. It is always at the worst time, too, when you have an important exam coming up, a term paper due, and a packed social calendar. Why are you sick yet again? What can you do about it?

If you ask your friends and relatives, you will hear the usual advice on how to prevent and treat colds: Take massive doses of vitamin C. Suck on zinc lozenges. Drink plenty of echinacea tea. Meditate. Get more rest. Exercise vigorously every day. Put that hat on when you go outside! You are left with an overwhelming list of options, often contradictory and some counter to common sense. If you keep up with health news, you may be even more confused. One website reports that a popular over-the-counter cold treatment is effective, while a local TV news story details the risks of using this remedy and highlights its ineffectiveness. How do you decide what to do?

Faced with this bewildering situation, most people follow the advice that makes the most sense to them, and if they find they still feel terrible, they try another remedy. Testing ideas and discarding ones that don't work is a kind of "everyday science." However, this technique has its limitations—for example, even if you feel better after trying a new cold treatment, you can't know if your recovery occurred because the treatment was effective or because the cold was ending anyway.

What professional scientists do is a more refined version of this everyday science, using strategies that help eliminate other possible explanations for a result. And while some fields of science may use unintelligible words or complicated and expensive equipment, the basic process

Take massive doses of Vitamin C?

Drink echinacea tea?

How would a scientist determine which advice is best?

3

for testing ideas is simple and universal to all areas of science. An understanding of this process can help you evaluate information about many issues that may concern and intrigue you—from health issues, to global warming, to the origin of life and the universe—with more confidence. In this chapter, we introduce you to the powerful process scientists use by asking the question we've considered here: Is there a cure for the common cold?

1.1 The Process of Science

The term *science* can refer to a body of knowledge—for example, the science of **biology** is the study of living organisms. You may believe that science requires near-perfect recall of specific sets of facts about the world. In reality, this goal is impossible and unnecessary—we do have reference books, after all. The real action in science is not memorizing what is already known but using the process of science to discover something new and unknown.

This process—making observations of the world, proposing ideas about how something works, testing those ideas, and discarding (or modifying) our ideas in response to the test results—is the essence of the **scientific method.** The scientific method allows us to solve problems and answer questions efficiently and effectively. Can we use the scientific method to solve the complicated problem of preventing and treating colds?

The Nature of Hypotheses

The statements our friends and family make about which actions will help us remain healthy (for example, the advice to wear a hat) are in some part based on the advice giver's understanding of how our bodies resist colds. Ideas about "how things work" are called **hypotheses.** Or, more formally, a hypothesis is a proposed explanation for one or more **observations.**

Hypotheses in biology come from knowledge about how the body and other biological systems work, experiences in similar situations, our understanding of other scientific research, and logical reasoning; they are also shaped by our creative mind (**Figure 1.1**). When your mom tells you to dress warmly to avoid colds, she is basing her advice on the following hypothesis: Becoming chilled makes you more susceptible to illness.

The hallmark of science is that hypotheses are subject to rigorous testing. Therefore, scientific hypotheses must be **testable**—it must be possible to evaluate a hypothesis through observations of the measurable universe. Not all hypotheses are testable. For instance, the statement that "colds are generated by disturbances in psychic energy" is not a scientific hypothesis because psychic energy does not have a material nature and thus cannot be seen or measured in a test.

In addition, hypotheses that require the intervention of a supernatural force cannot be tested scientifically. If something is **supernatural,** it is not constrained by any laws of nature, and its behavior cannot be predicted using our current understanding of the natural world. A scientific hypothesis must also be **falsifiable;** that is, an observation or set of observations could potentially prove it false. The hypothesis that exposure to cold temperatures increases your susceptibility to colds is falsifiable; we can imagine an observation that would cause us to reject this hypothesis (for instance, the observation that people exposed to cold temperatures do not catch more colds than people

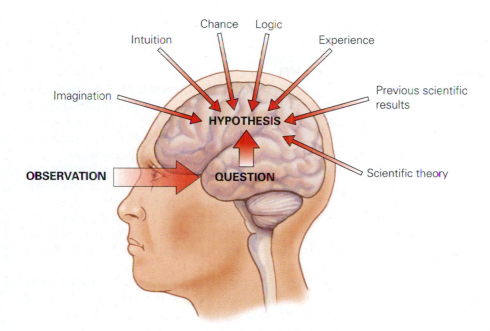

Intuition   Chance   Logic   Experience

Imagination

Previous scientific results

HYPOTHESIS

OBSERVATION   **QUESTION**

Scientific theory

Figure 1.1 Hypothesis generation.
All of us generate hypotheses. Many different factors, both logical and creative, influence the development of a hypothesis. Scientific hypotheses are both testable and falsifiable.

protected from chills). Of course, not all hypotheses are proved false, but it is essential in science that incorrect ideas be discarded, which occur only if it is *possible* to prove those ideas false. Lack of falsifiability is another reason supernatural hypotheses cannot be scientific. Because a supernatural force can cause any possible result of a test, hypotheses that rely on supernatural forces cannot be falsified.

Finally, statements that are value judgments, such as, "It is wrong to cheat on an exam," are not scientific because different people have different ideas about right and wrong. It is impossible to falsify these types of statements. To find answers to questions of morality, ethics, or justice, we turn to other methods of gaining understanding—such as philosophy and religion.

Scientific Theories

Most hypotheses fit into a larger picture of scientific understanding. We can see this relationship when examining how research upended a commonly held belief about diet and health—that chronic stomach and intestinal inflammation is caused by eating too much spicy food. This belief directed the standard medical practice for ulcer treatment for decades. Patients with ulcers were prescribed drugs that reduced stomach acid levels and advised to avoid eating acidic or highly spiced foods. These treatments were rarely successful, and ulcers were considered chronic, possibly lifelong, problems.

In 1982 Australian scientists Robin Warren and Barry Marshall discovered that a particular microscopic organism, specifically the bacterium *Helicobacter pylori*, was present in nearly all samples of ulcer tissue that they examined (**Figure 1.2**). From this observation, Warren and Marshall reasoned that *H. pylori* infection—invasion of the stomach wall by the bacteria—was the cause of most ulcers. Barry Marshall even tested this hypothesis on himself by consuming live *H. pylori*. He subsequently suffered from acute stomach pain.

Warren and Marshall's colleagues were at first unconvinced that ulcers could have such a simple cause. Today the hypothesis that *H. pylori* infection is responsible for most ulcers is accepted as fact. The primary reasons why this is the case? First, no reasonable alternative hypotheses about the causes of ulcers (for instance, consumption of spicy foods) has been consistently supported by hypothesis tests; and second, the hypothesis has not been rejected—that is,

(a)

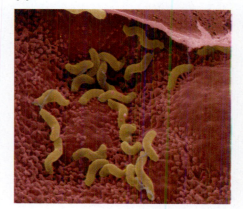

(b)

Figure 1.2 A scientific breakthrough.
(a) *Helicobacter pylori* on stomach lining (image from electron microscope). (b) Robin Warren and Barry Marshall won the 2005 Nobel Prize in Medicine for their discovery of the link between *H. pylori* and ulcers.

there have been no carefully designed experiments that show that *H. pylori* removal fails to cure most ulcers.

The third reason that the relationship between *H. pylori* and ulcers is considered fact is that it conforms to a well-accepted scientific principle, namely, the germ theory of disease. A **scientific theory** is an explanation for a set of related observations that is based on well-supported hypotheses from several different, independent lines of research. The basic premise of germ theory is that microorganisms (that is, organisms too small to be seen with the naked eye) are the cause, through infection, of some or all human diseases.

The biologist Louis Pasteur first observed that bacteria cause milk to become sour. From this observation, he reasoned that these same types of organisms could injure humans. Later, Robert Koch demonstrated a link between anthrax bacteria and a specific set of fatal symptoms in mice, providing additional evidence. Germ theory is further supported by the observation that antibiotic treatment that targets particular microorganisms can cure certain illnesses—as is the case with bacteria-caused ulcers.

In everyday speech, the word *theory* is synonymous with untested ideas based on little information. In contrast, scientists use the term when referring to well-supported ideas of how the natural world works. The supporting foundation of all scientific theories is multiple hypothesis tests.

The Logic of Hypothesis Tests

One common hypothesis about cold prevention is that taking vitamin C supplements keeps you healthy. This hypothesis is very appealing, especially given the following generally known facts:

1. Fruits and vegetables contain a lot of vitamin C.
2. People with diets rich in fruits and vegetables are generally healthier than people who skimp on these food items.
3. Vitamin C is known to be an anti-inflammatory agent, reducing throat and nose irritation.

With these facts in mind, we can state the following falsifiable hypothesis: Consuming vitamin C decreases the risk of catching a cold.

This hypothesis makes sense given the statements just listed and the experiences of the many people who insist that vitamin C keeps them healthy. The process used to construct this hypothesis is called **inductive reasoning**—combining a series of specific observations (here, statements 1–3) to discern a general principle. Inductive reasoning is an essential tool for understanding the world. However, a word of caution is in order: Just because the inductive reasoning that led to a hypothesis seems to make sense does not mean that the hypothesis is necessarily true. The example that follows demonstrates this point.

Consider the ancient hypothesis that the sun revolves around Earth. This hypothesis was induced based on the observations that the sun rose in the east every morning, traveled across the sky, and set in the west every night. For almost all of history, this hypothesis was considered to be a "fact" by nearly all of Western society. It wasn't until the early seventeenth century that this hypothesis was overturned—as the result of Galileo Galilei's observations of Venus. His observations proved false the hypothesis that the sun revolved around Earth. Galileo's work helped to confirm the more modern hypothesis, proposed by Nicolaus Copernicus, that Earth revolves around the sun.

So, even though the hypothesis about vitamin C is sensible, it needs to be tested to see if it can be proved false. Hypothesis testing is based on **deductive reasoning** or deduction. Deduction involves using a general principle to predict an expected observation. This **prediction** concerns the outcome of an action, test, or investigation. In other words, the prediction is the result we expect from a hypothesis test.

Deductive reasoning takes the form of "if/then" statements. That is, if our general principle is correct, then we expect to observe a specific outcome. A prediction based on the vitamin C hypothesis could be: *If* vitamin C decreases the risk of catching a cold, *then* people who take vitamin C supplements with their regular diets will experience fewer colds than will people who do not take supplements.

Stop & Stretch Consider the following scenario: Your coworker, Homer, is a notorious doughnut lover who has a nose for free food. You walk into the break room one morning to discover a box from the doughnut shop that is already completely empty. According to this information, what most likely happened to the doughnuts? Is this an inductive or deductive hypothesis?

Deductive reasoning, with its resulting predictions, is a powerful method for testing hypotheses. However, the structure of such a statement means that hypotheses can be clearly rejected if untrue but impossible to prove if true (**Figure 1.3**). This shortcoming is illustrated using the if/then statement concerning vitamin C and colds.

Consider the possible outcomes of a comparison between people who supplement with vitamin C and those who do not. People who take vitamin C supplements may suffer through more colds than people who do not; they may have the same number of colds as the people who do not supplement; or supplementers may in fact experience fewer colds. What does each of these results tell us about the hypothesis?

If, in a well-designed test, people who take vitamin C have more colds or the same number of colds as those who do not supplement, then the hypothesis that vitamin C provides protection against colds can be clearly rejected. But what if people who supplement with vitamin C do experience fewer colds? If this is the case, then we can only say that the hypothesis has been supported and not disproven.

Why is it impossible to say that the hypothesis that vitamin C prevents colds is true? Because there are **alternative hypotheses** that explain why people with different vitamin-taking habits vary in their cold susceptibility. In other words, demonstrating the truth of the *then* portion of a deductive statement does not prove that the *if* portion is true.

Stop & Stretch Consider this if/then statement: If your coworker Homer ate all the doughnuts, then there won't be any doughnuts left in the box that was placed in the break room 30 minutes ago. Imagine you find that the doughnuts are all gone—did you just prove that Homer ate them? Why or why not?

Consider the alternative hypothesis that frequent exercise reduces susceptibility to catching a cold. And suppose that people who take vitamin C supplements are more likely to engage in regular exercise. If both of these hypotheses

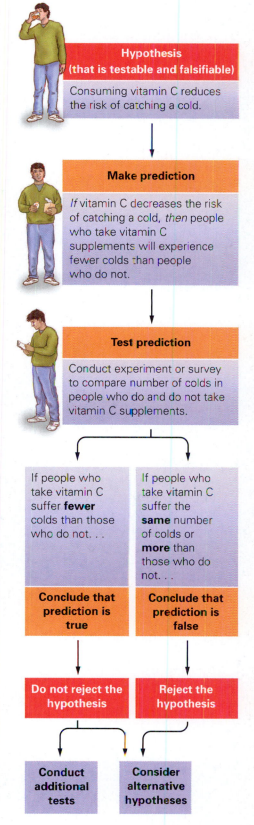

Figure 1.3 The scientific method. Tests of hypotheses follow a logical path. This flowchart illustrates the process of deduction as practiced by scientists.
Visualize This: According to this flowchart, scientists should consider alternative hypotheses even if their hypothesis is supported by their research. Explain why this is the case.

are true, then the prediction that vitamin C supplementers experience fewer colds than people who do not supplement would be true but not because the original hypothesis (vitamin C reduces the risk of colds) is true. Instead, people who take vitamin C supplements experience fewer colds because they are also more likely to exercise, and it is exercise that reduces cold susceptibility.

A hypothesis that seems to be true because it has not been rejected by an initial test may be rejected later because of a different test. This is what happened to the hypothesis that vitamin C consumption reduces susceptibility to colds. The argument for the power of vitamin C was popularized in 1970 by Nobel Prize–winning chemist Linus Pauling. Pauling based his assertion—that large doses of vitamin C reduce the incidence of colds by as much as 45%—on the results of a few studies that had been published between the 1930s and 1970s. However, repeated, careful tests of this hypothesis have since failed to support it. In many of the studies Pauling cited, it appears that alternative hypotheses explain the difference in cold incidence between vitamin C supplementers and nonsupplementers. Today, most health scientists agree that the hypothesis that vitamin C prevents colds has been convincingly falsified.

The example of the vitamin C hypothesis also highlights a challenge of communicating scientific information. You can see why the belief that vitamin C prevents colds is so widespread. If you don't know that scientific knowledge relies on rejecting incorrect ideas, a book by a Noble Prize–winning scientist may seem like the last word on the benefits of vitamin C. It took many years of careful research to show that this "last word" was, in fact, wrong.

1.2 Hypothesis Testing

The previous discussion may seem discouraging: How can scientists determine the truth of any hypothesis when there is always a chance that the hypothesis could be falsified? Even if one of the hypotheses about cold prevention is supported, does the difficulty of eliminating alternative hypotheses mean that we will never know which approach is truly best? The answer is yes—and no.

Hypotheses cannot be proven absolutely true; it is always possible that the true cause of a phenomenon may be found in a hypothesis that has not yet been tested. However, in a practical sense, a hypothesis can be proven beyond a reasonable doubt. That is, when one hypothesis has not been disproven through repeated testing and all reasonable alternative hypotheses have been eliminated, scientists accept that the well-supported hypothesis is, in a practical sense, true. "Truth" in science can therefore be defined as *what we know and understand based on all currently available information.* But scientists always leave open the possibility that what seems true now may someday be proven false.

An effective way to test many hypotheses is through rigorous scientific experiments. Experimentation has enabled scientists to prove beyond a reasonable doubt that the common cold is caused by a virus. A virus is a microscopic entity with a simple structure—it typically contains a short strand of genetic material and a few proteins encased in a relatively tough protein shell and sometimes surrounded by a membrane. A virus must infect a cell to reproduce. Of the over 200 types of viruses that are known to cause the common cold, most infect the cells in our noses and throats. The sneezing, coughing, congestion, and sore throat of a cold appear to result from the body's protective response—established by our immune system—to a viral invasion (**Figure 1.4**).

As you may know, if we survive a virus infection, we are unlikely to experience a recurrence of the disease the virus causes. For example, it is extremely rare to suffer from chicken pox twice because one exposure to the

(a) Cold-causing virus

(b) How the virus causes a cold

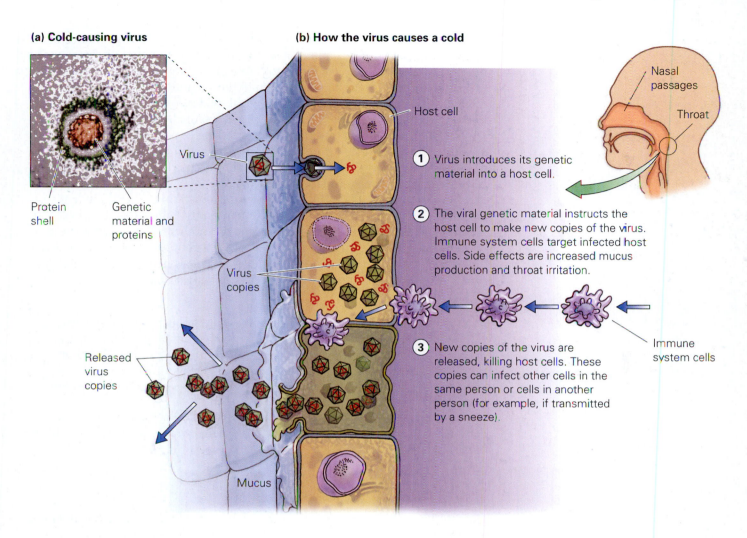

Protein shell

Genetic material and proteins

Virus

Host cell

Nasal passages

Throat

1 Virus introduces its genetic material into a host cell.

2 The viral genetic material instructs the host cell to make new copies of the virus. Immune system cells target infected host cells. Side effects are increased mucus production and throat irritation.

Virus copies

Released virus copies

3 New copies of the virus are released, killing host cells. These copies can infect other cells in the same person or cells in another person (for example, if transmitted by a sneeze).

Immune system cells

Mucus

chicken pox virus (through either infection or vaccination) usually provides lifelong immunity to future infection. The sheer number of cold viruses makes immunity to the common cold—and the development of a vaccine to prevent it—very improbable. Scientists thus focus their experimental research about common colds on methods of prevention and treatment.

The Experimental Method

Experiments are sets of actions or observations designed to test specific hypotheses. Generally, an experiment allows a scientist to control the conditions that may affect the subject of study. Manipulating the environment allows a scientist to eliminate some alternative hypotheses that may explain the result.

Experimentation in science is analogous to what a mechanic does when diagnosing a car problem. There are many reasons why a car engine might not start. If a mechanic begins by tinkering with numerous parts to apply all possible fixes before restarting the car, the mechanic will not know what exactly caused the problem (and will have an unhappy customer who is charged for unnecessary parts and labor). Instead, a mechanic begins by testing the battery for power; if the battery is charged, then she checks the starter motor; if the car still doesn't start, she looks over the fuel pump; and she continues in this manner until identifying the problem. Likewise, a scientist systematically attempts to eliminate hypotheses that do not explain a particular phenomenon.

Figure 1.4 A cold-causing virus. (a) An electron microscopic image of a typical rhinovirus, one of the many types of viruses that cause the common cold. (b) A rhinovirus causes illness by invading nose and throat cells and using them as "factories" to make virus copies. Cold symptoms result from immune system attempts to eliminate the virus.

Visualize This: Find two points in this process where intervention by drugs or other treatment could disrupt it and lead to reduced cold symptoms.

Not all scientific hypotheses can be tested through experimentation. For instance, hypotheses about how life on Earth originated or the cause of dinosaur extinction are usually not testable in this way. These hypotheses are instead tested using careful observation of the natural world. For instance, the examination of fossils and other geological evidence allows scientists to test hypotheses regarding the extinction of the dinosaurs (**Figure 1.5**).

The information collected by scientists during hypothesis testing is known as **data.** The data are collected on the **variables** of the test, that is, any factor that can change in value under different conditions. In an experimental test, scientists manipulate an **independent variable** (one whose value can be freely changed) to measure the effect on a **dependent variable.** The dependent variable may or may not be influenced by changes in the independent variable, but it cannot be systematically changed by the researchers. For example, to measure the effect of vitamin C on cold prevention, scientists can vary individuals' vitamin C intake (the independent variable) and measure their susceptibility to illness on exposure to a cold virus (the dependent variable).

Stop & Stretch Some of the experiments that established the link between *H. pylori* and stomach ulcer consisted of feeding gerbils *H. pylori* and measuring the rate of ulcer development compared to untreated animals. What are the independent and dependent variables in these experiments?

Data obtained from well-designed experiments should allow researchers to convincingly reject or support a hypothesis. This is more likely to occur if the experiment is controlled.

Controlled Experiments

Control has a specific meaning in science. A **control** for an experiment is a subject similar to an experimental subject except that the control is not exposed to experimental treatment. Controlled experiments are thus designed to eliminate as many alternative hypotheses as possible.

Figure 1.5 Testing hypotheses through observation. The fossil record provides a source of data to test hypotheses about evolutionary history.

Once subjects are enlisted in an experiment, they are assigned to a control or an experimental group. If members of the control and experimental groups differ at the end of a well-designed test, then the difference is likely due to the experimental treatment.

Our question about effective cold treatments lends itself to a variety of controlled experiments on possible drug therapies. For example, an extract of *Echinacea purpurea* (a common North American prairie plant) in the form of echinacea tea has been promoted as a treatment to reduce the likelihood as well as the severity and duration of colds (**Figure 1.6**). A recent scientific experiment on the efficacy of *Echinacea* involved asking individuals suffering from colds to rate the effectiveness of a tea in relieving their symptoms. In this study, people who used echinacea tea felt that it was 33% more effective. The "33% more effective" is in comparison to the rated effectiveness of a tea that did not contain *Echinacea* extract—that is, the results from the control group.

Control groups and experimental groups must be as similar as possible to each other to eliminate alternative hypotheses that could explain the results. In the case of cold treatments, the groups should not systematically differ in age, diet, stress level, or other factors that might affect cold susceptibility. One effective way to minimize differences between groups is the **random assignment** of individuals. For example, a researcher might put all of the volunteers' names in a hat, draw out half, and designate the people drawn as the experimental group and the remaining people as the control group. As a result, each group should be a rough cross-section of the population in the study. In the echinacea tea experiment just described, members of both the experimental and control groups were female employees of a nursing home who sought relief from their colds at their employer's clinic. The volunteers were randomly assigned into either the experimental or control group as they came into the clinic.

Figure 1.6 *Echinacea purpurea,* **an American coneflower.** Extracts from the leaves and roots of this plant are among the most popular herbal remedies sold in the United States.

Stop & Stretch In the echinacea tea experiment, a nonrandom assignment scheme might have put the first 25 visitors to the clinic in the control group and the next 25 in the experimental group. Imagine that in this version of the experiment, the experimental group did recover in fewer days than the control group. Describe an alternative hypothesis related to the nonrandom assignment of subjects that was not eliminated by this experimental design and that could explain these results.

The second step in designing a good controlled experiment is to treat all subjects identically during the course of the experiment. In this study, all participants, whether in the control or experimental group, received the same information about the supposed benefits of echinacea tea, and during the course of the experiment, all participants were given tea to drink five to six times daily until their symptoms subsided. However, individuals in the control group received "sham tea" that did not contain *Echinacea* extract. Treating all participants the same ensures that no factor related to the interaction between subject and researcher influences the results.

The sham tea in this experiment would be equivalent to the sugar pills that are given to control subjects during drug trials. Like other intentionally ineffective medical treatments, sham tea is a **placebo.** Employing a placebo generates only one consistent difference between individuals in the two groups—in this case, the type of tea they consumed.

In the echinacea tea study, the data indicated that cold severity was lower in the experimental group compared to those who received placebo. Because their study utilized controls, the researchers can be confident that the groups

(a)

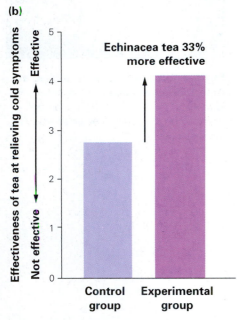

Control group	Experimental group
Experiencing early cold symptoms	Experiencing early cold symptoms
Sought treatment from clinic	Sought treatment from clinic
Received **placebo** tea	Received **echinacea** tea

(b)

Echinacea tea 33% more effective

Effectiveness of tea at relieving cold symptoms

Control group   Experimental group

Figure 1.7 A controlled experiment.
(a) In a controlled experiment testing echinacea tea as a treatment for colds, all 95 subjects were treated identically except for the type of tea they were given. (b) The results of the experiment indicated that echinacea tea was 33% more effective than the placebo.

differed because of the effect of *Echinacea*. By reducing the likelihood that alternative hypotheses could explain their results, the researchers could strongly infer that they were measuring a real, positive effect of echinacea tea on colds (**Figure 1.7**).

The study described here supports the hypothesis that echinacea tea reduces the severity of colds. However, it is extremely rare that a single experiment will cause the scientific community to accept a hypothesis beyond a reasonable doubt. Dozens of studies, each using different experimental designs and many using extracts from different parts of the plant, have investigated the effect of *Echinacea* on common colds and other illnesses. Some of these studies have shown a positive effect, but many others have shown none. In the medical community as a whole, there is significant skepticism and disagreement regarding the effectiveness of this popular herb as a cold treatment. Only through continued controlled tests of the hypothesis will we discern an accurate answer to the question, Is *Echinacea* an effective cold treatment?

Minimizing Bias in Experimental Design

Scientists and human research subjects may have strong opinions about the truth of a particular hypothesis even before it is tested. These opinions may cause participants to unfairly influence, or **bias,** the results of an experiment.

One potential source of bias is subject expectation. Individual experimental subjects may consciously or unconsciously model the behavior they feel the researcher expects from them. For example, an individual who knew she was receiving echinacea tea may have felt confident that she would recover more quickly. This might cause her to underreport her cold symptoms. This potential problem is avoided by designing a **blind experiment,** in which individual subjects are not aware of exactly what they are predicted to experience. In experiments on drug treatments, this means not telling participants whether they are receiving the drug or a placebo.

Another source of bias arises when a researcher makes consistent errors in the measurement and evaluation of results. This phenomenon is called observer bias. In the echinacea tea experiment, observer bias could take various forms. Expecting a particular outcome might lead a scientist to give slightly different instructions about which symptoms constituted a cold to subjects who received echinacea tea. Or, if the researcher expected people who drank echinacea tea to experience fewer colds, she might make small errors in the measurement of cold severity that influenced the final result.

To avoid the problem of experimenter bias, the data collectors themselves should be "blind." Ideally, the scientist, doctor, or technician applying the treatment does not know which group (experimental or control) any given subject is part of until after all data have been collected and analyzed (**Figure 1.8**). Blinding the data collector ensures that the data are **objective** or, in other words, without bias.

We call experiments **double-blind** when both the research subjects and the technicians performing the measurements are unaware of either the hypothesis or whether a subject is in the control or experimental group. Double-blind experiments nearly eliminate the effects of human bias on results. When both researcher and subject have few expectations about the outcome, the results obtained from an experiment are more credible.

Using Correlation to Test Hypotheses

Double-blind, placebo-controlled, randomized experiments represent the gold standard for medical research. However, well-controlled experiments can be difficult to perform when humans are the subjects. The requirement that both experimental and control groups be treated nearly identically means that some

Technician "blind"

Subject "blind"

What they know

- **Limited knowledge** of experimental hypothesis

- **No knowledge** of which group participants belong to

How they behave

- **No difference** in instructions to participants

- **No difference** in treatment of participants

- **No difference** in data collection

- **Limited knowledge** of experimental hypothesis

- **No knowledge** of which group he or she belongs to

- **Unbiased** reporting of symptoms or effects of treatment

Figure 1.8 Double-blind experiments. Double-blind experiments result in data that are more objective.

people unknowingly receive no treatment. In the case of healthy volunteers with head colds, the placebo treatment of sham tea did not hurt those who received it.

However, placebo treatments are impractical or unethical in many cases. For instance, imagine testing the effectiveness of a birth control drug using a controlled experiment. This would require asking women to take a pill that may or may not prevent pregnancy while not using any other form of birth control!

Experiments on Model Systems. Scientists can use **model systems** when testing hypotheses that would raise ethical or practical problems when tested on people. For basic research that helps us understand how cells and genes function, the model systems are easily grown and manipulated organisms, such as certain species of bacteria, nematodes, and fruit flies, or even isolated cells from larger organisms that reproduce in dishes in the laboratory. In the case of research on human health and disease, model systems are typically other mammals, including human cells. Mammals are especially useful as model organisms in medical research because they are closely related to us. Like us, they have hair and give birth to live young, and thus they also share with us similarities in anatomy and physiology (**Figure 1.9**).

The vast majority of animals used in biomedical research are rodents such as rats, mice, and guinea pigs, although some areas of research require animals that are more similar to humans in size, such as dogs or pigs, or share a closer evolutionary relationship, such as chimpanzees.

The use of model systems, especially animals, allows experimental testing on potential drugs and other therapies before these methods are employed on people. Research on model organisms such as lab rats has contributed to a better understanding of nearly every serious human health threat, including cancer, heart disease, Alzheimer's disease, and AIDS (acquired immunodeficiency syndrome). However, ethical concerns about the use of animals in research persist and can complicate such studies. In addition, the results of animal studies are not always directly applicable to humans—despite a shared evolutionary history, animals still can have important differences from humans. Testing hypotheses about human health in human beings still provides the clearest answer to these questions.

(a)

(b)

(c)

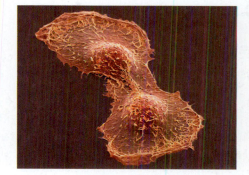

Figure 1.9 Model systems in science. (a) Research on the nematode *Caenorhabditis elegans* (also known as *C. elegans*) has led to major advances in our understanding of animal development and genetics. (b) The classic "lab rat" is easy to raise and care for and as a mammal is an important model system for testing drugs and treatments designed for humans. (c) HeLa cells are derived from a cancerous growth biopsied from Henrietta Lacks in 1951. These cells have the unique property of being able to divide indefinitely, making them valuable for research on a wide variety of subjects from HIV to human growth hormone.

Looking for Relationships Between Factors. Scientists can also test hypotheses using correlations when controlled experiments on humans are difficult or impossible to perform. A **correlation** is a relationship between two variables.

Suggestions about using meditation to reduce susceptibility to colds are based on a correlation between psychological stress and susceptibility to cold virus infections (**Figure 1.10**). This correlation was generated by researchers who collected data on subjects' psychological stress levels before giving them nasal drops that contained a cold virus. Doctors later reported on the incidence and severity of colds among participants in the study. Note that while the cold virus was applied to each participant in the study, the researchers had no influence on the stress level of the study participants—in other words, this was not a controlled experiment because people were not randomly assigned to different "treatments" of low or high stress.

Stop & Stretch Even though the researchers did not manipulate the independent variable in this study, there are still an independent variable and a dependent variable. What are they?

Let's examine the results presented in Figure 1.10. The horizontal axis of the graph, or **x-axis**, illustrates the independent variable. This variable is stress, and the graph ranks subjects along a scale of stress level—from low stress on the left edge of the scale to high stress on the right. The vertical axis of the graph, the **y-axis**, is the dependent variable—the percentage of study participants who developed colds as reported by their doctors. Each point on the graph represents a group of individuals and tells us what percentage of people in each stress category had clinical colds.

The line connecting the 5 points on the graph illustrates a correlation—the relationship between stress level and susceptibility to cold virus infections. Because the line rises to the right, it illustrates a positive correlation. These data tell us that people who had higher stress levels were more likely to come down with colds. But does this relationship mean that high stress causes increased cold susceptibility?

To conclude that stress causes illness, we need the same assurances that are given by a controlled experiment. In other words, we must assume that the

Figure 1.10 Correlation between stress level and illness. The graph indicates that people reporting higher levels of stress became infected after exposure to a cold virus more often than did people who reported low levels of stress.
Visualize This: This graph groups people with similar, but not identical, stress index measures. Why might this have been necessary? If people with stress indices 3 and 4 have the same susceptibility to colds, does this call into question the correlation?

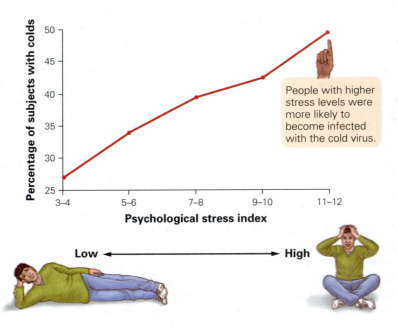

People with higher stress levels were more likely to become infected with the cold virus.

(a) Does high stress cause high cold frequency?

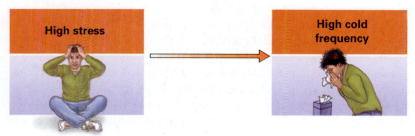

(b) Or does one of the causes of high stress also cause high cold frequency?

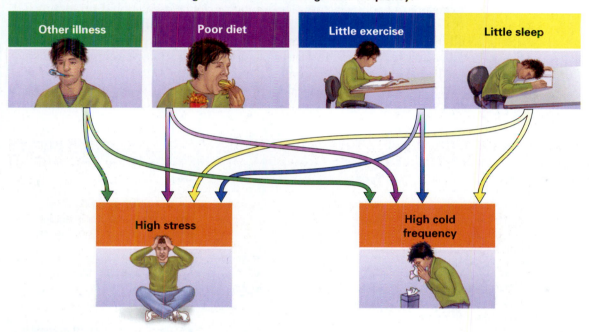

Figure 1.11 Correlation does not signify causation. A correlation typically cannot eliminate all alternative hypotheses.

individuals measured for the correlation are similar in every way except for their stress levels. Is this a good assumption? Not necessarily. Most correlations cannot control for all alternative hypotheses. People who feel more stressed may have poorer diets because they feel time-limited and rely on fast food more often. In addition, people who feel highly stressed may be in situations where they are exposed to more cold viruses. These differences among people who differ in stress level may also influence their cold susceptibility (**Figure 1.11**). Therefore, even with a strong correlational relationship between the two factors, we cannot strongly infer that stress causes decreased resistance to colds.

Researchers who use correlational studies can eliminate a number of alternative hypotheses by closely examining their subjects. For example, this study on stress and cold susceptibility collected data from subjects on age, weight, sex, education, and their exposure to infected individuals. None of these factors differed consistently among low-stress and high-stress groups. While this analysis increases our confidence in the hypothesis that high stress levels increase susceptibility to colds, people with high-stress lifestyles still may have important differences from those with low-stress lifestyles. It is possible that one of those differences is the real cause of disparities in cold frequency.

As you can see, it is difficult to demonstrate a cause-and-effect relationship between two factors simply by showing a correlation between them. In other words, correlation does not equal causation. For example, a commonly understood correlation exists between exposure to cold air and epidemics of the common cold. It is true that as outdoor temperatures drop, the incidence of colds increases. But numerous controlled experiments have indicated that chilling an individual does not increase his susceptibility to colds. Instead, cold outdoor temperatures mean increased close contact with other people (and their viruses), and this contact increases the number of colds we get during cold weather. Despite the correlation, cold air does not cause colds—exposure to viruses does.

Correlational studies are the main tool of epidemiology, the study of the distribution and causes of diseases. One commonly used epidemiological technique is a cross-sectional survey. In this type of survey, many individuals are both tested for the presence of a particular condition and asked about their exposure to various factors. The limitations of cross-sectional surveys include the effect of subject bias and poor recall by survey participants, in addition to all of the problems associated with interpreting correlations. **Table 1.1** provides an overview of the variety of correlational strategies employed to study the links between our environment and our health.

TABLE 1.1

Types of correlational studies.

Name	Description	Pros	Cons
Ecological studies	Examine specific human populations for unusually high levels of various diseases (e.g., documenting a "cancer cluster" around an industrial plant)	Inexpensive and relatively easy to do	Unsure whether exposure to environmental factor is actually related to onset of the disease
Cross-sectional surveys	Question individuals in a population to determine amount of exposure to an environmental factor and whether disease is present	More specific than ecological study	• Expensive • Subjects may not know exposure levels • Cannot control for other factors that may be different among individuals in survey • Cannot be used for rare diseases
Case-control studies	Compare exposures to specific environmental factors between individuals who have a disease and individuals matched in age and other factors who do not have the disease	• Relatively fast and inexpensive • Best method for rare diseases	• Does not measure absolute risk of disease as a result of exposure • Difficult to select appropriate controls to eliminate alternative hypotheses • Examines just one disease possibly associated with an environmental factor
Cohort studies	Follow a group of individuals, measuring exposure to environmental factors and disease prevalence	• Can determine risk of various diseases associated with exposure to particular environmental factor	• Expensive and time-consuming • Difficult to control for alternative hypotheses • Not feasible for rare diseases
Correlational experiment	Expose individuals who experience varying levels of an independent variable to an experimental treatment	• Can control the amount of exposure to at least one environmental factor of interest	• Cannot eliminate alternative hypotheses • Only feasible for hypotheses for which an experimental treatment can be applied

1.3 Understanding Statistics

During a review of scientific literature on cold prevention and treatment, you may come across statements about the "significance" of the effects of different cold-reducing measures. For instance, one report may state that factor A reduced cold severity, but that the results of the study were "not significant." Another study may state that factor B caused a "significant reduction" in illness. We might then assume that this statement means factor B will help us feel better, whereas factor A will have little effect. This is an incorrect assumption because in scientific studies, *significance* is defined a bit differently from its usual sense. To evaluate the scientific use of this term, we need a basic understanding of statistics.

An Overview: What Statistical Tests Can Tell Us

We often use the term *statistics* to refer to a summary of accumulated information. For instance, a baseball player's success at hitting is summarized by a statistic: his batting average, the total number of hits he made divided by "at bats," the number of opportunities he had to hit. The science of **statistics** is a bit different; it is a specialized branch of mathematics used to evaluate and compare data.

An experimental test utilizes a small subgroup, or **sample,** of a population. Statistical methods can summarize data from the sample—for instance, we can describe the average, also known as the **mean,** length of colds experienced by experimental and control groups. **Statistical tests** can then be used to extend the results from a sample to the entire population.

When scientists conduct an experiment, they hypothesize that there is a true, underlying effect of their experimental treatment on the entire population. An experiment on a sample of a population can only estimate this true effect because a sample is always an imperfect "snapshot" of an entire population.

Consider a hypothesis that the average hair length of women in a college class in 1963 was shorter than the average hair length at the same college today. To test this hypothesis, we could compare a sample of snapshots from college yearbooks. If hairstyles were very similar among the women in a snapshot, you could reasonably assume that the average hair length in the college class is close to the average length in the snapshot. However, what if you see that women in a snapshot have a variety of hairstyles, from short bobs to long braids? In this case, it is difficult to determine whether the average hair length in the snapshot is at all close to the average for the class. With so much variation, the snapshot could, by chance, contain a surprisingly high number of women with very long hair, causing the average length in the sample to be much longer than the average length for the entire class (**Figure 1.12**).

A statistical test calculates the likelihood, given the number of individuals sampled and the variation within samples, that the difference between two samples reflects a real, underlying difference between the populations from which these samples were drawn. A **statistically significant** result is one that is very unlikely to be due to chance differences between the experimental and control samples, so thus likely represents a true difference between the groups.

In the experiment with the echinacea tea, statistical tests indicated that the 33% reduction in cold severity observed by the researchers was statistically significant. In other words, there is a low probability that the difference in cold susceptibility between the two samples in the experiment is due to chance. If the experiment is properly designed, a statistically significant result allows researchers to infer that the treatment had an effect. *See the next section for A Closer Look at statistics.*

(a) Average hair length in this snapshot is shorter...

Class of 1963

- Little variability
- High probability of reflecting average of all women in the class

(b) ...than average hair length in this snapshot...

Class of 2013

- High variability
- Low probability of reflecting average of all women in the class

...so, is hair today longer than in 1963?

Figure 1.12 The role of statistics. Statistical tests calculate the variability within groups to determine the probability that two groups differ only by chance.

A Closer Look:
Statistics

We can explore the role that statistical tests play more closely by evaluating another study on cold treatments. This study examined the efficacy of lozenges containing zinc on reducing cold severity.

Some forms of zinc can block common cold viruses from invading cells in the nasal cavity. This observation led to the hypothesis that consuming zinc at the start of a cold could decrease the severity of cold symptoms by reducing the number of cells that become infected. Researchers at the Cleveland Clinic tested this hypothesis using a sample of 100 of their employees who volunteered for a study within 24 hours of developing cold symptoms. The researchers randomly assigned subjects to control or experimental groups. Members of the experimental group received lozenges containing zinc, while members of the control group received placebo lozenges. Members of both groups received the same instructions for using the lozenges and were asked to rate their symptoms until they had recovered. The experiment was double-blind.

When the data from the zinc lozenge experiment were summarized, the statistics indicated that the mean recovery time was more than 3 days shorter in the zinc group than in the placebo group (**Figure 1.13**). On the surface, this result appears to support the hypothesis. However, recall the example of the snapshot of women's hair length. A statistical test is necessary because of the effect of chance.

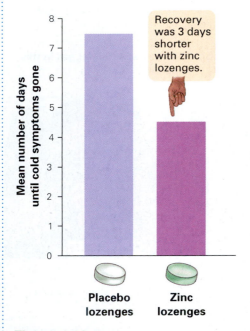

Figure 1.13 Zinc lozenges reduce the duration of colds. Fifty individuals taking zinc lozenges had colds lasting about 4½ days as opposed to approximately 7½ days for 50 individuals taking placebo.

The Problem of Sampling Error

The effect of chance on experimental results is known as **sampling error**—more specifically, sampling error is the difference between a sample and the population from which it was drawn. Similarly, in any experiment, individuals in the experimental group will differ from individuals in the control group in random ways. Even if there is *no* true effect of an experimental treatment, the data from the experimental group will never be identical to the data from the control group.

For example, we know that people differ in their ability to recover from a cold infection. If we give zinc lozenges to 1 volunteer and placebo lozenges to another, it is likely that the 2 volunteers will have colds of different lengths. But even if the zinc-taker had a shorter cold than the placebo-taker, you would probably say that the test did not tell us much about our hypothesis—the zinc-taker might just have had a less severe cold for other reasons.

Now imagine that we had 5 volunteers in each group and saw a difference, or that the difference was only 1 day instead of 3 days. How would we determine if the lozenges had an effect? Statistical tests allow researchers to look at their data and determine how likely it is—that is, the **probability**—that the result is due to sampling error. In this case, the statistical test distinguished between two possibilities for why the experimental group had shorter colds: Either the difference was due to the effectiveness of zinc as a cold treatment or it was due to a chance difference from the control group. The results indicated that there was a low probability, less than 1 in 10,000 (0.01%), that the experimental and control groups were so different simply by chance. In other words, the result is statistically significant.

A statistical measure of the amount of variability in a sample is often expressed as the **standard error.** The standard error is used to generate the **confidence interval**—the range of values that has a high probability (usually 95%) of containing the true population mean (**Figure 1.14**). Put simply, although the average of a sample is unlikely to be exactly the average of the population, the average plus the standard error represents the highest likely value for the population average, and the average minus the standard error represents the lowest likely average value. The confidence interval provides a way to express how much sampling error is influencing the results. A smaller confidence interval indicates that sampling error is likely to be small, and results with small confidence intervals are more likely to be statistically significant if the hypothesis is true.

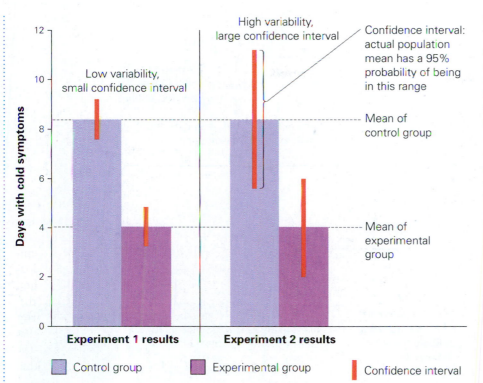

Figure 1.14 Confidence interval. The red lines on these bar graphs represent the confidence interval for each sample mean. A more variable sample has a larger confidence interval than a less variable sample. Even though the means are the same for both experimental and control groups in these two sets of bars, only the data summarized at left illustrate a significant difference because of the greater variability in the samples illustrated on the right side.

Stop & Stretch Opinion polling before a recent election indicated that candidate A was favored by 47% of a sample of likely voters, and candidate B was favored by 51% of the sample. The standard error was 3%. Why did reporters refer to this poll as a "statistical tie"?

Factors That Influence Statistical Significance

One characteristic of experiments that influences sampling error is **sample size**—the number of individuals in the experimental and control groups. If a treatment has no effect, a small sample size could return results that appear significantly different because of an unusually large and consistent sampling error. This was the case with the vitamin C hypothesis described at the beginning of the chapter. Subsequent tests with larger sample sizes, encompassing a wider variety of individuals with different underlying susceptibilities to colds, allowed scientists to reject the hypothesis that vitamin C prevents colds.

Conversely, if the effect of a treatment is real but the sample size of the experiment is small, a single experiment may not allow researchers to determine convincingly that their hypothesis has support. Several of the experiments that tested the efficacy of *Echinacea* extract demonstrate this phenomenon. For example, one experiment performed at a Wisconsin clinic with 48 participants indicated that echinacea tea drinkers were 30% less likely to experience any cold symptoms after virus exposure as compared to individuals who received a placebo tea. However,

the small sample size of the study meant that this result was not statistically significant.

The more participants there are in a study, the more likely it is that researchers will see a true effect of an experimental treatment even if it is very small. For example, a study of over 21,000 men over 6 years in Finland demonstrated that men who took vitamin E supplements had 5% fewer colds than the men who did not take these supplements. In this case, the large sample size allowed researchers to see that vitamin E has a real, but relatively tiny, effect on cold incidence. In other words, this statistically significant result has little real-world *practical* significance; a 5% effect probably won't convince people to begin supplementing with vitamin E to prevent colds. The relationship among hypotheses, experimental tests, sample size, and statistical and practical significance is summarized in **Figure 1.15** (on the next page).

There is one final caveat to this discussion. A statistically significant result is typically defined as one that has a *probability* of 5% or less of being due to chance alone. But probability is not certainty. If all scientific research uses this same standard, as many as 1 in every 20 statistically significant results (that is, 5% of the total) is actually reporting an effect that is *not real*. In other words, some statistically significant results are "false positives," representing a surprisingly large difference between experimental and control groups that occurred only as a result of sampling error. This potential error explains why one supportive experiment is not enough to convince all scientists that a particular hypothesis is accurate.

(continued on the next page)

(A Closer Look continued)

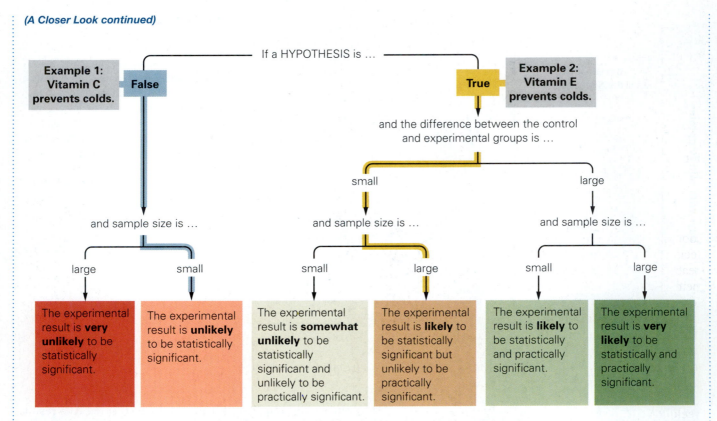

Figure 1.15 Factors that influence statistical significance. This flowchart summarizes the relationship between the true effect of a treatment and the sample size of an experiment on the likelihood of obtaining statistical significance.

Visualize This: The Nurse's Health Survey, which has followed nearly 80,000 women for over 20 years, has found that a diet low in refined carbohydrates such as white flour and sugar but relatively high in vegetable fat and protein cuts heart disease risk by 30% relative to women with a high refined carbohydrate diet. How does this result map out on the flowchart here?

Even with a statistical test indicating that the result had a probability of less than 0.01% of occurring by chance, we should begin to feel assured that taking zinc lozenges will reduce the duration of colds only after additional hypothesis tests give similar results. In fact, scientists continue to test this hypothesis, and there is still no consensus about the effectiveness of zinc as a cold treatment.

What Statistical Tests Cannot Tell Us

All statistical tests operate with the assumption that the experiment was designed and carried out correctly. In other words, a statistical test evaluates the chance of sampling error, not observer error, and a statistically significant result is not the last word on an experimentally tested hypothesis. An examination of the experiment itself is required.

In the test of the effectiveness of zinc lozenges, the experimental design—double-blind, randomized, and controlled—minimized the likelihood that alternative hypotheses could explain the results. Given such a well-designed experiment, this statistically significant result allows researchers to infer strongly that consuming zinc lozenges reduces the duration of colds.

As we saw with the experiment on vitamin E intake and cold prevention, statistical significance is also not equivalent to *significance* as we usually define the term, that is, as "meaningful or important." Unfortunately, experimental results reported in the news often use the term *significant* without clarifying this important distinction. Understanding that problem, as well as other misleading aspects of how science can be presented, will enable you to better use scientific information.

1.4 Evaluating Scientific Information

The previous sections should help you see why definitive scientific answers to our questions are slow in coming. However, a well-designed experiment can certainly allow us to approach the truth.

Primary Sources

Looking critically at reports of experiments can help us make well-informed decisions about actions to take. Most of the research on cold prevention and treatment is first published as **primary sources,** written by the researchers themselves and reviewed within the scientific community (**Figure 1.16** on the next page). The process of **peer review,** in which other scientists critique the results and conclusions of an experiment before it is published, helps increase confidence in scientific information. Peer-reviewed research articles in journals such as *Science, Nature,* the *Journal of the American Medical Association,* and hundreds of others represent the first and most reliable sources of current scientific knowledge.

However, evaluating the hundreds of scientific papers that are published weekly is a task no one of us can perform. Even if we focused only on a particular field of interest, the technical jargon used in many scientific papers may be a significant barrier to our understanding.

Instead of reading the primary literature, most of us receive our scientific information from **secondary sources** such as books, news reports, and advertisements. How can we evaluate information in this context? The following sections provide strategies for doing so, and a recurring Savvy Reader feature in this textbook will help you practice these evaluation skills.

Information from Anecdotes

Information about dietary supplements such as echinacea tea and zinc lozenges is often in the form of **anecdotal evidence**—meaning that the advice is based on one individual's personal experience. A friend's enthusiastic plug for vitamin C, because she felt it helped her, is an example of a testimonial—a common form of anecdote. Advertisements that use a celebrity to pitch a product "because it worked for them" are classic forms of testimonials.

You should be cautious about basing decisions on anecdotal evidence, which is not at all equivalent to well-designed scientific research. For example, many of us have heard anecdotes along the lines of the grandpa who was a pack-a-day smoker and lived to the age of 94. However, hundreds of studies have demonstrated the clear link between cigarette smoking and premature death. Although anecdotes may indicate that a product or treatment has merit, only well-designed tests of the hypothesis can determine its safety and efficacy.

Stop & Stretch One of the challenges of presenting science is that anecdotes make a much bigger impression on people than statistical summaries of data. Why do you think this is the case?

Figure 1.16 Primary sources: publishing scientific results. Most scientific journals require papers to go through stringent review before publication. **Visualize This:** How does having other anonymous experts review a draft of a scientist's paper make the final paper more reliable?

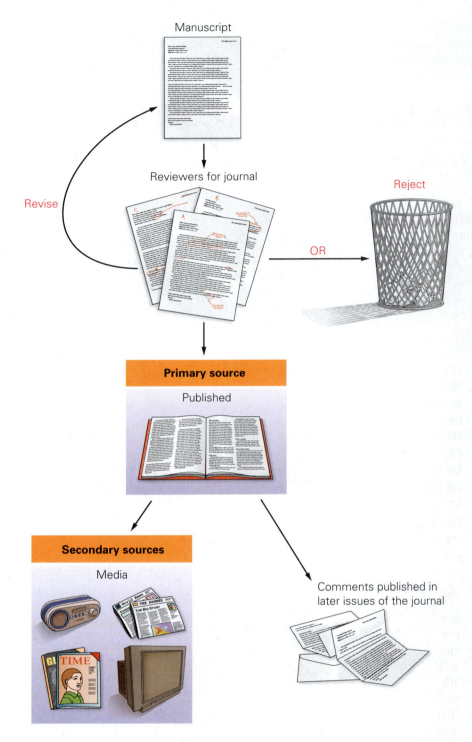

Science in the News

Popular news sources provide a steady stream of science information. However, stories about research results often do not contain information about the adequacy of experimental design or controls, the number of subjects, or the source of the scientist's funding. How can anyone evaluate the quality of research that supports dramatic headlines such as those describing risks to human health and the environment?

First, you must consider the source of media reports. Certainly news organizations will be more reliable reporters of fact than will entertainment tabloids, and news organizations with science writers should be better reporters of the substance of a study than those without. Television talk shows, which need to fill airtime, regularly have guests who promote a particular health claim. Too often these guests may be presenting information that is based on anecdotes or an incomplete summary of the primary literature.

Paid advertisements are a legitimate means of disseminating information. However, claims in advertising should be carefully evaluated. Advertisements of over-the-counter and prescription drugs must conform to rigorous government standards regarding the truth of their claims. However, lower standards apply to advertisements for herbal supplements, many health food products, and diet plans. Be sure to examine the fine print because advertisers often are required to clarify the statements made in their ads.

Another commonly used source for health information is the Internet. As you know, anyone can post information on this great resource. Typing in "common cold prevention" on a standard web search engine will return thousands of web pages—from highly respected academic and government sources to small companies trying to sell their products to individuals who have strong, sometimes completely unsupported, ideas about cures. Often it can be difficult to determine the reliability of a well-designed website, and even well-used sites such as Wikipedia may contain erroneous or misleading information. Here are some things to consider when using the web as a resource for health information:

1. Choose sites maintained by reputable medical establishments, such as the National Institutes of Health (NIH) or the Mayo Clinic.
2. It costs money to maintain a website. Consider whether the site seems to be promoting a product or agenda. Advertisements for a specific product should alert you to possible bias.
3. Check the date when the website was last updated and see whether the page has been updated since its original posting. Science and medicine are disciplines that must frequently evaluate new data. A reliable website will be updated often.
4. Determine whether unsubstantiated claims are being made. Look for references and be suspicious of any studies that are not from peer-reviewed journals.

Understanding Science from Secondary Sources

Once you are satisfied that a media source is relatively reliable, you can examine the scientific claim that it presents. Use your understanding of the process of science and of experimental design to evaluate the story and the science. Does the story about the claim present the results of a scientific study, or is it built around an untested hypothesis? Were the results obtained using the scientific method, with tests designed to reject false hypotheses? Is the story confusing correlation with causation? Does it seem that the information is applicable to nonlaboratory situations, or is it based on results from preliminary or animal studies?

Look for clues about how well the reporters did their homework. Scientists usually discuss the limitations of their research in their papers. Are these cautions noted in an article or television piece? If not, the reporter may be overemphasizing the applicability of the results.

Then, note if the scientific discovery itself is controversial. That is, does it reject a hypothesis that has long been supported? Does it concern a subject that is controversial? Might it lead to a change in social policy? In these cases, be extremely cautious. New and unexpected research results must be evaluated in light of other scientific evidence and understanding. Reports that lack comments from other experts may miss problems with a study or fail to place it in the context of other research. The Savvy Reader feature in this chapter provides a checklist of questions to ask and answer as you evaluate news reports.

Finally, the news media generally highlight only those science stories that editors find newsworthy. As we have seen, scientific understanding accumulates relatively slowly, with many tests of the same hypothesis finally leading to an accurate understanding of a natural phenomenon. News organizations are also more likely to report a study that supports a hypothesis rather than one that gives less-supportive results, even if both types of studies exist.

In addition, even the most respected media sources may not be as thorough as readers would like. For example, a recent review published in the *New England Journal of Medicine* evaluated the news media's coverage of new medications. Of 207 randomly selected news stories, only 40% that cited experts with financial ties to a drug told readers about this relationship. This potential conflict of interest calls into question the expert's objectivity. Another 40% of the news stories did not provide basic statistics about the drugs' benefits. Most of the news reports also failed to distinguish between absolute benefits (how many people were helped by the drug) and relative benefits (how many people were helped by the drug relative to other therapies for the condition). The journal's review is a vivid reminder that we need to be cautious when reading or viewing news reports on scientific topics.

Even after following all of these guidelines, you will still find well-researched news reports on several scientific studies that seem to give conflicting and confusing results. As you now know, such confusion is the nature of the scientific process—early in our search for understanding, many hypotheses are proposed and discussed; some are tested and rejected immediately, and some are supported by one experiment but later rejected by more thorough experiments. It is only by clearly understanding the process and pitfalls of scientific research that you can distinguish "what we know" from "what we don't know."

1.5 Is There a Cure for the Common Cold?

So where does our discussion leave us? Will we ever find the best way to prevent a cold or reduce its effects? In the United States alone, over 1 billion cases of the common cold are reported per year, costing many more billions of dollars in medical visits, treatment, and lost workdays. Consequently, there is an enormous effort to find effective protection from the different viruses that cause colds.

A search of medical publication databases indicates that every year nearly 100 scholarly articles regarding the biology, treatment, and consequences of common cold infection are published. This research has led to several important discoveries about the structure and biochemistry of common cold viruses, how they enter cells, and how the body reacts to these infections.

Despite all of the research and the emergence of some promising possibilities, the best prevention method known for common colds is still the old standby—keep your hands clean. Numerous studies have indicated that rates of common cold infection are 20% to 30% lower in populations who employ effective hand-washing procedures. Cold viruses can survive on surfaces for many hours; if you pick them up on your hands from a surface and transfer them to your mouth, eyes, or nose, you may inoculate yourself with a 7-day sniffle (**Figure 1.17**).

Figure 1.17 Preventing colds. According to the Centers for Disease Control and Prevention, the best defense against infection is effective hand washing. Scrub vigorously with a liquid soap for 20 seconds, then rinse under running water for 10 seconds. Alcohol-based hand gels (without water) containing at least 60% ethanol are effective if soap and water are unavailable but will not remove visible dirt.

Of course, not everyone gets sick when exposed to a cold virus. The reason one person has more colds than another might not be due to a difference in personal hygiene. The correlation that showed a relationship between stress and cold susceptibility appears to have some merit. Research indicates that among people exposed to viruses, the likelihood of ending up with an infection increases with high levels of psychological stress—something that many college students clearly experience.

Research also indicates that vitamin C intake, diet quality, exposure to cold temperatures, and exercise frequency appear to have no effect on cold susceptibility, although along with echinacea tea and zinc lozenges, there is some evidence that vitamin C may reduce cold symptoms a bit. Surprisingly, even though medical research has led to the elimination of killer viruses such as smallpox and polio, scientists are still a long way from "curing" the common cold.

SAVVY READER

A Toolkit for Evaluating Science in the News

The following checklist can be used as a guide to evaluating science information in the news. Although any issue that raises a red flag should cause you to be cautious about the conclusions of the story, no one issue should cause you to reject the story (and its associated research) out of hand. However, in general, the fewer red flags raised by a story, the more reliable is the report of the significance of the scientific study. Use **Table 1.2** to evaluate this extract from an article by Terry Gupta in the Nashua (New Hampshire) *Telegraph*:

> [Paula] Fortier, a licensed massage therapist, shared her special brew. To boiled water and tea, she adds cayenne pepper, honey and fresh squeezed lemon. Fortier's enthusiasm may be more contagious than the flu when she describes this heated, "unbeatable" combination. "Your throat opens up," she said. "You're less irritated and you sweat out some unwanted toxins." She varies the exact measures of the ingredients according to tolerance and taste. While on the topic of hot water, one of Fortier's favorite remedies for "an achy body that feels sore all over is a hot bath in a steeping tub with 2 cups of cider vinegar for 15 minutes. This draws off toxins, reduces inflammation and leaves you feeling better."

TABLE 1.2

A guide for evaluating science in the news. For each question, check the appropriate box.				
Question	Preferred answer		Raises a red flag	
1. What is the basis for the story?	Hypothesis test	☐	Untested assertion No data to support claims in the article.	☐
2. What is the affiliation of the scientist?	Independent (university or government agency)	☐	Employed by an industry or advocacy group Data and conclusions could be biased.	☐
3. What is the funding source for the study?	Government or nonpartisan foundation (without bias)	☐	Industry group or other partisan source (with bias) Data and conclusions could be biased.	☐

(continued)

TABLE 1.2 *(continued)*

A guide for evaluating science in the news. For each question, check the appropriate box.

Question	Preferred answer		Raises a red flag	
4. If the hypothesis test is a correlation: Did the researchers attempt to eliminate reasonable alternative hypotheses?	Yes	☐	No. Correlation does not equal causation. One hypothesis test provides poor support if alternatives are not examined.	☐
If the hypothesis test is an experiment: Is the experimental treatment the only difference between the control group and the experimental group?	Yes	☐	No. An experiment provides poor support if alternatives are not examined.	☐
5. Was the sample of individuals in the experiment a good cross section of the population?	Yes	☐	No. Results may not be applicable to the entire population.	☐
6. Was the data collected from a relatively large number of people?	Yes	☐	No. Study is prone to sampling error.	☐
7. Were participants blind to the group they belonged to or to the "expected outcome" of the study?	Yes	☐	No. Subject expectation can influence results.	☐
8. Were data collectors or analysts blinded to the group membership of participants in the study?	Yes	☐	No. Observer bias can influence results.	☐
9. Did the news reporter put the study in the context of other research on the same subject?	Yes	☐	No. Cannot determine if these results are unusual or fit into a broader pattern of results.	☐
10. Did the news story contain commentary from other independent scientists?	Yes	☐	No. Cannot determine if these results are unusual or if the study is considered questionable by others in the field.	☐
11. Did the reporter list the limitations of the study or studies on which he or she is reporting?	Yes	☐	No. Reporter may not be reading study critically and could be overstating the applicability of the results.	☐

Chapter Review

Learning Outcomes

MasteringBIOLOGY®

Go to the Study Area at www.masteringbiology.com for practice quizzes, myeBook, BioFlix™ 3-D animations, MP3Tutor sessions, videos, current events, and more.

LO1 Describe the characteristics of a scientific hypothesis (Section 1.1).

- Science is a process of testing hypotheses—statements about how the natural world works. Scientific hypotheses must be testable and falsifiable (p. 4).

LO2 Compare and contrast the terms *scientific hypothesis* and *scientific theory* (Section 1.1).

- A scientific theory is an explanation of a set of related observations based on well-supported hypotheses from several different, independent lines of research (p. 6).

LO3 Distinguish between inductive and deductive reasoning (Section 1.1).

- Hypotheses are often developed via inductive reasoning, which consists of making a number of observations and then inferring a general principle to explain them (p. 6).
- Hypotheses are tested via the process of deductive reasoning, which allows researchers to make specific predictions about expected observations (pp. 6–7).

LO4 Explain why the truth of a hypothesis cannot be proven conclusively via deductive reasoning (Section 1.1).

- Absolutely proving hypotheses is impossible because there may be other reasons besides the one hypothesized that could lead to the predicted result (pp. 7–8).

LO5 Describe the features of a controlled experiment, and explain how these experiments eliminate alternative hypotheses for the results (Section 1.2).

- Controlled experiments test hypotheses about the effect of experimental treatments by comparing a randomly assigned experimental group with a control group (pp. 10–11).
- Controls are individuals who are treated identically to the experimental group except for application of the treatment (p. 10).

LO6 List strategies for minimizing bias when designing experiments (Section 1.2).

- Bias in scientific results can be minimized with double-blind experiments that keep subjects and data collectors unaware of which individuals belong in the control or experimental group (p. 12).

LO7 Define correlation, and explain the benefits and limitations of using this technique to test hypotheses (Section 1.2).

- If performing controlled experiments on humans is considered unethical, scientists sometimes use correlations. The correlation of structure and function between humans and model organisms, such as other mammals, allows experimental tests of these hypotheses (pp. 12–13).
- Correlations cannot exclude alternative hypotheses. Thus, a correlation study can describe a relationship between two factors, but it does not strongly imply that one factor causes the other (p. 15).

LO8 Describe the information that statistical tests provide (Section 1.3).

- Statistics help scientists evaluate the results of their experiments by determining if results appear to reflect the true effect of an experimental treatment on a sample of a population (p. 17).
- A statistically significant result is one that is very unlikely to be due to chance differences between the experimental and control groups (p. 17).
- Even when an experimental result is highly significant, hypotheses are tested multiple times before scientists come to consensus on the true effect of a treatment (pp. 19–20).

LO9 Compare and contrast primary and secondary sources (Section 1.4).

- Primary sources of information are experimental results published in professional journals and peer reviewed by other scientists before publication (pp. 21–24).
- Secondary sources are typically news, books, or web sites that summarize the scientific research presented in primary sources. Secondary sources are not peer reviewed and sometimes contain poorly supported information (pp. 21–24).

LO10 Summarize the techniques you can use to evaluate scientific information from secondary sources (Section 1.4).

- Anecdotal evidence is an unreliable means of evaluating information, and media sources are of variable quality; distinguishing between news stories and advertisements is important when evaluating the reliability of information. The Internet is a rich source of information, but users should look for clues to a particular website's credibility (p. 21).
- Stories about science should be carefully evaluated for information on the actual study performed, the universality of the claims made by the researchers, and other studies on the same subject. Sometimes confusing stories about scientific information are a reflection of controversy within the scientific field itself (pp. 23–24).

Roots to Remember MB

The following roots of words come mainly from Latin and Greek and will help you to decipher terms:

bio- means life. Chapter term: *biology*

deduc- means to reason out, working from facts. Chapter term: *deductive reasoning*

induc- means to rely on reason to derive principles (also, to cause to happen). Chapter term: *inductive reasoning*

hypo- means under, below, or basis. Chapter term: *hypothesis*

-ology means the study of or branch of knowledge about. Chapter term: *biology*

Learning the Basics

1. **LO5 LO6** What is the value of a placebo in an experimental design?.

2. **LO3** Which of the following is an example of inductive reasoning?

   A. All cows eat grass; **B.** My cow eats grass and my neighbor's cow eats grass; therefore all cows probably eat grass; **C.** If all cows eat grass, when I examine a random sample of all the cows in Minnesota, I will find that they all eat grass; **D.** Cows may or may not eat grass, depending on the type of farm where they live.

3. **LO1** A scientific hypothesis is _____.

   A. an opinion; **B.** a proposed explanation for an observation; **C.** a fact; **D.** easily proved true; **E.** an idea proposed by a scientist.

4. **LO2** How is a scientific theory different from a scientific hypothesis?

   A. It is based on weaker evidence; **B.** It has not been proved true; **C.** It is not falsifiable; **D.** It can explain a large number of observations; **E.** It must be proposed by a professional scientist.

5. **LO3** One hypothesis states that eating chicken noodle soup is an effective treatment for colds. Which of the following results does this hypothesis predict?

   A. People who eat chicken noodle soup have shorter colds than do people who do not eat chicken noodle soup; **B.** People who do not eat chicken noodle soup experience unusually long and severe colds; **C.** Cold viruses cannot live in chicken noodle soup; **D.** People who eat chicken noodle soup feel healthier than do people who do not eat chicken noodle soup; **E.** Consuming chicken noodle soup causes people to sneeze.

6. **LO4** If I perform a hypothesis test in which I demonstrate that the prediction I made in question 5 is true, I have _____.

   A. proved the hypothesis; **B.** supported the hypothesis; **C.** not falsified the hypothesis; **D.** B and C are correct; **E.** A, B, and C are correct.

7. **LO5** Control subjects in an experiment

   A. should be similar in most ways to the experimental subjects; **B.** should not know whether they are in the control or experimental group; **C.** should have essentially the same interactions with the researchers as the experimental subjects; **D.** help eliminate alternative hypotheses that could explain experimental results; **E.** all of the above.

8. **LO6** An experiment in which neither the participants in the experiment nor the technicians collecting the data know which individuals are in the experimental group and which ones are in the control group is known as

   A. controlled; **B.** biased; **C.** double-blind; **D.** falsifiable; **E.** unpredictable.

9. **LO7** A relationship between two factors, for instance between outside temperature and the number of people with active colds in a population, is known as a(n)

   A. significant result; **B.** correlation; **C.** hypothesis; **D.** alternative hypothesis; **E.** experimental test.

10. **LO9** A primary source of scientific results is

    A. the news media; **B.** anecdotes from others; **C.** articles in peer-reviewed journals; **D.** the Internet; **E.** all of the above.

11. **LO8** A story on your local news station reports that eating a 1-ounce square of milk chocolate each day reduces the risk of heart disease in rats, and that this result is statistically significant. This means that

    A. People who eat milk chocolate are healthier than those who do not; **B.** The difference between chocolate- and non-chocolate-eating rats in heart disease rates was greater than expected by chance; **C.** Rats like milk chocolate; **D.** Milk chocolate reduces the risk of heart disease; **E.** Two ounces of milk chocolate per day is likely to be even better for heart health than 1 ounce.

12. **LO10** What features of the story on milk chocolate and heart health described in Question 11 should cause you to consider the results less convincing?

    A. The study was sponsored by a large milk chocolate manufacturer; **B.** A total of ten rats were used in the study; **C.** The only difference between the rats was that subjects of the experimental group received chocolate along with their regular diets, while subjects of the control group received no additional food; **D.** The reporter notes that other studies indicate that milk chocolate does not have beneficial effect on heart health; **E.** All of the above.

Analyzing and Applying the Basics

1. **LO7** There is a strong correlation between obesity and the occurrence of a disease known as type 2 diabetes—that is, obese individuals have a higher instance of diabetes than nonobese individuals do. Does this mean that obesity causes diabetes? Explain.

2. **LO5 LO6** In an experiment examining vitamin C as a cold treatment, students with cold symptoms who visited the campus medical center either received vitamin C or were treated with over-the-counter drugs. Students then reported on the length and severity of their colds. Both the students and the clinic health providers knew which treatment students were receiving. This study indicated that vitamin C significantly reduced the length and severity of colds. Which factors make this result somewhat unreliable?

3. **LO7** Brain-derived neurotrophic factor (BDNF) is a substance produced in the brain that helps nerve cells (neurons) to grow and survive. BDNF also increases the connectivity of neurons and improves learning and mental function. A 2002 study that examined the effects of intense wheel-running on rats and mice found a positive correlation between BDNF levels and running distance. What could you conclude from this result?

Connecting the Science

1. Much of the research on common cold prevention and treatment is performed by scientists employed or funded by drug companies. Often these companies do not allow scientists to publish the results of their research for fear that competitors at other drug companies will use this research to develop a new drug before they do. Should our society allow scientific research to be owned and controlled by private companies?

2. Should society restrict the kinds of research performed by government-funded scientists? For example, many people believe that research performed on tissues from human fetuses should be restricted; these people believe that such research would justify abortion. If most Americans feel this way, should the government avoid funding this research? Are there any risks associated with *not* funding research with public money?

Answers to **Stop & Stretch, Visualize This, Savvy Reader,** and **Chapter Review** questions can be found in the **Answers** section at the back of the book.

CHAPTER 2

Are We Alone in the Universe?

Water, Biochemistry, and Cells

Cartoon images aside, we know of no other intelligent life in our solar system.

Is there life on other planets? Scientists have found persuasive, if not conclusive, evidence of life on Mars. When the National Aeronautics and Space Administration (NASA) scientist David McKay first proposed that this cold, dry, rather harsh planet could harbor life, people were astounded.

The evidence of life found by Dr. McKay and his team did not in any way resemble the cartoon images often used to depict Martians. Instead, what these scientists found was evidence of life in a 3.6-billion-year-old, potato-sized rock. They believe that the rock broke off the surface of Mars after being hit by an asteroid. After floating through space for 16 million years, it fell to Earth, landing in Antarctica about 13,000 years ago and remaining there until discovered by scientists. This meteorite, drably named ALH84001, appeared to contain the same features that scientists use to demonstrate the existence of life in 3.6-billion-year-old Earth rocks—there were fossils, various minerals that are characteristic of life, and evidence of complex chemicals typically produced by living organisms.

While many scientists debate the assertion that this rock provides evidence of life on Mars, the announcement served to inject new energy into Mars exploration. Since then, multiple robotic rovers and mapping satellites have been sent to the planet, with the goal of someday sending astronauts there. While there are many reasons to explore Mars, the question that remains most intriguing—and is a significant focus of several of these missions—is whether life ever existed there.

The fascination about potential Martian life speaks to a fundamental question that many humans share: Are the creatures on Earth the only living organisms in the universe? Our galaxy is filled with countless stars

Some evidence suggests that Mars may harbor life.

A meteorite ejected from Mars and found in Antarctica may contain lifelike forms.

Will the Mars rover find evidence of life outside Earth?

and planets, and the universe teems with galaxies. Even if we find no convincing evidence of life on Mars, there are a seemingly infinite number of places to look for other living beings. In this chapter, we discuss the characteristics and requirements of life and examine techniques that scientists use to search the universe for other living creatures.

2.1 What Does Life Require?

Because the galaxy likely contains billions of planets, scientists looking for life elsewhere seek to identify the range of conditions under which they would expect life to arise. What is it that scientists look for when identifying a planet (or moon) as a candidate for hosting life?

A Definition of Life

In science-fiction movies, alien life-forms are often obviously alive and even somewhat familiar looking. But in reality, living organisms may be truly alien; that is, they may look nothing like organisms we are familiar with on Earth. So how would we determine whether an entity found on another planet was actually alive?

Surprisingly, biologists do not have a simple definition for a "living organism." A list of the attributes found in most earthly life-forms includes growth, movement, reproduction, response to external environmental stimuli, and **metabolism** (all of the chemical processes that occur in cells, including the breakdown of substances to produce energy, the synthesis of substances necessary for life, and the excretion of wastes generated by these processes). However, this definition could apply to things that no one considers to be living. For example, fire can grow, consume energy, give off waste, move, reproduce by sending off sparks, and change in response to environmental conditions. And some organisms that are clearly living do not conform to this definition. Male mules grow, metabolize, move, and respond to stimuli, but they are unable to reproduce.

Figure 2.1 Homeostasis. Black-capped chickadees can maintain a core body temperature of 108°F (~42°C) during the day, even when the air temperature is well below zero.

If we examine more closely the characteristics of living organisms on Earth, we will see that all organisms contain a common set of biological molecules, are composed of cells, and can maintain **homeostasis,** that is, a roughly constant internal environment despite an ever-changing external environment (**Figure 2.1**). The ability to maintain homeostasis requires complex feedback mechanisms between multiple sensory and physiological systems and is possible only in living organisms. In addition, populations of living organisms can evolve, that is, change in average physical characteristics over time. If we search the universe for planets that could support life similar to that found on Earth—and thus organisms that we would clearly identify as "living"—the list of planetary requirements becomes more stringent. In particular, an Earth-like planet should have abundant liquid **water** available.

The Properties of Water

Water is a requirement for life. Although Mars does not currently appear to have any liquid water, ice is found at its poles (**Figure 2.2a**), and features of its surface indicate that it once contained salty seas and flowing water (**Figure 2.2b**).

(a) Frozen water

(b) Running water

Figure 2.2 Water on Mars. (a) This image from the Mars rover indicates that frozen water exists on Mars. (b) This photograph, taken by the European Mars Express orbiter, shows a channel on Mars that may have been formed by running water.

The presence of liquid water on Mars would fulfill an essential prerequisite for the appearance of life. But why is water such an important feature?

Water is made up of two elements: hydrogen and oxygen. **Elements** are the fundamental forms of matter and are composed of atoms that cannot be broken down by normal physical means such as boiling.

Atoms are the smallest units that have the properties of any given element. Ninety-two natural elemental atoms have been described by chemists, and several more have been created in laboratories. Hydrogen, oxygen, and calcium are examples of elements commonly found in living organisms. Each element has a one- or two-letter symbol: H for hydrogen, O for oxygen, and Ca for calcium, for example.

Atoms are composed of subatomic particles called **protons, neutrons, and electrons.** Protons have a positive electric charge; these particles and the uncharged neutrons make up the **nucleus** of an atom. All atoms of a particular element have the same number of protons, giving the element its **atomic number.** The negatively charged electrons are found outside the nucleus in an "electron cloud." Electrons are attracted to the positively charged nucleus (**Figure 2.3**). A *neutral atom* has equal numbers of protons and electrons. Electrically charged **ions** do not have an equal number of protons and electrons. In this case, the atom is not neutral and is instead charged.

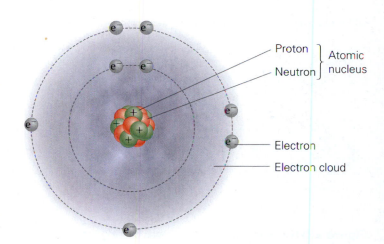

Figure 2.3 Atomic structure. An oxygen atom contains a nucleus made up of 8 protons and 8 neutrons. Orbiting electrons surround the nucleus. Although the number of particles within each atom differs, all atoms have the same basic structure.

The Structure of Water. The chemical formula for water is H_2O, indicating that it contains 2 hydrogen atoms for every 1 oxygen. Water, like other molecules, consists of 2 or more atoms joined by chemical bonds. A molecule can be composed of the same or different atoms. For example, a molecule of oxygen consists of 2 oxygen atoms joined to each other, while a molecule of carbon dioxide consists of 1 carbon and 2 oxygen atoms.

Water Is a Good Solvent. Water has the ability to dissolve a wide variety of substances. A substance that dissolves when mixed with another substance is called a solute. When a solute is dissolved in a liquid, such as water, the liquid is called a solvent. Once dissolved, components of a particular solute can pass freely throughout the water, making a chemical mixture or solution.

Water is a good solvent because it is **polar,** meaning that different regions, or poles, of the molecule have different charges. The polarity arises because oxygen is more attractive to electrons—that is, it is more **electronegative**—than most other atoms, including hydrogen. As a result of oxygen's electronegativity, electrons in a water molecule spend more time near the nucleus of the oxygen atom than near the nuclei of the hydrogen atoms. With more negatively charged electrons near it, the oxygen in water carries a partial negative charge, symbolized by the Greek letter delta, δ^-. The hydrogen atoms thus have a partial positive charge, symbolized by δ^+ (**Figure 2.4**). When atoms of a molecule carry no partial charge, they are said to be **nonpolar.** The carbon-hydrogen bonds, for example, share electrons equally and are nonpolar.

Water molecules tend to orient themselves so that the hydrogen atom (with its partial positive charge) of one molecule is near the oxygen atom (with its

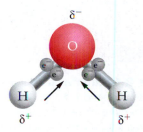

Figure 2.4 Polarity in water. Water is a polar molecule. Its atoms do not share electrons equally.
Visualize This: Toward which atom are the electrons of a water molecule pulled?

(a) Bonds between two water molecules   **(b) Bonds between many water molecules**

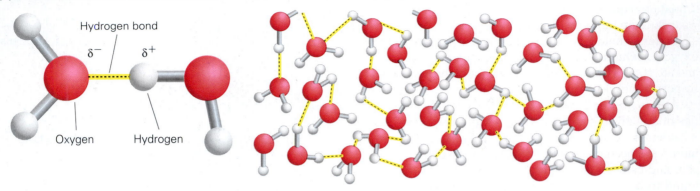

Figure 2.5 Hydrogen bonding. Hydrogen bonding can occur when there is a weak attraction between the hydrogen and oxygen atoms between (a) two or (b) many different water molecules.

partial negative charge) of another molecule (**Figure 2.5a**). The weak attraction between hydrogen atoms and oxygen atoms in adjacent molecules forms a **hydrogen bond.** Hydrogen bonding is a type of weak chemical bond that forms when a partially positive hydrogen atom is attracted to a partially negative atom. Hydrogen bonds can be intramolecular, involving different regions within the same molecule, or they can be intermolecular, between different molecules, as is the case in hydrogen bonding between different water molecules. **Figure 2.5b** shows the hydrogen bonding that occurs between water molecules in liquid form.

The ability of water to dissolve substances such as sodium chloride is a direct result of its polarity. Each molecule of sodium chloride is composed of 1 sodium ion (Na^+) and 1 chloride ion (Cl^-). In the case of sodium chloride, the negative pole of water molecules will be attracted to a positively charged sodium ion and separate it from a negatively charged chloride ion (**Figure 2.6**). Water can also dissolve other polar molecules, such as alcohol, in a similar manner. Polar molecules are called **hydrophilic** ("water-loving") because of their ability to dissolve in water. Nonpolar molecules, such as oil, do not contain charged atoms and are referred to as **hydrophobic** ("water-fearing") because they do not easily mix with water.

Water Facilitates Chemical Reactions. Because it is such a powerful solvent, water can facilitate **chemical reactions**, which are changes in the chemical composition of substances. Solutes in a mixture, called **reactants,** can come in contact with each other, permitting the modification of chemical bonds that occur during a reaction. The molecules formed as a result of a chemical reaction are known as **products.**

Water Is Cohesive. The tendency of like molecules to stick together is called **cohesion.** Cohesion is much stronger in water than in most liquids because of the sheer number of hydrogen bonds between water molecules. Cohesion is an important property of many biological systems. For instance, many plants depend on cohesion to help transport a continuous column of water from the roots to the leaves.

Water Moderates Temperature. When heat energy is added to water, its initial effect is to disrupt the hydrogen bonding among water molecules.

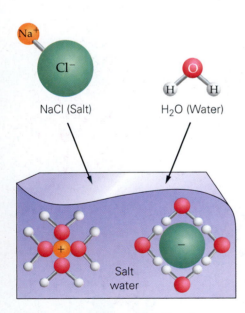

Figure 2.6 Water as a solvent. When salt is placed in water, the negatively charged regions of the water molecules surround the positively charged sodium ion, and the positively charged regions of the water molecules surround the negatively charged chlorine ion, breaking the bond holding sodium and chloride together and dissolving the salt.

Therefore, this heat energy can be absorbed without changing the temperature of water. Only after the hydrogen bonds have been broken can added heat increase the temperature. In other words, the initial input of energy is absorbed.

The flowing water that was once found on Mars is now only in the form of ice. Until scientists can land on Mars and collect ice samples for analysis, its actual composition is a matter of conjecture. However, images taken by a NASA rover have led scientists to believe that some of the rocks on Mars were probably produced from deposits at the bottom of a body of saltwater. We know from surveying Earth's oceans that saltwater is hospitable to millions of different life-forms. In fact, most hypotheses about the origin of life on Earth presume that our ancestors first arose in the salty oceans.

Acids, Bases, and Salts

Salts are produced by the reaction of an **acid** (a substance that donates H^+ ions to a solution) with a **base** (a substance that accepts H^+ ions). The **pH** scale is a measure of the relative amounts of these ions in a solution (**Figure 2.7**). The more acidic a solution is, the higher the H^+ concentration is relative to the OH^- ions. Hydrogen ion concentration is inversely related to pH, so the higher the H^+ concentration, the lower the pH. Basic solutions have fewer H^+ ions relative to OH^- ions and thus a higher pH. These ions can react with other charged molecules and help to bring them into the water solution. At any given time, a small percentage of water molecules in a pure solution will be dissociated. There are equal numbers of these ions in pure water, so it is neutral, which on the pH scale is 7. The pH of most cells is very close to 7.

When a European Space Agency probe landed on Titan, one of Saturn's moons and a place where the chemical composition of the atmosphere may be similar to that found on early Earth, the photos transmitted by the probe indicated that liquid was present on the surface of this bitterly cold place. At atmospheric temperatures of approximately 2292°F (2180°C), the liquid is obviously not water; instead, it is most likely a mixture of ethane and methane, both nonpolar molecules. As a result, oceans on Titan are much poorer solvents than are oceans on Earth, and conditions in these oceans are probably not suitable for the evolution of life.

Organic Chemistry

The Martian meteorite ALH84001 had one characteristic that provided some evidence that the rock once contained living organisms: the presence of complex molecules containing the element carbon.

All life on Earth is based on the chemistry of the element carbon. The branch of chemistry that is concerned with complex carbon-containing molecules is called **organic chemistry.**

An Overview: Chemical Bonds. Chemical bonds between atoms and molecules involve attractions that help stabilize various configurations. In general, this involves the sharing or transfer of electrons. One of the most important elements in biology is carbon, which is often involved in chemical bonding because of its ability to make bonds with up to four other elements.

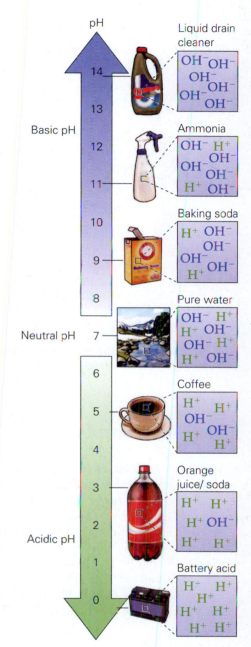

Figure 2.7 The pH scale. The pH scale is a measure of hydrogen ion concentration ranging from 0 (most acidic) to 14 (most basic). Each pH unit actually represents a 10-fold (10x) difference in the concentration of H^+ ions.

Visualize This: A substance with a pH of 5 would have how many times more H^+ ions than a substance with a pH of 7?

Like a Tinkertoy™ connector, carbon has multiple sites for connections that allow carbon-containing molecules to take an almost infinite variety of shapes (**Figure 2.8**). *See the next section for **A Closer Look** at chemical bonds.*

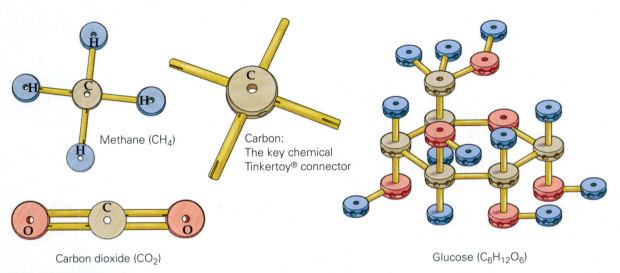

Methane (CH_4)

Carbon:
The key chemical
Tinkertoy® connector

Carbon dioxide (CO_2)

Glucose ($C_6H_{12}O_6$)

Figure 2.8 Carbon, the chemical Tinkertoy™ connector. Because carbon forms four covalent bonds at a time, carbon-containing compounds can have diverse shapes.

A Closer Look:
Chemical Bonds

The ability of elements to make chemical bonds depends on the atom's electron configuration. The electrons in the electron cloud that surrounds the atom's nucleus have different energy levels based on their distance from the nucleus. The first energy level, or **electron shell,** is closest to the nucleus, and the electrons located there have the lowest energy. The second energy level is a little farther away, and the electrons located in the second shell have a little more energy. The third energy level is even farther away, and its electrons have even more energy, and so on.

Each energy level can hold a specific maximum number of electrons. The first shell holds 2 electrons, and the second and third shells each hold a maximum of 8. Electrons fill the lowest energy shell before advancing to fill a higher energy-level shell. For example, hydrogen with its 1 electron needs only 1 more electron to fill its first shell.

Atoms with the same number of electrons in their outermost energy shell, called the **valence shell,** exhibit similar chemical behaviors. When the valence shell is full, the atom will not normally form chemical bonds with other atoms. Atoms whose valence shells are not full of electrons often combine via chemical bonds.

Atoms with 4 or 5 electrons in the outermost valence shell tend to share electrons to complete their valence shells. When atoms share electrons, a type of bond called a **covalent bond** is formed.

Carbon, with its 4 valence electrons, is said to be tetravalent (**Figure 2.9a**). In other words, it can form up to four bonds. Carbon can form 4 single bonds, 2 double bonds, 1 double bond and 2 single bonds, and so on, depending on the number of electrons needed by the atom that is its partner. **Figure 2.9b** shows carbon covalently bonded to 4 hydrogens to produce methane, an organic compound that has been found on Mars and Titan. Covalent bonds are symbolized by

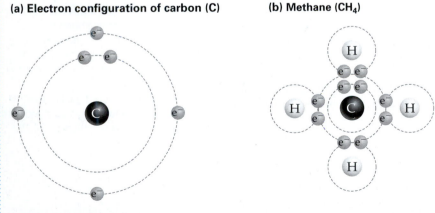

(a) Electron configuration of carbon (C)

(b) Methane (CH_4)

Figure 2.9 Covalent bonding. (a) Carbon has 4 unpaired electrons in its valence shell and is thus able to make up to 4 chemical bonds. (b) Methane consists of carbon covalently bonded to 4 hydrogens.

(a) Methane

H
|
H—C—H
|
H

(b) Ethylene

H   H
\   /
C=C
/   \
H   H

Figure 2.10 Single and double bonds. (a) Covalent bonds are symbolized by a short line indicating a shared pair of electrons. (b) Double covalent bonds involve two pairs of shared electrons, symbolized by two horizontal lines.

a short line indicating a shared pair of electrons (**Figure 2.10a**). When an element such as carbon enters into bonds involving two *pairs* of shared electrons, this is called a double bond. A carbon-to-carbon double bond is symbolized by two horizontal lines (**Figure 2.10b**).

Atoms with 1, 2, or 3 electrons in their valence shell tend to lose electrons and therefore become positively charged ions, while atoms with 6 or 7 electrons in the valence shell tend to gain electrons and become negatively charged ions. Positively and negatively charged ions associate into a type of bond called the ionic bond.

Stop & Stretch What does not make sense (chemically) about the structure below?

H–C=C–H

Ionic bonds form between charged atoms attracted to each other by similar, opposite charges. For example, the sodium atom forms an ionic bond with a chlorine atom to produce table salt (sodium chloride) when the sodium atom gives up an electron and the chlorine atom gains 1 (**Figure 2.11**). More than 2 atoms can be involved in an ionic bond. For instance, calcium will react with 2 chlorine atoms to produce calcium chloride ($CaCl_2$). This is because calcium has 2 electrons in its valence shell—when it loses these it has 2 more protons than electrons, giving it a double-positive charge. Each chlorine atom, with 7 electrons, picks only 1 more electron to have a stable outer shell and a single negative charge. Thus, 2 chlorine ions will be attracted to a single calcium ion.

Ionic bonds are about as strong as covalent bonds. They can be more easily disrupted, however, when mixed with certain liquids containing electrical charges. Water is one liquid that causes ions in molecules to dissociate or fall apart.

The simple organic molecules found in the Martian meteorite that appear to have formed on Mars are carbonates, molecules containing carbon and oxygen, and **hydrocarbons,** made up of chains and rings of carbon and hydrogen. Carbonates and hydrocarbons can form under certain natural conditions even without the presence of life. However, the meteorite lacked convincing evidence of **macromolecules,** large organic molecules made of many subunits, that are known to be produced only by living organisms.

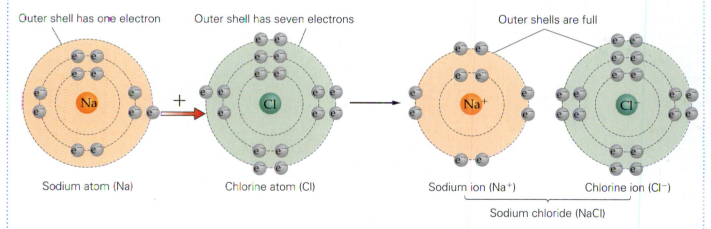

Outer shell has one electron   Outer shell has seven electrons   Outer shells are full

Sodium atom (Na)   Chlorine atom (Cl)   Sodium ion (Na⁺)   Chlorine ion (Cl⁻)

Sodium chloride (NaCl)

Figure 2.11 Ionic bonding. Ionic bonds form when electrons are transferred between charged atoms.

Structure and Function of Macromolecules

The macromolecules present in living organisms are carbohydrates, proteins, lipids, and nucleic acids. To date, every living Earth organism, whether bacteria, plant, or animal, has been found to contain these same macromolecules.

Carbohydrates. Sugars, or **carbohydrates,** provide the major source of energy for daily activities. Carbohydrates also play important structural roles in cells. The simplest carbohydrates are composed of carbon, hydrogen, and oxygen in the ratio (CH_2O). For example, the carbohydrate glucose is symbolized as $6(CH_2O)$ or $C_6H_{12}O_6$. Glucose is a simple sugar, or monosaccharide, that consists of a single ring-shaped structure. Disaccharides are two rings joined together. Table sugar, called sucrose, is a disaccharide composed of glucose and fructose, a sugar found in fruits.

Joining many individual subunits, or monomers, together produces polymers. Polymers of sugar monomers are called **polysaccharides** (**Figure 2.12**). Plants use tough polysaccharides in their cell walls as a sort of structural skeleton. The polysaccharide cellulose, found in plant cell walls, is the most abundant carbohydrate on Earth. The external skeletons of insects, spiders, and lobsters are composed of the polysaccharide chitin, and the cell walls that surround bacterial cells are rich in structural polysaccharides.

According to David McKay and his colleagues, the particular set of hydrocarbons found in the Martian meteorite is identical to the set formed when carbohydrates break down in certain bacteria on Earth. These trace remains of possible Martian carbohydrates are an important piece of evidence that scientists use to argue that Mars once harbored Earth-like life. Evidence of the presence of proteins on the meteorite is less convincing.

Proteins. Living organisms require **proteins** for a wide variety of processes. Proteins are important structural components of cells; they are integral to the

Figure 2.12 Carbohydrates.
Monosaccharides, which tend to form ring structures in aqueous solutions, are individual sugar molecules. Disaccharides are 2 monosaccharides joined together, and polysaccharides are long chains of sugars joined together by covalent bonds. The monosaccharide glucose and the disaccharide sucrose are important sources of energy, and cellulose plays a structural role in plant cell walls.

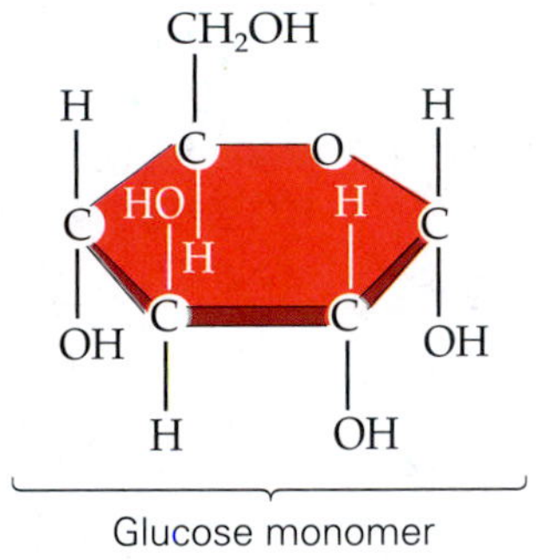

Glucose monomer

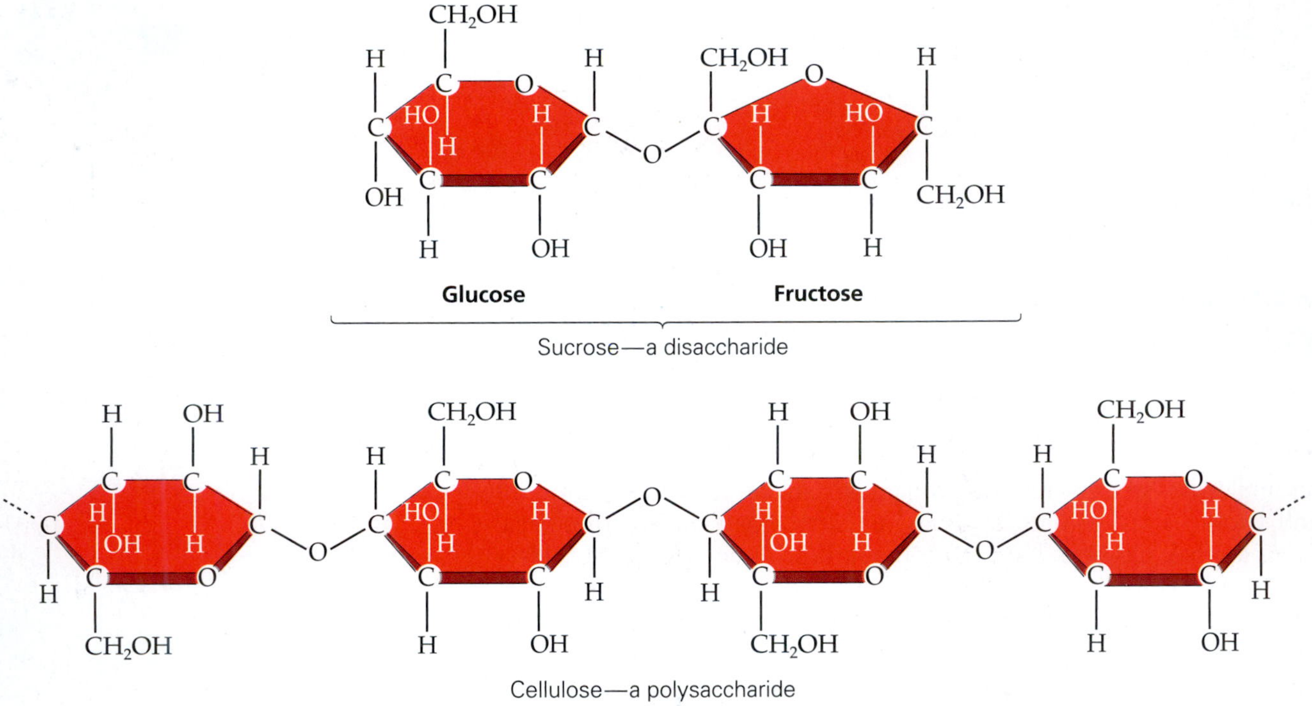

Sucrose—a disaccharide

Cellulose—a polysaccharide

structure of cell membranes and make up half the dry weight of most cells. Some cells, such as animal muscle cells, are largely composed of proteins. Proteins called **enzymes** accelerate and help regulate all the chemical reactions that build up and break down molecules inside cells. The catalytic power of enzymes (their ability to drastically increase reaction rates) allows metabolism to occur under normal cellular conditions. Proteins can also serve as channels through which substances are brought into cells, and they can function as hormones that send chemical messages throughout an organism's body.

Proteins are large molecules made of monomer subunits called **amino acids.** There are 20 commonly occurring amino acids. Like carbohydrates, amino acids are made of carbons, hydrogens, and oxygens; these form the amino acid's carboxyl group ($-COO^-$). In addition, amino acids have nitrogen as part of an amino (NH_2^+) group along with various side groups. Side groups are chemical groups that give amino acids different chemical properties (**Figure 2.13a**).

Polymers of amino acids can be joined together in various sequences called polypeptides. A covalent bond joins adjacent amino acids to each other. **Figure 2.13b** shows three amino acids—valine, alanine, and phenylalanine—joined by peptide bonds. Precisely folded polypeptides produce specific proteins in much the same manner that children can use differently shaped beads to produce a wide variety of structures (**Figure 2.13c**). Each amino acid side group has unique chemical properties, including being polar or nonpolar. Because each protein is composed of a particular sequence of amino acids, each protein has a unique shape and therefore specialized chemical properties.

Scientists have found no evidence of proteins in the Martian meteorite, although one group of investigators did report the presence of tiny amounts

(a) General formula for amino acid   **(b) Peptide bond formation**

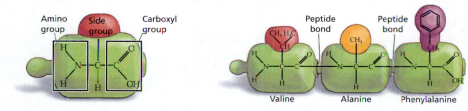

(c) Protein

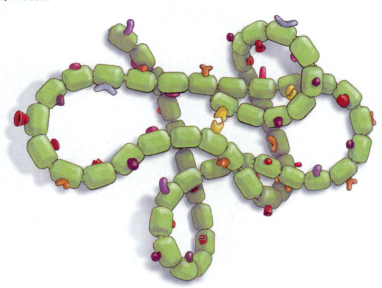

Figure 2.13 Amino acids, peptide bonds, and proteins. (a) All amino acids have the same backbone but different side groups. (b) Amino acids are joined together by peptide bonds. Long chains of these are called polypeptides. (c) Polypeptide chains fold upon themselves to produce proteins.

of three amino acids within the rock. However, it may be the case that these amino acids are contaminants; that is, they are present in the meteor because the meteor has been on protein-rich Earth for several thousand years. In addition, some amino acids are known to form under conditions where life is not present, so the presence of amino acids is not necessarily evidence of life.

Lipids.   One type of organic molecule, abundant in living organisms, that has not been found in the Martian meteorite is lipids. **Lipids** are partially or entirely hydrophobic organic molecules made primarily of hydrocarbons. Important lipids include fats, steroids, and phospholipids.

Fat.   The structure of a **fat** is that of a 3-carbon glycerol molecule with up to 3 long hydrocarbon chains attached to it (**Figure 2.14a**). Like the hydrocarbons present in gasoline, these can be burned to produce energy. The long hydrocarbon chains are called **fatty acid** tails of the fat. Fats are hydrophobic and function in storing energy within living organisms.

Steroids.   **Steroids** are composed of 4 fused carbon-containing rings. Cholesterol (**Figure 2.14b**) is one steroid that is probably familiar; its primary function in animal cells (plant cells do not contain cholesterol) is to help maintain the fluidity of membranes. Other steroids include the sex hormones testosterone, estrogen, and progesterone, which are produced by the sex organs and have effects throughout the body.

Phospholipids.   **Phospholipids** are similar to fats except that each glycerol molecule is attached to 2 fatty acid tails (not 3, as you would find in a dietary fat). The third bond in a phospholipid is to a phosphate head group. The phosphate head group consists of a phosphorous atom attached to four

Figure 2.14 Three types of lipids. (a) Fats are composed of a glycerol molecule with 3 hydrocarbon-rich fatty acid tails attached. (b) Cholesterol is a steroid common in animal cell membranes. (c) Phospholipids are composed of a glycerol backbone with 2 fatty acids attached and 1 phosphate head group. The cartoon drawing to the right shows how phospholipids are often depicted.

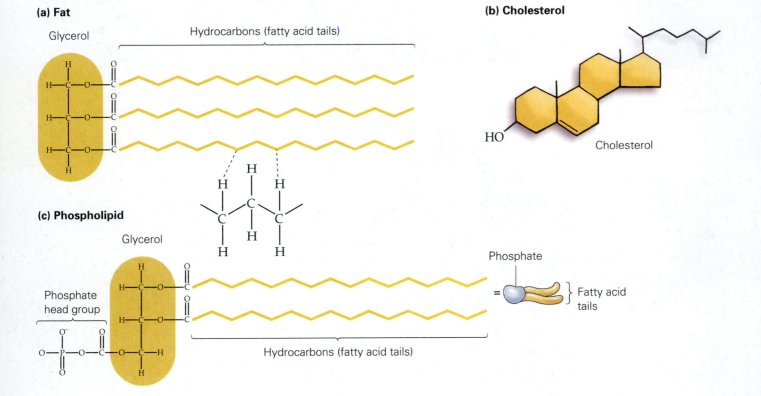

(a) Fat

Glycerol

Hydrocarbons (fatty acid tails)

(b) Cholesterol

Cholesterol

(c) Phospholipid

Glycerol

Phosphate head group

Hydrocarbons (fatty acid tails)

Phosphate

Fatty acid tails

oxygen atoms and is hydrophilic. Thus a phospholipid has a hydrophilic head and two hydrophobic tails (**Figure 2.14c**). Phospholipids often have an additional head group, attached to the phosphate, that also confers unique chemical properties on the individual phospholipid. Phospholipids are important constituents of the membranes that surround cells and that designate compartments within cells.

Even if the Martian meteorite contained unambiguous traces of carbohydrates, proteins, and lipids, the source of these molecules would not clearly be a living organism. Evidence of a mechanism for passing traits to the next generation would also be required. The hereditary, or genetic, information common to all life on Earth is in the form of nucleic acids.

Nucleic Acids. Nucleic acids are composed of long strings of monomers called **nucleotides.** A nucleotide is made up of a sugar, a phosphate, and a nitrogen-containing base. There are two classes of nucleic acids in living organisms. **Ribonucleic acid (RNA)** plays a key role in helping cells synthesize proteins (and is discussed in detail in further chapters). The nucleic acid that serves as the primary storage of genetic information in nearly all living organisms is **deoxyribonucleic acid (DNA). Figure 2.15** (on the next page) shows the three-dimensional structure of a DNA molecule and zooms inward to the chemical structure. You can see that DNA is composed of two curving strands that wind around each other to form a double helix. The sugar in DNA is the 5-carbon sugar deoxyribose. The nitrogen-containing bases, or **nitrogenous bases,** of DNA have one of four different chemical structures, each with a different name: **adenine (A), guanine (G), thymine (T),** and **cytosine (C).** Nucleotides are joined to each other along the length of the helix by covalent bonds.

Nitrogenous bases form hydrogen bonds with each other across the width of the helix. On a DNA molecule, an adenine (A) on one strand always pairs with a thymine (T) on the opposite strand. Likewise, guanine (G) always pairs with cytosine (C). The term **complementary** is used to describe these pairings. For example, A is complementary to T, and C is complementary to G. Therefore, the order of nucleotides on one strand of the DNA helix predicts the order of nucleotides on the other strand. Thus, if one strand of the DNA mole-cule is composed of nucleotides AACGATCCG, then we know that the order of nucleotides on the other strand is TTGCTAGGC.

As a result of this **base-pairing rule** (A pairs with T; G pairs with C), the width of the DNA helix is uniform. There are no bulges or dimples in the structure of the DNA helix because A and G, called **purines,** are structures composed of two rings; C and T are single-ring structures called **pyrimidines.** A purine always pairs with a pyrimidine and vice versa, so there are always 3 rings across the width of the helix. A-to-T base pairs have 2 hydrogen bonds holding them together. G-to-C pairs have 3 hydrogen bonds holding them together.

Each strand of the helix thus consists of a series of sugars and phosphates alternating along the length of the helix, the **sugar-phosphate backbone.** The strands of the helix align so that the nucleotides face "up" on one side of the helix and "down" on the other side of the helix. For this reason, the two strands of the helix are said to be antiparallel.

The overall structure of a DNA molecule can be likened to a rope ladder that is twisted, with the sides of the ladder composed of sugars and phosphates (the sugar-phosphate backbone) and the rungs of the ladder composed of the nitrogenous-base sequences A, C, G, and T. The structure of DNA was determined

(a) DNA double helix is made of two strands.

(b) Each strand is a chain of of antiparallel nucleotides.

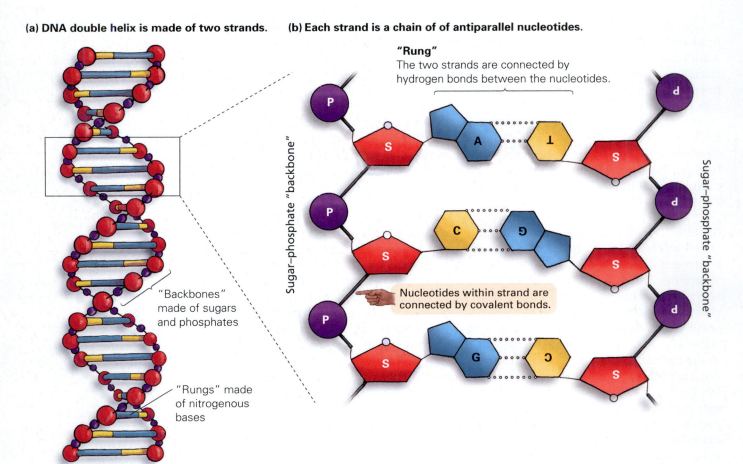

"Backbones" made of sugars and phosphates

"Rungs" made of nitrogenous bases

"Rung"
The two strands are connected by hydrogen bonds between the nucleotides.

Sugar-phosphate "backbone"

Sugar-phosphate "backbone"

Nucleotides within strand are connected by covalent bonds.

(c) Each nucleotide is composed of a phosphate, a sugar, and a nitrogenous base.

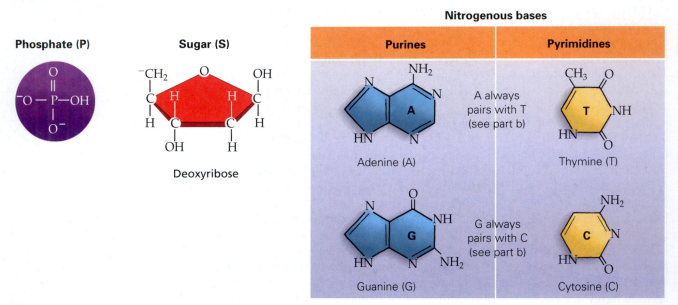

Phosphate (P)

Sugar (S)

Deoxyribose

Nitrogenous bases

Purines	Pyrimidines
Adenine (A)	A always pairs with T (see part b)	Thymine (T)
Guanine (G)	G always pairs with C (see part b)	Cytosine (C)

Figure 2.15 DNA structure. (a) DNA is a double-helical structure composed of nucleotides. (b) Each strand of the helix is composed of repeating units of sugars and phosphates, making the sugar-phosphate backbone, and of nitrogenous bases across the width of the helix. (c) Each nucleotide is composed of a phosphate, a sugar, and a nitrogenous base. Adenine and guanine are purines, which have a double-ring structure; cytosine and thymine are pyrimidines, which have a single-ring structure.

Visualize This: Look at part (a) of this figure. What do the red and purple spheres represent? Why are there two colors and sizes of nitrogenous bases?

by a group of scientists in the 1950s, most notably James Watson and Francis Crick (**Figure 2.16**).

How Might Macromolecules on Other Planets Differ? Many scientists argue that the fundamental constituents described here—carbohydrates, proteins, lipids, and nucleic acids—will be essentially similar wherever life is found. They will readily admit that the finer details are very likely to differ, however. For example, all proteins known on Earth contain only 20 different amino acids, despite an infinite number of possibilities. Presumably, proteins on other planets could contain completely different amino acids and many more than 20.

Not all scientists agree with the hypothesis that life on other planets will be based on carbon-containing organic compounds. Carbon is not the only chemical Tinkertoy™ connector; other elements, including silicon, can also make connections with four other atoms. Silicon is also relatively abundant in the universe and could theoretically form the backbone of an alternative organic chemistry. The basic constituents of silicon-based life could be very different from the chemical building blocks of life on Earth.

Even if all life in the universe is based on carbon chemistry, it is very unlikely that the suite of organisms found on another planet will look much like life on our planet. However, understanding the history of life on Earth also provides insight into the possible nature of life elsewhere in the universe.

Figure 2.16 The DNA model. American James Watson (left) and Englishman Francis Crick are shown with the three-dimensional model of DNA they devised while working at the University of Cambridge in England.

2.2 Life on Earth

One of the most dramatic features of the Martian meteorite is the presence of fossils that look remarkably like the tiniest living organisms known from Earth. The largest of these fossils is less than 1/100th of the diameter of a human hair, and most are about 1/1000th of the diameter of a human hair—small enough that it would take about 1000 laid end to end to span the dot at the end of this sentence. Some are egg-shaped, while others are tubular. These fossils appear similar to the simplest and most ancient of known organisms and are the strongest piece of evidence supporting the hypothesis that Mars once was home to living organisms.

David McKay and his colleagues argue that the fossil structures in the Martian meteorite are the remains of tiny cells. A **cell** is the fundamental structural unit of life on Earth, separated from its environment by a membrane and sometimes an external wall. Bacteria are single cells, which perform all of the activities required for life. More complex organisms can be composed of trillions of cells working together and do not have any cells that could survive and reproduce independently.

Prokaryotic and Eukaryotic Cells

All cells can be placed into one of two categories, prokaryotic or eukaryotic, based on the presence or absence of certain cellular structures. Scientists believe that the first prokaryotic cells appeared on Earth over 3.5 billion years ago and that the first eukaryotes appeared about 1.7 billion years later (Chapter 13).

Bacteria are **prokaryotic** cells, which are much smaller than eukaryotic cells (**Figure 2.17a** on the next page). Prokaryotes do not have a **nucleus,** a separate membrane-bound compartment to hold the genetic material. They also do not contain any membrane-bound internal compartments. They do however, have a **cell wall** that helps them maintain their shape (**Figure 2.17b**). According to the fossil record, prokaryotic cells predate the more complex eukaryotic cells (**Figure 2.17c**). The fossils in the Martian meteorite resemble modern prokaryotic cells known as nanobacteria and show no evidence of similar walls.

Eukaryotic cells have a nucleus and other internal structures with specialized functions, called **organelles,** that are surrounded by membranes.

(a) Different sizes: prokaryotic (red) vs. eukaryotic (white) cells

(c) Eukaryotic cell features

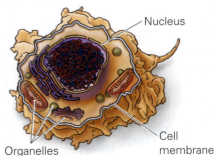

Nucleus

Cell
membrane

Organelles

(b) Prokaryotic cell features

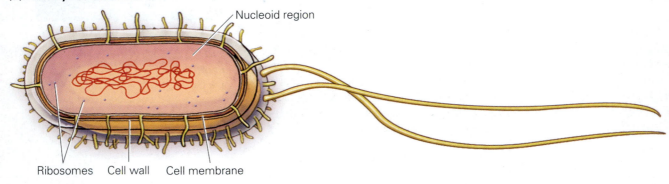

Nucleoid region

Ribosomes   Cell wall   Cell membrane

Figure 2.17 Prokaryotic and eukaryotic cells. (a) Prokaryotic cells are typically about 1/10 the diameter of a eukaryotic cell, as evidenced by the size of the two bacterial cells and a white blood cell. (b) Prokaryotic cells are structurally less complex than (c) eukaryotic cells.

Eukaryotic organisms include single-celled organisms such as amoebas and yeast as well as multicellular plants, fungi, and animals.

Many scientists dispute David McKay's interpretation that structures in ALH84001 are cellular and therefore evidence of life on Mars. In fact, similar structures can be formed in the absence of life by certain minerals under extremes of heat and pressure. If the Martian fossils are indeed cells, they should contain features similar to those found inside earthly cells.

Cell Structure

Each living cell can be considered a veritable factory, working to break down nutrients and to build required components. Let's look at the outside of the cell and then examine the structure and function of various internal cell components. **Table 2.1** describes the structures and functions of most cellular organelles in detail.

Plasma Membrane. All cells are enclosed by a structure called a **plasma membrane.** The plasma membrane defines the outer boundary of each cell, isolates the cell's contents from the environment, and serves as a semipermeable barrier that determines which nutrients are allowed into and out of the cell. Membranes that enclose structures inside the cell are usually referred to as cell membranes, while the outer boundary is the plasma membrane.

Internal and external membranes are composed in part of phospholipids. The chemical properties of these lipids make membranes flexible and self-sealing. When phospholipid molecules are placed in a watery solution, such as in a cell, they orient themselves so that their hydrophilic heads are exposed to the water and their hydrophobic tails are away from the water. They cluster into a form called a **phospholipid bilayer,** in which the tails of the phospholipids interact with themselves and exclude water, while the heads maximize their exposure to the surrounding water both inside and outside the membrane. The bilayer of phospholipids is stuffed with proteins that carry out enzymatic functions, serve as receptors for outside substances, and help transport substances throughout the cell.

A Fluid Mosaic of Lipids and Proteins. All of the lipids and most of the proteins in the plasma membrane are free to bob about, sliding from one location in

TABLE 2.1

Cell components. Illustrations and descriptions of cell components and their functions.	
Component	**Function**
Plasma membrane 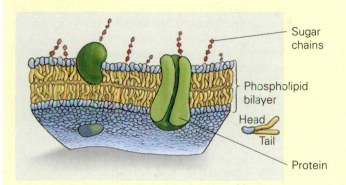	All cells are surrounded by a plasma membrane. It is composed of a bilayer of phospholipids perforated by proteins. Proteins in the bilayer help transport substances across the hydrophobic core of the membrane.
Cell wall 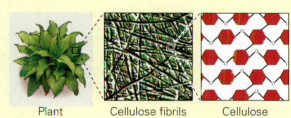	The cell wall is found outside the plasma membrane of plant and bacterial cells. The cell wall in plants is rich in the polysaccharide cellulose. Cellulose is assembled into strong fibrils and embedded in a matrix.
Nucleus 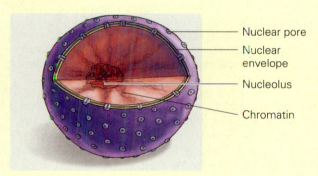	Eukaryotic cells contain a nucleus. The nucleus is a spherical structure surrounded by two membranes, together called the nuclear envelope. The nuclear envelope is studded with nuclear pores that regulate traffic into and out of the nucleus. Inside the nucleus is chromatin, composed of DNA and proteins. The nucleolus is where ribosomes are produced.
Mitochondrion 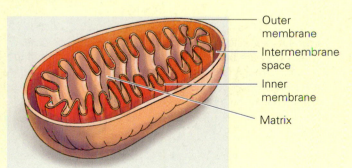	Plant and animal cells contain mitochondria, energy-producing organelles surrounded by two membranes. The inner and outer mitochondrial membranes are separated by the intermembrane space. The highly convoluted inner membrane carries many of the proteins involved in producing ATP. The matrix of the mitochondrion is the location of many of the reactions of cellular respiration.

In the Plasma membrane illustration: Sugar chains, Phospholipid bilayer, Head, Tail, Protein.

In the Cell wall illustration: Plant, Cellulose fibrils, Cellulose.

In the Nucleus illustration: Nuclear pore, Nuclear envelope, Nucleolus, Chromatin.

In the Mitochondrion illustration: Outer membrane, Intermembrane space, Inner membrane, Matrix.

(continued)

TABLE 2.1 *(continued)*

Cell components. Illustrations and descriptions of cell components and their functions.

Component	Function
Chloroplast 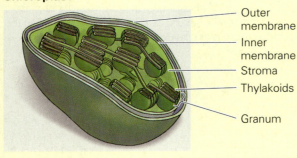 Outer membrane — Inner membrane — Stroma — Thylakoids — Granum	An important organelle present in plant cells, the chloroplast uses the sun's energy to convert carbon dioxide and water into sugars. Each chloroplast has an outer membrane, an inner membrane, a liquid interior called the stroma, and a network of membranous sacs called thylakoids that stack on one another to form structures called grana (singular: granum). Chloroplasts also contain pigment molecules that give green parts of plants their color.
Lysosome  Membrane — Digestive enzymes and digested material	A lysosome is a membrane-enclosed sac of digestive enzymes that degrade proteins, carbohydrates, and fats. Lysosomes roam around the cell and engulf targeted molecules and organelles for recycling.
Endoplasmic reticulum (ER) 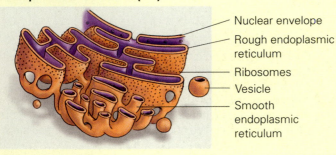 Nuclear envelope — Rough endoplasmic reticulum — Ribosomes — Vesicle — Smooth endoplasmic reticulum	The ER is a large network of membranes that begins at the nuclear envelope and extends into the cytoplasm of a eukaryotic cell. ER with ribosomes attached is called rough ER. Proteins synthesized on rough ER will be secreted from the cell or will become part of the plasma membrane. ER without ribosomes attached is called smooth ER. The function of the smooth ER depends on cell type but includes tasks such as detoxifying harmful substances and synthesizing lipids. Vesicles are pinched-off pieces of membrane that transport substances to the Golgi apparatus or plasma membrane.
Golgi apparatus  Vesicle from ER arriving at Golgi apparatus — Vesicle departing Golgi apparatus	The Golgi apparatus is a stack of membranous sacs. Vesicles from the ER fuse with the Golgi apparatus and empty their protein contents. The proteins are then modified, sorted, and sent to the correct destination in new transport vesicles that bud off from the sacs.
Ribosomes 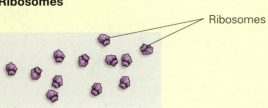 Ribosomes	Ribosomes are found in eukaryotic and prokaryotic cells. Ribosomes are built in the nucleolus and shipped out of the nucleus through nuclear pores to the cytoplasm, where they are used as workbenches for protein synthesis. They can be found floating in the cytoplasm or tethered to the ER.

TABLE 2.1

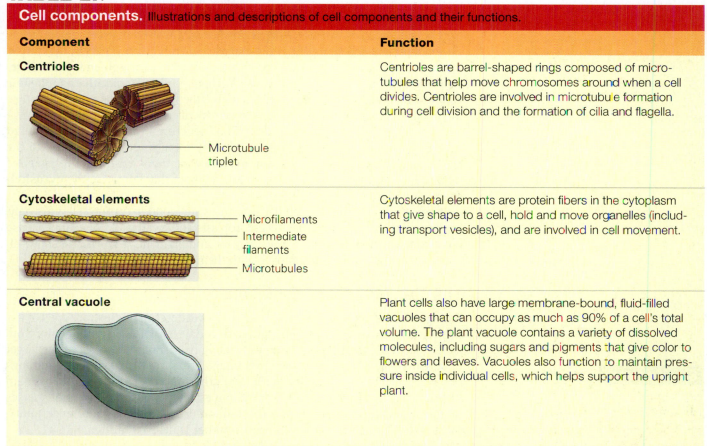

Cell components. Illustrations and descriptions of cell components and their functions.	
Component	**Function**
Centrioles	Centrioles are barrel-shaped rings composed of microtubules that help move chromosomes around when a cell divides. Centrioles are involved in microtubule formation during cell division and the formation of cilia and flagella.
Cytoskeletal elements	Cytoskeletal elements are protein fibers in the cytoplasm that give shape to a cell, hold and move organelles (including transport vesicles), and are involved in cell movement.
Central vacuole	Plant cells also have large membrane-bound, fluid-filled vacuoles that can occupy as much as 90% of a cell's total volume. The plant vacuole contains a variety of dissolved molecules, including sugars and pigments that give color to flowers and leaves. Vacuoles also function to maintain pressure inside individual cells, which helps support the upright plant.

the membrane to another. Because lipids and proteins move about laterally within the membrane, the membrane is a **fluid mosaic** of lipids and proteins. The membrane is fluid because the composition of any one location on the membrane can change. In the same manner that a patchwork quilt is a mosaic (different fabrics making up the whole quilt), so, too, is the membrane a mosaic with different regions of membrane composed of different types of phospholipids and proteins.

Cell membranes are **semipermeable** in the sense that they allow some substances to cross and prevent others from crossing. This characteristic allows cells to maintain a different internal composition from the surrounding solution.

Nucleus. All eukaryotic cells contain a nucleus surrounded by a double nuclear membrane, which houses the DNA. Inside the nucleus is the nucleolus, which is where ribosomes are assembled.

Ribosomes. Ribosomes are workbenches where proteins are assembled. They are composed of two subunits and can be found floating free in the cytosol or attached to certain membranes.

Cytosol. Between the nucleus and the plasma membrane lies the cytosol, a watery matrix containing water, salts, and many of the enzymes required for cellular reactions. The cytosol houses the organelles. The term **cytoplasm** includes the cytosol and organelles.

Organelles. Organelles are to cells as organs are to the body. Each organelle performs a specific job required by the cell, and all organelles work together to keep an individual cell healthy and to produce the raw materials that the cell needs to survive. Each organelle is enclosed in its own lipid bilayer. Some organelles are involved in metabolism. For example, organelles called **mitochondria** help the cells convert food energy into a form usable by cells, called ATP (adenosine triphosphate), while **chloroplasts** in plant cells use energy from sunlight to make sugars. **Lysosomes** help break down substances. Other organelles are involved in producing proteins. The endoplasmic reticulum (ER) is an extensive membranous organelle that can be studded with ribosomes and involved in protein synthesis (**rough endoplasmic reticulum**) or tubular in shape and involved in lipid synthesis (**smooth endoplasmic reticulum**). Proteins that are assembled on the membranes of the rough ER can be modified and sorted in a membranous structure called the **Golgi apparatus.**

There are other important subcellular structures that are not considered organelles because they are not bounded by membranes. In addition to the ribosomes described previously, **centrioles** are involved in moving genetic material around when a cell divides. Many fibers that compose the **cytoskeleton** help maintain the cell shape.

Some organelles and subcellular structures are found in certain cell types only. For instance, in addition to having chloroplasts and a cell wall, the plant cell also has a **central vacuole** to store water, sugars, and pigments. **Figure 2.18** shows an animal cell and a plant cell complete with their complement of organelles.

The Tree of Life and Evolutionary Theory

Biologists disagree about the total number of different **species,** or types of living organisms, that are present on Earth today. This uncertainty stems from lack of knowledge. Although scientists likely have identified most of the larger organisms—such as land plants, mammals, birds, reptiles, and fish—millions of species of insects, fungi, bacteria, and other microscopic organisms remain unknown to science. Amazingly, credible estimates of the number of species on Earth range from 5 million to 100 million. Despite this level of uncertainty, most biologists think that the likeliest number is near 10 million.

Theory of Evolution.   While the diversity of living organisms is tremendous, there exists remarkable similarity among all known species. All have the same basic biochemistry, including carbohydrates, lipids, proteins, and nucleic acids. All consist of cells surrounded by a plasma membrane. All eukaryotic organisms (including fungi, animals, and plants) contain nearly the same suite of cellular organelles. The best explanation for the shared characteristics of all species, what biologists refer to as

Stop & Stretch Would you expect prokaryotic cells to contain ribosomes? Why or why not?

Figure 2.18 Animal and plant cells. These drawings of a generalized (a) animal cell and (b) plant cell show the locations and sizes of organelles and other structures.
Visualize This: What three structures are present in plant but not animal cells?

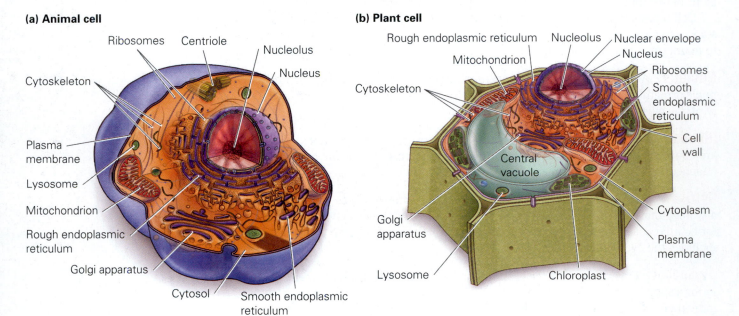

(a) Animal cell

Ribosomes   Centriole   Nucleolus   Nucleus   Cytoskeleton   Plasma membrane   Lysosome   Mitochondrion   Rough endoplasmic reticulum   Golgi apparatus   Cytosol   Smooth endoplasmic reticulum

(b) Plant cell

Rough endoplasmic reticulum   Nucleolus   Nuclear envelope   Mitochondrion   Nucleus   Ribosomes   Cytoskeleton   Smooth endoplasmic reticulum   Central vacuole   Cell wall   Golgi apparatus   Cytoplasm   Lysosome   Plasma membrane   Chloroplast "the unity of life," is that all living organisms share a common ancestor that arose on Earth nearly 4 billion years ago. The divergence and differences among modern species arose as a result of changes in the characteristics of populations, both in response to environmental change (a process called *natural selection*) and due to chance. These ideas underlie the entire science of biology and are known as the **theory of evolution.**

A major process in the diversification of life from a single ancestor was natural selection. The basic principle of the theory of natural selection (Chapter 11) is simple: Individual organisms vary from each other, and some of these variations increase their chances of survival and reproduction. A genetic trait that increases survival and reproduction should become more prevalent over time. In contrast, less successful variants should eventually be lost from the population.

The common ancestor can be thought of as the starting place for life on Earth, and the continual divergence among species and groups of species can be thought of as life's branching. Modern organisms can therefore be arranged on a tree of life that reflects their basic unity and relationships. According to current understanding, living organisms can be grouped into three large groups: two that are prokaryotic and one containing all eukaryotes. Eukaryotes can be further grouped into several categories made up primarily of free-living, single-celled organisms (such as amoebas and algae) and the three major multicellular groups—plants, fungi, and animals (**Figure 2.19**). (Chapter 13 provides a deeper exploration of the diversity of life on Earth.)

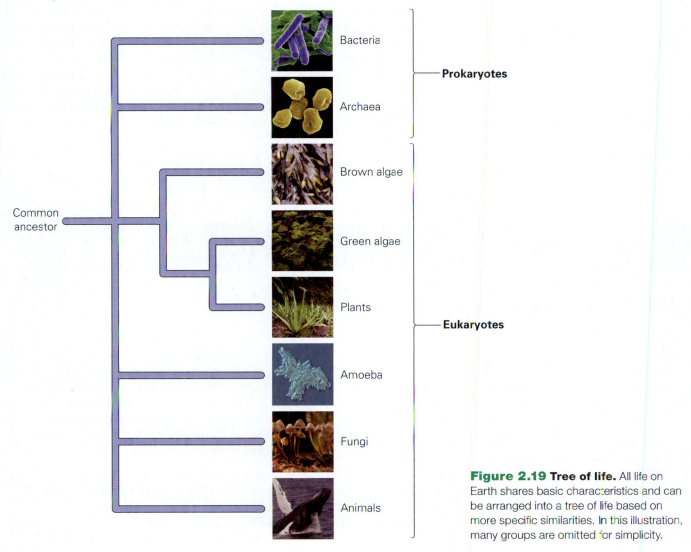

Figure 2.19 Tree of life. All life on Earth shares basic characteristics and can be arranged into a tree of life based on more specific similarities. In this illustration, many groups are omitted for simplicity.

Figure 2.20 Diversity of body form. Not all animals are bilaterally symmetric like us, with two eyes, two ears, two arms, two legs, an obvious head and "tail," and one central axis. A sea star is radially symmetric—it can be divided into two equal halves in any direction.

Visualize This: Is an earthworm bilaterally symmetric?

Because evolutionary change results from chance events and environmental changes (including the appearance of other species), the groups of organisms present on Earth today represent only one set of an infinite number of possibilities. Similarly, life on other planets need not look identical to life on Earth. For example, instead of the common body form found in animals, called bilateral symmetry, by which bodies can be visually divided into two mirror-image halves, life on other planets could be primarily radially symmetric and thus look very different (**Figure 2.20**). In fact, it is possible that life on other planets might not be based on carbon and may not be sustained by water. No one knows what living organisms would look like on a planet where the important molecules are based on the element silicon or where the sustaining liquid is not water.

Life in the Universe. Do other living organisms exist in the universe? Given the universe's sheer size and complexity, most scientists who study this question think that the existence of life on other planets is nearly certain.

While the evidence of life in the Martian meteorite is unconvincing to many scientists, we may in our lifetimes find out that life exists, or once existed, on our planetary neighbor. What about the existence of intelligent life that could communicate with us? Some scientists argue that as a result of natural selection, the evolution of intelligence is inevitable wherever life arises. Others point to the history of life on Earth—consisting of at least 2.5 billion years, during which all life was made up of single-celled organisms—to argue that most life in the universe must be "simple and dumb." It is clear from our explorations of the solar system that none of the sun's other planets hosts intelligent life. The nearest sun-like stars that could host an Earth-like planet, Alpha Centauri A and B, are over 4 light years away—nearly 40 trillion miles. With current technologies, it would take nearly 50,000 years to reach the Alpha Centauri stars, and there is certainly no guarantee that intelligent life would be found on any planets that circle them. For all practical purposes, at this time in human history, we are still alone in the universe.

SAVVY READER

Ionized Water

Several U. S. companies sell products called water ionizers. These are plastic-encased filtering systems about the size and shape of a small suitcase, which can be installed on or under your sink at a cost of close to $1000. What follows is an example of the kinds of claims made by these manufacturers' websites:

> Ionized water contains a negative electric charge that produces hydroxyl ions, and these charged particles help you to feel younger; have more energy; relieve chronic pain without medication; sleep better through the night, every night; rapidly skyrocket your energy and stamina to do all the things you love; enjoy a hugely improved quality of life every day. If you are looking for a natural solution that most conventional doctors don't know about and you want to get rid of cancer, chronic illness, headaches, high cholesterol, diabetes, arthritis, depression, or just not feeling well enough, you need to try this amazing healing water.

Research purporting to support the anticancer properties of ionized water is offered by these manufacturers. One of the claims made is that ionized water can clean up damaging oxygen free radicals (highly reactive molecules), the concentration of which is measured in various mouse tissues after exposure to ionized water or tap water.

1. Based on what you know about water, does the manufacturer's claim that ionized water contains a negative electric charge that produces hydroxyl ions make sense?

2. Evaluate this website by listing the first two red flags that are raised. (Use the questions in the checklist provided in Chapter 1, Table 1.2.)

3. Give two examples of language used on this website that differs from the way scientists usually describe results.

4. Do the authors provide evidence that a decrease in oxygen free radicals is associated with decreased cancer?

5. What key statistical element is missing from the graph?

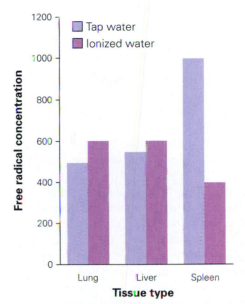

Anti-Cancer Effect of Ionized Water.

Chapter Review
Learning Outcomes

LO1 Describe the properties associated with living organisms (Section 2.1).

- Living organisms must be able to grow, metabolize substances, reproduce, and respond to external stimuli (p. 32).
- Living organisms contain a common set of biological molecules, are composed of cells, and can maintain homeostasis and evolve (p. 32).

LO2 List the components of water and some of the properties that make it important in living organisms (Section 2.1).

- Water consists of 1 oxygen and 2 hydrogen atoms (p. 33).
- Water is a good solvent in part because of its polarity and ability to form hydrogen bonds. Hydrogen bonding in water facilitates chemical reactions, promotes cohesion, and allows for heat absorption (pp. 33–34).
- The polarity of water also facilitates the dissolving of salts (p. 34).
- Water has a neutral pH. The pH scale is a measure of the relative percentages of ions in a solution and ranges from 0 (acidic or rich in hydrogen ions) to 14 (basic or rich in hydroxyl ions). Water has the same number of hydrogen as hydroxyl ions (p. 35).

LO3 Describe how atomic structure affects chemical bonding (Section 2.1).

- Chemical bonding depends on an element's electron configuration. Electrons closer to the nucleus have less energy than those that are farther away from the nucleus (p. 36).
- Atoms that have space in their valence shell form chemical bonds (p. 36).

LO4 Compare and contrast hydrogen, covalent, and ionic bonds (Section 2.1).

- Hydrogen bonds are weak attractions between hydrogen atoms and oxygen atoms in adjacent molecules (p. 34).
- Covalent bonds form when atoms share electrons. These tend to be strong bonds (p. 36).
- Ionic bonds form between positively and negatively charged ions. These tend to be weaker bonds (p. 37).

LO5 Discuss the importance of carbon in living organisms (Section 2.1).

- Life on Earth is based on the chemistry of the element carbon, which can make bonds with up to four other elements (pp. 35–36).

LO6 Describe the structure of carbohydrates, proteins, lipids, and nucleic acids and the roles these macromolecules play in cells (Section 2.1).

- Carbohydrates function in energy storage and play structural roles. They can be single-unit monosaccharides or multiple-unit polysaccharides with sugar monomers joined by covalent bonds (pp. 37–38).
- Proteins play structural, enzymatic, and transport roles in cells. They are composed of amino acid monomers joined by covalent bonds (pp. 38–40).
- Lipids are partially or entirely hydrophobic and come in three different forms. Fats are composed of glycerol covalently bonded to three fatty acids. Fats store energy. Phospholipids are composed of glycerol, 2 fatty acids, and a phosphate group. They are important structural components of cell membranes. Steroids are composed of 4 fused rings. Cholesterol is a steroid found in some animal cell membranes and helps maintain fluidity. Other steroids function as hormones (pp. 40–41).
- Nucleic acids are polymers of covalently bonded nucleotides, each of which is composed of a sugar, a phosphate, and a nitrogen-containing base (pp. 41–42).

LO7 Compare and contrast prokaryotic and eukaryotic cells (Section 2.2).

- There are two main categories of cells: Those with nuclei and membrane-bound organelles are eukaryotes; those lacking a nucleus and membrane-bound organelles are prokaryotes (pp. 43–44).

LO8 Describe the structure and function of cell membranes (Section 2.2).

- The plasma membrane that surrounds cells is a semipermeable boundary composed of a phospholipid bilayer that has embedded proteins and cholesterol. Cells also have internal membranous structures (p. 47).
- Lipids and proteins can move throughout membranes. This fluidity of the membrane allows changes in the protein and lipid composition (p. 47).
- Some organisms, such as plants, fungi, and bacterial cells, have a cell wall outside the plasma membrane that helps protect these cells and maintain their shape (p. 45).

LO9 Describe the structure and function of the subcellular organelles (Section 2.1).

- Subcellular organelles and structures perform many different functions within the cell. Mitochondria and chloroplasts are involved in energy conversions. Lysosomes are involved in the breakdown of macromolecules. Ribosomes serve as sites for protein synthesis. Proteins can be synthesized on ribosomes attached to rough endoplasmic reticulum. Smooth endoplasmic reticulum synthesizes lipids. The Golgi apparatus sorts proteins and sends them to their cellular destination. Centrioles help cells divide. The plant cell central vacuole stores water and other substances (pp. 45–48).

LO10 Provide a general summary of the theory of evolution (Section 2.2).

- There may be nearly 10 million unique life-forms on Earth. Despite all of this diversity, all life on Earth shares the same organic chemistry, genetic material, and basic cellular structures (p. 48).
- The similarities among living organisms on Earth provide support for the theory of evolution, which states that all life on Earth derives from a common ancestor. The process of evolutionary change since the origin of that ancestor led to the modern relationships among organisms, called the tree of life (pp. 49–50).

Roots to Remember

The following roots of words come mainly from Latin and Greek and will help you to decipher terms:

cyto- and **-cyte** mean cell or a kind of cell. Chapter terms: cytosol, cytoplasm, cytoskeleton

homeo- means like or similar. Chapter term: homeostasis

hydro- means water. Chapter terms: hydrophilic, hydrophobic

-mer means subunit. Chapter terms: polymer, monomer

macro- means large. Chapter term: macromolecule

micro- means small. Chapter term: microscopic

mono- means one. Chapter terms: monosaccharide, monomer

-philic means to love. Chapter term: hydrophilic

-phobic means to fear. Chapter term: hydrophobic

-plasm means a fluid. Chapter term: cytoplasm, plasma membrane

poly- means many. Chapter term: polymer

Learning the Basics

1. **LO6** List the four biological molecules commonly found in living organisms.

2. **LO9** Describe the structure and function of the subcellular organelles.

3. **LO8** Describe the structure and function of the plasma membrane.

4. **LO2** Water _____.

   A. is a good solute; B. dissociates into H^+ and OH^- ions; C. serves as an enzyme; D. makes strong covalent bonds with other molecules; E. has an acidic pH.

5. **LO3** Electrons _____.

   A. are negatively charged; B. along with neutrons comprise the nucleus; C. are attracted to the negatively charged nucleus; D. located closest to the nucleus have the most energy; E. all of the above are true.

6. **LO6** Which of the following terms is least like the others?

   A. monosaccharide; B. phospholipid; C. fat; D. steroid; E. lipid.

7. **LO6** Different proteins are composed of different sequences of _____.

   A. sugars; B. glycerols; C. fats; D. amino acids.

8. **LO6** Proteins may function as _____.

   A. genetic material; B. cholesterol molecules; C. fat reserves; D. enzymes; E. all of the above.

9. **LO6** A fat molecule consists of _____.

   A. carbohydrates and proteins; B. complex carbohydrates only; C. saturated oxygen atoms; D. glycerol and fatty acids.

10. **LO7** Eukaryotic cells differ from prokaryotic cells in that _____.

    A. only eukaryotic cells contain DNA; B. only eukaryotic cells have a plasma membrane; C. only eukaryotic cells are considered to be alive; D. only eukaryotic cells have a nucleus; E. only eukaryotic cells are found on Earth.

11. **LO4** Which of the following lists the chemical bonds from strongest to weakest?

    A. hydrogen, covalent, ionic; B. covalent, ionic, hydrogen; C. ionic, covalent, hydrogen D. covalent, hydrogen, ionic; E. hydrogen, ionic, covalent.

12. **LO10** Which of the following is not consistent with evolutionary theory?

    A. All living organisms share a common ancestor; B. The environment affects which organism survives to reproduce; C. Natural selection always favors the same traits, regardless of environment; D. Species change over time.

Analyzing and Applying the Basics

1. **LO1** A virus is made up of a protein coat surrounding a small segment of genetic material (either DNA or RNA) and a few proteins. Some viruses are also enveloped in membranes derived from the virus's host cell. Viruses cannot reproduce without taking over the genetic "machinery" of their host cell. Based on this description and biologists' definition of life, should a virus be considered a living organism?

2. **LO3** Recall that any molecule containing oxygen can be polar and that carbon-hydrogen bonds are nonpolar.

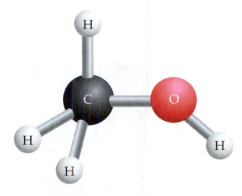

Figure 2.21 Methanol.

The structure of methanol (CH_3OH) is drawn in **Figure 2.21**. Which part of this molecule will have a partial negative charge, and which will have a partial positive charge?

3. **LO5** Some scientists have argued that silicon (Si) could also be an appropriate basis for organic chemistry because it is abundant and can form bonds with many other atoms. Carbon contains 6 electrons, and silicon contains 14. Recalling that the lowest electron shell contains 2 electrons, and the next 2 shells can contain a maximum of 8, how many "spaces" does silicon have in its valence shell? How does this compare to carbon?

Connecting the Science

1. Water's characteristic as an excellent solvent means that many human-created chemicals (including some that are quite toxic) can be found in water bodies around the globe. How would our use and manufacture of toxic chemicals be different if most of these chemicals could not be dissolved and diluted in water but instead accumulated where they were produced and used?

2. Do you believe that humans should expend considerable energy and resources looking for life, even intelligent life, elsewhere in the universe? Why or why not?

Answers to **Stop & Stretch, Visualize This, Savvy Reader,** and **Chapter Review** questions can be found in the **Answers** section at the back of the book.

CHAPTER **3**

Is It Possible to Supplement Your Way to Better Health?

Nutrients and Membrane Transport

Do supplemented waters improve hydration?

Nearly two-thirds of Americans take some sort of nutritional supplement, and virtually all of us have had a sports drink, vitamin water, or nutrition bar. A trip to the drug store will provide you access to hundreds of different vitamin, mineral, and plant-based herbal supplements, all claiming to enhance your health, looks, or longevity. Likewise, supermarket and convenience store aisles are filled with these conveniently packaged drinks and bars for eating on the go. Can these products help us make up for a poor diet? It seems that most Americans believe so—we spend around 6 billion dollars a year on these items. But are these products really necessary if we eat a healthy diet?

Do nutritional supplements improve health?

Are nutrition bars worth the cost?

Or is it better to just eat whole foods?

LEARNING OUTCOMES

LO1 Describe the role of nutrients in the body.

LO2 Explain the functions of water in the body.

LO3 Compare major dietary macronutrients, discussing the functions of each.

LO4 Compare dietary micronutrients, discussing the functions of each.

LO5 Describe the structure and function of the plasma membrane.

LO6 Distinguish between passive transport and active transport.

LO7 Compare the processes of endocytosis and exocytosis.

3.1 Nutrients

The food and drink that we ingest provide building-block molecules that can be broken down and used as raw materials for growth, for maintenance and repair, and as a source of energy. Another name for the substances in food that provide structural materials or energy is **nutrients.**

Macronutrients

Nutrients that are required in large amounts are called **macronutrients.** These include water, carbohydrates, proteins, and fats.

Water and Nutrition. Most animals can survive for several weeks with no nutrition other than water. However, survival without water is limited to just a few days. Besides helping the body disperse other nutrients, water helps dissolve and eliminate the waste products of digestion. Water helps to maintain blood pressure and is involved in virtually all cellular activities.

A decrease below the body's required water level, called **dehydration,** can lead to muscle cramps, fatigue, headaches, dizziness, nausea, confusion, and increased heart rate. Severe dehydration can result in hallucinations, heat stroke, and death. The body, including the brain, relies on water in the circulatory system to deliver nutrients as well as some oxygen. When water levels are low, the delivery system becomes less effective.

In addition, evaporation of water from the skin (sweating) helps maintain body temperature. When water is low and sweating decreases, the body temperature can rise to a harmful level.

Every day, humans lose about 3 liters of water as sweat, in urine, and in feces. To avoid dehydration, we must replace this water. We can replace some of it by consuming food that contains water. A typical adult obtains about 1.5 liters of water per day from food consumption, leaving a deficit of about 1.5 liters that must be replaced.

While it works quite well to replace this water with tap water, many people drink bottled water, in part because of concerns about the quality of tap water. However, the U.S. Food and Drug Administration (FDA), the government agency that sets the standards for bottled water, uses the same standards applied by the Environmental Protection Agency (EPA) to tap water. In other words, water from both sources should be equally clean. In fact, nearly 40% of bottled waters are actually derived directly from municipal tap water. Bottlers may, however, distill or filter this water to remove molecules that impart a smell or taste that consumers find objectionable.

Many people also chose to hydrate with sports drinks, most of which are high in simple sugars and calories. In fact, some of these drinks have so many additives that their solute concentration is higher than that of blood. When this happens, as you will learn more about in this chapter, water will leave the tissues, actually increasing the risk of dehydration.

Bottled waters and sports drinks are also chosen for convenience and portability, but this does come at a cost. The billions of bottles used each year require 1.5 million barrels of oil to produce. Although the bottles are recyclable, 86% of the bottles go to landfills each year. In addition, the energy required to transport water far exceeds the energy required to clean an equivalent amount of tap water.

Besides obtaining a healthy dose of water every day, people must consume foods that contain carbohydrates, proteins, and fats. (In Chapter 2, we explored the structure of these macromolecules, and now we focus on how they function in the body.)

Carbohydrates as Nutrients. Foods such as bread, cereal, rice, and pasta, as well as fruits and vegetables, are rich in sugars called carbohydrates. Carbohydrates are the major source of energy for cells. Energy is stored in the chemical bonds between the carbon, hydrogen, and oxygen atoms that comprise carbohydrate molecules (Chapter 2). Carbohydrates can exist as single-unit monosaccharides (individual sugar molecules) or can be bonded to each other to produce polysaccharides (long chains of sugar molecules).

The single-unit simple sugars are digested and enter the bloodstream quickly after ingestion. Sugars found in milk, juice, honey, and most refined foods are simple sugars. A sugar found in corn syrup, is shown in **Figure 3.1a**.

When multisubunit sugars are composed of many different branching chains of sugar monomers, they are called **complex carbohydrates.** Complex carbohydrates are found in vegetables, breads, legumes, and pasta.

When more sugar is present than needed, it can be stored for later use. Plants, such as potatoes, store their excess carbohydrates as the complex carbohydrate starch (**Figure 3.1b**). Animals store their excess carbohydrates as the complex carbohydrate glycogen in muscles and the liver (**Figure 3.1c**). Both starch and glycogen are polymers of glucose.

The body digests complex carbohydrates more slowly than it does simpler sugars because complex carbohydrates have more chemical bonds to break. Endurance athletes will load up on complex carbohydrates for several days before a race to increase the amount of stored energy they can draw on during competition.

Nutritionists agree that most of the carbohydrates in a healthful diet should be in the form of complex carbohydrates, and that it is best to consume only minimal amounts of processed sugars. A **processed food** is one that has undergone extensive refinement and, in doing so, has been stripped of much of its nutritive value. For example, unrefined raw brown sugar is made from the juice of the sugar cane plant and contains the minerals and nutrients found in the plant. Processing brown sugar to produce refined brown sugar results in the loss of these vitamins and minerals.

Fiber is an important part of a healthy diet. Also called *roughage,* fiber is composed mainly of those complex carbohydrates that humans cannot digest. For this reason, dietary fiber is passed into the large intestine, where some

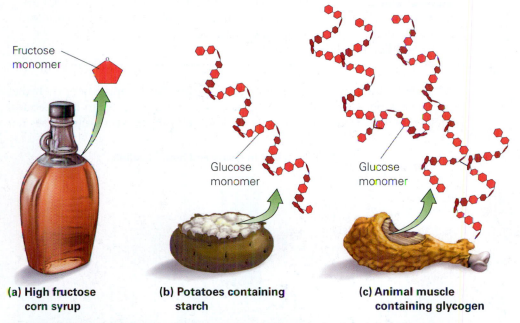

Fructose
monomer

Glucose
monomer

Glucose
monomer

(a) High fructose corn syrup　　　　**(b) Potatoes containing starch**　　　　**(c) Animal muscle containing glycogen**

Figure 3.1 Dietary carbohydrates. (a) The simple sugar fructose is found in corn syrup. Complex carbohydrates include (b) starch in potatoes and other vegetables and (c) glycogen found in animal muscle.

of it is digested by bacteria, and the remainder gives bulk to the feces. Foods that have not been stripped of their nutrition by processing are called **whole foods.** Whole grains, beans, and many fruits and vegetables are good sources of dietary fiber.

Although fiber is not a nutrient because it is not absorbed by the body, it is still an important part of a healthful diet. Fiber helps maintain healthy cholesterol levels. Fiber may also decrease your risk of various cancers.

One type of dietary supplement you can find on the shelves of grocery stores is fiber bars. While fiber bars can be rich in fiber, they are also often rich in processed sugars. In addition, many of these bars contain chocolate chips or other sugary sweeteners. Nutritionists recommend that such highly processed, sugar-rich foods occupy a very small portion of your diet because they contain more calories than nutrition.

Proteins as Nutrients.   Protein-rich foods include beef, poultry, fish, beans, eggs, nuts, and dairy products such as milk, yogurt, and cheese.

Proteins are composed of amino acids, which differ from each other based on their side groups (Chapter 2). Amino acids are bonded to each other in an infinite variety of combinations to produce a diverse array of proteins with many different functions.

Your body is able to synthesize many of the commonly occurring amino acids. Those your body cannot synthesize are called **essential amino acids** and must be supplied by the foods you eat. **Complete proteins** contain all the essential amino acids your body needs. Proteins obtained by eating meat are more likely to be complete than are those obtained by eating plants; plant proteins can often be missing one or more essential amino acids (**Figure 3.2**).

In the past, there was some concern that vegetarians might be at risk for deficiencies in certain amino acids. However, scientific studies have shown that there is little cause for concern. If a vegetarian's diet is rich in a wide variety of plant-based foods, the body will have little trouble obtaining all the amino acids it needs to build proteins.

Figure 3.2 Essential amino acids. Essential amino acids, such as lysine and valine, cannot be synthesized by the body and must be obtained from the diet. Some proteins are high in one amino acid but low in another. Eating a wide variety of proteins thus ensures that all the necessary amino acids are available for growth, development, and maintenance.

Stop & Stretch   Many cultures have characteristic meals that include protein from two different sources. For example, beans and rice, eggs and toast, and cereal and milk are common food pairings. Why might this practice of pairing different foods have evolved?

People looking to gain muscle mass will sometimes supplement their diet with protein powders that can be mixed with milk or water to help rebuild protein-rich muscle after a strenuous workout. However, extra protein does not

(a) Lentils are high in lysine and low in valine.   **(b) Rice is low in lysine and high in valine.**

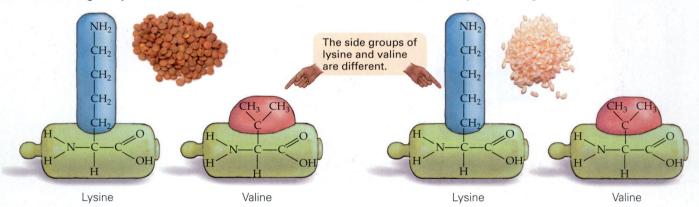

The side groups of lysine and valine are different.

Lysine   Valine   Lysine   Valine

always lead to extra muscle. Like any nutrient, excess protein is stored as fat. So if you eat too much protein, you will end up increasing body fat. In addition, diets that are too rich in protein can lead to health problems such as bone loss and kidney damage. If you are looking to build muscle, the most sensible thing to do is to combine your workout regimen with a healthful diet. Nutritionists recommend that a 100-pound person eat around 36 grams of protein a day, a 150-pound person around 54 grams of protein, and a 200-pound person around 72 grams of protein. Since the average chicken breast contains close to 50 grams of protein, a slice of cheese around 7 grams of protein, and one cup of broccoli around 5—you can see that it is easy to obtain all the protein you need through your diet. Even the most active people, endurance athletes during training, need a maximum of 50% more protein, still easily obtained through whole foods.

Fats as Nutrients. The body uses fat, a type of lipid, as a source of energy. Gram for gram, fat contains a little more than twice as much energy as carbohydrates or protein. This energy is stored in the chemical bonds of the fat molecule.

Foods that are rich in fat include meat, milk, cheese, vegetable oils, and nuts. Muscle is often surrounded by stored fat, but some animals store fat throughout muscle, leading to the marbled appearance of some red meat. Other animals—chickens, for example—store fat on the surface of the muscle, making it easy to remove for cooking (**Figure 3.3**). Most mammals, including humans, store fat just below the skin to help cushion and protect vital organs, to insulate the body from cold weather, and to store energy in case of famine. Some scientists believe that prehistoric humans often faced times of famine and may have evolved to crave fat.

Fats consist of a glycerol molecule with hydrogen- and carbon-rich fatty acid tails attached (Chapter 2). Your body can synthesize most of the fatty acids it requires. Those that cannot be synthesized are called **essential fatty acids.** Like essential amino acids, essential fatty acids must be obtained from the diet. Omega-3 and omega-6 fatty acids are essential fatty acids that can be obtained by eating fish. These fatty acids are thought to help protect against heart disease. Nutritionists recommend that people eat about 12 ounces of fish every week. Some people who aren't eating that much fish supplement their diets with fish oil capsules. However, fish are a good source of vitamin D and the mineral selenium, while fish oil capsules are not.

The fatty acid tails of a fat molecule can differ in the number and placement of double bonds (**Figure 3.4a** on the next page). When the carbons of a fatty acid are bound to as many hydrogens as possible, the fat is said to be a **saturated fat** (saturated in hydrogens). When there are carbon-to-carbon double bonds, the fat is not saturated in hydrogens, and it is therefore an **unsaturated fat** (**Figure 3.4b**). The more double bonds there are, the higher the degree of unsaturation will be. When it contains many unsaturated carbons, the fat is referred to as **polyunsaturated.** The double bonds in unsaturated fats make the structures kink instead of lying flat. This form prevents the adjacent fat molecules from packing tightly together, so unsaturated fat tends to be liquid at room temperature. Cooking oil is an example of an unsaturated fat. Unsaturated fats are more likely to come from plant sources, while the fats found in animals are typically saturated. Saturated fats, with their absence of carbon-to-carbon double bonds, pack tightly together to make a solid structure. This is why saturated fats, such as butter, are solid at room temperature.

Commercial food manufacturers sometimes add hydrogen atoms to unsaturated fats by combining hydrogen gas with vegetable oils under pressure. This process, called **hydrogenation,** increases the level of saturation of a fat. This process solidifies liquid oils, thereby making food seem less greasy and extending their shelf life. Margarine is vegetable oil that has undergone hydrogenation.

(a) Fat within muscle

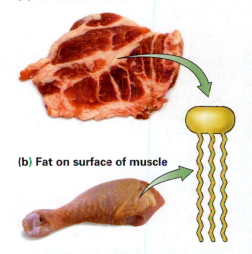

(b) Fat on surface of muscle

Figure 3.3 Fat storage. Fat can be intertwined with muscle tissue, as seen in this marbled piece of beef (a), or it can lie on the surface, as seen on this chicken leg (b).

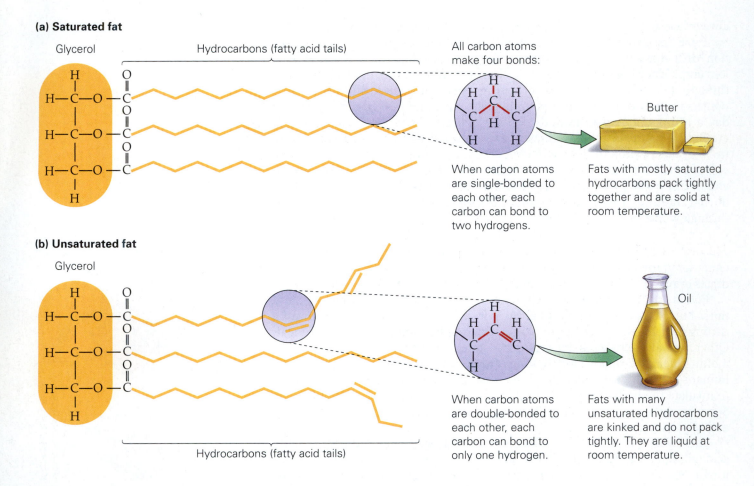

(a) Saturated fat

Glycerol

Hydrocarbons (fatty acid tails)

All carbon atoms make four bonds:

When carbon atoms are single-bonded to each other, each carbon can bond to two hydrogens.

Butter

Fats with mostly saturated hydrocarbons pack tightly together and are solid at room temperature.

(b) Unsaturated fat

Glycerol

Hydrocarbons (fatty acid tails)

When carbon atoms are double-bonded to each other, each carbon can bond to only one hydrogen.

Oil

Fats with many unsaturated hydrocarbons are kinked and do not pack tightly. They are liquid at room temperature.

Figure 3.4 Saturated and unsaturated fats. The types of bonds formed determine whether a fat will be (a) solid or (b) liquid at room temperature.
Visualize This: Which of the fatty acid tails in this figure is polyunsaturated?

Trans fats are produced by incomplete hydrogenation, which also changes the structure of the fatty acid tails in the fat so that, even though there are carbon-carbon double bonds, the fatty acids are flat and not kinked.

In contrast to most fats, trans fats are not required or beneficial. Fast food along with many of the energy bars, protein bars, fiber bars, and other nutrition bars on the market contain trans fats.

While definitive studies are currently under way, the potential health risks of consuming foods rich in trans-fatty acids, common in fast foods, include increased risk of clogged arteries, heart disease, and diabetes. Because fat contains more stored energy per gram than carbohydrate and protein do, and because excess fat intake is associated with several diseases, nutritionists recommend that you limit the amount of fat in your diet.

Micronutrients

Nutrients that are essential in minute amounts, such as vitamins and minerals, are called **micronutrients**. They are neither destroyed by the body during use nor burned for energy.

Vitamins. Vitamins are organic substances, most of which the body cannot synthesize. Most vitamins function as **coenzymes**, or molecules that help enzymes, and thus speed up the body's chemical reactions. When a vitamin is not present in sufficient quantities, deficiencies can affect every cell in the body because many different enzymes, all requiring the same vitamin, are involved in numerous different bodily functions. (The structure and function of enzymes will be covered in detail in Chapter 4.) Vitamins also help with the absorption of other nutrients; for example, vitamin C increases the absorption of iron from the intestine. Some vitamins may even help protect the body against cancer and heart disease and slow the aging process.

Vitamin D, also called calcitriol, is the only vitamin your cells can synthesize. Because sunlight is required for synthesis, people living in cold climates can develop deficiencies in vitamin D, and many health-care providers recommend vitamin D supplementation.

All other vitamins must be supplied by the foods you eat. Many vitamins, such as the B vitamins and vitamin C, are water soluble, so boiling causes them to leach out into the water—this is why fresh vegetables are more nutritious than cooked ones. Steaming vegetables or using the vitamin-rich broth of canned vegetables when making soup helps preserve the vitamin content. Because water-soluble vitamins are not stored by the body, a lack of water-soluble vitamins is more likely than a lack of fat-soluble vitamins to be the cause of dietary deficiencies. Vitamins A, D, E, and K are fat soluble and build up in stored fat; allowing an excess of these vitamins to accumulate in the body can be toxic. **Table 3.1** lists some vitamins and their roles in the body.

Many people take a daily vitamin supplement, and while this does not hurt, it is better, when possible, to obtain vitamins through eating whole foods.

TABLE 3.1

Vitamins. Water-soluble and fat-soluble vitamins.			
Water-Soluble Vitamins	• Small organic molecules • Will dissolve in water • Cannot be synthesized by body • Supplements packaged as pressed tablets • Excesses usually not a problem because water-soluble vitamins are excreted in urine, not stored		
Vitamin	**Sources**	**Functions**	**Effects of Deficiency**
Thiamin (B₁)	Pork, whole grains, leafy green vegetables	Required component of many enzymes	Water retention and heart failure
Riboflavin (B₂)	Milk, whole grains, leafy green vegetables	Required component of many enzymes	Skin lesions
Folic acid	Dark green vegetables, nuts, legumes (dried beans, peas, and lentils), whole grains	Required component of many enzymes	Neural tube defects, anemia, and gastrointestinal problems
B₁₂	Chicken, fish, red meat, dairy	Required component of many enzymes	Anemia and impaired nerve function
B₆	Red meat, poultry, fish, spinach, potatoes, and tomatoes	Required component of many enzymes	Anemia, nerve disorders, and muscular disorders
Pantothenic acid	Meat, vegetables, grains	Required component of many enzymes	Fatigue, numbness, headaches, and nausea
Biotin	Legumes, egg yolk	Required component of many enzymes	Dermatitis, sore tongue, and anemia
C	Citrus fruits, strawberries, tomatoes, broccoli, cabbage, green pepper	Collagen synthesis; improves iron absorption	Scurvy and poor wound healing
Niacin (B₃)	Nuts, leafy green vegetables, potatoes	Required component of many enzymes	Skin and nervous system damage

(continued)

TABLE 3.1 *(continued)*

Vitamins. Water-soluble and fat-soluble vitamins.

Fat-Soluble Vitamins	• Small organic molecules • Will not dissolve in water • Cannot be synthesized by body (except vitamin D) • Supplements packaged as oily gel caps • Excesses can cause problems since fat-soluble vitamins are not excreted readily	

Vitamin	Sources	Functions	Effects of Deficiency	Effects of Excess
A	Leafy green and yellow vegetables, liver, egg yolk	Component of eye pigment	Night blindness, scaly skin, skin sores, and blindness	Drowsiness, headache, hair loss, abdominal pain, and bone pain
D	Milk, egg yolk	Helps calcium be absorbed and increases bone growth	Bone deformities	Kidney damage, diarrhea, and vomiting
E	Dark green vegetables, nuts, legumes, whole grains	Required component of many enzymes	Neural tube defects, anemia, and gastrointestinal problems	Fatigue, weakness, nausea, headache, blurred vision, and diarrhea
K	Leafy green vegetables, cabbage, cauliflower	Helps blood clot	Bruising, abnormal clotting, and severe bleeding	Liver damage and anemia

Instead of taking a vitamin C supplement, it would be better to eat an orange because the orange, in addition to providing vitamin C, provides beta carotene (an antioxidant), calcium, fiber, and other nutrients. Vitamin C tablet supplements lack these other healthy components.

Minerals. Minerals are substances that do not contain carbon but are essential for many cell functions. Because they lack carbon, minerals are said to be inorganic. They are important for proper fluid balance, in muscle contraction and conduction of nerve impulses, and for building bones and teeth. Calcium, chloride, magnesium, phosphorus, potassium, sodium, and sulfur are all minerals. Like some vitamins, minerals are water soluble and can leach out into the water during boiling. Also like vitamins, minerals are not synthesized in the body and must be supplied through your diet. **Table 3.2** lists the various functions of minerals that your body requires and describes what happens when there is a deficiency or an excess in certain minerals.

Calcium is one of the more commonly supplemented minerals. Your body needs calcium to help blood clot, muscles contract, nerves fire, and maintain healthy bone structure. When dietary calcium is low, calcium is removed from the bones, weakening them. If you are not getting enough calcium in your diet, around 1000 mg/day, many health-care providers will recommend calcium supplements. Eight ounces of yogurt, 1.5 ounces of cheese, and one glass of milk together contain around 1000 mg of calcium.

Antioxidants

In addition to containing vitamins and minerals, many whole foods contain molecules called **antioxidants** that are thought to play a role in the prevention of many diseases, including cancer. Biologists are currently investigating antioxidants to see whether these substances can slow the aging process. Antioxidants protect cells from damage caused by highly reactive molecules that are

TABLE 3.2

Minerals. The minerals we require and their roles in the body.				
Minerals	• Will dissolve in water • Inorganic elements (do not contain carbon) • Cannot be synthesized by body • Supplements packaged as pressed tablets			
Mineral	**Sources**	**Functions**	**Effects of Deficiency**	**Effects of Excess**
Calcium	Milk, cheese, dark green vegetables, legumes	Bone strength, blood clotting	Stunted growth, osteoporosis	Kidney stones
Chloride	Table salt, processed foods	Formation of stomach acid	Muscle cramps, reduced appetite, poor growth	High blood pressure
Magnesium	Whole grains, leafy green vegetables, legumes, dairy, nuts	Required component of many enzymes	Muscle cramps	Neurological disturbances
Phosphorus	Dairy, red meat, poultry, grains	Bone and tooth formation	Weakness, bone damage	Impaired ability to absorb nutrients
Potassium	Meats, fruits, vegetables, whole grains	Water balance, muscle function	Muscle weakness	Muscle weakness, paralysis, and heart failure
Sodium	Table salt, processed foods	Water balance, nerve function	Muscle cramps, reduced appetite	High blood pressure
Sulfur	Meat, legumes, milk, eggs	Components of many proteins	None known	None known

generated by normal cell processes. These highly reactive molecules, called free radicals, have an incomplete electron shell, which makes them more chemically reactive than molecules with complete electron shells. Free radicals can damage cell membranes and DNA. Antioxidants inhibit the chemical reactions that involve free radicals and decrease the damage they can do in cells. Antioxidants are abundant in fruits and vegetables, nuts, grains, and some meats. **Table 3.3** (on the next page) describes food sources of common antioxidants.

Many different companies manufacture supplements claiming to be rich in antioxidants, and they even advertise that antioxidants have been shown to decrease risk of heart disease and stroke. While this may be true for people who eat diets rich in antioxidants, those findings have not yet been shown to hold true for people who take antioxidant supplements.

3.2 Transport Across Membranes

Once nutrients arrive at cells, they must traverse the membrane that surrounds cells, the plasma membrane. Molecules must cross the plasma membrane to gain access to the inside of the cell, where they can be used to synthesize cell components or be metabolized to provide energy for the cell. The chemistry of the membrane facilitates the transport of some substances and prevents the transport of others.

An Overview: Membrane Transport

The plasma membrane that surrounds cells is composed of a phospholipid bilayer (Chapter 2). The interior of the bilayer is hydrophobic. Hydrophobic substances can dissolve in the membrane and pass through it more easily than hydrophilic

TABLE 3.3

Antioxidants. Antioxidants are being investigated for their disease-preventing abilities.	
Antioxidants	• Present in whole foods • Protect cells from damage caused by free radicals • Thought to have a role in disease prevention
Antioxidant	**Source**
Beta-carotene	Foods rich in beta-carotene are orange in color; they include carrots, cantaloupe, squash, mangoes, pumpkin, and apricots. Beta-carotene is also found in some leafy green vegetables, such as collard greens, kale, and spinach.
Flavenoid	Cocoa and dark chocolate contain flavenoids.
Lutein	Lutein, which is known to help keep eyes healthy, is also found in leafy green vegetables such as collard greens, kale, and spinach.
Lycopene	Lycopene is a powerful antioxidant found in watermelon, papaya, apricots, guava, and tomatoes.
Selenium	Selenium is a mineral (not an antioxidant) that serves as a coenzyme for many antioxidant enzymes, thereby increasing their effectiveness. Rice, wheat, meats, bread, and Brazil nuts are major sources of dietary selenium.
Vitamin A	Foods rich in vitamin A include sweet potatoes, liver, milk, carrots, egg yolks, and mozzarella cheese.
Vitamin C	Foods rich in vitamin C include most fruits, vegetables, and meats.
Vitamin E	Vitamin E is found in almonds, many cooking oils, mangoes, broccoli, and nuts.

ones. In this sense, the membrane of the cell is differentially permeable to the transport of molecules, allowing some to pass and blocking others from passing.

Substances that can cross the membrane will do so until the concentration is equal on both sides of the membrane, a condition called equilibrium. Carbon dioxide, water, and oxygen move freely across the membrane. Larger molecules, charged molecules, and ions cannot cross the lipid bilayer on their own. If these substances need to be moved across the membrane, they must move through proteins embedded in the membrane. Proteins in the membrane

Plasma membrane

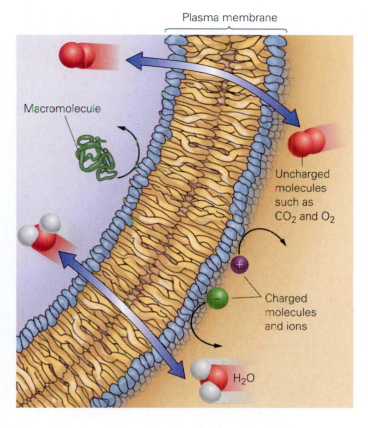

Macromolecule

Uncharged
molecules
such as
CO_2 and O_2

Charged
molecules
and ions

H_2O

Figure 3.5 **Transport of substances across membranes.**
The ability of a substance to cross a membrane is, in part, a function
of its size and charge.

can serve as channels to allow such molecules to cross until their concentration
is equal on both sides of the membrane. Proteins in the membrane can also
move substances across the membrane when more of a substance is required
on one side of the membrane than the other (**Figure 3.5**). *See the next section for*
A Closer Look at membrane transport.

A Closer Look:
Membrane Transport

The manner in which a substance is moved across a membrane depends on its particular chemistry and concentration
in the cell. Each type of transport has its own characteristics and energy requirements.

Passive Transport: Diffusion, Facilitated Diffusion, and Osmosis

All molecules contain energy that makes them vibrate
and bounce against each other, scattering around like bil-
liard balls during a game of pool. In fact, molecules will
bounce against each other until they are spread out over
all the available area. In other words, molecules will move from
their own high concentration to their own low concentration.
This movement of molecules from where they are in high
concentration to where they are in low concentration is called
diffusion. During diffusion, the net movement of molecules
is from their own high to their own low concentration or *down*

a concentration gradient. This movement does not require an
input of outside energy; it is spontaneous. Diffusion will continue until there is equilibrium, at which time no concentration gradient exists, and no net movement of molecules.

Diffusion also occurs in living organisms. When substances diffuse across the plasma membrane, we call the
movement **passive transport**. Passive transport is so
named because it does not require an input of energy. The
structure of the phospholipid bilayer that comprises the
plasma membrane prevents many substances from diffusing across it. Only very small, hydrophobic molecules are
able to cross the membrane by diffusion. In effect, these
molecules dissolve in the membrane to slip from one side of
the membrane to the other (**Figure 3.6a**).

Hydrophilic molecules and charged molecules such as
ions are unable simply to diffuse across the hydrophobic core
of the membrane. For example, when you have a meal of
chicken, rich in charged amino acids, and a green salad, rich

(continued on the next page)

(A Closer Look continued)

in hydrophilic carbohydrates and ions such as calcium (Ca^+), these amino acids, sugars, and ions cannot gain access to the inside of the cell on their own. Instead, these molecules are transported across membranes by proteins embedded in the lipid bilayer. This type of passive transport does not require an input of energy and is called **facilitated diffusion**.

Facilitated diffusion is so named because the specific membrane transport proteins are helping or "facilitating" the diffusion of substances across the plasma membrane

(**Figure 3.6b**). Although transport proteins are used to help the substance move across the plasma membrane, this form of transport is still considered to be diffusion because substances are traveling from high to low concentration.

The movement of water across a membrane is a type of passive transport called **osmosis**. Like other substances, water moves from its own high concentration to its own low concentration. Water can move through protein pores in the membrane, called aquaporins, but even without these, water can still cross the membrane. When an animal cell is placed in a solution of salt water, water leaves the cell, causing the cell

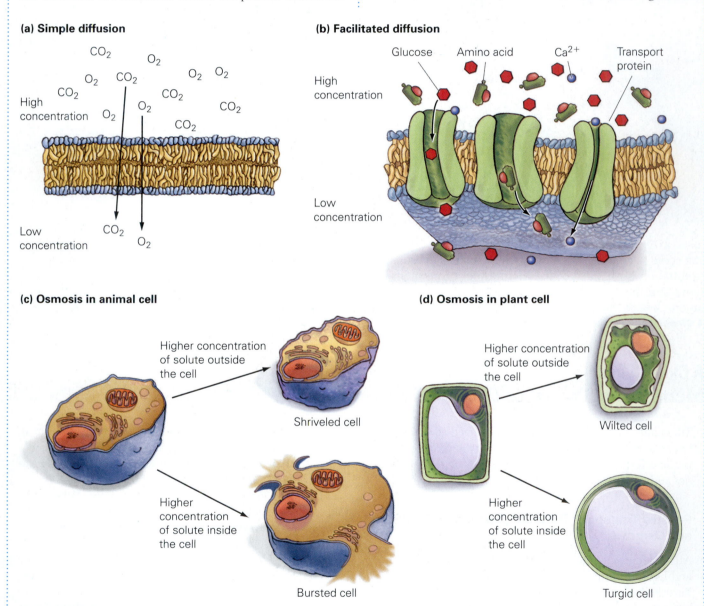

(a) Simple diffusion

(b) Facilitated diffusion

(c) Osmosis in animal cell

(d) Osmosis in plant cell

Figure 3.6 Three types of passive transport. (a) Simple diffusion of molecules across the plasma membrane occurs with the concentration gradient and does not require energy. Small hydrophobic molecules, carbon dioxide, and oxygen can diffuse across the membrane. (b) Facilitated diffusion is the diffusion of molecules assisted by substrate-specific proteins. Molecules move with their concentration gradient, which does not require energy. (c) Osmosis is a special type of diffusion that involves the movement of water in response to a concentration gradient. Water moves toward a region that has more dissolved solute. When water leaves an animal cell, it shrinks. (d) When water leaves a plant cell, the plant wilts instead of shrinks due to the presence of the cell wall.

to shrivel (**Figure 3.6c**). When an animal cell is placed in a solution with less dissolved solute than the cell, water will enter the cell, and it will expand and may even break open. Likewise, plants that are overfertilized or exposed to road salt wilt because water leaves the cells to equilibrate the concentration of water on either side of the plasma membrane (**Figure 3.6d**).

Stop & Stretch Suppose an endurance athlete was drinking a sports drink rich in substances that dissociate into ions called electrolytes, sugars, and other dissolved solutes. After drinking this drink, the athlete experiences dehydration. What might be happening?

Active Transport: Pumping Substances Across the Membrane

In some situations, a cell will need to maintain a concentration gradient. For example, nerve cells require a high concentration of certain ions inside the cell to transmit nerve impulses. To maintain this difference in concentration across the membrane requires the input of energy. Think of a hill with a steep incline or grade. Riding your

bike down the hill requires no energy, but riding your bike up the grade requires energy. At the cellular level, that energy is in the form of ATP (adenosine triphosphate, the main source of energy in cell reactions). **Active transport** is transport that uses proteins, powered by ATP, to move substances up a concentration gradient (**Figure 3.7**).

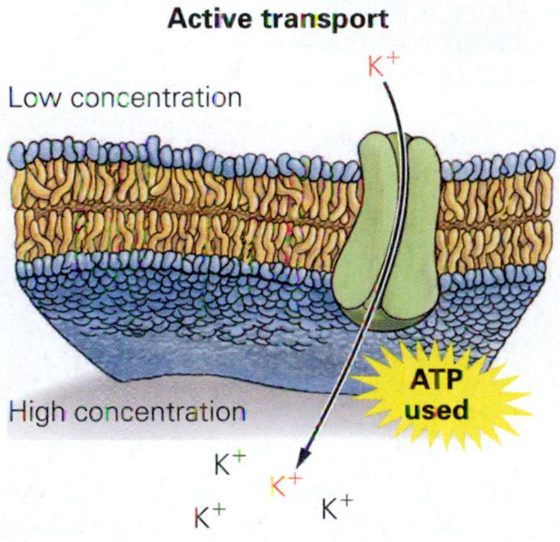

Stop & Stretch Muscle cells require a high internal concentration of potassium (K^+). Suppose you took a potassium supplement that needed to get inside a muscle cell. What type of transport would allow this ion to exist in a higher concentration inside than outside muscle cells?

Exocytosis and Endocytosis: Movement of Large Molecules Across the Membrane

Larger molecules are often too big to diffuse across the membrane or to be transported through a protein, regardless of whether they are hydrophobic or hydrophilic. Instead, they must be moved around inside membrane-bound vesicles (small sacs) that can fuse with membranes. **Exocytosis** (**Figure 3.8a**) occurs when a membrane-bound vesicle, carrying some substance, fuses with the plasma membrane and releases its contents into the exterior of the cell. **Endocytosis** (**Figure 3.8b**) occurs when a substance is brought into the cell by a vesicle pinching the plasma membrane inward.

Active transport

Low concentration

K$^+$

High concentration

K$^+$ K$^+$ K$^+$ K$^+$

ATP used

Figure 3.7 Active transport. Active transport moves substances against their concentration gradient and requires ATP energy to do so.

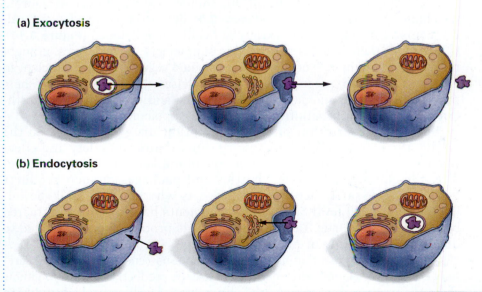

(a) Exocytosis

(b) Endocytosis

Figure 3.8 Movement of large substances. (a) Exocytosis is the movement of substances out of the cell. (b) Endocytosis is the movement of substances into the cell

(a) Protein breakdown

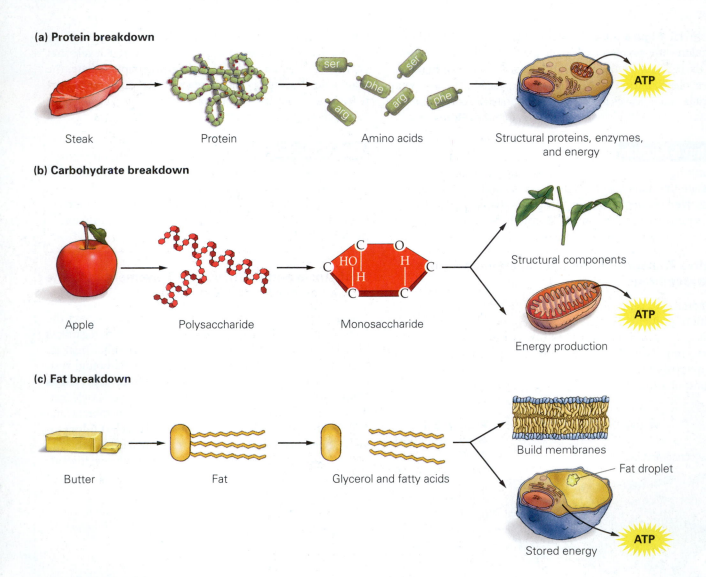

Steak   Protein   Amino acids   Structural proteins, enzymes, and energy

(b) Carbohydrate breakdown

Apple   Polysaccharide   Monosaccharide   Structural components

Energy production

(c) Fat breakdown

Butter   Fat   Glycerol and fatty acids   Build membranes

Fat droplet

Stored energy

Figure 3.9 You are what you eat. Food is digested into component molecules that are used to build cellular structures and generate ATP. (a) Proteins are broken down into amino acids and reassembled into proteins the cell requires and can be sources of energy. (b) Carbohydrates are broken down into sugars that can be reassembled into other sugars the cell requires or uses to produce energy. (c) Fats are broken down and reassembled to be used in membranes, burned to produce energy, or stored as fat.

You Are What You Eat

As you have seen, all the nutrients you consume and dismantle must find some way into your cells so that they can be used for energy and to build cellular components. Knowing all this should give heightened meaning to the phrase, "You are what you eat" (**Figure 3.9**). Because you are what you eat, it is important to eat a wide variety of healthy whole foods (**Figure 3.10**). If you do so, there is usually no need to supplement. In rare cases, it might make sense to supplement. For example, some vegetarians are deficient in vitamin B_{12} or iron, pregnant women are advised to take folic acid to help prevent spinal cord anomalies in their babies, and people living in northern climates might benefit from a vitamin D supplement.

Before a prescription or nonprescription drug is released to the public, the FDA requires scientific testing to prove its safety and effectiveness. However, the same is not true of dietary supplements. Therefore, the claims made on the bottles and packages of such products have not necessarily been proven. This is why you will find the following asterisked footnote: "These statements have not been evaluated by the Food and Drug Administration. This product is not intended to diagnose, treat, cure, or prevent any disease."

As with any product that makes claims about its benefits, you should carefully evaluate the claims before deciding to use the product. Use what you know about the process of science and the importance of well-controlled studies in

Figure 3.10 USDA Food Guide Pyramid. The newly designed pyramid stresses the importance of physical activity. The size of each triangle represents the relative proportion of your diet that should be composed of each food group.

Visualize This: The unlabeled yellow triangle represents the dietary nutrient you need in the smallest quantity. What does the unlabeled yellow triangle represent?

your evaluation. If you don't have the time or desire to do the research yourself, consult your physician or a trustworthy website. (Use the guidelines presented in Chapter 1.) The FDA also keeps a record on their website of dietary supplements that are under review for causing adverse effects.

Why pop high doses of expensive pills, powders, and potions when a healthy diet will get you all the nutrients you need, without the processing and costs?

SAVVY READER

The Açaí Bandwagon

It is often difficult to find trustworthy, research-based information about dietary supplements. This text was excerpted from a June 2009 Wellness Alert from the U.C. Berkeley Guide to Dietary Supplements (http://www.wellnessletter.com/html/ds/dsAcai.php).

In the nutraceutical or nutritional supplements market, there is never any shortage of bandwagons. One of the loudest and largest these days is the açaí bandwagon. Harvested from a Brazilian palm, açaí (ah-SAH-ee) berries are a dietary staple in Brazil and have also been used medicinally by Amazonian tribes. Açaí juice was introduced in the U.S. in 2001, and there are now more than 50 new food and drink products containing açaí. As a juice, pulp, powder, or capsule, it is marketed as a magic path to weight loss, a wrinkle remover, a way to cleanse the body of "toxins," and indeed just a plain old miracle cure. It is often combined with other

ingredients, such as glucosamine, so that the claims for benefits multiply exponentially.

Since açaí came on the market there have been a few studies pointing to potential benefits. Like many other fruits, açaí berries are high in antioxidants (molecules that quell cell-damaging free radicals) and other interesting compounds. But these were lab studies, and the results may not apply to humans. There is no scientific basis for weight-loss claims or any other health claims for açaí. The term "antioxidant" has become a sales tool.

Consumer protection groups such as the Center for Science in the Public Interest (CSPI) and the Better Business Bureau (BBB) have now come out against açaí marketers. "If Bernard Madoff were in the food business," said a CSPI nutritionist, "he'd be offering 'free' trials of açaí-based weight-loss products." Online ads regularly promise a free trial, saying that all

(continued on the next page)

you have to pay is shipping and handling. The catch is that you must supply your credit card number, and you'll automatically be signed up for $50 monthly shipments that will prove hard to cancel.

We urge you not to give your credit card number to anybody selling açaí products. Hundreds of complaints have been registered, and you may never get your money back. Beware of web-sites warning you of açaí scams—far from helping you get your money back, most turn out to be just sales pitches for more açaí.

There is no magic berry for weight loss or good health. Açaí berries are no doubt a good food, like other berries, but why pay a fortune for them or supplements containing them?

1. What are some indicators that this website might be a good source of information? (Use the checklist in Chapter 1 to evaluate this article.)

2. This article, and hundreds of other articles summarizing the research on dietary supplements, is put out by the University of California Berkeley. Should you be more skeptical of claims made by a university putting together a database of resources for the public or a private company that manufactures an herbal supplement? Justify your answer.

3. Many product websites will include some data that seem to support their claims. Private companies can hire their own scientists to perform studies that often have results that differ from those of government and university-sponsored scientists. Would you be more skeptical of results produced by scientists hired by the company whose product they are testing or scientists who work for the government or a university? Justify your answer.

Chapter Review

Learning Outcomes

Mastering BIOLOGY

Go to the Study Area at www.masteringbiology.com for practice quizzes, myeBook, BioFlix™ 3-D animations, MP3Tutor sessions, videos, current events, and more.

LO1 Describe the role of nutrients in the body (Section 3.1).

- Nutrients provide structural units and energy for cells (p. 56).

LO2 Describe the function of water in the body (Section 3.1).

- Water is an important dietary constituent that helps dissolve and eliminate wastes and maintain blood pressure and body temperature (p. 56).

LO3 Describe the major dietary macronutrients and discuss the functions of each (Section 3.1).

- Macronutrients are required in large amounts for proper growth and development. Macronutrients include carbohydrates, proteins, and fats. All of these molecules are composed of subunits that can be broken down for use by the cell (pp. 56–60).

LO4 List the major dietary micronutrients and describe their functions (Section 3.1).

- Micronutrients are dietary substances required in minute amounts for proper growth and development; they include vitamins and minerals (p. 60).

- Vitamins are organic substances, most of which the body cannot synthesize. Many vitamins serve as coenzymes to help enzymes function properly (pp. 60–61).
- Minerals are inorganic substances essential for many cell functions (p. 62).

LO5 Describe the structure and function of the plasma membrane (Section 3.2).

- To gain access to cells, nutrients move across the plasma membrane, which functions as a semipermeable barrier that allows some substances to pass and prevents others from crossing (p. 63).
- The plasma membrane is composed of two layers of phospholipids, in which are embedded proteins and cholesterol (pp. 63–65).

LO6 Distinguish between passive transport and active transport (Section 3.2).

- Passive transport mechanisms include simple diffusion and facilitated diffusion (diffusion through proteins). Passive transport always moves substances with their concentration gradient and does not require energy (pp. 65–67).

- Osmosis, the diffusion of water across a membrane, can involve the movement of water through protein pores in the membrane (pp. 66–67).
- Active transport is an energy-requiring process that requires proteins in cell membranes to move substances against their concentration gradients (p. 67).

LO7 Describe the processes of endocytosis and exocytosis (Section 3.2).

- Larger molecules move into (endocytosis) and out (exocytosis) of cells enclosed in membrane-bound vesicles (p. 67).

Roots to Remember

The following roots of words come mainly from Latin and Greek and will help you to decipher terms:

endo- means inside. Chapter term: endocytosis

exo- means outside. Chapter term: exocytosis

osmo- refers to water. Chapter term: osmosis

Learning the Basics

1. **LO1** What are the two main functions of nutrients?

2. **LO6** List three common cellular substances that can pass through cell membranes unaided.

3. **LO3** Macronutrients _____.

   A. include carbohydrates and vitamins; B. should comprise a small percentage of a healthful diet; C. are essential in minute amounts to help enzymes function; D. include carbohydrates, fats, and proteins; E. are synthesized by cells and not necessary to obtain from the diet.

4. **LO2** Which of the following is not a function of water?

   A. dispersing nutrients throughout the body; B. helping the body eliminate digestive wastes; C. helping to regulate body temperature; D. helping to regulate blood pressure.

5. **LO4** Micronutrients _____.

   A. include vitamins and carbohydrates; B. are not metabolized to produce energy; C. contain more energy than fatty acids; D. can be synthesized by most cells.

6. **LO5** The main constituents of the plasma membrane are _____.

   A. carbohydrates and lipids; B. proteins and phospholipids; C. fats and carbohydrates; D. fatty acids and nucleic acids.

7. **LO6** A substance moving across a membrane against a concentration gradient is moving by _____.

   A. passive transport; B. osmosis; C. facilitated diffusion; D. active transport; E. diffusion.

8. **LO6** A cell that is placed in salty seawater will _____.

   A. take sodium and chloride ions in by diffusion; B. move water out of the cell by active transport; C. use facilitated diffusion to break apart the sodium and chloride ions; D. lose water to the outside of the cell via osmosis.

9. **LO6 LO7** Which of the following forms of membrane transport require specific membrane proteins?

   A. diffusion; B. exocytosis; C. facilitated diffusion; D. active transport; E. facilitated diffusion and active transport.

10. **LO6** Water crosses cell membranes _____.

    A. by active transport; B. through protein pores called aquaporins; C. against its concentration gradient; D. in plant cells but not in animal cells.

Analyzing and Applying the Basics

1. **LO1 LO3** A friend of yours does not want to eat meat, so instead she consumes protein shakes that she buys at a nutrition store. Can you think of a dietary strategy that would allow her to be a vegetarian while not consuming protein shakes?

2. **LO1 LO2** The energy drink Red Bull is banned in several European countries due in part to concerns over its high caffeine content. Excess caffeine content can cause anxiety, heart palpitations, irritability, and difficulty sleeping. Many people use high-caffeine energy drinks as a mixer for alcohol, a practice that worries many health care practitioners. Why might mixing alcohol with energy drinks have even more negative health consequences than mixing alcohol with more conventional mixes such as pop, water, or juice?

3. **LO4** A vitamin water manufacturer claims that their product has all the vitamin C of an orange. What substances would you not be getting from the vitamin water that you would by eating an orange?

Connecting the Science

1. Select one fitness water or sports energy drink and research whether the claims made on its label are backed up by scientific evidence.

2. The New York City Board of Health recently adopted the nation's first major municipal ban on the use of trans fats in restaurant cooking. Do you think state health regulators should also ban trans fats in commercially prepared food like cookies and potato chips?

Answers to **Stop & Stretch, Visualize This, Savvy Reader,** and **Chapter Review** questions can be found in the **Answers** section at the back of the book

Fat: How Much Is Right for You?

Enzymes, Metabolism, and Cellular Respiration

Media images of attractive people tend to show the same body type.

The average college student spends a lot of time thinking about his or her body and ways to make it more attractive. While people come in all shapes and sizes, attractiveness tends to be more narrowly defined by the images of men and women we see in the popular media. Nearly all media images equate attractiveness and desirability with a limited range of body types.

For men, the ideal includes a tall, broad-shouldered, muscular physique with so little body fat that every muscle is visible. The standards for female beauty are equally unforgiving and include small hips, long and thin limbs, large breasts, and no body fat.

Because these ideals are difficult, if not impossible, to reach, we can end up feeling bad about our bodies and confused about what a healthy normal body looks like. How do you know if your body weight is a healthful one? Should you be dieting or trying to gain weight, or are you fine just as you are? This chapter will try to help you answer these questions.

Males are tall, thin, broad-shouldered, and muscular, while females are thin and large breasted.

For many people, trying to attain a body type that differs from their own can lead to anorexia ...

or obesity.

LEARNING OUTCOMES

LO1 Define the term metabolism.

LO2 Describe the structure and function of enzymes.

LO3 Explain how enzymes decrease a reaction's activation energy barrier.

LO4 List several reasons why metabolic rates differ between people.

LO5 Describe the structure and function of ATP.

LO6 Describe the process of cellular respiration from the breakdown of glucose through the production of ATP.

LO7 Explain how proteins and fats are broken down during cellular respiration.

LO8 Compare and contrast aerobic and anaerobic respiration.

LO9 Outline the health risks that occur with being overweight and underweight.

4.1 Enzymes and Metabolism

It is important to eat a well-balanced diet rich in unprocessed foods and the right amount of food (Chapter 3). All food, whether carbohydrate, protein, or fat, can be turned into fat when too much is consumed. In this manner, energy stored in the chemical bonds of food is converted into fat and stored for later use.

The amount of fat that a given individual will store depends partly on how quickly or slowly he or she breaks down food molecules into their component parts. **Metabolism** is a general term used to describe all of the chemical reactions occurring in the body.

Enzymes

All metabolic reactions are regulated by proteins called **enzymes** that speed up, or **catalyze,** the rate of biological reactions. Enzymes can help break down or build up substances or build more complex substances from simpler ones. The enzymes that help your body break down the foods you ingest liberate the energy stored in the food's chemical bonds. Enzymes are usually named for the reaction they catalyze and end in the suffix -*ase*. For example, sucrase is the enzyme that breaks down the table sugar sucrose.

To break chemical bonds, molecules must absorb energy from their surroundings, often by absorbing heat. This is why heating chemical reactants will speed up a reaction. However, heating cells to an excessively high temperature can damage or kill them, in part because proteins begin to break down. Enzymes do not require heat to catalyze the body's chemical reactions, so they break chemical bonds without damaging or killing cells. Furthermore, by eliminating the heat energy required to start a chemical reaction, enzymes allow the breakdown of chemical bonds to occur more quickly.

Activation Energy. The energy required to start the metabolic reaction serves as a barrier to catalysis and is called the **activation energy** (**Figure 4.1**). If not for the activation energy barrier, all of the chemical reactions in cells would occur relentlessly, whether the products of the reactions were needed or not. Because most metabolic reactions need to surpass the activation energy barrier before proceeding, they can be regulated by enzymes. In other words, a given chemical reaction will occur only if the correct enzyme is available and active. How do enzymes decrease the activation energy barrier?

Induced Fit. The chemicals that are metabolized by an enzyme-catalyzed reaction are called the enzyme's **substrate.** Enzymes decrease activation energy by binding to their substrate and placing stress on its chemical bonds, decreasing the amount of initial energy required to break the bonds. The region of the enzyme where the substrate binds is called the enzyme's **active site.** Each active site has its own shape and chemical climate. When the substrate binds to the active site, the enzyme changes shape slightly to envelop the substrate. This shape change by the enzyme in response to substrate binding results in stress being placed on the bonds of the substrate. This is called the **induced fit** model of enzyme catalysis. When the enzyme changes shape, it binds to the substrate more tightly, making it easier to break the substrate's chemical bonds. In this manner, the enzyme helps convert the substrate to a reaction product and then resumes its original shape so that it can perform the reaction again (**Figure 4.2**).

Different enzymes catalyze different reactions by a property called **specificity.** The specificity of an enzyme is the result of its shape and the shape of its active site. Different enzymes have unique shapes because they are composed of amino acids in varying sequences. The 20 amino acids, each with its own unique side group, are arranged in distinct numbers and orders for

(a) No enzyme present

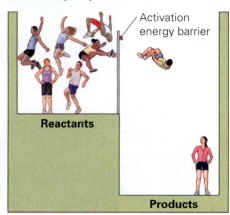

(b) Enzyme present

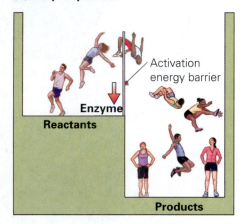

Figure 4.1 Activation energy.
(a) Few high jumpers can make it across the bar when the bar is high. Likewise, few reactants can break bonds to form products without surmounting an activation energy barrier. (b) Lowering the high-jump bar allows more jumpers to clear the bar. Enzymes lower the activation energy barrier in cells.

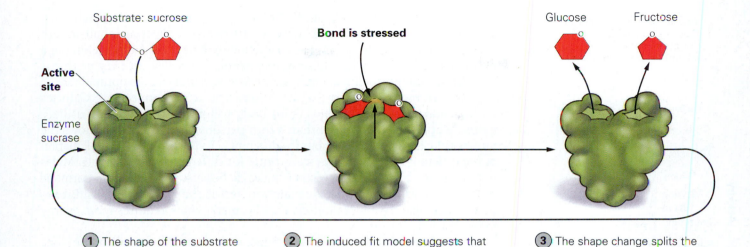

Substrate: sucrose

Active site

Enzyme sucrase

Bond is stressed

Glucose   Fructose

① The shape of the substrate matches the shape of the enzyme's active site.

② The induced fit model suggests that initial substrate binding to the active site changes the shape of the active site, inducing the substrate to fit even more snugly in the active site, stressing the bonds of the substrate.

③ The shape change splits the substrate and releases the two subunits. The enzyme is able to perform the reaction again.

each enzyme, producing enzymes of all shapes and sizes, each with an active site that can bind with its particular substrate. Although an infinite variety of enzymes could be produced, it is quite often the case that different organisms will utilize very similar enzymes, likely due to their evolution from a common ancestor.

Enzymes mediate all of the metabolic reactions occurring in an organism's cells. They even affect the rate of a particular individual's metabolism. This means that two similar-size individuals might need to consume different amounts of food to meet their daily energy requirements.

Figure 4.2 Enzymes. The enzyme sucrase is cleaving (splitting) the disaccharide sucrose into its monosaccharide subunits, fructose and glucose.
Visualize This: Is the enzyme itself permanently altered by the process of catalysis?

Calories and Metabolic Rate

Energy is measured in units called calories. A **calorie** is the amount of energy required to raise the temperature of 1 gram of water by 1 degree Celsius (1°C). In scientific literature, energy is usually reported in kilocalories, and 1 kilocalorie equals 1000 calories of energy. However, in physiology—and on nutritional labels—the prefix *kilo-* is dropped, and a kilocalorie is referred to as a **Calorie** (with a capital *C*). Calories are consumed to supply the body with energy to do work, which includes maintaining body temperature.

Balancing energy intake versus energy output means eating the correct amount of food to maintain health. When foods are eaten, they are broken down into their component subunits. The energy stored in the chemical bonds of food can be used to make a form of energy that the cell can use. When the supply of Calories is greater than the demand, the excess Calories can be stored as fat.

The speed and efficiency of many different enzymes will lead to an overall increase or decrease in the rate at which a person can break down food. Thus, when you say that your metabolism is slow or fast, you are actually referring to the speed at which enzymes catalyze chemical reactions in your body.

A person's **metabolic rate** is a measure of his or her energy use. This rate changes according to the person's activity level. For example, we require less energy when asleep than we do when exercising. The **basal metabolic rate** represents the resting energy use of an awake, resting, but alert person. The average basal metabolic rate is 70 Calories per hour, or 1680 Calories per day. However, this rate varies widely among individuals because many factors

To calculate the number of Calories you are burning per hour, multiply your weight by these numbers.

Activity	Calories/hour/pound of body weight
Walking	2.16
Hiking	2.52
Bicycling	2.70
Tennis	3.00
Mowing lawn	3.06
Swimming	3.48
Basketball	3.78
Jogging	4.15
Running	5.28

Figure 4.3 Energy expenditures for various activities. This bar graph can help you determine how many Calories you burn during certain activities.
Visualize This: How many Calories would a 160-pound person burn in 30 minutes of swimming?

influence each person's basal metabolic rate: exercise habits, body weight, sex, age, and genetics. Overall nutritional status can also affect metabolism. For example, vitamins can function as coenzymes (Chapter 3). A diet lacking in a particular vitamin may lead to slowed metabolism because an enzyme that is missing its vitamin coenzyme may not be able to perform at its optimal rate.

Exercise requires energy, which allows you to consume more Calories without having to store them. As for body weight, a heavy person utilizes more Calories during exercise than a thin person does. **Figure 4.3** shows the number of Calories used per hour for different activities, and allows calculation of the different rate of energy expenditure for different sizes of people. Males require more Calories per day than females do because testosterone, a hormone produced in larger quantities by males, increases the rate at which fat breaks down. Men also have more muscle than women, which requires more energy to maintain than fat does.

Age and genetics also play a role in metabolic rate. Two people of the same size and sex, who consume the same number of Calories and exercise the same amount, will not necessarily store the same amount of fat. The rate at which the foods you eat are metabolized slows as you age, and some people are simply born with lower basal metabolic rates. To obtain a rough measure of how many Calories you should consume per day, multiply the weight you wish to maintain by 11 and add the number of Calories you burn during exercise.

The properties of metabolic enzymes, like those of all proteins, are determined by the genes that encode them. Genes that influence a person's rate of fat storage and utilization are passed from parents to children. All of these variables help explain why some people seem to eat and eat and never gain an ounce, while others struggle with their weight for their entire lives.

Stop & Stretch Losing 1 pound of fat requires you to burn 3500 more Calories than you consume. If a person is trying to lose 1 pound per week and he decreases his caloric intake by 300 Calories per day, how many more Calories must he burn each day by exercising to reach his goal?

To be fully metabolized, food is broken down by the digestive system and then transported to individual cells via the bloodstream. Once inside cells, food energy can be converted into chemical energy by the process of cellular respiration.

4.2 Cellular Respiration

Cellular respiration is a series of metabolic reactions that converts the energy stored in chemical bonds of food into energy that cells can use while releasing waste products. Energy is stored in the electrons of chemical bonds, and when bonds are broken in a three-stage process, **adenosine triphosphate,** or **ATP,** is produced. ATP can supply energy to cells because it stores energy obtained from the movement of electrons that originated in food into its own bonds. Before trying to understand cellular respiration, it is important to have a better understanding of this chemical.

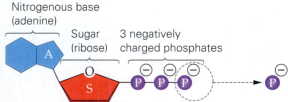

Figure 4.4 The structure of ATP. ATP is a nucleotide (sugar + phosphate + nitrogenous base) with a total of 3 negatively charged phosphates. Releasing a phosphate group from ATP releases energy.

Structure and Function of ATP

Structurally, ATP is a nucleotide triphosphate. It contains the nitrogenous base adenine, a sugar, and three phosphates (**Figure 4.4**). Each phosphate in the series of three is negatively charged. These negative charges repel each other, which contributes to the stored energy in this molecule.

Removal of the terminal phosphate group of ATP releases energy that can be used to perform cellular work. In this manner, ATP behaves much like a coiled spring. To think about this, imagine loading a dart gun. Pushing the dart into the gun requires energy from your arm muscles, and the energy you exert will be stored in the coiled spring inside the dart gun (**Figure 4.5**). When you shoot the dart gun, the energy is released from the gun and used to perform some work—in this case, sending a dart through the air.

The phosphate group that is removed from ATP can be transferred to another molecule. Thus, ATP can energize other compounds through **phosphorylation,** which means that it transfers a phosphate to another molecule. When a molecule, say an enzyme, needs energy, the phosphate group is transferred from ATP to the enzyme, and the enzyme undergoes a change in shape that allows the enzyme to perform its job. After removal of a phosphate group, ATP becomes adenosine diphosphate (ADP) (**Figure 4.6**). The energy released by the removal of the outermost phosphate of ATP can be used to help cells perform many different kinds of work. ATP helps power mechanical work such as the movement of cells, transport work such as the movement of substances across membranes during active transport, and chemical work such as the making of complex molecules from simpler ones (**Figure 4.7**).

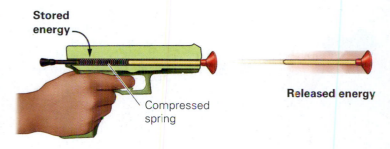

Figure 4.5 Stored energy. A dart gun uses energy stored in the coiled spring and supplied by the arm muscle to perform the work of propelling a dart.

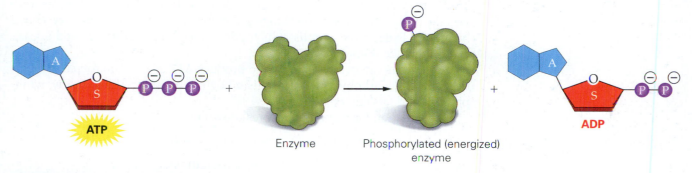

Figure 4.6 Phosphorylation. The terminal phosphate group of an ATP molecule can be transferred to another molecule, in this case an enzyme, to energize it. When ATP loses a phosphate, it becomes ADP.

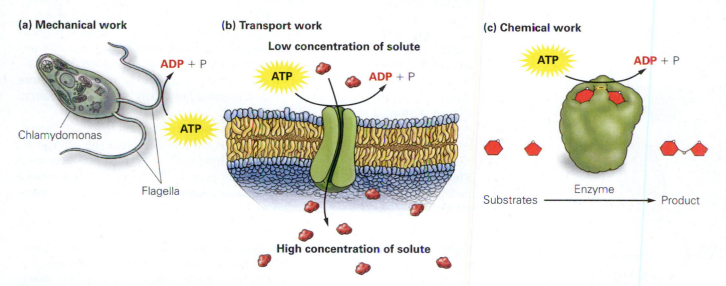

Figure 4.7 ATP and cellular work. ATP powers (a) mechanical work, such as the moving of the flagella of this single-celled green algae; (b) transport work, such as the active transport of a substance across a membrane from its own low to high concentration; and (c) chemical work, such as the enzymatic conversion of substrates to a product.

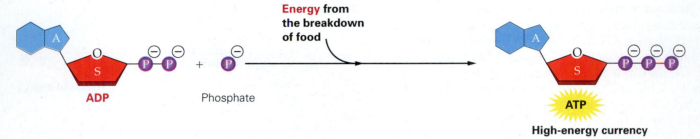

Figure 4.8 Regenerating ATP. ATP is regenerated from ADP and phosphate during the process of cellular respiration.

Cells are continuously using ATP. Exhausting the supply of ATP means that more ATP must be regenerated. ATP is synthesized by adding back a phosphate group to ADP during the process of cellular respiration (**Figure 4.8**). During this process, oxygen is consumed, and water and CO_2 are produced. Because some of the steps in cellular respiration require oxygen, they are said to be **aerobic** reactions, and this type of cellular respiration is called **aerobic respiration.**

An Overview: Cellular Respiration

The word *respiration* can also be used to describe breathing. When we breathe, we take oxygen in through our lungs and expel carbon dioxide. The oxygen we breathe in is delivered to cells, which undergo cellular respiration and release carbon dioxide (**Figure 4.9**).

Most foods can be broken down to produce ATP as they are routed through this process. Carbohydrate metabolism begins earliest in the pathway, while proteins and fats enter at later points.

The equation for carbohydrate breakdown is

$$C_6H_{12}O_6 + 6O_2 \rightarrow 6CO_2 + 6H_2O$$

Glucose + Oxygen → Carbon dioxide + Water

Glucose is an energy-rich sugar, but the products of its digestion—carbon dioxide and water—are energy poor. So where does the energy go? The energy released during the conversion of glucose to carbon dioxide and water is used to synthesize ATP. Many of the chemical reactions in this process occur in the mitochondria, organelles that are found in both plant and animal cells. Mitochondria are surrounded by an inner and an outer membrane. The space between the two membranes is called the **intermembrane space.** The semifluid medium inside the mitochondrion is called the matrix (**Figure 4.10a**).

Through a series of complex reactions in the mitochondrion, a glucose molecule breaks apart, and carbon and oxygen are released from the cell as carbon dioxide. Hydrogens from glucose combine with oxygen to produce water (**Figure 4.10b**). Gaining an appreciation for *how* this happens requires a more in-depth look. ***See the next section for A Closer Look at glycolysis, the citric acid cycle, and electron transport and ATP synthesis.***

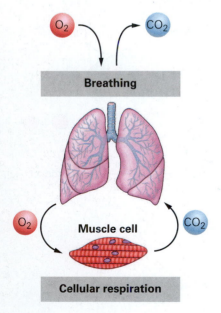

Figure 4.9 Breathing and cellular respiration. When you inhale, you bring oxygen from the atmosphere into your lungs. This oxygen is delivered through the bloodstream to tissues, such as the muscle tissue shown at the bottom of this figure, which use it to drive cellular respiration. The carbon dioxide produced by cellular respiration is released from cells and diffuses into the blood and to the lungs. Carbon dioxide is released from the lungs when you exhale.

(a) Cross section of a mitochondrion   **(b) Mitochondrion**

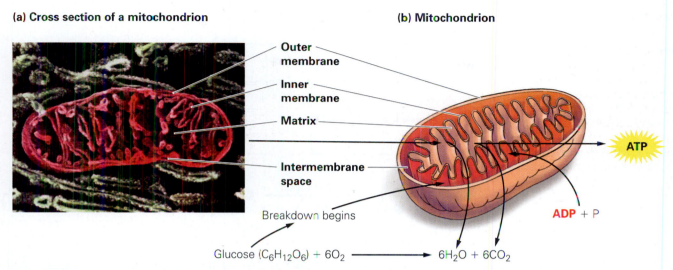

Outer
membrane

Inner
membrane

Matrix

Intermembrane
space

ATP

ADP + P

Breakdown begins

Glucose $(C_6H_{12}O_6)$ + $6O_2$ ⟶ $6H_2O$ + $6CO_2$

Figure 4.10 Overview of cellular respiration. (a) Much of this process of cellular respiration takes place within the mitochondrion. (b) The breakdown of glucose by cellular respiration requires oxygen and ADP plus phosphate. The energy stored in the bonds of glucose is harvested to produce ATP (from ADP and P), releasing carbon dioxide and water.

A Closer Look:
Glycolysis, the Citric Acid Cycle, and Electron Transport and ATP Synthesis

Cellular respiration occurs in three stages: glycolysis, the citric acid cycle, and electron transport and ATP synthesis (**Figure 4.11**).

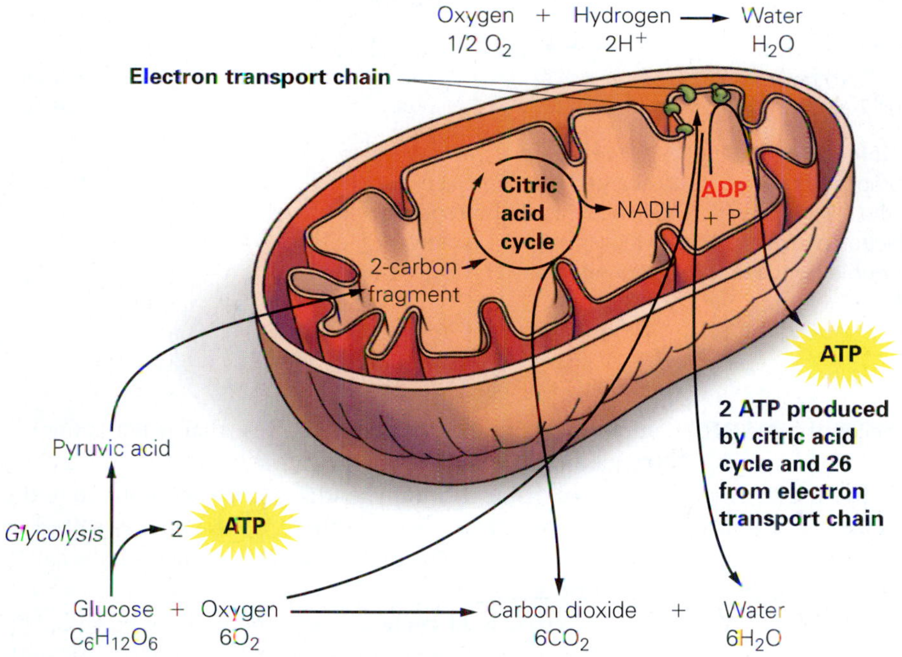

Stage 1: Glycolysis. To harvest energy from glucose, the 6-carbon glucose molecule is first broken down into two 3-carbon **pyruvic acid** molecules (**Figure 4.12** on the next page). This part of the process of cellular respiration actually occurs outside any organelle, in the fluid cytosol. Glycolysis does not require oxygen and produces 2 molecules of ATP.

In addition to producing ATP, glycolysis removes electrons from glucose for use in producing ATP during the final stage of cellular respiration. These electrons do not simply float around in a cell; this would damage the cell. Instead, they are carried by molecules called *electron carriers*. One of the electron carriers utilized by cellular respiration is a chemical called **nicotinamide adenine dinucleotide (NAD⁺).** NAD⁺ picks up 2 hydrogen atoms (along with their electrons) and releases

Oxygen + Hydrogen ⟶ Water
$1/2 O_2$ $2H^+$ H_2O

Electron transport chain

Citric
acid
cycle

NADH

ADP
+ P

2-carbon
fragment

ATP

Pyruvic acid

Glycolysis   2   ATP

2 ATP produced by citric acid cycle and 26 from electron transport chain

Glucose + Oxygen ⟶ Carbon dioxide + Water
$C_6H_{12}O_6$ $6O_2$ $6CO_2$ $6H_2O$

Figure 4.11 A more detailed look at cellular respiration. This figure diagrams the inputs and outputs of cellular respiration.

(continued on the next page)

(A Closer Look continued)

Figure 4.12 Glycolysis. Glyco-lysis is the enzymatic conversion of glucose into 2 pyruvic acid molecules. The pyruvic acid molecules are further broken down in the mitochondrion. Two ATP and 2 NADH are made during glycolysis.
Visualize This: Why is glycolysis an appropriate name for this process?

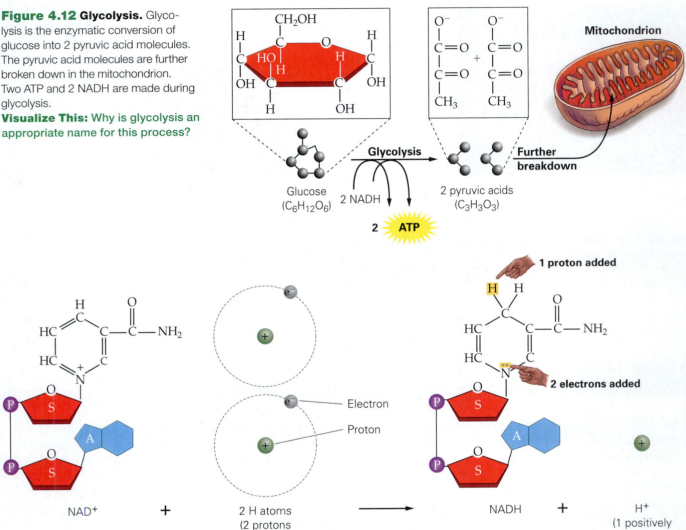

Figure 4.13 Nicotinamide adenine dinucleotide (NAD⁺). NAD⁺ can pick up a hydrogen atom along with its electron. Hydrogen atoms are composed of 1 negatively charged electron that circles around the 1 positively charged proton. When the electron carrier NAD⁺ encounters 2 hydrogen atoms (from food), it utilizes each hydrogen atom's electron and only 1 proton, thus releasing 1 proton.
Visualize This: Why does the number of double bonds in the nitrogenous base change when a proton is added to NAD⁺ to produce NADH?

1 positively charged hydrogen ion (H^+), becoming **NADH** (**Figure 4.13**).

 NADH serves as a sort of taxicab for electrons. The empty taxicab (NAD^+) picks up electrons. The full taxicab (NADH) carries electrons to their destination, where they are dropped off, and the empty taxicab returns for more electrons (**Figure 4.14**). NADH will deposit its electrons for use in the final step of cellular respiration.

 Stage 2: The Citric Acid cycle. After glycolysis, the pyruvic acid is decarboxylated (loses a carbon dioxide

molecule), and the 2-carbon fragment that is left is further metabolized inside the mitochondria.

 Once inside the mitochondrion, the energy stored in the remains of glucose is converted into the energy stored in the bonds of ATP. The first stage of this conversion is called the citric acid cycle.

 The **citric acid cycle** is a series of reactions catalyzed by 8 different enzymes, located in the matrix of each mitochondrion. This cycle breaks down the remains of a carbohydrate, harvesting its electrons and releasing carbon dioxide into the atmosphere (**Figure 4.15**). These reactions are a cycle because every trip around the pathway regenerates the first reactant, a 4-carbon molecule called oxaloacetate (OAA). OAA is always available to react with carbohydrate fragments entering the citric acid cycle.

Stage 3: Electron Transport and ATP Synthesis. Electrons harvested during the citric acid cycle are also carried by NADH to the final stage in cellular respiration. The **electron transport chain** is a series of proteins embedded in the inner mitochondrial membrane that functions as a sort of conveyer belt for electrons, moving them from one protein to another. The electrons, dropped off by NADH molecules generated during glycolysis and the citric acid cycle, move toward the bottom of the electron transport chain toward the matrix of the mitochondrion, where they combine with oxygen to produce water.

Each time an electron is picked up by a protein or handed off to another protein, the protein moving it changes shape. This shape change allows the movement of hydrogen ions (H⁺) from the matrix of the mitochondrion to the intermembrane space. So, while the proteins in the electron transport chain are moving electrons down the electron transport chain toward oxygen, they are also moving H⁺ ions across the inner mitochondrial membrane and into the intermembrane space. This decreases the concentration of H⁺ ions in the matrix and increases their concentration within the intermembrane space. Whenever a concentration gradient of a molecule exists, molecules will diffuse from an area of high concentration to an area of low concentration (Chapter 3). Because charged ions cannot diffuse across the hydrophobic core of the membrane, they escape through a protein channel in the membrane called **ATP synthase.** This enzyme uses the energy generated by the rushing H⁺ ions to synthesize 26 ATP from ADP and phosphate in the same manner that water rushing through a mechanical turbine can be used to generate electricity (**Figure 4.16** on the next page).

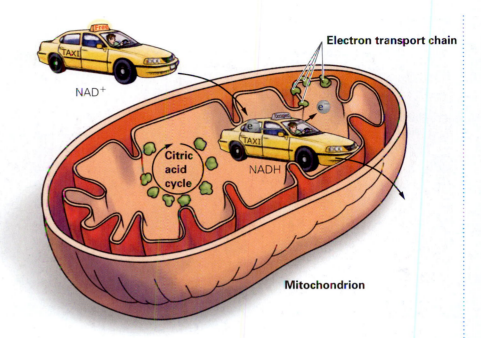

Figure 4.14 Electron carriers. NADH serves as an electron carrier, bringing electrons removed from the original glucose molecule to the electron transport chain. After dropping off its electrons, the electron carrier can be loaded up again and bring more electrons to the electron transport chain.

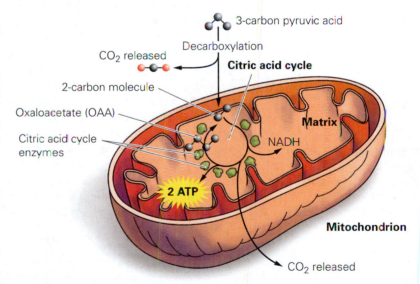

Figure 4.15 The citric acid cycle. The 3-carbon pyruvic acid molecules generated by glycolysis are decarboxylated, leaving a 2-carbon molecule that enters the citric acid cycle within the mitochondrial matrix. The 2-carbon fragment reacts with a 4-carbon OAA molecule and proceeds through a stepwise series of reactions that results in the production of more carbon dioxide and regenerates OAA. NADH and 2 ATP are also produced.

Overall, the two pyruvic acids produced by the breakdown of glucose during glycolysis are converted into carbon dioxide and water. Carbon dioxide is produced when it is removed from the pyruvic acid molecules during the citric acid cycle, and water is formed when oxygen combines with hydrogen ions at the bottom of the electron transport chain.

(continued on the next page)

(A Closer Look continued)

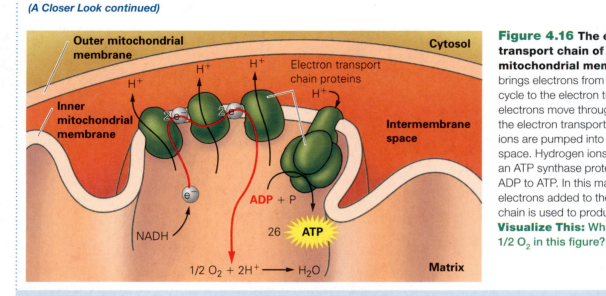

Figure 4.16 The electron transport chain of the inner mitochondrial membrane. NADH brings electrons from the citric acid cycle to the electron transport chain. As electrons move through the proteins of the electron transport chain, hydrogen ions are pumped into the intermembrane space. Hydrogen ions flow back through an ATP synthase protein, which converts ADP to ATP. In this manner, energy from electrons added to the electron transport chain is used to produce ATP.
Visualize This: What is meant by $1/2\ O_2$ in this figure?

Metabolism of Other Nutrients

Metabolism refers to all the chemical reactions that occur in an organism's cells. Most cells can break down not only carbohydrates but also proteins and fats. **Figure 4.17** shows the points of entry during cellular respiration for proteins and fats. Protein is broken down into component amino acids, which are then used to synthesize new proteins. Most organisms can also break down proteins to supply energy. However, this process takes place only when fats or carbohydrates are unavailable. In humans and other animals, the first step in producing energy from the amino acids of a protein is to remove the nitrogen-containing amino group of the amino acid. Amino groups are then converted to a compound called urea, which is excreted in the urine. The carbon, oxygen, and hydrogen remaining after the amino group is removed undergo further breakdown and eventually enter the mitochondria, where they are fed through the citric acid cycle and produce carbon dioxide, water, and ATP. The subunits of fats (glycerol and fatty acids) also go through the citric acid cycle and produce carbon dioxide, water, and ATP. Some cells will break down fat only when carbohydrate supplies are depleted.

Figure 4.17 Metabolism of other macromolecules. Carbohydrates, proteins, and fats can all undergo cellular respiration; they just feed into different parts of the metabolic pathway.

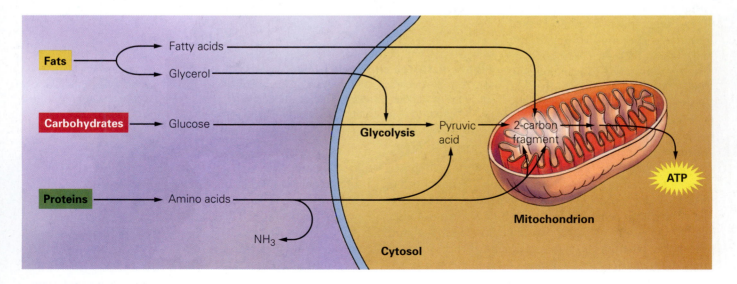

Metabolism Without Oxygen: Anaerobic Respiration and Fermentation

Aerobic cellular respiration is one way for organisms to generate energy. It is also possible for some cells to generate energy in the absence of oxygen, by a metabolic process called **anaerobic respiration.**

Muscle cells normally produce ATP by aerobic respiration. However, oxygen supplies diminish with intense exercise. When muscle cells run low on oxygen, they must get most of their ATP from glycolysis, the only stage in the cellular respiration process that does not require oxygen. When glycolysis happens without aerobic respiration, cells can run low on NAD⁺ because it is converted to NADH during glycolysis. When this happens, cells use a process called **fermentation** to regenerate NAD⁺.

Fermentation cannot, however, be used for very long because one of the by-products of this reaction leads to the buildup of a compound called lactic acid. Lactic acid is produced by the actions of the electron acceptor NADH, which has no place to dump its electrons during fermentation because there is no electron transport chain and no oxygen to accept the electrons. Instead, NADH deposits its electrons by giving them to the pyruvic acid produced by glycolysis (**Figure 4.18a**). Lactic acid is transported to the liver, where liver cells use oxygen to convert it back to pyruvic acid.

This requirement for oxygen, to convert lactic acid to pyruvic acid, explains why you continue to breathe heavily even after you have stopped working out. Your body needs to supply oxygen to your liver for this conversion, sometimes referred to as "paying back your oxygen debt." The accumulation of lactic acid also explains the phenomenon called "hitting the wall." Anyone who has ever felt as though their legs were turning to wood while running or biking knows this feeling. When your muscles are producing lactic acid by fermentation for a long time, the oxygen debt becomes too large, and muscles shut down until the rate of oxygen supply outpaces the rate of oxygen utilization, restoring proper feeling to your legs.

Stop & Stretch Aerobic exercise (such as running, swimming, and biking) strengthens the heart, allowing it to pump more blood per beat. Given the effects of anaerobic respiration on the body, how does aerobic exercise increase stamina?

Some fungi and bacteria also produce lactic acid during fermentation. Certain microbes placed in an anaerobic environment transform the sugars in milk into yogurt, sour cream, and cheese. It is the lactic acid present in these dairy products that gives them their sharp or sour flavor. Yeast in an anaerobic environment produces ethyl alcohol instead of lactic acid. Ethyl alcohol is formed when carbon dioxide is removed from pyruvic acid (**Figure 4.18b**). The yeast used to help make beer and wine converts sugars present in grains (beer) or grapes (wine) into ethyl alcohol and carbon dioxide. Carbon dioxide, produced by baker's yeast, helps bread to rise.

4.3 Body Fat and Health

You have seen how cells can use food to make energy. When more energy is consumed than is utilized, the excess is stored as fat. A clear understanding of how much body fat is healthful is hard to come by because cultural and biological definitions of the term "overweight" differ markedly. Cultural

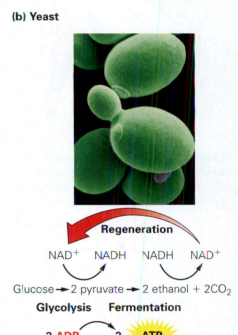

(a) Human muscle

Regeneration

NAD⁺ NADH   NADH NAD⁺

Glucose ⟶ 2 pyruvate ⟶ 2 lactate

Glycolysis   **Fermentation**

2 ADP   2   ATP

(b) Yeast

Regeneration

NAD⁺ NADH   NADH NAD⁺

Glucose ⟶ 2 pyruvate ⟶ 2 ethanol + 2CO₂

Glycolysis   **Fermentation**

2 ADP   2   ATP

Figure 4.18 Metabolism without oxygen. Glycolysis can be followed by (a) lactate fermentation to regenerate NAD⁺. This pathway also produces 2 ATP during glycolysis. Glycolysis followed by (b) alcohol fermentation also regenerates NAD⁺ and produces 2 ATP.

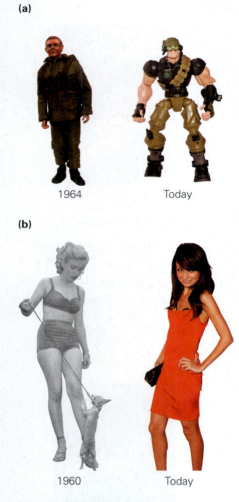

(a)

1964   Today

(b)

1960   Today

Figure 4.19 The perception of beauty. (a) GI Joe has become more muscular over time. (b) Female sex symbols have become thinner over time.

definitions of overweight have changed over the years. Men and women who were considered to be of normal weight in the past might not be seen as meeting today's standards. In the United States, the evolution of this trend has been paralleled by changes in children's action figures and dolls and in celebrities and movie stars over the last several decades. You need only compare the physiques of action figures from the 1960s and 1970s to the physiques seen on today's action figures to see how the standards have changed (**Figure 4.19a**). Standards for women have also changed. The 1960s sex symbol and movie star Marilyn Monroe was much curvier than more recent stars, such as Nicole Richie (**Figure 4.19b**).

The next time you read the newspaper, take note of advertisements for diets promoting diet products. It is often the case that "before" pictures show individuals of healthful weights, and "after" pictures show people who are too thin. Indeed, the average woman in the United States weighs 140 pounds and wears a size 12. The average model weighs 103 pounds and wears a size 4. With these distorted messages about body fat, it is difficult for average people to know how much body fat is right for them.

Evaluating How Much Body Fat Is Healthful

A person's sex, along with other factors, determines his or her ideal amount of body fat. Women need more body fat than men do to maintain their fertility. On average, healthy women have 22% body fat, and healthy men have 14%. To maintain essential body functions, women need at least 12% body fat but not more than 32%; for men, the range is between 3% and 29%. This difference between females and males, a so-called sex difference, exists because women store more fat on their breasts, hips, and thighs than men do. A difference in muscle mass leads to increased energy use by males because muscles use more energy than does fat.

Women also have an 8% thicker layer of tissue called the dermis under the outer epidermal layer of the skin as compared to men. This means that in a woman and a man of similar strength and body fat, the woman's muscles would look smoother and less defined than the man's muscles would.

A person's frame size also influences body fat—larger-boned people carry more fat. In addition, body fat tends to increase with age.

Unfortunately, it is a bit tricky to determine what any individual's ideal body weight should be. In the past, you simply weighed yourself and compared your weight to a chart showing a range of acceptable weights for given heights. The weight ranges on these tables were associated with the weights of a group of people who bought life insurance in the 1950s and whose health was monitored until they died. The problem with using these tables is that the subjects may not have been representative of the whole population. Generalizing results seen in one group to another group can lead to erroneous conclusions (Chapter 1). People who had the money to buy life insurance tended to have the other benefits of money as well, including easier access to health care, better nutrition, *and* lower body weight. Their longer lives may have had more to do with better health care and nutrition than with their weight.

To deal with some of the ambiguities associated with the insurance company's weight tables, a new measure of weight and health risk, the **body mass index (BMI),** has been developed. BMI is a calculation that uses both height and weight to determine a value that correlates an estimate of body fat with the risk of illness and death (**Table 4.1**).

Although the BMI measurement is a better approximation of ideal weight than are the insurance charts of the past, it is not perfect; BMI still does not account for differences in frame size, gender, or muscle mass. In fact, studies

TABLE 4.1

Body mass index (BMI). A chart based on height and weight correlations.

Height	Weight											
4'10"	91	96	100	105	110	115	119	124	129	134	138	143
4'11"	94	99	104	109	114	119	124	128	133	138	143	148
5'0"	97	102	107	112	118	123	128	133	138	143	148	153
5'1"	100	106	111	116	122	127	132	137	143	148	153	158
5'2"	103	109	115	120	126	131	136	142	148	153	158	164
5'3"	107	113	118	124	130	135	141	146	152	158	163	169
5'4"	110	116	122	128	134	140	145	151	157	163	169	174
5'5"	114	120	126	132	138	144	150	156	162	168	174	180
5'6"	117	124	130	136	142	148	155	161	167	173	179	186
5'7"	121	127	134	140	146	153	159	166	172	178	185	191
5'8"	125	131	138	144	151	158	164	171	177	184	190	197
5'9"	129	135	142	149	155	162	169	176	183	189	196	203
5'10"	132	139	146	153	160	167	174	181	188	195	202	207
5'11"	136	143	150	157	165	172	179	186	193	200	208	215
6'0"	140	147	154	162	169	177	184	191	198	206	213	221
6'1"	144	151	159	166	174	182	189	197	205	212	219	227
6'2"	148	155	163	171	179	186	194	202	210	218	225	233
6'3"	151	160	168	176	184	192	200	208	216	224	232	240
6'4"	156	164	172	180	189	197	205	213	221	230	238	246
BMI	19	20	21	22	23	24	25	26	27	28	29	30

16   20   25   30

Anorexic   Underweight and possibly anorexic   Healthy   Overweight   Obese

show that as many as 1 in 4 people may be misclassified by BMI tables because this measurement provides no means to distinguish between lean muscle mass and body fat. For example, an athlete with a lot of muscle will weigh more than a similar-size person with a lot of fat because muscle is heavier than fat.

If your BMI falls within the healthy range (BMI of 20–25), you probably have no reason to worry about health risks from excess weight. If your BMI is high, you may be at increased risk for diseases associated with obesity.

Obesity

Close to 1 in 3 Americans has a BMI of 30 or greater and is therefore considered to be obese. This crisis in **obesity** is the result you would expect when constant access to cheap, high-fat, energy-dense, unhealthful food is combined with lack of exercise. This relationship is clearly illustrated by the case of the Pima Indians.

Several hundred years ago, the ancestral population of Pima Indians split into 2 tribes. One branch moved to Arizona and adopted the American diet and lifestyle; the typical Pima of Arizona gets as much exercise as the average American and, like most Americans, eats a high-fat, low-fiber diet. Unfortunately, the health consequences for these people are more severe than they are for most other Americans—close to 60% of the Arizona Pima are obese and diabetic. In contrast, the Pima of Mexico maintained their ancestral farming life; their diet is rich in fruits, vegetables, and fiber. The Pima of Mexico also engage in physical labor for close to 22 hours per week and are on average 60 pounds lighter than their Arizona relatives. Consequently, diabetes is virtually unheard of in this group.

This example illustrates the impact of lifestyle on health: The Pima of Arizona share many genes with their Mexican relatives but have far less healthful lives due to their diet and lack of exercise. The example also shows that genes influence body weight because the Pima of Arizona have higher rates of obesity and diabetes than those of other Americans whose lifestyle they share.

Whether obesity is the result of genetics, diet, or lack of exercise, the health risks associated with obesity are the same. As your weight increases, so do your risks of diabetes, hypertension, heart disease, stroke, and joint problems.

Diabetes. **Diabetes** is a disorder of carbohydrate metabolism characterized by the impaired ability of the body to produce or respond to insulin. **Insulin** is a hormone secreted by beta cells, which are located within clusters of cells in the pancreas. Insulin's role in the body is to trigger cells to take up glucose so that they can convert the sugar into energy. People with diabetes are unable to metabolize glucose; as a result, the level of glucose in the blood rises.

There are two forms of the disease. Type 1, the insulin-dependent form of diabetes, usually arises in childhood. People with Type 1 diabetes cannot produce insulin because their immune systems mistakenly destroy their own beta cells. When the body is no longer able to produce insulin, daily injections of the hormone are required. Type 1 diabetes is not correlated with obesity.

Type 2, the non-insulin-dependent form of diabetes, usually occurs after 40 years of age and is more common in the obese. Type 2 diabetes arises either from decreased pancreatic secretion of insulin or from reduced responsiveness to secreted insulin in target cells. People with Type 2 diabetes are able to control blood glucose levels through diet and exercise and, if necessary, by insulin injections.

Hypertension. **Hypertension,** or high blood pressure, places increased stress on the circulatory system and causes the heart to work too hard. Compared to a person with normal blood pressure, a hypertensive person is six times more likely to have a heart attack.

Blood pressure is the force, originated by the pumping action of the heart, exerted by the blood against the walls of the blood vessels. Blood vessels expand and contract in response to this force. Blood pressure is reported as 2 numbers: The higher number, called the **systolic blood pressure,** represents the pressure exerted by the blood against the walls of the blood vessels as the heart contracts; the lower number, called the **diastolic blood pressure,** is the pressure that exists between contractions of the heart when the heart is relaxing. Normal blood pressure is around 120 over 80 (symbolized as 120/80). Blood pressure is considered to be high when it is persistently above 140/90.

Problematic weight gain is typically the result of increases in the amount of fatty tissue versus increases in muscle mass. Fat, like all tissues, relies on oxygen and other nutrients from food to produce energy. As the amount of fat on your body increases, so does the demand for these substances. Therefore, the amount of blood required to carry oxygen and nutrients also increases. Increased blood volume means that the heart has to work harder to keep the blood moving through the vessels, thus placing more pressure on blood vessel walls and leading to increased heart rate and blood pressure.

Heart Attack, Stroke, and Cholesterol.

A **heart attack** occurs when there is a sudden interruption in the supply of blood to the heart caused by the blockage of a vessel supplying the heart. A **stroke** is a sudden loss of brain function that results when blood vessels supplying the brain are blocked or ruptured. Heart attack and stroke are more likely in obese people because the elevated blood pressure caused by obesity also damages the lining of blood vessels and increases the likelihood that cholesterol will be deposited there. Cholesterol-lined vessels are said to be *atherosclerotic*.

Because lipids like cholesterol are not soluble in aqueous solutions, cholesterol is carried throughout the body, attached to proteins in structures called lipoproteins. **Low-density lipoproteins (LDLs)** have a high proportion of cholesterol (in other words, they are low in protein). LDLs distribute both the cholesterol synthesized by the liver and the cholesterol derived from diet throughout the body. LDLs are also important for carrying cholesterol to cells, where it is used to help make plasma membranes and hormones. **High-density lipoproteins (HDLs)** contain more protein than cholesterol. HDLs scavenge excess cholesterol from the body and return it to the liver, where it is used to make bile. The cholesterol-rich bile is then released into the small intestine, and from there much of it exits the body in the feces. The LDL/HDL ratio is an index of the rate at which cholesterol is leaving body cells and returning to the liver.

Your physician can measure your cholesterol level by determining the amounts of LDL and HDL in your blood. If your total cholesterol level is over 200 or your LDL level is above 100 or so, then your physician may recommend that you decrease the amount of cholesterol and saturated fat in your diet. This may mean eating more plant-based foods and less meat (because plants do not have cholesterol and meat does have cholesterol), as well as reducing the amount of saturated fats in your diet. Saturated fat is thought to raise cholesterol levels by stimulating the liver to step up its production of LDLs and slowing the rate at which LDLs are cleared from the blood.

Stop & Stretch   A friend of yours has his cholesterol level checked and tells you that he is really relieved because his cholesterol is 195. What other factors should your friend be considering before he decides his cholesterol level gives him no cause to be concerned?

Cholesterol is not all bad; in fact, some cholesterol is necessary—it is present in cell membranes to help maintain their fluidity, and it is the building block for steroid hormones such as estrogen and testosterone. You do, however, synthesize enough cholesterol so that you do not need to obtain much from your diet.

For some people, those with a genetic predisposition to high cholesterol, controlling cholesterol levels through diet is difficult because dietary cholesterol makes up only a fraction of the body's total cholesterol. People with high cholesterol who do not respond to dietary changes may have inherited genes that increase the liver's production of cholesterol. These people may require prescription medications to control their cholesterol levels.

Cholesterol-laden, atherosclerotic vessels increase your risk of heart disease and stroke. Fat deposits narrow your heart's arteries, so less blood can flow to your heart. Diminished blood flow to your heart can cause chest pain, or angina. A complete blockage can lead to a heart attack. Lack of blood flow to the heart during a heart attack can cause the oxygen-starved heart tissue to die, leading to irreversible heart damage.

The same buildup of fatty deposits also occurs in the arteries of the brain. If a blood clot forms in a narrowed artery in the brain, it can completely block blood flow to an area of the brain, resulting in a stroke. If oxygen-starved brain tissue dies, permanent brain damage can result.

Anorexia and Bulimia

Eating disorders that make you underweight cause health problems that are as severe as those caused by too much weight (**Table 4.2**). **Anorexia,** or self-starvation, is rampant on college campuses. Estimates suggest that 1 in 5 college women and 1 in 20 college men restrict their intake of Calories so severely that they are essentially starving themselves to death. Others allow themselves to eat—sometimes very large amounts of food (called binge eating)—but prevent the nutrients from being turned into fat by purging themselves, by vomiting or using laxatives. Binge eating followed by purging is called **bulimia.**

Anorexia has serious long-term health consequences. Anorexia can starve heart muscles to the point that altered rhythms develop. Blood flow is reduced, and blood pressure drops so much that the little nourishment present cannot get to the cells. The lack of body fat accompanying anorexia can also lead to the cessation of menstruation, a condition known as amenorrhea. Amenorrhea occurs when a protein called leptin, which is secreted by fat cells, signals the brain that there is not enough body fat to support a pregnancy. Hormones (such as estrogen) that regulate menstruation are blocked, and menstruation ceases. Amenorrhea can be permanent and causes sterility in a substantial percentage of people with anorexia.

The damage done by the lack of estrogen is not limited to the reproductive system; bones are affected as well. Estrogen secreted by the ovaries during the menstrual cycle acts on bone cells to help them maintain their strength and size. Anorexics reduce the development of dense bone and put themselves at a much higher risk of breaking their weakened bones, in a condition called **osteoporosis.**

Besides experiencing the same health problems that anorexics face, people with bulimia can rupture their stomachs through forced vomiting. They often have dental and gum problems caused by stomach acid being forced into their mouths during vomiting, and they can become fatally dehydrated.

Achieving Ideal Weight

As you have seen, the health problems associated with obesity, anorexia, and bulimia are severe. To avoid these problems it is best to focus more on fitness and healthy eating and less on body weight.

Working slowly toward being fit and eating healthfully rather than trying the latest fad diet are more realistic and attainable ways to achieve the positive health outcomes that we all desire. In fact, fitness may be more important than body weight in terms of health. Studies show that fit but overweight people have better health outcomes than unfit slender people. In other words, lack of fitness is associated with higher health risks than excess body weight. Therefore, it makes more sense to focus on eating right and exercising than it does to focus on the number on the scale.

TABLE 4.2

Obesity and anorexia or bulimia. Health problems result from being either overweight or underweight.	
Health Problems Resulting from Obesity	**Health Problems Resulting from Anorexia and Bulimia**
• Adult-onset diabetes	• Altered heart rhythms
• Hypertension (high blood pressure)	• Amenorrhea (cessation of menstruation)
• Heart attack	• Osteoporosis (weakened bones)
• Stroke	• Ruptured stomach
• Joint problems	• Dental/gum problems
	• Dehydration

SAVVY READER

Hoodia for Weight Loss

Several websites claim that celebrities, including Angelina Jolie, used a substance derived from the Hoodia Gordonii plant to help them lose weight. Summarized below are some of the claims made by these sites:

- Hoodia plants require over five years to grow to maturity. This cactus plant is also a fairly uncommon plant. These two factors limit its availability. Because the supply of Hoodia cannot meet the demand, Hoodia pills are more expensive than one would expect.

- Hoodia diet pills are natural. They contain no additives or preservatives, just parts of the plant. This natural substance acts on your brain to make you eat less.

- Hoodia has been used by indigenous people to help them traverse the Kalahari Desert.

- Drug companies have spent millions of dollars performing tests of Hoodia diet pills. No side effects have been found. Hoodia must be safe for everyone.

- If you don't believe Hoodia works, just look at these photos of Angelina Jolie (**Figure 4.20**). This is proof of the effectiveness of this natural appetite suppressant.

1. List several claims presented on this website. Is any evidence presented for these claims?

2. Many products are marketed as natural and many consumers let their guard down when they hear that a product is natural. Are all

Figure 4.20 Angelina Jolie

natural substances good for humans? Give an example of something that is natural that is not good for humans.

3. One site claims that a drug company performed tests on Hoodia and found that there were no side effects and that Hoodia appears to be safe for everyone. Is this the way scientists report data? What kinds of information should be here to substantiate this claim?

4. Look at the before (left) and after (right) photos of Angelina Jolie. Did she need to lose weight in the before picture? Is there any evidence presented that she actually took Hoodia?

Chapter Review

Learning Outcomes

Mastering BIOLOGY

Go to the Study Area at www.masteringbiology.com for practice quizzes, myeBook, BioFlix™ 3-D animations, MP3Tutor sessions, videos, current events, and more.

LO1 **Define the term metabolism (Section 4.1).**

- Metabolic reactions include all the chemical reactions that occur in cells to build up or break down macromolecules (p. 74).

LO2 **Describe the structure and function of enzymes (Section 4.1).**

- Enzymes are proteins that catalyze specific cellular reactions. The active site of an enzyme is composed of amino acids that affect its ability to bind to its substrate (pp. 74–75).

LO3 **Explain how enzymes decrease a reaction's activation energy barrier (Section 4.1).**

- The binding of a substrate to the enzyme's active site causes the enzyme to change shape (induced fit), placing more stress on the bonds of the substrate and thereby lowering the activation energy (pp. 74–75).

LO4 **List several reasons why metabolic rates differ between people (Section 4.1).**

- An individual's metabolic rate is affected by many factors, including age, sex, exercise level, body weight, and genetics (pp. 75–76).

LO5 **Describe the structure and function of ATP (Section 4.2).**

- ATP is a nucleotide triphosphate. The nucleotide found in ATP contains a sugar and the nitrogenous base adenine (p. 76).
- The energy stored in the chemical bonds of food can be released by metabolic reactions and stored in the bonds of ATP. Cells use ATP to power energy-requiring processes (pp. 76–77).
- Breaking the terminal phosphate bond of ATP releases energy that can be used to perform cellular work and produces ADP plus a phosphate (p. 77).
- ATP is generated in most organisms by the process of cellular respiration, which consumes carbohydrates and releases water and carbon dioxide as waste products (p. 78).

LO6 **Describe the process of cellular respiration from the breakdown of glucose through the production of ATP (Section 4.2).**

- Cellular respiration begins in the cytosol, where a 6-carbon sugar is broken down into two 3-carbon pyruvic acid molecules during the anaerobic process of glycolysis (pp. 79–80).

- The pyruvic acid molecules then move across the two mitochondrial membranes, where they are decarboxylated. The remaining 2-carbon fragment then moves into the matrix of the mitochondrion, where the citric acid cycle strips them of carbon dioxide and electrons (p. 80).
- Electrons removed from chemicals that are part of glycolysis and the citric acid cycle are carried by electron carriers, such as NADH, to the inner mitochondrial membrane; there they are added to a series of proteins called the electron transport chain. At the bottom of the electron transport chain, oxygen pulls the electrons toward itself. As the electrons move down the electron transport chain, the energy that they release is used to drive protons (H^+) into the intermembrane space. Once there, the protons rush through the enzyme ATP synthase and produce ATP from ADP and phosphate (pp. 81–82).
- When electrons reach the oxygen at the bottom of the electron transport chain, they combine with the oxygen and hydrogen ions to produce water (pp. 81–82).

LO7 **Explain how proteins and fats are broken down during cellular respiration (Section 4.2).**

- Proteins and fats are also broken down by cellular respiration, but they enter the pathway at later points than glucose (p. 82).

LO8 **Compare and contrast aerobic and anaerobic respiration (Section 4.2).**

- Anaerobic respiration is cellular respiration that does not use oxygen as the final electron acceptor (p. 83).

LO9 **Outline the health risks that occur with being overweight and underweight (Section 4.3).**

- Obesity is associated with many health problems, including hypertension, heart attack and stroke, diabetes, and joint problems (pp. 86–87).
- Anorexia and bulimia can cause altered heart rhythms, amenorrhea, osteoporosis, dehydration, and dental and stomach problems (pp. 88–89).

Roots to Remember

The following roots of words come mainly from Latin and Greek and will help you decipher terms:

an- means absence of. Chapter term: anaerobic

-ase is a common suffix in names of enzymes. Chapter term: sucrase glyco- means sugar. Chapter term: glycolysis

lipo- refers to fat or lipid. Chapter term: lipoprotein

osteo- refers to bone. Chapter term: osteoporosis

Learning the Basics

1. **LO1** Define the term metabolism.

2. **LO3** What is meant by the term induced fit?

3. **LO6** What are the reactants and products of cellular respiration?

4. **LO2** Which of the following is a *false* statement regarding enzymes?

   A. Enzymes are proteins that speed up metabolic reactions; **B.** Enzymes have specific substrates; **C.** Enzymes supply ATP to their substrates; **D.** An enzyme may be used many times.

5. **LO3** Enzymes speed up chemical reactions by _____.

   A. heating cells; **B.** binding to substrates and placing stress on their bonds; **C.** changing the shape of the cell; **D.** supplying energy to the substrate.

6. **LO6** Cellular respiration involves _____.

   A. the aerobic metabolism of sugars in the mitochondria by a process called glycolysis; **B.** an electron transport chain that releases carbon dioxide; **C.** the synthesis of ATP, which is driven by the rushing of protons through an ATP synthase; **D.** electron carriers that bring electrons to the citric acid cycle; **E.** the production of water during the citric acid cycle.

7. **LO6** The electron transport chain _____.

   A. is located in the matrix of the mitochondrion; **B.** has the electronegative carbon dioxide at its base; **C.** is a series of nucleotides located in the inner mitochondrial membrane; **D.** is a series of enzymes located in the intermembrane space; **E.** moves electrons from protein to protein and moves protons from the matrix into the intermembrane space.

8. **LO5** Most of the energy in an ATP molecule is released _____.

   A. during cellular respiration; **B.** when the terminal phosphate group is hydrolyzed; **C.** in the form of new nucleotides; **D.** when it is transferred to NADH.

9. **LO9** The function of low-density lipoproteins (LDLs) is to _____.

   A. break down proteins; **B.** digest starch; **C.** transport cholesterol from the liver; **D.** carry carbohydrates into the urine.

10. **LO8** Anaerobic respiration _____.

    A. generates proteins for muscles to use; **B.** occurs in yeast cells only; **C.** does not use oxygen as the final electron acceptor; **D.** utilizes glycolysis, the citric acid cycle, and the electron transport chain.

Analyzing and Applying the Basics

1. **LO7** A friend decides he will eat only carbohydrates because carbohydrates are burned to make energy during cellular respiration. He reasons that he will generate more energy by eating carbohydrates than by eating fats and proteins. Is this true?

2. **LO9** A friend thinks taking an oral enzyme supplement will speed up his metabolism. Do you think this is true? What will happen to the tablet supplement once it is in the stomach?

3. **LO4** What would you say to a friend who qualifies as obese on a BMI chart but who exercises regularly and eats a well-balanced diet?

Connecting the Science

1. Early Earth was devoid of oxygen. How might bacteria on early Earth have generated energy?

2. Why do you think that anorexia and bulimia are more common among women than men?

Answers to **Stop & Stretch, Visualize This, Savvy Reader,** and **Chapter Review** questions can be found in the **Answers** section at the back of the book.

Life in the Greenhouse

Photosynthesis and Global Warming

Massive flooding triggered by Hurricane Katrina devastated 80% of New Orleans.

Does the same fate await many of the other great cities of the world?

New Orleans. New York. Miami. Amsterdam. Alexandria. Mumbai. Ho Chi Minh City. Bangkok. Hong Kong. Shanghai. Tokyo. According to predictions of the effect of global warming, some of the great cities of the world may share a common future— portions of each may experience catastrophic flooding and become essentially uninhabitable. New Orleans provides a stark example. In August 2005, when Hurricane Katrina came onshore along the Gulf Coast of the United States, 80% of the city was inundated. A massive evacuation and relief effort still could not prevent the deaths of over 1500 residents or adequately attend to the needs of those impacted by the storm and flooding. Even today, significant portions of New Orleans— a major metropolis in one of the richest countries on Earth—remain essentially abandoned. Imagine the impact of such a disaster on cities and communities that lack the resources of the United States.

This future catastrophic flooding is possible because rising temperatures brought about by global warming are causing ice all over Earth's surface to melt. As this stored water runs off the land and into the oceans, sea levels rise. According to the U.S. National Climate Data Center, the entire planet has warmed by 0.25°C (0.5°F) each decade during the twentieth century, which has contributed to a sea-level rise of 10 to 20 cm (4 to 8 inches). The rising tides have already caused the widespread erosion of beaches, and even submerged islands. If global warming continues as predicted, by the end of the twenty-first century the sea level will be between 50 and 100 centimeters (20 to 36 inches) higher than today—a much faster and more devastating change than what we have already witnessed. Compounding this rapid raise in level, as ocean temperatures increase, the amount of

Global warming is causing Earth's ice to melt, raising sea levels by a meter over this century.

Rising sea levels are already threatening millions of people along ocean coasts.

energy available to be released in storms will rise, resulting in more powerful hurricanes and causing even more catastrophic flooding.

The currently occurring melt of Earth's surface ice is dramatic and sobering. In Antarctica, rising temperatures have led to the collapse of massive portions of the continent's ice cap. In recent years, Rhode Island-sized shelves of ice have fallen into the ocean. The Greenland ice cap is becoming thinner at its margins every year, and melt water collecting underneath the ice threatens to cause large portions to slide off the island. Meanwhile, the dramatic glaciers in Glacier National Park, located in the northwest corner of Montana, are disappearing. Several of the park's glaciers have already shrunk to half their original size, and the total number of glaciers has decreased from approximately 150 in 1850 to around 35 today. If this trend continues, scientists predict that by the year 2030 not a single glacier will be left in Glacier National Park. All of this melt water must go somewhere—and like an overflowing bathtub, it spills over the edges of ocean basins, inundating low-lying cities and coasts.

Why is Earth's ice melting and threatening to swamp our coastlines, and can anything be done to stop or slow the global warming that is causing the melt?

5.1 The Greenhouse Effect

Global warming is the progressive increase of Earth's average temperature that has been occurring over the past century. Although the general public seems to believe that there is debate among scientists and government-appointed panels about global warming, that is not really the case. Scientists who publish in peer-reviewed journals and respected scientific panels and societies, such as the Intergovernmental Panel on Climate Control (IPCC), National Academy of Sciences, and the American Association for the Advancement of Science (AAAS), all agree that the climate is warming and that most of the warming observed in the last century is attributable to human activities.

Global warming is caused by recent increases in the concentrations of particular gases in the atmosphere, including water vapor, carbon dioxide (CO_2), methane (CH_4), and ozone (O_3). The accumulation of many of these *greenhouse* gases is a direct result of human activity, namely, coal, oil, and natural gas combustion. The most abundant gas emitted by combustion of these fuels is carbon dioxide; for this reason, carbon dioxide is considered the most important greenhouse gas to control.

The presence of carbon dioxide and the other greenhouse gases in the atmosphere leads to a phenomenon called the **greenhouse effect.** Despite this name, the phenomenon caused by these gases is not exactly like that of a greenhouse, where panes of glass allow radiation from the sun to penetrate inside and then trap the heat that radiates from warmed-up surfaces. On Earth, the greenhouse effect works like this: Warmth from the sun heats Earth's surface, which then radiates the heat energy outward. Most of this heat is radiated back into space, but some of the heat warms up the greenhouse gases in the atmosphere and then is reradiated to Earth's surface. In effect, greenhouse gases act like a blanket (**Figure 5.1**). When you sleep under

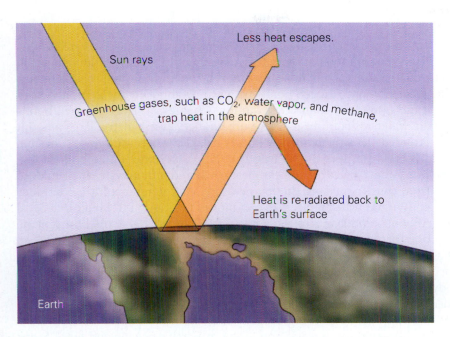

Figure 5.1 The greenhouse effect. Heat from the sun is absorbed in the atmosphere by water vapor, carbon dioxide, and other greenhouse gases and reradiated back to Earth.
Visualize This: What effect do you think increased levels of carbon dioxide have on the greenhouse effect?

a blanket at night, your body heat warms the blanket, which in turn keeps you warm. When the levels of greenhouse gases in the atmosphere increase, the effect is similar to sleeping under a thicker blanket—more heat is retained and reradiated, and therefore the temperature underneath the atmosphere is warmer.

The greenhouse effect is not in itself a dangerous phenomenon. If Earth's atmosphere did not have some greenhouse gases, too much heat would be lost to space, and Earth would be too cold to support life. At historical levels, greenhouse gases keep temperatures on Earth stable and hospitable to life. The danger posed by the greenhouse effect is in the *excess* warming caused by more and more carbon dioxide accumulation in the atmosphere as a result of coal, oil, and natural gas burning.

Stop & Stretch The moon's average daytime temperature is 107°C (225°F), and its average nighttime temperature is –153°C (–243°F). But the moon and Earth are the same distance from the sun. Think about the greenhouse gases on Earth to come up with a hypothesis for why temperatures on the moon might fluctuate so dramatically.

Water, Heat, and Temperature

Bodies of water absorb energy and help maintain stable temperatures on Earth. You have perhaps noticed that when you heat water on a stove, the metal pot becomes hot before the water. This is because water heats more slowly than metal and has a stronger resistance to temperature change than most substances.

Heat and temperature are measures of energy. **Heat** is the total amount of energy associated with the movement of atoms and molecules in a substance. **Temperature** is a measure of the intensity of heat—for example, how fast the molecules in the substance are moving. When you are swimming in a cool lake, your body has a higher temperature than the water; however, the lake contains more heat than your body because even though its molecules are moving more slowly, the sum total of molecular movement in its large volume is much greater than the sum total of molecular movements in your much smaller body.

The formation of hydrogen bonds between neighboring molecules of water makes it more cohesive than other liquids (Chapter 2). The hydrogen bonds also make water resistant to temperature change, even when a large amount of heat is added. This phenomenon occurs because when water is first heated, the heat energy disrupts the hydrogen bonds. Only after enough of the hydrogen bonds have been broken can heat cause individual water molecules to move faster, thus increasing the temperature. As the temperature continues to rise, individual water molecules can move fast enough to break free of all hydrogen bonds and rise into the air as water vapor. This is the basis for the water cycle that moves water from land, oceans, and lakes to clouds and then back again to Earth's surfaces (**Figure 5.2**). When water cools, hydrogen bonds re-form between adjacent molecules, releasing heat into the atmosphere. A body of water can release a large amount of heat into its surroundings while not decreasing its temperature much (**Figure 5.3**).

Water's high heat-absorbing capacity has important effects on Earth's climate. The vast amount of water contained in Earth's oceans and lakes keeps temperatures moderate by absorbing huge amounts of heat radiated by the sun and releasing that heat during less-sunny times, warming the air and preventing large temperature swings.

Figure 5.2 The water cycle. Water moves from the oceans and other surface water to the atmosphere and back, with stops in living organisms, underground pools and soil, and ice caps and glaciers on land.

Visualize This: What would happen to the amount of water in the ocean if all of the stored ice and snow melted? How would this melting affect the shoreline?

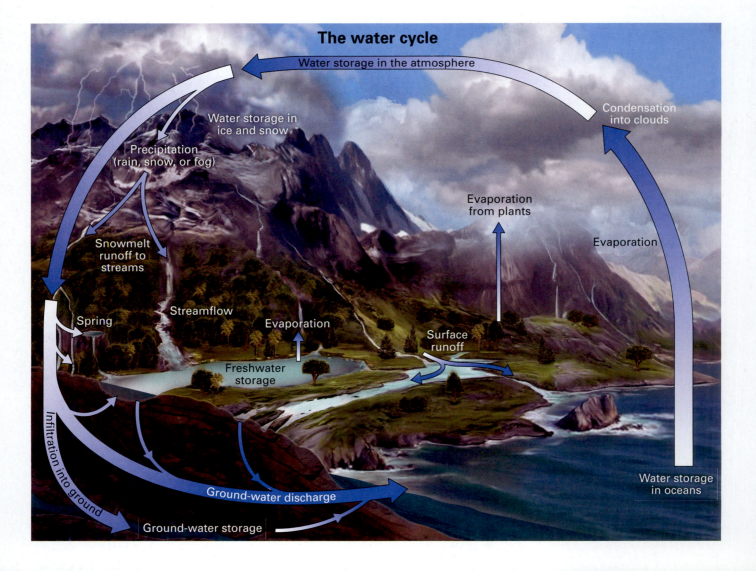

The water cycle

Water storage in the atmosphere

Water storage in ice and snow

Condensation into clouds

Precipitation (rain, snow, or fog)

Evaporation from plants

Evaporation

Snowmelt runoff to streams

Streamflow

Spring

Evaporation

Surface runoff

Freshwater storage

Infiltration into ground

Ground-water discharge

Ground-water storage

Water storage in oceans

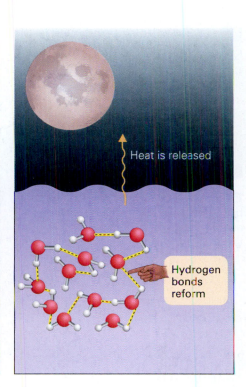

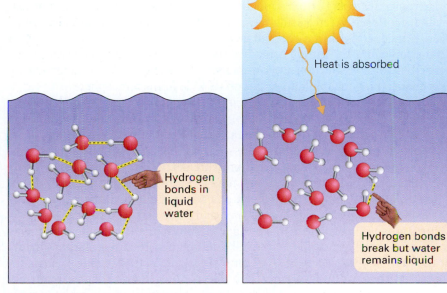

Figure 5.3 Hydrogen bonding in water. Hydrogen bonds break as they absorb heat and re-form as water releases heat.

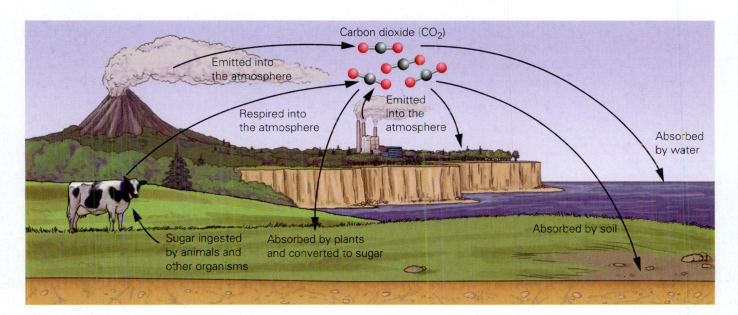

The balance between the release and storage of heat energy is vital to the maintenance of climate conditions on Earth. This balance can be disrupted when increasing levels of carbon dioxide cause more heat to be trapped in the atmosphere. Carbon dioxide in the atmosphere comes from many different sources, some of which are increasing in output.

5.2 The Flow of Carbon

The atoms that make up the complex molecules of living organisms move through the environment via biogeochemical cycles. **Figure 5.4** illustrates how carbon, like water, cycles back and forth between living organisms, the

Figure 5.4 The flow of carbon.
Living organisms, volcanoes, and fossil fuel emissions produce CO_2. Plants, oceans, and soil absorb CO_2 from the air.
Visualize This: Based on the observations that carbon dioxide levels have increased in the last 100 years and that volcanic activity has remained constant, which would you predict releases more carbon dioxide into the air: volcanic or human activity?

Figure 5.5 Burning fossil fuels. The burning of fossil fuels by industrial plants and automobiles adds more carbon dioxide to the environment.

atmosphere, bodies of water, and rock. The carbon dioxide you exhale enters the atmosphere, where it can absorb heat; these molecules can return to Earth's surface, where they can dissolve in water or be absorbed by plants. As you will learn in this chapter, carbon dioxide taken up by plants is converted into carbohydrates using the energy from sunlight. Other organisms use some of these carbohydrates for energy, rereleasing the carbon dioxide into the atmosphere in the process. Any unconsumed carbohydrates may become buried in the very rock of Earth for millennia; the carbon contained there can later be released through volcanic activity or by extraction and combustion by humans. It is the latter activity that is contributing to a buildup of carbon dioxide in the atmosphere.

The stored carbohydrates discussed in the previous paragraph are known as **fossil fuels** (**Figure 5.5**). These fuels—petroleum, coal, and natural gas—are "fossils" because they formed from the buried remains of ancient plants and microorganisms. Over a period of millions of years, the carbohydrates in these organisms were transformed by heat and pressure deep in Earth's crust into highly concentrated energy sources. Humans now tap these energy sources to power our homes, vehicles, and businesses, but as a result of our burning of these fuels, we have released millions of years of stored carbon as carbon dioxide.

Human use of fossil fuels is having a measurable effect; increases in carbon dioxide in the atmosphere are well documented by direct measurements over the past 50 years (**Figure 5.6**). Using "fossil air," scientists have also documented that the current level of carbon dioxide in the atmosphere greatly exceeds the levels present on Earth at any time in human history. Carbon dioxide concentrations in the past are measured by examining ice sheets that have existed for thousands of years. This measurement is possible because snow falling on an ice sheet surface traps air. As it accumulates, underlying snow is compressed into ice, and the trapped air becomes tiny ice-encased air bubbles. Thus, these bubbles are fossils—actual samples of the gases in the atmosphere at the time they formed. Cores can be removed from long-lived ice sheets and analyzed to determine the concentration of carbon dioxide trapped in fossil

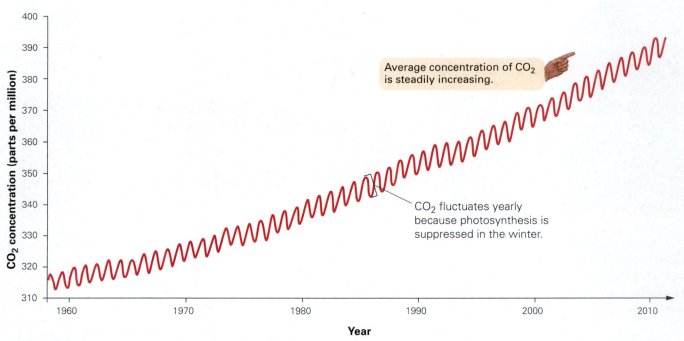

Figure 5.6 Increases in atmospheric carbon dioxide. Carbon dioxide levels from 1960 to present as measured by instruments at Mauna Loa observatory in Hawaii.

Visualize This: What evidence in the graph demonstrates the increased rate of carbon dioxide accumulation from 2000 to 2010 compared to the 1960s?

air (**Figure 5.7**). Other gases in the bubbles can provide indirect information about temperatures at the time the bubbles formed.

Ice core data from Antarctica (**Figure 5.8**) indicate that although Earth has gone through cycles of high carbon dioxide, the concentration of carbon dioxide in the atmosphere today is higher than at any other time in the last 400,000 years. The ice core data also demonstrate that increased levels of carbon dioxide occur at the same time as increased temperatures, suggesting that the carbon dioxide measurably warmed Earth. Taken together, these data are quite worrisome; they tell us that Earth may soon be facing temperatures well above those we experience in the current climate and warmer than Earth has seen in millennia.

5.3 Can Photosynthesis Reduce the Risk of Global Warming?

Modern plants and microorganisms do as their ancient predecessors did—absorb carbon dioxide from the atmosphere. Is it possible that we could depend on modern plants and microorganisms to reduce the amount of greenhouse gases released by fossil fuel burning and thus the threat of global warming?

Photosynthesis is the process by which plants and other microorganisms trap light energy from the sun and use it to convert carbon dioxide and water into sugar. In other words, photosynthesis transforms solar energy into the chemical energy required by all living things.

Chloroplasts: The Site of Photosynthesis

Chloroplasts are the specialized organelles in plant cells where photosynthesis takes place. Chloroplasts are surrounded by two membranes (**Figure 5.9** on the next page). The inner and outer membranes together are called the *chloroplast envelope* (Chapter 2). The chloroplast envelope encloses a compartment filled with **stroma,** the thick fluid that houses some of the enzymes of photosynthesis. Suspended in the stroma are disk-like membranous structures called **thylakoids,** which are typically stacked in piles like pancakes. The large amount of

Figure 5.7 Ice core. By analyzing ice cores, scientists can measure past atmospheric concentrations of carbon dioxide.

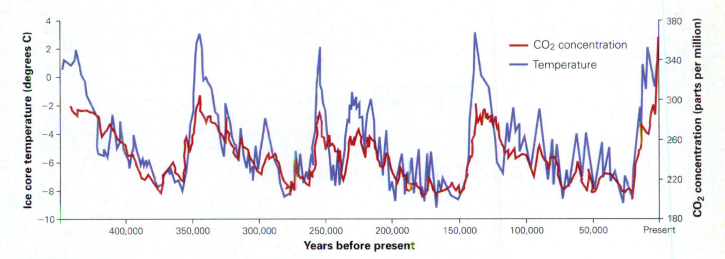

Figure 5.8 Records of temperature and atmospheric carbon dioxide concentration from Antarctic ice cores. These data indicate that increases in carbon dioxide levels are correlated with higher temperatures.

Visualize This: This graph has axes on both the left and right sides. Which line is indicated by the left axis and which by the right axis? Why are both lines included on the same graph?

(a)

(b)

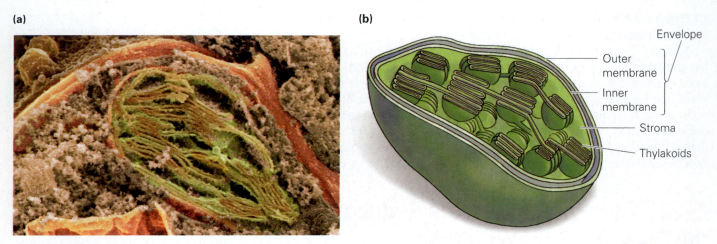

Envelope
Outer membrane
Inner membrane
Stroma
Thylakoids

Figure 5.9 Chloroplasts. The cross section (a) and drawing (b) of a chloroplast show the structures involved in photosynthesis.

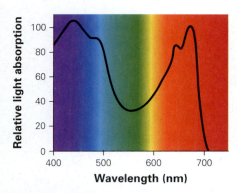

Figure 5.10 The absorption spectra of leaf pigments. Visible "white" light is actually made up of a series of wavelengths that appear as different colors to our eyes. The wavelengths absorbed by chlorophyll and other pigments in green plants are mainly red and blue and are thus removed from the spectrum of light that is reflected back to us from a leaf surface.

Visualize This: How would this graph appear for the pigments in the skin of a ripe red apple?

membrane inside the chloroplast provides abundant surface area on which some of the reactions of photosynthesis can occur.

On the surface of the thylakoid membrane are millions of molecules of **chlorophyll,** a pigment that absorbs energy from the sun. It is chlorophyll that gives leaves and other plant structures their green color. Like all pigments, chlorophyll absorbs light. Light is made up of rays with different colors, and each color corresponds to a different wavelength—to the human eye, shorter and middle wavelengths appear violet to green, and longer wavelengths appear yellow to red (**Figure 5.10**). Chlorophyll looks green to human eyes because it absorbs the shorter (blue) and longer (red) wavelengths of visible light and reflects the middle (green) range of wavelengths.

When a pigment such as chlorophyll absorbs sunlight, electrons associated with the pigment become excited, that is, increase in energy level. In effect, light energy is transferred to the chlorophyll and becomes chemical energy. For most pigments, the molecule remains excited for a very brief amount of time before this chemical energy is lost as heat. This is why a surface that looks black (that is, one composed of pigments that absorb all visible light wavelengths) heats quickly in the sun. (In contrast, a white surface, which does not absorb any light energy, remains relatively cool.) Inside a chloroplast, however, the chemical energy of the excited chlorophyll molecules is not released as heat; instead, the energy is captured.

Stop & Stretch In temperate areas, leaves change color in the fall. This occurs because chlorophyll is reabsorbed by the plant, making other pigments in the leaf visible. Why do you suppose the chlorophyll is reabsorbed?

An Overview: Photosynthesis

In plants and other photosynthetic organisms, solar energy is used to rearrange the atoms of carbon dioxide and water absorbed from the environment into carbohydrates (initially the sugar glucose). Photosynthesis produces oxygen as a waste product. The equation summarizing photosynthesis is as follows:

$$6\ CO_2 + 6\ H_2O + \text{Light energy} \rightarrow C_6H_{12}O_6 + 6\ O_2$$

Carbon dioxide + Water + Light energy → Glucose + Oxygen

Photosynthetic organisms use the carbohydrates that they produce to grow and supply energy to their cells. They, along with the organisms that eat them, lib-

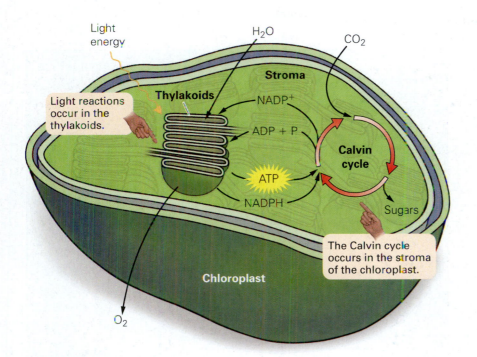

Figure 5.11 Overview of photosynthesis. Sunlight drives the synthesis of glucose and oxygen from carbon dioxide and water in plants and other photosynthetic organisms. The reaction requires two steps: the first that generates cellular energy (ATP) and harvests electrons from water (carried by NADPH); and the second that uses those products to convert carbon dioxide into sugar.

erate the energy stored in the chemical bonds of sugars via the process of cellular respiration. Any excess carbohydrates are stored in the body of the plant or other photosynthesizer and could become the raw material for fossil fuel production.

The process of photosynthesis can be broken down into two steps (**Figure 5.11**). The first, or "photo," step harvests energy from the sun during a series of reactions called the **light reactions,** which occur when there is sunlight. These reactions occur within the thylakoids of the chloroplast and produce the energy-carrying molecule ATP and energized electrons, which are transported around the cell by the molecule NADPH. The second, or "synthesis," step, called the **Calvin cycle,** uses the ATP and NADPH generated by the light reactions to synthesize sugars from carbon dioxide. This step can occur in either the presence or the absence of sunlight, and for this reason, the Calvin cycle is also sometimes referred to as the light-independent reactions. *See the next section for A Closer Look at photosynthesis.*

Stop & Stretch   In what ways are photosynthesis and cellular respiration (Chapter 4) related?

A Closer Look:
The Two Steps of Photosynthesis

The Light Reactions

When sunlight strikes chlorophyll molecules in a thylakoid membrane, the excited electrons are trapped by another molecule (**Figure 5.12**, step 1, on the next page). The energized electrons are then transferred along a specified path to other molecules in an electron transport chain located in the thylakoid membrane—a process very similar to the electron transport chain in mitochondria (Chapter 4). The electrons lose some of their energy with every step along the transport chain, but the chloroplast uses this energy to do work (**Figure 5.12**, step 2, on the next page). In particular, some of the proteins in the electron transport chain use the energy of the electrons to pump protons from the exterior to the interior of the thylakoid, setting up a gradient between the inside and outside of this structure. The stockpiled protons diffuse back out of the thylakoid through an enzyme in the thylakoid membrane, producing ATP in the same way that mitochondria

(continued on the next page)

(A Closer Look continued)

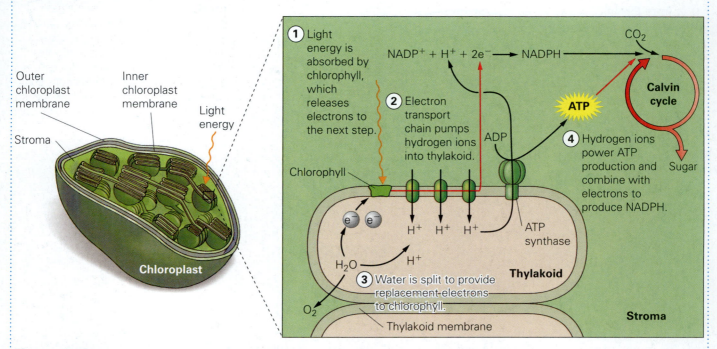

Figure 5.12 The light reactions of photosynthesis. (1) Sunlight strikes chlorophyll molecules located in the thylakoid membrane, exciting electrons, which then move to a higher energy level. (2) The electrons are captured by an electron transport chain, and their energy is used to pump hydrogen ions across the thylakoid membrane. (3) Water is split. Electrons removed from water are used to replace those lost from chlorophyll. Oxygen gas is released. (4) The movement of hydrogen ions out of the thylakoid power ATP production and generate NADPH. These molecules are produced in the stroma, where they will be available to the enzymes of the Calvin cycle.

Visualize This: How is this set of reactions similar to the electron transport chain in cellular respiration? How is it different? (To review cellular respiration, see Chapter 4.)

make ATP. The newly synthesized ATP is released into the stroma of the chloroplast, where it can be used by the enzymes of the Calvin cycle to produce sugars and other organic molecules.

Oxygen is produced during the light reactions when water (H_2O) is "split" into $2H^+$ ions and a single oxygen atom (O) (**Figure 5.12**, step 3). Two oxygen atoms combine to produce O_2, which is released as a waste product from the chloroplast. Since the hydrogen atom contains a single proton around which orbits a single electron, the splitting of water to produce two H^+ ions also releases two electrons. These two electrons are transferred back to chlorophyll molecules in the thylakoid membrane to replace those passed along the electron transport chain.

At the end of the electron transport chain, electrons are transferred to the electron carrier for plants, nicotinamide adenine dinucleotide phosphate, or NADP. Just like the NAD^+ involved in cellular respiration, $NADP^+$ functions as an electron taxicab. The $NADP^+$ used during photosynthesis picks up one H^+ ion and two electrons from the electron transport chain (**Figure 5.12**, step 4). The resulting NADPH ferries electrons to the stroma, where the enzymes of the Calvin cycle will use the electrons to assemble sugars. Thus, the

light reactions produce ATP and a source of electrons, both of which are used in the synthesis step, and release oxygen as a by-product.

Stop & Stretch ATP is used to power all activities in a cell that require energy input. Why can't plants just use the ATP produced during the light reactions of photosynthesis for all of their activities, instead of going through the step of producing glucose, which the plants will later break down to produce ATP?

The Calvin Cycle

The Calvin cycle consists of a series of reactions occurring in the stroma that use the ATP and NADPH produced during photosynthesis to convert carbon dioxide into sugars. CH_2O is the general formula for sugars. For example, glucose is $C_6H_{12}O_6$ or $6(CH_2O)$. A quick comparison of the formulas of these molecules makes it obvious that converting CO_2 into CH_2O requires the incorporation of hydrogen atoms and their associated electrons. These components are provided by the NADPH produced during the light reactions.

During the Calvin cycle, carbon dioxide from the environment reacts with 5-carbon molecules called ribulose bisphosphate, or RuBP (**Figure 5.13**). The enzyme that catalyzes this reaction is ribulose bisphosphate carboxylase oxygenase, or **rubisco,** the most abundant protein on the planet. Immediately, the resulting 6-carbon molecules break down into pairs of 3-carbon molecules. These 3-carbon molecules go through several rearrangements, with the help of energy released from ATP and electrons from NADPH, to produce the sugar glyceraldehyde 3-phosphate (or G3P). In the last step of the cycle, five G3P molecules are rearranged into three RuBP molecules. Excess G3P produced by this process is used by the cell to make glucose and other carbohydrate compounds. Because the first stable product of the Calvin cycle is a 3-carbon compound, this pathway is often called C_3 photosynthesis.

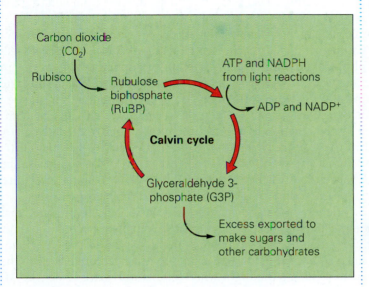

Figure 5.13 Reactions of the Calvin cycle. Carbon dioxide is incorporated into sugar in plants through a series of reactions that regenerate the initial carbon-containing starting product, the sugar ribulose bisphosphate (RuBP). The enzyme that attaches carbon dioxide to RuBP in the first step of the Calvin cycle is called rubisco. Glyceraldehyde 3-phosphate (G3P) is the simple sugar produced by these reactions. Excess G3P is exported to other pathways to produce organic molecules the plant needs.
Visualize This: How would an absence of sunlight affect the Calvin cycle?

Stop & Stretch   Plants can make all of the macromolecules they need to survive from simple, inorganic (non-carbon-containing) components. They accumulate carbon from the atmosphere and hydrogen and oxygen from water to make the overwhelming majority of their mass. The remaining elements they require are from the soil. Consider the atoms found in proteins and nucleic acids and list two or three soil nutrients that you think are most important to plant growth.

While photosynthesis does take carbon dioxide out of the atmosphere, it is important to realize that the fossil fuels that we have been using only for the last century or so took over 100 million years to form. In other words, right now, carbon dioxide is being released into the atmosphere many times faster than it can be absorbed via natural photosynthesis. We cannot simply rely on photosynthesis to remove excess greenhouse gases as a way to prevent global warming. In fact, rising temperatures may actually slow photosynthesis and reduce its effectiveness.

5.4 How Global Warming Might Reduce Photosynthesis

The high temperatures caused by global warming can reduce photosynthesis in a number of ways, but primarily because higher temperatures generally lead to drier conditions.

Land plants bring carbon dioxide for photosynthesis into their leaves through openings on their surface called **stomata** (**Figure 5.14**). Stomatal openings are surrounded by two kidney-bean-shaped cells called **guard cells.** When the guard cells are compressed against each other, the stomata are closed, thus restricting the flow of gases into or out of the plant. When the guard cells change shape to create a gap between them, the stomata are open, and carbon dioxide and oxygen gases can be exchanged. In addition to the exchange of gases, water moves out of the plant through the stomatal opening via a process called

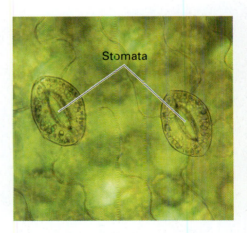

Figure 5.14 Stomata. Stomata are adjustable microscopic pores found on the surface of leaves that allow for gas exchange.

(a) Open

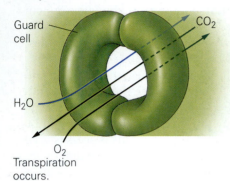

Guard cell

CO_2

H_2O

O_2

Transpiration occurs.

(b) Closed

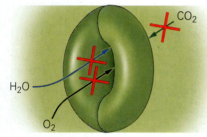

CO_2

H_2O

O_2

Transpiration does not occur.

Figure 5.15 Gas exchange and water loss. (a) When stomata are open, oxygen gases and carbon dioxide can be exchanged, but water can be lost from a plant through a process called transpiration. (b) When the guard cells change shape to block the opening, gas exchange and transpiration do not occur.

transpiration (**Figure 5.15**). The transpired water is replaced in the plant by water from the soil. On hot, dry days, plants cannot absorb enough water from the soil and will close their stomata to reduce the rate of water lost to transpiration. Closing stomatal openings also prevents carbon dioxide from entering the plant. Thus, with rising temperatures, the rate of photosynthesis declines.

Closing the stomatal openings may not just reduce the rate of photosynthesis; it may even counteract it. This occurs as the result of another series of reactions known as **photorespiration.** During photorespiration, the first enzyme in the Calvin cycle, rubisco, uses oxygen instead of carbon dioxide as a substrate for the reaction it catalyzes with RuBP. Oxygen is used when carbon dioxide levels are low inside the leaf, such as when the stomata are closed. The compound produced by the reaction between RuBP and oxygen, called glycolate, cannot be used in the Calvin cycle. In fact, glycolate must be destroyed by the plant because high levels of this acid will further inhibit photosynthesis. The breakdown of glycolate requires energy provided by cellular respiration—thus, instead of taking up carbon dioxide, plants that are photorespiring release it.

In warmer and drier environments, natural selection should favor plants that can minimize photorespiration despite having closed stomata much of the time. Two mechanisms for reducing photorespiration are known as C_4 and CAM photosynthesis, in which additional pathways that concentrate carbon dioxide in the plant (C_4) or capture it during the night when temperatures are cooler (CAM) occur before the C_3 photosynthetic pathway. C_4 and CAM, along with the C_3 process that occurs during the Calvin cycle, are briefly summarized in **Table 5.1**.

As Earth warms, it is possible that the abundance of C_4 and CAM plants will increase while the number of C_3 plants declines as a result of the increased burden of photorespiration. However, replacing C_3 plants, many of which are large trees, with C_4 plants, mostly grasses, will reduce the total amount of photosynthesis and the amount of carbon that can be taken out of the atmosphere. Because net rates of photosynthesis (as measured by grams of carbon dioxide removed from the atmosphere per acre per year) in grasslands are 30% to 60% less than rates in forests, the loss of trees significantly decreases the removal of carbon dioxide from the atmosphere.

The replacement of trees with grasses, **deforestation,** may happen naturally on a warming Earth as C_4 plants outperform C_3 types, but it is already happening at high rates thanks to human activities. The process occurs when forests are cleared for logging, farming, and ever-expanding human settlements. Deforestation also contributes directly to the increase in carbon dioxide within the atmosphere; current estimates are that up to 25% of the carbon dioxide introduced into the atmosphere originates from the cutting and burning of forests in the tropics alone. Clearly, one effective way to reduce global warming is to reduce deforestation and promote photosynthesis by planting trees through reforestation projects.

In the worst cases, rising temperatures due to global warming will cause vegetated land to become desert, completely eliminating photosynthesis in certain areas. Warming temperatures will expose more snow- and ice-covered regions, allowing additional photosynthesis there, but this additional carbon dioxide uptake is likely to be offset by carbon dioxide released from the more rapid decay of carbohydrates in the soil in these formerly frozen regions. As we have seen, photosynthesis is not likely to soak up much of the excess carbon dioxide released by fossil fuel combustion and, in fact, it may even decline on a warmer planet.

5.5 Decreasing the Effects of Global Warming

The devastation Hurricane Katrina caused in New Orleans is paralleled by environmental damage seen worldwide as a result of climate change. Sea-level rise is not the only effect; global warming will change weather patterns, resulting in droughts in some areas and torrential rains in others. It will

TABLE 5.1

C₃, C₄, and CAM plant photosynthesis. Plants have evolved adaptations that prevent water loss.		
Type of plant and example	**Stomata status**	**Description**
C₃ Soybean		The Calvin cycle converts two 3-carbon sugars into glucose.
C₄ Corn		Enzymes scavenge CO_2 to produce 4-carbon sugars, even when stomata are closed. The 4-carbon sugars "pump" carbon dioxide molecules to the Calvin cycle when they break down.
CAM Jade		Water loss is slowed by opening stomata only at night. Carbon dioxide is stored as an organic acid in the vacuole. The acid's breakdown releases this carbon dioxide to the Calvin cycle during the day.

cause the extinction of animals and plants that have narrow temperature requirements. It may contribute to the spread of infectious disease and dangerous pests that are now limited by temperature to the tropics. And the carbon dioxide absorbed by the ocean is causing it to acidify, threatening coral reefs, algae in the open ocean, and the organisms that depend on them. Clearly, excess carbon dioxide in the atmosphere is imposing a cost on people.

The United States has a disproportionate impact on the rate of carbon dioxide emissions into the atmosphere. Home to only 4% of the world's population, the United States produces close to 25% of the carbon dioxide emitted by fossil fuel burning. The emissions rate of carbon dioxide for an average American is twice that of a Japanese or German individual, three times that of the global average, four times that of a Swede, and 20 times that of the average Indian.

Most of the emissions for an individual country come from industry, followed by transportation and then by commercial, residential, and agricultural emissions. All of us can work to reduce our personal contribution to global warming by decreasing residential and transportation emissions. Most residential emissions are from energy used to heat and cool homes and to power electrical appliances. Transportation emissions are affected by our choice of the vehicles we use for transport, the fuel economy of cars, and the number of miles traveled. **Table 5.2**

TABLE 5.2

Decreasing your greenhouse gas emissions. Here are some ideas that you can use to help slow the rate of global warming.

Action		Annual decrease in carbon dioxide production
Drive an energy-efficient vehicle. SUVs average 16 miles per gallon, while smaller cars average 25 miles per gallon.		13,000 pounds
Carpool 2 days per week.		1,590 pounds
Recycle glass bottles, aluminum cans, plastic, newspapers, and cardboard.		850 pounds
Walk 10 miles per week instead of driving.		590 pounds
Buy high-efficiency appliances.		400 pounds per appliance
Buy food and other products with reusable or recyclable packaging, or reduced packaging, to save the energy required to manufacture new containers.		230 pounds
Use a push mower instead of a power mower.		80 pounds
Plant shade trees around your home to decrease energy consumption and to remove carbon dioxide by photosynthesis.		50 pounds

describes many ways that you can decrease your greenhouse gas emissions and indicates the number of pounds of carbon dioxide that each action would save annually. These reductions may seem trivial in comparison to the scope of the problem, but when they are multiplied by the more than 300 million people in the United States, the savings become significant.

Having an effect on industrial, commercial, and agricultural sectors is difficult for any one individual. Changes in these areas will instead take leadership from the policy makers who are committed to reducing emissions. Making these changes requires that our leaders, and all of us, understand that even though the implications of and solutions to global warming may be open to debate, the fact that it is occurring at an unprecedented rate is not.

SAVVY READER

Refugees from Global Warming

by Sean Mattson

CARTI SUGDUB, Panama—Rising seas from global warming, coming after years of coral reef destruction, are forcing thousands of indigenous Panamanians to leave their ancestral homes on low-lying Caribbean islands.

Seasonal winds, storms and high tides combine to submerge the tiny islands, crowded with huts of yellow cane and faded palm fronds, leaving them ankle-deep in emerald water for days on end.

Pablo Preciado, leader of the island of Carti Sugdub, remembers that in his childhood floods were rare, brief and barely wetted his toes. "Now it's something else. It's serious," he said.

The increase of a few inches in flood depth is consistent with a global sea level rise over Preciado's 64 years of life and has been made worse by coral mining by the islanders that reduced a buffer against the waves.

If the islanders abandon their homes as planned, the exodus will be one of the first blamed on rising sea levels and global warming.

"This is no longer about a scientist saying that climate change and the change in sea level will flood (a people) and affect them," said Hector Guzman, a marine biologist and coral specialist at the Smithsonian Tropical Research Institute in Panama. "This is happening now in the real world."

The fiercely independent Kuna, famed for rebellions against Spanish conquistadors, French pirates and Panamanian overlords, have accelerated their fate by mining coral, which they use to expand islands and build artificial islets and breakwaters.

Guzman, based at a Pacific island research center on the edge of Panama City, has warned of the risks of coral mining for a decade but says speaking out against a legally permitted traditional activity is "taboo."

"(The Kuna) have increased their vulnerability to storms, wave action, and above all, the action of the rise in sea level," he told Reuters.

1. What is the hypothesis of this excerpted article?

2. Did the author explore other hypotheses about the increased flooding in Carti Sugdub, even if only to refute them?

3. Is the islanders' plight definitively caused by global warming?

Source: Thomson Reuters, July 12, 2010. http://www .reuters.com/article/idUSTRE66B0PL20100712

Chapter Review

Learning Outcomes

Mastering BIOLOGY

Go to the Study Area at www.masteringbiology.com for practice quizzes, myeBook, BioFlix™ 3-D animations, MP3Tutor sessions, videos, current events, and more.

LO1 Describe the greenhouse effect (Section 5.1).

- Greenhouse gases, particularly carbon dioxide, increase the amount of heat retained in Earth's atmosphere, which then leads to increased surface temperatures (pp. 94–95).

LO2 Explain why water is slow to change temperature (Section 5.1).

- Water can absorb large amounts of heat without undergoing rapid or drastic changes in temperature because heat must first be used to break hydrogen bonds between adjacent water molecules. A high heat-absorbing capacity is a characteristic of water (pp. 95–97).

LO3 Compare and contrast the water cycle and the carbon cycle (Section 5.1, Section 5.2).

- Both water and carbon cycle between animals, plants, soil, oceans, and the atmosphere. Most of the movement of water occurs outside living organisms, while carbon is mostly cycled due to biological activity (pp. 96–98).

LO4 Discuss the origin of fossil fuels and their relationship to the carbon cycle (Section 5.2).

- Fossil fuels are the buried remains of ancient plants, which took carbon dioxide out of the atmosphere and transformed it into carbohydrates. The burning of fossil fuels is returning carbon to the atmosphere, leading to rising concentrations of carbon dioxide and global warming (pp. 98–99).

LO5 State the basic equation of photosynthesis (Section 5.3).

- Photosynthesis utilizes carbon dioxide from the atmosphere to make sugars and other substances. During photosynthesis, energy from sunlight is used to rearrange the atoms of carbon dioxide and water to produce sugars and oxygen (p. 100).

LO6 Describe the light reactions of photosynthesis (Section 5.3).

- Photosynthesis occurs in chloroplasts. Sunlight strikes the chlorophyll molecule within chloroplasts, boosting electrons to a higher energy level. These excited electrons are dropped down an electron transport chain located in the thylakoid membrane, and ATP is made (pp. 101–102).
- Electrons are also passed to electron carriers (NADPH). The electrons that are lost from chlorophyll become replaced by electrons acquired during the splitting of water, and oxygen is released (p. 102).

LO7 Explain the events that occur during the Calvin cycle, and describe the relationship between the light reactions and the Calvin cycle (Section 5.3).

- The Calvin cycle utilizes the products of the light reactions (ATP and the electron carrier NADPH) to incorporate carbon dioxide into sugars, regenerating the starting products of the cycle and exporting excess sugars to be used as chemical building blocks for plant compounds (pp. 102–103).

LO8 Explain the role of stomata in balancing photosynthesis and water loss (Section 5.3).

- Stomata on a plant's surface not only allow in carbon dioxide for photosynthesis but also allow water to escape from the plant. Guard cells surrounding the stomata can change shape to close the stomata and restrict water loss (pp. 103–104).

LO9 Define *photorespiration*, and explain why this process is detrimental to plants that experience it (Section 5.4).

- Photorespiration occurs when stomata are closed, carbon dioxide declines in the plant, and oxygen is incorporated into the first step of the Calvin cycle. The resulting product is poisonous to the plant, and energy must be expended to reduce it (p. 104).
- C_4 and CAM plants have evolved to perform photosynthesis while reducing the risk of photorespiration in dry conditions (pp. 104–105).

LO10 Discuss how our own activities contribute to or reduce the risk of global warming (Section 5.5).

- Humans are deforesting Earth's land surface, reducing the rate of photosynthesis and thus the uptake of atmospheric carbon dioxide, so reforestation is one strategy to reduce global warming (p. 104).
- Humans can reduce carbon dioxide emissions by increasing energy efficiency and reducing energy use (pp. 106–107).

Roots to Remember

The following roots of words come mainly from Latin and Greek and will help you decipher terms:

chloro- means green. Chapter terms: chloroplast, chlorophyll

photo- means light. Chapter terms: photosynthesis, photorespiration

phyll- means leaf. Chapter term: chlorophyll

trans- means across or to the other side. Chapter term: transpiration

Learning the Basics

1. **LO5** What are the reactants and products of photosynthesis?

2. **LO1** Carbon dioxide functions as a greenhouse gas by _____.

    A. interfering with water's ability to absorb heat; B. increasing the random molecular motions of oxygen; C. allowing radiation from the sun to reach Earth and absorbing the reradiated heat; D. splitting into carbon and oxygen and increasing the rate of cellular respiration

3. **LO2** Water has a high heat-absorbing capacity because _____.

    A. the sun's rays penetrate to the bottom of bodies of water, mainly heating the bottom surface; B. the strong covalent bonds that hold individual water molecules together require large inputs of heat to break; C. it has the ability to dissolve many heat-resistant solutes; D. initial energy inputs are first used to break hydrogen bonds between water molecules and only after these are broken, to raise the temperature; E. all of the above are true

4. **LO4** The burning of fossil fuels _____.

    A. releases carbon dioxide to the atmosphere; B. primarily occurs as a result of human activity. C. is contributing to global warming; D. is possible thanks to photosynthesis that occurred millions of years ago; E. all of the above are correct.

5. **LO8** Stomata on a plant's surface _____.

    A. prevent oxygen from escaping; B. produce water as a result of photosynthesis; C. cannot be regulated by the plant; D. allow carbon dioxide uptake into leaves; E. are found in stacks called thylakoids.

6. **LO6** Which of the following **does not** occur during the light reactions of photosynthesis?

    A. Oxygen is split, releasing water; B. Electrons from chlorophyll are added to an electron transport chain; C. An electron transport chain drives the synthesis of ATP for use by the Calvin cycle; D. NADPH is produced and will carry electrons to the Calvin cycle; E. Oxygen is produced when water is split.

7. **LO6 LO7** Which of the following is a **false** statement about photosynthesis?

    A. During the Calvin cycle, electrons and ATP from the light reactions are combined with atmospheric carbon dioxide to produce sugars; B. The enzymes of the Calvin cycle are located in the chloroplast stroma; C. Oxygen produced during the Calvin cycle is released into the atmosphere; D. Sunlight drives photosynthesis by boosting electrons found in chlorophyll to a higher energy level; E. Electrons released when sunlight strikes chlorophyll are replaced by electrons from water.

8. **LO10** Which human activity generates the most carbon dioxide?

    A. driving; B. cooking; C. bathing; D. using aerosol sprays.

9. **LO9** Photorespiration occurs _____.

    A. under hot and dry conditions; B. when oxygen is incorporated in the first step of the Calvin cycle; C. when carbon dioxide levels are high inside the plant; D. A and B are correct; E. A, B, and C are correct.

10. **LO5** Select the **true** statement regarding metabolism in plant and animal cells.

    A. Plant and animal cells both perform photosynthesis and aerobic respiration; B. Animal cells perform aerobic respiration only, and plant cells perform photosynthesis only; C. Plant cells perform aerobic respiration only, and animal cells perform photosynthesis only; D. Plant cells perform cellular respiration and photosynthesis, and animal cells perform aerobic respiration only.

Analyzing and Applying the Basics

1. **LO3** During the last Ice Age, global temperatures were 4 to 5°C lower than they are today. How would this temperature difference affect the water cycle?

2. **LO3** Imagine an Earth without living organisms on it. How does this difference change the water cycle and the carbon cycle?

3. **LO5** Before life evolved on Earth, there was very little oxygen in the air, mostly because oxygen is a very reactive molecule and tends to combine with other compounds. Explain why oxygen can be maintained at 16% in our atmosphere today.

Connecting the Science

1. A "carbon footprint" is the amount of carbon dioxide you produce as a result of your lifestyle. List five realistic actions you can take to reduce your carbon footprint. Which of these actions will have the greatest impact on your footprint? How difficult for you would it be to take this action?

2. The impact of global warming is not uniformly negative—some regions will benefit from a change in the climate. In addition, technological fixes exist that can help humans deal with some of the consequences of sea-level rise, and people can move from threatened coastlines and areas experiencing the negative effects of global warming. Given these facts, should humans commit to trying to stop global warming, or should we instead focus on mitigating its consequences?

Answers to **Stop & Stretch, Visualize This, Savvy Reader,** and **Chapter Review** questions can be found in the **Answers** section at the back of the book.

Cancer

DNA Synthesis, Mitosis, and Meiosis

Cancer cells divide when they should not.

Nicole's early college career was similar to that of most students. She enjoyed her independence and the wide variety of courses required for her double major in biology and psychology. She worried about her grades and finding ways to balance her coursework with her social life. She also tried to find time for lifting weights in the school's athletic center and snowboarding at a local ski hill. Some weekends, to take a break from school, she would ride the bus home to see her family.

Managing to get schoolwork done, see friends and family, and still have time left to exercise had been difficult, but possible, for Nicole during her first two years at school. That changed drastically during her third year of school.

One morning in October of her junior year, Nicole began having severe pains in her abdomen. The first time this happened, she was just beginning an experiment in her cell biology laboratory. Hunched over and sweating, she barely managed to make it through her two-hour biology lab. Over the next few days, the pain intensified so much that she was unable to walk from her apartment to her classes without stopping several times to rest.

Later that week, as she was preparing to leave for class, the pain was so severe that she had to lie down in the hallway of her apartment. When her roommate got home a few minutes later, she took Nicole to the student health center for an emergency visit. The physician at the health center first determined that Nicole's appendix had not burst and then made an appointment for Nicole to see a gynecologist the next day.

After hearing Nicole's symptoms, the gynecologist pressed on her abdomen and felt what he

Nicole got sick during her junior year of college.

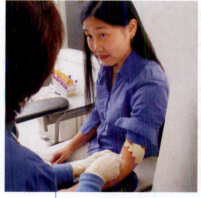

She had to undergo some procedures to see if she had cancer.

She wants to understand why she got cancer.

thought was a mass on her right ovary. He used a noninvasive procedure called ultrasound to try to get an image of her ovary. This procedure requires the use of high-frequency sound waves. These waves, which cannot be heard by humans, bounce off tissues and produce a pattern of echoes that can be used to create a picture called a sonogram. Healthy tissues, fluid-filled cysts, and tumors all look different on a sonogram.

Nicole's sonogram convinced her gynecologist that she had a large growth on her ovary. He told her that he suspected this growth was a *cyst,* or fluid-filled sac. Her gynecologist told her that cysts often go away without treatment, but this one seemed to be quite large and should be surgically removed.

Even though the idea of having an operation was scary for Nicole, she was relieved to know that the pain would stop. Her gynecologist also assured her that she had nothing to worry about because cysts are not cancerous. A week after the abdominal pain began, Nicole's gynecologist removed the cyst and her completely engulfed right ovary through an incision just below her navel.

After the operation, Nicole's gynecologist assured her that the remaining ovary would compensate for the missing ovary by ovulating (producing an egg cell) every month. He added that he would have to monitor her remaining ovary carefully to see that it did not become cystic or, even worse, cancerous. She could not afford to lose another ovary if she wanted to have children someday.

Monitoring her remaining ovary involved monthly visits to her gynecologist's office, where Nicole had her blood drawn and analyzed. The blood was tested for the level of a protein called CA125, which is produced by ovarian cells. Higher-than-normal CA125 levels usually indicate that the ovarian cells have increased in size or number and are thus associated with the presence of an ovarian tumor.

Nicole went to her scheduled checkups for five months after the original surgery. The day after her March checkup, Nicole received a message from her doctor asking that she come to see him the next day. Because she needed to study for an upcoming exam, Nicole tried to push aside her concerns about the appointment. By the time she arrived at her gynecologist's office, she had convinced herself that nothing serious could be wrong. She thought a mistake had probably been made and that he just wanted to perform another blood test.

The minute her gynecologist entered the exam room, Nicole could tell by his demeanor that something was wrong. As he started speaking to her, she began to feel very anxious—when he said that she might have a tumor on her remaining ovary, she could not believe her ears. When she heard the words *cancer* and *biopsy,* Nicole felt as though she was being pulled underwater. She could see that her doctor was still talking, but she could not hear or understand him. She felt too nauseated to think clearly, so she excused herself from the exam room, took the bus home, and immediately called her parents.

After speaking with her parents, Nicole realized that she had many questions to ask her doctor. She did not understand how it was possible for such a young woman to have lost one ovary to a cyst and then possibly to have a tumor on the other ovary. She wondered how this tumor would be treated and what her prognosis would be. Before seeing her gynecologist again, Nicole decided to do some research in order to make a list of questions for her doctor.

6.1 What Is Cancer?

Cancer is a disease that begins when a single cell replicates itself although it should not. **Cell division** is the process a cell undergoes to make copies of itself. This process is normally regulated so that a cell divides only when more cells are required and when conditions are favorable for division. A cancerous cell is a rebellious cell that divides without being given the go-ahead.

Tumors Can Be Cancerous

Unregulated cell division leads to a pileup of cells that form a lump or **tumor.** A tumor is a mass of cells; it has no apparent function in the body. Tumors that stay in one place and do not affect surrounding structures are said to be **benign.** Some benign tumors remain harmless; others become cancerous. Tumors that invade surrounding tissues are **malignant** or cancerous. The cells of a malignant tumor can break away and start new cancers at distant locations through a process called **metastasis** (**Figure 6.1**).

Cancer cells can travel virtually anywhere in the body via the lymphatic and circulatory systems. The lymphatic system collects fluids, or lymph, lost from blood vessels. The lymph is then returned to the blood vessels, thus allowing

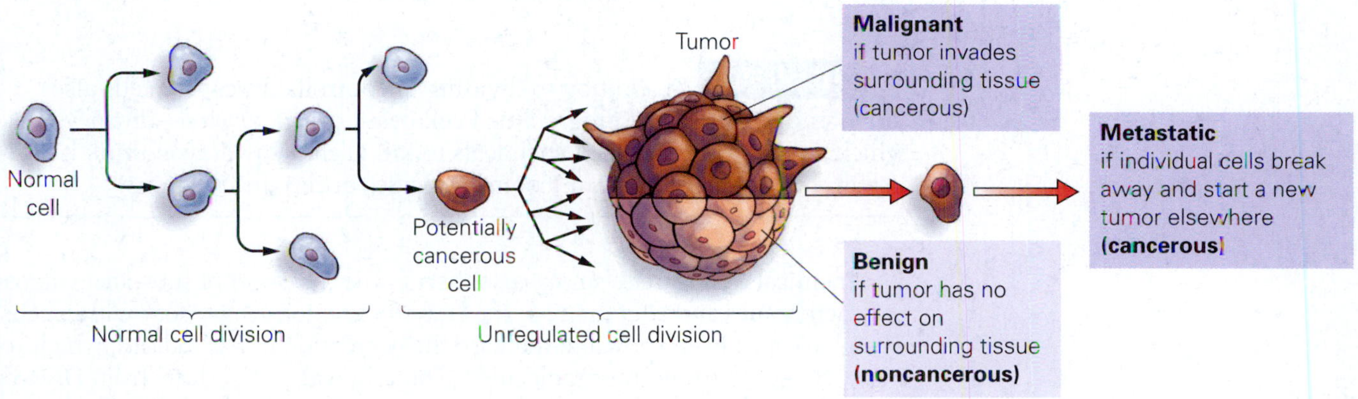

Figure 6.1 What is cancer? A tumor is a clump of cells with no function. Tumors may remain benign, or they can invade surrounding tissues and become malignant. Tumor cells may move, or metastasize, to other locations in the body. Malignant and metastatic tumors are cancerous.

cancer cells access to the bloodstream. Lymph nodes are structures that filter the lost fluids. When a cancer patient is undergoing surgery, the surgeon will often remove a few lymph nodes to see if any cancer cells are in the nodes. If cancer cells appear in the nodes, then some cells have left the original tumor and are moving through the bloodstream. If this has happened, cancerous cells likely have metastasized to other locations in the body.

When cancer cells metastasize, they can gain access not only to the **circulatory system,** which includes blood vessels to transport the blood, but also to the heart, which pumps the blood. Once inside a blood vessel, cancer cells can drift virtually anywhere in the body.

Cancer cells differ from normal cells in three ways: (1) They divide when they should not; (2) they invade surrounding tissues; and (3) they can move to other locations in the body. All tissues that undergo cell division are susceptible to becoming cancerous. However, there are ways to increase or decrease the probability of getting cancer.

Risk Factors for Cancer

Certain exposures and behaviors, called **risk factors,** increase a person's risk of obtaining a disease. General risk factors for virtually all cancers include tobacco use, a high-fat and low-fiber diet, lack of exercise, obesity, excess alcohol consumption, and increasing age. **Table 6.1** outlines other risk factors that are linked to particular cancers.

Tobacco Use. The use of tobacco of any type, whether via cigarettes, cigars, pipes, or chewing tobacco, increases your risk of many cancers. While smoking is the cause of 90% of all lung cancers, it is also the cause of about one-third of all cancer deaths. Cigar smokers have increased rates of lung, larynx, esophagus, and mouth cancers. Chewing tobacco increases the risk of cancers of the mouth, gums, and cheeks. People who do not smoke but who are exposed to secondhand smoke have increased lung cancer rates.

Tobacco smoke contains more than 20 known cancer-causing substances called **carcinogens.** For a substance to be considered carcinogenic, exposure to the substance must be correlated with an increased risk of cancer. Examples of carcinogens include cigarette smoke, radiation, ultraviolet light, asbestos, and some viruses.

The carcinogens that are inhaled during smoking come into contact with cells deep inside the lungs. Chemicals present in cigarettes and cigarette smoke have been shown to increase cell division, inhibit a cell's ability to repair damaged DNA, and prevent cells from dying when they should.

> **Stop & Stretch**   In addition to dividing uncontrollably, cancer cells also fail to undergo a type of programmed cell death called *apoptosis,* during which a cell uses specialized chemicals to kill itself. Why do you think it is useful that our own cells can "commit suicide" in certain situations?

Chemicals in cigarette smoke also disrupt the transport of substances across cell membranes and alter many of the enzyme reactions that occur within cells. They have also been shown to increase the generation of *free radicals,* which remove electrons from other molecules. The removal of electrons from DNA or other molecules causes damage to these molecules—damage that, over time, may lead to cancer. Cigarette smoking provides so many different opportunities for DNA damage and cell damage that tumor formation and metastasis are quite likely for smokers. In fact, people who smoke cigarettes increase their odds of developing almost every cancer.

TABLE 6.1

Cancer risk. Risk factors and detection methods for particular cancers are given.

Cancer Location	Risk Factors	Detection	Comments
Ovary Oviduct Ovary	• Smoking • Mutation to *BRCA2* gene • Advanced age • Oral contraceptive use and pregnancy decrease risk	• Blood test for elevated CA125 level • Rectovaginal exam	• Fifth leading cause of death among women in the United States
Breast 	• Smoking • Mutation to *BRCA1* gene • High-fat, low-fiber diet • Use of oral contraceptives may slightly increase risk	• Monthly self-exams, look and feel for lumps or changes in contour • Mammogram	• Only 5% of breast cancers are due to *BRCA1* mutations • Second-highest cause of cancer-related deaths • 1% of breast cancer occurs in males
Cervix Uterus Cervix Vagina	• Smoking • Exposure to sexually transmitted human papilloma virus (HPV)	• Annual Pap smear tests for the presence of precancerous cells	• Precancerous cells can be removed by laser surgery or cryotherapy (freezing) before they become cancerous
Skin 	• Smoking • Fair skin • Exposure to ultraviolet light from the sun or tanning beds	• Monthly self-exams, look for growths that change in size or shape	• Skin cancer is the most common of all cancers; usually curable if caught early
Blood (leukemia) 	• Exposure to high-energy radiation such as that produced by atomic bomb explosions in Japan during World War II	• A sample of blood is examined under a microscope	• Cancerous white blood cells cannot fight infection efficiently; people with leukemia often succumb to infections

(continued)

TABLE 6.1 *(continued)*

Cancer risk. Risk factors and detection methods for particular cancers are given.

Cancer Location	Risk Factors	Detection	Comments
Lung	• Smoking • Exposure to secondhand smoke • Asbestos inhalation	• X-ray	• Lung cancer is the most common cause of death from cancer, and the best prevention is to quit, or never start, smoking
Colon and rectum Small intestine Colon	• Smoking • Polyps in the colon • Advanced age • High-fat, low-fiber diet	• Change in bowel habits • Colonoscopy is an examination of the rectum and colon using a lighted instrument	• Benign buds called polyps can grow in the colon; removal prevents them from mutating and becoming cancerous
Prostate Bladder Prostate Rectum	• Smoking • Advanced age • High-fat, low-fiber diet	• Blood test for elevated level of prostate-specific antigen (PSA) • Physical exam by physician, via rectum	• More common in African American men than Asian, white, or Native American men
Testicle Testicle Scrotum	• Abnormal testicular development	• Monthly self-exam, inspect for lumps and changes in contour	• Testicular cancer accounts for only 1% of all cancers in men but is the most common form of cancer found in males between the ages of 15 and 35

A High-Fat, Low-Fiber Diet. Cancer risk may also be influenced by diet. The American Cancer Society recommends eating at least 5 servings of fruits and vegetables every day as well as 6 servings of food from other plant sources, such as breads, cereals, grains, rice, pasta, or beans. Plant foods are low in fat and high in fiber. A diet high in fat and low in fiber is associated with increased risk of cancer. Fruits and vegetables are also rich in *antioxidants*. These substances help to neutralize the electrical charge on free radicals and thereby prevent the free radicals from taking electrons from other molecules, including DNA. There is some evidence that antioxidants may help prevent certain cancers by minimizing the number of free radicals that may damage the DNA in our cells.

Lack of Exercise. Regular exercise decreases the risk of most cancers, partly because exercise keeps the immune system functioning effectively. The immune system helps destroy cancer cells when it can recognize them as foreign to the host body. Unfortunately, since cancer cells are actually your own body's cells run amok, the immune system cannot always differentiate between normal cells and cancer cells.

Obesity. Exercise also helps prevent obesity, which is associated with increased risk for many cancers, including cancers of the breast, uterus, ovary, colon, gallbladder, and prostate. Because fatty tissues can store hormones, the abundance of fatty tissue has been hypothesized to increase the odds of hormone-sensitive cancers such as breast, uterine, ovarian, and prostate cancer.

Excess Alcohol Consumption. Drinking alcohol is associated with increased risk of some types of cancer. Men who want to decrease their cancer risk should have no more than two alcoholic drinks a day, and women one or none. People who both drink and smoke increase their odds of cancer in a multiplicative rather than additive manner. In other words, if one type of cancer occurs in 10% of smokers and in 2% of drinkers, someone who smokes and drinks multiplies chances of developing cancer to a rate that is closer to 20% than 12%. The risk factor percentages are multiplied rather than added.

Increasing Age. As you age, your immune system weakens, and its ability to distinguish between cancer cells and normal cells decreases. This weakening is part of the reason many cancers are far more likely in elderly people. Additional factors that help explain the higher cancer risk with increasing age include cumulative damage. If we are all exposed to carcinogens during our lifetime, then the longer we are alive, the greater the probability that some of those carcinogens will mutate genes involved in regulating the cell cycle. Also, because multiple mutations are necessary for a cancer to develop, it often takes many years to progress from the initial mutation to a tumor and then to full-blown cancer. Scientists estimate that most cancers large enough to be detected have been growing for at least five years and are composed of close to one billion cells.

Nicole's cancer affected ovarian tissue. Why might ovarian cells be more likely to become cancerous than some other types of cells? Cells that divide frequently are more prone to cancer than those that don't divide often. When an egg cell is released from the ovary during ovulation, the tissue of the ovary becomes perforated. Cells near the perforation site undergo cell division to heal the damaged surface of the ovary. For Nicole, these cell divisions may have become uncontrolled, leading to the growth of a tumor.

6.2 Passing Genes and Chromosomes to Daughter Cells

Most cell division does not lead to cancer. Cell division produces new cells to heal wounds, replace damaged cells, and help organisms grow and reproduce themselves. Each of us begins life as a single fertilized egg cell that undergoes millions of rounds of cell division to produce all the cells that comprise the tissues and organs of our bodies.

(a) Amoeba

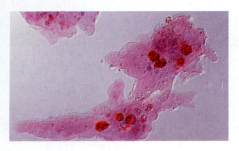

(b) English ivy

Figure 6.2 Asexual reproduction.
(a) This single-celled amoeba divides by copying its DNA and producing offspring that are genetically identical to the original, parent amoeba. (b) Some multicellular organisms, such as this English ivy plant, can reproduce asexually from cuttings.

(a) Uncondensed DNA

(b) DNA condensed into chromosomes

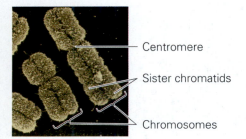

— Centromere

— Sister chromatids

— Chromosomes

Some organisms reproduce by producing exact copies of themselves via cell division. Reproduction of this type, called **asexual reproduction,** does not require genetic input from two parents and results in offspring that are genetically identical to the original parent cell. Single-celled organisms, such as bacteria and amoeba, reproduce in this manner (**Figure 6.2a**). Some multicellular organisms can reproduce asexually also. For example, most plants can grow from clippings of the stem, leaves, or roots and thereby reproduce asexually (**Figure 6.2b**). Organisms whose reproduction requires genetic information from two parents undergo **sexual reproduction.** Humans reproduce sexually when sperm and egg cells combine their genetic information at fertilization.

Genes and Chromosomes

Whether reproducing sexually or asexually, all dividing cells must first make a copy of their genetic material, the **DNA (deoxyribonucleic acid).** The DNA carries the instructions, called **genes,** for building all of the proteins that cells require. The DNA in the nucleus is wrapped around proteins to produce structures called **chromosomes.** Chromosomes can carry hundreds of genes along their length. Different organisms have different numbers of chromosomes in their cells. For example, dogs have 78 chromosomes in each cell, humans have 46, and dandelions have 24.

Chromosomes are in an uncondensed, string-like form prior to cell division (**Figure 6.3a**). Before cell division occurs, the DNA in each chromosome is condensed (compressed) in a short, linear form (**Figure 6.3b**). Condensed linear chromosomes are easier to maneuver during cell division and are less likely to become tangled or broken than are the uncondensed and string-like structures. When a chromosome is replicated, a copy is produced that carries the same genes. The copied chromosomes are called **sister chromatids,** and each sister chromatid is composed of one DNA molecule. Sister chromatids are attached to each other at a region toward the middle of the replicated chromosome, called the **centromere.**

The DNA molecule itself is double stranded and can be likened to a twisted rope ladder (Chapter 2). The backbone or "handrails" of each strand are composed of alternating sugar and phosphate groups. Across the width or "rungs" of the DNA helix are the nitrogenous bases, paired together via hydrogen bonds such that adenine (A) makes a base pair with thymine (T), and guanine (G) makes a base pair with cytosine (C).

You also learned that two of the people credited with determining DNA structure are James Watson and Francis Crick. Watson and Crick reported their hypothesis about the structure of the DNA molecule in a 1953 paper for the journal *Nature*. Although they did not go so far as to propose a detailed model for how the DNA molecule was replicated, they did say, "It has not escaped our notice that the specific pairing we have postulated immediately suggests a copying mechanism for the genetic material." The copying mechanism that Watson and Crick referred to is also called *DNA replication*.

Figure 6.3 DNA condenses during cell division. (a) DNA in its replicated but uncondensed form prior to cell division. (b) During cell division, each copy of DNA is wrapped neatly around many small proteins, forming the condensed structure of a chromosome. After DNA replication, two identical sister chromatids are produced and joined to each other at the centromere.

DNA Replication

During the process of **DNA replication** that precedes cell division, the double-stranded DNA molecule is copied, first by splitting the molecule in half up the middle of the helix. New nucleotides are added to each side of the original parent molecule, maintaining the A-to-T and G-to-C base pairings. This process results in two daughter DNA molecules, each composed of one strand of parental nucleotides and one newly synthesized strand (**Figure 6.4a**). Because each newly formed DNA molecule consists of one-half conserved parental DNA and one-half new daughter DNA, this method of DNA replication is referred to as *semiconservative replication*.

Replicating the DNA requires an enzyme that assists in DNA synthesis. This enzyme, called **DNA polymerase**, moves along the length of the unwound helix and helps bind incoming nucleotides to each other on the newly forming daughter strand (**Figure 6.4b**). When free nucleotides floating in the nucleus have an affinity for each other (A for T and G for C), they bind to each other across the width of the helix. Nucleotides that bind to each other are said to be *complementary* to each other.

(a) DNA replication **(b) The DNA polymerase enzyme facilitates replication.**

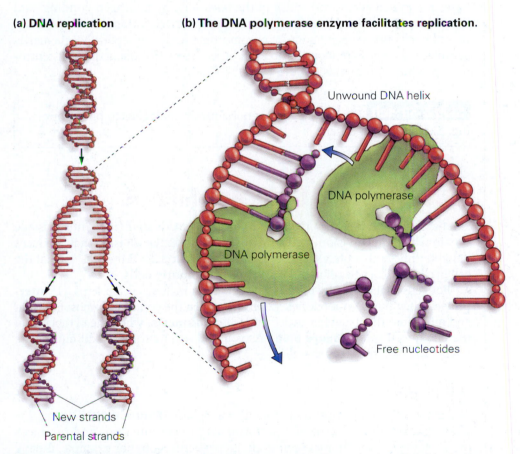

Unwound DNA helix

DNA polymerase

DNA polymerase

Free nucleotides

New strands

Parental strands

Figure 6.4 DNA replication. (a) DNA replication results in the production of two identical daughter DNA molecules from one parent molecule. Each daughter DNA molecule contains half of the parental DNA and half of the newly synthesized DNA.
Visualize This: Assume another round of replication were to occur with the incoming nucleotides still being purple in color. How many total DNA molecules would be produced, and what proportion of each DNA molecule would be purple?
(b) The DNA polymerase enzyme moves along the unwound helix, tying together adjacent nucleotides on the newly forming daughter DNA strand. Free nucleotides have three phosphate groups, two of which are cleaved to provide energy for this reaction before the nucleotide is added to the growing chain.

Figure 6.5 Unduplicated and duplicated chromosomes. An unreplicated chromosome is composed of one double-stranded DNA molecule. A replicated chromosome is X-shaped and composed of two identical double-stranded DNA molecules. Each DNA molecule of the duplicated chromosome is a copy of the original chromosome and is called a sister chromatid. In this illustration, the letters A, b, and C represent different genes along the length of the chromosome.

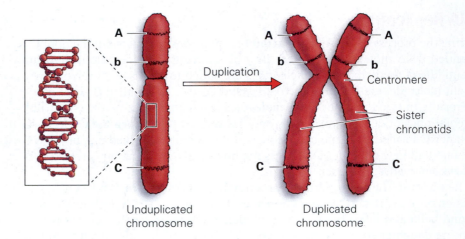

Unduplicated chromosome

Duplication

Centromere

Sister chromatids

Duplicated chromosome

The DNA polymerase enzyme catalyzes the formation of the covalent bond between nucleotides along the length of the helix. The paired nitrogenous bases are joined across the width of the backbone by hydrogen bonding, and the DNA polymerase advances along the parental DNA strand to the next unpaired nucleotide. When an entire chromosome has been replicated, the newly synthesized copies are identical to each other. They are attached at the centromere as sister chromatids (**Figure 6.5**).

Stop & Stretch DNA replication is not entirely analogous to photocopying. Explain how it differs.

6.3 The Cell Cycle and Mitosis

After a cell's DNA has been replicated, the cell is ready to divide. Mitosis is one way in which this division occurs. **Mitosis** is an asexual division that produces daughter nuclei that are exact copies of the parent nuclei. Mitosis is part of the cell cycle, or life cycle, of non-sex cells called **somatic cells.**

For cells that divide by mitosis, the cell cycle includes three steps: (1) **interphase,** when the DNA replicates; (2) *mitosis,* when the copied chromosomes split and move into the daughter nuclei; and (3) **cytokinesis,** when the cytoplasm of the parent cell splits (**Figure 6.6a**). As you will see, interphase and mitosis are further divided into steps as well.

Interphase

A normal cell spends most of its time in interphase (**Figure 6.6b**). During this phase of the cell cycle, the cell performs its typical functions and produces the proteins required for the cell to do its particular job. For example, during interphase, a muscle cell would be producing proteins required for muscle contraction, and a blood cell would be producing proteins required to transport oxygen. Different cell types spend varying amounts of time in interphase. Cells that frequently divide, like skin cells, spend less time in interphase than do those that seldom divide, such as some nerve cells. A cell that will divide also begins preparations for division during interphase. Interphase can be separated into three phases: G_1, S, and G_2.

During the G_1 (first gap or growth) phase, most of the cell's organelles duplicate. Consequently, the cell grows larger during this phase. During the

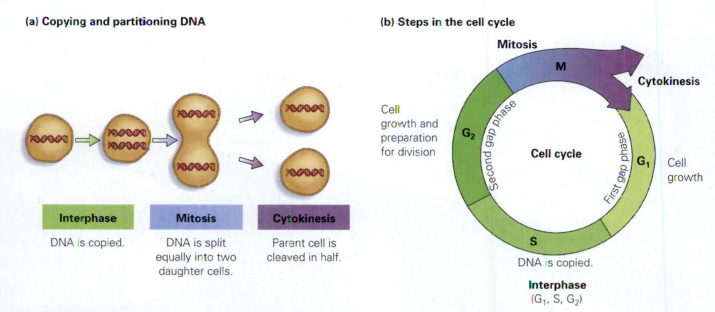

(a) Copying and partitioning DNA

Interphase	Mitosis	Cytokinesis
DNA is copied.	DNA is split equally into two daughter cells.	Parent cell is cleaved in half.

(b) Steps in the cell cycle

Mitosis

M

Cytokinesis

Cell growth and preparation for division — G_2 Second gap phase

Cell cycle

First gap phase — G_1

Cell growth

S

DNA is copied.

Interphase (G_1, S, G_2)

Figure 6.6 The cell cycle. (a) During interphase, the DNA is copied. Separation of the DNA into two daughter nuclei occurs during mitosis. Cytokinesis is the division of the cytoplasm, creating 2 daughter cells. (b) During interphase, there are two stages when the cell grows in preparation for cell division, G_1 and G_2 stages, and one stage where the DNA replicates, the S stage. The chromosomes are separated and two daughter cells are formed during the M phase.

S (synthesis) phase, the DNA in the chromosomes replicates. During the G_2 (second gap) phase of the cell cycle, proteins are synthesized that will help drive mitosis to completion. The cell continues to grow and prepare for the division of chromosomes that will take place during mitosis.

Mitosis

The movement of chromosomes into new cells occurs during mitosis. Mitosis takes place in all cells with a nucleus, although some of the specifics of cell division differ among kingdoms. Whether these phases occur in an animal or a plant, the outcome of mitosis and the next phase, cytokinesis, is the same: the production of genetically identical daughter cells. To achieve this outcome, the sister chromatids of a replicated chromosome are pulled apart, and one copy of each is placed into each newly forming nucleus. Mitosis is accomplished during 4 stages: prophase, metaphase, anaphase, and telophase. **Figure 6.7** (on the next page) summarizes the cell cycle in animal cells. The four stages of mitosis are nearly identical in plant cells.

During **prophase,** the replicated chromosomes condense, allowing them to move around in the cell without becoming entangled. Protein structures called **microtubules** also form and grow, ultimately radiating out from opposite ends, or **poles,** of the dividing cell. The growth of microtubules helps the cell to expand. Motor proteins attached to microtubules also help pull the chromosomes around during cell division. The membrane that surrounds the nucleus, called the **nuclear envelope,** breaks down so that the microtubules can gain access to the replicated chromosomes. At the poles of each dividing animal cell, structures called **centrioles** physically anchor one end of each forming microtubule. Plant cells do not contain centrioles, but microtubules in these cells do remain anchored at a pole.

During **metaphase,** the replicated chromosomes are aligned across the middle, or equator, of each cell. To do this, the microtubules, which are

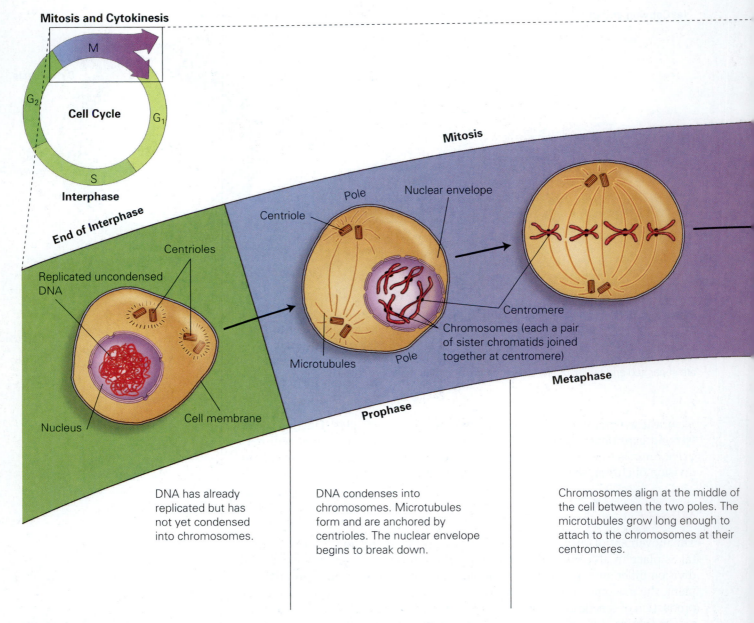

Figure 6.7 Cell division in animal cells. This diagram illustrates how cell division proceeds from interphase through mitosis and cytokinesis.

attached to each chromosome at the centromere, line up the chromosomes in single file across the middle of the cell.

During **anaphase,** the centromere splits, and the microtubules shorten to pull each sister chromatid of a chromosome to opposite poles of the cell.

In the last stage of mitosis, **telophase,** the nuclear envelopes re-form around the newly produced daughter nuclei, and the chromosomes revert to their uncondensed form.

Stop & Stretch   How does the similarity in the process of mitosis between animals and plants support the idea that all organisms share a common ancestor?

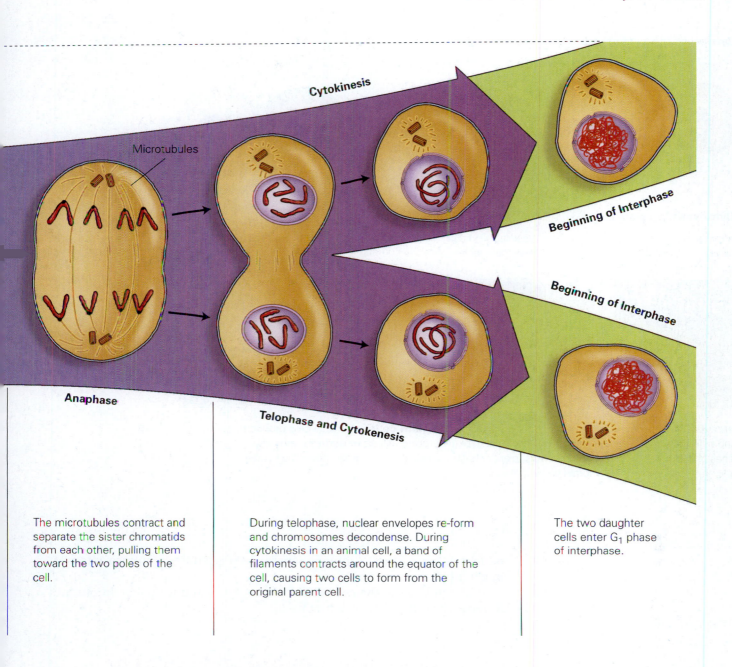

Cytokinesis

Microtubules

Anaphase

Telophase and Cytokenesis

Beginning of Interphase

Beginning of Interphase

The microtubules contract and separate the sister chromatids from each other, pulling them toward the two poles of the cell.

During telophase, nuclear envelopes re-form and chromosomes decondense. During cytokinesis in an animal cell, a band of filaments contracts around the equator of the cell, causing two cells to form from the original parent cell.

The two daughter cells enter G_1 phase of interphase.

Cytokinesis

Cytokinesis divides the cytoplasm, and daughter cells are produced. During cytokinesis in animal cells, a band of proteins encircles the cell at the equator and divides the cytoplasm. This band of proteins contracts to pinch apart the two nuclei and surrounding cytoplasm, creating two daughter cells from the original parent cell. Cytokinesis in plant cells requires that cells build a new **cell wall,** an inflexible structure surrounding the plant cells. **Figure 6.8** (on the next page) shows the difference between cytokinesis in animal and plant cells. During telophase of mitosis in a plant cell, membrane-bound vesicles from the Golgi apparatus (Chapter 2) deliver the materials required for building the cell wall to the center of the cell. The materials include a tough, fibrous carbohydrate called **cellulose** as well as some proteins. The membranes surrounding the vesicles gather in the center of the cell to form a structure called a **cell plate.** The cell plate and cell wall grow across the width of the cell

(a) Cytokinesis in an animal cell

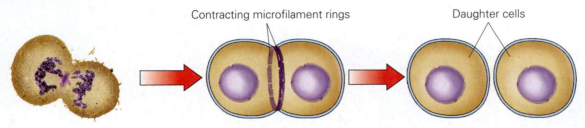

Contracting microfilament rings

Daughter cells

(b) Cytokinesis in a plant cell

Forming cell wall

Parent cell wall

Vesicles with cell wall material

Forming cell plate

New cell wall

Daughter cells

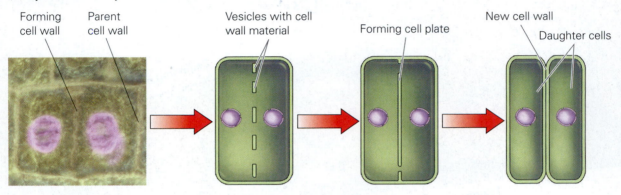

Figure 6.8 A comparison of cytokinesis in animal and plant cells. (a) Animal cells produce a band of filaments that divide the cell in half. (b) Plant cells undergoing mitosis must do so within the confines of a rigid cell wall. During cytokinesis, plant cells form a cell plate that grows down the middle of the parent cell and eventually forms a new cell wall.

and form a barrier that eventually separates the products of mitosis into two daughter cells.

After cytokinesis, the cell reenters interphase, and if the conditions are favorable, the cell cycle may repeat itself.

6.4 Cell-Cycle Control and Mutation

When cell division is working properly, it is a tightly controlled process. Cells are given signals for when and when not to divide. The normal cells in Nicole's ovary and the rest of her body were responding properly to the signals telling them when and how fast to divide. However, the cell that started her tumor was not responding properly to these signals.

An Overview: Controls in the Cell Cycle

Instead of proceeding in lockstep through the cell cycle, normal cells halt cell division at a series of **checkpoints.** At each checkpoint, proteins survey the cell to ensure that conditions for a favorable cellular division have been met. Three checkpoints must be passed before cell division can occur; one takes place during G_1, one during G_2, and the last during metaphase (**Figure 6.9**).

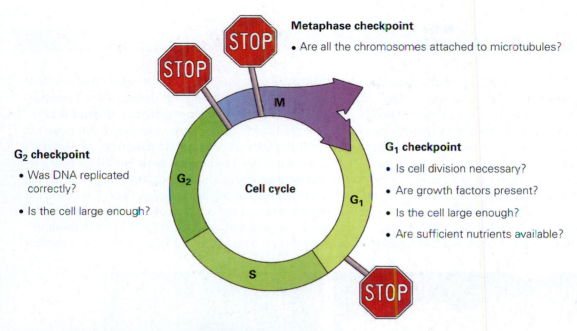

Metaphase checkpoint
- Are all the chromosomes attached to microtubules?

G$_2$ checkpoint
- Was DNA replicated correctly?
- Is the cell large enough?

Cell cycle

G$_1$ checkpoint
- Is cell division necessary?
- Are growth factors present?
- Is the cell large enough?
- Are sufficient nutrients available?

Figure 6.9 Controls of the cell cycle. Checkpoints at G$_1$, G$_2$, and metaphase determine whether a cell will continue to divide.
Visualize This: What would happen if a cell started M phase without having copied all of its chromosomes?

Proteins at the G$_1$ checkpoint determine whether it is necessary for a cell to divide. To do this, they survey the cell environment for the presence of other proteins called **growth factors** that stimulate cells to divide. Growth factors bind to cell membrane-bound proteins called **receptors** that elicit a response from the cell. If enough growth factors are present to trigger cell division, then other proteins check to see if the cell is large enough to divide and if all the nutrients required for cell division are available. When growth factors are limited in number, cell division does not occur. At the G$_2$ checkpoint, other proteins ensure that the DNA has replicated properly and double-check the cell size, again making sure that the cell is large enough to divide. The third and final checkpoint occurs during metaphase. Proteins present at metaphase verify that all the chromosomes have attached themselves to microtubules so that cell division can proceed properly.

If proteins surveying the cell at any of these three checkpoints determine that conditions are not favorable for cell division, the process is halted. When this happens, the cell may die.

Proteins that regulate the cell cycle, like all proteins, are coded by genes. When these proteins are normal, cell division is properly regulated. When these cycle-regulating proteins are unable to perform their jobs, unregulated cell division leads to large masses of cells called *tumors*. Mistakes in cell cycle regulation arise when the genes controlling the cell cycle are altered, or mutated, versions of the normal genes. A **mutation** is a change in the sequence of DNA. Changes to DNA can change a gene and in turn can alter the protein that the gene encodes, or provides instructions for. Mutant proteins do not perform their required cell functions in the same way that normal proteins do. If mutations occur to genes that encode the proteins regulating the cell cycle, cells can no longer regulate cell division properly. One or more cells in Nicole's ovary must have accumulated mutations in the cell-cycle control genes, leading to the development of cancer. *See the next section for A Closer Look at mutations to cell-cycle control genes.*

A Closer Look:
At Mutations to Cell-Cycle Control Genes

Those genes that encode the proteins regulating the cell cycle are called **proto-oncogenes** (*proto* means "before," and *onco* means "cancer"). Proto-oncogenes are normal genes located on many different chromosomes that enable organisms to regulate cell division. Cancer can develop when the normal proto-oncogenes undergo mutations and become **oncogenes.** A wide variety of organisms carry proto-oncogenes, which means that many different types of organisms can develop cancer (**Figure 6.10**).

Many proto-oncogenes provide the cell with instructions for building growth factors. A normal growth factor stimulates cell division only when the cellular environment is favorable and all conditions for division have been met. Oncogenes can overstimulate cell division (**Figure 6.11a**).

One gene involved in many cases of ovarian cancer is called *HER2*. (Names of genes are italicized, while names of the proteins they produce are not.) The *HER2* gene carries instructions for building a **receptor** protein. When the shape of the receptor on the surface of the cell is normal, it signals the inside of the cell to allow division to occur. Mutations to the gene that encodes this receptor can result in a receptor protein with a different shape from that of the nor-

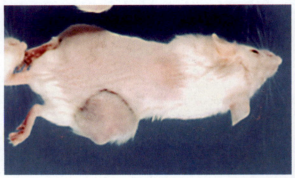

Figure 6.10 Cancer in nonhuman organisms. Many organisms carry proto-oncogenes, which can mutate into oncogenes and cause the development of tumors or cancer.

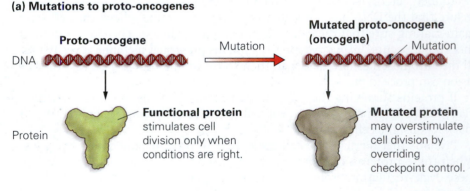

(a) Mutations to proto-oncogenes

Proto-oncogene — Mutation → Mutated proto-oncogene (oncogene) — Mutation

DNA

Protein — **Functional protein** stimulates cell division only when conditions are right.

Mutated protein may overstimulate cell division by overriding checkpoint control.

(b) Mutations to tumor-suppressor genes

Tumor suppressor — Mutation → Mutated tumor suppressor — Mutation

DNA

Protein — **Tumor-suppressor protein** stops tumor formation by suppressing cell division.

Mutated tumor-suppressor protein fails to stop tumor growth.

Figure 6.11 Mutations to proto-oncogenes and tumor-suppressor genes. (a) Mutations to proto-oncogenes and (b) tumor-suppressor genes can increase the likelihood of cancer developing.

mal receptor protein. When mutated or misshapen, the receptor protein functions as if many growth factors were present, even when there are actually few-to-no growth factors.

Another class of genes involved in cancer are **tumor suppressors.** These genes, also present in all humans and many other organisms, carry the instructions for producing proteins that suppress or stop cell division if conditions are not favorable. These proteins can also detect and repair damage to the DNA. For this reason, normal tumor suppressors serve as backups in case the proto-oncogenes undergo mutation. If a growth factor overstimulates cell division, the normal tumor suppressor impedes tumor formation by preventing the mutant cell from moving through a checkpoint (**Figure 6.11b**).

When a tumor-suppressor protein is not functioning properly, it does not force the cell to stop dividing even though conditions are not favorable. Mutated tumor suppressors also allow cells to override cell-cycle checkpoints. One well-studied tumor suppressor, named p53, helps to determine whether cells will repair damaged DNA or commit cellular suicide if the damage is too severe. Mutations to the gene that encodes p53 result in damaged DNA being allowed to proceed through mitosis, thereby passing on even more mutations. Over half of all human cancers involve mutations to the gene that encodes p53.

Mutations to a tumor-suppressor gene are common in cells that have become cancerous. Researchers believe that a normal *BRCA2* gene encodes a protein that is involved in helping to repair damaged DNA. The misshapen, mutant version of the protein cannot help to repair damaged DNA. This means that damaged DNA will be allowed to undergo mitosis, thus passing new mutations on to their daughter cells. As more and more mutations occur, the probability that a cell will become cancerous increases. **Figure 6.12** summarizes the roles of growth factors and tumor suppressors in the development of cancer.

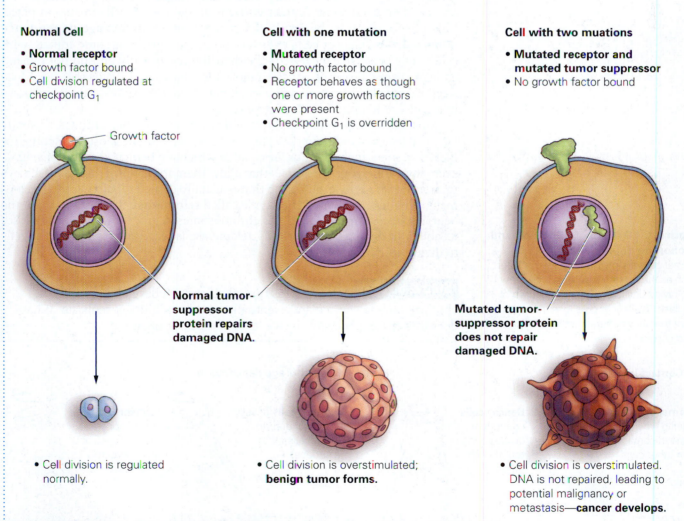

Normal Cell

- **Normal receptor**
- Growth factor bound
- Cell division regulated at checkpoint G_1

Cell with one mutation

- **Mutated receptor**
- No growth factor bound
- Receptor behaves as though one or more growth factors were present
- Checkpoint G_1 is overridden

Cell with two muations

- **Mutated receptor and mutated tumor suppressor**
- No growth factor bound

— Growth factor

Normal tumor-suppressor protein repairs damaged DNA.

Mutated tumor-suppressor protein does not repair damaged DNA.

- Cell division is regulated normally.

- Cell division is overstimulated; **benign tumor forms.**

- Cell division is overstimulated. DNA is not repaired, leading to potential malignancy or metastasis—**cancer develops.**

Figure 6.12 Growth factor receptors and tumor suppressors. When a cell has normal growth factors and tumor suppressors, cell division is properly regulated. Mutations to growth factor receptors can cause cell division to be overstimulated, and a benign tumor can form. Additional mutations to tumor-suppressor genes increase the likelihood of malignancy and metastasis.

Cancer Development Requires Many Mutations

Some mutations that occur as a result of damaged DNA being allowed to undergo mitosis are responsible for the progression of a tumor from a benign state, to a malignant state, to metastasis. For example, some cancer cells can stimulate the growth of surrounding blood vessels through a process called **angiogenesis.** These cancer cells secrete a substance that attracts and reroutes blood vessels so that they supply a developing tumor with oxygen (necessary for cellular respiration) and other nutrients. When a tumor has its own blood supply, it can grow at the expense of other, noncancerous cells. Because the growth of rapidly dividing cancer cells occurs more quickly than the growth of normal cells in this process, entire organs can eventually become filled with cancerous cells. When this occurs, an organ can no longer work properly, leading to compromised functioning or organ failure. Damage to organs also explains some of the pain associated with cancer.

Normal cells also display a property called **contact inhibition,** which prevents them from dividing when doing so would require them to pile up on each other. Cancer cells, conversely, have undergone mutations that allow them to continue to divide and form a tumor (**Figure 6.13a**). In addition, normal cells do need some contact with an underlayer of cells to stay in place. This phenomenon is the result of a process called **anchorage dependence** (**Figure 6.13b**). Cancer cells override this requirement for some contact with other cells because cancer cells are dividing too quickly and do not expend enough energy to secrete adhesion molecules that glue the cells together. Once a cell loses its anchorage dependence, it may leave the original tumor and move to the blood, lymph, or surrounding tissues.

Most cells are programmed to divide a certain number of times—usually 50 to 70 times—and then they stop dividing. This limits most developing tumors to a small mole, cyst, or lump, all of which are benign. Cancer cells, however, have undergone mutations that allow them to be immortal. They achieve immortality by activating a gene that is usually turned off after early development. This gene produces an enzyme called **telomerase** that helps prevent the degradation of chromosomes. As chromosomes degrade with age, a cell loses its ability to divide. In cancer cells, telomerase is reactivated, allowing the cells to divide without limit.

Stop & Stretch   Telomerase is turned off early in development, causing chromosomes to shorten and eventually lose their ability to replicate. What does this fact imply about the life span of tissue or organ?

Figure 6.13 Contact inhibition and anchorage dependence. (a) When normal cells are grown on a solid support such as the bottom of a flask, they grow and divide until they cover the bottom of the flask. (b) Cancer cells lose the requirement that they adhere to other cells or a solid support.

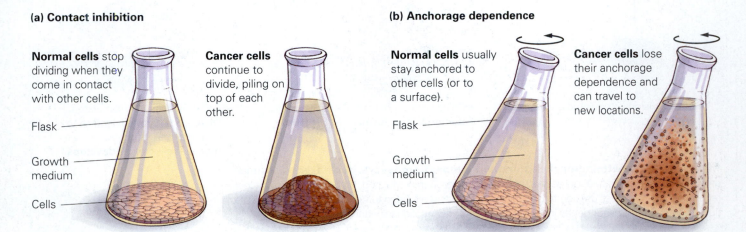

(a) Contact inhibition

Normal cells stop dividing when they come in contact with other cells.

Flask

Growth medium

Cells

Cancer cells continue to divide, piling on top of each other.

(b) Anchorage dependence

Normal cells usually stay anchored to other cells (or to a surface).

Flask

Growth medium

Cells

Cancer cells lose their anchorage dependence and can travel to new locations.

In Nicole's case, the progression from normal ovarian cells to cancerous cells may have occurred as follows: (1) One single cell in her ovary may have acquired a mutation to its *HER2* growth factor receptor gene. (2) The descendants of this cell would have been able to divide faster than neighboring cells, forming a small, benign tumor. (3) Next, a cell within the tumor may have undergone a mutation to its *BRCA2* tumor-suppressor gene, resulting in the inability of the BRCA2 protein to fix damaged DNA in the cancerous cells. (4) Cells produced by the mitosis of these doubly mutant cells would continue to divide even though their DNA is damaged, thereby enlarging the tumor and producing cells with more mutations. (5) Subsequent mutations could result in angiogenesis, lack of contact inhibition, reactivation of the telomerase enzyme, or overriding of anchorage dependence. If Nicole were very unlucky, the end result of these mutations could be that cells carrying many mutations would break away from the original ovarian tumor and set up a cancer at 1 or more new locations in her body.

Multiple-Hit Model. Because multiple mutations are required for the development and progression of cancer, scientists describe the process of cancer development using the phrase **multiple-hit model.** Nicole may have inherited some of these mutations, or they may have been induced by environmental exposures. Even though cancer is a disease caused by malfunctioning genes, most cancers are not caused only by the inheritance of mutant genes. In fact, scientists estimate that close to 70% of cancers are caused by mutations that occur during a person's lifetime.

Most of us will inherit few if any mutant cell-cycle control genes. Our level of exposure to risk factors will determine whether enough mutations will accumulate during our lifetime to cause cancer. Risk factors for ovarian cancer include smoking and uninterrupted ovulation. Ovarian cancer risk is thought to decrease for many years if ovulation is prevented for periods of time, as it is when a woman is pregnant, breastfeeding, or taking the birth control pill. When an egg cell is released from the ovary during ovulation, some tissue damage occurs. In the absence of ovulation, there is no damaged ovarian tissue and hence no need for extra cell divisions to repair damaged cells. Because mutations are most likely to occur when DNA is replicating before division, fewer divisions equal a lower likelihood of the appearance of cancer-causing mutations.

Regardless of its origin, once cancer is suspected, detection and treatment follow.

6.5 Cancer Detection and Treatment

Early detection and treatment of cancer increase the odds of survival dramatically. Being on the lookout for warning signs (**Figure 6.14**) can help alert individuals to developing cancers.

Detection Methods: Biopsy

Different cancers are detected using different methods. Some cancers are detected by the excess production of proteins that are normally produced by a particular cell type. Ovarian cancers often show high levels of CA125 protein in the blood. Nicole's level of CA125 led her gynecologist to think that a tumor might be forming on her remaining ovary. Once he suspected a tumor, Nicole's physician scheduled a biopsy. A **biopsy** is the surgical removal of cells, tissue, or fluid that will be analyzed to determine whether they are cancerous.

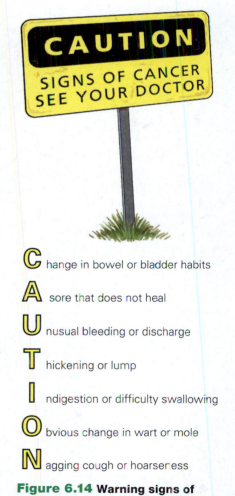

C hange in bowel or bladder habits

A sore that does not heal

U nusual bleeding or discharge

T hickening or lump

I ndigestion or difficulty swallowing

O bvious change in wart or mole

N agging cough or hoarseness

Figure 6.14 Warning signs of cancer. Self-screening for cancer can save your life. If you experience 1 or more of these warning signs, see your doctor.

When viewed under a microscope, benign tumors consist of orderly growths of cells that resemble the cells of the tissue from which thy were taken. Malignant or cancerous cells do not resemble other cells found in the same tissue; they are dividing so rapidly that they do not have time to produce all the proteins necessary to build normal cells. This leads to the often abnormal appearance of cancer cells as seen under a microscope.

A needle biopsy is usually performed if the cancer is located on or close to the surface of the patient's body. For example, breast lumps are often biopsied with a needle to determine whether the lump contains fluid and is a noncancerous cyst or whether it contains abnormal cells and is a tumor. When a cancer is diagnosed, surgery is often performed to remove as much of the cancerous growth as possible without damaging neighboring organs and tissues.

In Nicole's case, getting at the ovary to find tissue for a biopsy required the use of a surgical instrument called a **laparoscope.** For this operation, the surgeon inserted a small light and a scalpel-like instrument through a tiny incision above Nicole's navel.

Nicole's surgeon preferred to use the laparoscope because he knew Nicole would have a much easier recovery from laparoscopic surgery than she had from the surgery to remove her other, cystic ovary. Laparoscopy had not been possible when removing Nicole's other ovary—the cystic ovary had grown so large that her surgeon had to make a large abdominal incision to remove it.

A laparoscope has a small camera that projects images from the ovary onto a monitor that the surgeon views during surgery. These images showed that Nicole's tumor was a different shape, color, and texture from the rest of her ovary. They also showed that the tumor was not confined to the surface of the ovary; in fact, it appeared to have spread deeply into her ovary. Nicole's surgeon decided to shave off only the affected portion of the ovary and leave as much intact as possible, with the hope that the remaining ovarian tissue might still be able to produce egg cells. He then sent the tissue to a laboratory so that the pathologist could examine it. Unfortunately, when the pathologist looked through the microscope this time, she saw the disorderly appearance characteristic of cancer cells. Nicole's ovary was cancerous, and further treatment would be necessary.

Treatment Methods: Chemotherapy and Radiation

A treatment that works for one woman with ovarian cancer might not work for another ovarian cancer patient because a different suite of mutations may have led to the cancer in each woman's ovary. Luckily for Nicole, her ovarian cancer was diagnosed very early. Regrettably, this is not the case for most women with ovarian cancer, many of whom are diagnosed after the disease has progressed. The symptoms of ovarian cancer are often subtle and can be overlooked or ignored. They include abdominal swelling, pain, bloating, gas, constipation, indigestion, menstrual disorders, and fatigue. The difficulty of diagnosis is compounded because no routine screening tests are available. For instance, CA125 levels are checked only when ovarian cancer is suspected because (1) ovaries are not the only tissues that secrete this protein; (2) CA125 levels vary from individual to individual; and (3) these levels depend on the phase of the woman's menstrual cycle. Elevated CA125 levels usually mean that the cancer has been developing for a long time. Consequently, by the time the diagnosis is made, the cancer may have grown quite large and metastasized, making it much more difficult to treat.

Nicole's cancer was caught early. However, her physician was concerned that some of her cancerous ovarian cells may have spread through blood vessels or lymph ducts on or near the ovaries or spread into her abdominal cavity, so he started Nicole on chemotherapy after her surgery.

Chemotherapy. During **chemotherapy,** chemicals are injected into the bloodstream. These chemicals selectively kill dividing cells. A variety of chemotherapeutic agents act in different ways to interrupt cell division.

Chemotherapy involves many drugs because most chemotherapeutic agents affect only one type of cellular activity. Cancer cells are rapidly dividing and do not take the time to repair mistakes in replication that lead to mutations. These cells are allowed to proceed through the G_2 checkpoint with many mutations. Therefore, cancer cells can randomly undergo mutations, a few of which might allow them to evade the actions of a particular chemotherapeutic agent. Cells that are resistant to one drug proliferate when the chemotherapeutic agent clears away the other cells that compete for space and nutrients. Cells with a preexisting resistance to the drugs are selected for and produce more daughter cells with the same resistant characteristics, requiring the use of more than one chemotherapeutic agent.

Scientists estimate that cancer cells become resistant at a rate of approximately one cell per million. Because tumors contain about 1 billion cells, the average tumor will have close to 1000 resistant cells. Therefore, treating a cancer patient with a combination of chemotherapeutic agents aimed at different mechanisms increases the chances of destroying all the cancerous cells in a tumor.

Unfortunately, normal cells that divide rapidly are also affected by chemotherapy treatments. Hair follicles, cells that produce red and white blood cells, and cells that line the intestines and stomach are often damaged or destroyed. The effects of chemotherapy therefore include temporary hair loss, anemia (dizziness and fatigue due to decreased numbers of red blood cells), and lowered protection from infection due to decreases in the number of white blood cells. In addition, damage to the cells of the stomach and intestines can lead to nausea, vomiting, and diarrhea.

Several hours after each chemotherapy treatment, Nicole became nauseated; she often had diarrhea and vomited for a day or so after her treatments. Midway through her chemotherapy treatments, Nicole lost most of her hair.

Radiation Therapy. Cancer patients often undergo radiation treatments as well as chemotherapy. **Radiation therapy** uses high-energy particles to injure or destroy cells by damaging their DNA, making it impossible for these cells to continue to grow and divide. Radiation is applied directly to the tumor when possible. A typical course of radiation involves a series of 10 to 20 treatments performed after the surgical removal of the tumor, although sometimes radiation is used before surgery to decrease the size of the tumor. Radiation therapy is typically used only when cancers are located close to the surface of the body because it is difficult to focus a beam of radiation on internal organs such as an ovary. Therefore, Nicole's physician recommended chemotherapy only.

Stop & Stretch One risk of radiation therapy is an increased likelihood of secondary cancer emerging 5 to 15 years later. Why might this treatment increase cancer risk?

Nicole's treatments consisted of many different chemotherapeutic agents, spread over many months. The treatments took place at the local hospital on Wednesdays and Fridays. She usually had a friend drive her to the hospital

Figure 6.15 Chemotherapy. Many chemotherapeutic agents, such as Nicole's Taxol, are administered through an intravenous (IV) needle.

very early in the morning and return later in the day to pick her up. The drugs were administered through an intravenous (IV) needle into a vein in her arm (**Figure 6.15**). During the hour or so that she was undergoing chemotherapy, Nicole usually studied for her classes. She did not mind the actual chemotherapy treatments that much. The hospital personnel were kind to her, and she got some studying done. It was the aftermath of these treatments that she hated. Most days during her chemotherapy regimen, Nicole was so exhausted that she did not get out of bed until late morning, and on the day after her treatments, she often slept until late afternoon. Then she would get up and try to get some work done or make some phone calls before going back to bed early in the evening. After 6 weeks of chemotherapy, Nicole's CA125 levels started to drop. After another 2 months of chemotherapy, her CA125 levels were back down to their normal, precancerous level. If Nicole has normal CA125 levels for 5 years, she will be considered to be in **remission,** or no longer suffering negative impacts from cancer. After 10 years of normal CA125 levels, she will be considered cured of her cancer. Because Nicole's cancer responded to chemotherapy, she was spared from having to undergo any other, more experimental treatments.

Even though her treatments seemed to be going well, Nicole had other worries. She worried that her remaining ovary would not recover from the surgery and chemotherapy, which meant that she would never be able to have children. Nicole had always assumed that she would have children someday, and although she did not currently have a strong desire to have a child, she wondered if her feelings would change. Even though she was not planning to marry anytime soon, she also wondered how her future husband would feel if she were not able to become pregnant.

In addition to her concerns about being able to become pregnant, Nicole also became worried that she might pass on mutated, cancer-causing genes to her children. For Nicole, or anyone, to pass on genes to his or her children, reproductive cells must be produced by another type of nuclear division called meiosis.

6.6 Meiosis

Recall that mitosis is the division of a somatic cell's nucleus that ultimately results in two daughter nuclei that are exact copies of their parent. **Meiosis** is a form of cell division that *reduces* the number of chromosomes to produce specialized cells called **gametes** with only one-half the number of chromosomes of the parent cell. Gametes further differ from somatic cells in that gametes are produced only within the **gonads,** or sex organs.

In humans, and in most animals, the male gonads are the testes, and the female gonads are the ovaries. In animals, the male gametes are the sperm cells, while the gametes produced by the female are the egg cells. Because human somatic cells have 46 chromosomes and meiosis reduces the chromosome number by one-half, the gametes produced during meiosis contain 23 chromosomes each. (When an egg cell and a sperm cell combine their 23 chromosomes at fertilization, the developing embryo will then have the required 46 chromosomes.)

The placement of chromosomes into gametes is not random. For example, meiosis in humans does not simply place any 23 of the 46 human chromosomes into a gamete. Instead, meiosis apportions chromosomes in a very specific manner. Chromosomes in somatic cells occur in pairs. For example, the 46 chromosomes in human somatic cells are actually 23 different pairs of chromosomes. Meiosis produces gamete cells that contain one chromosome of every pair.

Autosomes (22 pairs)   **Sex chromosomes** (1 pair)

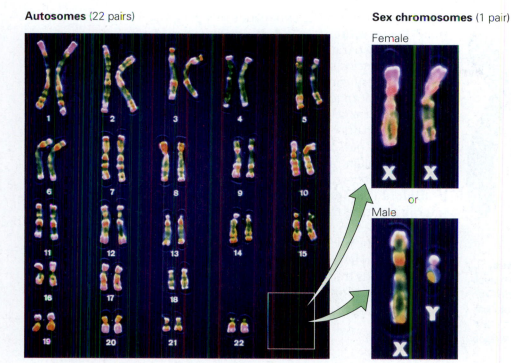

Female

X X

or

Male

Y

X

Figure 6.16 Karyotype. The pairs of chromosomes in this karyotype are arranged in order of decreasing size and numbered from 1 to 22. The X and Y sex chromosomes are the 23rd pair. The sex chromosomes from a female and a male are shown in the insets.
Visualize This: How do the X and Y chromosomes differ in terms of structure?

Homologous pair of chromosomes

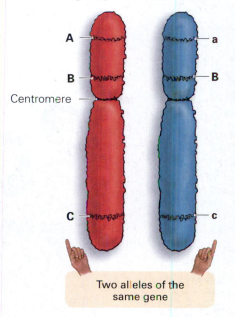

A — a

B — B

Centromere —

C — c

Two alleles of the same gene

Figure 6.17 Homologous and non-homologous pairs of chromosomes. Homologous pairs of chromosomes have the same genes (shown here as A, B, and C) but may have different alleles. The dominant allele is represented by an uppercase letter, while the recessive allele is shown with the same letter in lowercase. Note that the chromosomes of a homologous pair each have the same size, shape, and positioning of the centromere. One member of each pair is inherited from one's mother (and colored pink), while the other is inherited from one's father (and colored blue).

It is possible to visualize chromosome pairs by preparing a **karyotype,** a highly magnified photograph of the chromosomes arranged in pairs. A human karyotype is usually prepared from chromosomes that have been removed from the nuclei of white blood cells, which have been treated with chemicals to stop mitosis at metaphase (**Figure 6.16**). Because these chromosomes are at metaphase of mitosis, they are composed of replicated sister chromatids and are shaped like the letter X. The 46 human chromosomes can be arranged into 22 pairs of nonsex chromosomes, or **autosomes,** and one pair of **sex chromosomes** (the X and Y chromosomes) to make a total of 23 pairs. Human males have an X and a Y chromosome, while females have two X chromosomes.

Each chromosome is paired with a mate that is the same size and shape and has its centromere in the same position. These pairs of chromosomes are called **homologous pairs.** Each member of a homologous pair of chromosomes carries the same genes along its length, although not necessarily the same versions of those genes. Different versions of the same gene are called **alleles** of a gene (**Figure 6.17**). In your cells, one member of each pair was inherited from your mother and the other from your father.

There are normal and mutant alleles of the *BRCA2* gene. Note that there is a difference between the same type of information in the sense that both alleles of this gene code for a cell-cycle control protein, but they happen to code for different versions of the same protein. Alleles are alternate forms of a gene in the same way that chocolate and vanilla are alternate forms of ice cream.

When a chromosome is replicated, during the S phase of the cell cycle, the DNA is duplicated. Replication results in two copies, called *sister chromatids,* that are genetically identical. For this reason, we would find exactly the same

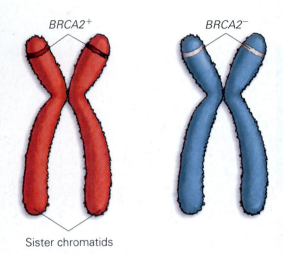

Sister chromatids

Figure 6.18 Duplicated chromosomes. This homologous pair of chromosomes has been duplicated. Note that the normal version of the *BRCA2* gene (symbolized *BRCA2⁺*) is present on both sister chromatids of the chromosome on the left. Its homologue on the right carries the mutant version of this allele (symbolized *BRCA2⁻*).

information on the sister chromatids that comprise a replicated chromosome (**Figure 6.18**).

Meiosis separates the members of a homologous pair from each other. Once meiosis is completed, there is one copy of each chromosome (1–23) in every gamete. When only one member of each homologous pair is present in a cell, we say that the cell is **haploid (*n*)**—both egg cells and sperm cells are haploid. All somatic cells in humans contain homologous pairs of chromosomes and are therefore diploid. For a diploid cell in a person's testes or ovary to become a haploid gamete, it must go through meiosis. After the sperm and egg fuse, the fertilized cell, or **zygote,** will contain two sets of chromosomes and is said to be **diploid (2*n*)** (**Figure 6.19**). Like mitosis, meiosis is preceded by an interphase stage that includes G₁, S, and G₂. Interphase is followed by two phases of meiosis, called meiosis I and meiosis II, in which divisions of the nucleus take place (**Figure 6.20**). Meiosis I separates the members of a homologous pair from each other. Meiosis II separates the chromatids from each other. Both meiotic divisions are followed by cytokinesis, during which the cytoplasm is divided between the resulting daughter cells.

Interphase

The interphase that precedes meiosis consists of G₁, S, and G₂. This interphase of meiosis is similar in most respects to the interphase that precedes mitosis. The centrioles from which the microtubules will originate are present. The G phases are times of cell growth and preparation for

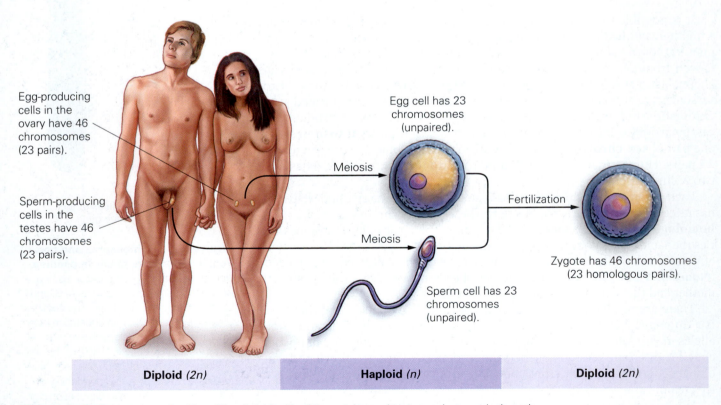

Egg-producing cells in the ovary have 46 chromosomes (23 pairs).

Sperm-producing cells in the testes have 46 chromosomes (23 pairs).

Meiosis

Meiosis

Egg cell has 23 chromosomes (unpaired).

Fertilization

Sperm cell has 23 chromosomes (unpaired).

Zygote has 46 chromosomes (23 homologous pairs).

| Diploid *(2n)* | Haploid *(n)* | Diploid *(2n)* |

Figure 6.19 Gamete production. The diploid cells of the ovaries and testes undergo meiosis and produce haploid gametes. At fertilization, the diploid condition is restored.

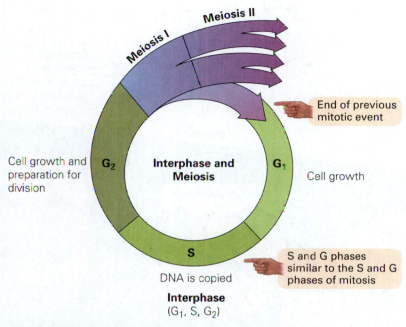

Figure 6.20 Interphase and meiosis. Interphase consists of G_1, S, and G_2 and is followed by 2 rounds of nuclear division, meiosis I and meiosis II.
Visualize This: Does DNA duplication occur during the interphase preceding meiosis II?

division. The S phase is when DNA replication occurs. Once the cell's DNA has been replicated, it can enter meiosis I.

Meiosis I

The first meiotic division, meiosis I, consists of prophase I, metaphase I, anaphase I, and telophase I (**Figure 6.21** on the next page).

During prophase I of meiosis, the nuclear envelope starts to break down, and the microtubules begin to assemble. The previously replicated chromosomes condense so that they can be moved around the cell without becoming entangled. The condensed chromosomes can be seen under a microscope. At this time, the homologous pairs of chromosomes exchange genetic information in a process called *crossing over*, which will be explained in a moment.

At metaphase I, the homologous pairs line up at the cell's equator, or middle of the cell. Microtubules bind to the metaphase chromosomes near the centromere. Homologous pairs are arranged arbitrarily regarding which member faces which pole. This process is called *random alignment*. At the end of this section, you will find detailed descriptions of crossing over and random alignment along with their impact on genetic diversity.

At anaphase I, the homologous pairs are separated from each other by the shortening of the microtubules, and at telophase I, nuclear envelopes re-form around the chromosomes. DNA is then partitioned into each of the two daughter cells by cytokinesis. Because each daughter cell contains only one copy of each member of a homologous pair, at this point the cells are haploid. Now both of these daughter cells are ready to undergo meiosis II.

Stop & Stretch How is meiosis I similar to mitosis? How is it different?

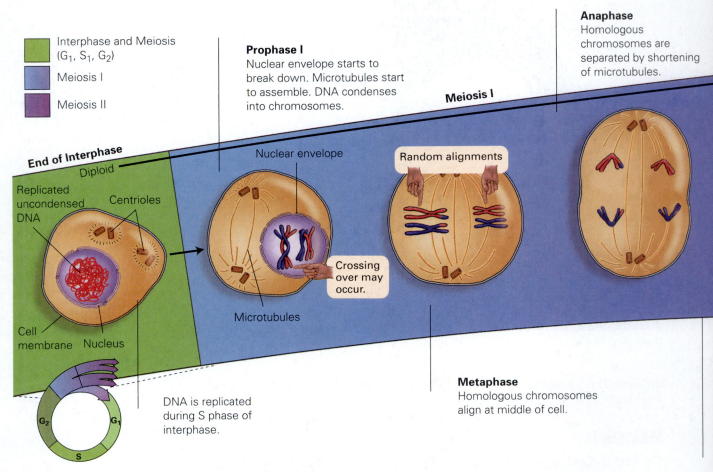

Figure 6.21 Sexual cell division. This diagram illustrates interphase, meiosis I, meiosis II, and cytokinesis in an animal cell.

Meiosis II

Meiosis II consists of prophase II, metaphase II, anaphase II, and telophase II. This second meiotic division is virtually identical to mitosis and serves to separate the sister chromatids of the replicated chromosome from each other.

At prophase II of meiosis, the cell is readying for another round of division, and the microtubules are lengthening again. At metaphase II, the chromosomes align in single file across the equator in much the same way that they do during mitosis—not as pairs, as was the case with metaphase I. At anaphase II, the sister chromatids separate from each other and move to opposite poles of the cell. At telophase II, the separated chromosomes each become enclosed in their own nucleus. In this fashion, half of a person's genes are physically placed into each gamete; thus, children carry one-half of each parent's genes.

Each parent can produce millions of different types of gametes due to two events that occur during meiosis I—crossing over and random alignment. Both of these processes greatly increase the number of different kinds of gametes that an individual can produce and therefore increase the variation in individuals that can be produced when gametes combine.

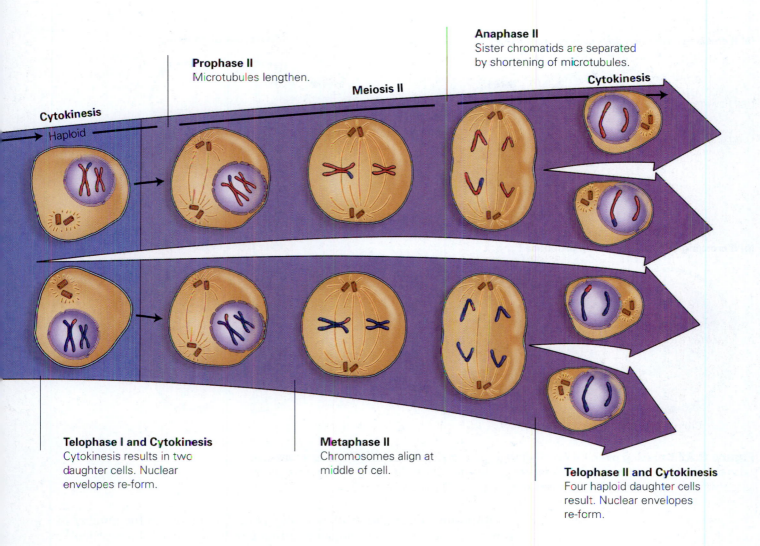

Prophase II
Microtubules lengthen.

Meiosis II

Anaphase II
Sister chromatids are separated
by shortening of microtubules.

Cytokinesis

Cytokinesis
Haploid

Telophase I and Cytokinesis
Cytokinesis results in two
daughter cells. Nuclear
envelopes re-form.

Metaphase II
Chromosomes align at
middle of cell.

Telophase II and Cytokinesis
Four haploid daughter cells
result. Nuclear envelopes
re-form.

Crossing Over and Random Alignment

Crossing over occurs during prophase I of meiosis I. It involves the exchange of portions of chromosomes from one member of a homologous pair to the other member. Crossing over is believed to occur several times on each homologous pair during each occurrence of meiosis.

To illustrate crossing over, consider an example using genes involved in the production of flower color and pollen shape in sweet pea plants. These two genes are on the same chromosome and are called **linked genes.** Linked genes move together on the same chromosome to a gamete, and they may or may not undergo crossing over.

If a pea plant has red flowers and long pollen grains, the chromosomes may appear as shown in **Figure 6.22** (on the next page). It is possible for this plant to produce four different types of gametes with respect to these two genes. Two types of gametes would result if no crossing over occurred between these genes—the gamete containing the red flower and long pollen chromosome and the gamete containing the white flower and short pollen chromosome. Two additional types of gametes could be produced if crossing over did occur—one type containing the red flower and short pollen grain chromosome and the other type containing the reciprocal white flower and long pollen grain chromosome. Therefore, crossing over increases genetic diversity by increasing the number of distinct combinations of genes that may be present in a gamete.

(a) If crossing over *does not* occur in prophase I

Two **types of gametes**

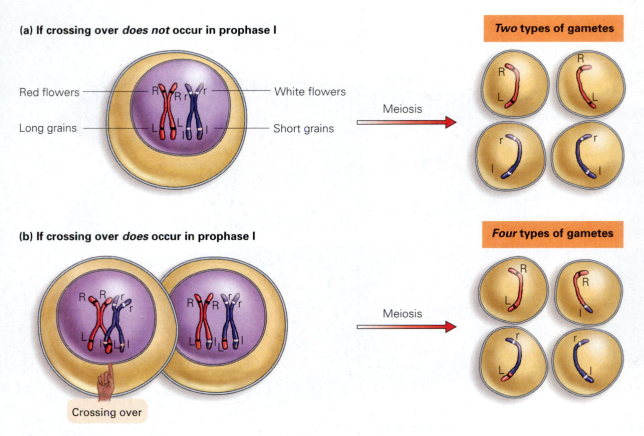

Red flowers

White flowers

Long grains

Short grains

Meiosis

(b) If crossing over *does* occur in prophase I

Four **types of gametes**

Meiosis

Crossing over

Figure 6.22 Crossing over. If a flower with the above arrangement of alleles undergoes meiosis, it can produce (a) two different types of gametes for these two genes if crossing over does not occur or (b) four different types of gametes for these two genes if crossing over occurs at L.

Random alignment of homologous pairs also increases the number of genetically distinct types of gametes that can be produced. Using Nicole's chromosomes as an example (**Figure 6.23**), let us assume that she did in fact inherit mutant versions of both the *BRCA2* and *HER2* genes and that these genes are located on different chromosomes. The arrangement of homologous pairs of chromosomes at metaphase I determines which chromosomes will end up together in a gamete. If we consider only these two homologous pairs of chromosomes, then two different alignments are possible, and four different gametes can be produced. For example, when Nicole produces egg cells, the two chromosomes that she inherited from her dad could move together to the gamete, leaving the two chromosomes she inherited from her mom to move to the other gamete. It is equally probable that Nicole could undergo meiosis in which one chromosome from each parent would align randomly together, resulting in two more types of gametes being produced. (In Chapters 7 and 8, you will see how random alignment and crossing over affect the inheritance of genes in greater detail.)

Mistakes in Meiosis

Sometimes mistakes occur during meiosis that result in the production of offspring with too many or too few chromosomes. Too many or too few chromosomes can result when there is a failure of the homologues (or sister chromatids) to separate during meiosis. This failure of chromosomes to separate is termed **nondisjunction** (**Figure 6.24**). The presence of an extra chromosome is termed **trisomy.** The absence of one chromosome of a homologous pair is termed **monosomy.** Nondisjunction can occur on autosomes or sex chromosomes.

Because the X and Y sex chromosomes do not carry the same genes and are not the same size and shape, they are not considered to be a homologous pair.

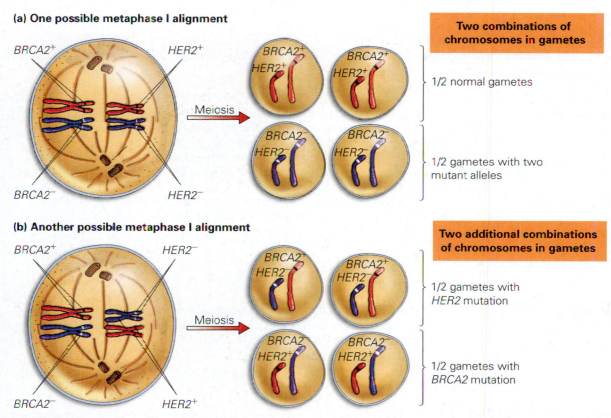

(a) One possible metaphase I alignment

$BRCA2^+$ $HER2^+$

$BRCA2^-$ $HER2^-$

Meiosis

Two combinations of chromosomes in gametes

$BRCA2^+$ $HER2^+$ $BRCA2^+$ $HER2^+$

1/2 normal gametes

$BRCA2^-$ $HER2^-$ $BRCA2^-$ $HER2^-$

1/2 gametes with two mutant alleles

(b) Another possible metaphase I alignment

$BRCA2^+$ $HER2^-$

$BRCA2^-$ $HER2^+$

Meiosis

Two additional combinations of chromosomes in gametes

$BRCA2^+$ $HER2^-$ $BRCA2^+$ $HER2^-$

1/2 gametes with *HER2* mutation

$BRCA2^-$ $HER2^+$ $BRCA2^-$ $HER2^+$

1/2 gametes with *BRCA2* mutation

Figure 6.23 Random alignment. Two possible alignments, (a) and (b), can occur when there are two homologous pairs of chromosomes. These different alignments can lead to novel combinations of genes in the gametes.
Visualize This: How many different alignments are possible with three homologous pairs of chromosomes?

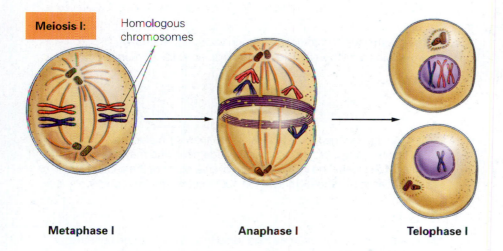

Meiosis I: Homologous chromosomes

Metaphase I   Anaphase I   Telophase I

Figure 6.24 Nondisjunction.
Nondisjunction during meiosis I produces gametes with too many and too few chromosomes.

There is, however, a region at the tip of each chromosome that is similar enough so that they can pair up during meiosis. Typically, early embryos with too many or too few chromosomes will die because they have too much or too little genetic information. However, in some situations, such as when the extra or missing chromosome is very small (such as in chromosomes 13, 18, and 21) or contains very little genetic information (such as the Y chromosome or an X chromosome that will later be inactivated), the embryo can survive. **Table 6.2** (on the next page) lists some chromosomal anomalies in humans that are compatible with life.

TABLE 6.2

Autosomal and sex-linked chromosomal anomalies.

Conditions Caused by Nondisjunction of Autosomes	Comments
Trisomy 21—Down syndrome	Affected individuals are mentally retarded and have abnormal skeletal development and heart defects. The frequency of Down syndrome increases with parental age. Down syndrome among the children of parents over 40 occurs in 6 per 1000 births, which is six times higher than the rate found among the children of couples under 35 years old.
Trisomy 13—Patau syndrome	Affected individuals are mentally retarded and deaf and have a cleft lip and palate. Approximately 1 in 5000 newborns is affected.
Trisomy 18—Edwards syndrome	Affected individuals have malformed organs, ears, mouth, and nose, leading to an elfin appearance. These babies usually die within 6 months of birth. Approximately 1 in 6000 newborns is affected.

Conditions Caused by Nondisjunction of Sex Chromosomes	Comments
XO—Turner Syndrome	Females with one X chromosome can be sterile if their ovaries fail to develop. Webbing of the neck, shorter stature, and hearing impairment are also common. Approximately 1 in 5000 female newborns is born with only one X chromosome. Turner syndrome is the only human monosomy that is viable.
Trisomy X: Meta female	Females with three X chromosomes tend to develop normally. Approximately 1 in 1000 females is born with an extra X chromosome.
XXY—Kleinfelter syndrome	Males with the XXY genotype are less fertile than XY males; have small testes, sparse body hair, some breast enlargement; and may have mental retardation. Testosterone injections can reverse some of the anatomical abnormalities in the approximately 1 in 1000 males with this condition.
XYY condition	Males with two Y chromosomes tend to be taller than average but have an otherwise normal male phenotype. Approximately 1 in 1000 newborn males has an extra Y chromosome.

From the previous discussions, you have learned that cells undergo mitosis for growth and repair and meiosis to produce gametes. **Figure 6.25** compares the significant features of mitosis and meiosis.

It is now possible to revisit the question of whether Nicole will pass on cancer-causing genes to any children she may have. Because Nicole developed cancer at such a young age, it seems likely that she may have inherited at least one mutant cell-cycle control gene; thus, she may or may not pass that gene on. If Nicole has both a normal and a mutant version of a cell-cycle control gene, then she will be able to make gametes with and without the mutant allele. Therefore, she could pass on the mutant allele if a gamete containing that allele is involved in fertilization. We have also seen that it takes many "hits" or mutations for a cancer to develop. Therefore, even if Nicole does pass

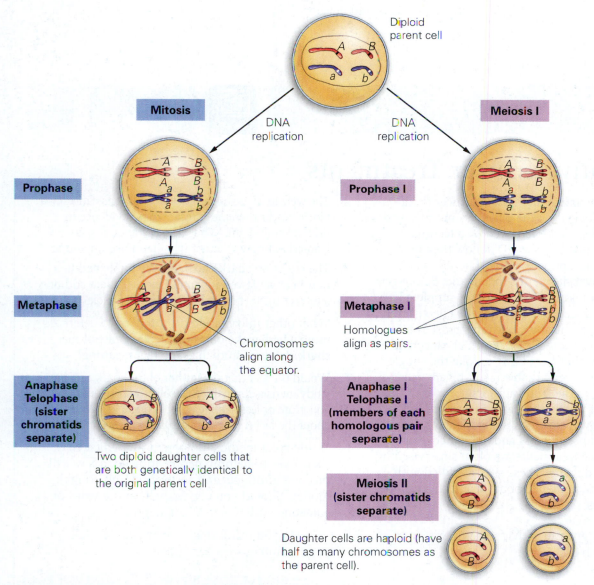

Figure 6.25 Comparing mitosis and meiosis. Mitosis is a type of cell division that occurs in somatic cells and gives rise to daughter cells that are exact genetic copies of the parent cell. Meiosis occurs in cells that will give rise to gametes and decreases the chromosome number by one-half. To do this and still ensure that each gamete receives one member of each homologous pair, the two members of each homologous pair align across the equator at the metaphase I of meiosis and are separated from each other during anaphase I.

on one or a few mutant versions of cell-cycle control genes to a child, environmental conditions will dictate whether enough other mutations will accumulate to allow a cancer to develop.

Mutations caused by environmental exposures are not passed from parents to children unless the mutation happens to occur in a cell of the gonads that will be used to produce a gamete. Nicole's cancer occurred in the ovary, the site of meiosis, but not all cells in the ovary undergo meiosis. Nicole's cancer originated in the outer covering of the ovary, a tissue that does not undergo meiosis. The cells involved in ovulation are located inside the ovary. A skin cancer that develops from exposure to ultraviolet light will not be passed on, and the same is true for most of the mutations that Nicole obtained from environmental exposures. Therefore, for any children that Nicole (or any of us) might have, it is the combined effects of inherited mutant alleles and any mutations induced by environmental exposures that will determine whether cancers will develop.

SAVVY READER

Alternative Cancer Treatments

One website claims that 80% of cancer therapies are blocked by the patient's emotions and that these blocks can be removed by a technique called tapping. Just buy the CD, follow the tapping instructions, and you will feel better in no time. Another website claims that sharks do not get cancer and that taking shark-cartilage supplements can help treat cancer. In a study of the effectiveness of shark cartilage, the study authors report that 15 of 29 patients diagnosed with terminal cancer were still alive 1 year after beginning to take the supplements, which is "a remarkable result by any measure," according to the study authors.

Websites for alternative cancer treatment centers offer unproven treatments to patients that cost tens of thousands of dollars and are not covered by insurance. These treatments are often supervised by actual medical doctors. These sites are filled with testimonials, but no data, about the effectiveness of such treatments.

1. Is there any guarantee that something written on a website is true?

2. The website that suggested that emotions can block cancer therapies presented no evidence to back up this claim. What would you do if you wanted to know whether that claim had any merit?

3. The study on shark cartilage was published in a non-peer reviewed journal. Does this fact add or subtract from the credibility of this article?

4. What other information do you need to determine whether the one-year survival rate of the shark cartilage study has any real meaning?

5. Would the fact that the author of the shark cartilage study owns a company that sells the product make you more or less skeptical of his findings? What about the fact that sharks actually do get cancer?

6. While most medical doctors have dedicated their lives to helping people in need, a few will promote products and services simply to make money. Should you always believe the word of someone with an advanced degree?

7. Why do you think that so many people fall prey to dubious cancer cures?

Chapter Review

Learning Outcomes

Mastering BIOLOGY®

Go to the Study Area at www.masteringbiology.com for practice quizzes, myeBook, BioFlix™ 3-D animations, MP3Tutor sessions, videos, current events, and more.

LO1 **Describe the cellular basis of cancer (Section 6.1).**

- Unregulated cell division can lead to the formation of a tumor (p. 113).

LO2 **Compare and contrast benign and malignant tumors (Section 6.1).**

- Benign or noncancerous tumors stay in one place and do not prevent surrounding organs from functioning. Malignant tumors are those that are invasive or those that metastasize to surrounding tissues, starting new cancers (p. 113).

LO3 **List several risk factors for cancer development (Section 6.1).**

- Risk factors for cancer include smoking, a poor diet, lack of exercise, obesity, alcohol use, and aging (pp. 114–117).

LO4 **List the normal functions of cell division (Section 6.2).**

- Cell division is a process required for growth and development (p. 118).

LO5 **Describe the structure and function of chromosomes (Section 6.2).**

- Chromosomes are composed of DNA wrapped around proteins. Chromosomes carry genes (p. 118).

LO6 **Outline the process of DNA replication (Section 6.2).**

- During DNA replication or synthesis, one strand of the double-stranded DNA molecule is used as a template for the synthesis of a new daughter strand of DNA. The newly synthesized DNA strand is complementary to the parent strand. The enzyme DNA polymerase ties together the nucleotides on the daughter strand. (pp. 119–120)

LO7 **Describe the events that occur during interphase of the cell cycle (Section 6.3).**

- Interphase consists of two gap phases of the cell cycle (G_1 and G_2), during which the cell grows and prepares to enter mitosis or meiosis, and the S (synthesis) phase, during which time the DNA replicates. The S phase of interphase occurs between G_1 and G_2 (pp. 120–121).

LO8 **Diagram two chromosomes as they proceed through mitosis of the cell cycle (Section 6.3).**

- During mitosis, the sister chromatids are separated from each other into daughter cells. During prophase, the replicated DNA condenses into linear chromosomes. At metaphase, these replicated chromosomes align across the middle of the cell. At anaphase, the sister chromatids separate from each other and align at opposite poles of the cells. At telophase, a nuclear envelope re-forms around the chromosomes lying at each pole (pp. 121–123).

LO9 **Describe the process of cytokinesis in animal and plant cells (Section 6.3).**

- Cytokinesis is the last phase of the cell cycle. During cytokinesis, the cytoplasm is divided into two portions, one for each daughter cell (pp. 123–124).

LO10 **Describe how the cell cycle is regulated and how dysregulation can lead to tumor formation (Section 6.4).**

- When cell division is working properly, it is a tightly controlled process. Normal cells divide only when conditions are favorable. Proteins survey the cell and its environment at checkpoints as the cell moves through G_1, G_2, and metaphase, and can halt cell division if conditions are not favorable. Mistakes in regulating the cell cycle arise when genes that control the cell cycle are mutated. Proto-oncogenes regulate the cell cycle. Oncogenes are mutated versions of these genes. Tumor suppressors are normal genes that can encode proteins to stop cell division if conditions are not favorable and can repair damage to the DNA. They serve as backups in case the proto-oncogenes undergo mutation (pp. 126–127).

LO11 **Explain how genes and environment both impact cancer risk (Section 6.4).**

- Mutated genes can be inherited, or mutations can be caused by exposure to carcinogens (pp. 128–129).

LO12 **Discuss the various methods of cancer detection and treatment (Section 6.5).**

- A biopsy is a common method for detecting cancer. It involves removing some cells or tissues suspected of being cancerous and analyzing them. Typical cancer treatments include chemotherapy, which involves injecting chemicals that kill rapidly dividing cells, and radiation, which involves killing tumor cells by exposing them to high-energy particles (pp. 129–132).

LO13 Explain what types of cells undergo meiosis, the end result of this process, and how meiosis increases genetic diversity (Section 6.6).

- Meiosis is a type of sexual cell division, occurring in cells, that gives rise to gametes. Gametes contain half as many chromosomes as somatic cells do. The reduction of chromosome number that occurs during meiosis begins with diploid cells and ends with haploid cells (pp. 132–134).
- Meiosis is preceded by an interphase stage in which the DNA is replicated. During meiosis I, the members of a homologous pair of chromosomes are separated from each other. During meiosis II, the sister chromatids are separated from each other (pp. 134–137).

LO14 Diagram four chromosomes from a diploid organism undergoing meiosis (Section 6.6).

- Homologues align in pairs during meiosis I and as individual chromosomes at metaphase II (pp. 135–137).

LO15 Explain the significance of crossing over and random alignment in terms of genetic diversity (Section 6.6).

- Homologous pairs of chromosomes exchange genetic information during crossing over at prophase I of meiosis, thereby increasing the number of genetically distinct gametes that an individual can produce. The alignment of members of a homologous pair at metaphase I is random with regard to which member of a pair faces which pole. This random alignment of homologous chromosomes increases the number of different kinds of gametes an individual can produce (pp. 137–138).

Roots to Remember

The following roots of words come mainly from Latin and Greek and will help you decipher terms:

cyto- and **-cyte** relate to cells. Chapter term: cytoplasm

-kinesis means motion. Chapter term: cytokinesis

meio- means to make smaller. Chapter term: meiosis

mito- means a thread. Chapter term: mitosis

onco- means cancer. Chapter term: oncogene

proto- means before. Chapter term: proto-oncogene

soma- and **-some** mean body. Chapter terms: somatic and chromosome

telo- means end or completion. Chapter term: telophase

Learning the Basics

1. **LO8 LO14** List the ways in which mitosis and meiosis differ.

2. **LO12** What property of cancer cells do chemotherapeutic agents attempt to exploit?

3. **LO8** A cell that begins mitosis with 46 chromosomes produces daughter cells with _____.

   A. 13 chromosomes; B. 23 chromosomes; C. 26 chromosomes; D. 46 chromosomes

4. **LO5** The centromere is a region at which _____.

   A. sister chromatids are attached to each other; B. metaphase chromosomes align; C. the tips of chromosomes are found; D. the nucleus is located

5. **LO4 LO8** Mitosis _____.

   A. occurs in cells that give rise to gametes; B. produces haploid cells from diploid cells; C. produces daughter cells that are exact genetic copies of the parent cell; D. consists of two separate divisions, mitosis I and mitosis II

6. **LO4 LO8** At metaphase of mitosis, _____.

   A. the chromosomes are condensed and found at the poles; B. the chromosomes are composed of one sister chromatid; C. cytokinesis begins; D. the chromosomes are composed of two sister chromatids and are lined up along the equator of the cell

7. **LO5** Sister chromatids _____.

   A. are two different chromosomes attached to each other; B. are exact copies of one chromosome that are attached to each other; C. arise from the centrioles; D. are broken down by mitosis; E. are chromosomes that carry different genes

8. **LO6** DNA polymerase _____.

   A. attaches sister chromatids at the centromere; B. synthesizes daughter DNA molecules from fats and phospholipids; C. is the enzyme that facilitates DNA synthesis; D. causes cancer cells to stop dividing

9. **LO13** After telophase I of meiosis, each daughter cell is _____.

   A. diploid, and the chromosomes are composed of one doublestranded DNA molecule; B. diploid, and the chromosomes are composed of two sister chromatids; C. haploid, and the chromosomes are composed of one double-stranded DNA molecule; D. haploid, and the chromosomes are composed of two sister chromatids

10. **LO15** List two things that happen during meiosis that increase genetic diversity.

11. **LO10** Define the terms *proto-oncogene* and *oncogene*.

12. **LO7 LO9** In what ways is the cell cycle similar in plant and animal cells, and in what ways does it differ?

13. **LO14** State whether the chromosomes depicted in parts (a)–(d) of **Figure 6.26** are haploid or diploid.

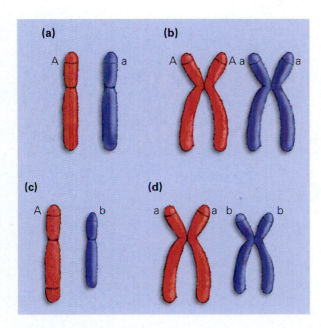

(a) A   a

(b) A   A a   a

(c) A   b

(d) a   a b   b

Figure 6.26 Haploid or diploid chromosomes?

Analyzing and Applying the Basics

1. **LO3 LO11** Would a skin cell mutation that your father obtained from using tanning beds make you more likely to get cancer? Why or why not?

2. **LO2** Will all tumors progress to cancers?

3. **LO12** Why are some cancers treated with radiation therapy while others are treated with chemotherapy?

Connecting the Science

1. Should members of society be forced to pay the medical bills of smokers when the cancer risk from smoking is so evident and publicized? Explain your reasoning.

2. Would you want to be tested for the presence of cell-cycle control mutations? How would knowing whether you had some mutated proto-oncogenes be of benefit or harm?

Answers to **Stop & Stretch, Visualize This, Savvy Reader,** and **Chapter Review** questions can be found in the **Answers** section at the back of the book

CHAPTER

7

Are You Only as Smart as Your Genes?

Mendelian and Quantitative Genetics

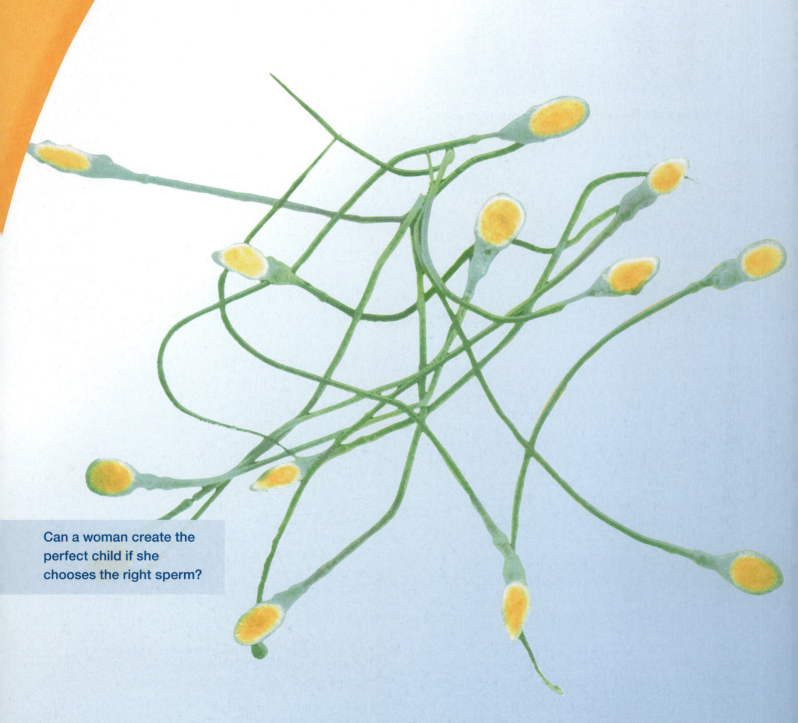

Can a woman create the
perfect child if she
chooses the right sperm?

The Fairfax Cryobank is a nondescript brick building located in a quiet, tree-lined suburb of Washington, DC. Stored inside this unremarkable edifice are the hopes and dreams of thousands of women and their partners. The Fairfax Cryobank is a sperm bank; inside its many freezers are vials containing sperm collected from hundreds of men. Women can order these sperm for a procedure called artificial insemination, which may allow them to conceive a child despite the lack of a fertile male partner.

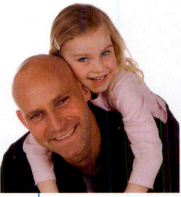

If a woman chooses a donor with the right genes, her child may look like her partner.

Women who purchase sperm from the Fairfax Cryobank can choose from hundreds of potential donors. The donors are categorized into three classes, and their sperm is priced accordingly. Most women who choose artificial insemination want detailed information about the donor before they purchase a sample; while all Fairfax Cryobank donors submit to comprehensive physical exams and disease testing and provide a detailed family health history, not all provide childhood pictures, audio CDs of their voices, or personal essays. Sperm samples from men who did not provide this additional information are sold at a discount because most women seek a donor who seems compatible in interests and aptitudes.

If she chooses a donor with the right genes, will her child be a genius?

However, in addition to the information-rich donors, there is also a set of premium sperm donors referred to by Fairfax Cryobank as its "Doctorate" category. These men either are in the process of earning or have completed a doctoral degree in medicine, law, or academia. A sperm sample from this donor category is 30% more expensive than sperm from the standard donor, and because several samples are typically needed, ensuring pregnancy from one of these donors is likely to cost hundreds of additional dollars.

Or is a child's intelligence more influenced by his environment?

Why would some women be willing to pay significantly more for sperm from a donor who has an advanced degree? Because academic achievement is associated with intelligence. These women want intelligent children, and they are willing to pay more to provide their offspring with "extra-smart" genes. But are these women putting their money in the right place? Is intelligence about genes, or is it a function of the environmental conditions in which a baby is raised? In other words, is who we are a result of our "nature" or our "nurture"? As you read this chapter, you will see that the answer to this question is not a simple one—our characteristics come from both our biological inheritance and the environment in which we developed.

7.1 The Inheritance of Traits

Most of us recognize similarities between our birth parents and ourselves. Family members also display resemblances—for instance, all the children of a single set of parents may have dimples. However, it is usually quite easy to tell siblings apart. Each child of a set of parents is unique, and none of us is simply the "average" of our parents' traits. We are each more of a combination—one child may be similar to her mother in eye color and face shape, another similar to mom in height and hair color.

To understand how your parents' traits were passed to you and your siblings, you need to understand the human life cycle. A **life cycle** is a description of the growth and reproduction of an individual (**Figure 7.1**).

A human baby is typically produced from the fusion of a single sperm cell produced by the male parent and a single egg cell produced by the female parent. Egg and sperm (called *gametes*) fuse at **fertilization,** and the resulting cell, called a zygote, duplicates all of the genetic information it contains and undergoes mitosis to produce two identical daughter cells. Each of these daughter cells

Figure 7.1 Instructions inside.
Both parents contribute genetic information to their offspring via their sperm and egg. The single cell that results from the fusion of one sperm with one egg contains all of the instructions necessary to produce an adult human.

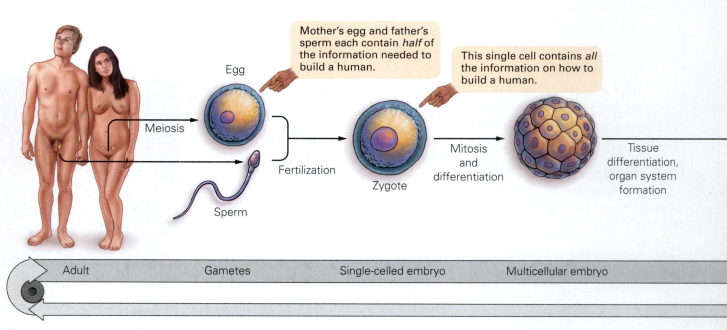

Mother's egg and father's sperm each contain *half* of the information needed to build a human.

This single cell contains *all* the information on how to build a human.

Egg

Sperm

Meiosis

Fertilization

Zygote

Mitosis and differentiation

Tissue differentiation, organ system formation

Adult   Gametes   Single-celled embryo   Multicellular embryo divides dozens of times in the same way. The cells in this resulting mass then differentiate into specialized cell types, which continue to divide and organize to produce the various structures of a developing human, called an embryo. Continued division of this single cell and its progeny leads to the production of a full-term infant and eventually an adult.

We are made up of trillions of individual cells, all of them the descendants of that first product of fertilization and nearly all containing exactly the same information originally found in the zygote. All of our traits are influenced by the information contained in that tiny cell.

Genes and Chromosomes

Each normal sperm and egg contains information about "how to build an organism." A large portion of that information is in the form of genes, segments of DNA that generally code for proteins.

Imagine genes as roughly equivalent to the words used in an instruction manual. These words are contained on chromosomes, which are roughly analogous to pages in the manual.

Prokaryotes such as bacteria typically contain a single, circular chromosome that floats freely inside the cell and is passed in its entirety to each offspring. In contrast, eukaryotes carry their genes on more than one linear chromosome. The number of chromosomes in eukaryotes can vary greatly, from 2 in the jumper ant (*Myrmecia pilosula*) to an incredible 1260 in the stalked adder's tongue, a species of fern (*Ophioglossum reticulatum*, **Figure 7.2**). Human cells contain 46 chromosomes, most of which carry thousands of genes. Thus, each cell has 46 pages of instructions, with each page containing thousands of words.

The instruction manual inside a cell is different from the instruction manual that comes with, for instance, a kit for building a model car. You would read the manual for building the car beginning at page 1 and follow an orderly set of steps to produce the final product. The human instruction manual is much more complicated—the pages and words to be read are different for different types of cell and may even change according to the situation. The "final product" of any given cell depends on the words used and the order in which the words are read from this common instruction manual.

For instance, eye cells and pancreas cells in mammals both carry instructions for the protein rhodopsin, which helps detect light, but rhodopsin is

(a) Stalked adder's tongue (a fern)

(b) Single fern cell

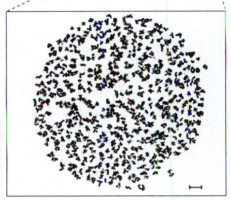

Figure 7.2 Variation in chromosome number. The amount of genetic information in an organism does not correlate to its complexity, as can be seen by examination of the 1260 chromosomes contained in a single cell of the stalked adder's tongue fern.

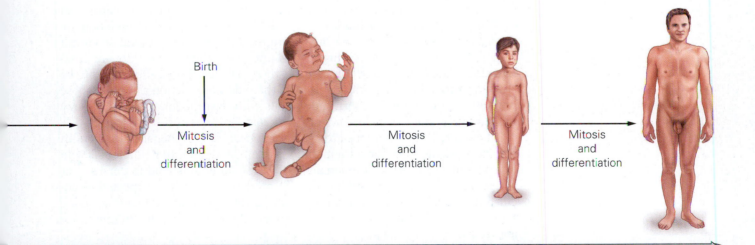

Birth

Mitosis and differentiation

Mitosis and differentiation

Mitosis and differentiation

Fetus | Baby | Child | Adult

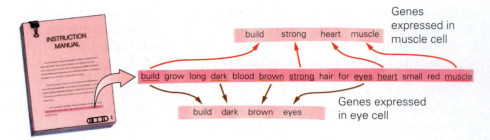

Figure 7.3 Genes as words in an instruction manual. Different words from the manual are used in different parts of the body, and identical words may be used in distinctive combinations in different cells.
Visualize This: Write several more instructions that can be extracted from this group of 14 words.

produced only in eye cells, not in pancreas cells. Rhodopsin requires assistance from another protein, called transducin, to translate the light that strikes it into the actions of the eye cell. Transducin is also produced in pancreas cells, but there it is combined with a third protein where it may help coordinate release of the hormone insulin.

Thus, a protein may serve two or more different functions depending on its context. Because genes, like words, can be used in many combinations, the instruction manual for building a living organism is very flexible (**Figure 7.3**).

Producing Diversity in Offspring

During reproduction, genes from both parents are copied and transmitted to the next generation. The copying and transmittal of genes from one generation to the next introduces variation among genes. It is this **genetic variation** that most interests individuals seeking sperm donors; the genes in the sperm they select will provide information about the traits of their offspring.

Gene Mutation Creates Genetic Diversity. Recall the process of DNA replication. Like a retyped instruction manual page, copies of chromosomes are rewritten rather than "photocopied." As a result, there is a chance of a typographical error, or mutation, every time a cell divides. Mutations in genes lead to different versions, or **alleles**, of the gene. The various types of mutation are described in **Figure 7.4**. As you can see from this figure, many mutations result in nonsensical instructions, that is, dysfunctional alleles, and thus are often harmful to the individual who possesses them. Dysfunctional alleles tend to be "lost" over time because individuals with them do not function as well as those without the mutation. In short, individuals with dysfunctional mutations may not survive or may reproduce at very low rates.

However, some mutations are neutral in effect, or even beneficial in certain situations, while some harmful mutations can be hidden if the individuals who carry them also carry a functional allele. These mutations tend to persist over generations. Because mutations occur at random and are not expected to occur in the same genes in different individuals, each of us should have a unique set of alleles reflecting the mutations passed along to us by our unique set of ancestors.

By contributing to differences among families over many generations, mutation creates genetic variation in a population in the form of new alleles. When a novel characteristic increases an individual's chance of survival and reproduction, the mutation contributes to a population's adaptation to its environment (Chapter 11). Genetic misspellings are thus the engine that drives evolution itself.

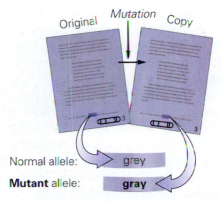

Normal allele: grey

Mutant allele: **gray**

(a) The mutant allele has the same meaning (mutant allele function the same as the original allele).

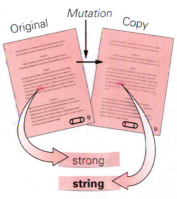

strong

string

(b) The mutant allele has a different meaning (mutant allele functions differently than the original allele).

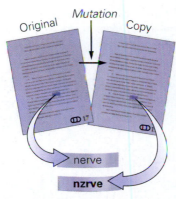

nerve

nzrve

(c) The mutant allele has no meaning (mutant allele is no longer functional).

Figure 7.4 The formation of different alleles. Different alleles for a gene may form as a result of copying errors. In this analogy, misspellings (mutations) do not change the meaning of the word (allele), but some may result in altered meanings (different allele function) or have no meaning at all (no allele function).

Stop & Stretch Not all of the DNA in a cell codes for proteins—some DNA functions as binding sites for gene promoters (proteins that "turn on" genes), other segments may provide structural support for chromosomes, and other parts may be meaningless. What do you think the effect of "misspellings" is on these nongene segments of DNA?

Segregation and Independent Assortment Create Gamete Diversity.

Both parents contribute genetic instructions to each child, but they do not contribute their entire manual. If they did, the genetic instructions carried in human cells would double every generation, making for a pretty crowded cell. Instead, the process of meiosis reduces the number of chromosomes carried in gametes by one-half (Chapter 6).

Although they are only transmitting half of their genetic information in a gamete, each parent actually gives a complete copy of the instruction manual to each child (**Figure 7.5**). This can occur because, in effect, our body cells each contain two copies of the manual—that is, each has two versions of each page, with each version containing essentially the same words. In other words, the 46 chromosomes each cell contains are actually 23 pairs of chromosomes, with each member of a pair containing essentially the same genes. Each set of two equivalent chromosomes is referred to as a **homologous pair.** The members of a homologous pair are equivalent, but not identical, because even though both have the same genes, each contains a unique set of alleles inherited from one or the other parent.

The process of meiosis separates homologous pairs of chromosomes and also places chromosomes independently into each gamete. These two events explain why siblings are not identical (with the exception of identical twins). For the most part, it is because parents do not give all of their offspring exactly the same set of alleles.

When homologous pairs are separated during meiosis, the alleles carried on the members of the pair are separated as well. The separation of pairs of alleles during the production of gametes is called **segregation.** Thus, a parent with two different alleles of a gene will produce gametes with a 50% probability of containing one version of the allele and a 50% probability of containing the other version.

The segregation of chromosomes during meiosis leads to **independent**

Figure 7.5 Equivalent information from parents. Each parent provides a complete set of instructions to each offspring.

Egg

Sperm

Zygote

The 23 pages of each instruction manual are roughly equivalent to the 23 chromosomes in each egg and sperm.

The zygote has 46 pages, equivalent to 46 chromosomes.

Parent cells have 2 copies of each chromosome — that is, 2 full sets of instruction manual pages, 1 from each parent.

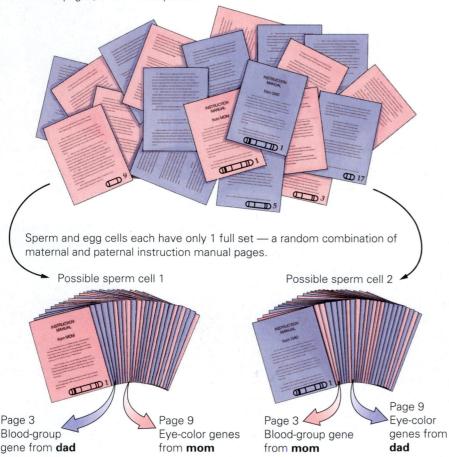

Sperm and egg cells each have only 1 full set — a random combination of maternal and paternal instruction manual pages.

Possible sperm cell 1   Possible sperm cell 2

Page 3
Blood-group
gene from **dad**

Page 9
Eye-color genes
from **mom**

Page 3
Blood-group gene
from **mom**

Page 9
Eye-color
genes from
dad

Figure 7.6 Each egg and sperm is unique. Because each sperm is produced independently, the set of pages in each sperm will be a unique combination of the pages that the man inherited from his mom and dad.
Visualize This: This man could produce four distinctly different kinds of sperm cell when considering these two genes and four alleles. List the other two possibilities not pictured here.

assortment, in which the alleles for each gene are inherited (mostly) independently of each other. Independent assortment arises from the *random alignment* of chromosomes during meiosis, which is the uncoordinated "lining up" of chromosome pairs before the first division of meiosis. As a result of random alignment, each homologous chromosome pair is segregated into daughter cells independently of all the other pairs during the production of gametes. Therefore, genes that are on different chromosomes are inherited independently of each other.

Due to independent assortment, the instruction manual contained in a single sperm cell is made up of a unique combination of pages from the manuals a man received from each of his parents. In fact, almost every sperm he makes will contain a unique subset of chromosomes—and thus a unique subset of his alleles. **Figure 7.6** illustrates this. In the figure, you can see that independent assortment causes an allele for an eye-color gene to end up in a sperm cell independently from an allele for the blood-group gene.

The independent assortment of segregated chromosomes into daughter cells is repeated every time a sperm is produced, and thus the set of alleles that a child receives from a father is different for all of his offspring. The sperm that contributed half of your genetic information might have carried an eye-color allele from your father's mom and a blood-group allele from his dad, while the sperm that produced your sister might have contained both the allele for eye color and the allele for blood group from your paternal grandmother. As a result of independent assortment, only about 50% of an individual's alleles are identical to those found in another offspring of the same parents—that is, for each gene, you have a 50% chance of being like your sister or brother.

Stop & Stretch In Figure 7.6, you can see that a man can produce four different combinations of alleles in his sperm when considering two genes on two different chromosomes. How many different allele combinations within gametes can be produced when considering three genes on three different chromosomes? What is the relationship between chromosome number (n) and the number of possible combinations that can be produced via independent assortment?

Random Fertilization Results in a Large Variety of Potential Offspring. As a result of the independent assortment of 23 pairs of chromosomes, each individual human can make at least 8 million different types of either egg or sperm. Consider now that each of your parents was able to produce such an enormous diversity of gametes. Further, any sperm produced by your father had an equal chance (in theory) of fertilizing any egg produced by your mother.

In other words, gametes combine without regard to the alleles they carry, a process known as **random fertilization.** Hence, the odds of your receiving your particular combination of chromosomes are 1 in 8 million times 1 in 8 million—or 1 in 64 trillion. Remarkably, your parents together could have made more than 64 trillion genetically different children, and you are only one of the possibilities.

Mutation creates new alleles, and independent assortment and random fertilization result in unique combinations of alleles in every generation. These processes help to produce the diversity of human beings.

A Special Case—Identical Twins.

Although the process of sexual reproduction can produce two siblings who are very different from each other, an event that may occur after fertilization can result in the birth of two children who share 100% of their genes.

Identical twins are referred to as **monozygotic twins** because they develop from one zygote—the product of a single egg and sperm. Recall that after fertilization the zygote grows and divides, producing an embryo made up of many daughter cells containing the same genetic information. Monozygotic twinning occurs when cells in an embryo separate from each other. If this happens early in development, each cell or clump of cells can develop into a complete individual, yielding twins who carry identical genetic information (**Figure 7.7a**).

In contrast to identical twins, nonidentical twins (also called fraternal twins) occur when two separate eggs fuse with different sperm. These twins are called **dizygotic,** and although they develop simultaneously, they are genetically no more similar than siblings born at different times (**Figure 7.7b**). In humans, about 1 in every 80 pregnancies produces dizygotic twins, while only approximately 1 of every 285 pregnancies results in identical twins.

Identical twins provide a unique opportunity to study the relative effects of our genes and environment in determining who we are. Because identical twins carry the same genetic information, researchers are able to study how important genes are in determining health, tastes, intelligence, and personality.

We will examine the results of some twin studies later in the chapter, as we continue to explore the question of predicting the heredity of the complex genetic traits possessed by a particular sperm donor. Our review of the relationship among parents, offspring, and genetic material has now prepared us to examine the inheritance of traits controlled by a single gene.

7.2 Mendelian Genetics: When the Role of Genes Is Clear

A few human genetic traits have easily identifiable patterns of inheritance. These traits are said to be "Mendelian" because Gregor Johann Mendel (**Figure 7.8**) was the first person to accurately describe their inheritance.

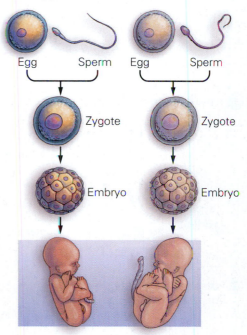

(a) Monozygotic (identical) twins

Egg · Sperm

Zygote

Embryo

Embryo splits

Two embryos

100% genetically identical

(b) Dizygotic (fraternal) twins

Egg · Sperm · Egg · Sperm

Zygote · Zygote

Embryo · Embryo

50% identical (no more similar than siblings born at different times)

Figure 7.7 The formation of twins. (a) Monozygotic twins form from one fertilization event and thus are genetically identical. (b) Dizygotic twins form from two independent fertilizations, resulting in two embryos who are as genetically similar as any other siblings.

Figure 7.8 Gregor Mendel. The father of the science of genetics.

① A pea flower normally self-pollinates.

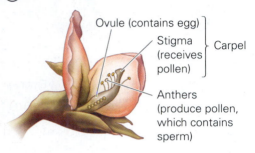

② Pollen containing structures can be removed to prevent self-fertilization.

Tweezers

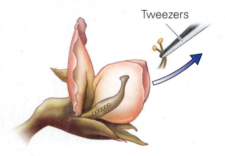

③ Pollen from another flower is dabbed on to stigma.

Paint brush

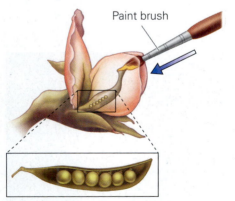

The resulting seeds will contain information on flower color, seed shape and color, and plant height from both parents.

Figure 7.9 Peas and genes. Pea plants were an ideal study organism for Mendel because their reproduction is easy to control, they complete their life cycle in a matter of weeks, and a single plant can produce thousands of offspring.

Mendel was born in Austria in 1822. Because his family was poor and could not afford private schooling, he entered a monastery to obtain an education. After completing his monastic studies, Mendel attended the University of Vienna. There he studied math and botany in addition to other sciences. Mendel attempted to become an accredited teacher but was unable to pass the examinations. After leaving the university, he returned to the monastery and began his experimental studies of inheritance in garden peas.

Mendel studied close to 30,000 pea plants over a 10-year period. His careful experiments consisted of controlled matings between plants with different traits. Mendel was able to control the types of mating that occurred by hand-pollinating the peas' flowers—that is, by taking pollen, which produces sperm, from the anthers of one pea plant and applying it to the stigma of the carpel (the egg-containing structure) of another pea plant. By growing the seeds that resulted from these controlled matings, he could evaluate the role of each parent in producing the traits of the offspring (**Figure 7.9**).

Although Mendel did not understand the chemical nature of genes, he was able to determine how traits were inherited by carefully analyzing the appearance of parent pea plants and their offspring. His patient, scientifically sound experiments demonstrated that both parents contribute equal amounts of genetic information to their offspring.

Mendel published the results of his studies in 1865, but his scientific contemporaries did not fully appreciate the significance of his work. Mendel eventually gave up his genetic studies and focused his attention on running the monastery until his death in 1884. His work was independently rediscovered by three scientists in 1900; only then did its significance to the new science of genetics become apparent.

The pattern of inheritance Mendel described occurs primarily in traits that are the result of a single gene with a few distinct alleles. **Table 7.1** lists some of the traits Mendel examined in peas; we will examine the principles he discovered, such as dominance and recessiveness, by looking at human disease genes that are of interest to prospective parents.

Genotype and Phenotype

We call the genetic composition of an individual his **genotype** and his physical traits his **phenotype.** The genotype is a description of the alleles for a particular gene carried on each member of a homologous pair of chromosomes (see Chapter 6, Figure 6.17). An individual who carries two different alleles for a gene has a **heterozygous** genotype. An individual who carries two copies of the same allele has a **homozygous** genotype.

The effect of an individual's genotype on her phenotype depends on the nature of the alleles she carries. Some alleles are **recessive,** meaning that their effects can be seen only if a copy of a dominant allele (described below) is not also present. For example, in pea plants the allele that codes for wrinkled seeds is recessive to the allele for round seeds. Wrinkled seeds will only appear when seeds carry only the wrinkled allele and no copies of the round allele.

A typical recessive allele is one that codes for a nonfunctional protein. Homozygotes having two copies of such an allele produce no functional protein. In contrast, heterozygotes carrying one copy of the functional allele have normal phenotypes because the normal protein is still produced. The functional allele in the case of the pea plant with round seeds prevents water from accumulating in the seed. When seeds containing one or two copies of this allele dry, they look much the same as when they first matured. However, wrinkled seeds result from two recessive, nonfunctional alleles (the

TABLE 7.1

Pea traits studied by Mendel.		
Character Studied	**Dominant Trait**	**Recessive Trait**
Seed shape	Round	Wrinkled
Seed color	Yellow	Green
Flower color	Purple	White
Stem length	Tall	Dwarf

absence of a functional allele). When the functional allele is missing from a seed, water flows into the seed, inflating it and causing its coat to increase in size. When this seed dries, it deflates and wrinkles, much like the surface of a balloon becomes wrinkly after it is blown up once and then deflated.

Dominant alleles are so named because their effects are seen even when a recessive allele is present. In the wrinkled seed example, the dominant allele is the one that produces a functional protein. The dominant allele does not always code for the "normal" condition of an organism, however. Sometimes mutations can create abnormal dominant alleles that essentially mask the effects of the recessive, normal allele. For example, "American Albino" horses—known for their snow-white coats, pink skin, and dark eyes—result from an allele that stops a horse's hair-color genes from being expressed during the horse's development. Because the allele prevents normal coat color development, it has its effect even if the animal carries only one copy—in other words, albinism in horses is dominant to the normal coat color (**Figure 7.10**). Dominant conditions in humans include cheek dimples and, surprisingly, the production of six fingers and toes.

Mendel worked with traits in peas that expressed only simple dominance and recessive relationships. However, for some genes, more than one dominant allele may be produced, and for others, a dominant allele may have different effects in a heterozygote than a homozygote. These situations are referred to as *codominance* and *incomplete dominance*, respectively (and are discussed in Chapter 8). For now, like Mendel, we will focus on simple dominant and recessive traits only.

Figure 7.10 A dominant allele.
American Albino horses occur when an individual carries an allele that prevents normal coat color development. Because the product of this allele actively interferes with a biochemical pathway, horses that have only one copy of the allele are albino. (Photo by Linda Gordon.)

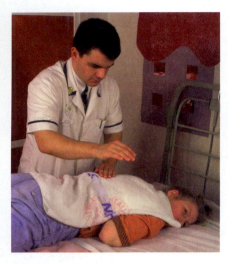

Figure 7.11 Treatment for cystic fibrosis. Percussive therapy consists of pounding the back of a patient with CF to loosen mucus inside the lungs. The mucus is then coughed up, reducing opportunities for bacterial infections to take hold.

Genetic Diseases in Humans

Most alleles in humans do not cause disease or dysfunction; they are simply alternative versions of genes. The diversity of alleles in the human population contributes to diversity among us in our appearance, physiology, and behaviors.

While women who use sperm banks are likely interested in a wide variety of traits in donors, sperm banks are primarily concerned with providing sperm that do not carry a risk of common genetic diseases. Some genetic diseases are produced by recessive alleles, while others are the result of dominant alleles.

Cystic Fibrosis Is a Recessive Condition. Individuals with cystic fibrosis (CF) cannot transport chloride ions into and out of cells lining the lungs, intestines, and other organs. As a result of this dysfunction, the balance between sodium and chloride in the cell is disrupted, and the cell produces a thick, sticky mucus layer instead of the thin, slick mucus produced by cells with the normal allele. Affected individuals suffer from progressive deterioration of their lungs and have difficulties absorbing nutrients across the lining of their intestines. Most children born with CF suffer from recurrent lung infections and have dramatically shortened life spans (**Figure 7.11**).

Cystic fibrosis results from the production of a mutant chloride ion transporter protein that cannot embed in a cell membrane. The CF allele is recessive because individuals who carry only one copy of the normal allele can still produce the functional chloride transporter protein. The disease affects only homozygous individuals with two mutant alleles and thus no functional proteins.

Cystic fibrosis is among the most common genetic diseases in European populations; nearly 1 in 2500 individuals in these populations is affected with the disease, and 1 in 25 is heterozygous for the allele. Heterozygotes for a recessive disease are called **carriers** because even though they are unaffected, these individuals can pass the trait to the next generation. Sperm banks can test donor sperm for several recessive disorders, including CF; any men who are carriers of CF are excluded from most donor programs, including Fairfax Cryobank. Other relatively well-known recessive conditions in humans include albinism—absence of pigment in skin and hair—and Tay-Sachs disease, a fatal condition that causes the relentless deterioration of mental and physical abilities of affected infants as a result of the accumulation of wastes in the brain.

Huntington's Disease Is Caused by a Dominant Allele. Early symptoms of Huntington's disease include restlessness, irritability, and difficulty in walking, thinking, and remembering. These symptoms typically begin to manifest in middle age. Huntington's disease is progressive and incurable—the nervous, mental, and muscular symptoms gradually become worse and eventually result in the death of the affected individual. Huntington's disease is an example of a fatal genetic condition caused by a dominant allele.

The Huntington's allele causes production of a protein that forms clumps inside the nuclei of cells. Nerve cells in areas of the brain that control movement are especially likely to contain these protein clumps, and these cells gradually die off over the course of the disease (**Figure 7.12**). Because this dysfunctional allele produces a mutant protein that damages cells, the presence of the normal allele cannot compensate or correct for this mutant version. An individual needs only one copy of the Huntington's allele to be affected by the disease; that is, even heterozygotes exhibit the symptoms of Huntington's.

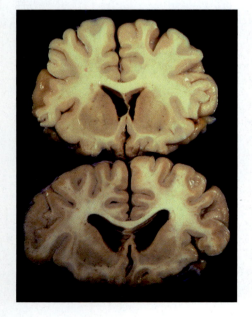

Figure 7.12 Huntington's and the brain. Cells containing mutant proteins die, shrinking and deforming the brain. This image compares the brain of a Huntington's sufferer (top) with an unaffected individual (bottom).

Stop & Stretch Typically, dominant mutations that result in death cannot be passed on from one generation to the next. What characteristic of Huntington's disease allows this allele to persist in the human population?

Only since the mid-1980s has genetic testing allowed people with a family history of Huntington's disease to learn whether they are affected before they show signs of the disease. Although most sperm banks do not test for the presence of the Huntington's allele because it is a rare condition, the detailed family medical histories required of sperm bank donors enable Fairfax Cryobank to exclude men with a family history of Huntington's disease from their donor list.

Using Punnett Squares to Predict Offspring Genotypes

Traits such as cystic fibrosis and Huntington's disease are the result of a mutation in a single gene, and the inheritance of these conditions and of other single-gene traits is relatively easy to understand. We can predict the likelihood of inheritance of small numbers of these single-gene traits by using a tool developed by Reginald Punnett, a British geneticist. A **Punnett square** is a table that lists the different kinds of sperm or eggs parents can produce relative to the gene or genes in question and then predicts the possible outcomes of a **cross**, or mating, between these parents (**Figure 7.13**).

Using a Punnett Square with a Single Gene. Imagine a woman and a sperm donor who are both carriers of the CF allele. Different alleles for a gene are symbolized with letters or number codes that refer to a trait that the gene affects. For instance, the CF gene is symbolized *CFTR* for *cystic fibrosis transmembrane regulator*. The dysfunctional *CFTR* allele is called *CFTR-ΔF508*, so both carriers would have the genotype *CFTR/CFTR-ΔF508*. However, to make this easier to follow, we will use a simpler key: the letters *F* and *f*, representing the dominant functional allele and recessive nonfunctional allele, respectively. A carrier for CF has the genotype *Ff*. A genetic cross between two carriers could then be symbolized as follows:

$$Ff \times Ff$$

We know that the female in this cross can produce eggs that carry either the *F* or *f* allele since the process of meiosis will segregate the two alleles from

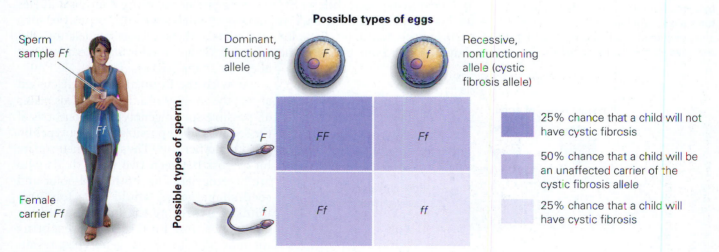

Figure 7.13 A cross between heterozygote individuals. What are the risks associated with sperm donated by a man who carries one copy of the cystic fibrosis allele? This Punnett square illustrates the likelihood that a woman who carries the cystic fibrosis allele would have a child with cystic fibrosis if the sperm donor were also a carrier.

each other. We place these two egg types across the horizontal axis of what will become a Punnett square. The male can also make two types of sperm, containing either the *F* or the *f* allele. We place these along the vertical axis. Thus, the letters on the horizontal and vertical axes represent all the possible types of eggs and sperm that the mother and father can produce by meiosis if we consider only the gene that codes for the chloride transport protein.

Inside the Punnett square are all the genotypes that can be produced from a cross between these two heterozygous individuals. The content of each box is determined by combining the alleles from the egg column and the sperm row.

Note that for a cross involving a single gene with two different alleles, there are three possible offspring types. The chance of this cross producing a child affected with CF is one in four, or 25%, because the *ff* combination of alleles occurs once out of the four possible outcomes. You can see why sperm banks would exclude carriers of CF from their donor rolls: The risk of an affected child born to a woman who might not know she is a carrier is too high. The *FF* genotype is also represented once out of four times, meaning that the probability of a homozygous unaffected child is also 25%. The probability of producing a child who is a **carrier** of cystic fibrosis is one in two, or 50%, since two of the possible outcomes inside the Punnett square are unaffected heterozygotes—one produced by an *F* sperm and an *f* egg and the other produced by an *f* sperm and an *F* egg.

When parents know which alleles they carry for a single-gene trait, they can easily determine the probability that a child they produce will have the disease phenotype (**Figure 7.14**). You should note that this probability is generated independently for each child. In other words, each offspring of two carriers has a 25% chance of being affected.

Punnett Squares for Crosses with More Genes. **Dihybrid crosses** are

genetic crosses involving two traits. Let's go back to Mendel's peas as an example. Seed color and seed shape are each determined by a single gene, and each is carried on different chromosomes. The two seed-color gene alleles Mendel studied are designated here as *Y*, which is dominant and codes for yellow color, and *y*, the recessive allele, which results in green seeds when homozygous. The two seed-shape alleles Mendel studied are designated as *R*, the dominant allele, which codes for a smooth, round shape, and *r*, which is recessive and codes for a wrinkled shape.

Because the genes for seed color and seed shape are on different chromosomes, they are placed in eggs and sperm independently of each other. In other words, a pea plant that is heterozygous for both genes (genotype *YyRr*) can make four different types of eggs: one carrying dominant alleles for both genes (*YR*), one carrying recessive alleles for both genes (*yr*), one carrying the dominant allele for seed color and the recessive allele for seed shape (*Yr*), and one carrying the recessive allele for color and the dominant allele for shape (*yR*).

As with the Punnett square discussed above, the analysis of a dihybrid cross places all possible sperm genotypes on one axis of the square and all possible egg genotypes on the other other axis. Thus, a Punnett square for a cross between two individuals who are heterozygous for both seed-color and seed-shape genes would have four columns representing the four possible egg genotypes and four rows representing the four possible sperm genotypes; resulting in 16 boxes within the square describing four different possible phenotypes (**Figure 7.15**).

Figure 7.14 A cross between a heterozygote individual and a homozygote individual. This Punnett square illustrates the outcome of a cross between a man who carries a single copy of the dominant Huntington's disease allele and an unaffected woman.
Visualize This: If one child produced by these parents carries the Huntington's allele, what is the likelihood that a second child produced by this couple will have Huntington's disease?

Mother is homozygous (*hh*), with two copies of the normal allele (*h*)

Possible types of eggs

Father is heterozygous (*Hh*), with one copy of the Huntington's allele (*H*)

Possible types of sperm

	h	*h*
H	*Hh*	*Hh*
h	*hh*	*hh*

50% chance the child will have Huntington's disease

50% chance the child be unaffected

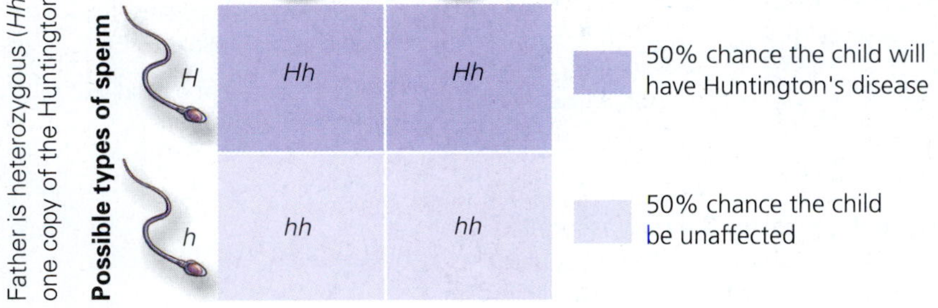

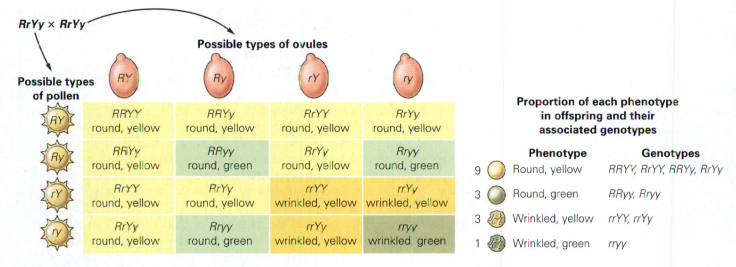

Figure 7.15 A dihybrid cross. Punnett squares can be used to predict the outcome of a cross involving two different genes. This cross involves two pea plants that are both heterozygous for the seed-color and seed-shape genes.

Visualize This: Curly hair (*CC*) is dominant over wavy (*Cc*) or straight hair (*cc*), and darkly pigmented eyes (*DD* or *Dd*) are dominant over blue eyes (*dd*). (Eye color is actually determined by several different genes, but we'll imagine it is controlled by only one in this example.) What fraction of the offspring of a mother and father who are heterozygous for both of these traits will have curly hair and dark eyes?

The phenotypes produced by a dihybrid cross result in a 9:3:3:1 **phenotypic ratio.** In this case, 9/16 include those genotypes produced by both dominant alleles (Y_R_ where the dashes indicate that the second allele for each gene could be either dominant or recessive); 3/16 include those produced by dominant alleles of one gene only (Y_rr); 3/16 include those produced by dominant alleles of the other gene (yyR_); and 1/16 have the phenotype produced by possessing recessive alleles only (yyrr).

Stop & Stretch Punnett squares can be produced for any number of genes, each of which has two alleles. Imagine a cross between two heterozygous tall, yellow, and round-seeded pea plants (TtYyRr). How many different types of gametes can each plant make? How many different cells would the resulting Punnett square table contain?

As you might imagine, as the number of genes in a Punnett square analysis increases, the number of boxes in the square increases, as does the number of possible genotypes. With two genes, each with two alleles, the number of unique gametes produced by a heterozygote is four, the number of boxes in the Punnett square is 16, and the number of unique genotypes of offspring that can be produced is nine. With three genes, each with two alleles, the Punnett square has 64 boxes and 22 different possible genotypes. With four genes, the square has 256 boxes, and with five genes, there are over 1000 boxes! Predicting the outcome of a cross becomes significantly more difficult as the number of genes we are following increases.

As scientists identify more genes and alleles, the amount of information about the genes of sperm donors—or any potential parent—will also increase. Identifying and testing for particular genes in potential parents will allow us to predict the *likelihood* of numerous genotypes in their offspring. Unfortunately, this increase in genetic testing is not necessarily equaled by an increase in our understanding of how more complex traits develop, as we shall see in the next section.

7.3 Quantitative Genetics: When Genes and Environment Interact

The single-gene traits discussed in the previous section have a distinct off-or-on character; individuals have either one phenotype (for example, the pea seed is round) or the other (the pea seed is wrinkled). Traits like this are known as *qualitative* traits. However, many of the traits that interest women who are choosing a sperm donor, such as height, weight, eye color, musical ability, susceptibility to cancer, and intelligence, do not have this off-or-on character. These traits are called **quantitative traits** and show **continuous variation;** that is, we can see a large range of phenotypes in a population—for instance, from very short people to very tall people. Wide variation in quantitative traits leads to the great diversity we see in the human population.

The distribution of phenotypes of a quantitative trait in a population can be displayed on a graph. These data often take the form of a curve called a *normal distribution*. **Figure 7.16a** illustrates the normal distribution of heights in a college class. Each individual is standing behind the label (at the bottom) that indicates their height. The curved line drawn across the photo summarizes these data—note the similarity of this curve to the outline of bell, leading to its common name, a bell curve.

A bell curve contains two important pieces of information. The first is the highest point on the curve, which generally corresponds to the average, or **mean,** value for data. The mean is calculated by adding all of the values for a trait in a population and dividing by the number of individuals in that population. The second is in the width of the bell itself, which illustrates the variability of a population. The variability is described with a mathematical measure called **variance,** which is essentially the average distance any one individual in the population is from the mean. If a low variance for a trait indicates a small amount of variability in the population, a high variance indicates a large amount of variability (**Figure 7.16b**).

(a) Normal distribution of student height in one college class

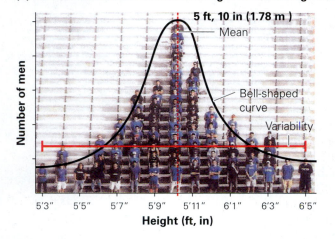

(b) Variance describes the variability around the mean.

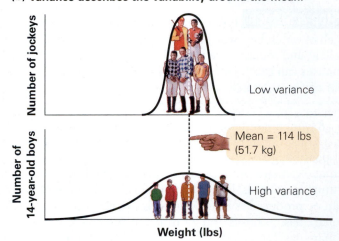

Figure 7.16 A quantitative trait. (a) This photo of men arranged by height illustrates a normal distribution. The highest point of the bell curve is also the mean height of 5 feet, 10 inches. (b) Fourteen-year-old boys and professional jockeys have the same average weight—approximately 114 pounds. However, to be a jockey, you must be within about 4 pounds of this average. Thus, the variance among jockeys in weight is much smaller than the variance among 14-year-olds.

Visualize This: Examine Figure 7.16a closely: Does an average height of 5 feet, 10 inches in this particular population imply that most men were this height? Were most men in this population close to the mean, or was there a wide range of heights?

An Overview: Why Traits Are Quantitative

A range of phenotypes may exist for a trait because numerous genotypes are found among the individuals in the population. This happens when a trait is influenced by more than one gene; traits influenced by many genes are called **polygenic traits.**

As we saw above, when a single gene with two alleles determines a trait, only three possible genotypes are present: *FF, Ff,* and *ff,* for example. But when more than one gene, each with more than one allele, influences a trait, many genotypes are possible. For example, eye color in humans is a polygenic trait influenced by at least three genes, each with more than one allele. These genes help produce and distribute the pigment melanin, a brown color, to the iris. When different alleles for the genes for eye pigment production and distribution interact, a range of eye colors, from dark brown (lots of melanin produced) to pale blue (very little melanin produced), is found in humans. The continuous variation in eye color among people is a result of several genes, each with several alleles, influencing the phenotype.

Continuous variation also may occur in a quantitative trait due to the influence of environmental factors. In this case, each genotype is capable of producing a range of phenotypes depending on outside influences. For a clear example of the effect of the environment on phenotype, see **Figure 7.17**. These identical twins share 100% of their genes but are quite different in appearance. Their difference is entirely due to variations in their environment—one twin smoked cigarettes and had high levels of sun exposure, whereas the other did not smoke and spent less time in the sun.

Most traits that show continuous variation are influenced by both genes and the environment. Skin color in humans is an example of this type of trait. The shade of an individual's skin is dependent on the amount of melanin present near the skin's surface. As with eye color, a number of genes have an effect on skin-color phenotype—those that influence melanin production and those that affect melanin distribution. However, the environment, particularly the amount of exposure to the sun during a season or lifetime, also influences the skin color of individuals (**Figure 7.18**). Melanin production increases, and any melanin that is present darkens in sun-exposed skin. In fact, after many years of intensive sun exposure, skin may become permanently darker.

Stop & Stretch The average height of American men has stayed the same over the past 50 years, while average heights in other countries (for example, Denmark) have increased. Height is a classic quantitative trait. Provide both a genetic and an environmental hypothesis for why the height of the American population is not increasing along with other countries.

Women choosing Doctorate category sperm donors from Fairfax Cryobank are presumably interested in having smart, successful children, but intelligence has both a genetic and an environmental component. With an important role for both influences, how can we predict if the child of a father with a doctorate will also be capable of earning a doctorate?

To determine the role of genes in determining quantitative traits, scientists must calculate the **heritability** of the trait. To estimate heritability for quantitative traits in most populations, researchers use correlations between individuals with varying degrees of genetic similarity. A correlation determines how accurately one can predict the measure of a trait in an

Figure 7.17 The effect of the environment on phenotype. These identical twins have exactly the same genotype, but they are quite different in appearance due to environmental factors. The twin on the right was a life-long smoker, while the one on the left never smoked.

(a) Genes

(b) Environment

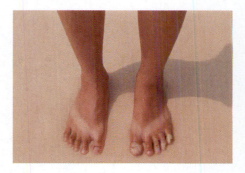

Figure 7.18 Skin color is influenced by genes and environment. (a) The difference in skin color between these two women is due primarily to variations in several alleles that control skin pigment production. (b) The difference in color between the sun-protected and sun-exposed portions of the individual in this picture is entirely due to environmental effects.

Points represent parent-offspring pairs with matching immunity levels.
• Weak • Average • Strong

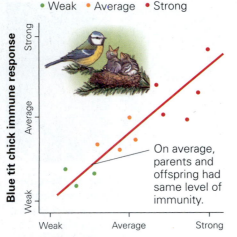

Blue tit parent immune response

On average, parents and offspring had same level of immunity.

Figure 7.19 Using correlation to calculate heritability. The close correlation of immune system response between parents and offspring in the blue tit, a European bird, indicates that immune response is highly heritable.

Visualize This: How would a graph that showed a low correlation between parents and offspring differ from this graph?

individual when its measure in a related individual is known. For example, **Figure 7.19** shows a correlation between parent birds and their offspring in the strength of their response to tetanus vaccine. An individual that responded strongly produced a large number of antitetanus proteins, called antibodies, while ones that responded weakly produced a lower number of antibodies.

As you can see from the graph, parents with weak responses tended to have offspring with weak responses, and parents with strong responses had offspring with strong responses. This strong correlation indicates that the ability to respond to tetanus is highly heritable—most of the difference between birds in their immune system response results from genetic differences.

Heritability in humans is typically measured by examining correlations between groups. These studies calculate how similar or different parents are to their children, or siblings are to each other, in the value of a particular trait. When examined across an entire population, the strength of a correlation provides a measure of heritability (**Figure 7.20**). The uses and limitations of heritability are addressed more specifically in Section 7.4. *See the next section for A Closer Look at using correlations to calculate heritability of intelligence.*

For most traits, such as body size, people come in a wide variety of types.

How much of our differences is due to the environment, and how much is a result of different genes?

Correlations between relatives, (for instance, here between fathers and sons) can provide information about the importance of genes in determining variation among individuals.

Figure 7.20 Determining heritability in humans. Comparisons between parents and children can help us estimate the genetic component of a quantitative trait.

A Closer Look:
Using Correlations to Calculate the Heritability of Intelligence

Intelligence is often measured by performance on an IQ test. The French psychologist Alfred Binet developed the intelligence quotient (or IQ) in the early 1900s to identify schoolchildren who were in need of remedial help. Binet's IQ test was not based on any theory of intelligence and was not meant to comprehensively measure mental ability, but the tests remain a commonly used way to measure innate or "natural" intelligence. There is still significant controversy over the use of IQ tests in this way.

Even if IQ tests do not really measure general intelligence, IQ scores have been correlated with academic success—meaning that individuals at higher academic levels usually have higher IQs. So, even without knowing their IQ scores, we can reasonably expect that donors in the Doctorate category have higher IQs than do other available sperm donors. However, the question of whether the high IQ of a prospective sperm donor has a genetic basis still remains.

The average correlation between IQs of parents and their children—in other words, the estimated heritability of IQ by this method—is 0.42. However, parents and the children who live with them are typically raised in a similar social and economic environment. As a result, correlations of IQ between the two groups cannot distinguish the relative importance of genes from the importance of the environment. This is the problem found in most arguments about "nature versus nurture"—do children resemble their parents because they are "born that way" or because they are "raised that way"?

To avoid the problem of overlap in environment and genes between parents and children, researchers seek situations that remove one or the other overlap. These situations are called **natural experiments** because one factor is "naturally" controlled, even without researcher intervention. Human twins are one source of a natural experiment to test hypotheses about the heritability of quantitative traits in humans.

By comparing monozygotic twins, who share all of their alleles, to dizygotic twins, who share, on average, 50% of their alleles, researchers can begin to separate the effects of shared genes from the effects of shared environments. Because twins raised in the same family have similar childhood experiences, one would expect that the only real difference between monozygotic and dizygotic twins is their genetic similarity. The average heritability of IQ calculated from a number of twin studies of this type is about 0.52. According to these studies, 52% of the variability in IQ among humans is due to differences in genotypes. Surprisingly, this value is even higher than the 42% calculated from the correlation between parents and children.

However, monozygotic twins and dizygotic twins likely *do* differ in more than just genotype. In particular, identical twins are treated more similarly than nonidentical twins. This occurs both because they look very similar and because other people may presume that they are identical in all other respects. If monozygotic twins are *expected* to be more alike than dizygotic twins, their IQ scores may be similar because they are encouraged to have the same experiences and to achieve at the same level.

There is one natural experiment that can address this problem, however. By comparing identical and nonidentical twins raised apart, the problem of differential treatment of the two *types* of twins is minimized because no one would know that the individual members of a pair have a twin (**Figure 7.21**). If variation in genes does not explain much of the variation among peoples' IQ scores, then identical twins raised apart should be no more similar than any two unrelated people of the same age and sex.

Unfortunately for scientists (but perhaps fortunately for children), the frequency of early twin separation is extremely low. Researchers have estimated the heritability of IQ at a remarkable 0.72 in a small sample of twins raised apart. This study and the other correlations appear to support the hypothesis that differences in our genes explain the majority of the variation in IQ among people. **Table 7.2** summarizes the estimates of IQ heritability but previews the cautions discussed in the next section of this chapter.

Figure 7.21 Twins separated at birth. Tamara Rabi (left) and Adriana Scott were reunited at age 20 when a mutual friend noticed their remarkable resemblance. Tamara was raised on the Upper West Side of Manhattan and Adriana in suburban Long Island, but they both have similar outgoing personalities, love to dance, and even prefer the same brand of shampoo.

(continued on the next page)

(A Closer Look continued)

TABLE 7.2

To what extent is IQ heritable? A summary of various estimates of IQ heritability, their shortcomings, and the problems with using them to understand the role of genes in determining an individual's potential intelligence.

Method of Measurement	Estimated Percentage of Genetic Influence	Warnings When Interpreting This Result	Warnings That Apply to All Measurements of Heritability
Correlation between parents' IQ and children's IQ in a population	42%	Because parents and children are similar in genes and environment, a correlation cannot be used to indicate the relative importance of genes and environment in determining IQ.	• Heritability values are specific to the populations for which they were measured. • High heritability for a trait does not mean that it is not heavily influenced by environmental conditions; we cannot predict how the trait will respond to a change in the environment. • Heritability is a measure of a population, not an individual.
Natural experiment comparing IQ in pairs of identical twins versus nonidentical twins	52%	Identical twins are treated as more alike than nonidentical twins. Therefore, their environment is different from that of nonidentical twins—the heritability value could be an overestimate.	
Natural experiment comparing IQ of identical twins raised apart versus nonidentical twins raised apart	72%	Small sample size may skew results.	

7.4 Genes, Environment, and the Individual

Perhaps we can now determine the importance of a sperm donor father who has earned a doctorate to his child's intellectual development. We know that a sperm donor will definitely influence some of his child's traits—eye and skin color and perhaps even susceptibility to certain diseases. In addition, according to the studies discussed above, the donor will probably pass on some intellectual traits to the child. In fact, with a heritability of IQ at above 50%, it appears that genes are primary in determining an individual's intelligence. Perhaps it is a good idea to pay a premium price for Doctorate category sperm after all.

However, we need to be very careful when applying the results of twin studies to questions about the individual sperm donors. To understand why, we will take a closer look at the practical significance of heritability.

The Use and Misuse of Heritability

A calculated heritability value is unique to the population in which it was measured and to the environment of that population. We should be very cautious when using heritability to measure the *general* importance of genes to the development of a trait. The following sections illustrate why.

Differences Between Groups May Be Entirely Environmental. A "thought experiment" can help illustrate this point. Body weight in laboratory mice has a strong genetic component, with a calculated heritability of about 0.90. In a population of mice in which weight is variable, bigger mice have bigger offspring, and smaller mice have smaller offspring.

Imagine that we randomly divide a population of variable mice into two groups—one group is fed a rich diet, and the other group is fed a poor diet. Otherwise, the mice are treated identically. As you might predict, regardless of their genetic predispositions, the well-fed mice become fat, while the poorly fed mice become thin. Consider the outcome if we were to keep the mice in these same conditions and allowed the two groups to reproduce. Not surprisingly, the second generation of well-fed mice is likely to be much heavier than the second generation of poorly fed mice. Now imagine that another researcher came along and examined these two populations of mice without knowing their diets. Knowing that body weight is highly heritable, the researcher might logically conclude that the groups are genetically different. However, we know this is not the case—both are grandchildren of the same original source population. It is the environment of the two populations that differs (**Figure 7.22**).

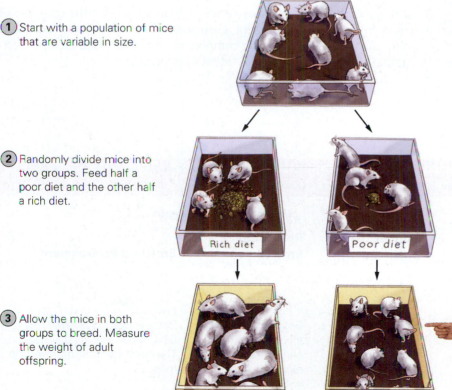

① Start with a population of mice that are variable in size.

② Randomly divide mice into two groups. Feed half a poor diet and the other half a rich diet.

Rich diet

Poor diet

③ Allow the mice in both groups to breed. Measure the weight of adult offspring.

Rich diet

Poor diet

Figure 7.22 The environment can have powerful effects on highly heritable traits. If genetically similar populations of mice are raised in radically diverse environments, then differences between the populations are entirely due to environment.

Average weight of the mice in the rich-diet environment is twice the average weight of the population in the poor-diet environment. However, there is no genetic difference between the two groups.

Now extend the same thought experiment to human groups. Imagine that we have two groups of humans, and we have determined that IQ had high heritability. In this case, people in one group were affluent, and their average IQ was higher. The other group was impoverished, and their average IQ was lower. What conclusions could you draw about the genetic differences between these two populations? The answer to the question above is none—as with the laboratory mice, these differences could be entirely due to environment. The high heritability of IQ cannot tell us if two human groups in differing social environments vary in IQ because of variations in genes or because of differences in environment.

A Highly Heritable Trait Can Still Respond to Environmental Change. A high heritability for IQ might seem to imply that IQ is not strongly influenced by environmental conditions. However, intelligence in other animals can be demonstrated to be both highly heritable and strongly influenced by the environment.

Rats can be bred for maze-running ability, and researchers have produced rats that are "maze bright" and rats that are "maze dull." Maze-running ability is highly heritable in the laboratory environment; that is, bright rats have bright offspring, and dull rats have dull offspring. The results of an experiment that measured the number of mistakes made by maze-bright and maze-dull rats raised in different environments are presented in **Table 7.3**.

In the typical lab environment, bright rats were much better at maze running than dull rats. But in both a very boring or restricted environment and a very enriched environment, the two groups of rats did about the same. In fact, no rats excelled in a restricted environment, and all rats did better at maze running in enriched environments, with the duller rats improving most dramatically.

What this example demonstrates is that we cannot predict the response of a trait to a change in the environment, even when that trait is highly heritable. Thus, even if IQ has a strong genetic component, environmental factors affecting IQ can have big effects on an individual's intelligence.

TABLE 7.3

A highly heritable trait is not identical in all environments. This table describes the average number of mistakes made by rats of two different genotypes in three different environments.

		Number of Mistakes in ...		
	Phenotype	**Normal Environment**	**Restricted Environment**	**Enriched Environment**
	Maze-bright rats	115	170	112
	Maze-dull rats	165	170	122
Explanation of Results		Maze-dull rats made more mistakes than maze-bright rats when running a maze.	Both groups made the same number of mistakes when running a maze.	Both groups made fewer mistakes when running the maze. The maze-dull rats improved the most.

Stop & Stretch Some commentators have argued that given IQ's high heritability, policies that increase financial resources to failing schools will ultimately fail to increase achievement because such a predominantly genetic trait will not respond well to environmental change. Use your understanding of the proper application of heritability to refute this argument.

Heritability Does Not Tell Us Why Two Individuals Differ. High heritability of a trait is often presumed to mean that the difference between two individuals is mostly due to differences in their genes. However, even if genes explain 90% of the population variability in a particular environment, the reason one individual differs from another may be entirely a function of environment (as an example of this, look back at the identical twins in Figure 7.18).

Currently, there is no way to determine if a particular child is a poor student because of genes, a poor environment, or a combination of both factors. There is also no way to predict whether a child produced from the sperm of a man with a doctorate will be an accomplished scholar. All we can say is that given our current understanding of the heritability of IQ and the current social environment, the alleles in Doctorate category sperm *may* increase the probability of having a child with a high IQ.

How Do Genes Matter?

We know that genes can have a strong influence on eye color, risk of genetic diseases such as CF, and even the structure of the brain. But what really determines who we are—nature or nurture?

Even with single-gene traits, the outcome of a cross between a woman and a sperm donor is not a certainty; it is only a probability. Couple this with traits being influenced by more than one gene, and independent assortment greatly increases the offspring types possible from a single mating (recall how the number of cells in a Punnett square increased exponentially when more genes were added). Knowing the phenotype of potential parents gives you relatively little information about the phenotype of their children. So, even if genes have a strong effect on traits, we cannot "program" the traits of children by selecting the traits of their parents (**Figure 7.23**).

In truth, we are really asking ourselves the wrong question when we wonder if nature or nurture has a more powerful influence on who we are. Both nature and nurture play an important role. Our cells carry instructions for all the essential characteristics of humanity, but the process of developing from embryo to adult takes place in a physical and social environment that influences how these genes are expressed. Scientists are still a long way from answering questions about how all of these complex, interacting circumstances result in who we are.

What is the lesson for women and couples who are searching for a sperm donor from Fairfax Cryobank? Donors in the Doctorate category may indeed have higher IQs than donors in the cryobank's other categories, but there is no real way to predict if a particular child of one of these donors will be smarter than average. According to the current data on the heritability of IQ, sperm from high-IQ donors may increase the odds of having an offspring with a high IQ, but only if parents provide them with a stimulating, healthy, and challenging environment in which to mature. This, of course, would be good for children with any alleles.

Figure 7.23 Genes are not destiny. Even when the traits of both parents are known, the children they produce may be very different in interests and aptitudes.

SAVVY READER

Who Picks Your Friends?

Researchers [have] found genetic links among groups of friends that suggest people tend to form friendships with others who share at least some genetic markers with them.

"People's friends may not only have similar traits, but actually resemble each other on a genotypic level," write researcher James H. Fowler, of the division of medical genetics at the University of California, San Diego and colleagues, in the *Proceedings of the National Academy of Sciences*.

In two separate groups, one comprised of adolescents and the other adults, researchers found correlations in two of six genetic markers tested in those who forged friendships.

The study showed those who carried the DRD2 genetic marker, which has been previously associated with alcoholism, tended to form friendships with others who shared the same marker. Similarly, those who lacked this gene formed friendships with others who also did not have the gene.

Researchers say this finding supports the notion that people are attracted to others with similar traits—the "birds of feather flock together" theory.

In contrast, a second genetic correlation provided evidence to support the belief that "opposites attract." The results showed that people who had a gene associated with an open personality, CYAP26, tended to have friends who did not share this gene.

Researchers say together these results could have important implications in explaining the role genes play in shaping human behavior.

"For example, a person with a genotype that makes her susceptible to alcoholism may be directly influenced to drink," write the researchers. "However, she may also be indirectly influenced to drink because she chooses friends with the same genotype."

1. Decode this article. What do the researchers' results seem to indicate about the role of heritable influence versus environmental influence in determining our choices of friends?

2. How might these results inform studies of alcoholism prevention?

3. Put this study in context with what you learned about heritability in this chapter. Are your friendships determined by your genes, your environment, or both?

Source: Web MD Health News: "Genes May Link Friends, Not Just Family" by Jennifer Warner. January 18, 2011. (http://www.webmd.com/news/20110118/genes-may-link-friends-not-just-family: Accessed Jan 25, 2011)

Chapter Review

Go to the Study Area at www.masteringbiology.com for practice quizzes, myeBook, BioFlix™ 3-D animations, MP3Tutor sessions, videos, current events, and more.

Learning Outcomes

LO1 **Describe the relationship between genes, chromosomes, and alleles (Section 7.1).**

- Children resemble their parents in part because they inherit their parents' genes, segments of DNA that contain information about how to make proteins (pp. 148–149).

- Chromosomes contain genes. Different versions of a gene are called alleles (p. 150).

- Mutations in genes generate a variety of alleles. Each allele typically results in a slightly different protein product (p. 151).

LO2 **Explain why cells containing identical genetic information can look different from each other (Section 7.1).**

- Although nearly all cells in an individual contain the exact same genetic information, the genes that are "read" in those cells differ. Even when the same gene is expressed in two different cells, their products may have very different effects (p. 150).

LO3 **Define *segregation* and *independent assortment* and explain how these processes contribute to genetic diversity (Section 7.1).**

- Segregation separates alleles during the production of eggs and sperm, meaning that a parent contributes only half of its genetic information to an offspring (p. 151).
- In independent assortment, individual chromosomes are sorted into eggs and sperm independent of each other, meaning that the subset of information passed on by parents is unique for each gamete (pp. 151–152).

LO4 **Distinguish between homozygous and heterozygous genotypes and describe how recessive and dominant alleles produce particular phenotypes when expressed in these genotypes (Section 7.2).**

- An individual heterozygous for a particular gene carries two different alleles for the gene, while one who is homozygous carries two identical alleles (p. 154).
- A dominant allele is expressed even when the genotype is heterozygous, while a recessive allele is only expressed when the individual carries no copies of the dominant allele—that is, when it is homozygous recessive (pp. 154–155).

LO5 **Demonstrate how to use a Punnett square to predict the likelihood of a particular offspring genotype and phenotype from a cross of two individuals with known genotype (Section 7.2).**

- A Punnett square helps us determine the probability that two parents of known genotype will produce a child with a particular genotype (pp. 157–159).

LO6 **Define *quantitative trait* and describe the genetic and environmental factors that cause this pattern of inheritance (Section 7.3).**

- Many traits—such as height, IQ, and musical ability—show quantitative variation, which results in a range of values for the trait within a given population (p. 160).
- Quantitative variation in a trait may be generated because the trait is influenced by several genes, because the trait can be influenced by environmental factors, or due to a combination of both factors (p. 161).

LO7 **Describe how heritability is calculated and what it tells us about the genetic component of quantitative traits (Section 7.4).**

- The role of genes in determining the phenotype for a quantitative trait is estimated by calculating the heritability of the trait (pp. 161–162).
- Heritability is calculated by examining the correlation between parents and offspring or by comparing pairs of monozygotic twins to pairs of dizygotic twins (pp. 163–164).

LO8 **Explain why a high heritability still does not always mean that a given trait is determined mostly by the genes an individual carries (Section 7.4).**

- Calculated heritability values are unique to a particular population in a particular environment. The environment may cause large differences among individuals, even if a trait has high heritability (pp. 165–167).
- Our current understanding of the relationship between genes and complex traits does not allow us to predict the phenotype of a particular offspring from the phenotype of its parents (p. 167).

Roots to Remember

The following roots of words come mainly from Latin and Greek and will help you decipher terms:

di-	means two. Chapter term: dizygotic
hetero-	means the other, another, or different. Chapter term: heterozygous
homo-	means the same. Chapter term: homozygous
mono-	means one. Chapter term: monozygotic
pheno-	comes from a verb meaning "to show." Chapter term: phenotype
poly-	means many. Chapter term: polygenic
-zygous	derives from *zygote*, the "yoked" cell resulting from the union of an egg and sperm. Chapter terms: monozygous, heterozygote

Learning the Basics

1. **LO2** What is the relationship between genotype and phenotype?

2. **LO6** What factors cause quantitative variation in a trait within a population?

3. **LO1** Which of the following statements correctly describes the relationship between genes and chromosomes?

   A. Genes are chromosomes; B. Chromosomes contain many genes; C. Genes are made up of hundreds or thousands of chromosomes; D. Genes are assorted independently during meiosis, but chromosomes are not; **E.** More than one of the above is correct.

4. **LO1** An allele is a _____.

A. version of a gene; **B.** dysfunctional gene; **C.** protein; **D.** spare copy of a gene; **E.** phenotype

5. **LO3** Sperm or eggs in humans always _____.

A. each have 2 copies of every gene; **B.** each have 1 copy of every gene; **C.** each contain either all recessive alleles or all dominant alleles; **D.** are genetically identical to all other sperm or eggs produced by that person; **E.** each contain all of the genetic information from their producer

6. **LO2** Scientists have recently developed a process by which a skin cell from a human can be triggered to develop into a human heart muscle cell. This is possible because _____.

A. most cells in the human body contain the genetic instructions for making all types of human cells; **B.** a skin cell is produced when all genes in the cell are expressed; turning off some genes in the cell results in a heart cell; **C.** scientists can add new genes to old cells to make them take different forms; **D.** a skin cell expresses only recessive alleles, so it can be triggered to produce dominant heart cell alleles; **E.** it is easy to mutate the genes in skin cells to produce the alleles required for other cell types

7. **LO3** What is the physical basis for the independent assortment of alleles into offspring?

A. There are chromosome divisions during gamete production; **B.** Homologous chromosome pairs are separated during gamete production; **C.** Sperm and eggs are produced by different sexes; **D.** Each gene codes for more than one protein; **E.** The instruction manual for producing a human is incomplete.

8. **LO4** Among heritable diseases, which genotype can be present in an individual without causing a disease phenotype in that individual?

A. heterozygous for a dominant disease; **B.** homozygous for a dominant disease; **C.** heterozygous for recessive disease; **D.** homozygous for a recessive disease; **E.** all of the above

9. **LO6** A quantitative trait _____.

A. may be one that is strongly influenced by the environment; **B.** varies continuously in a population; **C.** may be influenced by many genes; **D.** has more than a few values in a population; **E.** all of the above

10. **LO7 LO8** When a trait is highly heritable, _____.

A. it is influenced by genes; **B.** it is not influenced by the environment; **C.** the variance of the trait in a population can be explained primarily by variance in genotypes; **D.** A and C are correct; **E.** A, B, and C are correct

Genetics Problems

1. **LO1** A single gene in pea plants has a strong influence on plant height. The gene has two alleles: tall (*T*), which is dominant, and short (*t*), which is recessive. What are the genotypes and phenotypes of the offspring of a cross between a *TT* and a *tt* plant?

2. **LO5** What are the genotypes and phenotypes of the offspring of *Tt* × *Tt*?

3. **LO4** The "*P*" gene controls flower color in pea plants. A plant with either the *PP* or *Pp* genotype has purple flowers, while a plant with the *pp* genotype has white flowers. What is the relationship between *P* and *p*?

4. **LO4 LO5** Albinism occurs when individuals carry 2 recessive alleles (*aa*) that interfere with the production of melanin, the pigment that colors hair, skin, and eyes. If an albino child is born to two individuals with normal pigment, what is the genotype of each parent?

5. **LO5** Pfeiffer syndrome is a dominant genetic disease that occurs when certain bones in the skull fuse too early in the development of a child, leading to distorted head and face shape. If a man heterozygous for the allele that causes Pfeiffer syndrome marries a woman who is homozygous for the nonmutant allele, what is the chance that their first child will have this syndrome?

6. **LO4 LO5** A cross between a pea plant that produces yellow peas and a pea plant that produces green peas results in 100% yellow pea offspring.

A. Which allele is dominant in this situation? **B.** What are the likely genotypes of the yellow pea and green pea plants in the initial cross?

7. **LO5** A cross between a pea plant that produces round (*R*), yellow (*Y*) peas and a pea plant that produces wrinkled (*r*), green (*y*) peas results in 50% yellow, round pea offspring and 50% green, wrinkled pea offspring. What are the genotypes of the plants in the initial cross?

8. **LO5** A woman who is a carrier for the cystic fibrosis allele marries a man who is also a carrier.

A. What percentage of the woman's eggs will carry the cystic fibrosis allele? **B.** What percentage of the man's sperm will carry the cystic fibrosis allele? **C.** The probability that this couple will have a child who carries two copies of the cystic fibrosis allele is equal to the percentage of eggs that carry the allele times the percentage of sperm that carry the allele. What is this probability? **D.** Is this the same result you would generate when doing a Punnett square of this cross?

9. **LO5 LO6** The allele *BRCA2* was identified in families with unusually high rates of breast and ovarian cancer. Up to 80% of women with one copy of the *BRCA2* allele develop one of these cancers in their lifetime.

A. Is *BRCA2* a dominant or a recessive allele? **B.** How is *BRCA2* different from the typical pattern of Mendelian inheritance?

Analyzing and Applying the Basics

1. **LO5 LO6** Two parents both have brown eyes, but they have two children with brown eyes and two with blue eyes. How is it possible that two people with the same eye color can have children with different eye color? If eye color in this family is determined by differences in genotype for a single gene with two alleles, what percentage of the children are expected to have blue eyes? If the ratio of brown to blue eyes in this family does not conform to expectations, why does this result not refute Mendelian genetics?

2. **LO7 LO8** Does a high value of heritability for a trait indicate that the average value of the trait in a population will not change if the environment changes? Explain your answer.

3. **LO8** The heritability of IQ has been estimated at about 72%. If John's IQ is 120 and Jerry's IQ is 90, does John have stronger "intelligence" genes than Jerry does? Explain your answer.

Connecting the Science

1. If scientists find a gene that is associated with a particular "undesirable" personality trait (for instance, a tendency toward aggressive outbursts), will it mean greater or lesser tolerance toward people with that trait? Will it lead to proposals that those affected by the "disorder" should undergo treatment to be "cured" and that measures should be taken to prevent the birth of other individuals who are also afflicted?

2. The higher price for Doctorate sperm at the Fairfax Cryobank seems to imply that these donors are rare and highly desirable. If you were a woman who was looking for a sperm donor, would you focus your selection process on the Doctorate donors? What might you miss by focusing only on these donors?

3. Down syndrome is caused by a mistake during meiosis and results in physical characteristics such as a short stature and distinct facial features as well as cognitive impairment (also known as mental retardation). Does the fact that Down syndrome is a genetic condition that results in low IQ mean that we should put fewer resources into education for people with Down syndrome? How does your answer to this question relate to questions about how we should treat individuals with other genetic conditions?

Answers to **Stop & Stretch, Visualize This, Savvy Reader,** and **Chapter Review** questions can be found in the **Answers** section at the back of the book.

DNA Detective

Complex Patterns of Inheritance and DNA Fingerprinting

The Romanov family ruled Russia until their overthrow, exile, and 1918 execution.

On the night of July 16, 1918, the tsar of Russia, Nicholas II, his wife, Alexandra Romanov, their five children, and four family servants were executed in a small room in the basement of the house to which they had been exiled. These murders ended three centuries of rule by the Romanov family over the Russian Empire.

In February 1917, in the wake of protests throughout Russia, Nicholas II had relinquished his power by abdicating for both himself and on behalf of his only son, Alexis, then 13 years old. The tsar hoped that these abdications would protect his son, the heir to the throne, as well as the rest of the family from harm.

The political climate in Russia at that time was explosive. During the summer of 1914, Russia and other European countries became embroiled in World War I. This war proved to be a disaster for the imperial government. Russia faced severe food shortages, and the poverty of the common people contrasted starkly with the luxurious lives of their leaders. The Russian people felt deep resentment toward the tsar's family. This sentiment sparked the first Russian Revolution in February 1917. Following Nicholas's abdication, the imperial family was kept under guard at one of their palaces outside St. Petersburg.

In November 1917, the Bolshevik Revolution brought the communist regime, led by Vladimir Lenin, to power. Ridding the country of the last vestige of Romanov rule became a priority for Lenin and his political party. Lenin believed that doing so would solidify his regime as well as garner support among people who felt that the exiled Romanovs and their opulent lifestyle represented all that was wrong with Imperial Russia.

The fall of the communist Soviet Union prompted the desire for a proper burial of the Romanov family.

Photo by Dr. Sergey Nikitin. People believed that bones found in a grave in Ekaterinburg were those of the slain Romanovs.

Forensic evidence, like that used on popular crime shows, helped confirm that the bones buried in the shallow grave belonged to the Romanovs.

Fearing any attempt by pro-Romanov forces to save the family, Lenin ordered them to the town of Ekaterinburg in Siberia.

Shortly after midnight on July 16, the family was awakened and asked to dress. Nicholas, Alexandra, and their children—Olga, Tatiana, Maria, Anastasia, and Alexis—along with the family physician, cook, maid, and valet, were escorted to a room in the basement of the house in which they had been kept. Believing they were to be moved, the family waited. A soldier entered the room and read a short statement indicating they were to be killed. Armed men stormed into the room, and after a hail of bullets, the royal family and their entourage lay dead.

After the murders, the men loaded the bodies of the Romanovs and their servants into a truck and drove to a remote, wooded area in Ekaterinburg. Historical accounts differ regarding whether the bodies were dumped down a mineshaft, later to be removed, or were immediately buried. There is also some disagreement regarding the burial of two of the people who were executed. Some reports indicated that all eleven people were buried together, and two of them either were badly decomposed by acid placed on the ground of the burial site or were burned to ash. Other reports indicated that two members of the family were buried separately. Some people even believe that two victims escaped the execution. In any case, the bodies of at least nine people were buried in a shallow grave, where they lay undisturbed until 1991.

The bodies were not all that remained buried. For decades, details of the family's murder were hidden in the Communist Party archives in Moscow. However, after the dissolution of the Soviet Union, postcommunist leaders allowed the bones to be exhumed so that they could be given a proper burial. This exhumation took on intense political meaning because the people of Russia hoped to do more than just give the family a proper burial. The event took on the symbolic significance of laying to rest the brutality of the communist regime that took power after the murders of the Romanov family.

Because all that remained of the bodies when they were exhumed was a pile of bones, it was difficult to know if these were the remains of the royal family. A great deal of circumstantial evidence provided by forensic science pointed to that conclusion.

Forensic science is a branch of science that helps answer questions of interest to the legal system. Many of the techniques typically used by forensic scientists were not possible to use in the case of the Romanovs due to the decay of the bodies and the crime scene. For instance, no fingerprint, toxicology, footprint, or ballistic (firearm) evidence remained at the scene. Evidence that did remain at the scene included dental and bone structure. Dental evidence showed that five of the bodies had gold, porcelain, and platinum dental work, which had been available only to aristocrats.

The bones seemed to indicate that they belonged to six adults and three children. Investigators electronically superimposed the photographs of the skulls on archived photographs of the family. They compared the skeletons' measurements with clothing known to have belonged to the family. These and other data were consistent with the hypothesis that the bodies could be those of the tsar, the tsarina, three of their five children, and the four servants.

Karyotype analysis (Chapter 6) requires dividing cells, which also did not exist at the crime scene. Therefore, scientists tried to determine the sex of the buried individuals based on pelvic bone structure. Females have evolved to have wider pelvic openings to accommodate the passage of a child through the birth canal. Russian scientists thought that all three of the children's skeletons and two of the adults' skeletons were probably female (and four of the adult skeletons were male). However, the pelvises had decayed, so it was impossible to be certain.

By using the forensic evidence available to them, Russian scientists had shown only that these skeletons might be the Romanovs. They had not yet shown with any degree of certainty that these bodies *did* belong to the slain royals. The new Russian leaders did not want to make a mistake when symbolically burying a former regime. Unassailable proof was necessary because so much was at stake politically.

Before more sophisticated techniques became available to scientists, solving the puzzle of the Ekaterinburg bones required that scientists be able to show relatedness between the tsar and tsarina's skeletons and their children's skeletons.

8.1 Extensions of Mendelian Genetics

Patterns of inheritance are fairly simple to predict when genes are inherited in a straightforward manner. Predictions about traits that are controlled by one or two genes with dominant and recessive alleles can be made by using Punnett squares (Chapter 7). Patterns of inheritance that are a little more complex are said to be *extensions* of Mendelian genetics.

For some genes, two identical copies of a dominant allele are required for expression of the full effect of a phenotype. In this case, the phenotype of the heterozygote is intermediate between both homozygotes—a situation called **incomplete dominance.** For example, the alleles that determine flower color in snapdragons are incompletely dominant: One homozygote produces red flowers, the other, presumably carrying two nonfunctional copies of a color gene, produces white flowers; the heterozygote, carrying one "red" and one "white" allele, produces pink flowers (**Figure 8.1** on the next page).

Alternatively, the phenotype of a heterozygote in which neither allele is dominant to the other may be a combination of both fully expressed traits. This situation, by which two different alleles of a gene are both expressed in an individual, is known as **codominance.** In cattle, for example, the allele

Flower color in snapdragons

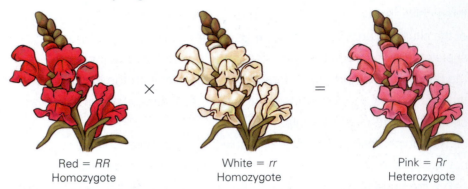

Red = *RR*
Homozygote

White = *rr*
Homozygote

Pink = *Rr*
Heterozygote

Figure 8.1 Incomplete dominance. Snapdragons show incomplete dominance in the inheritance of flower color. The heterozygous flower has a phenotype that is in between that of the two homozygotes.

Coat color in cattle

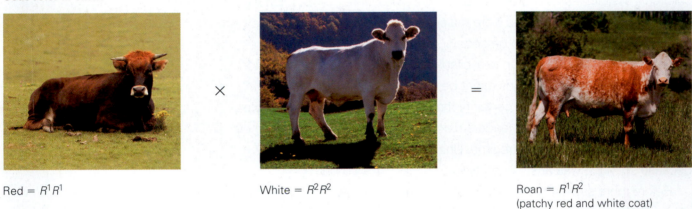

Red = R^1R^1

White = R^2R^2

Roan = R^1R^2
(patchy red and white coat)

Figure 8.2 Codominance. Roan coat color in cattle is an example of codominance. Both alleles are equally expressed in the heterozygote, so the conventional uppercase and lowercase nomenclature for alleles no longer applies.

that codes for red hair color and the allele that codes for white hair color are both expressed in a heterozygote. These individuals have patchy coats that consist of an approximately equal mixture of white hairs and red hairs (**Figure 8.2**).

Because all that was left of the Romanovs was a pile of bones, the scientists could study only a few genetic traits to show the relatedness of the adult skeletons to two of the four children's skeletons. Genetic traits that were obvious, such as bone size and structure, are controlled by many genes and affected by environmental components like nutrition and physical activity level. Traits that are affected by the interactions between many are called **polygenic traits.** This made using bone size and structure to predict which of the adults' skeletons were related to the children's skeletons a matter of guesswork. The scientists had to use more sophisticated analyses.

A technique that scientists use to help determine relatedness of people is blood typing, which involves determining if certain carbohydrates are located on the surface of red blood cells. These surface markers are part of the **ABO blood system.** The ABO blood system displays two extensions of Mendelism—codominance and **multiple allelism,** which occurs when there are more than two alleles of a gene in the population. In fact, three distinct alleles of one blood-group gene code for the enzymes that synthesize the sugars found on the surface of red blood cells. Two of the three alleles display codominance to each other, and one allele is recessive to the other two.

Figure 8.3 summarizes the possible genotypes and phenotypes for the ABO blood system. The three alleles of this blood-type gene are I^A, I^B, and i. A given individual will carry only two alleles, one on each of his or her homologous pairs of chromosomes, even though three alleles are being passed on in the entire population. In other words, one person may carry the I^A and I^B alleles, and another might carry the I^A and i alleles. There are three different alleles, but each individual can carry only two alleles.

The symbols used to represent these alternate forms of the blood-type gene tell us something about their effects. The lowercase i allele is recessive to both the I^A and I^B alleles. Therefore, a person with the genotype $I^A i$ has type A blood, and a person with the genotype $I^B i$ has type B blood. A person with both recessive alleles, genotype ii, has type O blood. The uppercase I^A and I^B alleles display codominance in that neither masks the expression of the other. Both of these alleles are expressed. Thus, a person with the genotype $I^A I^B$ has type AB blood.

Stop & Stretch Imagine a new allele I^C that is codominant with I^A and I^B and dominant to i. List all of the possible blood-group genotypes and phenotypes in a population containing all four alleles.

Another molecule on the surface of red blood cells is called the **Rh factor.** Someone who is positive (+) for this trait has the Rh factor on his or her red blood cells, while someone who is negative (−) does not. This trait, unlike the ABO blood system, is inherited in a straightforward two-allele, completely dominant manner with Rh⁺ dominant to Rh⁻. Persons who are Rh-positive can have the genotype Rh⁺Rh⁺ or Rh⁺Rh⁻. An Rh-negative individual has the genotype Rh⁻Rh⁻.

Blood typing is often used to help establish whether a given set of parents could have produced a particular child. For example, a child with type AB blood and parents who are type A and type B could be related, but a child with type O blood could not have a parent with type AB blood. Likewise, if a child has blood type B and the known mother has blood type AB, then the father of that child could have type AB, A, B, or O blood, which does not help to establish parentage.

If a child has a blood type consistent with alleles that he or she may have inherited from a man who might be his or her father, this finding does not mean that the man is the father. Instead, it is only an indication that the man could be. In fact, many other men would also have that blood type (**Table 8.1**). Therefore, blood-type analysis can be used only to eliminate people from consideration. Blood typing cannot be used to positively identify someone as the parent of a particular child.

Clinicians must take ABO blood groups into account when performing blood transfusions. Persons receiving transfusions from incompatible blood

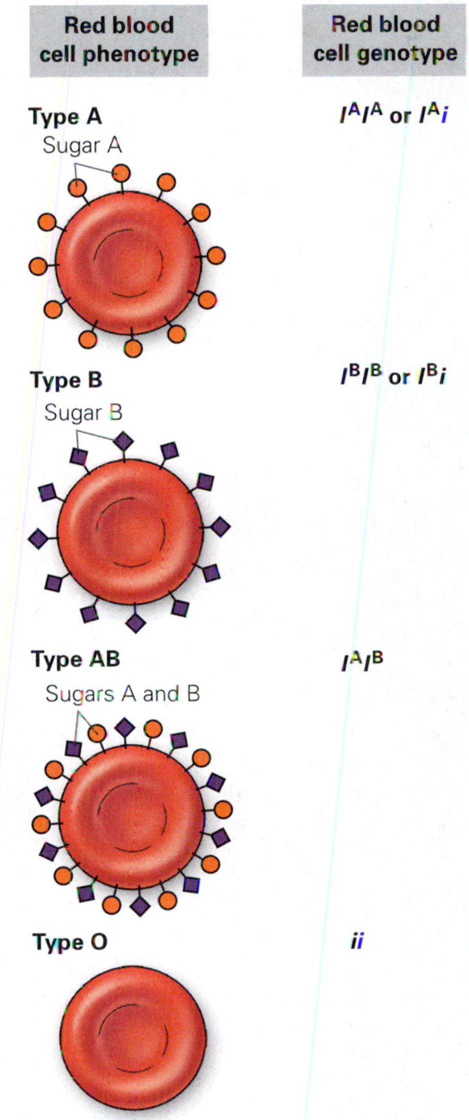

Red blood cell phenotype	Red blood cell genotype
Type A Sugar A	$I^A I^A$ or $I^A i$
Type B Sugar B	$I^B I^B$ or $I^B i$
Type AB Sugars A and B	$I^A I^B$
Type O	ii

Figure 8.3 ABO blood system. Red blood cell phenotypes and corresponding genotypes. Alleles I^A and I^B are codominant, and both are dominant to i.

TABLE 8.1

Frequency of blood types in U.S. population. Percentages of the population with a given blood type are listed from most to least common. The negative and positive superscripts refer to the absence or presence of the Rh factor.

Blood Type	O⁺	A⁺	B⁺	O⁻	A⁻	AB⁺	B⁻	AB⁻
Frequency in U.S. Population (%)	40	32	11	7	5	3	1.5	0.5

TABLE 8.2

Blood transfusion compatibilities.		
Recipient	**Recipient Can Receive**	**Recipient Cannot Receive**
Type O	Type O	Type A
		Type B
		Type AB
Type A	Type O	Type B
	Type A	Type AB
Type B	Type O	Type A
	Type B	Type AB
Type AB	Type O	None
	Type A	
	Type B	
	Type AB	

groups will mount an immune response against those sugars that they do not carry on their own red blood cells. The presence of these foreign red blood cell sugars causes a severe reaction in which the donated, incompatible red blood cells form clumps. This clumping can block blood vessels and kill the recipient. **Table 8.2** shows the types of blood transfusions individuals of various blood types can receive.

Blood-typing analysis can sometimes provide scientists with some information about potential relatedness of victims. However, it was not an option in the case of the Romanovs. The very old remains contained no blood, so blood typing was not possible.

The Romanov family was known to carry a trait that demonstrates another extension of Mendelism: pleiotropy. **Pleiotropy** is the ability of a single gene to cause multiple effects on an individual's phenotype. Alexis was the last of several Romanovs to have **hemophilia,** a pleiotropic blood-clotting disorder. A person with the most common form of hemophilia cannot produce a protein called clotting factor VIII. When this protein is absent, blood does not form clots to stop bleeding from a cut or internal blood vessel damage. Affected individuals bleed excessively, even from small cuts.

Due to the direct effects of excessive bleeding, hemophilia can lead to excessive bruising, pain and swelling in the joints, vision loss from bleeding into the eye, and anemia, resulting in fatigue. In addition, neurological problems may occur if bleeding or blood loss occurs in the brain.

Historical records indicate that Alexis, heir to the throne, was so ill with hemophilia that his father actually had to carry him to the basement room where he was executed.

8.2 Sex Determination and Sex Linkage

It appears that Alexis inherited the hemophilia allele from his mother. We can deduce this pattern of inheritance because we now know that the clotting factor gene is inherited in a sex-specific manner. The clotting factor VIII gene (the gene that, when mutated, causes hemophilia) is located in the X chromosome. Of the 23 pairs of chromosomes present in the cells of human males, 22 pairs are **autosomes**, or nonsex chromosomes, and one pair, X and Y, are the **sex chromosomes.** Males have 22 pairs of autosomes and one X and one Y sex chromosome. Females also have 22 pairs of autosomes, but their sex chromosomes are comprised of two X chromosomes.

Chromosomes and Sex Determination

The X and Y chromosomes are involved in producing the sex of an individual through a process called **sex determination.** When men produce sperm and the chromosome number is divided in half through meiosis, their sperm cells contain one member of each autosome and either an X or a Y chromosome. Females produce gametes with 22 unpaired autosomes and one of their two X chromosomes. Therefore, human egg cells normally contain one copy of an X chromosome, but sperm cells can contain either an X or a Y chromosome.

The sperm cell determines the sex of the offspring resulting from a particular fertilization. If an X-bearing sperm unites with an egg cell, the resulting offspring will be female (XX). If a sperm bearing a Y chromosome unites with an egg cell, the resulting offspring will be male (XY). **Figure 8.4** summarizes the process of sex determination in humans, and **Table 8.3** (on the next page) outlines some mechanisms of sex determination in nonhumans.

Sex Linkage

Genes located on the X or Y chromosome are called **sex-linked genes** because biological sex is inherited along with, or "linked to," the X or Y chromosome. Sex-linked genes found on the X chromosomes are said to be X linked, while those on the Y chromosome are Y linked. The X chromosome is much larger than the Y chromosome, which carries very little genetic information (**Figure 8.5**).

X-Linked Genes. **X-linked genes** are located on the X chromosome. The fact that males have only one X chromosome leads to some peculiarities in inheritance of sex-linked genes. Males always inherit their X chromosome from their mother because they must inherit the Y chromosome from their father to be male. Thus, males will inherit X-linked genes only from their mothers. Males are more likely to suffer from diseases caused by recessive alleles on the X chromosome because they have only one copy of any X-linked gene. Females are less likely to suffer from these diseases because they carry two copies of the X chromosome and thus have a greater likelihood of carrying at least one functional version of each X-linked gene.

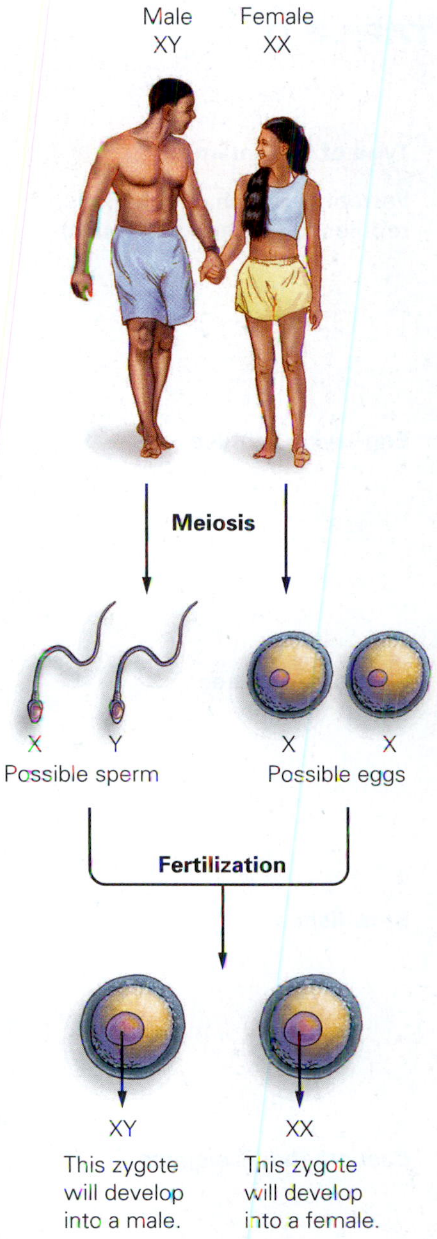

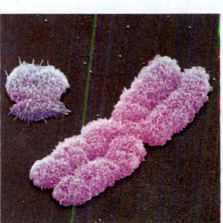

Figure 8.5 The X and Y chromosomes. The Y chromosome (left) is smaller than the X chromosome and carries fewer genes.

Male
XY

Female
XX

Meiosis

X Y
Possible sperm

X X
Possible eggs

Fertilization

XY
This zygote will develop into a male.

XX
This zygote will develop into a female.

Figure 8.4 Sex determination in humans. In humans, sex is determined by the male because males produce sperm that carry either an X or a Y chromosome (in addition to 22 autosomes). The egg cell always carries an X chromosome along with the 22 unpaired autosomes. When an X-bearing sperm fertilizes the egg cell, a female (XX) results. When a Y-bearing sperm fertilizes an egg cell, an XY male results. **Visualize This:** What is the probability that a couple will have a boy? What is the probability that a couple with four boys will have a girl for their fifth child?

TABLE 8.3

Sex determination in nonhuman organisms.		
Type of Organism		**Mechanism of Sex Determination**
Vertebrates (fish, amphibians, reptiles, birds, and mammals)		The male may have two of the same chromosomes and the female two different chromosomes. In these cases, the female determines the sex of the offspring.
Egg-laying reptiles		In many egg-laying species, two organisms with the same suite of sex chromosomes could become different sexes. Sex depends on which genes are activated during embryonic development. For example, the sex of some reptiles is determined by the incubation temperature of the egg.
Wasps, ants, bees		Sex is determined by the presence or absence of fertilization. In bees, males (drones) develop from unfertilized eggs. Females (workers or queens) develop from fertilized eggs.
Bony fishes		Some species of bony fishes change their sex after maturation. All individuals will become females unless they are deflected from that pathway by social signals such as dominance interactions.
Caenorhabditis elegans		The nematode *C. elegans* can either be male or have both male and female reproductive organs. Such individuals are called hermaphrodites.

A **carrier** of a recessively inherited trait has one copy of the recessive allele and one copy of the normal allele and will not exhibit symptoms of the disease. Only females can be carriers of X-linked recessive traits because males with a copy of the recessive allele will have the trait. Both males and females can be carriers of non-sex-linked, autosomal traits.

Even though female carriers of an X-linked recessive trait will not display the recessive trait, they can pass the trait on to their offspring. For this reason, most women carrying the hemophilia allele will not even realize that they are a carrier until their son becomes ill. **Figure 8.6a** illustrates that a cross between a male who does not have hemophilia and a female carrier can produce unaffected females, carrier females, unaffected males, and affected males. **Figure 8.6b** illustrates that no male children produced by a cross between an affected male and an unaffected

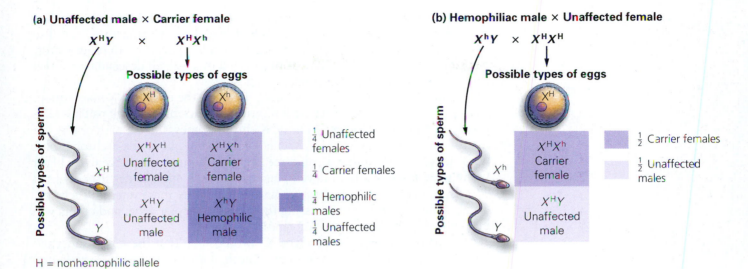

(a) Unaffected male × Carrier female

X^HY × X^HX^h

Possible types of eggs

	X^H	X^h
X^H	X^HX^H Unaffected female	X^HX^h Carrier female
Y	X^HY Unaffected male	X^hY Hemophilic male

Possible types of sperm

- $\frac{1}{4}$ Unaffected females
- $\frac{1}{4}$ Carrier females
- $\frac{1}{4}$ Hemophilic males
- $\frac{1}{4}$ Unaffected males

(b) Hemophiliac male × Unaffected female

X^hY × X^HX^H

Possible types of eggs

	X^H
X^h	X^HX^h Carrier female
Y	X^HY Unaffected male

Possible types of sperm

- $\frac{1}{2}$ Carrier females
- $\frac{1}{2}$ Unaffected males

H = nonhemophilic allele
h = hemophilic allele

Figure 8.6 Genetic crosses involving the X-linked hemophilia trait. Cross (a) shows the possible outcomes and associated probabilities of a mating between a nonhemophilic male and a female carrier of hemophilia. Cross (b) shows the possible outcomes and associated probabilities of a cross between a hemophilic male and an unaffected female. **Visualize This:** What genetic cross would result in the highest frequency of affected males?

female would have hemophilia. All daughters produced by this cross would be carriers of the trait. In the United States, there are over 20,000 hemophiliacs, nearly all of whom are male.

Stop & Stretch What must be true of the parents of a female who has an X-linked recessive disease?

Red-green colorblindness is another X-linked trait. This trait affects approximately 4% of males. Red blindness is the inability to see red as a distinct color. Green blindness is the inability to see green as a distinct color. When normal (in this case, the dominant alleles are normal), these genes code for the production of proteins called opsins that help absorb different wavelengths of light. A lack of opsins causes insensitivity to light of red and green wavelengths.

Duchenne muscular dystrophy is a progressive, fatal X-linked disease of muscle wasting that affects approximately 1 in 3500 males. The onset of muscle wasting occurs between 1 and 12 years of age, and by age 12, affected boys are often confined to a wheelchair. The affected gene is one that normally codes for the dystrophin protein. When at least one allele is normal, dystrophin stabilizes cell membranes during muscle contraction. It is thought that the absence of normal dystrophin protein causes muscle cells to break down and muscle tissue to die.

Most of the protein products of over 100 genes on the X chromosome have nothing at all to do with the production of biological sex differences. Accordingly, females and males should require equal doses of the products of X-linked genes. How can we account for the fact that females, with their two X chromosomes, could receive two doses of X-linked genes, while males receive only one? The answer comes from a phenomenon called **X inactivation** that occurs in all of the cells of a developing female embryo. This inactivation guarantees that all females actually receive only one dose of the proteins produced by genes on the X chromosomes. Inactivation of the genes on one of the two X chromosomes takes place in the embryo at about the time that the embryo implants in the uterus. One chromosome is inactivated when a string of RNA is wrapped around it (**Figure 8.7**).

Active X chromosome

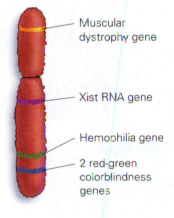

- Muscular dystrophy gene
- Xist RNA gene
- Hemophilia gene
- 2 red-green colorblindness genes

Inactive X chromosome

- Xist RNA

Figure 8.7 X inactivation. One of a female's two X chromosomes is inactivated when strands of RNA wrap around the chromosome.

This inactivation is random with respect to the parental source—either of the two X chromosomes can be inactivated in a given cell. Inactivation is also irreversible and, as such, is inherited during cell replication. In other words, once a particular X chromosome is inactivated in a cell, all descendants of that cell continue inactivating the same chromosome.

Cats with tortoiseshell coat coloring illustrate the effects of X inactivation. The coats of tortoiseshell cats are a mixture of black and orange patches. The genes for fur color are located on the X chromosomes. If a cat with orange fur mates with a cat with black fur, a female kitten could have one X chromosome with the gene for orange fur, and one X chromosome with the gene for black fur (**Figure 8.8a**). Early in development, when the embryo consists of about 16 cells, one of the two X chromosomes is randomly inactivated in each cell. Thus, some cells will be expressing the orange fur-color gene, and others will be expressing the black fur-color gene.

The pattern of inactivation (the X chromosome that the kitten inherited from its mother or the X chromosome that the kitten inherited from its father) is passed on to the daughter cells of the 16-celled embryo, resulting in the patches of orange and black fur color seen in tortoiseshell cats (**Figure 8.8b**). Because this pattern of coat coloration requires the expression of both alleles of

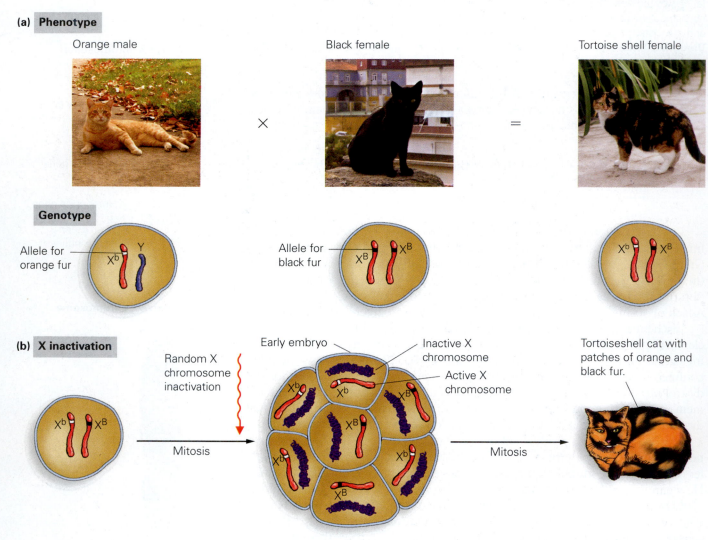

(a) Phenotype

Orange male   Black female   Tortoise shell female

Genotype

Allele for orange fur — X^b Y

Allele for black fur — X^B X^B

X^b X^B

(b) X inactivation

Random X chromosome inactivation

Early embryo
Inactive X chromosome
Active X chromosome

X^b X^B Mitosis X^b X^b X^B X^B X^b X^b X^B Mitosis   Tortoiseshell cat with patches of orange and black fur.

Figure 8.8 X inactivation and patchy gene expression. (a) An orange male cat and a black female cat can produce female offspring with tortoiseshell coats. (b) Random inactivation of one X chromosome early in development leads to patches of coat color.

the color gene in different patches of cells, a cat must have two different alleles of this X-linked gene. Therefore, these cats are almost always female. On rare occasions, male cats can have this pattern of coloring. This can happen only if a male cat has two X chromosomes and one Y chromosome—a situation that does occur, though infrequently, via nondisjunction (Chapter 6)—and results in sterility.

Y-Linked Genes. **Y-linked genes** are located on the Y chromosome and are passed from fathers to sons. Although this distinctive pattern of inheritance should make Y-linked genes easy to identify, very few genes have been localized to the Y chromosomes. One gene known to be located exclusively on the Y chromosome is called the *SRY* gene (for sex-determining region of the Y chromosome). The expression of this gene triggers a series of events leading to development of the testes and some of the specialized cells required for male sexual characteristics. Genes other than *SRY,* on chromosomes other than the Y, code for proteins that are unique to males but are not expressed unless testes develop.

Stop & Stretch Sometimes chromosomal sex does not match biological sex. For instance, males can sometimes have two X chromosomes and no Y chromosome. Most of these males carry a copy of the SRY gene on one X chromosome. What event during meiosis can explain this phenomenon? Some genes on the Y chromosome code for sperm production. What does this imply about XX males?

In the case of the bones thought to belong to the Romanovs, because karyotype (review Figure 6.18) and pelvic bone analyses were difficult to perform on the decayed bones, scientists analyzed DNA from the bones for sequences known to be present only on the Y chromosome. When DNA that was isolated from the children's remains was analyzed, it became clear that the children's bones all belonged to girls. If these bones did belong to the Romanovs, one of the two missing children was Alexis, the Romanovs' only son.

8.3 Pedigrees

Another line of evidence was provided by the extensive family trees of the Romanovs and their relatives. Because the hemophilia gene is X linked, Alexis Romanov inherited the disease from his mother, who must have been a carrier of the disease. We can trace the lineage of this disease through the Romanov family by using a chart called a pedigree. A **pedigree** is a family tree that follows the inheritance of a genetic trait for many generations of relatives. Pedigrees are often used in studying human genetics because it is impossible and unethical to set up controlled matings between humans the way one can with fruit flies or plants. Pedigrees allow scientists to study inheritance by analyzing matings that have already occurred. **Figure 8.9** identifies some of the symbols used in pedigrees, and **Figure 8.10** (on the next page) shows how scientists can use pedigrees to determine whether a trait is inherited as autosomal dominant or recessive or as sex-linked recessive.

Information is available about the Romanovs' ancestors because they were royalty and because scientists interested in hemophilia had kept very good records of the inheritance of that trait. Hemophilia was common among European royal families but rare among the rest of the population. This was because members of the royal families intermarried to preserve the royal bloodlines. The tsarina must have been a carrier of the hemophilia allele because her son had

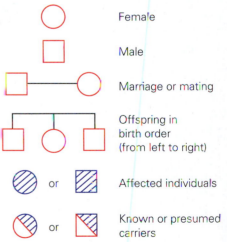

Pedigree analysis symbols

Female

Male

Marriage or mating

Offspring in birth order (from left to right)

Affected individuals

Known or presumed carriers

Figure 8.9 Pedigree analysis.
Symbols used in pedigrees.
Visualize This: Draw a pedigree showing a female who has a boy with her first husband and a girl with her second husband.

(a) Dominant trait: Polydactyly

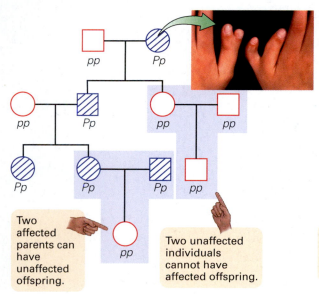

Two affected parents can have unaffected offspring.

Two unaffected individuals cannot have affected offspring.

(b) Recessive trait: Attached earlobes

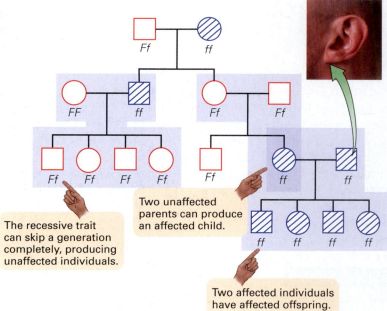

The recessive trait can skip a generation completely, producing unaffected individuals.

Two unaffected parents can produce an affected child.

Two affected individuals have affected offspring.

(c) Sex-linked trait: Muscular dystrophy

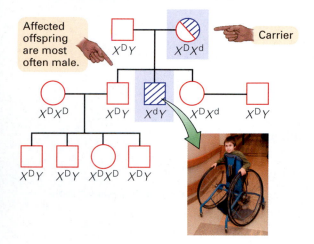

Affected offspring are most often male.

Carrier

Figure 8.10 Pedigrees showing different modes of inheritance.
(a) Polydactyly is a dominantly inherited trait. People with this condition have extra fingers or toes. (b) Having attached earlobes is a recessively inherited trait. (c) Muscular dystrophy is inherited as an X-linked recessive trait.

the trait. Her mother, Alice, must also have been a carrier because the tsarina's brother Fred had the disease, as did two of her sister Irene's sons, Waldemar and Henry. The tsarina's grandmother, Queen Victoria, seems to have been the first carrier of this allele in the family, because there is no evidence of this disease before her eighth child, Leopold, was affected.

Queen Victoria's mother, Princess Victoria, most likely incurred a mutation to the clotting factor VIII gene while the cells of her ovaries were undergoing DNA synthesis to produce egg cells. The egg cell that carried the mutant clotting factor VIII gene was passed from Princess Victoria to her daughter, who came to be crowned Queen Victoria. When the inherited mutant cell divided by mitosis to produce Queen Victoria's body (somatic) cells, the mutant clotting factor VIII gene was passed on to each of those cells. When the cells in her ovaries underwent meiosis to produce gametes, the Queen passed the mutant version on to three of her nine children (**Figure 8.11**). The extensive pedigree available to the scientists working on the Romanov case, in concert with a powerful technique called DNA fingerprinting, would provide the key data in solving the mystery of the buried bones.

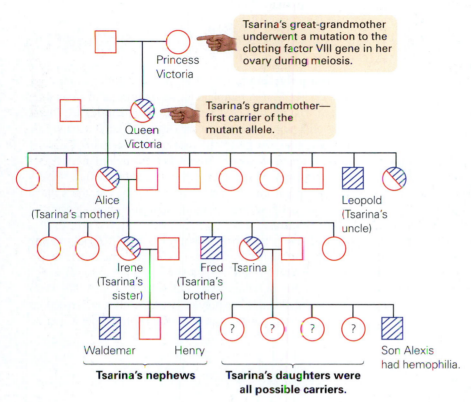

Tsarina's great-grandmother underwent a mutation to the clotting factor VIII gene in her ovary during meiosis.

Princess Victoria

Tsarina's grandmother— first carrier of the mutant allele.

Queen Victoria

Alice (Tsarina's mother)

Leopold (Tsarina's uncle)

Irene (Tsarina's sister)

Fred (Tsarina's brother)

Tsarina

Waldemar   Henry

Tsarina's nephews

Tsarina's daughters were all possible carriers.

Son Alexis had hemophilia.

Figure 8.11 Origin and inheritance of the hemophilia allele. This abbreviated pedigree shows the origin of the hemophilia allele and its inheritance among the tsarina's family. It appears that the tsarina's great-grandmother underwent a mutation that she passed on to her daughter, who then passed it on to three of her nine children—one of whom was the tsarina's mother, Alice.

8.4 DNA Fingerprinting

Limits on the power of conventional forensic techniques such as blood typing and karyotyping to identify the bones found in the Ekaterinburg grave necessitated the use of more sophisticated techniques. To do so, scientists took advantage of the fact that any two individuals who are not identical twins have small differences in the sequences of nucleotides that comprise their DNA. To test the hypothesis that the bones buried in the Ekaterinburg grave belonged to the Romanov family, the scientists had to answer the following questions:

1. Which of the bones from the pile are actually different bones from the same individuals?
2. Which of the adult bones could have been from the Romanovs, and which bones could have belonged to their servants?
3. Are these bones actually from the Romanovs, not some other related set of individuals?

An Overview: DNA Fingerprinting

All of these questions were answered using **DNA fingerprinting**. This technique allows unambiguous identification of people in the same manner that traditional fingerprinting has been used in the past. To begin this process, it is necessary to isolate the DNA to be fingerprinted. Scientists can isolate DNA from blood, semen, vaginal fluids, a hair root, skin, and even (as was the case in Ekaterinburg) degraded skeletal remains.

Because often there is not much DNA present, it must first be copied to generate larger quantities for analysis. Since each person has a unique sequence of DNA, particular regions of DNA, which vary in size, are selected for copying. When the copied DNA fragments are separated and stained, a unique pattern is produced. *See the next section for A Closer Look at DNA Fingerprinting.*

A Closer Look:
DNA Fingerprinting

When very small amounts of DNA are available, as is often the case, scientists can make many copies of the DNA by first performing a DNA-amplifying reaction.

Polymerase Chain Reaction (PCR)

The **polymerase chain reaction (PCR)** is used to amplify, or produce many copies of, a particular region of DNA (**Figure 8.12**). To perform PCR, scientists place four essential components into a tube: (1) the **template,** which is the double-stranded DNA to be copied; (2) the **nucleotides** adenine (A), cytosine (C), guanine (G), and thymine (T), which are the individual building-block subunits of DNA; (3) short single-stranded pieces of DNA called **primers,** which are complementary to the ends of the region of DNA to be copied; and (4) an enzyme called *Taq* polymerase.

Taq **polymerase** is a DNA polymerase that uses one strand of DNA as a template for the synthesis of a daughter strand that carries complementary nucleotides (A:T base pairs are complementary, as are G:C base pairs) (Chapter 6). This enzyme was given the first part of its name (*Taq*) because it was first isolated from *Thermus aquaticus*, a bacterium that lives in hydrothermal vents and can withstand very high temperatures. The second part of the enzyme's name (polymerase) describes its synthesizing activity—it acts as a DNA polymerase.

The main difference between human DNA polymerase and *Taq* polymerase is that the *Taq* polymerase is resistant to extremely high temperatures, temperatures at which human DNA polymerase would be inactivated. The heat-resistant qualities of *Taq* polymerase thus allow PCR reactions to be run at very high temperatures. High temperatures are

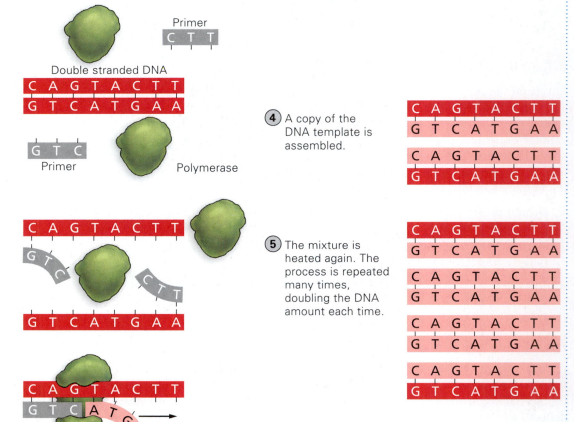

1 PCR is used to amplify, or make copies of, a specific region of DNA. During a PCR reaction, primers (short stretches of single-strand DNA), free nucleotides, and template DNA are mixed with heat-tolerant *Taq* polymerase.

2 The DNA is heated to separate, or denature, the two strands. As the mixture cools, the primers bond to the region of the DNA template they complement.

3 The *Taq* polymerase uses the primers to initiate synthesis.

4 A copy of the DNA template is assembled.

5 The mixture is heated again. The process is repeated many times, doubling the DNA amount each time.

Figure 8.12 The polymerase chain reaction (PCR). PCR is used to amplify, or make copies of, DNA. Each round of PCR doubles the number of DNA molecules. This type of exponential growth can yield millions of copies of DNA.

necessary because the DNA molecule being amplified must first be **denatured,** or split up the middle of the double helix, to produce single strands. After heating, the DNA solution is allowed to cool, and the primers bond to the region of the DNA template they complement. *Taq* polymerase then uses the primers to initiate DNA synthesis, producing double-stranded DNA molecules. This cycle of heating and cooling the tube is repeated many times, with each round of PCR doubling the amount of double-stranded DNA present in the tube.

Once scientists have produced enough DNA by PCR, they can analyze the differences between individuals. Because each individual has distinct nucleotide sequences, copying different people's DNA with the same enzyme can produce fragments of different sizes.

Analysis of Variable Number Tandem Repeats (VNTRs)

During PCR, the copied regions of DNA have different lengths due to the presence of DNA sequences that vary in number, called variable number tandem repeats (VNTRs) (**Figure 8.13**). These are nucleotide sequences that all of us carry, but in different numbers. For example, one person may have four copies of the following sequence (CGATCGA) on one chromosome

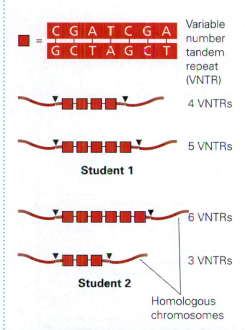

Figure 8.13 Variable number tandem repeats (VNTRs). Student 1 has four repeat sequences comprising the VNTRs on one of his chromosomes and five on the other. Student 2 has six copies of the same repeat sequence on one of her chromosomes and three on the other member of the homologous pair. The repeat sequence is represented as a box. PCR amplifies the region of DNA containing the VNTRs, generating fragments of different sizes. PCR primer sites around the VNTRs are shown as arrowheads.

and five copies of the same sequence on the other, homologous, chromosome. Another person may have six repeating copies of that sequence on one chromosome and three copies on the other member of the homologous pair. Within a population, there are variable numbers of these known tandem repeat sequences. When PCR is used to copy the region of DNA containing a VNTR, those segments of DNA that carry more repeats will be longer than those that carry fewer repeats. Because it is impossible to determine the size of DNA fragments by simply looking at the DNA in the tube, techniques that allow for the separation and visualization of the DNA are required.

Stop & Stretch VNTRs do not code for proteins. If two individuals have different numbers of VNTRs around a particular gene, does that mean that they carry different alleles for the gene? Explain your answer.

Gel Electrophoresis

The amplified fragments of DNA generated by PCR can be separated from each other by allowing the fragments to migrate through a solid support called an **agarose gel,** which resembles a thin slab of gelatin. When an electric current is applied, the gel impedes the progress of the larger DNA fragments more than it does the smaller ones, facilitating the size-based separation. The size-based separation of molecules when an electric current is applied to a gel is a technique called **gel electrophoresis.** Segments of DNA with more repeats would be longer than those with fewer repeats. Longer DNA segments will not migrate as far in the gel as shorter DNA fragments. Thus, agarose gel electrophoresis separates the DNA fragments on the basis of their size (**Figure 8.14**).

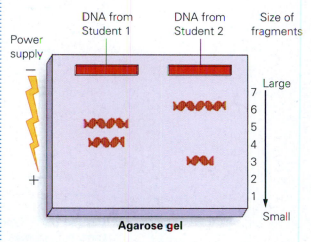

Figure 8.14 DNA fingerprinting. Fragments of DNA that were amplified by PCR can be separated by gel electrophoresis. The DNA itself is not visible to the unaided eye and must be stained to produce the DNA fingerprint.

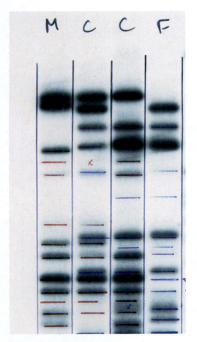

Figure 8.15 DNA fingerprint. This photograph of a DNA fingerprint shows a mother (M), a father (F), and their two children (C). Note that every band present in a child must also be present in one of the parents.

DNA Fingerprinting Evidence Helped Solve the Romanov Mystery

In 1992, a team of Russian and English scientists used the DNA fingerprinting technique to determine which of the bones discovered at Ekaterinburg belonged to the same skeleton. They used a slightly different DNA fingerprinting technique in that they isolated DNA from mitochondria located in the bone cells to confirm that the pile of decomposed bones in the Ekaterinburg grave belonged to nine different individuals.

Once scientists had established that the bones from nine different people were buried in the grave, they tried to determine which bones might belong to the adult Romanovs and which belonged to the servants. For the answer to these questions, scientists took advantage of the fact that Romanov family members would have more DNA sequences in common with each other than they would with the servants.

In fact, each DNA region that a child carries is inherited from one of his or her parents. Therefore, each band produced in a DNA fingerprint of a child must be present in the DNA fingerprint of one of that child's parents (**Figure 8.15**). By comparing DNA fingerprints made from the smaller skeletons, scientists were able to determine which of the six adult skeletons could have been the tsar and tsarina. **Figure 8.16** shows a hypothetical DNA fingerprint that illustrates how the banding patterns produced can be used to determine which of the bones belonged to the parents of the smaller skeletons and which bones may have belonged to the unrelated servants. Notice that Figure 8.16 has more than two bands in each gel lane. This is because the PCR reaction was done in a way that amplified many different VNTR sites at the same time.

Stop & Stretch Why doesn't a child possess all of the same fragments found in one parent?

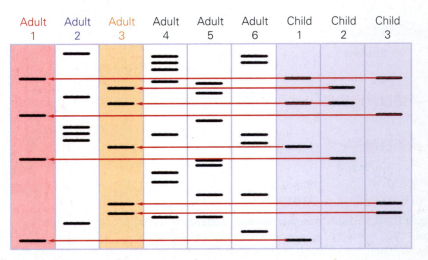

Figure 8.16 Hypothetical fingerprint of adult- and child-sized skeletons. A hypothetical DNA fingerprint of DNA from the bone cells of individuals found in the Ekaterinburg grave is shown. From the results of this fingerprint, it is evident that children 1, 2, and 3 are the offspring of adults 1 and 3. Note that each band from each child has a corresponding band in either adult 1 or adult 3. The remaining DNA from adults does not match any of the children, so these adults are not the parents of any of these children.

DNA evidence was further used to help put to rest claims made by many pretenders to the throne. People from all over the world had alleged that they were either a Romanov who had escaped execution or a descendant of an escapee.

The most compelling of these claims was made by a young woman who was rescued from a canal in Berlin, Germany, two years after the murders. This young woman suffered from amnesia and was cared for in a mental hospital, where the staff named her Anna Anderson (**Figure 8.17**). She later came to believe that she was Anastasia Romanov, a claim she made until her death in 1984. The 1956 Hollywood film *Anastasia,* starring Ingrid Bergman, made Anna Anderson's claim seem plausible. A more recent animated version of the story of an escaped princess, also titled *Anastasia,* convinced many young viewers that Anna Anderson was indeed the Romanov heiress.

Because the sex-typing analysis showed only that one daughter was missing from the grave, but not which daughter, scientists again looked to the fingerprinting data. DNA fingerprinting had been done in the early 1990s on intestinal tissue removed during a surgery performed before Anna Anderson's death. The analysis showed that Anna was not related to anyone buried in the Ekaterinburg grave. She could not be Anastasia.

Thus far in our narrative, scientists have answered two of the questions posed. They have determined (1) that nine different individuals were buried in the Ekaterinburg grave and (2) that two of the adult skeletons were the parents of the three children. The last question is still unanswered. How did the scientists show that these were bones from the Romanov family, not just some other set of related individuals?

To answer this question, the scientists turned to living relatives of the Romanovs. DNA testing was performed on England's Prince Philip, who is a grandnephew of Tsarina Alexandra. In addition, Nicholas II's dead brother

(a) Anna Anderson

(b) Anastasia Romanov

Figure 8.17 Anna Anderson and Anastasia Romanov. Photos were not useful in determining whether Anna Anderson was Anastasia Romanov, as she claimed. DNA evidence was.

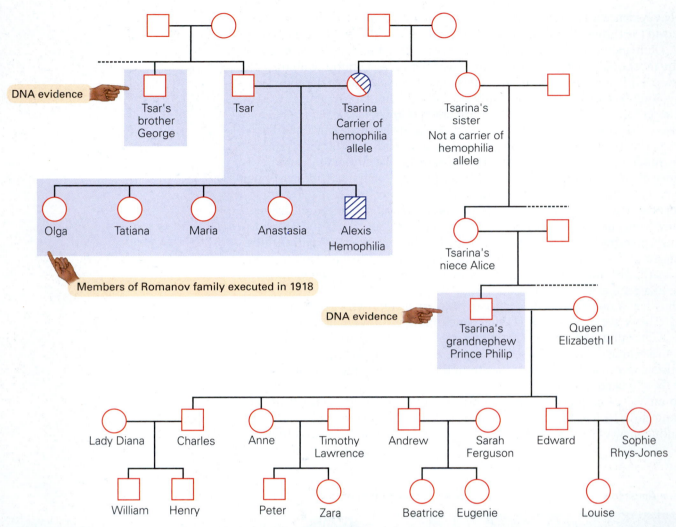

Figure 8.18 Romanov family pedigree. This pedigree shows only the pertinent family members. DNA from the tsar's brother George showed that he was related to the tsar. Note that Prince Philip is the tsarina's grandnephew. Prince Philip married Queen Elizabeth II. Together they had four children, Charles, Anne, Andrew, and Edward, the current British royal family. The tsarina's sister does not appear to have been a carrier of the hemophilia allele because none of her descendants has been affected by the disease.

George was exhumed, and his DNA was tested. **Figure 8.18** shows the Romanov family pedigree; you can see how these individuals are related to each other. The DNA testing performed on these individuals showed that George was genetically related to the adult male skeleton related to the children's skeletons, and that Prince Philip was genetically related to the adult female skeleton shown to be related to the children's skeletons. This evidence strongly supported the hypothesis that the adult skeletons were indeed those of the tsar and tsarina. The process of elimination suggests that the remaining four skeletons had to be the servants.

Table 8.4 summarizes how scientists used the scientific method to test the hypothesis that the remains were indeed those of the Romanovs.

In 1998, some 80 years after their execution, the Romanov family was finally laid to rest, and the people of postcommunist Russia symbolically laid to rest this part of their country's political history. Ten years after that, all cases of pretenders to the throne were dismissed when bone fragments unearthed in a forest near where the family was killed were found to belong to Alexis and his sister.

TABLE 8.4

The scientific method. A summary of tests and the conclusions that were drawn from them. **Hypothesis:** The bones found in the Ekaterinburg grave belonged to the Romanov family and their servants.	
Test	**Description of Results**
Analyze teeth	Expensive dental work was typically seen only in royalty.
Measure skeletons	The skeletons are those of six adults and three children.
Sex typing	The two children missing from the grave are a male and a female.
DNA fingerprinting	Children in the grave are related to two adults in the grave.
DNA fingerprinting	Claims to be one of the missing Romanov children or their descendants are disproved.
DNA fingerprinting	The two adults related to the children are related to known Romanov relatives.

Conclusion: When you look at each result individually, the evidence is less compelling than when you look at all the evidence together. As a whole, the evidence strongly supports the hypothesis that it was indeed the Romanovs who were buried in the Ekaterinburg grave.

SAVVY READER

Determining the Sex of Your Child

You are killing time in a pharmacy while your prescription is filled. You come across a product that claims to help you conceive a child of a particular sex. This product claims to have a success rate of 60%.

In a different aisle of the pharmacy, there is a magazine with a cover story claiming that more attractive parents have more girls. This article claims that Americans who are rated as attractive have close to 60% chance of having a daughter.

1. What are the odds of having a child of a particular gender by chance? Can we make any decisions about the validity of these assertions without data and information about the significance of any reported differences?

2. Can you think of any way your gametes could be directed to produce a male or female offspring?

3. A friend of yours believes that parents have children of the gender that the parents are best able to rear. Couples who have daughters have traits that are better for daughters to have and vice versa. Your friend believes this is consistent with the theory of natural selection. Based on your understanding of natural selection and gametogenesis, does this seem possible?

Chapter Review

Learning Outcomes

Mastering**BIOLOGY**®

Go to the Study Area at www.masteringbiology.com for practice quizzes, myeBook, BioFlix™ 3-D animations, MP3Tutor sessions, videos, current events, and more.

LO1 Differentiate incomplete dominance from codominance (Section 8.1).

- Incomplete dominance is an extension of Mendelian genetics in which the phenotype of the progeny is intermediate to that of both parents (pp. 175–176).
- Codominance occurs when both alleles of a given gene are expressed (pp. 175–176).

LO2 Explain how polygenic and pleiotropic traits differ (Section 8.1).

- Polygenic inheritance occurs when many genes control one trait (p. 176).
- Pleiotropy occurs when a single gene leads to multiple effects (p. 178).

LO3 Outline the inheritance of the multiple allelic ABO blood system (Section 8.1).

- Genes that have more than two alleles segregating in a population are said to have multiple alleles (p. 176).
- The ABO blood system displays both multiple allelism (alleles I^A, I^B, and i) and codominance because both I^A and I^B are expressed in the heterozygote (pp. 176–178).

LO4 Describe the mechanism of sex determination in humans (Section 8.2).

- One mechanism of sex determination involves the suite of sex chromosomes present (p. 179).
- In humans, males have an X and a Y chromosome and can produce gametes containing either sex chromosome, and females have two X chromosomes and always produce gametes containing an X chromosome. When an X-bearing sperm fertilizes an egg cell, a female baby will result. When a Y-bearing sperm fertilizes an egg cell, a male baby will result (pp. 179–180).

LO5 Explain the pattern of inheritance exhibited by sex-linked genes (Section 8.2).

- Genes linked to the X and Y chromosomes show characteristic patterns of inheritance. Males need only one recessive X-linked allele to display the associated phenotype. Females can be carriers of an X-linked recessive allele and may pass an X-linked disease on to their sons (pp. 179–181).
- Y-linked genes are passed from fathers to sons (p. 183).

LO6 Describe how X-inactivation changes patterns of inheritance (Section 8.2).

- One of the two X chromosomes in females is inactivated, and the genes residing on it are not expressed. This inactivation is faithfully propagated to all daughter cells (pp. 181–183).

LO7 Explain how scientists use pedigrees (Section 8.3).

- Pedigrees are charts that scientists use to study the transmission of genetic traits among related individuals (pp. 183–185).

LO8 Explain the significance of DNA fingerprinting and how the process works (Section 8.4).

- DNA fingerprinting is used to show the relatedness of individuals based on similarities in their DNA sequences (p. 185).
- When small amounts of DNA are available, the polymerase chain reaction (PCR) can be used to make millions of copies of the DNA (p. 186).
- The lengths of repeated sequences are characteristic of a given individual (p. 187).
- DNA samples in an agarose gel subjected to an electric current will separate according to their size (p. 187).

Roots to Remember

The following roots of words come mainly from Latin and Greek and will help you decipher terms:

forensic comes from a Latin word, *forum*, which was the place where legal disputes and lawsuits were heard.

hemo- means blood. Chapter term: hemophilia

pleio- and **poly-** both come from Greek words for many. Chapter terms: pleiotropy and polygenic

Learning the Basics

1. **LO3** What does it mean for a trait to have multiple alleles? Can an individual carry more than two alleles of one gene?

2. **LO4** How is sex determined in humans?

3. **LO8** Describe the technique of DNA fingerprinting.

4. **LO3** If a man with blood type A and a woman with blood type B have a child with type O blood, what are the genotypes of each parent?

5. **LO5** A man with type A⁺ blood whose father had type O⁻ blood and a woman with type AB⁻ blood could produce children with which phenotypes relative to these blood-type genes?

6. **LO8** Which of the following is *not* part of the procedure used to make a DNA fingerprint?

   A. DNA is amplified by PCR; **B.** DNA is placed in a gel and subjected to an electric current; **C.** The genes that encode fingerprint patterns are cloned into bacteria; **D.** DNA from blood, semen, vaginal fluids, or hair root cells can be used for analysis.

7. **LO8** Which of the following statements is consistent with the DNA fingerprint shown in **Figure 8.19**?

   A. B is the child of A and C; **B.** C is the child of A and B; **C.** D is the child of B and C; **D.** A is the child of B and C; **E.** A is the child of C and D.

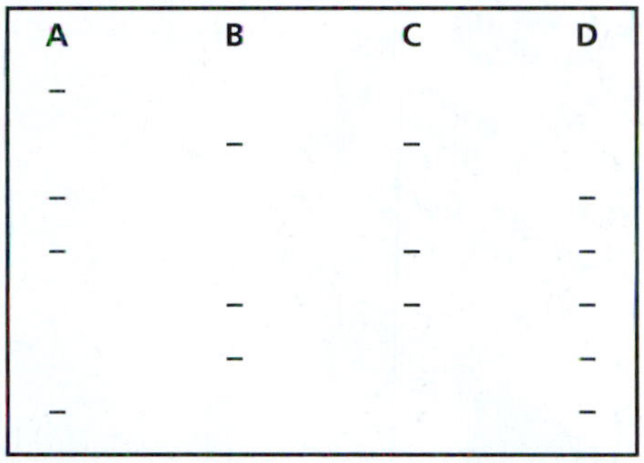

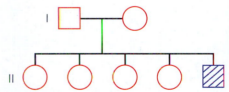

Figure 8.19 DNA fingerprint.

8. **LO4** What is the probability that a family with two children will have one boy and one girl?

   A. 100%; **B.** 75%, **C.** 50%; **D.** 25%

9. **LO3** The pedigree in **Figure 8.20** illustrates the inheritance of hemophilia (sex-linked recessive trait) in the royal family. What is the genotype of individual II-5 (Alexis)?

   A. $X^H X^H$; **B.** $X^H X^h$; **C.** $X^h X^h$; **D.** $X^H Y$; **E.** $X^h Y$

Figure 8.20 Pedigree.

10. **LO5** A woman is a carrier of the X-linked recessive color-blindness gene. She mates with a man with normal color vision. Which of the following is true of their offspring?

    A. All the males will be colorblind; **B.** All the females will be carriers; **C.** Half the females will be colorblind; **D.** Half the males will be colorblind.

11. **LO2** Which of the following traits is most likely to be a polygenic trait?

    A. a blood disease caused by one mutation; **B.** a complex trait like intelligence; **C.** a disease you get when exposed to radiation; **D.** having type AB blood

12. **LO6** What causes tortoiseshell cats to have patches of color?

    A. multiple alleles; **B.** environmental influences; **C.** pleiotropy; **D.** random X-inactivation

Analyzing and Applying the Basics

1. **LO1** Compare and contrast codominance and incomplete dominance.

2. **LO8** Draw a DNA fingerprint that might be generated by two sisters and their parents.

3. **LO7** Draw a pedigree of a mating between first cousins and use the pedigree to explain why matings between relatives can lead to an increased likelihood of offspring with rare recessive diseases.

Connecting the Science

1. Science helped solve the riddle of who was buried in the Ekaterinburg grave, leading to a church burial for some of the Romanov family. In this manner, science played a role in helping the people of Russia come to terms with the brutal communist regime that followed the deaths of the royal family. Can you think of other examples for which science has been used to help answer a question with great social implications?

2. The Innocence Project is a nonprofit legal clinic at the Benjamin Cardozo School of Law in New York City that attempts to help prisoners whose claims of innocence can be verified by DNA testing because biological evidence from their cases still exists. Over 250 inmates have been exonerated since the project started in 1992. Thousands of inmates await the opportunity for testing. Does the fact that the Innocence Project has shown that the criminal justice system convicts innocent people change your opinion about the death penalty? Why or why not?

Answers to **Stop & Stretch, Visualize This, Savvy Reader,** and **Chapter Review** questions can be found in the **Answers** section at the back of the book.

Genetically Modified Organisms

Gene Expression, Mutation, and Cloning

Many people are concerned about genetic engineering.

Genetic engineering is the use of laboratory techniques to alter hereditary traits in an organism. Genetically modified foods, also called GM foods, have had one or more of their genes modified or a gene from another organism inserted alongside their normal complement of genes. Emotions run high in the debate over the altering of our food supply, with people on both sides of the issue making their cases dramatically.

Demonstrators dressed in biohazard suits toss genetically modified crackers, cereal, and pasta into a garbage can in front of a supermarket while shareholders inside vote on whether to remove such foods from the store shelves. Protesters hack down plots of genetically modified corn and coffee. Some parents refuse to send their kids to day-care centers that use milk from cows treated with growth hormone.

On the other side of the issue are proponents of genetically modifying foods who believe the technology holds tremendous promise. Advocates claim that genetically modified foods have the potential to help wipe out hunger. They see a future in which all children will be vaccinated through the use of edible vaccines, and common human diseases like heart disease will be prevented when people begin to eat genetically modified animals such as pigs that produce omega-3 fatty acids. Proponents even see environmental benefits because some genetically modified crops allow farmers to use far smaller amounts of chemicals that damage the environment.

Even more controversy surrounds the idea of genetically modifying humans. Scientists have already cloned many different animals; can humans be far behind?

Can you tell which of these foods have been genetically modified?

Is it acceptable to modify animals such as these pigs that can produce an essential nutrient for human consumption?

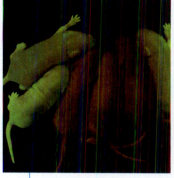

Or animals such as these luminescent mice?

How can the average citizen determine whether the benefits of genetic engineering outweigh the drawbacks? To answer these questions for your-self, you must learn about how genes typically produce their products and then how they can be modified to suit human desires.

9.1 Protein Synthesis and Gene Expression

To genetically modify foods, scientists can move a gene known to produce a certain protein from one organism to another. Alternatively, the scientists can change the amount of protein a gene produces. Regulating the amount of protein produced by a cell is also referred to as *regulating gene expression*.

One of the first examples of scientists controlling gene expression occurred in the early 1980s when genetic engineers began to produce **recombinant bovine growth hormone (rBGH)** in their laboratories. Recombinant (r) bo-vine growth hormone is a protein that has been made by genetically engineered bacteria. These bacterial cells have had their DNA manipulated so that it carries the instructions for, or encodes, a cow growth hormone that can be produced in the laboratory. Hormones are substances that are secreted from specialized glands and travel through the bloodstream to affect their target organs. Growth hormones act on many different organs to increase the overall size of the body. Bovine growth hormone that is produced in a laboratory can be injected into dairy cows to increase their milk production.

Production of growth hormone protein, or any protein, in the lab or in a cell, requires the use of the genetic information coded in the DNA.

From Gene to Protein

Protein synthesis involves using the instructions carried by a gene to build a particular protein. Genes do not build proteins directly; instead, they carry the instructions that dictate how a protein should be built. Understanding protein synthesis requires that we review a few basics about DNA, genes, and RNA. First, DNA is a polymer of nucleotides that make chemical bonds with each other based on their complementarity: adenine (A) to thymine (T) and cytosine (C) to guanine (G). Second, a gene is a sequence of DNA that encodes a pro-tein. Proteins are large molecules composed of amino acids. Each protein has a unique function that is dictated by its particular structure. The structure of a protein is the result of the order of amino acids that constitute it because the chemical properties of amino acids cause a protein to fold in a particular man-ner. Before a protein can be built, the instructions carried by a gene are first copied. When the gene is copied, the copy is made up not of DNA (deoxyribo-nucleic acid) but of **RNA (ribonucleic acid).**

RNA, like DNA, is a polymer of nucleotides. A nucleotide is composed of a sugar, a phosphate group, and a nitrogen-containing base. Whereas the sugar in DNA is deoxyribose, the sugar in RNA is ribose. RNA has the nitrogenous base uracil (U) in place of thymine. RNA is usually single stranded, not double stranded like DNA (**Figure 9.1**).

When a cell requires a particular protein, a strand of RNA is produced using DNA as a guide or template. RNA nucleotides are able to make base pairs

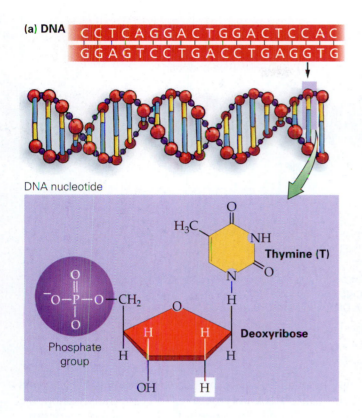

(a) DNA

C	C	T	C	A	G	G	A	C	T	G	G	A	C	T	C	C	A	C
G	G	A	G	T	C	C	T	G	A	C	C	T	G	A	G	G	T	G

DNA nucleotide

Thymine (T)

Phosphate group

Deoxyribose

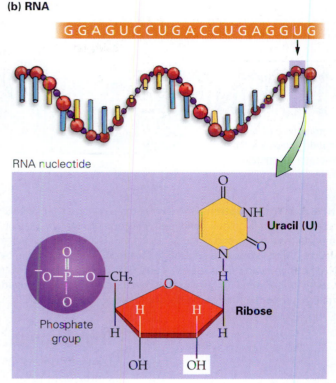

(b) RNA

G	G	A	G	U	C	C	U	G	A	C	C	U	G	A	G	G	U	G

RNA nucleotide

Uracil (U)

Phosphate group

Ribose

with DNA nucleotides. C pairs with G, and A pairs with U. The RNA copy then serves as a blueprint that tells the cell which amino acids to join together to produce a protein. Thus, the flow of genetic information in a eukaryotic cell is from DNA to RNA to protein (**Figure 9.2**).

How does this flow of information actually take place in a cell? Going from gene to protein involves two steps. The first step, called **transcription,** involves producing the copy of the required gene. In the same way that a transcript of a speech is a written version of the oral presentation, transcription inside a cell produces a transcript of the original gene, with the RNA nucleotides substituted for DNA nucleotides. The second step, called **translation,** involves decoding the copied RNA sequence and producing the protein for which it codes. In the same way that a translator deciphers one language into another, translation in a cell involves moving from the language of nucleotides (DNA and RNA) to the language of amino acids and proteins.

Figure 9.1 DNA and RNA. (a) DNA is double stranded. Each DNA nucleotide is composed of the sugar deoxyribose, a phosphate group, and a nitrogen-containing base (A, G, C, or T). (b) RNA is single stranded. RNA nucleotides are composed of the sugar ribose, a phosphate group, and a nitrogen-containing base (A, G, C, or U). **Visualize This:** Point out the chemical difference between the sugar in DNA and the one in RNA and the difference between the nitrogenous bases thymine and uracil.

Figure 9.2 The flow of genetic information. Genetic information flows from a DNA to an RNA copy of the DNA gene, to the amino acids that are joined together to produce the protein coded for by the gene.

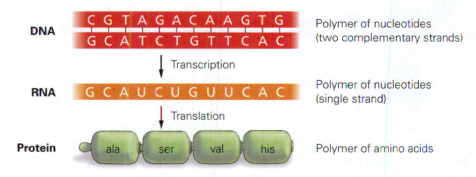

DNA

C	G	T	A	G	A	C	A	A	G	T	G
G	C	A	T	C	T	G	T	T	C	A	C

Polymer of nucleotides (two complementary strands)

↓ Transcription

RNA

G	C	A	U	C	U	G	U	U	C	A	C

Polymer of nucleotides (single strand)

↓ Translation

Protein ala ser val his

Polymer of amino acids

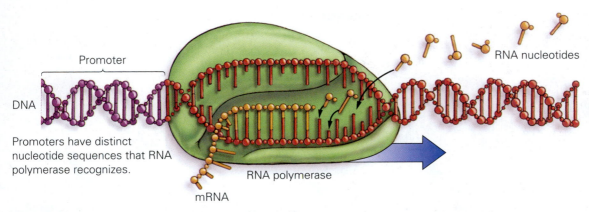

Promoter

DNA

Promoters have distinct nucleotide sequences that RNA polymerase recognizes.

RNA nucleotides

RNA polymerase

mRNA

Figure 9.3 Transcription. RNA polymerase ties together nucleotides within the growing RNA strand as they form hydrogen bonds with their complementary base on the DNA. When the RNA polymerase reaches the end of the gene, the mRNA transcript is released.
Visualize This: Propose a sequence for the DNA strand that is being used to produce the mRNA. Keep in mind that purines (A and G) are composed of two rings and thus are represented by longer pegs in this illustration. Once you have proposed a DNA sequence, determine the mRNA sequence that would be produced by transcription.

Transcription

Transcription is the copying of a DNA gene into RNA (**Figure 9.3**). The copy is synthesized by an enzyme called **RNA polymerase.** To begin transcription, the RNA polymerase binds to a nucleotide sequence at the beginning of every gene, called the **promoter.** Once the RNA polymerase has located the beginning of the gene by binding to the promoter, it then rides along the strand of the DNA helix that comprises the gene. As it is traveling along the gene, the RNA polymerase unzips the DNA double helix and ties together RNA nucleotides that are complementary to the DNA strand it is using as a template. This results in the production of a single-stranded RNA molecule that is complementary to the DNA sequence of the gene. This complementary RNA copy of the DNA gene is called **messenger RNA (mRNA)** because it carries the message of the gene that is to be expressed.

Translation

The second step in moving from gene to protein, translation, requires that the mRNA be used to produce the actual protein for which the gene encodes. For this process to occur, a cell needs mRNA, a supply of amino acids to join in the proper order, and some energy in the form of ATP. Translation also requires structures called ribosomes and transfer RNA molecules.

Ribosomes. **Ribosomes** are subcellular, globular structures (**Figure 9.4**) that are composed of another kind of RNA called **ribosomal RNA (rRNA),** which is wrapped around many different proteins. Each ribosome is composed of two subunits—one large and one small. When the large and small subunits of the ribosome come together, the mRNA can be threaded between them. In addition, the ribosome can bind to structures called **transfer RNA (tRNA)** that carry amino acids.

Transfer RNA (tRNA). Transfer RNA (**Figure 9.5**) is yet another type of RNA found in cells. An individual transfer RNA molecule carries one specific amino acid and interacts with mRNA to place the amino acid in the correct location of the growing polypeptide.

As mRNA moves through the ribosome, small sequences of nucleotides are sequentially exposed. These sequences of mRNA, called **codons,** are three nucleotides long and encode a particular amino acid. Transfer RNAs

Large subunit

Small subunit

Figure 9.4 Ribosome. Ribosomes are composed of two subunits. Each subunit in turn is composed of rRNA and protein.

also have a set of three nucleotides, which will bind to the codon if the right sequence is present. These three nucleotides at the base of the tRNA are called the **anticodon** because they complement a codon on mRNA. The anticodon on a particular tRNA binds to the complementary mRNA codon. In this way, the codon calls for the incorporation of a specific amino acid. The ribosome moves along the mRNA sequentially exposing codons for tRNA binding.

When a tRNA anticodon binds to the mRNA codon, a peptide bond is formed. The ribosome adds the amino acid that the tRNA is carrying to the growing chain of amino acids that will eventually constitute the finished protein. The transfer RNA functions as a sort of cellular translator, fluent in both the language of nucleotides (its own language) and the language of amino acids (the target language).

To help you understand protein synthesis, let us consider its similarity to an everyday activity such as baking a cake (**Figure 9.6**). To bake a cake, you would consult a recipe book (genome) for the specific recipe (gene) to make your cake (protein). You may copy the recipe (mRNA) out of the book so that the original recipe (gene) does not become stained or damaged. The original recipe (gene) is left in the book (genome) on a shelf (nucleus) so that you can make another copy when you need it. The original recipe (gene) can be copied again and again. The copy of the recipe (mRNA) is placed on the kitchen counter (ribosome) while you assemble the ingredients (amino acids). The ingredients (amino acids) for your cake (protein) include flour, sugar, butter, milk, and eggs. The ingredients are measured in measuring spoons and cups (tRNAs). (While in baking you might use the same cups and spoons for several ingredients, in protein synthesis we use tRNAs that are dedicated to one specific ingredient.) The measuring spoons and cups bring the ingredients to the kitchen counter. Like the ingredients in a cake that can be used in many ways to produce a variety of foods, amino acids can be combined in different orders to produce different proteins. The ingredients (amino acids) are always added according to the instructions specified by the original recipe (gene).

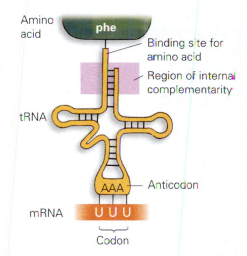

Figure 9.5 Transfer RNA (tRNA).
Transfer RNAs translate the language of nucleotides into the language of amino acids. The tRNA that binds to UUU carries only one amino acid (phe, the three letter abbreviation for phenylalanine).
Visualize This: The structure of a tRNA molecule involves regions where the RNA strand forms complementary bonds with itself, causing the RNA to fold up on itself. What nitrogenous bases might be involved in bonding in such regions of internal complementarity?

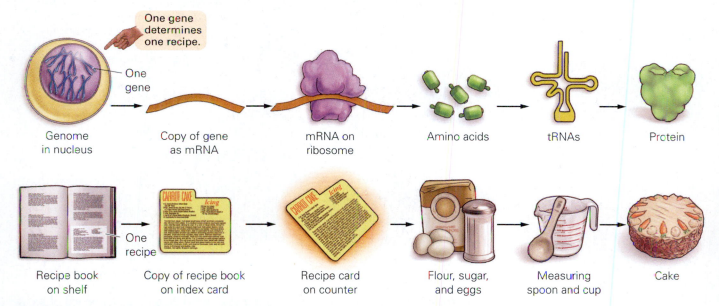

Figure 9.6 Protein synthesis and cake baking. A recipe book on a shelf resembles the genome in the nucleus of a cell. Copying one recipe onto an index card is similar to copying the gene to produce mRNA. The recipe card is placed on the kitchen counter, and the mRNA is placed on the ribosome. The ingredients for a cake are flour, sugar, butter, milk, and eggs. The ingredients to make a protein are various amino acids. The ingredients are measured in measuring spoons and cups (tRNAs) that are dedicated to one specific ingredient. The measuring spoons and cups bring the ingredients to the kitchen counter. The ingredients (amino acids) are always added according to the instructions specified by the original recipe (gene).

Figure 9.7 Translation. During translation, mRNA directs the synthesis of a protein.

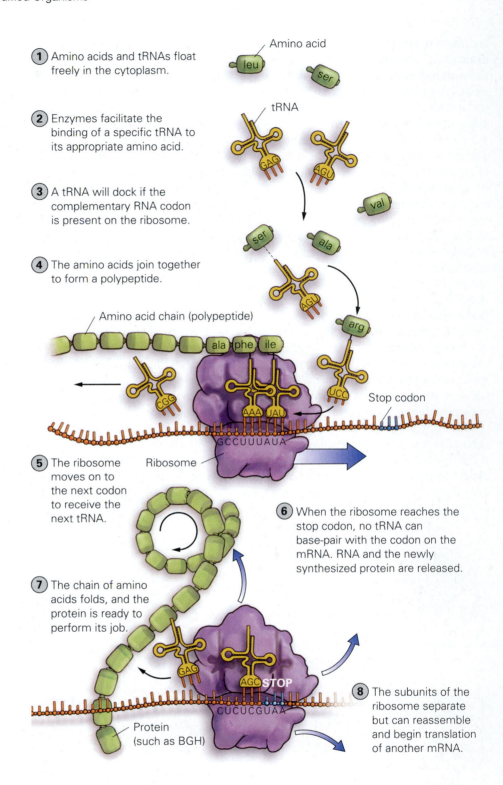

1. Amino acids and tRNAs float freely in the cytoplasm.

2. Enzymes facilitate the binding of a specific tRNA to its appropriate amino acid.

3. A tRNA will dock if the complementary RNA codon is present on the ribosome.

4. The amino acids join together to form a polypeptide.

5. The ribosome moves on to the next codon to receive the next tRNA.

6. When the ribosome reaches the stop codon, no tRNA can base-pair with the codon on the mRNA. RNA and the newly synthesized protein are released.

7. The chain of amino acids folds, and the protein is ready to perform its job.

8. The subunits of the ribosome separate but can reassemble and begin translation of another mRNA.

Amino acid

tRNA

Amino acid chain (polypeptide)

Stop codon

Ribosome

Protein (such as BGH)

Within cells, the sequence of bases in the DNA dictates the sequence of bases in the RNA, which in turn dictates the order of amino acids that will be joined together to produce a protein. Protein synthesis ends when a codon that does not code for an amino acid, called a **stop codon,** moves through the ribosome. When a stop codon is present in the ribosome, no new amino acid can be added, and the growing protein is released. Once released, the protein folds up on itself and moves to where it is required in the cell. A summary of the process of translation is shown in **Figure 9.7.**

The process of translation allows cells to join amino acids in the sequence coded by the gene. Scientists can determine the sequence of amino acids that a gene calls for by looking at the **genetic code.**

Genetic Code. The genetic code determines which mRNA codons code for which amino acids (**Table 9.1**). As Table 9.1 shows, there are 64 codons, 61 of which code for amino acids. Three of the codons are stop codons that occur near the end of an mRNA. Because stop codons do not code for an amino acid, protein synthesis ends when a stop codon enters the ribosome. In the table, you can see that the codon AUG functions both as a start codon (and thus is found near the beginning of each mRNA) and as a codon dictating that the amino acid methionine (met) be incorporated into the protein being synthesized. This initial methionine is often removed later. Notice also that the same amino acid can be coded for by more than one codon. For example, the amino acid threonine (thr) is incorporated into a protein in response to the codons ACU, ACC, ACA, and ACG. The fact that more than one codon can code for the same amino acid is referred to as *redundancy* in the genetic code. There is, however, no situation where a given codon can call for more than one amino acid. For example, AGU codes for serine (ser) and nothing else. Therefore, there is no *ambiguity* in the genetic code regarding which amino acid any codon will call for. The genetic code is also *universal* in the sense that organisms typically decode the same gene to produce the same protein. This is why genes can be moved from one organism to another, as was the case with the gene for luminescence in the mice of the chapter opener photo, which is a gene normally expressed in jellyfish.

TABLE 9.1

The genetic code.

It is possible to determine which amino acid is coded for by each mRNA codon using a chart called the genetic code. Look at the left-hand side of the chart for the first-base nucleotide in the codon; there are 4 rows, 1 for each possible RNA nucleotide—A, C, G, or U. By then looking at the intersection of the second-base columns at the top of the chart and the first-base rows, you can narrow your search for the codon to 4 different codons. Finally, the third-base nucleotide in the codon on the right-hand side of the chart determines the amino acid that a given mRNA codon codes for. Note the 3 codons UAA, UAG, and UGA that do not code for an amino acid; these are stop codons. The codon AUG is a start codon, found at the beginning of most protein-coding sequences.

First base	Second base: U	C	A	G	Third base
U	UUU UUC Phenylalanine (phe); UUA UUG Leucine (leu)	UCU UCC UCA UCG Serine (ser)	UAU UAC Tyrosine (tyr); UAA Stop codon; UAG Stop codon	UGU UGC Cysteine (cys); UGA Stop codon; UGG Tryptophan (trp)	U C A G
C	CUU CUC CUA CUG Leucine (leu)	CCU CCC CCA CCG Proline (pro)	CAU CAC Histidine (his); CAA CAG Glutamine (gln)	CGU CGC CGA CGG Arginine (arg)	U C A G
A	AUU AUC AUA Isoleucine (ile); AUG Methionine (met) Start codon	ACU ACC ACA ACG Threonine (thr)	AAU AAC Asparagine (asn); AAA AAG Lysine (lys)	AGU AGC Serine (ser); AGA AGG Arginine (arg)	U C A G
G	GUU GUC GUA GUG Valine (val)	GCU GCC GCA GCG Alanine (ala)	GAU GAC Aspartic acid (asp); GAA GAG Glutamic acid (glu)	GGU GGC GGA GGG Glycine (gly)	U C A G

Figure 9.8 Substitution mutation.
A single nucleotide change from the normal DNA sequence (a) to the mutated sequence (b) can result in the incorporation of a different amino acid. If the substituted amino acid has chemical properties different from those of the original amino acid, then the protein may assume a different shape and thus lose its ability to perform its job.

(a) Normal DNA sequence

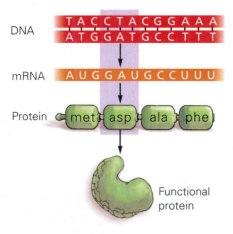

(b) Mutated DNA sequence

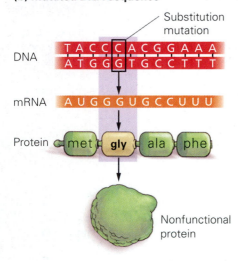

Figure 9.9 Neutral and frameshift mutations. (a) Neutral mutations result in the incorporation of the same amino acid as was originally called for. (b) The insertion (or deletion) of a nucleotide can result in a frameshift mutation.
Visualize This: Is the top or bottom strand of the DNA serving as the template for RNA synthesis in this figure?

Stop & Stretch What amino acids would be coded for by the mRNA CCU-AAU?

Mutations

Changes to the DNA sequence, called **mutations,** can affect the order or types of amino acids incorporated into a protein during translation. Mutations to a gene can result in the production of different forms, or alleles, of a gene. Different alleles result from changes in the DNA that alter the amino acid order of the encoded protein. Mutations can result in the production of either a nonfunctional protein or a protein different from the one previously required. If this protein does not have the same amino acid composition, it may not be able to perform the same job (**Figure 9.8**). For instance, a substitution of a single nucleotide results in the incorporation of a new amino acid in the hemoglobin protein and compromises the ability of cells to carry oxygen, producing sickle-cell disease.

There are also cases when a mutation has no effect on a protein. These cases may occur when changes to the DNA result in the production of a mRNA codon that codes for the same amino acid as was originally required. Due to the redundancy of the genetic code, a mutation that changes the mRNA codon from say ACU to ACC will have no impact because both of these codons code for the amino acid threonine. This is called a **neutral mutation** (**Figure 9.9a**). In addition, mutations can result in the substitution of one amino acid for another with similar chemical properties, which may have little or no effect on the protein.

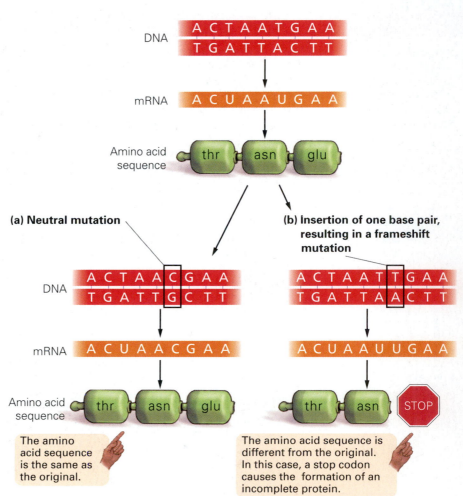

(a) Eukaryotic protein synthesis

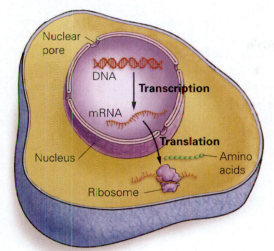

(b) Prokaryotic protein synthesis

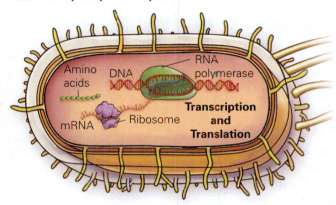

Figure 9.10 Protein synthesis in eukaryotic and prokaryotic cells. (a) In eukaryotes, transcription occurs in the nucleus and translation in the cytoplasm. (b) In prokaryotic cells, which lack nuclei, transcription and translation occur simultaneously.

Inserting or deleting a single nucleotide can have a severe impact because the addition (or deletion) of a nucleotide can change the groupings of nucleotides in every codon that follows (**Figure 9.9b**). Changing the triplet groupings is called altering the **reading frame.** All nucleotides located after an insertion or deletion will be regrouped into different codons, producing a **frameshift mutation.** For example, inserting an extra letter *H* after the fourth letter of the sentence, "The dog ate the cat," could change the reading frame to the nonsensical statement, "The dHo gat eth eca t." Inside cells, this often results in the incorporation of a stop codon and the production of a shortened, nonfunctional protein.

Stop & Stretch What effect does the formation of new alleles have on natural selection?

Cells in all organisms undergo this process of protein synthesis, with different cell types selecting different genes from which to produce proteins. **Figure 9.10a** shows the coordination of these two processes as they occur in cells with nuclei, that is, eukaryotic cells. In eukaryotic cells, transcription and translation are spatially separate, with transcription occurring in the nucleus and translation occurring in the cytoplasm. Cells lacking a membrane-bound nucleus and organelles are called prokaryotic cells. Prokaryotic cells (such as bacterial cells) also undergo protein synthesis, but transcription and translation occur at the same time and in the same location instead of occurring in separate places. As an mRNA is being transcribed, ribosomes attach and begin translating (**Figure 9.10b**).

An Overview: Gene Expression

Different cell types transcribe and translate different genes. Each cell in your body, except sperm or egg cells, has the same complement of genes you inherited from your parents but expresses only a small percentage of those genes. For example, because your muscles and nerves each perform a specialized suite of jobs, muscle cells turn on or express one suite of genes and nerve cells another (**Figure 9.11**). Turning a gene on or off, or modulating it more subtly, is called **regulating gene expression.** The expression of a given gene is regulated so that it is turned on and turned off in response to the cell's needs. *See the next section for **A Closer Look** at regulating gene expression.*

(a) Muscle cells

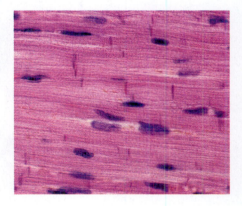

(b) Nerve cells

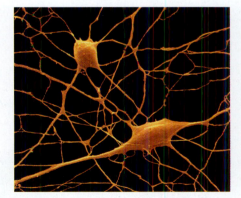

Figure 9.11 Gene expression differs from cell to cell. (a) A muscle cell performs different functions and expresses different genes than (b) a nerve cell. Both of these cells have the same suite of genes in their nucleus but express different subsets of genes.

A Closer Look:
Regulating Gene Expression

Gene expression can be regulated through the rate of transcription or translation. It is also possible to regulate how long an mRNA or protein remains functional in a cell.

Regulation of Transcription

Gene expression is most commonly regulated by controlling the rate of transcription. Regulation of transcription can occur at the promoter, the sequence of nucleotides adjacent to a gene to which the RNA polymerase binds to initiate transcription. When a cell requires a particular protein, the RNA polymerase enzyme binds to the promoter for that particular gene and transcribes the gene.

Prokaryotic and eukaryotic cells both regulate gene expression by regulating transcription but have different strategies for doing so. Prokaryotic cells typically regulate gene expression by blocking transcription via proteins called **repressors** that bind to the promoter and prevent the RNA polymerase from binding. When the gene needs to be expressed, the repressor will be released from the promoter so that the RNA polymerase can bind (**Figure 9.12a**). This is the main mechanism by which simple single-celled prokaryotes regulate gene expression.

The more complex eukaryotic cells have evolved more complex mechanisms to control gene expression. To control transcription, eukaryotic cells more commonly enhance gene expression using proteins called **activators** that help the RNA polymerase bind to the promoter, thus facilitating gene expression (**Figure 9.12b**). The rate at which the polymerase binds to the promoter is also affected by substances that are present in the cell. For example, the presence of alcohol in a liver cell might result in increased transcription of a gene involved in the breakdown of alcohol.

(a) Repression of transcription

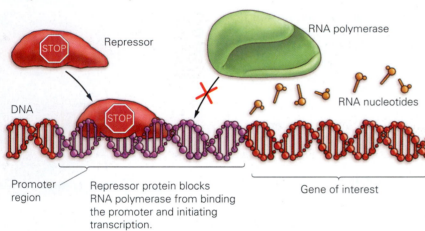

Repressor

RNA polymerase

DNA

RNA nucleotides

Promoter region

Repressor protein blocks RNA polymerase from binding the promoter and initiating transcription.

Gene of interest

(b) Activation of transcription

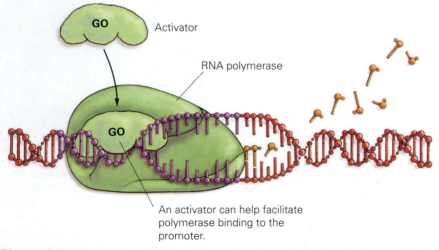

Activator

RNA polymerase

An activator can help facilitate polymerase binding to the promoter.

Figure 9.12 Regulation of gene expression. Gene expression can be regulated by (a) repression or (b) activation. Prokaryotes more typically use repression and eukaryotes activation of transcription to regulate gene expression.

Stop & Stretch How would a mutation in a repressor that makes it nonfunctional likely affect the individual with this mutation?

Regulation by Chromosome Condensation

It is also possible to regulate gene expression by condensing all or part of a chromosome. This prevents RNA polymerase from being able to access genes. The inactivation of an X chromosome turns off the expression of X-linked genes in organisms that have two X chromosomes (Chapter 8). Entire chromosomes are also inactivated when they condense during mitosis.

Regulation by mRNA Degradation

Eukaryotic cells can also regulate the expression of a gene by regulating how long a messenger RNA is present in the cytoplasm. Enzymes called *nucleases* roam the cytoplasm, cutting RNA molecules by binding to one end and breaking the bonds between nucleotides. If a particular mRNA has a long adenosine nucleotide "tail," it will survive longer in the cytoplasm and be translated more times. All mRNAs are eventually degraded in this manner; otherwise, once a gene had been transcribed one time, it would be expressed forever.

Regulation of Translation

It is also possible to regulate many of the steps of translation. For example, the binding of the mRNA to the ribosome can be slowed or hastened, as can the movement of the mRNA through the ribosome.

Regulation of Protein Degradation

Once a protein is synthesized, it will persist in the cell for a characteristic amount of time. Like the mRNA that provided the instructions for its synthesis, the life of a protein can be affected by cellular enzymes called *proteases* that degrade the protein. Speeding up or slowing down the activities of these enzymes can change the amount of time that a protein is able to be active inside a cell.

Genetic engineering permits precise control of gene expression in many different circumstances. For example, the problem of regulating gene expression is easily solved in the case of rBGH. Farmers can simply decide how much protein to inject into the bloodstream of a cow. However, they must first synthesize the protein.

9.2 Producing Recombinant Proteins

The first step in the production of the rBGH protein is to transfer the BGH gene from the nucleus of a cow cell into a bacterial cell. Bacteria are single-celled prokaryotes that copy themselves rapidly. They can thrive in the laboratory if they are allowed to grow in a liquid broth containing the nutrients necessary for survival. Bacteria with the BGH gene can serve as factories to produce millions of copies of this gene and its protein product. Making many copies of a gene is called **cloning** the gene.

Cloning a Gene Using Bacteria

The following three steps are involved in moving a BGH gene into a bacterial cell (**Figure 9.13** on the next page).

Step 1. **Remove the Gene from the Cow Chromosome.** The gene is sliced out of the cow chromosome on which it resides by exposing the cow DNA to enzymes that cut DNA. These enzymes, called **restriction enzymes,** act like highly specific molecular scissors. Most restriction enzymes cut DNA only at specific sequences, called *palindromes,* such as

Note that the bottom middle sequence is the reverse of the top middle sequence. Many restriction enzymes cut the DNA in a staggered pattern, leaving "sticky ends," such as

The unpaired bases form bonds with any complementary bases with which they come in contact. The enzyme selected by the scientist cuts on both ends of the *BGH* gene but not inside the gene.

| TATC | GTACG | BGH GENE | C | GTACG | AAC |
| ATAG | TACGC | BGH GENE | G | CATGC | TTG |

A particular restriction enzyme cuts DNA at a specific sequence. Therefore, scientists need some information about the entire suite of genes present in a particular organism, called the **genome,** to determine which restriction enzyme cutting sites surround the gene of interest. Cutting the DNA generates many different fragments, only one of which will carry the gene of interest.

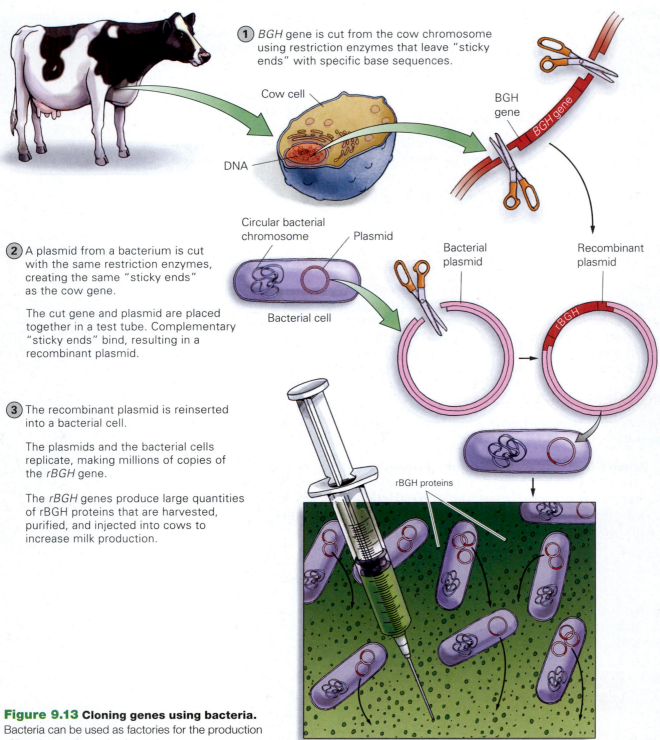

① *BGH* gene is cut from the cow chromosome using restriction enzymes that leave "sticky ends" with specific base sequences.

Cow cell

DNA

BGH gene

BGH gene

Recombinant plasmid

② A plasmid from a bacterium is cut with the same restriction enzymes, creating the same "sticky ends" as the cow gene.

The cut gene and plasmid are placed together in a test tube. Complementary "sticky ends" bind, resulting in a recombinant plasmid.

Circular bacterial chromosome

Plasmid

Bacterial cell

Bacterial plasmid

rBGH

③ The recombinant plasmid is reinserted into a bacterial cell.

The plasmids and the bacterial cells replicate, making millions of copies of the *rBGH* gene.

The *rBGH* genes produce large quantities of rBGH proteins that are harvested, purified, and injected into cows to increase milk production.

rBGH proteins

Figure 9.13 Cloning genes using bacteria. Bacteria can be used as factories for the production of human or other animal proteins.

Step 2. **Insert the BGH Gene into the Bacterial Plasmid.** Once the gene is removed from the cow genome, it is inserted into a bacterial structure called a plasmid. A plasmid is a circular piece of DNA that normally exists separate from the bacterial chromosome and can replicate independently of the bacterial chromosome. Think of the plasmid as a ferry that carries the gene into the bacterial cell, where it can be replicated. To incorporate the BGH gene into the plasmid, the plasmid is also cut with the same restriction enzyme used to cut the gene. Cutting both the plasmid and gene with the same enzyme allows the sticky ends that are generated to base-pair with each other

(A to T and G to C). When the cut plasmid and the cut gene are placed together in a test tube, they re-form into a circular plasmid with the extra gene incorporated.

The bacterial plasmid has now been genetically engineered to carry a cow gene. At this juncture, the BGH gene is referred to as the rBGH gene, with the r indicating that this product is genetically engineered, or recombinant, because it has been removed from its original location in the cow genome and recombined with the plasmid DNA.

Step 3. Insert the Recombinant Plasmid into a Bacterial Cell. The recombinant plasmid is now inserted into a bacterial cell. Bacteria can be treated so that their cell membranes become porous. When they are placed into a suspension of plasmids, the bacterial cells allow the plasmids back into the cytoplasm of the cell. Once inside the cell, the plasmids replicate themselves, as does the bacterial cell, making thousands of copies of the *rBGH* gene. Using this procedure, scientists can grow large amounts of bacteria capable of producing BGH.

Once scientists successfully clone the BGH gene into bacterial cells, the bacteria produce the protein encoded by the gene. Then the scientists are able to break open the bacterial cells, isolate the BGH protein, and inject it into cows. Bacteria can be genetically engineered to produce many proteins of importance to humans. For example, bacteria are now used to produce the clotting protein missing from people with hemophilia, as well as human insulin for people with diabetes.

Close to one-third of all dairy cows in the United States now undergo daily injections with rBGH. These injections increase the volume of milk that each cow produces by around 20%. Prior to marketing the recombinant protein to dairy farmers, scientists had to demonstrate that its product would not be harmful to cows or to humans who consume the cows' milk. This involved obtaining approval from the U.S. Food and Drug Administration (FDA).

The FDA is the governmental organization charged with ensuring the safety of all domestic and imported foods and food ingredients (except for meat and poultry, which are regulated by the U.S. Department of Agriculture). The manufacturer of any new food that is not **generally recognized as safe (GRAS)** must obtain FDA approval before marketing its product. Adding substances to foods also requires FDA approval unless the additive is GRAS.

According to the FDA, there is no detectable difference between milk from treated and untreated cows and no way to distinguish between the two. In 1993, the FDA deemed the milk from rBGH-treated cows as safe for human consumption.

Stop & Stretch Not all anti-GM organism activists are convinced that milk produced by rBGH-treated cows is identical to that from untreated cows. Given the effects of rBGH, how might the milk produced by these cows be different, even if it is safe?

In addition, because the milk from treated and untreated cows is indistinguishable, the FDA does not require that milk obtained from rBGH-treated cows be labeled in any manner. However, many distributors of milk from untreated cows label their milk as "hormone free," even though there is no evidence of the hormone in milk from treated cows (**Figure 9.14**).

But what about the welfare of the cows? There is some evidence that cows treated with rBGH are more susceptible to certain infections than untreated cows. Due in part to concerns about the health of these animals, Europe and Canada have banned rBGH use.

The rBGH story is a little different from that of genetically modified crop foods because rBGH protein is produced by bacteria and then administered to cows. When foods are genetically modified, the genome of the food itself is altered.

Figure 9.14 Hormone-free milk. Some manufacturers label their products as hormone free.

Figure 9.15 Artificial selection in corn. Selective-breeding techniques resulted in the production of modern corn (right) from ancient teosinite corn.

Figure 9.16 Golden rice. Golden rice has been genetically engineered to produce beta-carotene, which causes the rice to look more gold in color than the unmodified rice. This picture shows a mixture of modified and unmodified rice.

9.3 Genetically Modified Foods

Whether you realize it or not, you have probably been eating genetically modified foods for your entire life. Some genetic modifications involve moving genes between organisms in labs. Other modifications have occurred over the last several thousand years due to farmers' use of selective breeding techniques—breeding those cattle that produce the most milk or crossing crop plants that are easiest to harvest (**Figure 9.15**). While this artificial selection does not involve moving a gene from one organism to another, it does change the overall frequency of certain alleles for a gene in the population.

Unless you eat only certified organic foods, you have been eating food that has been modified. This may lead you to wonder why and how plants are genetically modified, what impact eating them has on your health, and whether growing them affects the environment.

Why Genetically Modify Crop Plants?

Crop plants are genetically modified to increase their shelf life, yield, and nutritive value. For example, tomatoes have been engineered to soften and ripen more slowly. The longer ripening time means that tomatoes stay on the vine longer, thus making them taste better. The slower ripening also increases the amount of time that tomatoes can be left on grocery store shelves without becoming overripe and mushy.

Genetic engineering techniques increase crop yield when plants are manipulated to be resistant to pesticides, herbicides, drought, and freezing. The nutritive value of crops can also be increased through genetic engineering. A gene that regulates the synthesis of beta-carotene has been inserted into rice, a staple food for many of the world's people. Scientists hope that the engineered rice will help decrease the number of people in underdeveloped nations who become blind due to a deficiency of beta-carotene, which is necessary for the synthesis of vitamin A. Eating this genetically modified rice, called *golden rice*, increases a person's ability to synthesize the vitamin (**Figure 9.16**). However, golden rice has not yet been approved for human consumption.

Modifying Crop Plants with the Ti Plasmid and Gene Gun

To modify crop plants, the gene must be able to gain access to the plant cell, which means it must be able to move through the plant's rigid outer cell wall. One "ferry" for moving genes into flowering plants is a naturally occurring plasmid of the bacterium *Agrobacterium tumefaciens*. In nature, this bacterium infects plants and causes tumors called **galls** (**Figure 9.17a**). The tumors are induced by a plasmid, called **Ti plasmid** (Ti for tumor inducing).

(a) Gall caused by *A. tumefaciens*

(b) Using the Ti plasmid

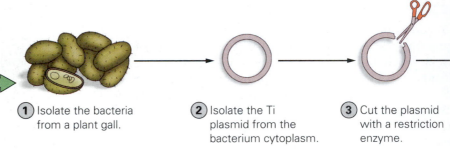

1 Isolate the bacteria from a plant gall.

2 Isolate the Ti plasmid from the bacterium cytoplasm.

3 Cut the plasmid with a restriction enzyme.

Figure 9.17 Genetically modifying plants using the Ti plasmid. (a) Plants infected by *Agrobacterium tumefaciens* in nature show evidence of the infection by producing tumorous galls. (b) The Ti plasmid from *A. tumefaciens* serves as a shuttle for incorporating genes into plant cells. The recombinant plasmid is then used to infect developing plant cells, producing a genetically modified plant. When the plant cell reproduces, it may pass on the engineered gene to its offspring.

Genes from different organisms can be inserted into the Ti plasmid by using the same restriction enzyme to cut the Ti plasmid and the gene, resulting in identical sticky ends, and then connecting the gene and plasmid. The recombinant plasmid can then be inserted into a bacterium. The *A. tumefaciens* bacterium with the recombinant Ti plasmid is then used to infect plant cells. During infection, the recombinant plasmid is transferred into the host-plant cell (**Figure 9.17b**). For genetic engineering purposes, scientists use only the portion of a plasmid that does not cause tumor formation.

Moving genes into other agricultural crops such as corn, barley, and rice can also be accomplished by using a device called a **gene gun**. A gene gun shoots metal-coated pellets covered with foreign DNA into plant cells (**Figure 9.18**). A small percentage of these genes may be incorporated into the plant's genome. When a gene from one species is incorporated into the genome of another species, a **transgenic organism** is produced. A transgenic organism is more commonly referred to as a **genetically modified organism (GMO).**

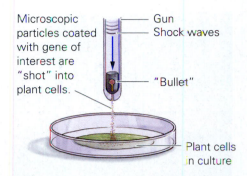

Figure 9.18 Genetically modifying plants using a gene gun. A gene gun shoots a plastic bullet loaded with tiny DNA-coated pellets into a plant cell. The bullet shells are prevented from leaving the gun, but the DNA-covered pellets penetrate the cell wall, cell membrane, and nuclear membrane of some cells.

Effect of GMOs on Health

Concerns about the potential negative health effects of consuming GM crops have led some citizens to fight for legislation requiring that genetically modified foods be labeled so that consumers can make informed decisions about the foods they eat. Manufacturers of GM crops argue that labeling foods is expensive and that the labels will be viewed by consumers as a warning, even in the absence of any proven risk. Those manufacturers believe that GM food labeling will decrease sales and curtail further innovation.

As the labeling controversy continues, most of us are already eating GM foods. Scientists estimate that over half of all foods in U.S. markets, including virtually all processed foods, contain at least small amounts of GM foods. Products that do not contain GMOs are often labeled to promote that fact. Concern about GM foods is not limited to their consumption. Many people are also concerned about the effects of GM crop plants on the environment.

GM Crops and the Environment

Many genetically modified crops have been engineered to increase their yield. For centuries, farmers have tried to increase yields by killing the pests that damage crops and by controlling the growth of weeds that compete for nutrients, rain, and sunlight. In the United States, farmers typically spray high volumes of chemical pesticides and herbicides directly onto their fields

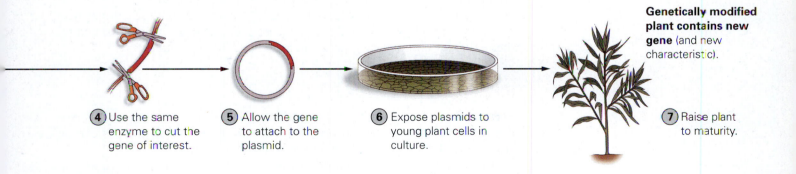

Genetically modified plant contains new gene (and new characteristic).

4 Use the same enzyme to cut the gene of interest.

5 Allow the gene to attach to the plasmid.

6 Expose plasmids to young plant cells in culture.

7 Raise plant to maturity.

Figure 9.19 Application of chemicals. It remains to be seen how genetic technologies might affect the use of herbicides and pesticides.

(**Figure 9.19**). This practice concerns people worried about the health effects of eating foods that have been treated by these often toxic or cancer-causing chemicals. In addition, both pesticides and herbicides may leach through the soil and contaminate drinking water.

To help decrease farmers' reliance on pesticides, agribusiness companies have engineered plants that are genetically resistant to pests. Unfortunately, this is a short-term solution because the pest populations will eventually evolve resistance. Application of a pesticide does not always kill all of the targeted organisms. The few pests that have preexisting resistance genes and are not susceptible survive and produce resistant offspring. Eventually, widespread resistance develops, and a new pesticide must be developed and applied.

The continued need for the development of new pesticides in farming is paralleled by farmers' reliance on herbicides. Roundup®, or glyophosphate, is a herbicide used to control weeds in soybean fields. Glyophosphate inhibits the synthesis of certain amino acids required by growing plants. Herbicide-resistant crop plants, such as Roundup Ready® soybeans, have been engineered to be resistant to Roundup. Roundup Ready plants are engineered to synthesize more of these amino acids than unengineered plants and thus can grow in the presence of the herbicide while nearby plants and weeds will die. Therefore, farmers can spray their fields of genetically engineered soybeans with herbicides that will kill everything but the crop plant. Some people worry that this resistance gene will allow farmers to spray more herbicide on their crops because there is no chance of killing the GM plant, thereby exposing consumers and the environment to even more herbicide.

There is also concern that GM crop plants may transfer engineered genes from modified crop plants to their wild or weedy relatives. Wind, rain, birds, and bees carry genetically modified pollen to related plants near fields containing GM crops (or even to farms where no GM crops are being grown). Many cultivated crops have retained the ability to interbreed with their wild relatives; in these cases, genes from farm crops can mix with genes from the wild crops.

While there is hope that genetic engineers will be able to help solve hunger problems by making farming more productive, there are also concerns about possible negative health and environmental effects of GM foods. It remains to be seen whether genetic engineering will constitute a lasting improvement to agriculture.

9.4 Genetically Modified Humans

Humans can also be genetically modified. Scientists may someday be able to cure or alleviate the symptoms of diseases by manipulating genes. It may also be possible for physicians to diagnose genetic defects in early embryos and fix them, thereby eliminating genetic disease in children and adults.

Stem Cells

Stem cells are unspecialized or **undifferentiated** precursor cells that have not yet been programmed to perform a specific function. Stem cells can be embryonic in origin or can be found in some adult tissues.

Imagine that you are remodeling an old home, and you have a type of material that you can mold into anything you might need for the remodeling job—brick, tile, pipe, plaster, and so forth. Having a supply of this material would help you fix many different kinds of damage. Scientists believe that stem cells may serve as this type of all-purpose repair material in the body. If cells are nudged in a particular developmental direction in the laboratory, they can be directed to become a particular tissue or organ. Using stem cells to produce healthy tissues as replacements for damaged tissues is a type of **therapeutic cloning.** Tissues and organs grown from stem cells in the laboratory may someday be used to replace organs damaged in accidents or that are gradually failing due to **degenerative diseases.** Degenerative diseases start with the slow breakdown of an organ and progress to organ failure. In addition, when one organ is not working properly, other organs are affected. Degenerative diseases include diabetes, liver and lung diseases, heart disease, multiple sclerosis, and Alzheimer's and Parkinson's diseases.

Stem cells could provide healthy tissue to replace those tissues damaged by spinal cord injury or burns. New heart muscle could be produced to replace muscle damaged during a heart attack. A diabetic could have a new pancreas, and people suffering from some types of arthritis could have replacement cartilage to cushion their joints. Thousands of people waiting for organ transplants might be saved if new organs were grown in the lab.

Stem cells can be isolated from early embryos that are left over after fertility treatments. *In vitro* (Latin, meaning "in glass") fertilization procedures often result in the production of excess embryos because many egg cells are harvested from a woman who wishes to become pregnant. These egg cells are then mixed with her partner's sperm in a petri dish, resulting in the production of many fertilized eggs that grow into embryos. A few of the embryos are then implanted into the woman's uterus. The remaining embryos are stored so that more attempts can be made if pregnancy does not result or if the couple desires more children. When the couple achieves the desired number of pregnancies, the remaining embryos are discarded or, with the couple's consent, used for stem cell research. The crux of the controversy surrounding stem cells involves the idea of some that early embryos constitute life, and that using them for research is unethical. Others argue that early embryos will not become life unless they are implanted in a woman's uterus.

Early embryonic cells are harvested because stem cells are **totipotent**, meaning they can become any other cell type immediately after fertilization (**Figure 9.20**). As the embryo develops, its cells become less and less able to produce other cell types. As a human embryo grows, the early cells start dividing and forming different, specialized cells such as heart cells, bone cells, and muscle cells. Once formed, specialized non-stem cells can divide only to produce replicas of themselves. They cannot backtrack and become a different type of cell.

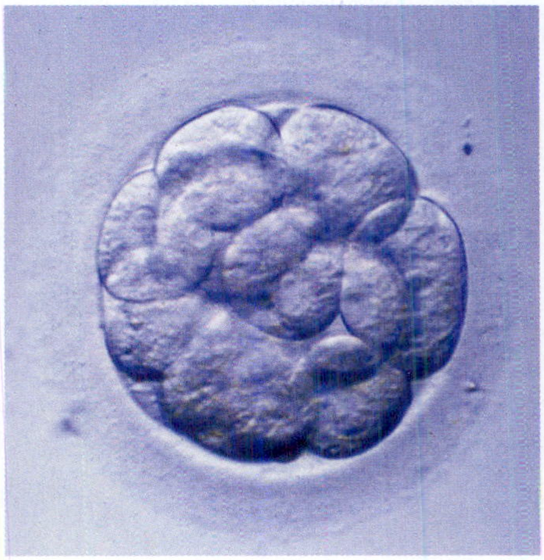

Figure 9.20 Human embryo. Stem cells can be obtained from an early embryo such as the one in this petri dish.

Stop & Stretch For parents known to carry a harmful genetic mutation, fertility clinics will take a cell from an 8-cell preembryo to test it for the presence of the defective gene. Taking a single cell causes no apparent harm to the developing embryo, and the remaining cells are allowed to continue undergoing development. Once a preembryo is found that does not carry the mutation, it can be implanted into the female parent's uterus. What does the fact that you can remove a cell from an embryo without harming subsequent development tell you about the capabilities of preembryonic cells?

There are also stem cells present in adult tissues, probably so that these tissues can repair themselves. Scientists have found adult stem cells in many different tissues, and work is under way to determine whether these cells can be used for transplants and treatment of degenerative diseases.

Genetic modifications may one day include replacing defective or non-functional alleles of a gene with a functional copy of the gene. This requires detailed information about the human genome, much of which has been obtained.

The Human Genome Project

The **Human Genome Project** involves determining the nucleotide-base sequence (A, C, G, or T) of the entire human genome and the location of each of the 20,000 to 25,000 human genes.

The scientists involved in this multinational effort also sequenced the genomes of the mouse, the fruit fly, the roundworm, baker's yeast, and a common intestinal bacterium named *Escherichia coli* (*E. coli*). Scientists thought it was important to sequence the genomes of organisms other than humans because these **model organisms** are easy to manipulate in genetic studies and because important genes are often found in many different organisms. In fact, 90% of human genes are also present in mice; 50% are in fruit flies, and 31% are in baker's yeast. Therefore, understanding how a certain gene functions in a model organism helps us understand how the same gene functions in humans.

To sequence the human genome, scientists first isolated DNA from white blood cells. They then cleaved the chromosomes into more manageable sizes using restriction enzymes, cloned them into plasmids, and determined the base sequence using automated DNA sequencers. These sequencing machines distinguish between nucleotides based on structural differences in the nitrogenous bases. Sequence information was then uploaded to the Internet. Scientists working on this, or any other project, could search for regions of sequence information that overlapped with known sequences, a process called *chromosome walking* (**Figure 9.21**). Using overlapping regions, scientists in laboratories all over the world worked together to patch together DNA sequence information. DNA sequence information obtained by means of the Human Genome Project may someday enable medical doctors to take blood samples from patients and determine which genetic diseases are likely to affect them.

Many people worry about having these types of tests performed because of the potential for insurance companies or employers to obtain the personal information and use it against them. Yet there is a positive side to having genetic information available. Once the genetic basis of a disease has been worked out—that is, how the gene of a healthy person differs from the gene of a person with a genetic disease—the information can be used to develop treatments or cures.

Figure 9.21 Chromosome walking. Scientists from many different labs worked together to sequence the human genome. Sequence information is uploaded to a common database, and scientists search for overlapping regions to fill in gaps in the sequence, much like assembling a jigsaw puzzle.

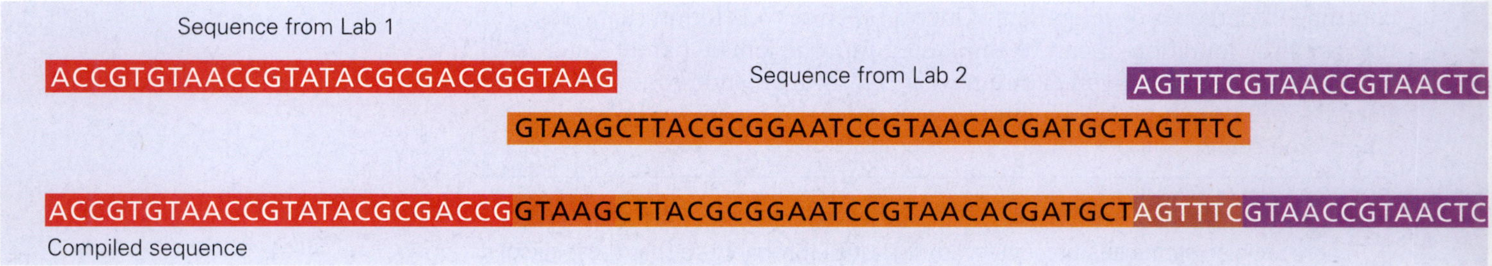

Sequence from Lab 1

ACCGTGTAACCGTATACGCGACCGGTAAG

Sequence from Lab 2

AGTTTCGTAACCGTAACTC

GTAAGCTTACGCGGAATCCGTAACACGATGCTAGTTTC

ACCGTGTAACCGTATACGCGACCGGTAAGCTTACGCGGAATCCGTAACACGATGCTAGTTTCGTAACCGTAACTC

Compiled sequence

Gene Therapy

Scientists who try to replace defective human genes (or their protein products) with functional genes are performing **gene therapy.** Gene therapy may someday enable scientists to fix genetic diseases in an embryo. To do so, the scientists would supply the embryo with a normal version of a defective gene; this so-called **germ-line gene therapy** would ensure that the embryo and any cells produced by cell division would replicate the new, functional version of the gene. Thus, most of the cells would have the corrected version of the gene, and when these genetically modified individuals had children, they would pass on the corrected version of the gene.

Another type of gene therapy, called **somatic cell gene therapy,** can be performed on body cells to fix or replace the defective protein in only the affected cells. Using this method, scientists introduce a functional version of a defective gene into an affected individual cell in the laboratory, allow the cell to reproduce, and then place the copies of the cell bearing the corrected gene into the diseased person.

This treatment may seem like science fiction, but it is likely that this method of treating genetic diseases will be considered a normal procedure in the not-too-distant future. In fact, genetic engineers already have successfully treated a genetic disorder called **severe combined immunodeficiency (SCID),** a disease caused by a genetic mutation that results in the absence of an important enzyme and severely weakens the individual's immune system. Because their immune systems are compromised, people with SCID are incapable of fighting off any infection, and they often suffer severe brain damage from the high temperatures associated with unabated infection. Any exposure to infection can kill or disable someone with SCID, so most patients must stay inside their homes and often live inside protective bubbles that separate them from everyone, even family members.

To devise a successful treatment for SCID, or any disease treated with gene therapy, scientists had to overcome a major obstacle—getting the therapeutic gene to the right place.

Proteins break down easily and are difficult to deliver to the proper cells, so it is more effective to replace a defective gene than to continually replace a defective protein. Delivering a normal copy of a defective gene only to the cell type that requires it is a difficult task. SCID, a disorder that has been treated successfully, was chosen by early gene therapists in part because defective immune system cells could be removed from the body, treated, and returned to the body.

Immune system cells that require the enzyme missing in SCID patients circulate in the bloodstream. Blood removed from a child with SCID is infected with nonpathogenic (non-disease-causing) versions of a virus. This virus is first engineered to carry a normal copy of the defective gene in SCID patients. After the immune system cells are infected with the virus, these recombinant

Sequence from Lab 3

`TACTACCGTTACGGATATGCTTACTGTAC`

Sequence from Lab 4

`TATGCTTACTGTACCTTCAAAACTGACTTTGGCTAACCGTACTCTGTACCTTCAAAACTGACTTTT`

`TACTACCGTTACGGATATGCTTACTGTACCTTCAAAACTGACTTTGGCTAACCGTACTCTGTACCTTCAAAACTGACTTTT`

(a) Gene therapy for SCID patients

(b) SCID survivor

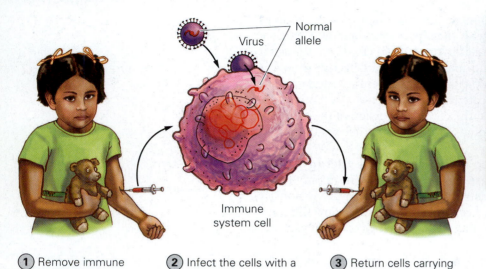

Virus · Normal allele

Immune system cell

1 Remove immune system cells from patient.

2 Infect the cells with a virus carrying the normal allele.

3 Return cells carrying the normal allele.

Figure 9.22 Gene therapy in an SCID patient. (a) A virus carrying the normal gene is allowed to infect immune system cells that have been removed from a person with SCID. The virus inserts the normal copy of the gene into some of the cells, and these cells are then injected into the SCID patient. (b) Ashi DiSilva, the first gene therapy patient.

cells, which now bear copies of the functional gene, are returned to the SCID patient (**Figure 9.22a**).

In 1990, a 4-year-old girl named Ashi DiSilva (**Figure 9.22b**) was the first patient to receive gene therapy for SCID. Ashi's parents were willing to face the unknown risks to their daughter because they were already far too familiar with the risks of SCID—the couple's two other children also had SCID and were severely disabled. Ashi is now a healthy adult with an immune system that is able to fight off most infections.

However, Ashi must continue to receive treatments because blood cells, whether genetically engineered or not, have limited life spans. When most of Ashi's engineered blood cells have broken down, she must be treated again; thus, she undergoes this gene therapy a few times each year. Since Ashi's gene therapy turned out well, many other SCID patients have been successfully treated and can live normal lives. Unfortunately, Ashi's gene therapy does not prevent her from passing on the defective allele to her biological children because this therapy is not "fixing" the allele in cells of her ovaries.

Although things worked out well for Ashi, successful gene therapy is far from routine. Two of 11 French boys treated with gene therapy for SCID developed leukemia that is thought to be related to their treatment, and an American teenager, Jesse Gelsinger, died from complications of experimental gene therapy meant to cure his relatively mild genetic disorder.

In addition to the risks involved in conducting any experimental therapy, not many genetic diseases can be treated with gene therapy. Gene therapy to date has focused on diseases caused by single genes for which defective cells can be removed from the body, treated, and reintroduced to the body. Most genetic diseases are caused by many genes, affect cells that cannot be removed and replaced, and are influenced by the environment.

Most people support the research of genetic engineers in their attempts to find better methods for delivering gene sequences to the required locations and for regulating the genes once they are in place. A far more controversial type of genetic engineering involves making an exact copy of an entire organism by a process called reproductive **cloning.**

Cloning Humans

Human cloning occurs commonly in nature via the spontaneous production of identical twins. These clones arise when an embryo subdivides itself into 2 separate embryos early in development. This is not the type of cloning that many people find objectionable; people are more likely to be upset by cloning that involves selecting which traits an individual will possess.

Natural cloning of an early embryo to make identical twins does not allow any more selection for specific traits than does fertilization. However, in the future it may be possible to select adult humans who possess desired traits and clone them. Because cloning does not actually alter an individual's genes, it is more of a reproductive technology than a genetic engineering technology. However, it may someday be possible to alter the genes of a cloned embryo.

Cloning offspring from adults with desirable traits has been successfully performed on cattle, goats, mice, cats, pigs, rabbits, and sheep. In fact, the animal that brought cloning to the attention of the public was a ewe named Dolly (**Figure 9.23**).

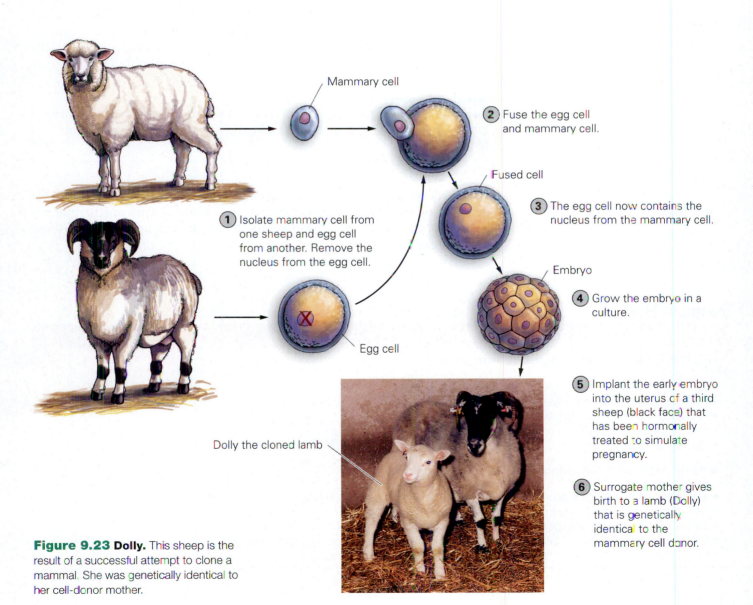

Mammary cell

2 Fuse the egg cell and mammary cell.

Fused cell

1 Isolate mammary cell from one sheep and egg cell from another. Remove the nucleus from the egg cell.

3 The egg cell now contains the nucleus from the mammary cell.

Egg cell

Embryo

4 Grow the embryo in a culture.

5 Implant the early embryo into the uterus of a third sheep (black face) that has been hormonally treated to simulate pregnancy.

Dolly the cloned lamb

6 Surrogate mother gives birth to a lamb (Dolly) that is genetically identical to the mammary cell donor.

Figure 9.23 Dolly. This sheep is the result of a successful attempt to clone a mammal. She was genetically identical to her cell-donor mother.

Dolly was cloned when Scottish scientists took the nucleus from a mammary gland cell of an adult female sheep and fused it with an egg cell that had previously had its nucleus removed. Treated egg cells were then placed in the uterus of an adult ewe that had been hormonally treated to support a pregnancy. Scientists had to try many times before this **nuclear transfer** technique worked. In all, 277 embryos were constructed before one was able to develop into a live lamb. Dolly was born in 1997.

The research that led to Dolly's birth was designed to provide a method of ensuring that cloned livestock would have the genetic traits that made them most beneficial to farmers. Sheep that produced the most high-quality wool and cattle that produced the best beef would be cloned. This technique is more efficient than allowing two prize animals to breed because each animal gives only half of its genes to the offspring. There is no guarantee that the offspring of 2 prize animals will have the desired traits. Even when a genetic clone is produced, there is no guarantee that the clone produced will be identical in the appearance and behavior to the original because of environmental differences.

No one knows if nuclear transfer will work in humans—or if cloning is safe. If Dolly is a representative example, cloning animals may not be safe. Dolly was euthanized at age six to relieve her from the discomfort of arthritis and a progressive lung disease, conditions usually found only in older sheep. The fact that Dolly developed these conditions has led scientists to hypothesize that she had aged prematurely.

Stop & Stretch Devise a hypothesis about why Dolly may have aged more rapidly than a sheep that was produced naturally. How could you test this hypothesis?

The debate about human cloning mimics the larger debate about genetic engineering (**Table 9.2**). As a society, we need to determine whether the potential for good outweighs the potential harm for each application of these technologies.

TABLE 9.2

Some pros and cons of genetic engineering.

Why the work of genetic engineers is important

- GM crops may make farms more productive.
- GM crops may be made to taste better.
- GM crops can be made to have longer shelf lives.
- GM crops can be made to contain more nutrients.
- Genetic engineers hope to cure diseases and save lives.

Why the work of genetic engineers is controversial

- GM crops encourage agribusiness, which may close down some small farms.
- GM animals and crops may cause health problems in consumers.
- GM crops might have unexpected adverse effects on the environment.
- Lack of genetic diversity of GM crops could lead to destruction of food supply worldwide by pest or environmental change.
- Present research might lead to the unethical genetic modification of humans.

SAVVY READER

Political Cartoons

Both of the political cartoons shown are about stem cells.

1. Can we trust the scientific accuracy of cartoons?

2. Is the question of whether an embryo constitutes life one that science can answer?

3. Do you think an early embryo should have the same rights as a fetus or a baby?

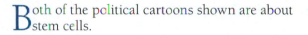

Chapter Review

Learning Outcomes

Mastering BIOLOGY®

Go to the Study Area at www.masteringbiology.com for practice quizzes, myeBook, BioFlix™ 3-D animations, MP3Tutor sessions, videos, current events, and more.

LO1 **Define the term *gene expression* (Section 9.1).**

- Genes carry instructions for synthesizing proteins. Gene expression involves turning a gene on that includes the processes of transcription and translation (pp. 196–197).

LO2 **Describe how messenger RNA (mRNA) is synthesized during the process of transcription (Section 9.1).**

- Transcription occurs in the nucleus of eukaryotic cells when an RNA polymerase enzyme binds to the promoter, located at the start site of a gene, and makes an mRNA that is complementary to the DNA gene (pp. 197–198).

LO3 **Describe how proteins are synthesized during the process of translation (Section 9.1).**

- Translation occurs in the cytoplasm of eukaryotic cells and involves mRNA, ribosomes, and transfer RNA (tRNA). Messenger RNA carries the code from the DNA, and ribosomes are the site where amino acids are assembled to synthesize proteins. Transfer RNA carries amino acids, which bind to triplet nucleotide sequences on the mRNA called codons. A particular tRNA carries a specific amino acid. Each tRNA has its unique anticodon that binds to the codon and carries instructions for its particular amino acid (pp. 198–201).

LO4 Use the genetic code to identify the amino acid for which a DNA segment or a codon codes (Section 9.1).

- The amino acid coded for by a particular codon can be determined using the genetic code (p. 201).

LO5 Define the term *mutation*, and explain how mutations can affect protein structure and function (Section 9.1).

- Mutations are changes to DNA sequences that can affect protein structure and function (p. 202).
- Neutral mutations are changes to the DNA that do not result in a different amino acid being incorporated. Insertions or deletions of nucleotides can result in frameshift mutations that change the protein drastically (pp. 202–203).

LO6 Explain why and how some genes are expressed in some cell types but not others (Section 9.1).

- A given cell type expresses only a small percentage of the genes that an organism possesses (p. 203).
- Turning the expression of a gene up or down is accomplished in different ways in prokaryotes and eukaryotes. Prokaryotes typically block the promoter with a repressor protein to keep gene expression turned off. Eukaryotes regulate gene expression in any of 5 ways: (1) increasing transcription through the use of proteins that stimulate RNA polymerase binding; (2) varying the time that DNA spends in the uncondensed, active form; (3) altering the mRNA life span; (4) slowing down or speeding up translation; and (5) affecting the protein life span (pp. 204–205).

LO7 Outline the process of cloning a gene using bacteria (Section 9.2).

- Bacteria can be used to clone genes by placing the gene of interest into a plasmid, which then makes millions of copies of the gene as the plasmid replicates itself inside its bacterial host. Bacteria can then express the gene by transcribing an mRNA copy and translating the mRNA into a protein (pp. 205–207).

LO8 Describe how foods are genetically modified, and list some of the health and environmental concerns surrounding this practice (Section 9.3).

- A Ti plasmid or gene gun can be used to insert a particular gene into plant cells (pp. 208–209).
- Although there have been no documented incidents of negative health effects from GM food consumption, there is concern that GM foods may be unhealthy. Concerns about GM crops include their impacts on surrounding organisms, the evolution of resistances, and transfer of modified genes to wild and weedy relatives (pp. 209–210).

LO9 Describe the science behind and significance of stem cells, the Human Genome Project, gene therapy, and cloning (Section 9.4).

- Stem cells are undifferentiated cells that can be reprogrammed to act as a variety of cell types. Stem cells may someday allow scientists to treat degenerative diseases (pp. 210–212).
- The Human Genome Project resulted in the production of the complete DNA sequence of humans and several other organisms (p. 212).
- Gene therapy involves replacing defective genes or their products in an embryo or in the affected adult tissue. Gene therapy is considered experimental but may hold promise once scientists determine how to target genes to the right locations and express them in the proper amounts (pp. 213–215).
- Nuclear transfer has been used to clone animals with desirable agricultural traits. It may someday be possible to clone humans, but it is unclear if these humans would be healthy (pp. 215–216).

Roots to Remember

The following roots of words come mainly from Latin and Greek and will help you decipher terms:

chromo- means color. Chapter term: chromosome

-ic is a common ending of acids. Chapter term: nucleic acid

nucleo- or **nucl-** refers to a nucleus. Chapter term: nucleotide

-ose is a common ending for sugars. Chapter terms: ribose and dexyribose

-some or **-somal** relates to a body. Chapter terms: chromosome and ribosome

Learning the Basics

1. **LO2** List the order of nucleotides on the mRNA that would be transcribed from the following DNA sequence: CGATTACTTA.

2. **LO4** Using the genetic code (Table 9.1 on p. 201), list the order of amino acids encoded by the following mRNA nucleotides: CAACGCAUUUUG.

3. **LO3** List the subcellular structures that participate in translation.

4. **LO2** Transcription _____.

   A. synthesizes new daughter DNA molecules from an existing DNA molecule; **B.** results in the synthesis of an RNA copy of a gene; **C.** pairs thymines (T) with adenines (A); **D.** occurs on ribosomes

5. **LO3** Transfer RNA (tRNA) _____.

A. carries monosaccharides to the ribosome for synthesis; **B.** is made of messenger RNA; **C.** has an anticodon region that is complementary to the mRNA codon; **D.** is the site of protein synthesis

6. **LO2** During the process of transcription, _____.

A. DNA serves as a template for the synthesis of more DNA; **B.** DNA serves as a template for the synthesis of RNA; **C.** DNA serves as a template for the synthesis of proteins; **D.** RNA serves as a template for the synthesis of proteins

7. **LO3** Translation results in the production of _____.

A. RNA; **B.** DNA; **C.** protein; **D.** individual amino acids; **E.** transfer RNA molecules

8. **LO2** The RNA polymerase enzyme binds to _____, initiating transcription.

A. amino acids; **B.** tRNA; **C.** the promoter sequence; **D.** the ribosome

9. **LO3** A particular triplet of bases in the coding sequence of DNA is TGA. The anticodon on the tRNA that binds to the mRNA codon is _____.

A. TGA; **B.** UGA; **C.** UCU; **D.** ACU

10. **LO2** RNA and DNA are similar because _____.

A. they are both double-stranded helices; **B.** uracil is found in both of them; **C.** both contain the sugar deoxyribose; **D.** both are made up of nucleotides consisting of a sugar, a phosphate, and a base

11. **LO6** **True/False** A kidney cell and a heart cell from the same cat would have the same genes.

12. **LO8** **True/False** Scientists know that genetically modified crops cannot hurt the environment.

13. **LO9** **True/False** Gene therapy is used commonly and with 100% success.

Analyzing and Applying the Basics

1. **LO5** Take another look at the genetic code (Table 9.1, p. 201). Do you see any similarities between codons that code for the same amino acid? Based on this difference, why might a mutation that affects the nucleotide in the third position of the codon be less likely to affect the structure of the protein than a mutation that affects the codon in the first position?

2. **LO5** Genes encode RNA polymerase molecules. What would happen to a cell that has undergone a mutation to its RNA polymerase gene?

3. **LO7** Draw a box around the 6-base-pair site at which a restriction enzyme would most likely be cut:

ATGAATTCCGTCCG
TACTTAAGGCAGGC

Connecting the Science

1. The first "test-tube baby," Louise Brown, was born over 30 years ago. Sperm from her father was combined with an egg cell from her mother. The fertilized egg cell was then placed into the mother's uterus for the period of gestation. At the time of Louise's conception, many people were very concerned about the ethics of scientists performing these in vitro fertilizations. Do you think human cloning will eventually be as commonplace as in vitro fertilizations are now? Why or why not?

2. Do you think it is acceptable to grow genetically modified foods if health risks turn out to be low but environmental effects are high?

Answers to **Stop & Stretch, Visualize This, Savvy Reader,** and **Chapter Review** questions can be found in the **Answers** section at the back of the book.

CHAPTER

10

Where Did We Come From?

The Evidence for Evolution

Should public school students be taught about alternative hypotheses to evolution?

ess than a decade ago, the journal *Science* published an article in which the authors compared the results of a simple survey among populations around the world. The survey asked whether the statement, "Human beings, as we know them, developed from earlier species of animals," is true or false, or whether the respondent is not sure or does not know. In the United States, 39% of respondents thought the statement was false, and another 21% were unsure—leaving only 40% who thought the statement was true. In similarly developed countries, such as Iceland, Denmark, Sweden, Great Britain, and France, up to 80% of the population agree with the statement. In fact, compared with the 32 countries discussed in the article, acceptance of this statement is lower only in Turkey.

Public opinion in the United States has not changed much since this result was announced. In 2009, a similar poll in the United States, Great Britain, and Canada asked respondents to choose a statement that best expressed their point of view regarding the origin and development of human beings, offering the choices: "Human beings evolved from less advanced life forms over millions of years" and "God created human beings in their present form in the last 10,000 years." In the United States, 35% chose the evolution statement, while 47% chose the creation statement. In Canada, acceptance of the evolution statement was 61%, and in Great Britain, 68%.

To biologists in the United States, these poll results are puzzling. The theory of evolution, including the concept that humans evolved from nonhuman ancestors, forms the bedrock of biological science. The United States has a world-class higher education system in the sciences, one that draws students from all over the world. So, why

One idea about the origin of humans: special creation.

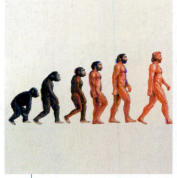

Another idea: evolution.

Why do biologists insist that evolution is the correct explanation?

is the American public seemingly so skeptical of an idea that is dogma in the scientific establishment?

Could it be that Americans, expressing the national tradition of questioning authority, reflexively reject any idea that seems to be proclaimed as truth when it seems counter to their traditions and experience? The majority of American citizens are practicing Christians, and many of these believe that the words in the Bible are absolute truth. And it seems clear that humans are set apart from all other animals, in our domination of nearly all terrestrial environments and their native organisms. Some would argue that the idea of evolution is as much a matter of faith as their own religious doctrine. In this chapter, we examine this question by exploring the theory of evolution and the origin of humans as a matter of science.

10.1 What Is Evolution?

Evolution really has two different meanings to biologists. The term can refer to either a process or an organizing principle, that is, a theory.

The Process of Evolution

Generally, the word *evolution* means change, and the process of evolution reflects this definition as it applies to populations of organisms. **Biological populations** are groups of individuals of the same species that are subdivided from other populations by geography and are somewhat independent of other groups. **Biological evolution,** then, is a change in the characteristics of a biological population that occurs over the course of generations. The changes in populations that are considered evolutionary are those that are inherited via genes. Changes that may take place in populations due only to short-term changes in their environment are not evolutionary. For example, the average dress size for women in the United States has increased from 8 to 14 over the past 50 years. This change is not genetic, but occurred because of an increase in average calorie intake and average age in the population, thus it is not an evolutionary change.

As an example of the process of biological evolution, consider the species of organism commonly known as head lice. Some populations of head lice in the United States have become resistant to the pesticide permethrin, found in over-the-counter delousing shampoos. Initially, lice infections were readily controlled through treatment with these products; however, over time, populations of lice evolved to become less susceptible to the effects of these chemicals. Unlike the example of the change in women's dress sizes resulting from increased food availability, this change in the susceptibility of the lice population to permethrin is a result of a change in their genes. The evolution of resistance can occur rapidly—a study in Israel demonstrated that populations of head lice were four times less susceptible to permethrin only 30 months (or 40 lice generations) after the pesticide was introduced in that country. Note that in this example, individual head lice did not "evolve" or change; instead, the population as a whole changed from one in which most lice were susceptible to the pesticide to one in which most lice were resistant to it.

The differences in resistance to permethrin among individuals in the population resulted from genetic variation. In particular, some lice carried gene variants (that is, alleles) that conferred resistance, and others carried nonresistant alleles. Because individuals with the resistant alleles survived permethrin treatment, they passed

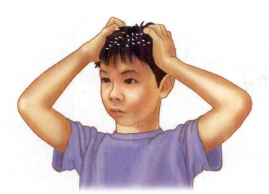

Initial lice infestation consists of both susceptible and resistant lice.

After permethrin treatment, most lice are dead, but a few that are resistant to the pesticide survive.

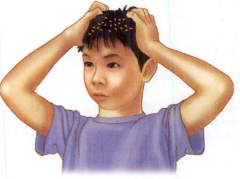

Reinfestation with the offspring of the resistant lice. The population of lice is now more resistant to permethrin.

these resistant alleles on to their offspring. As a result, a population made up primarily of individual lice that carried the susceptible alleles changed into one in which most of the lice carried the resistant alleles. This change in the characteristics of the population took many generations (**Figure 10.1**).

According to the definition of evolution presented in this chapter, the population of lice has evolved. In this case, the process of **natural selection,** the differential survival and reproduction of individuals in a population, brought about the evolutionary change. Natural selection is the process by which populations adapt to their changing environment (Chapter 11). Other forces, including chance, can cause evolutionary changes in the genetic makeup of populations as well (Chapter 12).

Most people accept that traits in populations can evolve. Evolutionary change in biological populations, such as the development of pesticide resistance in insects and antibiotic resistance in bacteria, has been observed multiple times (Chapter 11). Changes that occur within a biological population are referred to as **microevolution.** The controversy about teaching evolution instead comes from discussions of **macroevolution,** the changes that result in the origin of new species.

The Theory of Evolution

Although some Americans do not believe that microevolutionary changes within a species occur, theirs is a minority view. However, as the surveys described at the beginning of chapter illustrate, many dispute whether the theory of evolution, which includes both the process of microevolution and its macroevolutionary results, is "scientific truth."

Some of the controversy is generated by the use of the word *theory*. When people use *theory* in everyday conversation, they often are referring to a tentative explanation with little systematic support; for example, a sports fan might have a theory about why her team is losing, or a gardener might have a theory about why his roses fail to bloom. Usually, these explanations amount to a "best guess" regarding the cause of some phenomenon. A **scientific theory** is much more substantial—it is a body of scientifically acceptable general principles that provide the current best explanation of how the universe works (Chapter 1). Scientific theories are supported by numerous lines of evidence and have withstood repeated experimental tests. For instance, the theory of gravity explains the motion of the planets, and germ theory explains the relationship between microorganisms and human disease. The theory of evolution is a principle for understanding how species originate and why they have the characteristics that they exhibit. The **theory of evolution** can thus be stated:

All species present on Earth today are descendants of a single common ancestor, and all species represent the product of millions of years of accumulated microevolutionary changes.

Figure 10.1 The process of evolution. The evolution of pesticide resistance in lice occurs as a result of natural selection, one of the mechanisms by which traits in a population can change over time.
Visualize This: How would this figure be different if no lice in this child's hair were resistant to pesticide? Could evolution occur?

Figure 10.2 The theory of common descent. This theory states that all modern organisms are descended from a single common ancestor. Each branching point on the tree represents the origin of new species from an ancestral form.

Visualize This: Why do some branches on the tree stop short of the right side of the tree?

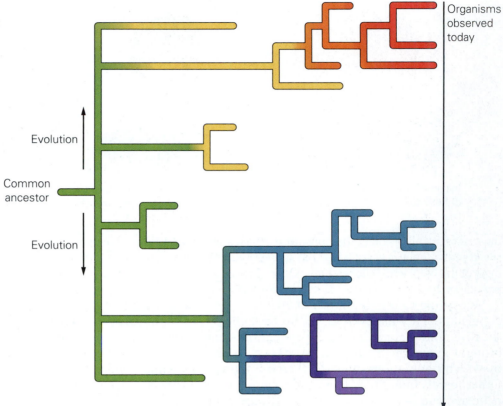

In other words, modern animals, plants, fungi, bacteria, and other living things are related to each other and have been diverging from their common ancestor, by various processes, since the origin of life on this planet. While the processes that cause evolution—such as natural selection—are not the subject of controversy among the general public, the part of the theory of evolution that states that all life shares a common ancestor, called the theory of **common descent**, is (**Figure 10.2**). It is this controversy that underlies the opposition to what is often termed *Darwinism*.

10.2 Charles Darwin and the Theory of Evolution

The theory of evolution is sometimes called "Darwinism" because Charles Darwin is credited with bringing it into the mainstream of modern science (**Figure 10.3**).

The youngest son of a wealthy physician, Darwin spent much of his early life as a lackluster student. After dropping out of medical school, Darwin entered Cambridge University to study for the ministry at the urging of his father. Darwin hated most of his classes but did strike up friendships with several scientists at the college. One of his closest companions was Professor John Henslow, an influential botanist who nurtured Darwin's deep curiosity about the natural world. It was Henslow who secured Darwin his first "job" after graduation. In 1831, at age 22, Darwin set out on what would become his life-defining journey—the voyage of the *HMS Beagle*.

The *Beagle*'s mission was to chart the coasts and harbors of South America. The ship's personnel were to include a naturalist for "collecting, observing, and noting anything worthy to be noted in natural history." Henslow recommended Darwin to be the unpaid assistant naturalist (and socially appropriate dinner

Figure 10.3 Charles Darwin. Beginning when he a young naturalist, Charles Darwin conceived of and developed a theory of evolution that remains one of the most well-supported ideas in biology.

companion) to the *Beagle*'s aristocratic captain after two other candidates had turned it down.

Early Views of Evolution

The hypothesis that organisms change over time was not new when Darwin embarked on his voyage. The Greek poet Anaximander (611–546 B.C.) seems to have been the first Western philosopher to postulate that humans evolved from fish that had moved onto land.

The first modern evolutionist, Jean Baptiste Lamarck, had published his ideas about evolution in 1809, the year of Darwin's birth. Lamarck was the first scientist to state clearly that organisms adapted to their environments. He proposed that all individuals of every species had an innate, inner drive for perfection, and that the traits they acquired over their lifetimes could be passed on to their offspring. Lamarck used this principle in an attempt to explain the long legs of wading birds, which he argued arose when the ancestors of those animals attempted to catch fish. As they waded in deeper water, they would stretch their legs to their full extent, resulting in gradual lengthening of the limbs. These longer legs would be passed on to the next generation, which would in turn stretch their legs while fishing and pass on even longer legs to the next generation.

Lamarck's contemporaries were unconvinced by his proposed mechanism for species change—for instance, it was easily seen that the children of muscular blacksmiths had similar-sized biceps to those of bankers' children and had not inherited their fathers' highly developed muscles. Lamarck's critics were also unwilling to question the more socially acceptable alternative hypothesis that Earth and its organisms were created in their current forms by God and did not change over time. It was this more acceptable hypothesis that Darwin found himself questioning as a result of his around-the-world voyage.

The Voyage of the Beagle

In the 5 years that he spent on the expedition of the *HMS Beagle,* Darwin spent most of his time on land—luckily for him as it turns out, because he was nearly constantly seasick on board the ship. The trip was a dramatic awakening for the young man, who was awed at the sight of the Brazilian rain forest, amazed by the scantily clothed natives in the chilly climate of Tierra del Fuego, and intrigued by the diversity of animals and plants he collected (**Figure 10.4**).

On the ship, Darwin had ample time to read, including Charles Lyell's book *Principles of Geology,* which put forth the hypothesis that the geological processes working today are no different from those working in the past, and that large geological features result from the accumulated effect of these geological forces. In supporting this idea, Lyell argued that deep canyons resulted from the gradual erosion of rock by rivers and streams over enormous timescales. Lyell's hypothesis called into question the belief that Earth was less than 10,000 years old, an age that was deduced from a literal reading of the Bible.

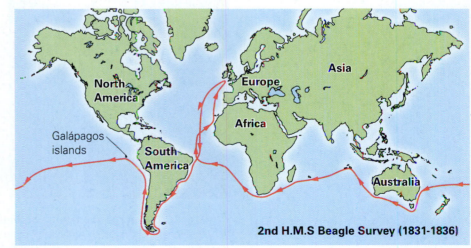

2nd H.M.S Beagle Survey (1831-1836)

Figure 10.4 The voyage of *HMS Beagle*. Darwin's expedition took him to tropical locales from South America to Tahiti.

(a) Dome-shelled tortoise from Santa Cruz Island, an island with abundant ground-level vegetation

(b) Flat-shelled tortoise from Española Island, an island with tall vegetation

Figure 10.5 Giant tortoises of the Galápagos. The subspecies of giant tortoises on the Galápagos Islands from different environments look distinct from each other.

Darwin was also strongly influenced by a stop in the Galápagos, a small archipelago of volcanic islands off the coast of Ecuador. While at first look the islands seemed nearly lifeless, during the month that the *Beagle* spent sailing them, Darwin collected an astonishing variety of organisms. Many of the birds and reptiles he observed appeared to be unique to each island. For instance, while all islands had populations of tortoises, the type of tortoise found on one island was different from the types found on other islands (**Figure 10.5**).

Darwin wondered why God would place different, unique subtypes of tortoise on islands in the same small archipelago. On his return to England, Darwin reflected on his observations and concluded that the populations of tortoises on the different islands must have descended from a single ancestral tortoise population. He noted a similar pattern of divergence between other groups of species on the closest mainland and the Galápagos islands—for instance, prickly pear cacti in mainland Ecuador have the ground-hugging shape familiar to many of us, whereas on the Galápagos, these plants are tree size, but both the mainland and island cacti are clearly related (**Figure 10.6**).

Developing the Hypothesis of Common Descent

Darwin's observations and portions of his fossil collection, sent periodically back to Henslow via other ships, made Darwin a scientific celebrity even before the *Beagle* returned to England. A journal of his travels was a bestseller, and on returning home to England, he settled into a comfortable life with his wife Emma, heiress to a large fortune. After the voyage and the publication of his journal, Darwin reflected on his journal entries and recognized that the observations he had made on living and fossilized plants and animals supported the hypothesis of common descent. However, knowing that this hypothesis was still considered radical, Darwin shared his ideas with only a few close friends.

(a) Prickly pear cactus in mainland South America

(b) Prickly pear cactus in Galápagos

Figure 10.6 Divergence from a common ancestor. Prickly pear cactuses have very different forms in South America compared to the Galápagos, but they clearly share ancestry, as evidenced by similar pad and flower structures.

Visualize This: What environmental factors might have caused the difference in evolution between prickly pears on the mainland and prickly pears on the islands?

Instead of publishing his ideas about evolution, Darwin spent the next two decades carefully collecting evidence and further developing his theory. He was finally spurred into publishing his ideas in 1858, after receiving a letter from Alfred Russel Wallace. With Wallace's letter was a manuscript detailing a mechanism for evolutionary change—nearly identical to Darwin's theory of natural selection.

Concerned that his years of scholarship would be forgotten if Wallace published his ideas first, Darwin had excerpts of both his and Wallace's work presented in July 1858 at a scientific meeting in London, and the next year he published the book, *On the Origin of Species by Means of Natural Selection, or the Preservation of Favoured Races in the Struggle for Life*. The main point of this text was to put forward the hypothesis of natural selection. But Darwin devoted the last several chapters of *The Origin of Species* to describing the evidence he had accumulated supporting common descent.

The evidence put forth in *The Origin of Species* was so complete, and from so many different areas of biology, that the hypothesis of common descent no longer appeared to be a tentative explanation. In response, scientists began to refer to this idea and its supporting evidence as the *theory* of common descent. Indeed, most biologists today would agree that common descent is a scientific fact. Darwin's careful catalogue of evidence in his book had revolutionized the science of biology.

Alternative Ideas on the Origins and Relationships among Organisms

Let us explore the statement, "common descent (or evolution) is a fact," more closely. When *The Origin of Species* was published, most Europeans believed that special creation explained how organisms came into being. According to this belief, God created organisms during the 6 days of creation described in the first book of the Bible, Genesis. This belief also states that organisms, including humans, have not changed significantly since creation. According to some biblical scholars, the Genesis story indicates that creation also occurred fairly recently, within the last 10,000 years.

For a hypothesis to be testable by science, we must be able to evaluate it through observations or measurements made within the material universe (Chapter 1). Because a supernatural creator is not observable or measurable, there is no way to determine the existence or predict the actions of such an entity through the scientific method. Therefore, as it is stated, special creation is not a scientific hypothesis. In fact, any statement that supposes a supernatural cause—including intelligent design, which argues that while evolution is possible, some specific features of organisms must have been designed by a creator—cannot be considered science.

Stop & Stretch *Faith* can be defined as the acceptance of ideals or beliefs that are not necessarily demonstrable through experimentation or direct observation of nature. How does faith differ from science? Can statements of faith be tested scientifically? Explain.

The idea of special creation does provide *some* scientifically testable hypotheses, however. For instance, the assertion that organisms came into being within the last 10,000 years and that they have not changed substantially since their creation is testable through observations of the natural world. We can call this hypothesis about the origin and relationships among living organisms the *static model hypothesis,* indicating that organisms are unchanging, in addition to being recently derived (**Figure 10.7a** on the next page). There are also several intermediate hypotheses between the static model and common descent. One intermediate hypothesis is that all living organisms were created, perhaps even millions of years ago, and that changes

(a) Static model
Species arise separately
and do not change over time.

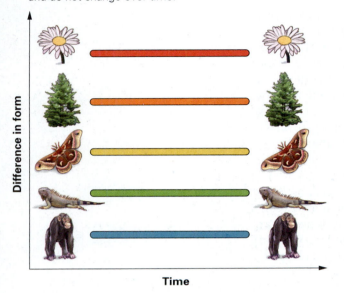

(b) Transformation
Species arise separately and change over time in
order to adapt to the changing environment.

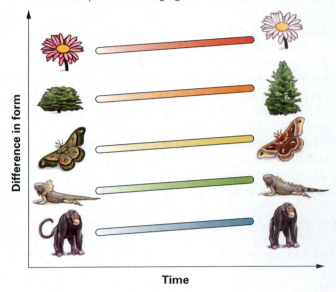

(c) Separate types
Species change over time, and new species can arise, but not
from a common ancestor. Each group of species derives from
a separate ancestor that arose independently.

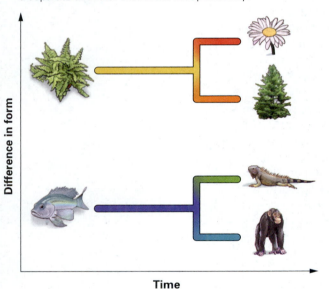

(d) Common descent
Species do change over time, and new species
can arise. All species derive from a common ancestor.

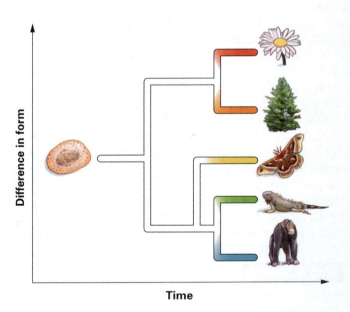

**Figure 10.7 Four hypotheses about
the origin of modern organisms.**
A graphical representation of the four
hypotheses.

have occurred by microevolution in these species, but that brand-new species have not
arisen. We will call this the *transformation hypothesis* (**Figure 10.7b**). Another intermediate hypothesis is that different *types* of organisms (for example, plants, animals
with backbones, or insects) arose separately and since their origin have diversified
into numerous species. We will call this the *separate types hypothesis* (**Figure 10.7c**).
Are these three alternatives to the theory of common descent (**Figure 10.7d**) equally
likely and reasonable explanations for the origin of biological diversity?

It appears from the polls that many Americans think so. But what about
all those scientists who insist that the theory of common descent is fact? Why
would so many maintain this position? As you will soon see, the three alternative hypotheses are not equivalent to the theory of common descent. To understand why, we must examine the observations that help us test these hypotheses.

10.3 Examining the Evidence for Common Descent

The evidence that all organisms share a common ancestor comes from several areas of biology and geology. According to many critics of evolution, it seems highly unlikely that humans, with our highly developed brains, could be "just another branch" in the animal family tree. Because this criticism is so common, we will address it head on—by examining the hypothesis that humans share a common ancestor with apes, including the chimpanzee and gorilla.

Any zookeeper will tell you that the primate house is their most popular exhibit. People love apes and monkeys. It is easy to see why—primates are curious, playful, and agile. But, something else drives our fascination with primates: We see ourselves reflected in them. The forward placement of their eyes and their reduced noses appear human-like. They have hands with fingernails instead of paws with claws. Some can stand and walk on two legs for short periods. They can finely manipulate objects with their fingers and opposable thumbs. They exhibit extensive parental care, and even their social relations are similar to ours—they tickle, caress, kiss, and pout (**Figure 10.8**).

Why are primates, particularly the great apes (gorillas, orangutans, chimpanzees, and bonobos) so similar to humans? Scientists contend that it is because humans and apes are recent descendants of a common biological ancestor.

Figure 10.8 Are humans related to apes? Biologists contend that apes and humans are similar in appearance and behavior because we share a relatively recent common ancestor.

Linnaean Classification

As the modern scientific community was developing in the sixteenth and seventeenth centuries, various methods for organizing biological diversity were proposed. Many of these classification systems grouped organisms by similarities in habitat, diet, or behavior; some of these classifications placed humans with the great apes, and others did not.

Into the classification debate stepped Carl von Linné, a Swedish physician and botanist. Von Linné gave all species of organisms a two-part, or binomial, name in Latin, which was the common language of science at the time. In fact, he adopted a Latinized name for himself—Carolus Linnaeus. The Latin names that Linnaeus assigned to other organisms typically contained information about the species' traits—for instance, *Acer saccarhum* is Latin for "maple tree that produces sugar," the sugar maple, whereas *Acer rubrum* is Latin for "red maple." The scientific name of a species contains information about its classification as well. For example, all species with the generic name *Acer* are all maples, and all species with the generic name *Ursus* are bears.

Linnaeus also created a logical and useful system for organizing diversity—grouping organisms in a hierarchy. From broadest to narrowest groupings, the classification levels he designated are

KINGDOM
 PHYLUM
  CLASS
   ORDER
    FAMILY
     GENUS
      SPECIES

Since Linnaeus's time, biologists have added a broader level—the Domain—as the largest grouping of organisms.

Although Linnaeus himself believed in special creation and preceded the theory of evolution, Darwin noted that Linnaeus's hierarchal classification system

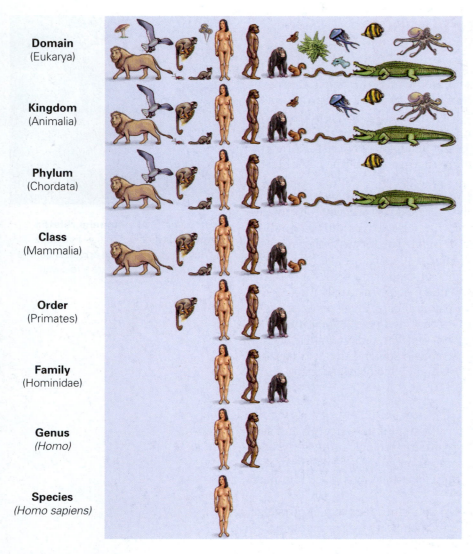

Domain
(Eukarya)

Kingdom
(Animalia)

Phylum
(Chordata)

Class
(Mammalia)

Order
(Primates)

Family
(Hominidae)

Genus
(*Homo*)

Species
(*Homo sapiens*)

Figure 10.9 The Linnaean classification of humans. All organisms within a category share basic characteristics, and as the groups become smaller subdivisions toward the bottom of the figure, the organisms have more similarities.

Visualize This: List one or more traits that all members of the order Primates share.

(**Figure 10.9**) implied evolutionary relationships among organisms. Linnaeus's classification system was so effective at capturing these relationships that, even after the widespread acceptance of evolutionary theory, it has only been modified slightly, with the addition of "sub" and "super" levels between his categories—such as superfamily between family and order—to better represent the complex relationships among groups of organisms (Chapter 13).

While his hierarchical categories are still in use, many of the actual groupings Linnaeus proposed in the eighteenth century have been overturned or radically altered by the accumulation of additional data. However, when it comes to humans, his original classification remains largely supported by the data. Linnaeus placed humans, monkeys, and apes in the same order, which he called Primates. Among the primates, humans are most similar to apes. Humans and apes share a number of characteristics, including relatively large brains, erect posture, lack of a tail, and increased flexibility of the thumb. Scientists now place humans and apes in the same family, Hominidae.

Humans and the African great apes share even more characteristics, including elongated skulls, short canine teeth, and reduced hairiness; they are placed together in the same *sub*family, Homininae. When the classification of humans, great apes, and other primates is shown in the form of a tree diagram (**Figure 10.10**), it is easy to see why Darwin concluded that humans and modern apes probably evolved from the same ancestor.

Anatomical Homology

The tree of relationships implied by Linnaeus's classification forms a hypothesis that can be tested. If modern species represent the descendants of ancestors that also gave rise to other species, we should be able to observe other, less-obvious similarities between humans and apes in anatomy, behavior, and genes. These similarities are referred to as **homology.** Homology in skeletal anatomy among a variety of organisms can be seen in comparing mammalian forelimbs, including the presence of a single bone in the upper part of the limb, paired bones in the lower portion, multiple small bones in the wrist, and a number of elongated "finger" bones (**Figure 10.11**). The similarities between human and chimpanzee forelimbs are striking, especially the presence of a distinct, opposable thumb.

Note, however, that not all superficial similarities are evidence of evolutionary relationship. Birds and bats both have wings and can fly but do not share a recent common ancestor. Their similarities are due to **convergence,**

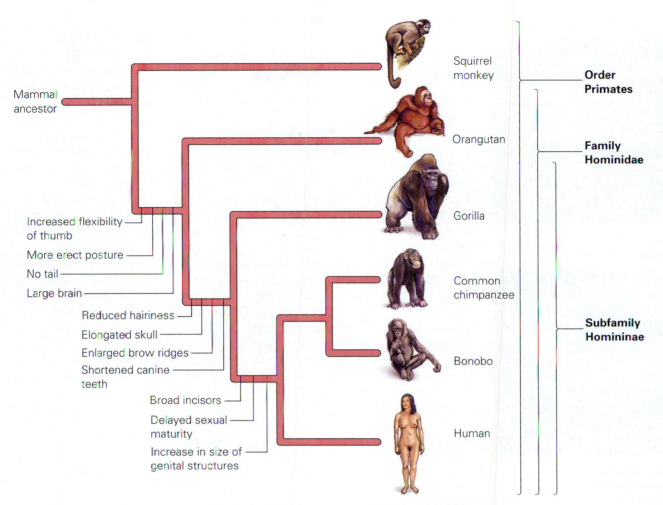

Figure 10.10 Shared characteristics among humans and apes imply shared ancestry.
This tree diagram represents the current classification of humans and apes. Characteristics noted on the side of the evolutionary tree are shared by all of the species on that branch and those to the right.
Visualize This: Add "squirrel" to this tree. How can you indicate the shared ancestor of squirrels and other members of this group?

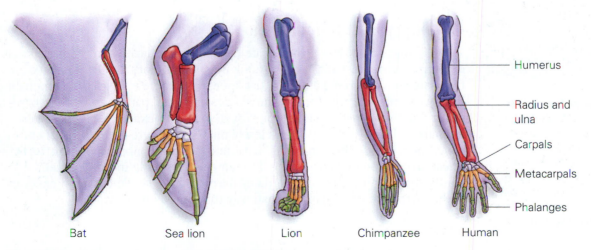

Figure 10.11 Homology of mammal forelimbs. The bones in the forelimbs of these mammals are very similar; equivalent bones in each organism are shaded the same color. The similarity in underlying structure despite great differences in function is evidence of shared ancestry.

(a) Fern sporophyte

**(b) Fern gametophyte
(an independent plant)**

**(c) Flowering plant sporophyte
containing microscopic
gametophyte**

Gametophyte
generation
is found here.

Figure 10.12 Vestigial structures in plants. The ancestors of flowering plants were similar to modern ferns. Ferns have two independent stages in their life cycle: (a) the familiar form, which is known as the sporophyte stage and (b) a tiny, independent, gametophyte stage. (c) Flowering plants no longer have two independent stages but do produce a tiny gametophyte within the flower itself.

the evolution of similar structures in unrelated organisms with similar lifestyles. Another example of convergence is the football-like shape found among animals who spend a significant amount of time hunting food in the water—from tuna fish, to penguins, to otters and dolphins. One challenge associated with the study of evolution is describing which similarities result from shared ancestry and distinguishing them from similarities that evolved in parallel in different groups.

Even more compelling evidence of shared ancestry comes from similarities between functional traits in one organism and seemingly nonfunctional or greatly reduced features, or **vestigial traits,** in another. In other words, vestigial traits represent a vestige, or remainder, of biological heritage. For example, flightless birds such as ostriches produce functionless wings, and flowering plants still produce a tiny "second generation" (gametophyte) within a developing flower ovule, a vestige of their relationship with ferns (**Figure 10.12**).

At least two vestigial traits in humans link us to other primates and great apes (**Figure 10.13**). Great apes and humans have a tailbone like other primates, yet neither great apes nor humans have a tail. In addition, all mammals possess tiny muscles called arrector pili at the base of each hair. When the arrector pili contract under conditions of emotional stress or cold temperatures, the hair is elevated. In furry mammals, the arrector pili help to increase the perceived size of the animal, and they increase the insulating value of the hair coat. In humans, the same emotional or physical conditions produce only goose bumps, which provide neither benefit.

Darwin maintained that the hypothesis of evolution provided a better explanation for vestigial structures than did the hypothesis of special creation represented by the static model. A useless trait such as goose bumps is better explained as the result of inheritance from our biological ancestors than as a feature that appeared—or was created—independently in our species.

Stop & Stretch Some vestigial structures persist because they may still have a function or did in the very recent past. Wisdom teeth, which erupt in early adulthood and often have to be removed because they do not "fit" in a modern person's jaw, are sometimes considered vestigial. However, they may have had a function in the recent past. Why might wisdom teeth have persisted over the course of human evolution?

Developmental Homologies

Multicellular organisms demonstrate numerous similarities in the process of development from fertilized egg to adult organism. Genes that control development are similar among animals that appear as different as humans and fruit flies. As a consequence of these shared developmental pathways, early embryos of very different species often look very similar. For example, all chordates—animals that have a backbone or closely related structure—produce structures called pharyngeal slits, and most have tails as early embryos (**Figure 10.14**). These structures are even seen in human embryos early in development. The similarities suggest that all chordates derive from a single common ancestor with a particular developmental pathway that they all inherited.

Molecular Homology

Scientists now understand that differences among individuals arise largely from differences in their genes. It stands to reason that differences among species must also derive from differences in their genes. If the hypothesis

"Useful" trait in primate relative	Vestigial trait in human

(a) Tail bone

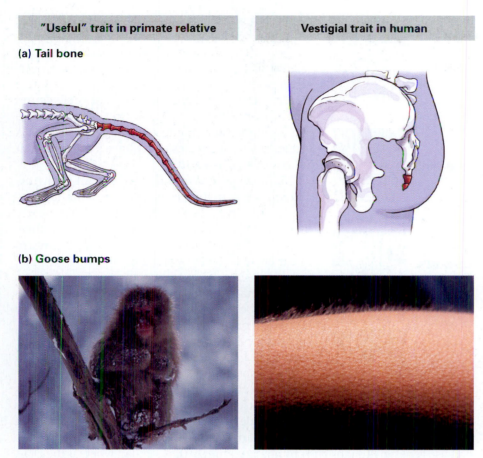

(b) Goose bumps

Figure 10.13 Vestigial traits reflect our evolutionary heritage. (a) Humans and other great apes do not have tails, but they do have a vestigial tailbone. (b) Goose bumps are reminders of our relatives' hairier bodies.

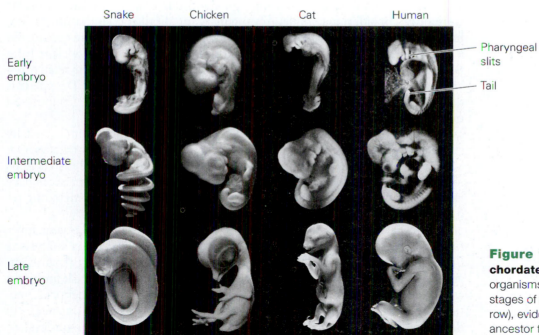

Figure 10.14 Similarity among chordate embryos. These diverse organisms appear very similar in the first stages of development (shown in the top row), evidence that they share a common ancestor that developed along the same pathway.

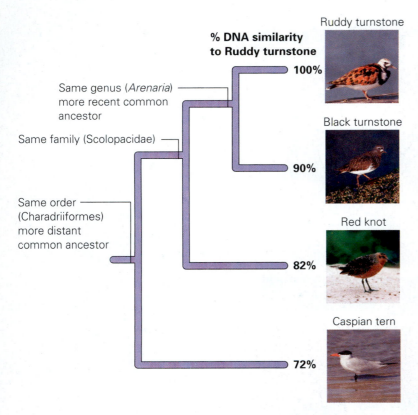

% DNA similarity
to Ruddy turnstone

Ruddy turnstone

100%

Same genus (*Arenaria*) more recent common ancestor

Same family (Scolopacidae)

Black turnstone

90%

Same order (Charadriiformes) more distant common ancestor

Red knot

82%

Caspian tern

72%

Figure 10.15 DNA evidence of relationship. Scientists used the physical characteristics of these four bird species to put them into different taxonomic categories, shown here in tree form. DNA studies later supported this hypothesis, since species that share a more recent common ancestor have more similar DNA sequences than species that share a more distant ancestor.

of common descent is correct, then species that appear to be closely related must have more similar genes than do species that are more distantly related. The most direct way to measure the overall similarity of two species' genes is to evaluate similarities in the DNA sequences of genes found in both organisms. Species that share a more common ancestor should have more similar DNA sequences than species that share a more distant ancestor (**Figure 10.15**).

Many genes are found in nearly all living organisms. For instance, genes that code for histones, the proteins that help store DNA neatly inside cells, are found in algae, fungi, fruit flies, humans, and all other organisms that contain linear chromosomes. Among organisms that share many aspects of structure and function, such as humans and chimpanzees, many genes are shared. However, because amino acids can be coded for by more than one DNA codon, the sequences of genes, even those that produce proteins with the same amino acid sequence, are typically not identical from one species to the next.

A comparison of the sequences of dozens of genes that are found in humans and other primates demonstrates the relationship between classification and gene sequence similarity (**Figure 10.16**). The DNA sequences of these genes in humans and chimpanzees are 99.01% similar, whereas the DNA sequences of humans and gorillas are identical over 98.9% of their length. More distantly related primates are less similar to humans in DNA sequence. This pattern of similarity in DNA sequence exactly matches the biological relationships implied by physical similarity in this group. This result supports the hypothesis of common descent among the primates.

Interestingly, DNA evidence has called into question groupings that once seemed obvious based on morphological similarity. For instance, reptiles were once considered a single group consisting of all the land animals that have scales as skin covering and lay shelled eggs, while birds were considered a distinct group based on their production of feathers. It is now clear from DNA evidence that some living reptiles—the crocodiles and their relatives—are more closely related to birds than they are to other reptiles. Despite this dramatic exception, DNA evidence generally confirms the broad groupings of organisms based on similar morphology and development.

At first, a finding of similarities in DNA sequence may not seem especially surprising. If genes are instructions, then you would expect the instructions for building a human and a chimpanzee to be more similar than the instructions for building a human and a monkey. After all, humans and chimpanzees have many more physical similarities than humans and monkeys do, including reduced hairiness and lack of a tail.

However, remember that the genes being compared perform the same function in all of these species. For example, one of the genes in this analysis is *BRCA1*, which in all organisms has the general function of helping repair damage to DNA. (In humans, *BRCA1* is also associated with increased risk for breast cancer, which explains the source of the gene's name.) Given the identical function of *BRCA1* in all organisms, there is no reason to expect that differences in *BRCA1* sequences among different species should display a pattern—unless the

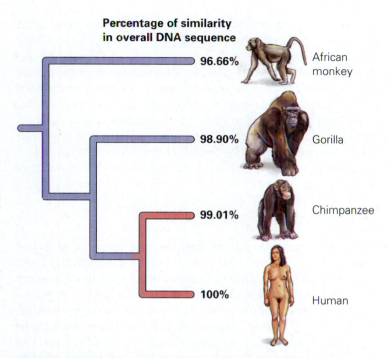

Percentage of similarity
in overall DNA sequence

96.66% — African monkey

98.90% — Gorilla

99.01% — Chimpanzee

100% — Human

Figure 10.16 DNA evidence for human ancestry. Similarities in DNA sequences among the primates parallels the hypothesis of relationship implied by similar morphology and development.

organisms are related by descent. But there is a pattern; the *BRCA1* gene of humans is more similar to the *BRCA1* gene of chimpanzees than to the same gene in monkeys. The best explanation for this observation is that humans and chimpanzees share a more recent common ancestor than either species shares with monkeys.

Differences in DNA sequence between humans and chimpanzees can also allow us to estimate when these two species diverged from their common ancestor. The estimate is based on a **molecular clock.** The principle behind a molecular clock is that the rate of change in certain DNA sequences, due to the accumulation of mutations that affect the DNA sequence but not the protein sequence, seems to be relatively constant within a species. According to one application of a molecular clock, the amount of time it takes for a 1% difference in DNA sequence (about the difference between humans and chimpanzees) to accumulate in diverging species is 5 to 6 million years. Scientists have examined the fossil record in an effort to confirm this hypothetical date.

Biogeography

The distribution of species on Earth is known as the **biogeography** of life. As discussed in this chapter, Darwin observed during his travels on the *Beagle* that each island in the Galápagos had a unique species of tortoise, and that the islands had species of prickly pear cactus that resembled, but were not identical to, the prickly pear found on the mainland. He also noted that the species on the Galápagos were very different from species found on other, similar tropical islands that he visited. If the alternative hypotheses that species appeared independently were true, we would predict either that all tropical islands have the same suite of species or that all tropical islands have completely unique sets of species. The observations of biogeography therefore suggest that species in a geographic location are generally descended from ancestors in nearby geographic locations, supporting the theory of branching descent from common ancestors.

By the time Darwin made his observations of tortoise and prickly pear biogeography, humans were distributed over the entire Earth. The modern biogeography of our species cannot help us determine our relationship to other organisms. However, Darwin reasoned that if humans and apes share a common ancestor, then we highly mobile humans must have first appeared where our less-mobile relatives can still be found. He predicted that evidence of early human ancestors would be found in Africa, the home of chimpanzees and other great apes.

The Fossil Record

Evidence of human ancestors comes from **fossils,** the remains of living organisms left in soil or rock. These fossils and those of other species form a record of ancient life and provide direct evidence of change in organisms over time. There are many examples of fossil series that show a progression from more ancient forms to more modern forms, as in the transition between ancient and modern horses (**Figure 10.17**). The chronological appearance of other groups of fossils also supports the theory of evolution. For example, evidence from anatomy and developmental biology indicates that modern mammals and other four-legged land animals evolved from fish ancestors. Correspondingly, we find the first fossil fish in very ancient rocks and the first fossil mammals in much younger layers of rock.

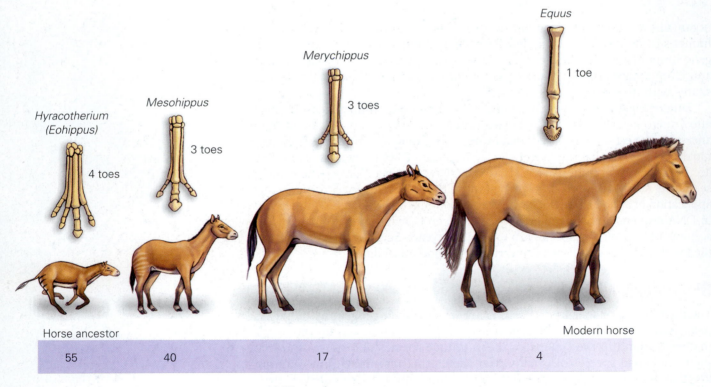

Figure 10.17 The fossil record of horses. Horse fossils provide a fairly complete sequence of evolutionary change from small, catlike animals with four toes to the modern horse with one massive toe.

Fossils typically form when the organic material decomposes and minerals fill the space left behind (**Figure 10.18**). However, fossilized impressions can also form—for instance, of shells, animal burrows, the soft tissues surrounding bones, or footprints. Fossilization is more likely to occur when organisms or their traces are quickly buried by sediment. Fortunately for scientists looking for fossils of **hominins**—humans and human ancestors—there is a relatively good record.

Hominin fossils can be distinguished from other primate fossils by some key characteristics. One essential difference between humans and other apes is our mode of locomotion. While chimpanzees and gorillas use all four limbs to move, humans are bipedal; that is, they walk upright on only two limbs. This difference in locomotion results in several anatomical differences between humans and other apes (**Figure 10.19**). In hominins, the face

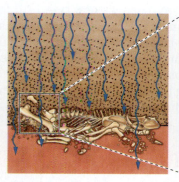

Figure 10.18 Fossilization. A fossil is commonly a rock "model" of an organic structure.

(1) An organism is rapidly buried in water, mud, sand, or volcanic ash. The tissues begin to decompose very slowly.

(2) Water seeping through the sediment picks up minerals from the soil and deposits them in the spaces left by the decaying tissue.

(3) After thousands of years, most or all of the original tissue is replaced by very hard minerals, resulting in a rock model of the original bone.

(4) When erosion or human disturbance removes the overlying sediment, the fossil is exposed.

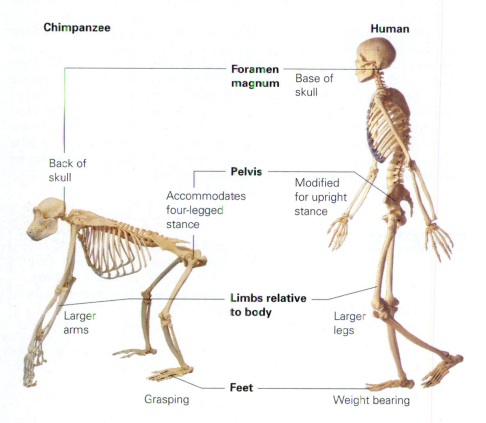

Chimpanzee

Human

Foramen magnum — Base of skull

Back of skull

Pelvis — Modified for upright stance

Accommodates four-legged stance

Larger arms

Limbs relative to body

Larger legs

Feet

Grasping   Weight bearing

Figure 10.19 Anatomical differences between humans and chimpanzees. Humans are bipedal animals, while chimpanzees typically travel on all fours. If any bipedal features are present in a fossil primate, the fossil is classified as a hominin.

is on the same plane as the back instead of at a right angle to it; thus, the foramen magnum, the hole in the skull through which the spinal cord passes, is found on the back of the skull in other apes but at the base of the skull in humans. In addition, the structures of the pelvis and knee are modified for an upright stance; the foot is changed from being grasping to weight bearing; and the lower limbs are elongated relative to the front limbs.

The first hominin fossils were found not in Africa, but in Europe. Remains of *Homo neanderthalensis* (Neanderthal man) were discovered in 1856 in a small cave within the Neander Valley of Germany. In 1891, fossils of older, human-like creatures now called *Homo erectus* (standing man) were discovered in Java (Indonesia) in 1891. It was not until 1924 that the first *African* hominin fossil, the Taung child, was discovered in South Africa. This fossil was later determined to be a much older species than Neanderthals and *Homo erectus* and was placed in a new genus, *Australopithecus*. Paleontologists continue to discover new hominin fossils in southern and eastern Africa, including the famous "Lucy," a remarkably complete skeleton of the species *Australopithecus afarensis*, discovered in 1974 in Ethiopia. Lucy's fossil skeleton included a large section of her pelvis, which clearly indicated that she walked upright.

By determining the age of these fossils and many other hominin species, scientists have confirmed Darwin's predictions—the earliest human ancestors arose in Africa.

An Overview: Radiometric Dating

Scientists can determine the date when an ancient fossil organism lived by estimating the age of the rock that surrounds the fossil. **Radiometric dating** relies on radioactive decay, which occurs as radioactive elements in rock spontaneously break down into different, unique elements known as *daughter products*. Each radioactive element decays at its own unique rate. The rate of decay is measured by the element's **half-life**—the amount of time required for one-half of the amount of the element originally present to decay into the daughter product.

Scientists have used radiometric dating to estimate the age of Earth and the time of origin of various groups of organisms. Using this technique, scientists have also determined that the most ancient hominin fossil, the species *Ardepithecus ramidus*, is 5.2 to 5.8 million years old. (Two even older fossil species, *Orrorin tugenensis* and *Sahelanthropus tchadensis*, were recently described as 6- and 7-million-year-old human ancestors, respectively. However, most scientists are reserving judgment about these fossil specimens until more examples are found.) These very early fossils probably represent hominins similar to the common ancestor of humans and chimpanzees. *See the next section for A Closer Look at radiometric dating and the human pedigree.*

Stop & Stretch In the chapter, you learned how scientists have used a molecular clock to estimate when humans and chimpanzees diverged. How does the age estimate for the fossils of our oldest human ancestors compare to this estimated divergence date? Do these two pieces of evidence about the timing of human evolution agree or disagree?

A Closer Look:
Radiometric Dating and the Human Pedigree

When rock is newly formed from the liquid underlying Earth's crust, it contains a fixed amount of any radioactive element. When the rock hardens, some of these radioactive elements become trapped. As a trapped element decays over time, the amount of radioactive material in the rock declines, and correspondingly, the amount of daughter product increases. By determining the ratio of radioactive element to daughter product in a rock sample and knowing the half-life of the radioactive element, scientists can estimate the number of years that have passed since the rock formed (**Figure 10.20**).

As the number of described hominin fossils has increased, a tentative genealogy of humans has emerged. These fossil species can be arranged in a pedigree of relationships, indicating ancient and more modern species determined by radiometric dating and grouping similar species and dividing unique ones by comparing the organisms' anatomy (**Figure 10.21** on the next page). What the pedigree indicates is that modern humans are the last remaining branch of a once-diverse group of hominins.

Like the fossil record of horses or the transition from fish to land animals, the hominin fossil record shows a clear progression—in this case, from more "apelike" to more "human-like" ancestors over time. Besides being bipedal, humans differ from other apes in having a relatively large brain, a flatter face, and a more extensive culture. The oldest hominins are bipedal but are otherwise similar to other apes in skull shape, brain size, and probable lifestyle. More modern hominins show greater similarity to modern humans, with flattened faces and increased brain size (**Figure 10.22** on the next page). Evidence collected with fossils of the most recent human ancestors indicates the existence of symbolic culture and extensive tool use, trademarks of modern humans.

The pedigree also illustrates a common theme seen in the fossil record of other groups: When a new lifestyle (in this case, bipedal ape) evolves, many different types appear, but only a small number of lineages survive over the long term. Biologists often refer to the lineages that die out as "evolutionary experiments." But does the pedigree provide

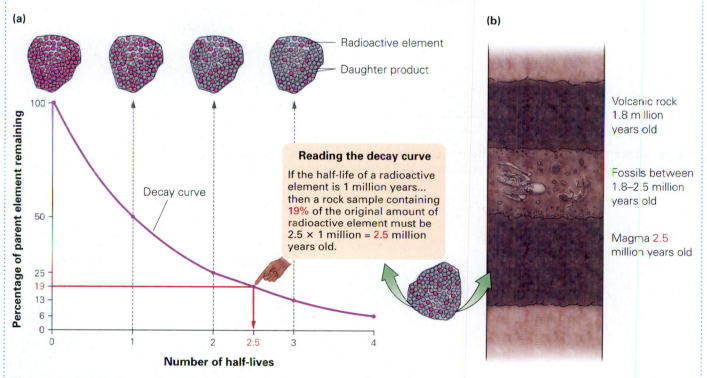

(a)

Radioactive element

Daughter product

Reading the decay curve

If the half-life of a radioactive element is 1 million years... then a rock sample containing **19%** of the original amount of radioactive element must be **2.5 × 1 million = 2.5 million years old.**

Decay curve

Percentage of parent element remaining

Number of half-lives

(b)

Volcanic rock 1.8 million years old

Fossils between 1.8–2.5 million years old

Magma 2.5 million years old

Figure 10.20 Radiometric dating. (a) The age of a fossil can be estimated when it is found between two layers of magma-formed rock. (b) The age of rocks can be estimated by measuring the amount of radioactive material (designated by dark purple circles) with a known half-life and the amount of daughter material (designated by light blue circles) in a sample of rock.
Visualize This: How old is a rock that contains 12.5% of the original parent element?

(continued on the next page)

(A Closer Look continued)

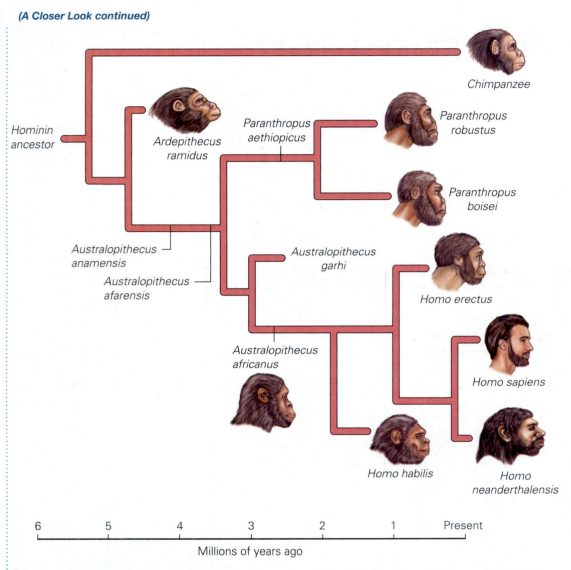

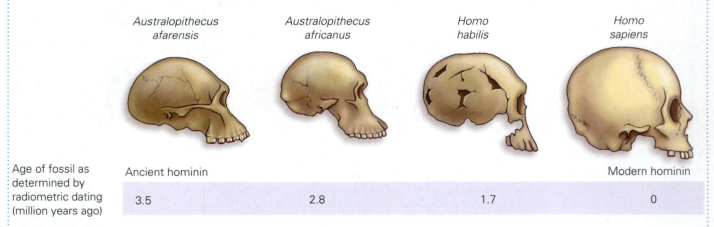

Figure 10.21 The evolutionary relationships among hominin species. This tree represents the current consensus among scientists who are attempting to uncover human evolutionary history.

Figure 10.22 The ape-to-human transition. Ancient hominins display numerous ape-like characteristics, including a large jaw, small braincase, and receding forehead. More recent hominins have a reduced jaw, larger braincase, and smaller brow ridge, much like modern humans.

convincing evidence that modern humans evolved from a common ancestor with other apes?

The common ancestor of humans and chimpanzees is often called the "missing link" because it has not been identified. However, finding the fossilized common ancestor between chimpanzees and humans, or between any two species for that matter, is extremely difficult, if not impossible. To identify a common ancestor, the evolutionary history of both species since their divergence must be clear. Like humans, modern chimpanzees have been evolving over the 5 million years since they diverged from humans. In other words, a missing link would not look like a modern chimpanzee with some human features or a cross between the two species—as suggested by nineteenth-century cartoons depicting Charles Darwin as an "ape man" (**Figure 10.23**). The fact that scientists cannot conclusively identify a fossil species that was the common ancestor of modern humans and modern chimpanzees is not evidence that these two species are not related. The vast majority of the evidence supports the hypothesis that chimpanzees are our closest living relatives.

Figure 10.23 The common ancestor of humans and chimpanzees? A "missing link" between humans and chimpanzees would not look half human and half ape, as this cartoon suggests.

10.4 Are Alternatives to the Theory of Evolution Equally Valid?

Observations of anatomical and genetic similarities among modern organisms and biogeographical patterns provide good evidence to support the theory of evolution. As with nearly all evidence in science, these observations allow us to infer the accuracy of the hypothesis but do not prove the hypothesis correct. This type of evidence is similar to the "circumstantial evidence" presented in a murder trial, such as finding the murder weapon in a car belonging to a suspect or the presence of the suspect's fingerprints on the door of the victim's home. The European scientific community in the nineteenth century was convinced by the indirect evidence that Darwin had catalogued and embraced the theory of common descent as the best explanation for the origin of species. Since Darwin's time, scientists have accumulated additional indirect evidence, such as the DNA sequence similarities discussed earlier, to support this theory.

But, as in a murder trial, direct evidence is always preferred to establish the truth—for instance, the testimony of an eyewitness or a recording of the crime by a security camera. Of course, there are no human eyewitnesses to the evolution of humans, but we have a type of "recording" in the form of the fossil record. Because fossilization requires specific conditions, the fossil record is not a complete recording of the history of life—it is more like a security video that captures only a small portion of the action, with many blank segments. Just as the blank segments in a security video do not make the video an incorrect record of events, "gaps" in the fossil record do not diminish its value. The evidence present in the fossil record provides convincing support for the theory of common descent. All of the lines of evidence for common descent are summarized in **Table 10.1** (beginning on the next page).

TABLE 10.1

The evidence for evolution.

Observation	Example	Why It Suggests Common Descent
Biological classification. The most logical and useful system groups organisms hierarchically.	The levels of classification can identify degrees of relationship.	All species in the same family share a relatively recent common ancestor, while all families in the same class share a more distant common ancestor.
Anatomical homology. Organisms that look quite different have surprisingly similar structures.	Mammalian forelimbs share a common set of bones organized in the same way, despite their very different functions.	The simplest explanation is that each species inherited the basic structure from the same common ancestor, and evolution led to their modification in each group.
Vestigial traits. Some species display traits that are nonfunctional but have a functional equivalent in other species.	Flightless birds such as ostriches produce functionless wings.	The simplest explanation is that the trait was functional in an ancestral species but lost its function over time in one branch of the evolutionary line.

TABLE 10.1

The evidence for evolution.

Observation	Example	Why It Suggests Common Descent
Biogeography. The distribution of organisms on Earth corresponds in part to the relationship implied by biological classification.	The different species of tortoises on the Galápagos Islands are clearly related. Round backed Flat backed	Similarities among species in a geographic location imply divergence from ancestors in that geographic location.
Homology in development. Early embryos of different species often look similar.	All chordates—animals that have a backbone or closely related structure—produce structures called pharyngeal slits, and most have tails as early embryos. Chicken Cat Human	Similarities in early development suggest that these organisms derived from a single common ancestor that developed along a similar pathway.
Homology of DNA. The DNA sequences of species that are closely related in a taxonomic grouping are more similar than those from more distantly related groups.	The closer two organisms are in classification, the more similar is the DNA sequence for the same genes, and vice versa. 90% Black turnstone Caspian tern 72% % DNA sequence similarity to Ruddy turnstone	Similar DNA sequences in different species imply that the species evolved from a common ancestor with a particular sequence.
The fossil record. The remains of extinct organisms show progression from more ancient forms to more modern forms.	The transition between ancient and modern horses 4 toes 3 toes 1 toe	Fossils provide direct evidence of change in organisms over time and suggest relationships among modern species.

Now we return to the three competing hypotheses: static model, transformation, and separate types (**Figure 10.24**). Do the observations described in the previous section allow us to reject any of these hypotheses?

Weighing the Alternatives

The physical evidence we have discussed thus far allows us to clearly reject only one of the hypotheses—the static model. The fossil record provides unambiguous evidence that the species that have inhabited this planet have changed over time, and radiometric dating indicates that Earth is far older than 10,000 years.

Of the remaining three hypotheses, transformation is the poorest explanation of the observations. If organisms arose separately and each changed on its own path, there is no reason to expect that different species would share structures—especially if these structures are vestigial in some of the organisms. There is also no reason to expect similarities among species in DNA sequence. The hypothesis of transformation predicts that we will find little evidence of biological relationships among living organisms. As our observations have indicated, evidence of relationships abounds.

Both the hypothesis of common descent and the hypothesis of separate types contain a process by which we can explain observations of relationships. That is, both hypothesize that modern species are descendants of common ancestors. The difference between the two theories is that common descent hypothesizes a single common ancestor for all living things, whereas separate types hypothesizes that ancestors of different groups arose separately and then gave rise to different types of organisms. Separate types seems more

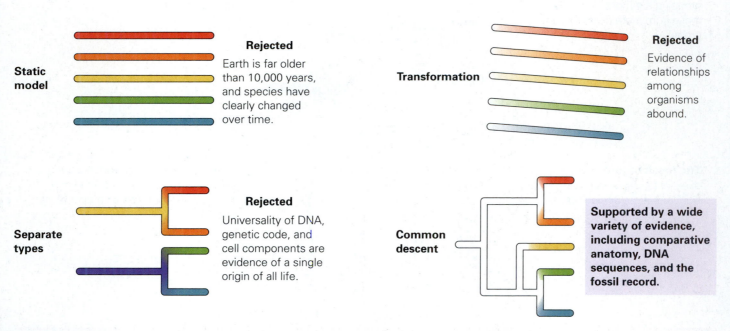

Figure 10.24 Four hypotheses about the origin of modern organisms. A scientific evaluation of these hypotheses leads to the rejection of all of them except for the theory of common descent.

reasonable than common descent to many people. It seems impossible that organisms as different as pine trees, mold, ladybugs, and humans share a common ancestor. However, several observations indicate that these disparate organisms are all related.

The most compelling evidence for the single origin of all life is the universality of both DNA and of the relationship between DNA and proteins. For example, genes from bacteria can be transferred to plants, and the plants will make a functional bacterial protein (Chapter 9). This is possible only because both bacteria and plants translate genetic material into functional proteins in a similar manner. If bacteria and plants arose separately, we could not expect them to translate genetic information similarly.

The fact that organisms as different as pine trees, mold, ladybugs, and humans contain cells with nearly all of the same components and biochemistry is also evidence of shared ancestry (**Figure 10.25**). A mitochondrion could have many different possible structural forms and still perform the same function; the fact that the mitochondria in a plant cell and an animal cell are essentially identical implies that both groups of organisms inherited these mitochondria from a common ancestor.

Pine trees, mold, ladybugs, and humans *are* very different. Proponents of the hypothesis of separate types argue that the differences among these organisms could not have evolved in the time since they shared a common ancestor. But the length of time during which these organisms have been diverging is immense—nearly 2 billion years. The remaining basic similarities among all living organisms serve as evidence of their ancient relationship.

Figure 10.25 The unity and diversity of life. The theory of evolution, including the theory of common descent, provides the best explanation for how organisms as distinct as pine trees, mold, ladybugs, and humans can look very different while sharing a genetic code and many aspects of cell structure and cell division.

The Best Scientific Explanation for the Diversity of Life

Scientists favor the theory of common descent because it is the best explanation for how modern organisms came about. The theory of evolution—including the theory of common descent—is robust, meaning that it is a good explanation for a variety of observations and is well supported by a wide variety of evidence from anatomy, geology, molecular biology, and genetics. Evidence for the theory of common descent demonstrates **consilience,** meaning that there is concurrence among observations derived from different sources. Consilience is a feature of all strongly supported scientific theories.

The theory of common descent is no more tentative than is atomic theory; few scientists disagree with the models that describe the basic structures of atoms, and few disagree that the evidence for the theory of common descent is overwhelming. Most scientists would say that both of these theories are so well supported that we can call them fact.

Evolutionary theory helps us understand the functions of human genes, comprehend the interactions among species, and predict the consequences of a changing global environment for modern species. Describing evolution as "*just a theory*" vastly understates the importance of evolutionary theory as a foundation of modern biology. People who do not have a grasp of this fundamental biological principle may lack an appreciation of the basic unity and diversity of life and fail to understand the effects of evolutionary history and change on the natural world and on ourselves. (For example, we will explore why evolution is important for understanding and treating human disease in Chapter 11.)

SAVVY READER

Do Scientists Doubt Evolution?

Another 100 scientists have joined the ranks of scientists from around the world publicly stating their doubts about the adequacy of Darwin's theory of evolution. "Darwinism is a trivial idea that has been elevated to the status of the scientific theory that governs modern biology," says dissent list signer Dr. Michael Egnor. Egnor is a professor of neurosurgery and pediatrics at State University of New York, Stony Brook, and an award-winning brain surgeon named one of New York's best doctors by *New York Magazine*. Discovery Institute's Center for Science and Culture today announced that over 700 scientists from around the world have now signed a statement expressing their skepticism about the contemporary theory of Darwinian evolution. The statement, located online at www.dissentfromdarwin.org, reads: "We are skeptical of claims for the ability of random mutation and natural selection to account for the complexity of life. Careful examination of the evidence for Darwinian theory should be encouraged."

1. The "dissent from Darwin" project can be thought of as an argument against the theory of evolution. Describe the essence of this argument.

2. Critique the argument. What counterpoints can be made?

3. The Internet contains volumes of information on topics that promote controversy, such as evolution. How would you determine if the website this article was drawn from was credible? If a group that sponsors a website has a clear agenda on an issue, how might that have an impact on the material posted there?

(*Note:* Discovery Institute is a prominent antievolution organization.)
Discovery Institute, February 8, 2007.

Chapter Review
Learning Outcomes

Mastering**BIOLOGY**®

Go to the Study Area at www.masteringbiology.com for practice quizzes, myeBook, BioFlix™ 3-D animations, MP3Tutor sessions, videos, current events, and more.

LO1 Define *biological evolution*, and distinguish it from other forms of nonevolutionary change in organisms (Section 10.1).

- The process of evolution is the change that occurs in the characteristics of organisms in a population over time (p. 222).
- The theory of evolution, as described by Charles Darwin, is that all modern organisms are related to each other and arose from a single common ancestor (pp. 223–224).

LO2 Illustrate the theory of common descent using a tree diagram (Section 10.1).

- Modern organisms can be arranged on a "tree" of relationships based on similarities in morphology, development, and genes (p. 224).

LO3 Summarize how Darwin's experiences led him to develop the outline of the theory of evolution (Section 10.2).

- Scientists before Charles Darwin had hypothesized that species could change over time (p. 225).
- Darwin's voyage on the *HMS Beagle* led him to suspect that this hypothesis was correct. Over the course of 20 years, he was able to gather enough evidence to support this hypothesis and his hypothesis of common descent so that most scientists accepted these theories as the best explanation for the diversity of life on Earth (pp. 225–227).

LO4 Detail the modern biological classification system, and explain how it supports the theory of evolution (Section 10.3).

- Linnaeus classified organisms based on physical similarities between them, for instance placing humans and monkeys in the same order of animals. Darwin argued that the pattern of biological relationships illustrated by Linnaeus's classification provided strong support for the theory of common descent (pp. 229–230).

LO5 Explain how homologies in anatomy and genetics, even in useless traits, support the theory of evolution (Section 10.3).

- Similarities in the underlying structures of a variety of organisms and the existence of vestigial structures, such as the appendix or goose bumps in humans, are difficult to explain except through the theory of common descent (pp. 230–232).
- Modern data on similarities of DNA sequences among organisms match the hypothesized evolutionary relationships suggested by anatomical similarities and provide an independent line of evidence supporting the hypothesis of common descent (pp. 233–235).

LO6 Describe how details of embryonic development support the theory of evolution (Section 10.3).

- Similarities in embryonic development and embryonic structures among diverse organisms—for instance, the presence of pharyngeal slits in the embryos of all chordates—are best explained as a result of their common ancestry (p. 233).

LO7 Define *biogeography*, and explain how it supports the theory of evolution (Section 10.3).

- Biogeography is the study of the geographical distribution of organisms (p. 235).
- Biogeographical patterns support the hypothesis of common descent because species that appear related physically are also often close to each other geographically: for example, the tortoises and mockingbirds on the Galápagos Islands (p. 236).

LO8 Explain how the fossil record provides direct evidence of evolutionary change in species over time (Section 10.3).

- The fossilized remains of extinct species in many groups of organisms demonstrate a progression of forms from more ancient to more modern types. For example, early hominins appear much more ape-like than more recent hominin fossils (pp. 236–241).

LO9 Articulate why the theory of evolution is considered the best explanation for the origin of humans and other organisms (Section 10.4).

- Evidence strongly supports the hypothesis that organisms have changed over time and are related to each other (pp. 241–244).
- Shared characteristics of all life, especially the universality of DNA and the relationship between DNA and proteins, provide evidence that all organisms on Earth descended from a single common ancestor rather than from multiple ancestors (p. 245).

Roots to Remember (MB)

These roots come from Greek or Latin and will help you decode the meaning of words:

evol- means to unroll. Chapter term: evolution

homini- means human-like. Chapter terms: hominid, hominin

homolog- indicates similar or shared origin; from a word meaning "in agreement." Chapter terms: homologus, homology

radio- is the combining form of radiation. Chapter term: radiometric

-metric means to measure. Chapter term: radiometric

Learning the Basics

1. **LO2** Describe the theory of common descent.

2. **LO3** What observations did Charles Darwin make on the Galápagos Islands that helped convince him that evolution occurs?

   A. the existence of animals that did not fit into Linnaeus's classification system; **B.** the similarities and differences among mockingbirds and tortoises on the different islands; **C.** the presence of species he had seen on other tropical islands far from the Galápagos; **D.** the radioactive age of the rocks of the islands; **E.** fossils of human ancestors

3. **LO1 LO9** The process of biological evolution _____.

   A. is not supported by scientific evidence; **B.** results in a change in the features of individuals in a population; **C.** takes place over the course of generations; **D.** B and C are correct; **E.** A, B, and C are correct

4. **LO1** In science, a theory is a(n) _____.

   A. educated guess; **B.** inference based on a lack of scientific evidence; **C.** idea with little experimental support; **D.** a body of scientifically acceptable general principles; **E.** statement of fact

5. **LO2** The theory of common descent states that all modern organisms _____.

   A. can change in response to environmental change; **B.** descended from a single common ancestor; **C.** descended from one of many ancestors that originally arose on Earth; **D.** have not evolved; **E.** can be arranged in a hierarchy from "least evolved" to "most evolved"

6. **LO5** The DNA sequence for the same gene found in several species of mammals _____.

   A. is identical among all species; **B.** is equally different between all pairs of mammal species; **C.** is more similar between closely related species than between distantly

related species; **D.** provides evidence for the hypothesis of common descent; **E.** more than one of the above is correct

7. **LO7** Marsupial mammals give birth to young that complete their development in a pouch on the mother's abdomen. All the native animals of Australia are marsupials, while these types of animals are absent or uncommon on other continents. This observation is an example of _____.

   A. developmental evidence for evolution; **B.** biogeographic evidence for evolution; **C.** genetic evidence for evolution; **D.** fossil evidence for evolution; **E.** not useful evidence for evolution

8. **LO6** Even though marsupial mammals give birth to live young, an eggshell forms briefly early in their development. This is evidence that _____.

   A. marsupials share a common ancestor with some egg-laying species; **B.** marsupials are not really mammals; **C.** all animals arose from a common ancestor; **D.** marsupial mammals were separately created by God; **E.** the fossil record of marsupial mammals is incorrect

9. **LO5** A species of crayfish that lives in caves produces eyestalks like its above-ground relatives, but no eyes. Eyestalks in cave-dwelling crayfish are thus _____.

   A. an evolutionary error; **B.** a dominant mutation; **C.** biogeographical evidence of evolution; **D.** a vestigial trait; **E.** evidence that evolutionary theory may be incorrect

10. **LO4** Which of the following taxonomic levels contains organisms that share the most recent common ancestor?

    A. family; **B.** order; **C.** phylum; **D.** genus; **E.** class

Analyzing and Applying the Basics

1. **LO2 LO4** The classification system devised by Linnaeus can be "rewritten" in the form of an evolutionary tree. Draw a tree that illustrates the relationship among the flowering species listed, given their classification (note that *subclass* is a grouping between class and order):

   Pasture rose (*Rosa carolina*, Family Rosaceae, Order Rosales, Subclass Rosidae)
   Live forever (*Sedum purpureum*, Family Crassulaceae, Order Rosales, Subclass Rosidae)
   Spring avens (*Geum vernum*, Family Rosaceae, Order Rosales, Subclass Rosidae)
   Spring vetch (*Vicia lathyroides*, Family Fabaceae, Order Fabales, Subclass Rosidae)
   Multiflora rose (*Rosa multiflora*, Family Rosaceae, Order Rosales, Subclass Rosidae)

2. **LO5** DNA is not the only molecule that is used to test for evolutionary relationships among organisms. Proteins can also be used, and the sequences of their building blocks

(called amino acids) can be compared in much the same way that DNA sequences are compared. *Cytochrome c* is a protein found in nearly all living organisms; it functions in the transformation of energy within cells. The percentage difference in amino acid sequence between humans and other organisms can be summarized as follows:

Cytochrome c	*Percentage difference from sequence of human sequence*
Chimpanzee	0.0
Mouse	8.7
Donkey	10.6
Carp	21.4
Yeast	32.7
Corn	33.3
Green algae	43.4

Draw the evolutionary tree implied by these data that illustrates the relationship between humans and the other organisms listed.

3. **LO2** Look at the tree you generated for Question 2. It does not imply that yeast is more closely related to corn than it is to green algae. Why not?

Connecting the Science

1. Humans and chimpanzees are more similar to each other genetically than many very similar-looking species of fruit fly are to each other. What does this similarity imply regarding the usefulness of chimpanzees as stand-ins for humans during scientific research? What do you think it implies regarding our moral obligations to these animals?

2. Creationists have argued that if students learn that humans descended from animals and are, in fact, a type of animal, these impressionable youngsters will take this fact as permission to act on their "animal instincts." What do you think of this claim?

Answers to **Stop & Stretch, Visualize This, Savvy Reader,** and **Chapter Review** questions can be found in the **Answers** section at the back of the book.

An Evolving Enemy

Natural Selection

What should we do with someone who has a contagious, deadly disease?

In the early summer of 2007, the U.S. Centers for Disease Control and Prevention (CDC), the federal agency responsible for monitoring and protecting public health, did something that hadn't occurred for over 40 years. They ordered a man to be placed under involuntary quarantine in a hospital.

The man was Andrew Speaker, a 31-year-old personal injury lawyer from Atlanta. In spring of that year, Speaker had been diagnosed with a disease that is relatively harmless in healthy Americans—tuberculosis. He had no apparent symptoms, and the diagnosis and subsequent drug treatment didn't interfere with his plans for later that spring, which were to fly from Atlanta to a Greek island to get married and then travel to Rome, Italy, for his honeymoon. But while he was in Rome, the CDC contacted him to inform him that on further study, they had determined that he had a very dangerous form of the disease.

Depending on which side tells the tale, the CDC either ordered or asked Speaker and his wife to remain in Rome until they could arrange travel back to the United States. Speaker instead booked a flight to Prague, Czechoslovakia, and from there to Montreal, Canada, before renting a car and driving it back to Atlanta. In his travels, everyone he had contact with incurred a risk of infection by the bacteria that causes this dangerous variety of tuberculosis.

People around the world were outraged at Andrew Speaker's seemingly reckless behavior. How could he put so many people at risk of a devastating and often-deadly disease? But others were outraged by the treatment he received from the CDC. Why would they pluck a seemingly healthy man from his honeymoon and place him in involuntary confinement? As it turns out, the CDC has some cause to be very concerned about this form of

Newlywed Andrew Speaker learned one answer to that question.

He may have knowingly exposed thousands of fellow airline passengers to his disease.

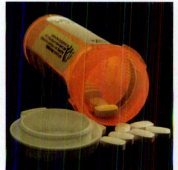

But it is the unconsciously reckless behavior of thousands of us that is the ultimate cause of the problem.

tuberculosis—but few of us are threatened by the actions of Speaker. Instead, it is the recklessness of millions of ordinary people around the world that put us at risk of this, and other, deadly microbes.

11.1 Return of a Killer

Tuberculosis (TB) has been plaguing humans for thousands of years. Tubercular decay has been found in the spines of Egyptian mummies from as long ago as 3000 B.C. In 460 B.C., the Greek physician Hippocrates identified a condition that appears to be tuberculosis as the most widespread disease of the times. As recently as 1906 in the United States, 2 of every 1000 deaths were due to TB infection.

Thanks to relatively recent advances in science and medicine, TB now accounts for only 1.5 of every 100,000 deaths in the United States. In the 1980s, the dramatic drops in infection and death due to TB around the world had many public health professionals convinced that this ancient scourge could be completely eliminated from the human population. Today, those hopes have largely faded as a result of both our own missteps and the powerful force of natural selection.

What Is Tuberculosis?

Tuberculosis is caused by the single-celled bacterium *Mycobacterium tuberculosis* (**Figure 11.1**). Tuberculosis is a public health problem not because it is especially deadly, but instead because it affects so many people. Two billion people, over one-third of the world's population, carry *M. tuberculosis*, and new infections are estimated to occur at a rate of 1 per second. Ninety percent of *M. tuberculosis* infections are symptomless, and most of these resolve as the bacteria are destroyed by the infected individual's immune system. However, the remaining 10% of infected individuals develop active disease, and more than one-half of these individuals will die without treatment. Tuberculosis now causes approximately 2 million deaths worldwide every year.

The symptoms of tuberculosis include a cough that produces blood, fever, fatigue, and a long, relentless wasting in which the patient gradually becomes weaker and thinner. These symptoms explain the antiquated name for this disease, consumption, because the infection seemed to consume people from within.

We now understand that the consumptive symptoms of TB arise from destruction of lung tissue caused by the body's reaction to active *M. tuberculosis* infection. Colonies of the bacteria in the lung are walled off by immune system cells, creating structures called tubercles (**Figure 11.2**); while this may

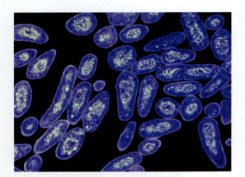

Figure 11.1 The organism that causes tuberculosis. The bacterial species *Mycobacterium tuberculosis*.

Figure 11.2 Lung tubercles.
(a) The white spots on this lung X-ray are nodules, or tubercles, within the lung tissue. (b) The tubercles contain *M. tuberculosis* encased by immune system cells and other tissues.

(a)

Tubercles

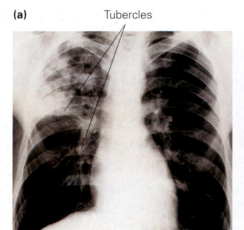

(b)

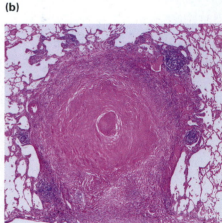

slow the spread of the disease in the body, it permanently damages the lung tissue. Physical wasting is a result of a reduction in the capacity of the lungs to provide oxygen to the body. *Mycobacterium tuberculosis* can lie dormant for months inside these tubercles, which may degrade and then release bacteria back into the lung tissue. Once released from the tubercle, the disease is considered active again, and these bacteria can cause additional infections within the lungs and be spread to other individuals.

Tuberculosis infection is a challenge to diagnose, primarily because *M. tuberculosis* is difficult to grow in the lab from saliva or lung tissue. Typically, infection is confirmed by an X-ray examination for tubercles in the lungs and a skin test that looks for the presence of an immune response to the bacteria (**Figure 11.3**). A significant reaction to the skin test indicates that the individual was infected with *M. tuberculosis* but cannot determine whether the infection is current or if the individual has cleared the bacterium from his body.

Although you can be infected without symptoms, transmission of tuberculosis is almost entirely from people with active disease. When these individuals cough, sneeze, speak, or spit, they expel infectious droplets. A single sneeze can release about 40,000 of these droplets (**Figure 11.4**). People with prolonged, frequent, or intense contact with infected individuals are at highest risk of becoming infected themselves.

Most individuals can fight off the infection, but many cannot. Those at highest risk of active TB are young children; the elderly; individuals in poor overall health due to poor nutrition, other illnesses, or drug abuse; and people with AIDS. Until about 60 years ago, these individuals had little chance of surviving tuberculosis. Today, their prognosis—especially in Western countries where high-quality health care is readily available—is much better. But that may be changing.

Treatment—and Treatment Failure

The treatments for tuberculosis in the nineteenth and early twentieth centuries, at least among the wealthy, consisted primarily of long stays in rural facilities where the air was fresh and unpolluted. These tuberculosis sanatoriums (**Figure 11.5**) were useful for two reasons. By moving patients from areas where the air was thick with lung-damaging particles and chemicals to sanatoriums, doctors were able to preserve patients' lung function for longer periods of time than would otherwise be possible. And because sanatoriums isolated patients, they reduced the spread of TB to the rest of the community. In poorer communities, individuals with active TB were often forcibly isolated in much more grim conditions.

The discovery of **antibiotics,** drugs that kill microbes, including bacteria, revolutionized tuberculosis treatment in the 1940s. Since then, infected patients with active TB are typically kept in isolation for only 2 weeks until antibiotics kill off most of the *M. tuberculosis* in the lungs. At this point, the patient is no longer contagious and can return to the community. However, because the *M. tuberculosis* can hide inside immune system cells for long periods, antibiotic treatment must be maintained for 6 to 12 months to completely eliminate the organism.

Since the 1980s, however, scientists have chronicled a disturbing rise in the number of **antibiotic-resistant** tuberculosis infections—ones that cannot be cured by the standard drug treatment. According to the CDC, 20% of

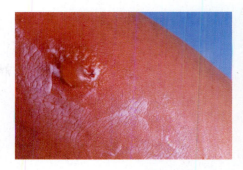

Figure 11.3 Tuberculosis test. The red spot on this individual's forearm marks the site where a small amount of disassembled *M. tuberculosis* was injected just under the surface of the skin. The cell fragments cannot cause disease, but if the individual has been exposed to the bacteria in the past, the spot will become inflamed, as seen here.

Figure 11.4 Transmission through the air. Like cold and flu viruses, the tuberculosis bacteria can be transmitted in droplets that are emitted in the sneeze or cough of an infected individual.

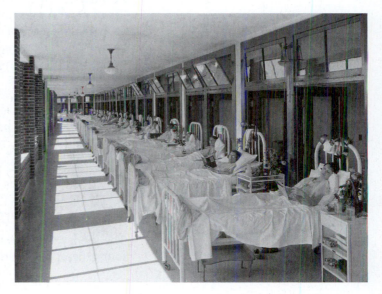

Figure 11.5 A tuberculosis sanatorium. Typically, patients at sanatoriums spent many hours every day exposed to fresh air.

reported TB cases from 2000 to 2004 did not respond to standard treatments (such cases are called multidrug-resistant TB, or MDR-TB), and 2% were resistant to treatment with second-line drugs (such cases are called extensively drug-resistant TB, or XDR-TB). Even in the United States, with abundant access to resources and drugs, one-third of individuals diagnosed with active XDR-TB have died of the disease.

As control has become less effective, the number of cases of TB in both developed and developing countries has begun to rise. In countries with fewer resources, the toll of XDR-TB could be much greater. In a recent XDR-TB outbreak in South Africa, 52 of 53 individuals diagnosed with the strain died within a month of showing signs of active disease. The resurgence of tuberculosis has now been declared a global health emergency by the World Health Organization.

Stop & Stretch   If the first antibiotic treatment provided to a person with active TB is ineffective, the individual likely is infected with an antibiotic-resistant variety of the bacterium. Even if they later receive effective treatment, why is the delay caused by the initial ineffective treatment a concern?

Why are we losing the battle against tuberculosis? And what can be done to stop it? Answering these questions requires an understanding of an important force for evolutionary change: natural selection.

11.2 Natural Selection Causes Evolution

In *The Origin of Species,* Charles Darwin put forth two major ideas: the theory of common descent (Chapter 10) and the theory of **natural selection.** Darwin's presentation of the theory of common descent—that all species living today appear to have descended from a single ancestor—was thorough and convincing. Within 20 years of the publication of his book, the theory of common descent had been accepted by most scientists. However, it was another 60 years before the scientific community accepted Darwin's theory of natural selection, which explains *how* organisms evolved from a common ancestor to become the great variety we see today. Darwin proposed that through the process of natural selection, the physical or behavioral traits of organisms that lead to increased survival or reproduction in a particular environment become common within their population, while less favorable traits are lost. The changes accumulating within populations via natural selection can lead to the development of new species.

Darwin reasoned that the process of natural selection is an inevitable consequence of the competition for survival among variable individuals in a population. Today, natural selection is considered one of the most important causes of evolution (although others, such as the processes of genetic drift and sexual selection as described in Chapter 12, also cause populations to change over time).

Four Observations and an Inference

The theory of natural selection is elegantly simple. It is an inference based on four general observations:

1. Individuals within Populations Vary. Observations of groups of humans support this statement—people do come in an enormous variety of

(a) Variation in coat color

(b) Variation in blooming time

Figure 11.6 Observation 1: Individuals within populations vary. (a) Gray wolves vary in coat color, even within a single litter of animals. (b) Flowers may vary in blooming time, with some individual plants blooming much earlier than others of the same species.

shapes, sizes, colors, and facial features. It may be less obvious that there is variation in nonhuman populations as well. For example, within a litter of gray wolves born to a single female, some individuals may be black, tawny, or reddish in color (**Figure 11.6**). We can add all kinds of less obvious differences to this visible variation; for example, the amount of caffeine produced in the seeds of a coffee plant varies among individuals in a wild population. Each different type of individual in a population is called a **variant.**

2. Some of the Variation among Individuals Can Be Passed on to Their Offspring. Although Darwin did not understand how it occurred, he observed many examples of the general resemblance between parents and offspring. He also noticed that people took advantage of the inheritance of variation in other species. Pigeon fanciers in Darwin's time clearly recognized the inheritance of variation; they could see that, for instance, pigeons with fan-shaped tails were more likely to produce offspring with fan-shaped tails than were pigeons with straight tails. Thus, when they wanted to produce a fan-tailed variety of pigeon, they encouraged breeding among the birds with this trait (**Figure 11.7**). Darwin hypothesized that offspring tend to have the same characteristics as their parents in natural populations as well.

For several decades after *The Origin of Species* was published, the observation that some variations were inherited was the most controversial part of the theory of natural selection. Because scientists could not adequately explain the origin and inheritance of variation, many were unwilling to accept that natural selection could be a mechanism for evolutionary change. When Gregor Mendel's work on inheritance in pea plants (Chapter 7) was rediscovered in the 1900s, the mechanism for this observation became clear—natural selection operates on genetic variation that is passed from one generation to the next.

Figure 11.7 Observation 2: Some of the variation among individuals can be passed on to their offspring. Darwin noted that breeders could create flocks of pigeons with fantastic traits by using as parents of the next generation only those individuals that displayed these traits.

3. Populations of Organisms Produce More Offspring Than Will Survive. This observation is clear to most of us—the trees in the local park make literally millions of seeds every summer, but a small fraction of these survive to germinate, and only a few of the seedlings live for more than a year or two.

If a female elephant (colored pink) lives a full fertile lifetime, she will bear about six calves in about 90 years. On average, half of her calves will be female.

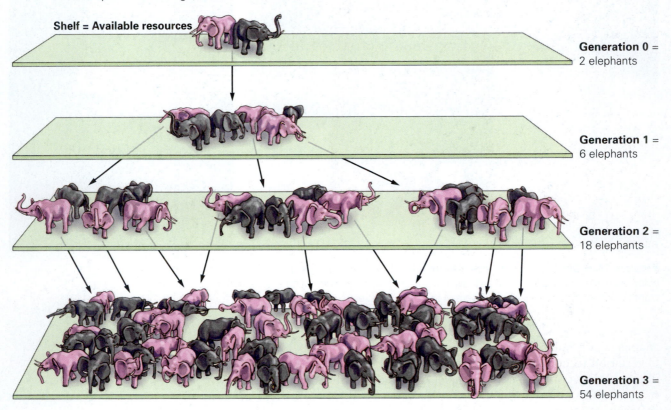

Shelf = Available resources

Generation 0 = 2 elephants

Generation 1 = 6 elephants

Generation 2 = 18 elephants

Generation 3 = 54 elephants

Figure 11.8 Observation 3: Populations of organisms produce more offspring than will survive. Even slow-breeding animals like elephants are capable of producing huge populations relatively quickly. **Visualize This:** What will happen to the elephants as a result of limited resource availability when Generation 4 is produced?

In *The Origin of Species*, Darwin gave a graphic illustration of the difference between offspring production and survival. In his example, he used elephants, animals that live long lives and are very slow breeders. A female elephant does not begin breeding until age 30, and she produces about 1 calf every 10 years until around age 90. Darwin calculated that even at this very low rate of reproduction, if all the descendants of a single pair of African elephants survived and lived full, fertile lives, after about 500 years their family would have more than 15 million members (**Figure 11.8**)—many more than can be supported by all the available food resources on the African continent!

4. Survival and Reproduction Are Not Random. In other words, the subset of individuals that survives long enough to reproduce is not an arbitrary group. Some variants in a population have a higher likelihood of survival and reproduction than other variants do; that is, there is differential survival and reproduction among individuals in the population. The survival and reproduction of one variant compared to others in the same population is referred to as its relative **fitness.** Traits that increase an individual's relative fitness in a particular environment are called **adaptations.** Individuals with adaptations to a particular environment are more likely to survive and reproduce than are individuals lacking such adaptations; in other words, these individuals have higher relative fitness.

Darwin referred to the results of differential survival and reproduction as natural selection. Adaptations are "naturally selected" in the sense that individuals possessing them survive and contribute offspring to the next generation. Although Darwin used the word *selection,* which implies some active

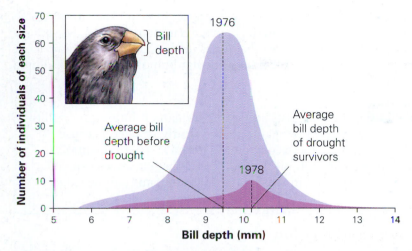

Figure 11.9 Observation 4: Survival and reproduction are not random.
The pale purple curve summarizes bill depth in ground finches on Daphne Island in the Galápagos in 1976. The dark purple curve below it represents the population in 1978, after the drought of 1977. These data indicate that survivors of the drought had an average bill depth 0.6 mm greater than that of the predrought population. The change in the population's average bill size occurred because finches with larger-than-average bills had higher fitness than did small-billed birds during the drought.
Visualize This: How did total population size of ground finches change between 1976 and 1978?

choice, natural selection is a passive process that is simply determined by differences among individuals and their success in their particular environment. For example, among the birds called medium ground finches living on an island in the Galápagos archipelago, scientists have observed that when rainfall is scarce, a large bill is an adaptation. This is because birds with larger bills are able to crack open large, tough seeds—the only food available during severe droughts. As shown in **Figure 11.9**, the 90 survivors of a 1977 drought had an average bill depth that was 6% greater than the average bill depth of the original population of 751 birds. In these environmental conditions, a large bill increases survival.

Adaptations are not only traits that increase survival. Any trait possessed by an individual that increases the number of offspring it produces relative to other individuals in a population is also an adaptation. For example, flowers in a crowded mountain meadow may have a relatively limited number of potential insect pollinators. More pollinator visits generally result in more seeds being produced by a single flower, so any trait that increases a flower's attractiveness to pollinators, such as a brighter color or greater nectar production, should be favored by natural selection (**Figure 11.10**).

Darwin's Inference: Natural Selection Causes Evolution

The result of natural selection is that favorable inherited variations tend to increase in frequency in a population over time, while unfavorable variations tend to be lost. In other words, adaptations become more common in a population as the individuals who possess them contribute larger numbers of their offspring to the succeeding generation. Natural selection results in a change in the traits of individuals in a population over the course of generations—that is, evolution. It is a testament to the power of the theory of natural selection that today it seems self-evident to us—we might even wonder why Darwin is considered a brilliant scientist for explaining something that seems so obvious.

Figure 11.10 Adaptations are not about survival only. Variations that increase a flower's attractiveness to a pollinator can increase its reproductive success by increasing the number of seeds it produces.

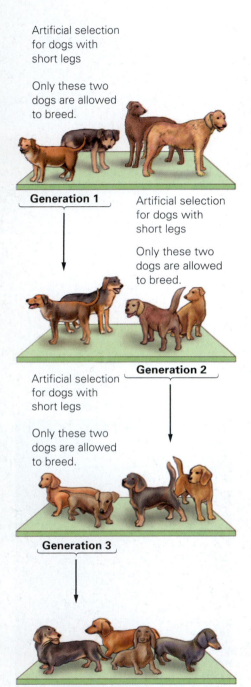

Artificial selection for dogs with short legs

Only these two dogs are allowed to breed.

Generation 1

Artificial selection for dogs with short legs

Only these two dogs are allowed to breed.

Generation 2

Artificial selection for dogs with short legs

Only these two dogs are allowed to breed.

Generation 3

Dachshunds

Figure 11.11 Artificial selection can cause evolution. When breeders select dogs with certain traits to produce the next generation of animals, they increase the frequency of that trait in the population. Over generations, the trait can become quite exaggerated. Dachshunds are descendants of dogs that were selected for the production of very short legs.

Visualize This: What would this sequence of changes look like if the selection was for dogs with long legs and coats, as in Afghan hounds?

But the theory of natural selection only became so powerful after it was tested and shown to work—in nature—in the manner Darwin described. Natural selection proved such a powerful idea that it has influenced how we think about many phenomena, from the success of particular brands of soft drinks to the relationships among nations.

Stop & Stretch Some biologists argue that human evolution as a result of natural selection on our physical traits has nearly stopped because of our ability to use technology to compensate for poor environmental conditions or to overcome physical handicaps. What examples can you give that support this view? In what ways might humans still be subject to natural selection?

Testing Natural Selection

Darwin proposed a scientific explanation of how evolution occurs, and like all good hypotheses, it needed to be tested. All of the tests described next illustrate that natural selection is an effective mechanism for evolutionary change.

Artificial Selection. Selection imposed by human choice is called *artificial selection*. It is artificial in the sense that humans deliberately control the survival and reproduction of individual plants and animals to change the characteristics of the population. Individuals with preferred traits are permitted to breed, whereas those that lack preferred traits are not allowed to breed.

The fancy pigeons that Darwin studied arose by artificial selection, and the great variety of domestic dogs also resulted from this process. In each case, different breeds evolved through selection by breeders for various traits (**Figure 11.11**). These examples demonstrate that differential survival and reproduction change the characteristics of populations. However, due to the direct intervention of humans on the survival and reproduction of these organisms, artificial selection is not exactly equivalent to natural selection. Can change in populations occur without direct human intervention?

Natural Selection in the Lab. Another test of the effectiveness of natural selection is to examine whether populations living in artificially manipulated laboratory environments change over time. An example of this kind of experiment is one performed on fruit flies placed in environments containing different concentrations of alcohol.

High concentrations of alcohol cause cell death. Many organisms, including fruit flies and humans, produce enzymes that metabolize alcohol—that is, they break it down, extract energy from it, and modify it into less-toxic chemicals. There is variation among fruit flies in the rate at which they metabolize alcohol. In a typical laboratory environment, most flies process alcohol relatively slowly, but about 10% of the population possesses an enzyme variant that allows those flies to metabolize alcohol twice as rapidly as the more common variant.

In an experiment (**Figure 11.12**), scientists divided a population of fruit flies into two randomized groups. Initially, these two groups had the same percentage of fast and slow alcohol metabolizers. One group of flies was placed in an environment containing typical food sources; the other group was placed in an environment containing the same food spiked with alcohol. After 57 generations, the percentage of fast alcohol-metabolizing flies in the environment with only typical food sources was the same as at the beginning of the experiment—10%. But after the same number of generations, the percentage of fast alcohol-metabolizing flies in the alcohol-spiked environment was 100%. Because all of the flies in this environment were now of the fast alcohol-metabolizing variety, the *average* rate of alcohol metabolism in

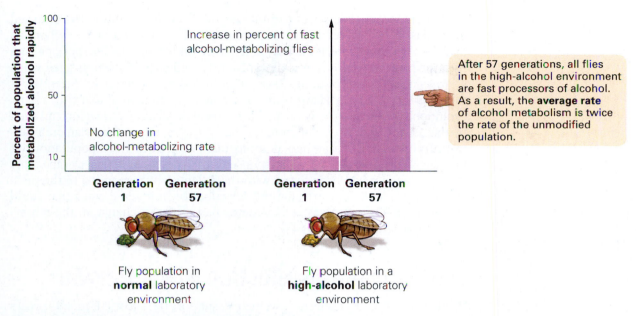

Figure 11.12 Natural selection in laboratory conditions. When fruit flies are placed in an alcohol-spiked environment, the percentage of flies that can rapidly metabolize alcohol increases over many generations because of natural selection. In the normal laboratory environment, there is no selection for faster alcohol processing, so the average rate of alcohol metabolism does not change.

the population in this environment was much higher in generation 57 than in generation 1. The population had evolved.

The evolution of the fruit flies in this experiment was a result of natural selection. In an environment where alcohol concentrations were high, individuals that were able to metabolize alcohol relatively rapidly had higher fitness. Because they lived longer and were less affected by alcohol, the fast alcohol-metabolizing flies left more offspring than the slow alcohol-metabolizing flies did. Thus, each generation had a higher frequency of fast alcohol-metabolizing individuals than the previous generation did. After many generations, flies that could rapidly metabolize alcohol predominated in the population.

Selection can change populations in highly regulated laboratory environments. But does it have an effect in natural wild populations?

Natural Selection in Wild Populations. The evolution of *M. tuberculosis* from susceptible to antibiotics to resistant is one example of natural selection in a wild population; clearly, a change in the environment (that is, the introduction of antibiotics) caused a change in the bacteria population. Dozens of other pathogens have become resistant to drugs and pesticides in the last 50 years as well. But even these changes may seem suspect because the adaptation is to a human-imposed environmental change. Although studying adaptation to natural environmental changes in the field is a significant challenge, the effects of natural selection have been observed in dozens of wild populations.

A classic example of natural selection in a natural setting is the evolution of bill size in Galápagos finches in response to drought (review Figure 11.9). The survivors of the drought tended to be those with the largest bills, which could more easily handle the tough seeds that were available in the dry environment.

The survival of this nonrandom subset of birds resulted in a true change in the next generation. The population of birds that hatched from eggs in 1978—the descendants of the drought survivors—had an average bill depth 4% to 5% larger than that of the predrought population.

A more recent example of natural selection causing evolution has occurred in the last few decades on the eastern coast of the United States, where an invasive Asian crab species is wreaking havoc on native mussels. But one species, the blue mussel, has quickly evolved the ability to thicken its shell when it grows in the presence of the Asian crab, thwarting their attacks. Scientists at the University of New Hampshire were able to demonstrate that this was an evolutionary change by comparing blue mussel populations in regions invaded by the Asian crab to those in more northerly waters, where the Asian crab cannot survive. These researchers demonstrated that while both populations of mussels thicken their shells in response to the presence of native crabs, only the mussels that had been living with the Asian crabs responded to the presence of this species. Clearly, natural selection for individual crabs that could recognize this new species as a predator had caused a change in the mussel population.

11.3 Natural Selection Since Darwin

While tests of natural selection seemed to support the theory, it took 60 years for the rest of biological science to catch up to Darwin's insight and adequately explain it. Scientists working in the new field of genetics in the 1920s began to recognize that genetic principles could explain how natural selection causes the evolution of populations.

The Modern Synthesis

The union between genetics and evolution was termed "the modern synthesis" during its development in the 1930s and 1940s. The modern synthesis outlines the model of evolutionary change accepted by the vast majority of biologists.

The modern synthesis is predicated on a number of genetic principles:

- Genes are segments of genetic material (typically DNA) that contain information about the structure of molecules called proteins (Chapter 9).
- The actions of proteins within an organism help determine its physical traits.
- Different versions of the same gene are called alleles, and variation in physical traits among individuals in a population is often due to variation in the alleles they carry.
- Half of the alleles carried by a parent are passed to their offspring via their eggs or sperm (Chapter 7).
- Different alleles for the same gene arise through mutation—changes in the DNA sequence.

We can apply these genetic principles to the fruit flies exposed to a high-alcohol environment. In this population there are two variants, or alleles, of the gene that controls alcohol processing. One allele produces a fast alcohol-metabolizing enzyme, and the other produces a slow alcohol-metabolizing enzyme. In the high-alcohol environment, flies that made mostly fast alcohol-metabolizing enzyme had more offspring than did flies that made primarily the slow alcohol-metabolizing enzyme. Therefore, in the next generation, produced primarily by fast-alcohol metabolizers, a higher percentage of individuals in the population inherited and thus carried at least one allele for the fast alcohol-metabolizing enzyme. This exemplifies why we can now describe the evolution of a population as an increase or decrease in the frequency of an allele for a particular gene.

The existence of two different alleles for alcohol metabolism in fruit flies suggests that one of these alleles is a mutated version of the other. In the normal laboratory environment, neither of these alleles appears to have a strong effect on fitness. Because the slow alcohol metabolizers are more numerous than

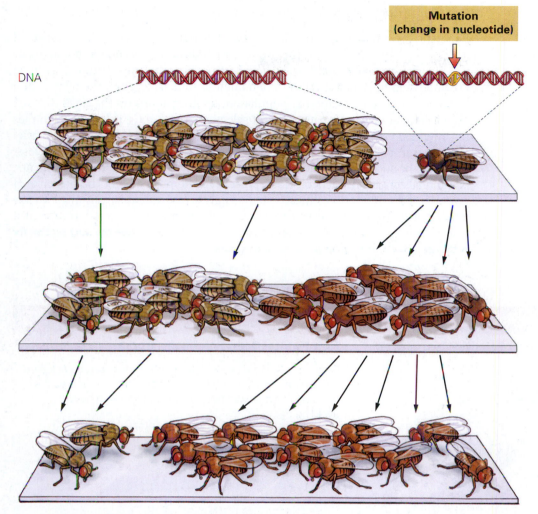

**Mutation
(change in nucleotide)**

DNA

1. Mutation arises in the gene for alcohol metabolizing enzyme, creating a new, mutant allele.

2. The mutation results in a protein that increases the flies' alcohol-metabolism rate.

3. In an alcohol-spiked environment, the flies with the fastest alcohol-metabolism rate produce the most offspring, many of which carry the mutant allele.

4. Over several generations, the frequency of the mutant allele increases within the population.

Figure 11.13 Mutation and natural selection. When a gene has mutated, its product may have a slightly different activity. If this new activity leads to increased fitness in individuals carrying the mutated gene, it will become more common in the population through the process of natural selection. **Visualize This:** How would the ratio of fast to slow metabolizers differ on each "shelf" of this figure if the flies were not in a high-alcohol environment?

the fast alcohol metabolizers, it appears that there might be a slight disadvantage to carrying the fast alcohol-metabolizing enzyme in a low-alcohol environment. However, in the high-alcohol environment, the mutation resulting in the fast alcohol-metabolizing allele gives a strong advantage, and its presence in the population allows for the population's evolution (**Figure 11.13**).

Scientists now understand that the random process of gene mutation generates the raw material—variations—for evolution, and that natural selection acts as a filter that selects for or against new alleles produced by mutation.

Stop & Stretch   The allele responsible for cystic fibrosis (discussed in Chapter 7) appears to protect carriers (individuals with one copy of the allele) from *M. tuberculosis* infection. In addition, it appears that the cystic fibrosis allele is more common in humans with ancestors from crowded, urban environments in Northern Europe. Use the modern understanding of evolution via natural selection to develop a hypothesis about why the cystic fibrosis allele is more common in these populations.

An Overview: The Subtleties of Natural Selection

Because the idea of natural selection has been applied to the realm of human society—such as to the success or failure of a particular company or technology—misunderstandings of how it works in nature are common. Common misunderstandings of natural selection fall into three categories: the relationship between the individual and the population, the limitations on the traits that can be selected, and the ultimate result of selection. **Table 11.1** provides examples of statements that you might hear in reference to natural selection and a brief explanation of how each demonstrates a misunderstanding of this process. In short, the subtleties of natural selection can be described in three statements. Natural selection only acts on traits present in the population. The presence of any adaptation sometimes means that a slightly less valuable trait must be forgone. Natural selection results in the fit of an organism to the environment it is currently experiencing, not some future environment. *See the next section for A Closer Look at the subtleties of natural selection.*

TABLE 11.1

Misunderstandings about natural selection.		
A Misunderstanding of Natural Selection		**How Natural Selection Really Works**
"The Dodo was too stupid to adapt to human hunters, so it had to go extinct."		Only traits that are present in the population can be selected for. The Dodo was not "stupid" or unworthy of survival; the population simply did not contain variants with hunter-avoiding traits.
"Some animals are not adapted to their environments very well. For instance, if swallows were well adapted to North America, they wouldn't have to migrate to the tropics every winter."		Natural selection does not lead to some sort of idealized perfect state. There are always trade-offs between traits. A swallow is very well adapted to catching flying insects. Any swallow variant with physical changes that allow it to eat the seeds that are available in winter would be less able to catch flying insects. These variants would thus lose in the competition for food during the summer.
"If natural selection improves populations, why aren't chimpanzees evolving into humans?"		Natural selection adapts organisms to their current environment; it is not a process that moves organisms from simpler to more complex. Evolution in chimpanzees fits these animals to their current situation. In their environment, bipedal humans already exist and fully exploit the resources in which our ancestors were able to first specialize. Traits that make chimpanzees more "human-like" would not be successful because the resources humans use are already taken.

A Closer Look:
Subtleties of Natural Selection

Natural selection does not result in the evolution of "perfect" organisms. To make sense of that statement, you need to have a good understanding of not only the power but also the limitations of this important evolutionary process.

Natural Selection Cannot Cause New Traits to Arise

One mistaken idea about natural selection is that certain species are doomed to extinction because they "will not" adapt to a new environment. Polar bears are threatened with extinction because they rely on sea ice for hunting, and sea ice in the Arctic is rapidly diminishing as a result of global warming. Why don't they just switch to another food? After all, grizzly bears eat a wide variety of foods.

This idea that species "should" change to survive results from our understanding of the selection process in business. A car company will fail if it does not change its designs to meet the changing needs of its customers. But change in the company occurs because businesses can *invent* new "adaptations" that respond to rapid changes, say in tastes or gas prices. In contrast, in living populations, evolution can occur only when traits that influence survival are already present in a population and have a genetic basis.

The example of the alcohol-metabolizing fruit flies illustrates this point. Selection did not cause change in individual flies; either a fly could rapidly metabolize alcohol or it could not. It was the differential survival and reproduction of these types of flies in an alcohol-laced environment that caused the population to change.

Natural Selection Does Not Result in Perfection

Natural selection does cause populations to become better fit to their environment, but the result of that process is not necessarily "better" organisms that can survive over a wider range of environments. The result of natural selection is simply organisms that are better adapted to their current environment.

Changes in traits that increase survival and reproduction in one environment may be liabilities in another environment. In other words, most adaptations are trade-offs between success in one situation versus another. For example, Richard Lenski and his coworkers at Michigan State University found that certain bacteria would adapt to an environment where food levels were low by evolving chemicals that were deadly to other bacteria. Individual bacteria that produced this chemical had an advantage over those that did not; by killing off their nearby competitors, the poisonous bacteria had more of the very limited food available to them and thus became

prevalent in the population. However, when the poisonous bacteria were grown in a food-rich environment, they did not grow as well as nonpoisonous bacteria did and were outcompeted for the available food. This occurs because the poisonous bacteria use energy to produce their toxin—a trade-off that reduces the amount of energy that can be used for growth and reproduction. In a food-rich environment, individual bacteria that expended the maximum amount of energy on growth had more offspring and thus were better adapted.

Adaptations of organisms are also constrained by their underlying biology. This is apparent throughout nature—adaptations are modifications of existing structures and pathways. The evolutionary biologist Stephen Jay Gould described one of the most famous examples of this kind of modification—the thumb found on the front paws of a giant panda (**Figure 11.14**). These animals apparently have six digits: five fingers composed of the same bones as our fingers and a thumb constructed from an enlarged bone equivalent to one found in our wrist. The muscles that operate this opposable thumb are rerouted hand muscles. This structure in pandas appears to be an adaptation that increases their ability to strip leaves from bamboo shoots, their primary food source. A more effective design for an opposable thumb is our own, adapted from one of the basic five digits. However, in the panda population, this variation did not exist. Individuals with enlarged wrist bones did exist, so what evolved in giant pandas was a thumb that does its limited job but is not as flexible as our own thumb. Unlike other bears, pandas may have a type of thumb, but they will never be able to text with it.

Another, more familiar, example of jury-rigged design is our own upright posture—judging by the number of lower

"Thumb"

Figure 11.14 The panda's thumb.
In addition to the five digits on its paw, the giant panda has a partially opposable "thumb" made up of elongated wrist bones.

(continued on the next page)

(A Closer Look continued)

back and knee injuries reported every year, humans are not perfectly designed for bipedal walking. This is because our skeleton evolved from a four-legged ancestor, and the changes that make us upright are adaptations of that ancestral design.

Natural Selection Does Not Cause Progression Toward a Goal

You may have heard natural selection described as "survival of the fittest"; however, it is important to recognize that natural selection favors those variants with the most appropriate adaptations to the current environment. The survivors are not fittest in an absolute sense, only relative to others in the same population at a given time. And natural selection is a passive process; in no sense are individual organisms "choosing" to change or adapt to environment. Instead, the fittest individuals are simply born with particular traits that increase their chances of survival or reproduction in a particular situation. Natural selection thus does not result in the progress of a population toward a particular predetermined goal; instead, it depends on the particular environment in which a population lives.

The example of the alcohol-metabolizing flies again helps to illustrate this point. Only the population of flies in the high-alcohol environment evolved a faster rate of alcohol metabolism. Without a change in the environment, the alcohol-metabolizing rate of the population of flies in the normal environment did not evolve.

The situational nature of natural selection can lead to evolutionary patterns that defy our sense that species are evolving toward a "more perfect" condition. For example, flowering plants evolved from nonflowering plants, and the flower is, in part, an adaptation to attract bees and other pollinators to help increase seed production. However, some species of flowering plants, especially grasses, have adapted to environments where wind pollination is particularly effective and the need to attract insect pollinators is much reduced. In these plants, natural selection has favored individuals that have very reduced flower parts and generate primarily pollen-producing and egg-producing structures—much like the reproductive structures of their distant nonflowering ancestors (**Figure 11.15**). Grasses have not regressed but instead have simply adapted to the environment they experience.

Figure 11.15 Natural selection does not imply progression. Grasses are flowering plants that evolved from ancestors with much showier flowers.

Patterns of Selection

As Darwin noted, natural selection is a force that causes the traits in a population to change over time. A more modern understanding of the process has helped scientists recognize that different environmental conditions may lead to no change in the population or even cause it to split into two species.

The type of natural selection experienced by the flies in the alcohol-laden environment and by bacteria responding to the use of antibiotics is called **directional selection** because it causes the population traits to move in a particular direction (**Figure 11.16a**). Directional selection is typically the type of selection that leads to change in a population over time—in the case of the flies, to a more alcohol-tolerant population, and in the case of bacteria, to a more antibiotic-resistant strain.

In certain environments, however, the average variant in the population may have the highest fitness. This results in **stabilizing selection,** in which the extreme variants in a population are selected against, and the traits of the population stay the same (**Figure 11.16b**). For example, in humans, the survival of newborns is correlated to birth weight—both extremely small and extremely large babies have lower survival, causing the average birth weight of babies to be relatively stable over time. Stabilizing selection causes populations to tend to resist change in unchanging environments.

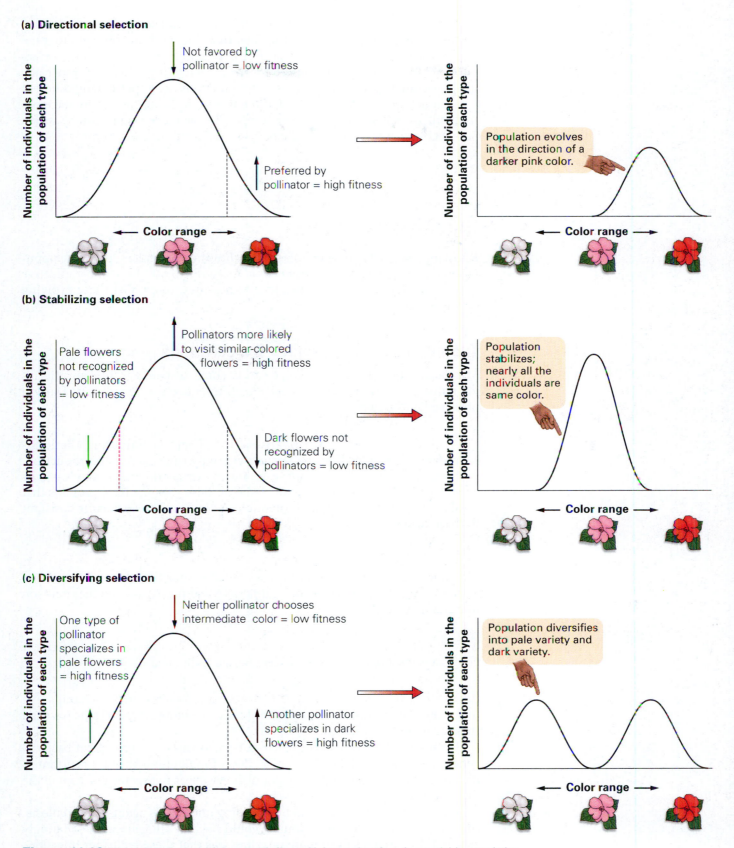

Figure 11.16 Directional, stabilizing, and diversifying selection. In a variable population, different environmental conditions cause different types of selection.

Finally, in some situations, the most common variant may have the lowest fitness, resulting in **diversifying selection,** known sometimes as *disruptive selection*. Diversifying selection causes the evolution of a population consisting of two or more variants (**Figure 11.16c**). Diversifying selection is especially likely within a species if different subpopulations are experiencing different environmental conditions, such that the traits that bring success in one environment are not so successful in another. The primary mechanism by which new species arise (explored in Chapter 12) is diversifying selection.

11.4 Natural Selection and Human Health

Knowing how natural selection works allows us to understand how populations of *M. tuberculosis* have become resistant to all of our most effective antibiotics. It also provides insight into how best to combat these evolving enemies.

Tuberculosis Fits Darwin's Observations

Mycobacterium tuberculosis evolved to become resistant to our antibiotics via natural selection because it fulfills all the necessary conditions Darwin observed and noted.

1. Organisms in the Population Vary.
Bacteria hidden in the lungs reproduce as they feed on the tissue there; any time there is reproduction, mutation can occur. As a result, even during drug treatment, new variants of *M. tuberculosis* continually arise. Some of these variants have proteins that disable or counteract certain antibiotics, making the bacteria more resistant to these drugs.

2. The Variation among Organisms Can Be Passed on to Offspring.
The traits that cause drug resistance are coded for in a bacterium's DNA. When a cell divides to reproduce, it copies this DNA and passes it—and the trait—on to its daughter cells.

3. More Organisms Are Produced than Survive.
The antibiotic treatment eliminates most of the bacteria in the infected individual's body.

4. An Organism's Survival Is Not Random.
Bacterial cells with traits that make them more resistant to the antibiotic are more likely to survive than those that are less resistant.

Because of the increased fitness of particular variants, subsequent bacterial generations consist of a greater percentage of individuals that are resistant to the antibiotic. In other words, the population evolves to become resistant to the drug treatment.

Immediately after scientists began using antibiotics against tuberculosis, they noticed that some individuals would become ill again after seemingly successful treatment. Even more puzzling was that these recurrent infections were much more difficult to treat than the initial one. It was only until scientists incorporated their understanding of natural selection into tuberculosis treatments that effective, long-term therapies were developed.

Two characteristics of the early treatment strategies for tuberculosis actually sped up the development of drug resistance. The first was using the drugs for too short a time period. The second was using only one type of antibiotic at a time in patients with active tuberculosis disease.

Selecting for Drug Resistance

Antibiotics are effective because they buy a patient time to control a bacterial infection with their own immune system. By keeping the bacterial population low, antibiotics allow the body to devote energy to developing an immune response instead of losing energy to the effects of the infection. Because most of the *M. tuberculosis* cells in an infected individual are susceptible to antibiotic, a few days of treatment will eliminate the majority of these organisms. Once most of the bacteria are dead, the patient feels much better as the most debilitating aspects of the disease (e.g., fever, severe cough) are reduced or eliminated. However, a small number of bacteria that are more resistant to the antibiotic take longer to kill. If drug treatment stops as soon as the patient feels better—the typical pattern in early treatment protocols—any of the more resistant bacteria that remain can multiply and restart the infection. A drug-resistant strain is born. And because this new population is more resistant, the resurgent infection is much more difficult to control (**Figure 11.17**). As the number of resistant bacteria increase in the patient, the chances increase that one will appear with a mutation that makes it strongly resistant to the antibiotic. If the patient returns to the community at this point, a disease that is highly resistant to the most commonly effective drugs could begin to spread.

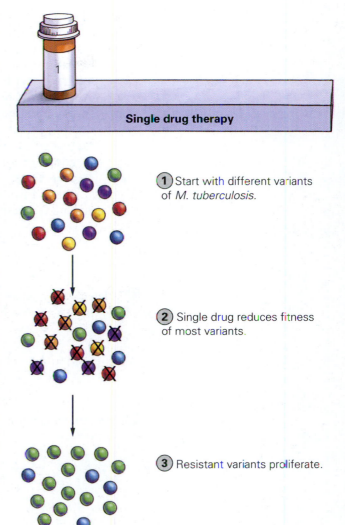

Single drug therapy

1 Start with different variants of *M. tuberculosis.*

2 Single drug reduces fitness of most variants.

3 Resistant variants proliferate.

Stop & Stretch Why would some individual bacteria carry a mutation that makes them more resistant to an antibiotic even though they have never been exposed to this antibiotic?

Figure 11.17 Directional selection in tuberculosis. In a variable population of *M. tuberculosis* bacteria, some individuals may be more resistant to a particular antibiotic. When this antibiotic is used, these resistant variants survive and continue to reproduce. The result is a population that is resistant to the antibiotic.

Stopping Drug Resistance

Combating the development of drug resistance in *M. tuberculosis* required removing the factors that promoted its development in the first place. One strategy is to maintain drug therapy for months, until all signs of the bacterial infection are cleared from the body. The other is to use multiple drugs on active infections to avoid selecting for strongly drug-resistant variants that already exist in the population.

Combination drug therapy, also called drug cocktail therapy, is commonly used on diseases for which resistance to a single drug can develop rapidly. HIV, the virus that causes AIDS, is one example. The effectiveness of combination therapy is based on the following fact: The greater the number of drugs used, the greater the number of changes that are required in the bacterial genome for resistance to develop.

The likelihood of a bacterial variant arising that is resistant to a single drug is relatively small but still very possible in a patient with 1 billion different bacterial variants. However, the likelihood of a bacterial variant arising with resistance to two or three drugs in a cocktail is extremely small. Put another way, the chance that any bacterium exists that is resistant to a single drug is analogous to the likelihood that in 1 billion lottery ticket holders, one person will hold the winning combination—in other words, relatively likely. The likelihood of a variant being resistant to several different drugs is analogous to that same ticket holder winning the lottery several times in a row—incredibly unlikely. Just as it is exceedingly rare to win the lottery twice in a row, it is very difficult for *M. tuberculosis* to adapt to an environment where it faces two "killer drugs" at once (**Figure 11.18**).

If scientists learned these lessons about drug resistance in the 1940s, why is drug-resistant *M. tuberculosis* making a comeback only now? Primarily, it is

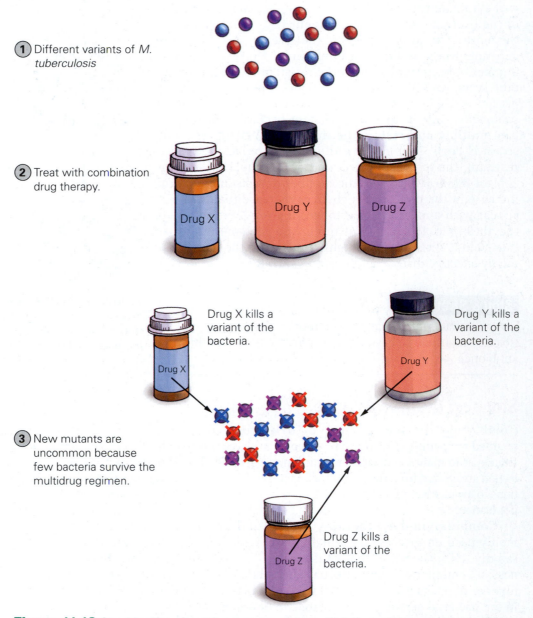

1 Different variants of *M. tuberculosis*

2 Treat with combination drug therapy.

Drug X

Drug Y

Drug Z

Drug X kills a variant of the bacteria.

Drug Y kills a variant of the bacteria.

Drug X

Drug Y

3 New mutants are uncommon because few bacteria survive the multidrug regimen.

Drug Z kills a variant of the bacteria.

Drug Z

Figure 11.18 Combination drug therapy prevents antibiotic resistance. Using multiple antibiotics makes the environment much harsher for *M. tuberculosis* and decreases the likelihood that a variant with multiple resistances will evolve.

because public health researchers let down their guard and didn't follow TB patients to ensure that they continued taking their drugs for the 6 to 12 months required to clear out all of the invading bacteria. It wasn't just Andrew Speaker's recklessness that exposed his fellow airline passengers to XDR-TB. It was thousands of ordinary TB patients who failed to complete their treatment regimen who led to the development of this dangerous variant. With this understanding, public health strategies that focus on maintaining the long-term treatment of individuals, and bringing that treatment to underserved communities, can go a long way in stemming the development of more cases of XDR-TB.

Not surprisingly, tuberculosis is not the only bacterial disease in which resistance has developed, largely as a result of inadequate treatment with antibiotics. MRSA, methicillin-resistant *Staphylococcus aureus,* is another formerly easily treated bacteria that has evolved into a more dangerous and deadly type (**Figure 11.19**).

The rise of antibiotic resistance stems not only from failure to follow drug treatment but also from the overuse of these drugs. Antibiotics are often erroneously prescribed to individuals with viral infections, such as the common cold, on which they have no effect and only serve to increase the likelihood that formerly easily controlled bacteria may develop resistance. Antibiotics are also commonly and liberally used in animal agriculture; they are given to poultry, cows, and pigs to prevent bacterial disease outbreaks in crowded feedlots. Drug resistance is also not confined to bacteria; the same evolutionary process has occurred in viruses, like HIV, and other pathogens, such as the protozoan that causes malaria.

To stop the development of these antibiotic-resistant "superbugs," patients and physicians must use antibiotics judiciously and wisely. But perhaps evolution has another trick up its sleeve that gives us some hope as well.

Can Natural Selection Save Us from Superbugs?

Bacteria can rapidly evolve resistance to our antibiotics. Why can't we evolve resistance to the bacteria? Can't natural selection save us from these pathogens?

There are clearly differences among healthy individuals in their susceptibility to long-term tuberculosis infection and even to active disease. Given the existence of this heritable variation and differences in survival among those exposed to this killer virus, can we expect natural selection to cause the human population to evolve resistance to *M. tuberculosis*?

Eventually, perhaps—but remember that natural selection occurs because of differential survival and reproduction of individuals over time. In short, for the human population as a whole to become resistant to this bacterium, nonresistant variants must die out of the population. This process has not occurred in the more than 6000 years since tuberculosis has been in human populations. Because a majority of nonresistant people are not even exposed to *M. tuberculosis* and thus will continue to survive and reproduce, nonresistant variants will probably never be lost from the population. Clearly, future *human* evolution is not a likely solution to the problem of antibiotic-resistant superbugs.

However, recall that natural selection does not result in perfect organisms, only those with traits that are effective in the current environmental conditions. In *M. tuberculosis*, variants that are resistant to multiple antibiotics are also much less likely to spread to other individuals. In other words, antibiotic resistance is a trade-off that reduces a bacterial cell's ability to survive and reproduce under normal conditions. This raises hopes that quick response to TB outbreaks in vulnerable populations can contain the spread of this dangerous pathogen.

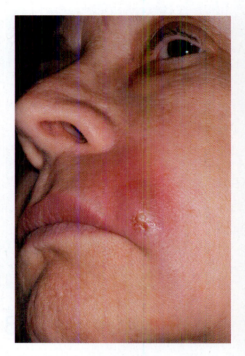

Figure 11.19 MRSA. Staph infections of the skin are relatively common, causing boils like these. However, if untreatable by antibiotics, staph can become a systemic disease that may be deadly.

Natural selection has assisted in our fight against tuberculosis in another way as well. By shaping human brains so that we are able to make predictions and test hypotheses, it has provided us with a powerful tool for fighting this disease—our own ingenuity. Understanding how the disease is spread has helped prevent thousands of new infections, and the development of antibiotics has fought off millions more. A vaccine that can stimulate the immune system to prevent initial infection is in development that may eventually eliminate the disease entirely. Before that happens, we need to apply our ingenuity fully to address the alarming spread of this resistant disease.

The good news for Andrew Speaker is that later tests of the *M. tuberculosis* variant in his lungs confirmed that it was the more treatable MDR-TB and not XDR-TB. He received antibiotic treatment and lung surgery and was relatively quickly released from the hospital. This outcome was good news for the people on Andrew Speaker's airplane flights, as was the fact that in the absence of active disease, his ability to infect other individuals was very low. Tuberculosis does not need to return to the unbeatable enemies list. And with attention to the process of natural selection guiding our use and application of antibiotics, it shouldn't become one.

SAVVY READER

Blogging the TB Story

What gall!

Who does Andrew Speaker think he is? Everyone knows that tuberculosis is a pretty bad disease and is spread through the air. Everyone knows that the air on an airplane is pretty polluted on the best days, and that one is more likely than not to come off an airplane with the virus du jour. So he knows he has TB, gets on a plane anyway, or make that several planes, long international flights. Runs from the CDC when they track him down—on another plane, no less. And his comment now is that he really didn't think he was putting anyone at risk?

Right. Maybe he should have stopped at "I didn't think." or maybe changed the sentence to, "I didn't care." What incredible disregard for other people. What arrogance! I'm afraid it rather sums up the way too many people have become these days. The thought is ME, ME, ME. Speaker is perhaps the quintessential ME guy. He is even willing to put his bride at risk, for goodness sakes! Not to mention all the hundreds of people he came in contact with. Shame on him.

1. List the facts presented within this opinion piece. Does the author provide any supporting evidence for them?

2. No opinion piece presents evidence for all the statements contained within it. What resources do you have to evaluate the credibility of this author?

3. This blog posting is a reasonable summary of the reactions many people had to initial reports about Andrew Speaker. How do these initial reactions appear now that more of the facts of the case have been revealed?

From the Wordpress blog site of "fwtandy" (no further identification) posted June 8, 2007.

Chapter Review

Learning Outcomes

Mastering**BIOLOGY**®

Go to the Study Area at www.masteringbiology.com for practice quizzes, myeBook, BioFlix™ 3-D animations, MP3Tutor sessions, videos, current events, and more.

L01 **Describe the history of tuberculosis in human populations, and explain why current treatments are ineffective against some strains of the disease (Section 1).**

- Tuberculosis is an ancient and widespread disease caused by bacterial infection and can result in lung disease and death in about 10% of cases (pp. 252–254).
- The advent of antibiotics turned tuberculosis from a possible death sentence to a readily curable condition. However, beginning about 25 years ago, variants of the tuberculosis bacteria appeared that are resistant to most antibiotics (pp. 253–254).

L02 **List the four observations that led to the inference of natural selection (Section 2).**

- Individuals in a population vary, and some of this variation can be passed on to offspring (pp. 254–255).
- Not all individuals born in a population survive to adulthood, and not all adults produce the maximum number of offspring possible (pp. 255–257).
- Survival and reproduction is not random. Advantageous traits, called adaptations, increase an individual's fitness, which is his or her chance of survival or reproduction (pp. 256–257).

L03 **Explain how natural selection causes evolutionary change (Section 2).**

- The increased fitness of individuals with particular adaptations causes the adaptation to become more prevalent in a population over generations (pp. 257–258).

L04 **Provide examples of evidence that supports the hypothesis that natural selection leads to the evolution of populations (Section 2).**

- Artificial selection, when humans deliberately control an organism's fitness, causes the evolution of different breeds of animals and varieties of plants (p. 258).
- Populations exposed to environmental changes, both in the lab and in nature, have been shown to evolve traits that make them better fitted to the environment (pp. 258–260).

L05 **Describe how natural selection works on allele frequencies in a population (Section 3).**

- The modern definition of evolution is a genetic change in a population of organisms (p. 260).
- Alleles that code for adaptations become more common in a population over generations as a result of natural selection (pp 261–262).

L06 **Discuss why natural selection does not result in "perfectly adapted" organisms or drive organisms toward some ideal state (Section 3).**

- Natural selection can act only on the variants currently available in the population. Natural selection results in a population that is better adapted to its environment but usually not perfectly adapted as a result of trade-offs. Natural selection does not push a population in the direction of a predetermined "goal" (pp. 262–264).

L07 **List the three patterns of selection, provide examples of each, and explain how they lead to different outcomes (Section 3).**

- Selection can cause the traits in a population to change in a particular direction. However, in some environments it may cause certain traits to resist change and in other environments cause multiple variants to evolve (pp. 264–266).

L08 **Using your understanding of natural selection, explain why combination drug therapy is an effective tool to combat drug resistance (Section 4).**

- Disease-causing organisms can evolve resistance to antibiotics because they consist of multiple variants that have differential survival when exposed to various drugs; thus, a population of disease-causing organisms can evolve drug resistance via natural selection (pp. 266–267).
- A mutant organism that is resistant to several different antibiotics is relatively unlikely, so combination drug therapy can reduce the risk of antibiotic resistance evolving (pp. 267–269).
- As a result of trade-offs, varieties of disease-causing organisms that are multiple-drug resistant are less likely to survive and reproduce in normal conditions than

are non-drug-resistant varieties. Resistant organisms, therefore, are less transmissible than are nonresistant varieties (pp. 269–270).

Roots to Remember

These roots come mainly from Latin and Greek and will help you understand terms:

adapt-	means to fit in. Chapter term: adaptation
anti-	means in opposition to. Chapter term: antibiotic
-biotic	indicates pertaining to life. Chapter term: antibiotic
-osis	indicates a condition. Chapter term: tuberculosis
patho-	means disease. Chapter term: pathogen
tubercul-	means a small swelling. Chapter term: tuberculosis

Learning the Basics

1. **LO1** What types of drugs have helped reduce the death rate due to tuberculosis infection, and why have they become less effective more recently?

2. **LO4** Define *artificial selection*, and compare and contrast it with natural selection.

3. **LO8** Describe how *Mycobacterium tuberculosis* evolves when it is exposed to an antibiotic.

4. **LO2** Which of the following observations is not part of the theory of natural selection?

   A. Populations of organisms have more offspring than will survive; **B.** There is variation among individuals in a population; **C.** Modern organisms are unrelated; **D.** Traits can be passed on from parent to offspring; **E.** Some variants in a population have a higher probability of survival and reproduction than other variants do.

5. **LO2** The best definition of *evolutionary fitness* is _____.

   A. physical health; **B.** the ability to attract members of the opposite sex; **C.** the ability to adapt to the environment; **D.** survival and reproduction relative to other members of the population; **E.** overall strength

6. **LO3** An adaptation is a trait of an organism that increases _____.

   A. its fitness; **B.** its ability to survive and replicate; **C.** in frequency in a population over many generations; **D.** A and B are correct; **E.** A, B, and C are correct

7. **LO5** The heritable differences among organisms are a result of _____.

   A. differences in their DNA; **B.** mutation; **C.** differences in alleles; **D.** A and B are correct; **E.** A, B, and C are correct

8. **LO5** Since the modern synthesis, the technical definition of evolution is a change in _____ in a _____ over the course of generations.

   A. traits, species; **B.** allele frequency, population; **C.** natural selection, natural environment; **D.** adaptations, single organism; **E.** fitness, population

9. **LO7** Ivory from elephant tusks is a valuable commodity on the world market. As a result, male African elephants with large tusks have been heavily hunted for the past few centuries. Today, male elephants have significantly shorter tusks at full adulthood than male elephants in the early 1900s. This is an example of _____.

   A. disruptive selection; **B.** stabilizing selection; **C.** directional selection; **D.** evolutionary regression; **E.** more than one of the above is correct

10. **LO8** Antibiotic resistance is becoming common among organisms that cause a variety of human diseases. All of the following strategies help reduce the risk of antibiotic resistance evolving in a susceptible bacterial population except _____.

   A. using antibiotics only when appropriate, for bacterial infections that are not clearing up naturally; **B.** using the drugs as directed, taking all the antibiotic over the course of days prescribed; **C.** using more than one antibiotic at a time for difficult-to-treat organisms; **D.** preventing natural selection by reducing the amount of evolution the organisms can perform; **E.** reducing the use of antibiotics in non-health care settings, such as agriculture

Analyzing and Applying the Basics

1. **LO3 LO6** Most domestic fruits and vegetables are a result of artificial selection from wild ancestors. Use your understanding of artificial selection to describe how domesticated strawberries must have evolved from their smaller wild relatives. What trade-offs do domesticated strawberries exhibit relative to their wild ancestors?

2. **LO3 LO5** The striped pattern on zebras' coats is considered to be an adaptation that helps reduce the

likelihood of a lion or other predator identifying and preying on an individual animal. The ancestors of zebras were probably not striped. Using your understanding of the processes of mutation and natural selection, describe how a population of striped zebras might have evolved from a population of zebras without stripes.

3. **LO2 LO3 LO6** Are all features of living organisms adaptations? How could you determine if a trait in an organism is a product of evolution by natural selection?

Connecting the Science

1. The theory of natural selection has been applied to human culture in many different realms. For instance, there is a general belief in the United States that "survival of the fittest" determines which people are rich and which are poor. How are the forces that produce differences in wealth among individuals like natural selection? How are they different?

2. Ninety-five percent of worldwide tuberculosis cases occur in developing countries, where most of the population cannot afford effective antibiotic therapy. Does the United States have an obligation to provide people in the developing world with low-cost, effective anti-TB therapy? In countries where the needs of daily survival often overshadow the requirement to take the drugs for an adequate amount of time, drug-resistant strains of TB are more likely to develop. How might governments and funding agencies combat this problem?

Answers to **Stop & Stretch, Visualize This, Savvy Reader,** and **Chapter Review** questions can be found in the **Answers** section at the back of the book.

CHAPTER 12

Who Am I?

Species and Races

This African American scholar is honoring the election of the first African American president with a poem.

January 20, 2009, represented a watershed moment in American history, the inauguration of the first African American president, Barack Obama. Regardless of how you feel about his presidency, everyone can appreciate how Obama's election represented the culmination of a major transformation in American society, less than six generations removed from the legal enslavement of Africans in this country. When the African American scholar and poet Elizabeth Alexander read a verse she penned for the inauguration entitled "Praise Song for the Day," she explicitly referenced how far the nation had come toward viewing people of her race as worthy of equal rights and opportunities:

> Say it plain: that many have died for
>   this day.
> Sing the names of the dead who
>   brought us here,
> who laid the train tracks, raised the
>   bridges,
>
> picked the cotton and the lettuce, built
> brick by brick the glittering edifices
> they would then keep clean and work inside of.
>
> Praise song for struggle, praise song for
>   the day.
> Praise song for every hand-lettered
>   sign,
> the figuring-it-out at kitchen tables.

African slavery was in part justified by a belief that black Africans were somehow biologically inferior to white Europeans. Even after slavery ended, the racism that helped excuse it lived on in social policies that kept blacks confined to segregated facilities and communities. Dr. Alexander's professional career has been dedicated to improving race relations and helping reduce the lingering

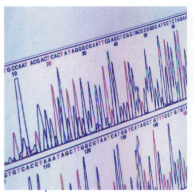

Less than a year after the inauguration, she learned that she was genetically more European than African.

In fact, she shares a recent ancestor with the European American comedian, Stephen Colbert.

How does this new genetic information fit into our experience of human races?

275

effects of racism that she has experienced and that impinge on other African Americans. Imagine Dr. Alexander's surprise just one year after Obama's inauguration when she learned that according to genetic analysis, her ancestry is actually only 27% African and a remarkable 66% European. The analysis was performed as part of a documentary series examining the diversity of America, and her bemused reaction to the results was caught on film.

Moments later, the host of the documentary, Harvard professor of African American studies Henry Louis Gates, Jr., revealed to Alexander that his analysis identified a distant cousin whose DNA was also analyzed for the series: the comedian Stephen Colbert, whose genetic ancestry is 100% European. Alexander was more closely related to Colbert than to Gates—and given these results, she likely shares little genetic history with many other African Americans to whom her work seeks to give voice.

But what does a genetic analysis that places a noted African American poet closer to a white entertainer than to other notable African Americans tell us about race? If she is 66% European, is she more white than black? Or is race more a social construction than a biological reality? This chapter examines the biological definition of race as a step on the road to the formation of new species and addresses the question of whether humans can be biologically grouped into distinct races.

(a) Lion

(b) Leopard

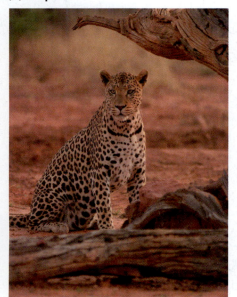

Figure 12.1 Same genus, different species. (a) *Panthera leo*, the lion, and (b) *Panthera pardus*, the leopard.

12.1 What Is a Species?

All humans belong to the same species. Before we can understand the concept of race, we first need to understand both what is meant by this statement and what is known about how species originate.

In the mid-1700s, the Swedish scientist Carolus Linnaeus began the task of cataloging all of nature. Linnaeus developed a classification scheme that grouped organisms according to shared traits (Chapter 10). The primary category in his classification system was the **species,** a group whose members have the greatest resemblance. Linnaeus assigned a two-part name to each species—the first part of the name indicates the **genus,** or broader group; the second part is specific to a particular species within that genus. For example, lions, the species *Panthera leo*, are classified in the same genus with other species of roaring cats, such as the leopard, *Panthera pardus* (**Figure 12.1**).

Linnaeus coined the binomial name *Homo sapiens* (*Homo* meaning "man" and *sapiens* meaning "knowing or wise") to describe the human species. Although Linnaeus recognized the variability among humans, by placing all of us in the same species he acknowledged our basic unity.

Modern biologists have kept the basic Linnaean classification, although they have added a **subspecies** name, *Homo sapiens sapiens*, to distinguish modern humans from earlier humans who appeared approximately 250,000 years ago. Other subspecies of humans include the Neanderthals, known as *Homo sapiens neanderthalensis*.

The Biological Species Concept

While most people intuitively grasp the differences between most species—lions and leopards are both definitely cats but not the same species—biologists have had difficulty finding a single definition that can be consistently applied. Several useful concepts have been proposed and used. The definition most commonly used is the biological species concept.

Biological Species Are Reproductively Isolated.
According to the **biological species concept,** a species is defined as a group of individuals that, in nature, can interbreed and produce fertile offspring but cannot reproduce with members of other species. In practice, this definition can be difficult to apply. For example, species that reproduce asexually (such as most bacteria) and species known only via fossils do not easily fit into this species concept. However, the biological species concept does help us understand why species are distinct from each other.

Recall that differences in traits among individuals arise partly from differences in their genes, and that the different forms of a gene that exist are known as *alleles* of the gene. By the process of evolution, a particular allele can become more common in a species. If interbreeding does not occur, then this allele cannot spread from one species to the other. In this way, two species can evolve differences from each other. For example, various lines of evidence suggest that the common ancestor of lions and leopards had a spotted coat. The allele that eliminated the spots from lions arose and spread within this species, but the allele is not found in leopards because lions and leopards cannot interbreed.

> **Stop & Stretch** Why might the allele for spotlessness have become common in the lion population?

Scientists refer to the sum total of the alleles found in all the individuals of a species as the species' **gene pool.** A single species theoretically makes up an impermeable container for its gene pool—a change in the frequency of an allele in a gene pool can take place only within a biological species.

The Nature of Reproductive Isolation.
The spread of an allele throughout a species' gene pool is called **gene flow.** Gene flow cannot occur between different biological species because pairing between them fails to produce fertile offspring. This reproductive isolation can take two general forms: prefertilization barriers or postfertilization barriers, summarized in **Table 12.1** (on the next page).

As you can see in the table, most mechanisms of reproductive isolation are invisible, occurring before mating or fertilization can occur or soon after fertilization. In the instances when a mating between two different species

TABLE 12.1

Mechanisms of reproductive isolation.

Type	Effect	Example
Prefertilization barriers prevent fertilization from occurring.		
Spatial isolation	Individuals from different species do not come in contact with each other.	Polar bear (Arctic) and spectacled bear (South America) never encounter each other in natural settings.
Behavioral	Ritual behaviors that prepare partners for mating are different in different species.	Many birds with premating songs or "dances" will not mate with individuals who do not know the ritual.
Mechanical	Sex organs are incompatible between different species, so sperm cannot reach egg.	Many insects with "lock-and-key" type genitals physically prevent sperm from contacting eggs of a different species.
Temporal	Timing of readiness to reproduce is different in different species.	Plants with different flowering periods cannot fertilize each other.
Gamete incompatibility	Proteins on egg that allow sperm binding do not bind with sperm from another species.	Animals with external reproduction, such as sponges, have specific proteins on their eggs that will only bind to sperm from the same species.
Postfertilization barriers: Fertilization occurs, but hybrid cannot reproduce.		
Hybrid inviability	Zygote cannot complete development because genetic instructions are incomplete.	A sheep crossed with a goat can produce an embryo, but the embryo dies in the early developmental stages.
Hybrid sterility	Hybrid organism cannot produce offspring because chromosome number is odd.	Mules (see Figure 12.2)

results in live offspring—a **hybrid** descendant of the two parent species—the offspring are often sterile. A well-known example of an interspecies hybrid is the mule, resulting from a cross between a horse and a donkey. Mules have a well-earned reputation as tough and sturdy farm animals, but they cannot produce their own offspring.

Hybrid sterility often occurs because hybrid individuals cannot produce viable sperm or egg cells. During the production of eggs and sperm, homologous chromosomes pair up and separate during the first cell division of meiosis (Chapter 6). Because a hybrid forms from the chromosome sets of two different species, its chromosomes are not homologous and thus cannot pair up correctly during this process.

In the case of mules, the horse parent has 64 chromosomes and therefore produces eggs or sperm with 32; the donkey parent has 62 chromosomes, producing eggs or sperm with 31. The mule will therefore have 63 chromosomes and no way to effectively sort these into pairs during the first division of meiosis (**Figure 12.2**). While a tiny number of female mules, surprisingly, have produced offspring, this event is so rare that the gene pools of donkeys and horses have remained separate.

Stop & Stretch Hybrid corn is produced when two different varieties of the species *Zea mays* (for example, a short variety and a sweet variety) are crossed. Do you expect that hybrid corn is sterile? Why or why not?

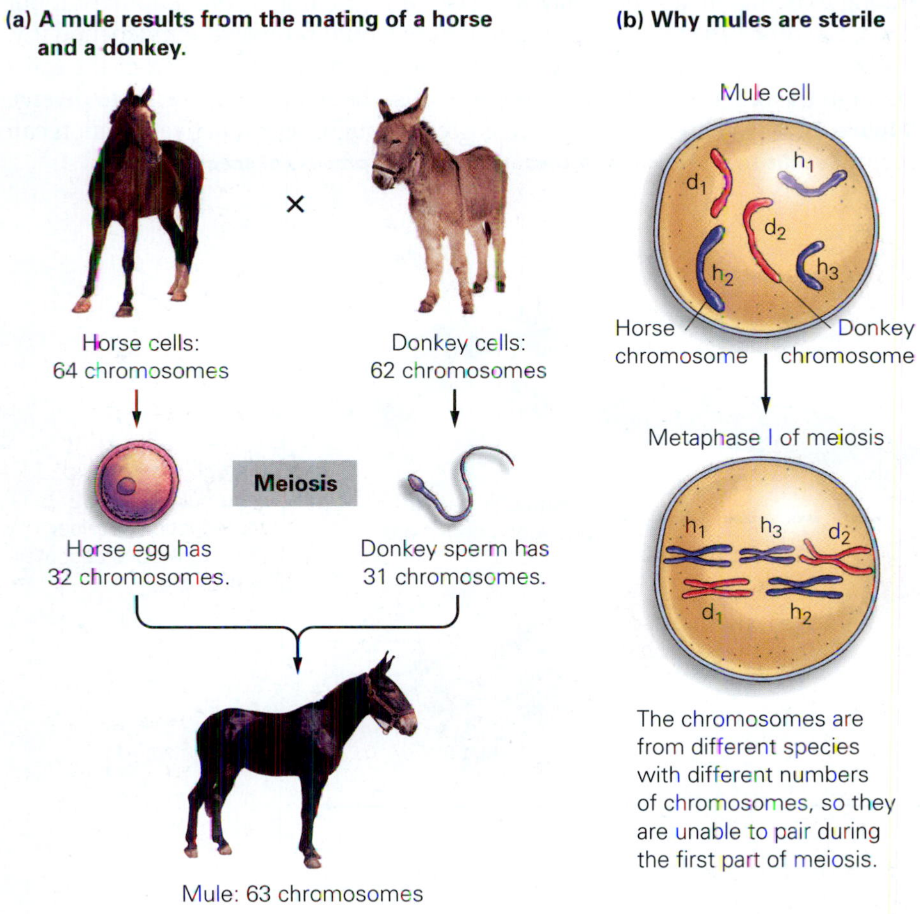

(a) A mule results from the mating of a horse and a donkey.

Horse cells:
64 chromosomes

×

Donkey cells:
62 chromosomes

Meiosis

Horse egg has
32 chromosomes.

Donkey sperm has
31 chromosomes.

Mule: 63 chromosomes

(b) Why mules are sterile

Mule cell

d_1 h_1

d_2

h_2 h_3

Horse
chromosome

Donkey
chromosome

Metaphase I of meiosis

h_1 h_3 d_2

d_1 h_2

The chromosomes are from different species with different numbers of chromosomes, so they are unable to pair during the first part of meiosis.

Figure 12.2 Reproductive isolation between horses and donkeys.
(a) A cross between a female horse and a male donkey produces a mule with 63 chromosomes. (b) Mules produce only very few eggs or sperm because their chromosomes cannot pair properly during meiosis. Only a small number of chromosomes are illustrated to simplify the drawing.

Visualize This: If donkeys had 60 chromosomes so that their sperm contained 30, a mule would have 62 chromosomes. Would having an even number of chromosomes permit mules to be fertile?

Given the definition of a biological species, it is clear that all humans belong to the same biological species. To understand the concept of races within a species, however, we must first examine how species form.

An Overview: Speciation

According to the theory of common descent, all modern organisms descended from a common ancestral species. The evolution of one or more species from an ancestral form is called **speciation.**

For one species to give rise to a new species, most biologists agree that three steps are necessary:

1. Isolation of the gene pools of subgroups, or **populations,** of the species;
2. Evolutionary changes in the gene pools of one or both of the isolated populations; and
3. The evolution of reproductive isolation between these populations, preventing any future gene flow.

Recall that gene flow occurs when reproduction is occurring within a species. Now imagine what would happen if two populations of a species became physically isolated from each other, so that the movement of individuals between these two populations was impossible. Even without genetic or behavioral barriers to mating between these two populations, gene flow between them would cease.

What is the consequence of eliminating gene flow between two populations? It is identical to what occurs in separate biological species. New alleles that arise in one population may not arise in the other. Thus, a new allele may become common in one population, but it may not exist in the other. Even among existing alleles, one may increase in frequency in one population but not in the other. In this way, each population would be evolving independently.

Over time, the traits found in one population begin to differ from the traits found in the other population. In other words, the populations begin to diverge (**Figure 12.3**). When the divergence is great enough, reproductive isolation can occur. *See the next section for **A Closer Look** at the process of speciation.*

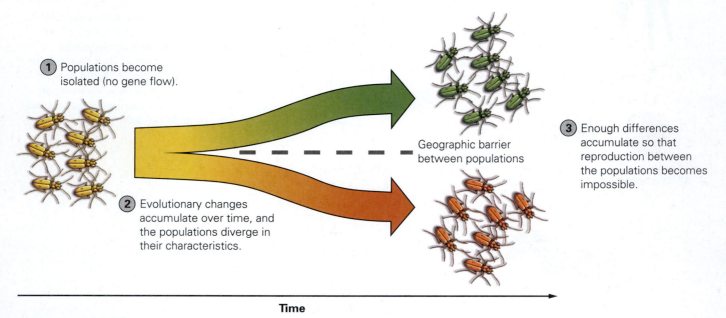

1. Populations become isolated (no gene flow).

2. Evolutionary changes accumulate over time, and the populations diverge in their characteristics.

Geographic barrier between populations

3. Enough differences accumulate so that reproduction between the populations becomes impossible.

Time

Figure 12.3 Speciation. Isolated populations diverge in traits. Divergence can lead to reproductive isolation and thus the formation of new species.

A Closer Look:
Speciation

Speciation is a historical process and thus likely to be different and somewhat unique in every group of organisms. However, there are patterns in nature that give us some insight into how and why speciation may happen.

Isolation and Divergence of Gene Pools

The gene pools of populations may become isolated from each other for several reasons. Often, a small population becomes isolated when it migrates to a location far from the main population. This is the case on many oceanic islands, including the Galápagos and Hawaiian islands. Bird, reptile, plant, and insect species on these islands appear to be the descendants of species from the nearest mainland. The original ancestral migrants arrived on the islands by chance.

Because it is rare for organisms to find their way across hundreds of miles of open ocean to these islands, populations at each site are nearly completely isolated from their ancestral population (**Figure 12.4**). In addition, because migrant populations are small, their gene pools can change rapidly via the process of genetic drift, as described in section 12.3.

The establishment of a new population far from the original population may lead to the evolution of several new species—an idea known as the **founder hypothesis.** According to this hypothesis, the diversity of unique species on oceanic islands, as well as in isolated bogs, caves, and lakes, resulted from colonization of these once "empty" environments by one founding species that rapidly diversified into many species by taking advantage of many different resources. For example, more than 50 species of silversword are now found on the Hawaiian islands, all descendants of one original founding plant population.

Populations may also become isolated from each other by the intrusion of a geologic barrier. This could be an event as slow as the rise of a mountain range or as rapid as a sudden change in the course of a river. The emergence of the Isthmus of Panama between 3 and 6 million years ago represents one such intrusion event. This land bridge ultimately connected South and North America but divided the Pacific Ocean and the Caribbean Sea. Scientists have described dozens of pairs of aquatic species—each made up of one member living in the Caribbean and one in the Pacific Ocean—whose members have diverged from each other since the rise of the isthmus. Populations that are isolated from each other by distance or a barrier are known as **allopatric.**

However, separation between the gene pools of two populations may occur even if the populations are living near each other, that is, if they are **sympatric.** This appears to be the case in populations of the apple maggot fly. Apple maggot flies are notorious pests of apples grown in northeastern North America. However, apple trees are not native to North America; they were first introduced to this continent less than 300 years ago. Apple maggot flies also infest the fruit of hawthorn shrubs, a group of species that are native to North America. Apples and hawthorns live in close proximity, and apple maggot flies clearly have the ability to

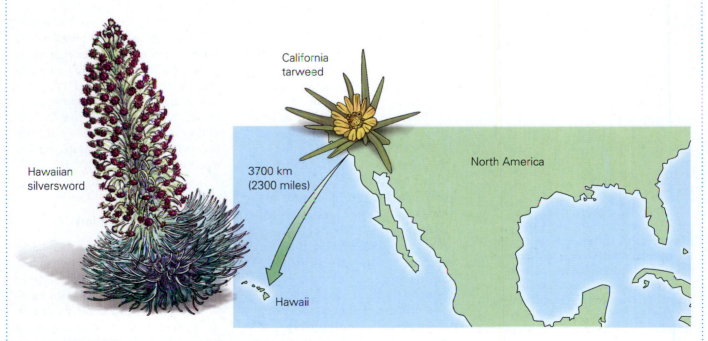

Figure 12.4 Migration leads to speciation. California tarweed seeds were blown or carried by birds to the Hawaiian Islands, creating an isolated population. With no gene flow between the two populations, a very different group of species, Hawaiian silverswords, evolved from the tarweed colonists.

(continued on the next page)

(A Closer Look continued)

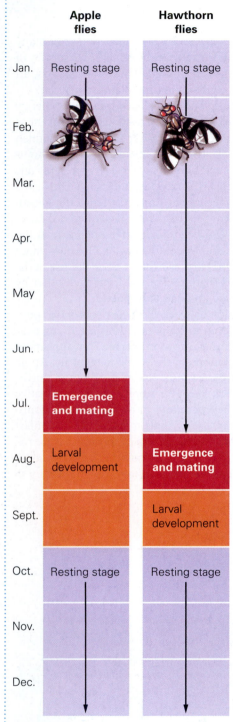

Figure 12.5 Differences in the timing of reproduction can lead to speciation. This graph illustrates the life cycle of two populations of the apple maggot fly: one that lives on apple trees and one that lives on hawthorn shrubs. The mating period for these two populations differs by a month, resulting in little gene flow between them.

fly between apple orchards and hawthorn shrubs. At first glance, it does not appear that the apple maggot flies that eat apples and those that eat hawthorn fruit could be isolated from each other.

However, on closer inspection, it is clear that flies on apples and those on hawthorns actually have little opportunity to mate. Flies mate on the fruit where they will lay their eggs, and hawthorns produce fruit approximately 1 month after apples do. Each population of fly has a strong preference regarding the fruit on which it will mate, and flies that lay eggs on hawthorns develop much faster than flies that lay eggs on apples. There appears to be little gene flow between the apple-preferring and hawthorn-preferring populations.

The gene pools of the two groups of apple maggot flies differ strongly in the frequency of some alleles. Thus it appears that divergence of two populations can occur even if those populations are in contact with each other as long as some other factor—in this case, the timing of reproduction resulting from variation in fruit preference—is keeping their gene pools isolated (**Figure 12.5**). While still not considered separate biological species, these two populations may continue to diverge and eventually become reproductively incompatible.

In plants, isolation of gene pools can occur instantaneously and without any barriers between populations. A simple hybrid between two plant species is typically infertile because it cannot make gametes (the same problem that occurs in mules; review Figure 12.2). However, some hybrid plants can become fertile again—if a mistake during mitosis produces a cell containing duplicated chromosomes. The process of chromosome duplication is called **polyploidy**, and it results in a cell that contains two copies of each chromosome from each parent species. If polyploidy

occurs inside a plant bud, all the cells of the branch that arises from that bud will be polyploid, containing two identical copies of every chromosome (**Figure 12.6**).

Because polyploid cells now contain pairs of identical chromosomes, meiosis can proceed, and thus flowers produced on the branch can produce eggs and sperm. Since most plants produce both types of gametes in a flower, a polyploid flower can self-fertilize and give rise to hundreds of offspring, representing a brand new species that is isolated from its parent plants. One example of species that formed via polyploidy is canola (*Brassica rapa*), an important agricultural crop grown for the oil produced in its seeds. Canola developed as a result of chromosome duplication in a hybrid of kale (*Brassica oleracea*) and turnip (*Brassica campestris*).

Polyploidy can result in speciation even in nonhybrid plants. For example, by chance, the geneticist Hugo de Vries discovered individual evening primrose plants in his research garden that had 28 chromosomes—twice as many as other plants in the garden. On investigation, he found that these plants were unable to produce viable offspring with evening primrose plants having the normal number of 14 chromosomes. Recent research suggests that this process of "instantaneous speciation" may have been a key factor in the evolution of as many as 50% of flowering plant species. Polyploidy occurs in some animal groups, such as insects and frogs, as well.

The Evolution of Reproductive Isolation

To become truly distinct biological species, diverging populations must become reproductively isolated either by their behavior or by genetic incompatibility. In the case of canola, genetic incompatibility occurs immediately—a cross between canola and kale does not result in offspring. In most animals, the process may be more gradual, occurring

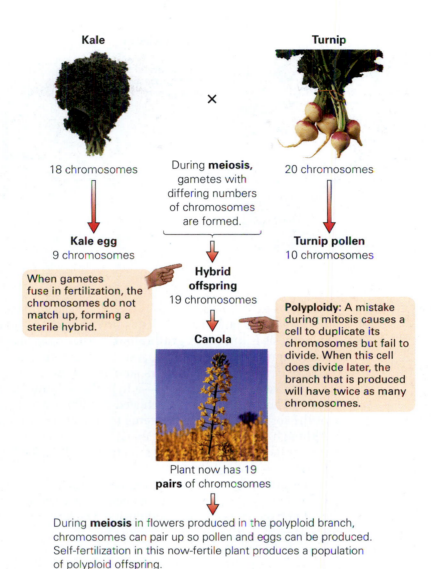

Kale

18 chromosomes

×

During **meiosis,** gametes with differing numbers of chromosomes are formed.

Turnip

20 chromosomes

Kale egg
9 chromosomes

Turnip pollen
10 chromosomes

When gametes fuse in fertilization, the chromosomes do not match up, forming a sterile hybrid.

Hybrid offspring
19 chromosomes

Canola

Polyploidy: A mistake during mitosis causes a cell to duplicate its chromosomes but fail to divide. When this cell does divide later, the branch that is produced will have twice as many chromosomes.

Plant now has 19 **pairs** of chromosomes

During **meiosis** in flowers produced in the polyploid branch, chromosomes can pair up so pollen and eggs can be produced. Self-fertilization in this now-fertile plant produces a population of polyploid offspring.

Figure 12.6 Instantaneous speciation. Canola evolved from a hybrid of kale and turnip via chromosome duplication. Because it has a different number of chromosomes from either of its parents, canola pollen cannot fertilize kale or turnip plant eggs and vice versa.

when the amount of divergence has caused numerous genetic differences between two populations.

There is no hard-and-fast rule about how much divergence is required; sometimes, a difference in a single gene can lead to incompatibility, while at other times, populations demonstrating great physical differences can produce healthy and fertile hybrids (**Figure 12.7**). Exactly how reproductive isolation evolves on a genetic level is still unknown and is an actively researched question in biology.

Once reproductively isolated, species that derived from a common ancestor can accumulate many differences, even completely new genes.

(a)

(b)

Figure 12.7 How different are two species? (a) These two species of dragonfly look alike but cannot interbreed. (b) Dog breeds provide a dramatic example of how the evolution of large physical differences does not always result in reproductive incompatibility.

(a) Gradualism

Gradual natural selection (and other processes) leads to divergence.

Time

(b) Punctualism

Rapid natural selection (and other processes) leads to divergence.

Time

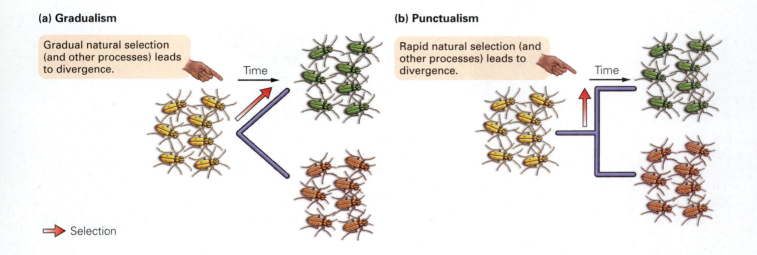

➡ Selection

Figure 12.8 Punctuated equilibrium. The pattern of evolutionary change in groups of species may be (a) smooth, representing a constant level of small changes, or (b) more punctuated, with hundreds of thousands of years of stasis followed unpredictably by rapid, large changes occurring within a few thousand years.

Is Speciation Gradual or Sudden? Darwin assumed that speciation occurred over millions of years as tiny changes gradually accumulated. This hypothesis is known as **gradualism.** Other biologists have argued that most speciation events are sudden, result in dramatic changes in form within the course of a few thousand years, and are followed by many thousands or millions of years of little change—a hypothesis known as **punctuated equilibrium.** The hypothesis of punctuated equilibrium is supported by observations of the fossil record, which seems to reflect just this pattern (**Figure 12.8**). Although the tempo of evolutionary change may not match Darwin's predictions, the process of natural selection he described can still explain many instances of divergence.

Regardless of the tempo of speciation, the period after the separation of the gene pools of two populations but before the evolution of reproductive isolation could be thought of as a period during which races of a species may form.

12.2 Are Human Races Biological?

Biologists do not agree on a standard definition of *biological race*. In fact, not all biologists feel that *race* is a useful term; many prefer to use the term *subspecies* to describe subgroups within a species, and others feel that race is not a useful biological concept at all. When the term is applied, it is often inconsistent. For example, populations of birds with slightly different colorations might be called different races by some bird biologists, while other biologists would argue that the same variations in color are meaningless.

However, our question about how race matters in practical terms leads us to a specific definition of race. What we want to know is: If an individual is identified as a member of a particular race, does that mean she is more closely related and thus biologically *more similar* to individuals of the *same* race than she is to individuals of *other* races? The definition of **biological race** that addresses this question is the following: Races are populations of a single species that have diverged from each other as a result of isolation of their gene pools. Biologists may alternatively refer to this as the **genealogical species** concept because it reflects closer shared ancestry—a genealogy—among certain individuals within a biological species. With little gene flow among races, evolutionary changes that occur in one race may not occur in a different race. However, to understand if the racial categories people in the United States identify with have a biological basis—that is, are true genealogical species—we must first understand how these racial categories came to be identified in American society.

The History of Human Races

Until the height of the European colonial period in the seventeenth and eighteenth centuries, few cultures distinguished between groups of humans according to shared physical characteristics. People primarily identified themselves and others as belonging to particular cultural groups with different customs, diets, and languages.

As northern Europeans began to contact people from other parts of the world, being able to set groups of people "apart" made colonization and slavery less morally troublesome. Thus, when Linnaeus classified all humans as a species, he was careful to distinguish definitive varieties (what we would now call races) of humans. Linnaeus recognized five races of *Homo sapiens*. Not only did Linnaeus describe physical characteristics, but also he ascribed particular behaviors and aptitudes to each race; reflecting his Western bias, he set the European race as the superior form (**Figure 12.9**).

The classification shown in Figure 12.9 is one of dozens of examples of how scientists' work has been used to legitimize cultural practices. In this case, hundreds of years of injustice were supported by a scientific classification that made the lower status of nonwhite groups in European and American society appear "natural."

Scientists since Linnaeus have also proposed hypotheses about the number of races of the human species. The most common number is 6 (white, black, Pacific Islander, Asian, Australian Aborigine, and Native American), although some scientists have described as many as 26 different races of the human species. Personal genotyping services, companies that analyze the genes of individuals to determine their ancestry, cut human diversity even more finely—some claim to be able to distinguish more than 50 population groups. Does each of these groups represent a biological race?

To answer our question about the biology of race, we can try to determine if the physical characteristics used to delineate supposed human races—skin color, eye shape, and hair texture, for instance—developed because these groups evolved independently of each other. We can test this hypothesis by

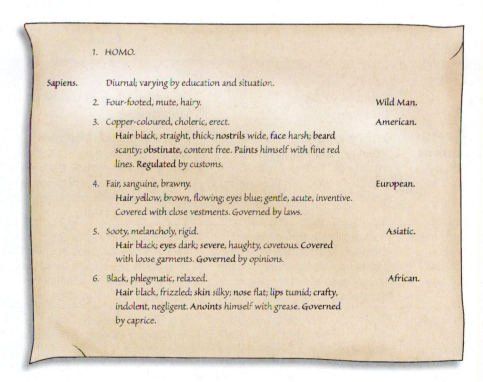

Figure 12.9 Linnaean classification of human variety. Linnaeus published this classification of the varieties of humans in the 10th edition of *Systema Naturae* in 1758. The text reflects a widespread bias among his contemporaries that the European race was superior to other races. Interestingly, one variety of humans recognized by Linnaeus—the "Wild Man"—is the chimpanzee.

looking at the fossil record for evidence of isolation during human evolution and by looking at the gene pools of modern populations of these supposed races for consistent differences among groups.

The Morphological Species Concept

The ancestors of humans are known only through the fossil record. We cannot delineate fossil species using either the biological or genealogical species concepts. Instead, paleontologists, scientists who study fossils, use a more practical definition: A species is defined as a group of individuals with some reliable physical characteristics distinguishing them from all other species. In other words, individuals in the same species have similar morphology—they look alike in some key feature. This is known as the **morphological species concept.** The morphological differences among species are assumed to correlate with isolation of gene pools. **Table 12.2** compares and contrasts the three species concepts.

Stop & Stretch   Which species concept, biological, genealogical, or morphological, do you think is more likely to be used by biologists who are trying to count the number of bird species in a local park? Why?

Using the morphological species concept, scientists have identified the fossils of our direct human ancestors. This has allowed them to reconstruct the movement of humans since our species' first appearance.

Modern Humans: A History

The immediate predecessor of *Homo sapiens* was *Homo erectus*, a species that first appeared in east Africa about 1.8 million years ago and spread to Asia and Europe over the next 1.65 million years. Fossils identified as early *H. sapiens* appear in Africa in rocks that are approximately 250,000 years old. The fossil record shows that these early humans rapidly replaced *H. erectus* populations in Africa, Europe, and Asia.

TABLE 12.2

Comparison of three species concepts.

Species Concept	Definition	Benefits of Using This Concept	Disadvantages of Using This Concept
Biological	Species consist of organisms that can interbreed and produce fertile offspring and are reproductively isolated from other species.	Useful in identifying boundaries between populations of similar organisms. Relatively easy to evaluate for sexually reproducing species.	Cannot be applied to organisms that reproduce asexually or to fossil organisms. May not be meaningful when two populations of the same species are separated by large geographical distances.
Genealogical	Species consist of organisms that can interbreed, are all descendants of a common ancestor, and represent independent evolutionary lineages.	Most evolutionarily meaningful because each species has its own unique evolutionary history. Can be used with asexually reproducing species.	Difficult to apply in practice. Requires detailed knowledge of gene pools of populations within a biological species. Cannot be applied to fossil organisms.
Morphological	Species consist of organisms that share a set of unique physical characteristics that is not found in other groups of organisms.	Easy to use in practice on both living and fossil organisms. Only a few key features are needed for identification.	Does not necessarily reflect evolutionary independence from other groups.

Most data support the hypothesis that all modern human populations descended from these African *H. sapiens* ancestors within the last few hundred thousand years. One line of evidence supporting this is that humans have much less genetic diversity (measured by the number of different alleles that have been identified for any gene) than any other great ape, indicating that there has been little time to accumulate many different gene variants. Using the same reasoning, we know that African populations must be the oldest human populations because they are more genetically diverse than others around the world. Thus, African populations are likely the source of all other human populations.

Given the evidence of recent African ancestry, the physical differences we see among human populations must have arisen in the last 150,000 to 200,000 years or in about 10,000 human generations. In evolutionary terms, this is not much time. All humans shared a common ancestor very recently; thus, the defined human races cannot be very different from each other.

Genetic Evidence of Divergence

Even with little genetic difference among populations, our question about the meaning of race is still relevant. After all, even if two races differ from one another only slightly, if the difference is consistent, then people are biologically more similar to members of their own race than to people of a different race. The ancestry results produced by personal genotyping services seem to indicate that a person can measure what percentage of their genome comes from any particular racial group. Do these results indicate that someone who has 74% European ancestry has more biological similarities to other Europeans than to Africans or Asians?

Researchers can examine the gene pool of populations described as a single race to determine if the race was once truly isolated from other races. Remember that when populations are isolated from each other, little gene flow occurs between them. If an allele appears in one population, it cannot spread to another. As a result, isolated populations should contain some unique alleles.

In addition to finding unique alleles, researchers should be able to observe gene pool differences among isolated populations in the frequency of particular alleles. When a trait becomes more common in a population due to evolution, it is because the allele for that trait has become more common (Chapter 10). In other words, evolution results in a change in **allele frequency** in a population, that is, in the percentage of copies of any given gene that are a particular allele. For example, consider a population of 50 people. Because every person carries 2 copies of each gene, there are actually 100 copies of any gene in this population. Many genes have several different forms; for example, the blood group gene has at least three alleles—O, A, and B (discussed in Chapter 8). Now, imagine that nearly everyone in a population carries two copies of the allele that codes for blood type O, but that two individuals carry one copy of the O allele and one copy of the allele that codes for blood type B. The frequency of the B allele is thus 2 out of 100, or 2%, in this population. An evolutionary change in blood type in this population would be seen as an increase or decrease in the frequency of the B allele in the next generation. If populations are isolated from each other, changes in allele frequency that occur in one population will not necessarily occur in another.

We can now make two predictions to test a hypothesis of whether biological races exist within a species. If a race has been isolated from other populations of the species for many generations, it should have these two traits:

1. Some unique alleles
2. Differences in allele frequency, relative to other races, for some genes

Before we answer the question of whether the proposed human races demonstrate these two traits, we need to learn a bit about how allele frequencies are calculated.

Using the Hardy-Weinberg Theorem to Calculate Allele Frequencies

Allele frequencies in a population can be determined using the mathematical relationship independently discovered by Godfrey Hardy of Cambridge University and German physician Wilhelm Weinberg. The **Hardy-Weinberg theorem** states that allele frequencies will remain stable in populations that are large in size, randomly mating, and experiencing no migration or natural selection. This rule provides a baseline for predicting how allele frequencies will change if any of these conditions are violated. In other words, the Hardy-Weinberg theorem enables scientists to quantify the effect of evolutionary change on allele frequencies. Today, the Hardy-Weinberg theorem forms the basis of the modern science of **population genetics.**

In the simplest case, the Hardy-Weinberg theorem (which we abbreviate to HW) describes the relationship between allele frequency and genotype frequency for a gene with two alleles in a stable population. The frequencies of these two alleles are written as p and q.

Let us say that 70% of the alleles in the population are dominant (A), and 30% are recessive (a) (**Figure 12.10a**). Thus, $p = 0.7$ and $q = 0.3$. Each gamete produced by members of the population carries one copy of the gene. Therefore, 70% of the gametes produced by this entire population will carry the dominant allele, and 30% will carry the recessive allele. The frequency of gametes produced of each type is equal to the frequency of alleles of each type.

HW assumes that every member of the population has an equal chance of mating with any member of the opposite sex. The fertilizations that occur in this situation are analogous to the result of a lottery drawing. In this analogy, we can imagine individuals in a population each contributing an equal number of gametes to a "bucket." Fertilizations result when a gamete drawn from the sperm bucket fuses with another drawn from the egg bucket. Because the frequency of gametes carrying the dominant allele in the bucket is equal to the frequency of the dominant allele in the population, the chance of drawing an egg that carries the dominant allele is 70%.

In **Figure 12.10b**, a modified Punnett square illustrates the relationship between allele frequency and genotype frequency in a stable population. On the horizontal axis of the square, we place the two types of gametes that can be produced by females in the population (A and a), while on the vertical axis we place the two types of gametes that can be produced by males. In addition, on each axis is an indication of the frequency of these types of egg and sperm in the population: 0.7 for A eggs and A sperm, 0.3 for a eggs and a sperm.

Used like the typical Punnett square (Chapter 7), the grid of the square also shows the frequency of each genotype in this population. The frequency of the AA genotype in the next generation will be equal to the frequency of A sperm being drawn (0.7) times the frequency that A eggs will be drawn (0.7), or 0.49. The frequency of the AA genotype is thus $p \times p$ ($=p^2$). This calculation can be repeated for each genotype. The frequency of the aa genotype $q \times q$ ($=q^2$), and the Aa genotype $p \times q \times 2$ ($=2pq$). By this theorem, Hardy and Weinberg mathematically proved that the frequency of genotypes in one generation of a population depends on the frequency of genotypes in the previous generation.

Once this relationship between allele and phenotype frequency became clear, scientists could use the known frequency of a phenotype produced by a recessive allele to calculate the allele frequency. For instance, if the frequency of individuals with blood type O is 1 in 100 births (0.01), then q^2—the frequency of homozygous recessive individuals in the population—is equal to 0.01. Therefore, q is simply the square root of this number, or 0.1. In this population, the frequency of the type O allele is 0.1, or 10%.

(a) Parental generation: 70% **A** ($p = 0.7$) and 30% **a** ($q = 0.3$)

Gametes: 14 out of 20 are **A** ($p = 0.7$) and 6 out of 20 are **a** ($q = 0.3$)

(b)

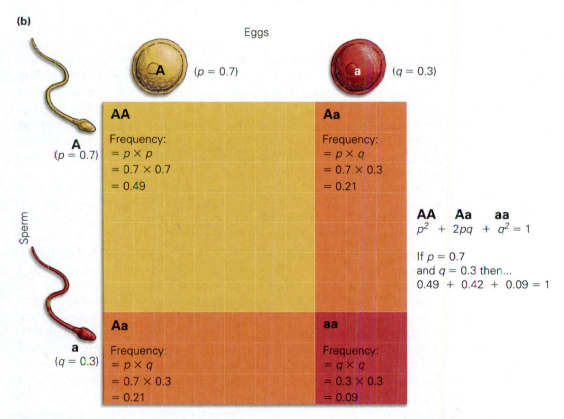

Figure 12.10 The relationship between allele frequency and gamete frequency.
(a) The frequency of any allele in a population of adults is equal to the frequency of that allele in gametes produced by that population. (b) Knowing the allele frequency of gametes allows us to predict the frequency of various genotypes in the next generation.
Visualize This: Why is the frequency of heterozygotes *2pq* instead of just *pq*?

The Human Genome Project has made the HW theorem obsolete for calculating allele frequencies in a population from genotype frequencies—now scientists can use DNA technology to determine how common an allele is in a population. However, the Hardy-Weinberg theorem is still useful in studies of human variation for identifying populations that are surprisingly different in allele frequency. Recall that one of the assumptions of the theorem is that the population is randomly mating. If groups of the population have been (or are) isolated, this assumption is not true—it is impossible to mate randomly if certain individuals can never come in contact with certain other individuals. A statistical test can identify when an allele does not conform to HW expectations and is thus useful for tracing human ancestry.

Human Races Are Not Isolated Biological Groups

Recall the six major human races described by many authors: white, black, Pacific Islander, Asian, Australian Aborigine, and Native American. Do these groups show the predicted pattern of race-specific alleles and unique patterns of allele frequency? Yes, but not in a way that provides convincing evidence of consistent difference among these groups. Instead, most evidence indicates that the genetic differences between unrelated individuals within a race are much greater than the average genetic differences between races.

No Alleles Are Found in All Members of a Race. The Human Genome Project has allowed researchers to scan the genome of thousands of individuals looking for evidence of alleles that are found in a single human group and not in other groups. The types of alleles that are most commonly used in this analysis are called **single nucleotide polymorphisms,** abbreviated SNPs ("snips"). A SNP is a single base pair in the DNA sequence of humans that can differ from one individual to another. Humans are remarkably similar, having identical gene sequences for 99% of our genome. The 1% of the genome where there is variability is primarily made up of SNPs. We can think of the different DNA bases found at a particular SNP site as different alleles.

The primary reason scientists are interested in identifying SNPs in the human genome is to understand the source of human diversity in disease susceptibility and other traits that affect health and well-being. However, because 99% of SNPs appear to be in parts of the genome that do not produce proteins, they may have little or no effect on evolutionary fitness and thus can be readily passed on to future generations. You can imagine that when a nonharmful SNP allele arises via mutation, it can readily spread throughout a human population as a result of evolution. For this reason, it is the more "neutral" SNPs that are most useful for understanding ancestry.

Researchers have identified a number of SNP alleles that are unique to particular human populations. Most genetic ancestry testing companies identify three major groups: African, European, and Asian (including Native American). When Elizabeth Alexander learned that she is genetically 27% African, the results were based on a search of her DNA for African-specific SNP alleles. Within those larger groups, certain populations display unique SNP alleles that may help identify an individual's ancestry more specifically. However, it is important to note that no SNP allele is found in every individual in any population, and among groups classified in the same major race, some populations may have no individuals who have a particular SNP allele considered unique to a race.

What is true of SNP alleles may be more clearly illustrated with alleles associated with readily apparent phenotypes. For example, sickle-cell anemia has long been thought of as a "black" disease. This illness occurs in individuals who carry two copies of the sickle-cell allele, resulting in red blood cells that deform into a sickle shape under certain conditions. The consequences of these sickling attacks include severe pain, heart, kidney, lung, and brain damage. Many individuals with sickle-cell anemia do not live past childhood.

Nearly 10% of African Americans and 20% of Africans carry one copy of the sickle-cell allele, whereas the allele is almost completely absent in European Americans. However, if we examine the distribution of the sickle-cell allele more closely, we see that this seemingly race-specific pattern is not so straightforward. Not all human populations classified as black have a high frequency of the sickle-cell allele. In fact, in populations from southern and north-central Africa, which are traditionally classified as black, this allele is very rare or absent. Among populations who are classified as white or Asian, there are some in which the sickle-cell allele is relatively common, such as white populations in the Middle East and Asian populations in northeast India (**Figure 12.11**). Thus, the sickle-cell allele is not a characteristic of all black populations or unique to a supposed "black race."

Similarly, cystic fibrosis, a disease that results in respiratory and digestive problems and early death, was often thought of as a "white" disease. Cystic fibrosis occurs in individuals who carry two copies of the cystic fibrosis allele. As with sickle-cell anemia, it has become clear that the allele that causes cystic fibrosis is not found in all white populations and is found, in low frequency, in some black and Asian populations.

These examples of the sickle-cell allele and cystic fibrosis allele demonstrate the typical pattern of gene distribution. Scientists have not identified a single allele that is found in all (or even most) populations of a commonly described race but not found in other races. Only a tiny number of SNPs have been identified as unique to a particular human racial group, and these are never found in all populations of the race or within every individual in a population.

The hypothesis that human races represent independent evolutionary groups is not supported by these observations.

Populations Within a Race Are Often as Different as Populations Compared Across Races.

Certain SNP alleles are useful in that they link individuals to particular population groups. This is not surprising since an ancestral population tends to be associated with a particular geography, and

Figure 12.11 The sickle-cell allele: Not a "black gene." The map illustrates where the sickle-cell allele is found in human populations. Note that it is not found in all African populations but is found in some European and Asian populations.

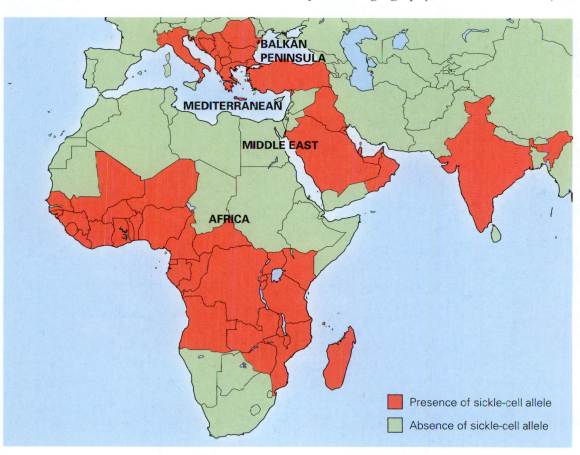

Presence of sickle-cell allele

Absence of sickle-cell allele

people living close to each other are likely to share more ancestors (and thus be more genetically similar) than people living far apart. However, if race is to be biologically meaningful, then the allele frequency for *many different* SNPs and genes should be more similar among populations within a race than between populations of different races.

Again, the pattern of SNP allele frequency among populations can be illustrated by more obvious alleles. The human populations in each part of **Figure 12.12** are listed by increasing frequency of an allele in the population. The color coding of each population group in each of the graphs corresponds to the racial category in which the population is typically placed. For example, at the top of 12.12a, the Taiwanese population is categorized as Asian, while the Cree Indian population of eastern Canada is categorized as Native American. If the hypothesis that human racial groups have a biological basis is correct, then populations from the same racial group should be clustered together on each bar graph.

Figure 12.12a shows the frequency of the allele that interferes with an individual's ability to taste the chemical phenylthiocarbamide (PTC) in several populations. People who carry two copies of this recessive allele cannot detect PTC, which tastes bitter to people who carry one or no copies of the allele. Note that members of a single "race," such as Asian, vary widely in frequency of this allele.

Figure 12.12b lists the frequency of one allele for the gene *haptoglobin 1* in a number of different human populations. Haptoglobin 1 is a protein that helps scavenge the blood protein hemoglobin from old, dying red blood cells. Again, we see a wide distribution in allele frequency within the race categories.

Figure 12.12 Do human races show genetic evidence of isolation? The bars on each of these histograms illustrate the frequency of the described allele in many different human populations. These histograms illustrate that populations within these "races" are not necessarily more similar to each other than they are to populations in different races.

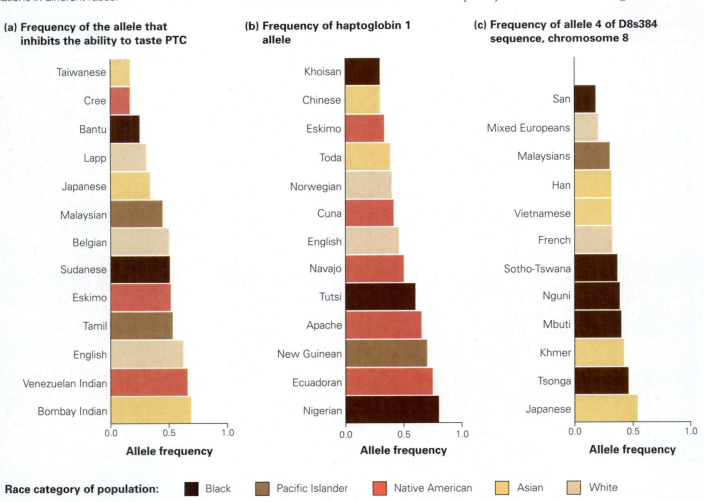

Race category of population: Black — Pacific Islander — Native American — Asian — White

Figure 12.12c illustrates variation among human populations in the frequency of a repeating DNA sequence on chromosome 8. Repeating sequences are common in the human genome, and differences among individuals in the number of repeats create the unique signatures of DNA fingerprints. The frequency of one pattern of repeating sequence in a segment of chromosome 8, called allele 4 of the D8s384 sequence, is illustrated for a number of populations.

What we see in the three graphs in Figure 12.12 is that allele frequencies for these genes are *not* more similar within racial groups than between racial groups. In fact, in two of the three graphs, Figure 12.12a and 12.12b, the populations with the highest and lowest allele frequencies belong to the same race—for these genes, there is more variability *within* a race than there are average differences *among* races. Even though certain SNPs can help us track an individual to a particular ancestral population, the pattern of diversity in other genes tells us that these "ancestral" patterns are not evidence of deep biological similarities within a racial group.

Stop & Stretch   We are accustomed to looking at differences in skin color and eye shape and presuming they signify underlying biological relationships. Can you think of another physical trait that varies greatly in human populations but that is rarely used as evidence for biological relationship? How would including this trait affect the classification of races?

Human races fail to meet the criteria for identifying populations as consistently isolated from each other. Both the fossil evidence and genetic evidence indicate that the six commonly listed human racial groups do *not* represent biological races.

Human Races Have Never Been Truly Isolated

The results of genetic ancestry tests on individuals in the United States typically reflect the mixing that has occurred as a result of the past 300 years of European and Asian immigration and the history of the African slave trade. Recent research indicates that most African Americans have at least 20% European SNPs, and that 30% of white U.S. college students have less than 90% European SNPs. But the "melting pot" of the United States is not unique in human history. In fact, the evidence that human populations have been "mixing" since modern humans first evolved is contained within the gene pool of human populations. For instance, the frequency of the B blood group decreases from east to west across Europe (**Figure 12.13** on the next page). The allele that codes for this blood type apparently evolved in Asia, and the pattern of blood group distribution seen in Figure 12.13 corresponds to the movement of Asians into Europe beginning about 2000 years ago. As the Asian immigrants mixed with the European residents, their alleles became a part of the European gene pool. Populations closest to Asia experienced a large change in their gene pools, while populations who were more distant encountered a more "diluted" immigrant gene pool made up of the offspring between the Asian immigrants and their European neighbors.

Other genetic analyses have led to similar maps. For example, one indicates that populations that practiced agriculture arose in the Middle East and migrated throughout Europe and Asia about 10,000 years ago. These data indicate that there are no clear boundaries within the human gene pool. Interbreeding of human populations over hundreds of generations has prevented the isolation required for the formation of distinct biological races.

Figure 12.13 Genetic mixing in humans. The decline in frequency of type B blood from west to east in Europe reflects the movement of alleles from Asian populations into European populations over the past 2000 years.

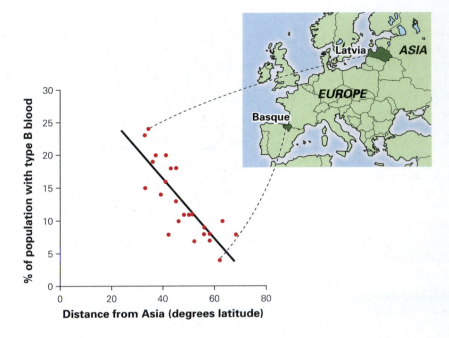

12.3 Why Human Groups Differ

Human races are not true biological races. However, as is clear to all of us, human populations do differ from each other in many traits. In this section, we explore what is known about why populations share certain superficial traits and differ in others.

Natural Selection

Recall the distribution of the sickle-cell allele in human populations shown in Figure 12.11. This allele is found in some populations of at least three of the typically described races. Why is it higher in certain populations?

The sickle-cell allele is higher in certain populations because in particular environments natural selection favors individuals who carry one copy of it. The sickle-cell anemia allele is an *adaptation,* a feature that increases fitness, within populations in malaria-prone areas. Malaria is a disease caused by a parasitic, single-celled organism that spends part of its life cycle feeding on red blood cells, eventually killing the cells. Because their red blood cells are depleted, people with severe malaria suffer from anemia, which may result in death. When individuals carry a single copy of the sickle-cell allele, their blood cells deform when infected by a malaria parasite. These deformed cells quickly die, reducing the ability of the parasite to reproduce and infect more red blood cells and therefore reducing a carrier's risk of anemia.

The sickle-cell allele reduces the likelihood of severe malaria, so natural selection has caused it to increase in frequency in susceptible populations. The protection that the sickle-cell allele provides to heterozygote carriers is demonstrated by the overlap between the distribution of malaria and the distribution of sickle-cell anemia (**Figure 12.14**).

Another physical trait that has been affected by natural selection is nose form. The pattern of nose shape in populations generally correlates to climate factors—populations in dry climates tend to have narrower noses than do populations in moist climates. A long, narrow nose has a greater internal surface area, exposing inhaled air to more moisture, reducing lung damage, and increasing the fitness of individuals in dry environments. Among tropical Africans, people living at drier high altitudes have much narrower noses than do those living in humid rain-forest areas (**Figure 12.15**).

Malaria sickle-cell overlap

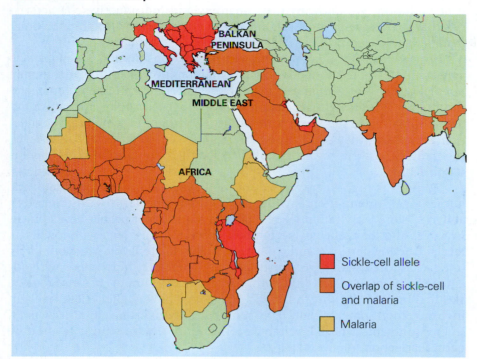

Figure 12.14 The sickle-cell allele is common in malarial environments. This map shows the distributions of the sickle-cell allele and malaria in human populations.

- Sickle-cell allele
- Overlap of sickle-cell and malaria
- Malaria

(a) Ethiopian with a narrow nose

(b) Bantu with a broad nose

Figure 12.15 Nose shape is affected by natural selection. (a) Long, narrow noses are more common among populations in cold, dry environments because they conserve heat and water. (b) Broad, flattened noses are more common in warm, wet environments because they effectively radiate body heat.

Interestingly, our preconception puts two populations of Africans (represented by the images in Figure 12.15) in the same race and explains differences in their nose shape as a result of natural selection, but we place white and black populations into different races and explain their skin color differences as evidence of long isolation from each other. However, like nose shape, skin color is a trait that is strongly influenced by natural selection.

Convergent Evolution

Traits that are shared by unrelated populations because they share similar environmental conditions are termed *convergent*. **Convergent evolution** occurs when natural selection for similar environmental factors causes unrelated organisms to resemble each other. For example, the similarity in

Figure 12.16 Convergence. The similarity in shape between dolphins and sharks results from similar adaptations to life as an oceanic predator of fish, not shared ancestry.

shape between white-sided dolphins and reef sharks is a result of convergent evolution. We know by their anatomy and reproductive characteristics that sharks are most closely related to other fish and dolphins to other mammals (**Figure 12.16**).

The pattern of skin color in human populations around the globe also appears to be the result of convergent evolution, in which unrelated human populations appear similar as a result of evolution in similar environmental conditions. When scientists compare the average skin color in a native human population to the level of ultraviolet (UV) light to which that population is exposed, they see a close correlation—the lower the UV light level, the lighter the skin, regardless of the race the population is classified as (**Figure 12.17**).

UV light is high-energy radiation in a range that is not visible to the human eye. Among its many effects, UV light interferes with the body's ability to store the vitamin folate. Folate is required for proper development in babies and for adequate sperm production in males. Men with low folate levels have low fertility, and women with low folate levels are more likely to have children with severe birth defects. Therefore, individuals with adequate folate have higher fitness than individuals without. Because darker-skinned individuals absorb less UV light, they have higher folate levels in high-UV environments than light-skinned individuals do. In other words, in environments where UV light levels are high, dark skin is favored by natural selection (**Figure 12.18**).

Human populations in low-UV environments face a different challenge. Absorption of UV light is essential for the synthesis of vitamin D. Vitamin D is crucial for the proper development of bones. Women are especially harmed by low vitamin D levels—inadequate development of the pelvic bones can make giving birth deadly. There is no risk of not making enough vitamin D when UV light levels are high, regardless of skin color. However, in areas where levels of UV light are low, individuals with lighter skin absorb a larger fraction

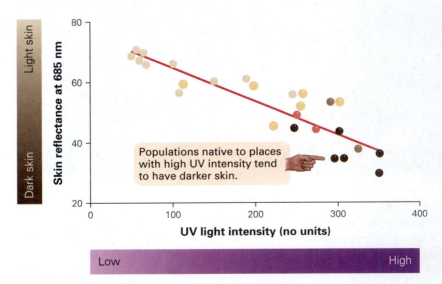

Figure 12.17 There is a strong correlation between skin color and exposure to UV light. Reflectance is an indication of color—higher reflectance indicates lighter skin. The color of the dots on the graph specifies the racial category of each population.
Visualize This: Find two populations classified as the same race that have very different skin color. Find two populations classified in different races that have the same skin color.

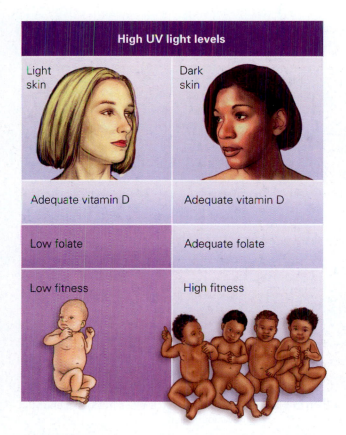

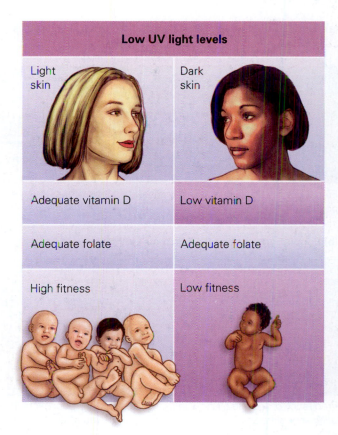

Figure 12.18 The relationship between UV light levels, folate, vitamin D, and skin color. Populations in regions where UV light levels are high experience selection for darker, UV-resistant skin. Populations in regions where UV light levels are low experience selection for lighter, UV-transparent skin.
Visualize This: What could an individual with dark skin in a low-UV environment do to compensate? What about an individual with light skin in a high-UV environment?

of UV light and thus have higher levels of vitamin D than do individuals with darker skin. Thus, in darker environments, light skin has been favored by natural selection.

Because UV light has important effects on human physiology, it has driven the evolution of skin color in human populations. Where UV light levels are high, dark skin is an adaptation, and populations become dark skinned. Where UV light levels are low, light skin is usually an adaptation, and populations evolve to become light skinned. The pattern of skin color in human populations is a result of the convergent evolution of different populations in similar environments, not necessarily evidence of separate races of humans.

Stop & Stretch   The word for mother in many different languages is a variation of the sound "ma." This appears to be an example of convergent evolution in language. Think about the first word-like sounds babies make. Why might many different cultures share this characteristic sound for mother?

Natural selection has caused differences among human populations, but it has also resulted in some populations superficially appearing more similar to some other human populations. In contrast, populations who appear on the surface to be similar may be quite different, simply by chance.

(a) Founder effect: A small sample of a large population establishes a new population.

  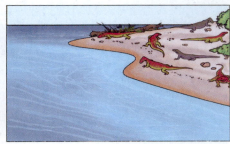

Frequency of red allele is low in original population.

Several of the travelers happen to carry the red allele.

Frequency of red allele is much higher in new population.

(b) Population bottleneck: A dramatic but short-lived reduction occurs in population size.

  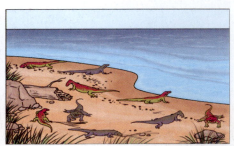

Frequency of red allele is low in original population.

Many survivors of tsunami happen to carry red allele.

Frequency of red allele is much higher in new population.

(c) Chance events in small populations: The carrier of a rare allele does not reproduce.

Frequency of red allele is low in original population.

The only lizard with red allele happens to fall victim to an eagle and dies.

Red allele is lost.

Figure 12.19 The effects of genetic drift. A population may contain a different set of alleles because (a) its founders were not representative of the original population; (b) a short-lived drop in population size caused a change in allele frequency in one of the populations; (c) one population is so small that low-frequency alleles are lost by chance.

Genetic Drift

A change in allele frequency that occurs due to chance is called **genetic drift.** Human populations tend to travel and colonize new areas, so we seem to be especially prone to evolution via genetic drift. Genetic drift occurs in three different types of situations (**Figure 12.19**).

Founder Effect. Genetic differences can occur when a small sample of a larger population establishes a new population. The gene pool of the immigrants is rarely an exact reflection of the gene pool of the source population. This difference leads to the **founder effect.**

Rare genetic diseases that are more common in certain populations may be a result of the founder effect. For example, the Amish of Pennsylvania are descended from a population of 200 German founders who immigrated to the United States over 200 years ago. Ellis-van Creveld syndrome, a recessive disease that causes dwarfism (among other effects), is 5000 times more common in the Pennsylvania Amish population than in other German American populations. This difference is a result of a single founder in that original population who carried the allele. Since the Pennsylvania Amish usually marry others within their small religious community, the allele has stayed at a high level—1 in 8 Pennsylvania Amish are carriers of the Ellis-van Creveld allele, compared to fewer than 1 in 100 non-Amish Americans of German ancestry.

Plants with animal-dispersed seeds appear to be especially prone to the founder effect. For example, cocklebur, a widespread weed that produces hitchhiker fruit designed to grab onto the fur of a passing mammal (**Figure 12.20**), consists of populations that are quite variable in size and shape. The variation among populations appears to have been caused by differences in the hitchhikers that happen to be carried to a new colony.

A variant of the founder effect is the bottleneck effect, in which a small subset of a population makes up the only survivors of a once-larger population. A sixteenth-century bottleneck on the island of Puka Puka in the South Pacific resulted in a human population that is clearly different from other Pacific island populations: The 17 survivors of a tsunami on Puka Puka were all relatively petite, and modern Puka Pukans are significantly shorter in stature compared to people native to nearby islands. Bottlenecks are experienced by nonhuman populations as well. For example, the genetic similarity among individuals in a large population of Galápagos tortoises on the island of Isabela seems to suggest a bottleneck. Geological evidence indicates that most of the tortoise population was wiped out during a volcanic eruption about 88,000 years ago. The current population apparently descended from a tiny group of survivors and thus has little genetic diversity.

Genetic Drift in Small Populations.

Even without a population bottleneck, allele frequencies may change in a population due to chance events. When an allele is in low frequency within a small population, only a few individuals carry a copy of it. If one of these individuals fails to reproduce, or if it passes on only the more common allele to surviving offspring, the frequency of the rare allele may drop in the next generation. If the population is small enough, even relatively high-frequency alleles may be lost after a few generations by this process.

A human population that illustrates the effects of genetic drift in small populations is the Hutterites, a religious sect with communities in South Dakota and Canada. Modern Hutterite populations trace their ancestry back to 442 people who migrated from Russia to North America between 1874 and 1877. Hutterites tend to marry other members of their sect, so the gene pool of this population is small and isolated from other populations. Genetic drift in this population over the last century has resulted in a near absence of type B blood among the Hutterites, as compared to a frequency of 15% to 30% in other European migrants in North America.

Genetic drift in populations that remain small for many generations can lead to a rapid loss of many different alleles. While this problem is uncommon in humans, the effects of genetic drift on small populations of endangered species can lead to extinction (Chapter 15).

Humans are a highly mobile species, and we have been founding new populations for millennia. Most early human populations were also probably quite small. These factors make human populations especially susceptible to the founder effect, genetic bottlenecks, and genetic drift and have contributed to the differences among modern human groups. However, in addition to natural selection and random genetic change, humans' highly social nature has contributed to superficial differences in the appearance of different populations.

Figure 12.20 An example of the founder effect in plants. Cocklebur is a hitchhiker plant, dispersing by hooking the spikes on its fruits to the fur (or sock) of a passing animal. When these burrs are removed at a distant location, the plant can found a new population that might be quite different in size and shape from the source population.

(a) Peacock

(b) Lion

(c) Blue morpho butterfly

Figure 12.21 The effects of sexual selection. Sexual selection is responsible for many unique and fantastic characteristics of organisms from (a) the peacock's tail to (b) the male lion's mane to (c) the bright colors of butterflies.

Sexual Selection

Men and women within a population may have preferences for particular physical features in their mates. These preferences can cause populations to differ in appearance. When a trait influences the likelihood of mating, that trait is under the influence of a form of natural selection called **sexual selection.**

Darwin hypothesized sexual selection in 1871 as an explanation for differences between males and females within a species. For instance, the enormous tail on a male peacock results from female peahens that choose mates with showier tails. Because large tails require so much energy to display and are more conspicuous to their predators, peacocks with the largest tails must be both physically strong and smart to survive. Peahens can use the size of the tail, therefore, as a measure of the "quality" of the male. Tail length does appear to be a good measure of overall fitness in peacocks; the offspring of well-endowed males are more likely to survive to adulthood than are the offspring of males with scanty tails. When a peahen chooses a male with a large tail, she is ensuring that her offspring will receive high-quality genes. Sexual selection explains the differences between males and females in many species (**Figure 12.21**).

Stop & Stretch In most cases of sexual selection in animals, it is female choice that drives the evolution of traits in the male. Why do you think there is more incentive for females to choose the right male than vice versa?

In humans, there is some evidence that the difference in overall body size between men and women is a result of sexual selection—namely, a widespread female preference for larger males—perhaps again because size may be an indication of overall fitness. However, some apparently sexually selected traits seem to have little or no relationship to fitness and reflect simply a "social preference." In our highly social species, this type of sexual selection may be common.

For example, some scientists have suggested that lack of thick facial and body hair in many Native American and Asian populations resulted from selection by both men and women for less hairy mates. Some scientists even hypothesize that many physical features that are unique to particular human populations evolved as a result of these socially derived preferences. While intriguing, there is as yet little evidence to support this idea and no simple way to test it.

Assortative Mating

Differences between human populations may be reinforced by the ways in which people choose their mates. Individuals usually prefer to mate with someone who is like themselves, by a process called positive **assortative mating.** For example, there is a tendency for people to mate assortatively by height—that is, tall women tend to marry tall men—and by skin color (**Figure 12.22**).

When two populations differ in obvious physical characteristics, the number of matings between them may be small if the traits of one population are considered culturally unattractive to members of the other population. Assortative mating has been observed in other organisms as well; for instance, sea horses choose mates that are similar in size to themselves, and in some species of fruit flies, females will mate only with males who have the same body color.

Figure 12.22 Assortative mating.
People tend to choose mates who look like them in socially important characteristics.

Positive assortative mating tends to maintain and even exaggerate physical differences between populations. In highly social humans, assortative mating may be an important amplifier of differences between groups.

While human populations may show superficial differences due to natural selection in certain environments, genetic drift, sexual selection, and assortative mating, the genetic evidence indicates that many of these differences are literally no more than skin deep. Beneath a veneer of physical differences, humans are basically the same.

SAVVY READER

Race and Medicine

Excerpted from an article by Gregory M. Lamb

Race-based medicine has gained momentum:

- The Food and Drug Administration is expected to approve the drug BiDil in June [2005], making it the first "ethnic drug" on the market. After failing in a broader study, BiDil was shown to be effective in treating heart failure in a clinical study that included only African-Americans.

- African-Americans need higher doses of one medication used to treat asthma than Caucasians, suggesting "an inherent predisposition" in blacks not to absorb the medicine as easily, says a study in the February [2005] issue of the journal *Chest*.

Much of what we're seeing in human illness can be explained by "issues as prosaic and mundane as access to healthy water and good nutrition," says Professor [Troy] Duster, who teaches at New York University and the University of California at Berkeley. For example: A 2002 report by the Institute of Medicine showed that racial and ethnic minorities tend to receive lower-quality healthcare than whites, even if factors such as insurance status, income, age, and severity of condition are similar.

Many in the field are calling for broader international studies to make sure the bigger picture emerges. Studies comparing white and black Americans, for example, have shown that blacks have higher rates of hypertension, suggesting a genetic difference. But earlier this year research by Dr. [Richard] Cooper and a team at Loyola showed that although African-Americans do show higher levels than North American whites, whites have higher levels than Nigerians and Jamaicans, who are "ethnically" black. Overall, the range of levels of hypertension among blacks in the study ranged from 14 to 44 percent while in whites it was higher, 27 to 55 percent.

"'Race' and 'ethnicity' are poorly defined terms that serve as flawed surrogates for multiple environmental and genetic factors in disease causation, including ancestral geographic origins, socioeconomic status, education, and access to health care," wrote Francis Collins, the head of the National Human Genome Research Institute (NHGRI), last fall. "Research must move beyond these weak and imperfect proxy relationships to define the more proximate factors that influence health."

But other researchers say they expect to keep finding genetic racial differences and that society must learn to become comfortable with the idea. "New forms of scientific knowledge will point out more and more ways in which we are diverse," wrote geneticist James Crow, a professor emeritus at the University of Wisconsin, in a 2002 article. "I hope that differences will be welcomed, rather than accepted grudgingly. Who wants a world of identical people, even if they are Mozarts or Jordans?"

1. In this article, what is the evidence that racial categories may be important to medical practice?
2. What evidence is presented that argues against this idea?
3. What more would you need to know to understand whether race has a place in medicine?

The Christian Science Monitor, March 3, 2005.

Chapter Review

Learning Outcomes

MasteringBIOLOGY®

Go to the Study Area at www.masteringbiology.com for practice quizzes, myeBook, BioFlix™ 3-D animations, MP3Tutor sessions, videos, current events, and more.

L01 Define *biological species*, and list the mechanisms by which reproductive isolation is maintained by biological species (Section 12.1).

- All humans belong to the same biological species, *Homo sapiens sapiens*. A biological species is defined as a group of individuals that can interbreed and produce fertile offspring. Biological species are reproductively isolated from each other, thus separating the gene pools of species (pp. 276–277).
- Reproductive isolation is maintained by prefertilization factors, such as differences in mating behavior or timing, or postfertilization factors, such as hybrid inviability or sterility (pp. 277–280).

L02 Describe the three steps in the process of speciation (Section 12.1).

- Speciation occurs when populations of a species become isolated from each other. These populations diverge from each other, and reproductive isolation between the populations evolves (pp. 280–284).

L03 Explain how a "race" within a biological species can be defined using the genealogical species concept (Section 12.2).

- Biological races are populations of a single species that have diverged from each other but have not become reproductively isolated. This definition of race corresponds to the genealogical species concept (pp. 284–286).

L04 List the evidence that modern humans are a young species that arose in Africa (Section 12.2).

- The morphology of human ancestors in the fossil record provides evidence that the modern human species is approximately 200,000 years old (p. 287).

- Genetic evidence indicates that modern humans have limited genetic diversity, a characteristic of young species, and that the oldest populations (containing the most genetic diversity) are found in Africa (pp. 287–288).

LO5 List the genetic evidence expected when a species can be divided into unique races (Section 12.2).

- The genetic evidence for biological race include (1) alleles that are unique to a particular race, (2) similar allele frequencies for a number of genes among populations within races, and (3) differences in allele frequencies among populations in different races (pp. 287–288).

LO6 Describe how the Hardy-Weinberg theorem is used in studies of population genetics (Section 12.2).

- The Hardy-Weinberg theorem provides a rule for calculating allele frequency from genotype frequency (pp. 288–290).
- If the assumptions of Hardy-Weinberg (no natural selection, no migration, and random mating of individuals within a population) are violated, the allele frequency will not match expectations. Scientists look for alleles that are not typical to identify those that might provide information about ancestry (pp. 288–289).

LO7 Summarize the evidence that indicates that human races are not deep biological divisions within the human species (Section 12.2).

- Although some single nucleotide polymorphism (SNP) alleles are more common in certain human population groups, modern human groups do not show evidence that they have been isolated from each other long enough to have formed distinct races (pp. 290–294).
- Other genetic evidence indicates that human groups have been mixing for thousands of years (pp. 293–294).

LO8 Provide examples of traits that have become common in certain human populations due to the natural selection these populations have undergone (Section 12.3).

- Similarities among human populations may evolve as a result of natural selection. The sickle-cell allele is selected for in populations in which malaria incidence is high, and light skin is selected for in areas where the UV light level is low (pp. 294–297).

LO9 Define *genetic drift*, and provide examples of how it results in the evolution of a population (Section 12.3).

- Human populations may show differences due to genetic drift, which is defined as changes in allele frequency due to chance events such as founder effects or population bottlenecks (pp. 298–300).

LO10 Describe how human and animal behavior can cause evolution via sexual selection and assortative mating (Section 12.3).

- Sexual selection, by which individuals—typically females—choose mates that display some "attractive" quality, may also be responsible for creating differences among human populations (p. 300).
- Positive assortative mating, in which individuals choose mates who are like themselves, can reinforce differences between human populations (pp. 300–301).

Roots to Remember

The following roots come mainly from Latin and Greek and will help you decipher terms:

allo- means different. Chapter term: allopatric

morpho- means shape or form. Chapter term: morphological

paleo- means ancient. Chapter term: paleontologists

-patric means country or place of origin. Chapter term: allopatric, sympatric

-ploid means the number of different types of chromosomes. Chapter term: polyploid

poly- means many. Chapter term: polyploid

sym- means same or united. Chapter term: sympatric

Learning the Basics

1. **LO2** Describe the three steps of speciation.

2. **LO4** Why is a relative lack of genetic diversity in humans evidence that we are a young species?

3. **LO9** Describe three ways that evolution can occur via genetic drift.

4. **LO1** Which of the following is an example of a prefertilization barrier to reproduction?

   A. A female mammal is unable to carry a hybrid offspring to term; B. Hybrid plants produce only sterile pollen; C. A hybrid between two bird species cannot perform a mating display; D. A male fly of one species performs a "wing-waving" display that does not convince a female of another species to mate with him; E. A hybrid embryo is not able to complete development.

5. LO2 According to the most accepted scientific hypothesis about the origin of two new species from a single common ancestor, most new species arise when _____.

A. many mutations occur; **B.** populations of the ancestral species are isolated from each other; **C.** there is no natural selection; **D.** a Creator decides that two new species would be preferable to the old one; **E.** the ancestral species decides to evolve

6. LO1 For two populations of organisms to be considered separate biological species, they must be _____.

A. reproductively isolated from each other; **B.** unable to produce living offspring; **C.** physically very different from each other; **D.** a and c are correct; **E.** a, b, and c are correct

7. LO6 Cystic fibrosis, a recessive disease, affects 1 of every 2500 Caucasian babies born in the United States, a frequency of 0.0004. Use the Hardy-Weinberg theorem to calculate the frequency of the cystic fibrosis allele in this population.

A. 0.0004; **B.** 0.04; **C.** 0.02; **D.** 0.00000016; **E.** There is not enough information in the question to calculate the answer.

8. LO7 All of the following statements support the hypothesis that humans cannot be classified into biological races *except*:

A. There is more genetic diversity within a racial group than average differences between racial groups; **B.** Alleles that are common in one population in a racial group may be uncommon in other populations of the same race; **C.** Geneticists can use particular SNP alleles to identify the ancestral group(s) of any individual human; **D.** There are no alleles found in all members of a given racial group; **E.** There is genetic evidence of mixing among human populations occurring thousands of years ago until the present.

9. LO8 Similarity in skin color among different human populations appears to be primarily the result of _____.

A. natural selection; **B.** convergent evolution; **C.** which biological race they belong to; **D.** a and b are correct; **E.** a, b, and c are correct

10. LO10 The tendency for individuals to choose mates who are like themselves is called _____.

A. natural selection; **B.** sexual selection; **C.** assortative mating; **D.** the founder effect; **E.** random mating

Analyzing and Applying the Basics

1. LO3 LO5 Wolf populations in Alaska are separated by thousands of miles from wolf populations in the northern Great Lakes of the lower 48 states. Wolves in both populations look similar and have similar behaviors. However, the U.S. government has treated these two populations quite differently, listing the Great Lakes populations as endangered until recently but allowing hunting of wolves in Alaska. Some opponents of wolf protection have argued that the "wolf" should not be considered endangered at all in the United States because of the large population in Alaska, while supporters of wolf protection state that the Great Lakes population represents a unique population that deserves special status. Should these two populations be considered different races or species? What information would you need to test your answer?

2. LO6 Phenylketonuria (PKU) is the inability to metabolize the amino acid phenylalanine. The frequency of PKU in Irish populations is 1 in every 7000 births, while the frequency in urban British populations is 1 in 18,000 and only 1 in 36,000 in Scandinavian populations. PKU is found only in individuals who are homozygous recessive for the disease allele. Use the Hardy-Weinberg theorem to calculate the frequency of the disease allele in each population. Give two reasons that this allele, which can result in severe mental retardation in homozygous individuals, may be found in different frequencies in these populations.

3. LO1 LO2 A species of aster normally blooms between mid-August and late September. The range of the aster is located in a climate that usually produces freezing temperatures in late October. The pollinators (bees, ants, and other insects) of the aster are active between late April and early October. Mutations can occur in the gene that controls flowering time for some individuals of this aster, causing them to flower earlier or longer than normal. The table shows possible flowering times.

Aster Type	Flowering Time
Normal	August 15–September 30
Early mutation	July 1–August 15
Expanded mutation	July 15–September 30

Describe how this population of asters might split into two species.

Connecting the Science

1. African Americans have higher rates of hypertension (high blood pressure), heart disease, and stroke than do whites in the United States. Is this difference likely to be biological? How could you test your hypothesis?

2. The only information that was collected from every household in the United States in the 2010 census was the names of residents of the household, their relationship to each other, sex, age, and race and ethnicity. Do you think it is important to collect race and ethnicity data from all citizens? Why or why not? Is there some other piece of information that you think is more useful to the government?

Answers to **Stop & Stretch**, **Visualize This, Savvy Reader**, and **Chapter Review** questions can be found in the **Answers** section at the back of the book.

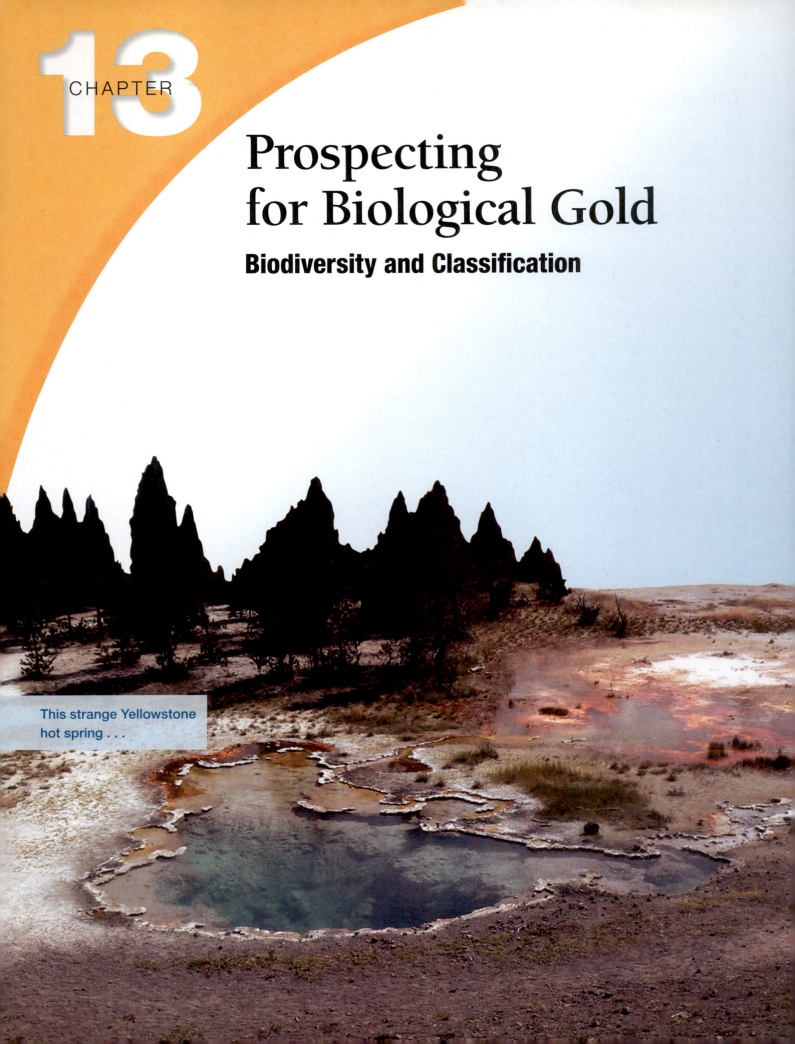

CHAPTER

13

Prospecting for Biological Gold

Biodiversity and Classification

This strange Yellowstone hot spring . . .

At the end of a narrow foot trail in Yellowstone National Park lies a natural curiosity—Octopus Spring. The boiling hot water of the spring is colored an otherworldly blue. A gooey white crust encircles its main pool, and along the banks of the drainage streams radiating from it are brightly colored mats and streamers of pink, yellow, green, and orange. Although Octopus Spring is certainly not the most beautiful or dramatic feature of Yellowstone, this relatively small spring looms large in the history of biological discovery.

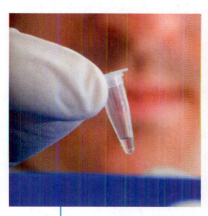

. . . is a source of this extremely valuable chemical, produced by microorganisms.

The brilliant colors of Octopus Spring result from microscopic organisms living in the water and on nearby surfaces. In the 1960s, Dr. Thomas Brock of the University of Wisconsin was the first to describe this biological community. One of the species he discovered, which he named *Themus aquaticus* for its affinity for hot water, contained a protein new to science. This protein, an enzyme called *Taq* polymerase, allows *T. aquaticus* to replicate its DNA at much higher temperatures than other organisms can. *Taq* polymerase is now an essential part of the polymerase chain reaction (PCR), a high-temperature process used to prepare DNA samples for research or for the process of DNA fingerprinting. PCR has revolutionized DNA research and has made many of the advances in genetic technology possible.

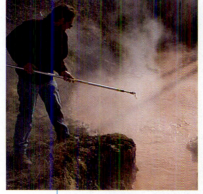

Scientists are interested in finding other useful products in Yellowstone.

Until it was ruled invalid in 1999, the patent for *Taq* polymerase was held by the Swiss pharmaceutical firm Hoffman-LaRoche.

Are they likely to succeed? Is there much left to discover about life?

Licensing agreements with users and producers of *Taq* polymerase netted the company over $100 million every year. Of this substantial sum, Yellowstone National Park, the National Park Service, and the U.S. Treasury received nothing in royalty payments. Even a small share of the royalties would have helped improve and manage this heavily used national park.

The managers of Yellowstone do not want to miss out on the financial rewards that may come from other valuable species protected by their borders. To capitalize on future discoveries in the park, Yellowstone entered into an agreement in 1997 with Diversa Corporation to identify and describe some of the microscopic species in the park. In return, Diversa agreed to make a one-time payment of $100,000 to Yellowstone National Park and to provide several thousand dollars of research services. Diversa also agreed to share a percentage of royalties from any profitable products that result.

The announcement of this deal set off a flurry of criticisms—from environmentalists who fear the exploitation of the park, to government watchdogs concerned about a few private stockholders profiting from a park maintained for the entire public. The National Parks Service and Diversa, which has since merged with another company and changed its name to Verenium, finally signed a benefits-sharing agreement in the summer of 2010, paving the way for exploration.

The agreement between the managers of Yellowstone and Verenium is a calculated risk by both parties. The company is investing millions of dollars in this venture, and it will need to conform to strict permitting requirements that may interfere with its success. Yellowstone risks damage to the wildlife and scenery that the park was designed to protect. Why are the parties to this agreement willing to take these risks? What is the likelihood of success in the search for valuable species within Yellowstone National Park? Can there be many organisms that humankind has yet to discover? What do we know about the organisms we have identified? And how can we screen newly discovered species for their valuable traits? We can answer many of these questions by applying evolutionary theory to investigations of the amazing variety of life on Earth.

13.1 Biological Classification

Verenium Corporation's proposed hunt for new organisms and new uses of known organisms is called **bioprospecting.** Bioprospectors seek to strike biological "gold" by finding the next penicillin (originally discovered in a fungus), aspirin (produced by willow trees), or *Taq* polymerase in the living world.

Yellowstone isn't the only potentially rich source of biological gold—drug companies are also bioprospecting in vast Amazonian rain forests, strange hydrothermal vents of the deep ocean (**Figure 13.1**), and bleak expanses of Antarctic ice. Other scientists are surveying more commonly encountered organisms such as airborne molds and the bacteria that cause tooth decay. To understand the challenges associated with this survey, we must know something about the diversity of life on Earth.

Figure 13.1 A source of biological riches? The organisms like these tube worms surrounding a deep-sea volcanic site have been known to science for fewer than 30 years.

How Many Species Exist?

A characteristic of life on Earth is that it is full of variety. Scientists refer to the variety within and among living species as **biodiversity.** Studies of biodiversity don't just provide drugs and chemicals to solve human problems. More important to many biologists, these studies help us understand the evolutionary origins of species and discover their role in healthy biological systems.

To bioprospectors, life's great variety presents great opportunity but is also a source of challenge. The number of species described by science is between 1.4 and 1.8 million; even if prospectors spent only a single day screening each species for valuable products, examining them would take more than 5000 years.

Listing the number of different species also greatly underestimates the biodiversity within a species. Just as a single species of tomato can come in many shapes, sizes, and flavors, there are species in which individuals differ greatly in the amount and potency of a particular biological molecule (**Figure 13.2**). Bioprospectors could miss a valuable molecule if they test only a handful of individuals from a single population.

Biologists disagree about the total number of species that have already been identified and described. This uncertainty stems from the method of storing and cataloguing known species. **Systematists** are scientists who specialize in describing and categorizing a particular group of organisms. For an organism to be considered a new species by the scientific community, a systematist must create a description of the species that clearly distinguishes it from similar species, and he or she must publish this description in a professional journal. The scientist also must collect individual specimens for storage in a specialized museum. Most animal collections are found in natural history museums, while plant repositories are called *herbaria* (**Figure 13.3** on the next page); microbes and fungi are kept in facilities called *type-collection centers*.

Because there are numerous large natural history museums and herbaria all over the world, along with many different professional journals, it can be unclear whether a species has already been described. Systematists regularly evaluate collections at different centers to see if there is overlap, but this process is slow and further complicated by the continual discovery and description of new species. Because there is no central resource for species collections and descriptions, the total number of described species is still only an

Figure 13.2 Biological diversity.
These varieties of the garden tomato illustrate diversity within a species.

(a) Natural history museum

(b) Herbarium

Figure 13.3 Biological collections. Most collections contain several examples of each species to show the range of variation within the species. (a) A collection of birds in a natural history museum. (b) Plant specimens are stored in herbaria.

estimate. The number of known species also represents only a fraction of the total number on Earth. Some estimates of the actual number are as large as 100 million unique species, but a typical estimate is around 10 million. Scientists also know that the diversity of life today is very different from the diversity in the past.

Paleontologists have been able to piece together the history of life on Earth by examining fossils and other ancient evidence. Early in this reconstruction, they recognized distinct "dynasties" of groups of organisms that appeared during different periods. The rise and fall of these dynasties allowed scientists to subdivide life's history into **geologic periods.** Each period is defined by a particular set of fossils. **Table 13.1** gives the names of major geologic periods, their age and length, and the major biological events that occurred during each period.

Despite the long history of diversity, it is very rare for scientists to describe either a modern or fossil organism that appears unrelated to all other species. In fact, species can be grouped into a few broad categories based on shared characteristics. The most general categories are kingdoms and domains.

An Overview: Kingdoms and Domains

Modern species are divided into three large groups, called **domains,** which are unified by similarities in cell structure. The three domains are named Bacteria, Archaea, and Eukarya. Within each of these domains are smaller groups, called **kingdoms.** Because most identified living organisms are in the domain Eukarya, we focus on the kingdoms only within this group. The other domains contain kingdoms as well, but these classifications are generally of interest only to specialists. **Table 13.2** (on page 312) summarizes the kingdoms and domains discussed in section 13.2. *Take **A Closer Look** at kingdoms and domains on page 313.*

TABLE 13.1

Geological periods. The history of life is divided into four major eras, with all but the first era divided into several periods. Periods are marked by major changes in the dominant organisms present on Earth.				

Era	Period	Millions of Years Ago	Features of Life on Earth	
Pre-Cambrian		4500	Formation of Earth. The first organisms, primitive bacteria, appear within 1000 million years.	
		543	Life is dominated by single-celled organisms in the ocean. Ediacaran fauna appear at the end of the era.	
Paleozoic	Cambrian	495	All modern animal groups appear in the oceans. Algae are abundant.	
	Ordovician	439	Life is diverse in the oceans. Cephalopods appear, and trilobites are common.	
	Silurian	408	Life begins to invade land. The first colonists are small seedless plants, primitive insects, and soft-bodied animals.	
	Devonian	354	Known as the age of fishes. Sharks and bony fish appear. Large trilobites are abundant in the oceans.	
	Carboniferous	290	Land is dominated by dense forests of seedless plants. Insects become abundant. Large amphibians appear.	
	Permian	251	Early reptiles appear on land. Seedless plants abundant. Coral and trilobites abundant in oceans. Permian ends with extinction of 95% of living organisms.	
Mesozoic	Triassic	206	Early dinosaurs, mammals, and cycads appear on land. Life "restarts" in the oceans.	
	Jurassic	144	Huge plant-eating dinosaurs evolve. Forests are dominated by cycads and tree ferns.	
	Cretaceous	65	Massive carnivorous and flying dinosaurs are abundant. Large cone-bearing plants dominate forests. Flowering plants appear.	
Cenozoic	Tertiary	1.8	After the extinction of the dinosaurs, mammals, birds, and flowering plants diversify.	
	Quaternary	0	Most modern organisms present.	

TABLE 13.2

The classification of life. Until recently, most biologists used the five-kingdom system to organize life's diversity. Now many use a six-category system, which better reflects evolutionary relationships by acknowledging the existence of three major domains as well as four of the kingdoms.

Kingdom Name		Kingdom Characteristics	Examples	Approximate Number of Known Species	Domain Name and Characteristics
Plantae		Eukaryotic, multicellular, make own food, largely stationary	Pines, wheat, moss, ferns	300,000	**Eukarya** All organisms contain eukaryotic cells.
Animalia		Eukaryotic, multicellular, rely on other organisms for food, mobile for at least part of life cycle	Mammals, birds, fish, insects, spiders, sponges	1,000,000	
Fungi		Eukaryotic, multicellular, rely on other organisms for food, reproduce by spores, body made up of thin filaments called hyphae	Mildew, mushrooms, yeast, *Penicillium*, rusts	100,000	
Protista		Eukaryotic, mostly single-celled forms, wide diversity of lifestyles, including plant-like, fungus-like, and animal-like types	Green algae, *Amoeba, Paramecium*, diatoms, chytrids	15,000	
Bacteria		Prokaryotic, mostly single-celled forms, although some form permanent aggregates of cells	*Escherichia coli, Salmonella, Bacillus anthracis, Anabena*, sulfur bacteria	4,000	**Bacteria** Prokaryotes with cell wall containing peptidoglycan. Wide diversity of lifestyles, including many that can make their own food.
Archaea			*Thermus aquaticus, Halobacteria halobium*, methanogens	1,000	**Archaea** Prokaryotes without peptidoglycan and with similarities to Eukarya in genome organization and control. Many known species live in extreme environments.

A Closer Look:
Kingdoms and Domains

Systematists work in the field of **biological classification,** in which they attempt to organize biodiversity into discrete and logical categories. The task of classifying life is much like categorizing books in a library—books can be divided into "fiction" or "nonfiction," and within each of these divisions, more precise categories can be made (for example, nonfiction can be divided into biography, history, science, etc.). The book-cataloguing system used in most public libraries, the Dewey decimal system, is only one way of shelving books. For instance, academic and research libraries use a different system, developed by the U.S. Library of Congress. Librarians use the cataloguing system that is appropriate to the collection of books they work with and the needs of the library's users. Just as there are alternative methods of organizing books, there is more than one way to organize biodiversity to meet differing needs.

Biologists have traditionally subdivided living organisms into large groups that share some basic characteristic. Sixty years ago, most biologists divided life into two categories: plants, for organisms that were immobile and apparently made their own food; and animals, for organisms that could move about and relied on other organisms for food.

Many organisms did not fit easily into this neat division of life, so scientists began to use a system of five **kingdoms,** which categorized organisms according to cell type and method of obtaining energy: the Monera, a group consisting of single-celled organisms without a nucleus (consisting of what we now know as the Archea and Bacteria); Protista, a group containing organisms with nuclei that spend some of their life cycle as single, mobile cells; and three kingdoms of multicellular organisms: Plantae, which make their own food; Animalia, which rely on other organisms for food; and Fungi, which digest dead organisms. The five-kingdom system is not perfect either; for instance, the Protista kingdom contains a wide diversity of organisms, from amoebas to seaweeds, with only superficial similarities.

More recently, many biologists have argued that the most appropriate way to classify life is according to evolutionary relationships among organisms. Recall that the theory of evolution states that all modern organisms represent the descendants of a single common ancestor. The process of divergence from early ancestors into the diversity of modern species has resulted in the modern "tree of life" (**Figure 13.4**).

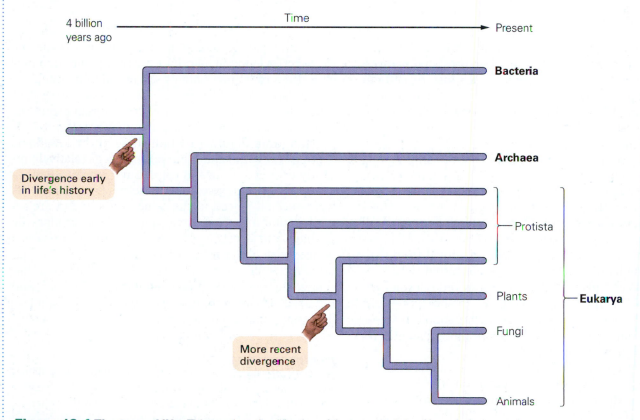

Figure 13.4 The tree of life. This tree is a simplification of the current state of knowledge regarding evolutionary relationships among living organisms. Living organisms represent a small remnant of all the species that have appeared over Earth's history.

(continued on the next page)

(A Closer Look continued)

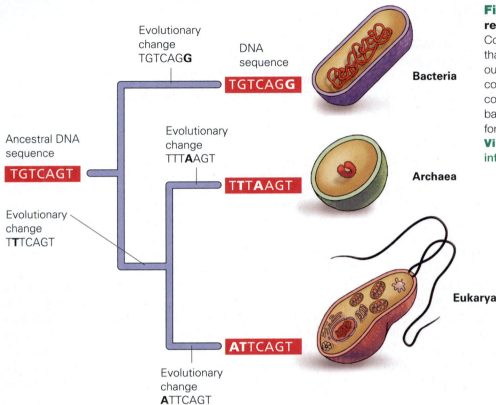

Evolutionary change TGTCAG**G**

DNA sequence

TGTCAGG

Bacteria

Ancestral DNA sequence

TGTCAGT

Evolutionary change TT**T**AAGT

TTTAAGT

Archaea

Evolutionary change T**T**TCAGT

ATTCAGT

Eukarya

Evolutionary change **A**TTCAGT

Figure 13.5 Determining the relationships among organisms. Comparisons of the DNA sequences that code for ribosomal RNA in various organisms have helped scientists construct the tree of life. The actual comparison consisted of thousands of base pairs; only a few are shown here for simplicity.
Visualize This: How can scientists infer the ancestral sequence?

When life is classified according to the relationships among organisms, major groupings correspond to divergences that occurred very early in life's history, and minor groupings correspond to more recent divergences. Classifying life according to evolutionary relationships may be useful to bioprospectors if relationships provide clues to chemical similarity.

Determining the evolutionary relationship among *all* living organisms requires comparisons of their DNA. Because each species is unique, the sequence of nucleotides within DNA of each species is unique. However, because all species share a common ancestor, all organisms also have basic similarities in their DNA sequences. As evolutionary lineages diverged from each other, mutations in DNA sequences occurred independently in each lineage and are now a record of evolutionary relationship among organisms. In short, the DNA sequences of closely related organisms should be more similar than the DNA sequences of more distantly related organisms (**Figure 13.5**).

When comparing all modern species, scientists must examine genes that perform a similar function among organisms as diverse as humans, willow trees, and *Thermus aquaticus*. The DNA sequence that best fits this requirement contains instructions for making ribosomal RNA (rRNA),

a structural part of ribosomes. Recall that ribo-somes are the factories found in all cells, translating genes into proteins. Each ribosome contains several rRNA molecules in both its large and small subunits. A comparison of the DNA coding for small-subunit rRNAs from myriad organisms yielded a tree diagram similar to that shown in Figure 13.4.

Note from Figure 13.4 that three of the kingdoms (Fungi, Animalia, and Plantae) represent relatively recently diverged groups of organisms. Single-celled organisms that do not contain a nucleus for storing DNA were once all placed in the same kingdom but are actually found in two different, quite distinct groups, the Archaea and Bacteria. And the kingdom Protista is a hodgepodge of many very different organisms. To better reflect such biological relationships, biologists categorize life into three **domains:** Bacteria, Archaea, and Eukarya. These domains represent the most ancient divergence of living organisms.

Stop & Stretch Can you think of a situation for which an evolutionary classification may be less helpful to scientists than a classification that groups organisms by other shared traits, such as similarity in color or habitat?

13.2 The Diversity of Life

Dividing life into six categories—those described in Table 13.2—simplifies our discussion of biodiversity and its bioprospecting potential. In this section we describe the six categories. In addition, we discuss entities that have many characteristics of living organisms but are not considered *alive*—the viruses.

The Domains Bacteria and Archaea

Life on Earth arose at least 3.6 billion years ago, according to the fossil record. The most ancient fossilized cells (**Figure 13.6**) are remarkably similar in external appearance to modern bacteria and archaea. Both **bacteria** and **archaea** are **prokaryotes;** this means they do not contain a nucleus, which provides a membrane-bound, separate compartment for the DNA in other cells. Prokaryotes also lack other internal structures bounded by membranes, such as mitochondria and chloroplasts, which are found in more complex **eukaryotes.**

Although some species may be found in chains or small colonies as in **Figure 13.7**, most prokaryotes are **unicellular,** meaning that each cell is an individual organism. Individual prokaryotic cells are hundreds of times smaller than the cells that make up our bodies; for this reason, they are often called microorganisms or **microbes,** and biologists who study these organisms (as well as unicellular eukaryotes) are known as **microbiologists.** Their small size, easily accessible DNA, and simple structure make prokaryotes very attractive to bioprospectors because they are generally easier to grow, study, and manipulate than eukaryotes.

The relatively simple structure of prokaryotes belies their incredible chemical complexity and diversity. Some prokaryotes can live on petroleum, others on hydrogen sulfide emitted by hydrothermal vents deep below the surface of the ocean, and some simply on water, sunlight, and air (Figure 13.7). Prokaryotes are found in and on nearly every square centimeter of Earth's surface, including very hot and very salty places, and even thousands of feet below ground. Prokaryotes are also incredibly numerous; for instance, there are more prokaryotes living in your mouth right now than the total number of humans who have ever lived!

Although most are likely harmless to humans, the majority of bacteria that have been identified are known because they are pathogenic, that is, they cause disease in humans or our crops. While enormous efforts are expended to control these organisms, the fact that they can live in and on other living creatures is remarkable.

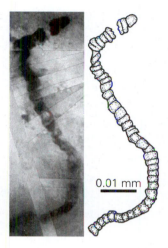

Figure 13.6 The oldest form of life. This photograph of a fossil is accompanied by an interpretive drawing showing the fossil's living form. The fossil was found in rocks dated at 3.465 billion years old.

Figure 13.7 A diversity of prokaryotes. (a) *Escherichia coli*, the "lab rat" of basic genetic studies, lives on the partially digested food in our intestines. (b) *Streptomyces venezuelae*, source of the antibiotic chloramphenicol and a member of the group of bacteria that live in soil and decompose dead plant matter. (c) *Halobacterium*, a salt-loving archaean, is found in high populations in these salty ponds. The red pigment in this bacterium's cells is used in photosynthesis.

(a) *Escherichia coli*

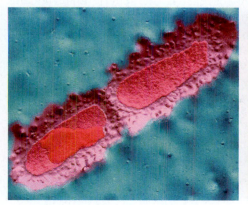

(b) *Streptomyces griseus*

(c) *Halobacterium*

To survive within a host organism, bacteria must escape eradication by the host's infection-fighting system. The molecules that allow bacteria to effectively colonize living humans could be useful in treating immune system diseases. Disease-causing organisms thus, surprisingly, represent a source of bacterial biological gold.

Many of the known bacteria obtain nutrients by decomposing dead organisms. Bacterial species that function as decomposers often have competitors—other species that also consume the same food source. When many individuals compete for the same resource, natural selection will favor traits that provide an edge over the competition. With competing bacterial decomposers, this edge is often in the form of **antibiotics** that kill or disable other bacteria. More than half of the antibiotic drugs we use to treat infections are derived from bacteria.

Bacterial competition has resulted in another class of valuable molecules called *restriction enzymes,* proteins that can chop up DNA at specific sequence sites and thus interfere with, or restrict, the growth of other organisms. Restriction enzymes are used in the production of DNA fingerprints. Bioprospectors are very interested in finding more of both of these extremely valuable compounds—antibiotics and restriction enzymes—within the domain Bacteria.

Like bacteria, archaea cells are prokaryotic. However, the domain Archaea differs from Bacteria in many fundamental ways, including the very structure of its cell membranes. The known Archaea are typically found in extreme environments, including high-salt, high-sulfur, and high-temperature habitats. *Thermus aquaticus*, the source of *Taq* polymerase and a hot-spring dweller, belongs to Archaea.

Taq polymerase is valuable because it operates at a high temperature—making it, along with compounds from other hot-spring archaeans, potentially useful in industrial settings. Natural selection acting on Archaea in extreme environments has likely caused the evolution of other biological molecules that can operate at high temperatures, at high pressures, or in extremely salty conditions.

Scientists and bioprospectors still have much to learn about bacteria and archaea, beginning with an understanding of exactly how diverse they are. For instance, while most archaeans are known from extreme environments, species in this domain are found everywhere they are sought. Some scientists estimate that the number of undescribed prokaryotic species could range up to 100 million.

The Origin of the Domain Eukarya

The third domain of life contains all of the organisms that keep their genetic material within a nucleus inside their cells—that is, the **eukaryotes.** The most ancient fossils of eukaryotic cells are approximately 2 billion years old, nearly 1.5 billion years younger than the oldest prokaryotic fossils.

The earliest eukaryotic cells likely developed from prokaryotes that produced excess cell membrane that folded into the cell itself. In some cells, these internal membranes may have segregated the genetic material into a primitive nucleus and formed channels for translating, rearranging, and packaging proteins, much like the modern endoplasmic reticulum and Golgi bodies. According to the **endosymbiotic theory,** the mitochondria and chloroplasts found in eukaryotic cells appear to have descended from bacteria that took up residence inside larger primitive eukaryotes. When organisms live together, the relationship is known as a *symbiosis.* In this case, the symbiosis was mutually beneficial, and over time the cells became inextricably tied together (**Figure 13.8**)

When biologist Lynn Margulis first popularized the endosymbiotic hypothesis in the United States in 1981, many of her colleagues were skeptical. But an examination of the membranes, reproduction, and ribosomes of

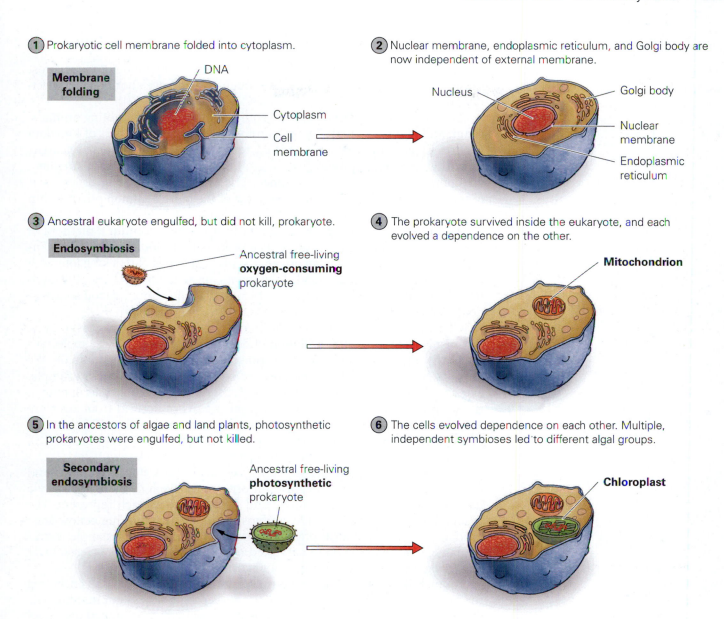

① Prokaryotic cell membrane folded into cytoplasm.

Membrane folding

DNA

Cytoplasm

Cell membrane

② Nuclear membrane, endoplasmic reticulum, and Golgi body are now independent of external membrane.

Nucleus

Golgi body

Nuclear membrane

Endoplasmic reticulum

③ Ancestral eukaryote engulfed, but did not kill, prokaryote.

Endosymbiosis

Ancestral free-living **oxygen-consuming** prokaryote

④ The prokaryote survived inside the eukaryote, and each evolved a dependence on the other.

Mitochondrion

⑤ In the ancestors of algae and land plants, photosynthetic prokaryotes were engulfed, but not killed.

Secondary endosymbiosis

Ancestral free-living **photosynthetic** prokaryote

⑥ The cells evolved dependence on each other. Multiple, independent symbioses led to different algal groups.

Chloroplast

Figure 13.8 The evolution of eukaryotes. Mitochondria and chloroplasts appear to be descendants of once free-living bacteria that took up residence within an ancient nucleated cell.

mitochondria and chloroplasts shows clear similarities to the same features in certain bacteria. Even more convincingly, the sequence of DNA found in mitochondria (mtDNA) is most similar to the DNA sequence found in species of the purple sulfur bacteria. After the evolution of mitochondria, independent endosymbioses between eukaryotic cells and various species of photosynthetic bacteria appeared. These relationships led to the evolution of chloroplasts—one line led to green algae and land plants, while other lines led to the unique chloroplasts in other modern groups, including red algae and brown algae.

Stop & Stretch   Mitochondria are thought to have evolved before chloroplasts. What is the evidence that supports this hypothesis?

Kingdom Protista

The kingdom **Protista** is made up of the simplest known eukaryotes. Most protists are single-celled creatures, although several have enormous multicellular (many-celled) forms.

The modern kingdom Protista contains organisms resembling animals, fungi, and plants. As with Bacteria and Archaea, most members of kingdom Protista remain unknown. There is currently no agreement among scientists regarding how many **phyla,** that is, groups just below the level of kingdom, are contained within Protista. Some argue as few as eight, and others propose as many as 80. **Table 13.3** lists a few of the more common phyla within the kingdom.

As a result of its diversity, kingdom Protista may contain a plethora of useful organisms and compounds. The protists that bioprospectors have investigated in the most detail are **algae,** the only members of this kingdom with the ability to manufacture food. Like plants, algae make food via photosynthesis. And also like plants, algae represent a rich and tempting food source to nonphotosynthetic predators, organisms that will eat them.

Consequently, natural selection in algae and most other photosynthetic organisms has favored the evolution of defensive chemicals. These molecules make the algae distasteful or even poisonous to a potential predator. We humans can often use these chemicals to control *our* predators. For example, extracts of red algae stop the reproduction of several different viruses in human tissues grown in laboratory dishes. As yet, no drugs derived from algae are available to consumers, but with great interest in treatments for HIV (the virus that causes AIDS), influenza, and severe acute respiratory syndrome (SARS), these organisms remain a focus for bioprospectors.

The group of organisms we call algae is actually made up of several distinct, quite divergent, categories of organisms. Each of these algal phyla has its own methods of producing and storing food as each represents the descendants of a separate endosymbiosis. Some of the unique compounds produced by different algal phyla are useful to humans. For example, carrageenan, a slimy carbohydrate harvested from red algae, is used as a stabilizer and thickener in foods, medicines, and cosmetics.

Animal-like protists include many notorious parasites of humans, including the organisms that cause malaria, African sleeping sickness (an insect-borne disease that affects the brain), trichomoniasis (a sexually transmitted disease of the urogenital tract), and giardiasis (a water-borne diarrheal illness). Fungus-like protists include slime molds and some important crop pests, most notably potato blight, the organism responsible for the Irish famine of the late 1840s.

The animal-like and fungus-like protists are less interesting to most bioprospectors, presumably because animals and fungi have not been as rich a source of useful biological products as photosynthesizers have. However, as we shall see, both the kingdom Animalia and the kingdom Fungi have some intriguing characteristics and can be a source of useful biochemicals. It may be that within these categories of protists, bioprospectors also will find natural products that humans can use.

Kingdom Animalia

From the origin of the first prokaryote until approximately 1.2 billion years ago, life on Earth consisted only of single-celled creatures. Then multicellular organisms first began to appear in the fossil record. The ancient, many-celled creatures of 600 million years ago, called the Ediacaran fauna, were organisms unlike any modern species and included giant fronds and ornamented disks (**Figure 13.9**).

Figure 13.9 Ediacaran fauna.
This reconstruction of multicellular organisms that lived before the Cambrian explosion is based on 580-million-year-old fossil remains.

TABLE 13.3

The diversity of Protista. Protista contains animal-like, fungus-like, and plant-like organisms. A sampling of protistan phyla is described here.

Kingdom Protista: Common Names and Characteristics of Select Phyla	Example
Animal-like protists **Ciliates** Free-living, single-celled organisms that use hair-like structures to move.	*Paramecium*
Flagellates Use one or more long whip-like tail for locomotion. Most are free living, but some cause disease by infecting human organs.	*Giardia*
Amoebas Flexible cells that can take any shape and move by extending pseudopodia ("false feet").	*Amoeba*
Fungus-like protists **Slime molds** Feed on dead and decaying material by growing net-like bodies over a surface or by moving about as single amoeba-like cells.	*Physarum*
Plant-like protists **Diatoms** Single cells encased in a silica (glass) shell.	Diatom
Brown algae Large multicellular seaweeds.	Kelp
Green algae Closest relatives to land plants. Single-celled to multicellular forms.	*Volvox*

Biologists are unsure which of these ancient species is the common ancestor of modern **animals**—defined as multicellular organisms that make their living by ingesting other organisms and that are motile (have the ability to move) during at least one stage of their life cycle. What is clear from the fossil record is that by about 530 million years ago, *all* modern animal groups had dramatically emerged.

The remarkably sudden appearance of the modern forms of animals—a period comprising little more than 1% of the history of life on Earth—is referred to as the **Cambrian explosion,** named for the geologic period during which it occurred. Some scientists hypothesize that the evolution of the animal lifestyle itself—that is, as predators of other organisms—led to the Cambrian explosion.

It can be difficult to conceive of an animal as complex as a human evolving from a simple eukaryotic ancestor. However, humans are not very different from other eukaryotes. When the first cell containing a nucleus appeared, all of the complicated processes that take place in modern cells, such as cell division and cellular respiration, must have evolved. When the first multicellular animals appeared, many of the processes required to maintain these larger organisms, such as communication systems among cells and the formation of organs and organ systems, arose. Although a human and a starfish appear to be very different, the way we develop and the structures and functions of our cells and common organs are nearly identical.

In fact, there appears to be surprisingly little genetic difference between humans and starfish; most of that difference occurs in a group of genes that control **development,** the process of transforming from a fertilized egg into an adult creature. In addition, the amount of time since the divergence of the major evolutionary lineages of animals—530 million years—is quite a long time for differences among phyla to evolve. Put in more familiar terms, if the time since the Cambrian explosion were one 24-hour day, all of human history would fit in only the last 2 seconds.

Zoologists, scientists who study the kingdom **Animalia,** have described more than 25 phyla in this kingdom. Some of the more well-known groups are illustrated in **Table 13.4**.

Most people typically picture mammals, birds, and reptiles when they think of animals, but species with backbones (including mammals, birds, reptiles, fish, and amphibians) represent only 4% of the total species in the kingdom. A small number of these **vertebrates** have traits interesting to a bioprospector. For instance, poison dart frogs (**Figure 13.10**) secrete high levels of toxins onto their skin. These toxins are nerve poisons that cause convulsions, paralysis, and even death to their potential predators. In fact, the name of these frogs derives from the use of their excretions to coat the tips of hunting darts. Poison dart frog toxins are potentially valuable as sources of potent, nonaddictive painkillers—in low doses, of course.

Most of the bioprospecting work in kingdom Animalia focuses on the remaining 96% of known organisms—the **invertebrates** (animals without backbones, as shown in **Figure 13.11** on page 323. The vast majority of multicellular organisms on Earth are invertebrates, and most of these animals are insects. Many invertebrates contain chemical compounds not found elsewhere in nature. Species of beetles, ants, bees, wasps, and spiders produce venom to kill prey or to repel predators and competitors; these venoms are sources of potential drugs. For example, the tropical ant *Pseudomyrmex triplarinus* produces venom that may be useful for treating the joint swelling and pain associated with arthritis. In addition, ants, bees, wasps, and termites that live in crowded colonies have evolved protective molecules that reduce disease spread in these environments. These compounds may have the potential to reduce the spread of disease in human populations as well.

Figure 13.10 Poison dart frog. This brightly colored frog contains glands in its skin that release poison when the frog is handled.

Visualize This: What is the advantage to the frog of having brightly colored skin?

TABLE 13.4

Phyla in the kingdom Animalia. A sampling of the diversity of animals. The rows are arranged generally in order of appearance in evolutionary time—from the more ancient sponges to the more recent chordates.

Kingdom Animalia: Major Phyla	Description	Example	
Porifera	Fixed to underwater surface and filter bacteria from water that is drawn into their loosely organized body cavity.	Sponge	
Cnidaria	Radially symmetric (like a wheel) with tentacles. Some are fixed to a surface as adults (e.g., corals), while others are free floating in marine environments.	Anenome	
Platyhelminthes	Flatworms with a ribbon-like form. Live in a variety of environments on land and sea or as parasites of other animals.	Tapeworm	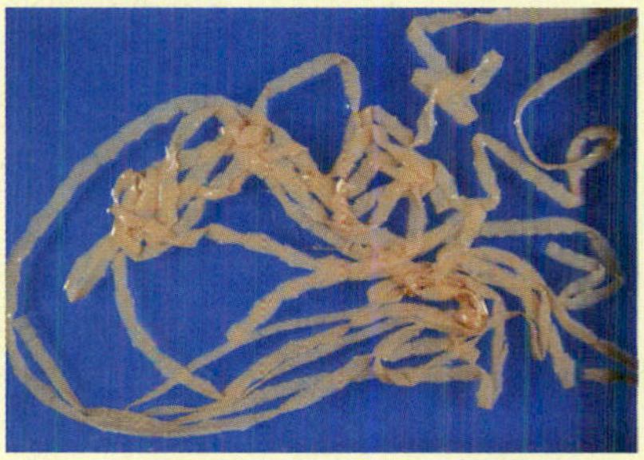
Mollusca	Soft-bodied animals often protected by a hard shell. Body plan consists of a single muscular foot and body cavity enclosed in a fleshy mantle. Phylum includes snails, clams, and squid.	Octopus	
Annelida	Segmented worms. Body divided into a set of repeated segments.	Earthworm	

(continued)

TABLE 13.4 *(continued)*

Phyla in the kingdom Animalia. A sampling of the diversity of animals. The rows are arranged generally in order of appearance in evolutionary time—from the more ancient sponges to the more recent chordates.

Kingdom Animalia: Major Phyla	Description	Example	
Nematoda	Roundworms with a cylindrical body shape. Very diverse and widespread in many environments.	Roundworm	
Arthropoda	Segmented animals in which the segments have become specialized into different roles (such as legs, mouthparts, and antennae). Body completely enclosed in an external skeleton that molts as the animal grows. Phylum includes insects and spiders as well as crabs and lobsters.	Shrimp	
Echinodermata	Slow-moving or immobile animals without segmentation and with radial symmetry. Internal skeleton with projections gives the animal a spiny or armored surface.	Sea urchin	
Chordata	Animals with a spinal cord (or spinal cord-like structure). Includes all large land animals, as well as fish, aquatic mammals, and salamanders.	Duck-billed platypus	

(a) Ant (*Pseudomyrmex triplarinus*)

Potential source of arthritis treatment

(b) Jellyfish (*Aequorea victoria*)

Source of fluorescent protein, a useful labeling tool in microbiology

(c) Horseshoe crab (*Limulus polyphemus*)

Source of blood proteins used to test for pathogens in humans

The animals that inhabit the oceans' incredibly diverse coral reefs are especially interesting to bioprospectors (**Figure 13.12**). These biological communities are crowded with life, and the individuals within them continually interact with predators and competitors. As a result of these challenges, many successful coral-reef organisms contain defensive chemicals that might be useful as drugs. The number of unknown invertebrate species, especially in the oceans, is estimated to be anywhere from 6 to 30 million.

Figure 13.11 Examples of invertebrates. Most animals are invertebrates, including these examples.

Stop & Stretch One mantra of reef bioprospectors is, "If it is bright red, slow-moving, and alive, we want it." Why is such an animal probably a good source of defensive compounds?

While our ignorance about the diversity of animals is great, another kingdom of multicellular eukaryotes is even less well known, although no less important to the functioning of ecosystems and as a source of molecules useful to humans—the fungi.

Kingdom Fungi

Like plants, fungi are immobile, and many produce fruit-like organs that disperse **spores,** cells that are analogous to plant seeds in that they germinate into new individuals. In fact, for generations biologists classified fungi in the plant kingdom. However, while plants make their own food via photosynthesis, fungi feed on other organisms by secreting digestive chemicals into their environment; these chemicals break down complex organic food sources into small molecules, which the fungi then absorb. Because they rely on other organisms for their food sources, fungi are more like animals than plants. In fact, DNA sequence analysis by **mycologists,** biologists who study fungi, indicates that Fungi and Animalia are more closely related to each other than either kingdom is to the plants.

Figure 13.12 A coral reef. This extremely diverse biological community is a rich source of interesting biological chemicals.

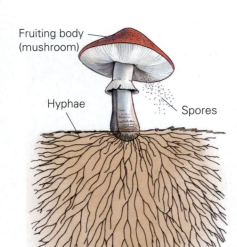

Fruiting body (mushroom)

Hyphae

Spores

Figure 13.13 Fungi. Hyphae can extend over a large area. The familiar mushroom, as well as the fruiting structures of less-familiar fungi, primarily function only as methods of dispersing spores.
Visualize This: Why is a mushroom produced outside the food resource for the hyphae?

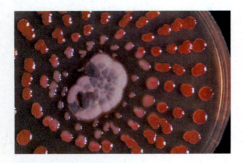

Figure 13.14 Antibiotics from fungi. The large white-and-gray mold in the center of this petri dish is a colony of *Penicillium*. The red dots on the edges of the dish are colonies of bacteria. The pink dots surrounding the *Penicillium* colony are bacterial colonies that are dying from contact with antibiotic secretions from the fungi.

The mushroom you think of when you imagine a fungus is a misleading image of the kingdom. Most of the functional part of fungi is made up of very thin, stringy material called **hyphae,** which grows over and within its food source (**Figure 13.13**). The string-like form of hyphae maximizes the surface over which feeding takes place. Some fungi feed on living tissue, while others decompose dead organisms. This latter role of fungi is key to recycling nutrients in biological systems.

The phyla of fungi are distinguished from each other by their method of spore formation (**Table 13.5**). However, convergent evolution, in which unrelated species take similar forms because of similar environments, has led to the appearance of similar body shapes and lifestyles—which we call "fungal forms"—among these different phyla. One of the most commercially important fungal forms is **yeast,** a single-celled type of fungi. Yeast forms are found in at least two different fungal phyla and live in liquids such as plant sap, soaked grains, or fruit juices. The activity of yeasts in oxygen-poor but sugar-rich environments results in the formation of alcohol. The metabolism of yeasts within flour batter leads to the production of carbon dioxide bubbles, which are trapped by wheat protein fibers and thus allow the dough to "rise" during bread making. Another fungal form known as **mold** is found in all phyla and is also commercially important. This quickly reproducing, fast-growing form can spoil fruits and other foods, although some produce certain types of flavorful cheese, including blue and Camembert, as they are "spoiling" milk.

About one-third of the bacteria-killing antibiotics in widespread use today are derived from fungi. Fungi produce antibiotics because many types of bacteria compete with them for their food sources. Penicillin, the first commercial antibiotic, is produced by a fungus (**Figure 13.14**). Its discovery is one of the great examples of good fortune in science.

Before he went on vacation during the summer of 1928, British bacteriologist Alexander Fleming left a dish containing the bacteria *Staphylococcus aureus* on his lab bench. While he was away, this culture was contaminated by a spore from a *Penicillium* fungus, probably from a different laboratory in the same building. When Fleming returned to his laboratory, he noticed that the growth of *S. aureus* had been inhibited on the fungus-contaminated culture dish. Fleming inferred that some antibiotic chemical substance had diffused from the fungus; he named it penicillin, after the fungus itself. The first batches of this bacteria-slaying drug became available during World War II and greatly reduced the number of deaths of wounded soldiers from infection. Since the discovery of penicillin, hundreds of other antibiotics have been isolated from different fungus species.

Some fungi infect living animals and thus also have potential as sources of drugs. Cyclosporin is a molecule produced by a number of different parasitic fungi and has the effect of suppressing a host's immune response. Humans have used cyclosporin to prevent the immune systems of organ transplant recipients from attacking foreign transplanted organs. Fungi are also the source of a powerful class of anticholesterol drugs, called statins, which help treat and prevent heart disease.

The fungus *Claviceps purpurea,* also known as ergot, has long been known to have powerful effects on the human body. Midwives throughout the nineteenth century used this pest of rye and wheat to stimulate uterine contractions and speed labor, and farmers throughout Europe knew that consuming grain infected with ergot could lead to neurological effects, including burning pain in the limbs, hallucinations, and convulsions. The symptoms of "demonic possession" that led to the Salem witch trials in 1691–1692 may have been caused by consumption of ergot-contaminated grain. More recently, the drug LSD has been derived from this fungus, with similar effects on recreational users.

While fungi have been the third most important source of molecules useful to humans—right after bacteria in terms of numbers and impacts of derived compounds—the source of most naturally derived drugs has been the plant kingdom.

TABLE 13.5

Fungal diversity. Fungi are classified into phyla based on their mode of spore production. The most common phyla are listed here.

Kingdom Fungi: Major Phyla	Description	Example	
Zygomycota	Sexual reproduction occurs in a small resistant structure called a zygospore. Most reproduction is asexual—directly via mitosis.	*Rhizopus stolonifera*—bread mold	
Ascomycota	Spores are produced in sacs on the tips of hyphae in fruiting structures.	Morel	
Basidiomycota	Spores are produced in specialized club-shaped appendages on the tips of hyphae in fruiting structures.	*Amanita muscaria*—the poisonous fly agaric	

Kingdom Plantae

The kingdom **Plantae** consists of multicellular eukaryotic organisms that make their own food via photosynthesis. Plants have been present on land for over 400 million years, and their evolution is marked by increasingly effective adaptations to the terrestrial environment (**Table 13.6**).

The first plants to colonize land were small and close to the ground. Their small size was necessary, for they had no way to transport water from where it is available in the soil to where it is needed in the leaves. The evolution of **vascular tissue,** made up of specialized cells that can transport water and other substances, allowed plants to reach tree-sized proportions and to colonize much drier areas. The evolution of **seeds,** structures that protect and provide a food source for young plants, represented another adaptation to dry conditions on land. However, most modern plants belong to a group that appeared only about 140 million years ago, the **flowering plants.** Like their ancestors, flowering plants possess vascular tissue and produce seeds; in addition, these plants evolved a specialized reproductive organ, the flower. Over 90% of the known plant species are flowering plants (**Figure 13.15**).

TABLE 13.6

Plant diversity. The four major phyla of plants are listed here in order of their appearance in evolutionary history.

Kingdom Plantae: Major Phyla	Description	Example	
Bryophyta	Mosses. Lacking vascular tissue, these plants are very short and typically confined to moist areas. Reproduce via spores.	Moss	
Pteridophyta	Ferns and similar plants. Contain vascular tissue and can reach tree size. Reproduce via spores.	Fern	
Coniferophyta	Cone-bearing plants resembling the first seed producers	Cycad	
Anthophyta	Flowering plants. Seeds produced within fruits, which develop from flowers. Advances in vascular tissues and chemical defenses contribute to their current dominance on Earth.	Orchid	

(a) Foxglove (*Digitalis purpurea*) **(b) Aloe (*Aloe barbadensis*)** **(c) Willow (*Salix alba*)**

Source of heart drug digitalis

Source of aloe vera, used to treat burns and dry skin

Source of aspirin

Figure 13.15 Diversity of flowering plants. A few of the enormous variety of plants that provide important medicines.

From about 100 million to 80 million years ago, the number of distinct groups, or families, of flowering plants increased from around 20 to over 150. During this time, flowering plants became the most abundant plant type in nearly every habitat. The rapid expansion of flowering plants is called **adaptive radiation**—the diversification of one or a few species into a large and varied group of descendant species. Adaptive radiation typically occurs either after the appearance of an evolutionary breakthrough in a group of organisms or after the extinction of a competing group. For example, the radiation of animals during the Cambrian explosion may be a result of the evolution of the predatory lifestyle, and the radiation of mammals beginning about 65 million years ago occurred after the extinction of the dinosaurs. The radiation of flowering plants must be due to an evolutionary breakthrough—some advantage they had over other plants allowed them to assume roles that were already occupied by other species.

Plant biologists, or **botanists,** still debate which traits of flowering plants give them an advantage over nonflowering types. Some botanists believe that the unique reproductive characteristics of flowering plants led to their radiation (**Figure 13.16**).

One trait leading to diversity is that flowering plants often rely on the assistance of animals in transferring gametes—as an insect or other animal becomes a reliable visitor to a particular flower type, the plant evolves to become more effective at "loading" the visitor with pollen (including highly attractive **petals** on the flower), leading to an exclusive relationship and thus divergence from other plants. The pattern of both species in a relationship evolving adaptations to each other is known as **coevolution.** Coevolution has also occurred between animals and flowering plants related to seed dispersal—flowering plants make fruit, and different types of fruit attract or depend on different types of animals.

1. **Flower petals** attract insects that move pollen from one flower to another, helping fertilization to occur.

2. **Double fertilization** occurs. The pollen tube carries two sperm. One fertilizes the embryo, and the other fuses with two nuclei in another cell to produce the endosperm.

3. **Fruit** consists of seeds packaged in a structure that aids their dispersal, such as tasty flesh or a parachute.

4. **Seeds** contain an embryo and endosperm, and are highly resistant to drying. The endosperm is a tissue that nourishes the embryo.

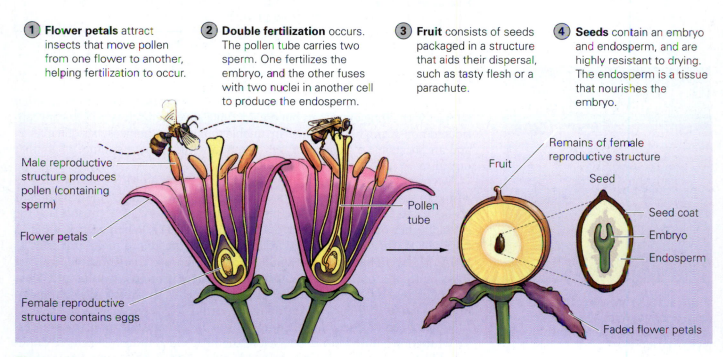

Male reproductive structure produces pollen (containing sperm)

Flower petals

Female reproductive structure contains eggs

Pollen tube

Fruit

Remains of female reproductive structure

Seed

Seed coat

Embryo

Endosperm

Faded flower petals

Figure 13.16 Sexual reproduction in flowering plants. The reproductive differences between flowering plants and other plants, including the production of fruit for dispersal, may have led to their adaptive radiation.

Another trait that may have led to success among flowering plants is **double fertilization,** wherein sperm from a single pollen grain fertilize both the egg and a specialized food-producing tissue. Double fertilization ensures that a plant only commits energy in the form of food resources to an existing embryo.

Other botanists think that chemical defenses in flowering plants contributed to their success by reducing their susceptibility to predators. The diversity of chemical defenses in flowering plants makes them particularly interesting to bioprospectors. Because plants cannot physically escape from their predators, natural selection has favored the production of predator-deterring toxins via side reactions to primary biochemical pathways. These chemicals are known as **secondary compounds.** For instance, curare vines produce toxins that block the connection between nerves and muscles. Organisms that get this toxin, called *curarine,* in their bodies become paralyzed and can do little damage to the vine. Curarine is produced via a side reaction of the process of amino acid synthesis. In this case, natural selection must have favored genetic variations leading to production of not only normal amino acids but also this toxic secondary compound. Today, doctors use curarine as a muscle relaxant during surgery.

The kingdom Plantae is the source of many other well-known, naturally derived drugs. Aspirin from willow, the heart drug digitalis from foxglove, the anticancer chemical vincristine from the rosy periwinkle, morphine from opium poppies, caffeine from coffee, and dozens of other pharmaceutical products are derived directly from plants. Pharmaceutical manufacturers also now reproduce hundreds of other drugs that were first identified in plants. Many botanists believe that the number of unknown plant species is relatively small—probably a few thousand—but the potential drugs available in even the known species are mostly unknown.

Stop & Stretch Why do you think most of the known nature-derived drugs used by humans come from plants?

Not Quite Living: Viruses

Any accounting of biodiversity must also consider viruses, organic entities that interact with living organisms but are not quite alive themselves. Viruses are considered non-living because they are unable to maintain homeostasis (Chapter 2) and are incapable of growth or reproduction without the assistance of another organism.

A virus typically consists simply of a strand of DNA or RNA surrounded by a protein shell, called a *capsid*. Some viruses are also enveloped in membranous coat. Viruses are basically rogue pieces of genetic material that must hijack the cells of other organisms to reproduce (**Figure 13.17**). When a virus enters a host cell, it uses the transcription machinery of the cell to make copies of its DNA or RNA. The replicated viral genomes then use the ribosome, transfer RNA, and amino acids of the host to make new protein capsules and other polypeptides. If the virus is enveloped, it even uses the host cell's own membrane to make envelopes for the daughter viruses.

Once a cell has been hijacked by a virus, it cannot perform its own necessary functions and will die. But before the hijacked cell dies, dozens of duplicated viruses can be released and move on to infect other cells. Some of the most devastating pathogens of humans are viruses, including polio, smallpox,

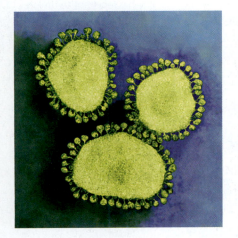

Figure 13.17 Viruses. The severe acute respiratory virus (the cause of severe acute respiratory syndrome, SARS), viewed under very high magnification. Visible here is the protein capsid, studded with receptors that bind to host cells.

influenza, and HIV. HIV is especially troublesome because it destroys immune system cells, which are essential for fighting off infection. An HIV-infected individual gradually loses the ability to combat other diseases, including the HIV infection itself.

13.3 Learning About Species

The living world is amazingly diverse, and our knowledge of it is only fragmentary. Many biologists are attracted to the study of biological diversity for just this reason—because the variety of life is remarkable, fascinating, and largely an unexplored frontier in human knowledge. And as we observed in the previous section, nonhuman species can represent an enormous resource for humans.

Our survey of life's diversity illustrates the essential problem for a bioprospector—there are many more potentially valuable species than there are resources to find and evaluate them. This challenge has led many drug companies to abandon bioprospecting in favor of "rational drug design," which uses an understanding of the causes of a particular illness to create synthetic drugs used to treat it. However, some of the most effective plant-based drugs are so strange in structure that a comparable synthetic drug would likely never have been developed using rational design.

Bioprospectors continue to hope to find more of these strange natural compounds. What tools can biologists offer to bioprospectors who seek to mine biological gold?

Fishing for Useful Products

The National Cancer Institute (NCI) has taken a brute-force approach to screening species for cancer-suppressing chemicals. NCI scientists receive frozen samples of organisms from around the world, chop them up, and separate them into a number of extracts, each probably containing hundreds of components. These extracts are tested against up to 60 different types of cancer cells for their efficacy in stopping or slowing growth of the cancer. Promising extracts are then further analyzed to determine their chemical nature, and chemicals in the extract are tested singly to find the effective compound. This approach is often referred to as the "grind-'em-and-find-'em" strategy.

The most well-known anticancer chemical discovered using this approach is paclitaxel, also known by the trade name Taxol, from the Pacific yew (**Figure 13.18**). Paclitaxel continues to be produced from the needles of yew trees and is effective against ovarian cancer, advanced breast cancers, malignant skin cancer, and some lung cancers. This screening method has identified dozens of other less-well-known anticancer drugs as well.

Understanding Ecology

The grind-'em-and-find-'em approach works best when researchers are seeking treatments for a specific disease or set of diseases, such as cancer. But most bioprospectors are much more speculative—they are interested in determining whether an organism contains a chemical that is useful against *any* disease. Doing this effectively requires a more thoughtful approach, taking into account the biology of the species.

Figure 13.18 The Pacific yew. The source of the anticancer drug paclitaxel was overharvested after the drug was discovered. Fortunately, the drug can be synthesized from other common yew species.

Figure 13.19 Induced defenses.
Linden viburnum damaged by deer is subsequently more resistant to insect attack because the plant increases production of defensive chemicals after being damaged once.

One aspect of an organism's biology that can be illuminating is its **ecology**—that is, its relationship to the environment and other living organisms. Our survey of diversity illustrated some ecological characteristics that increase the likelihood that a species contains valuable chemicals. These characteristics include survival in extreme environments, high levels of competition with bacteria and fungi, susceptibility to predation, ability to live in and on other living organisms, and high population density. In these cases, natural selection can lead to the evolution of biochemicals that function at high temperature or salinity; antibacterial, antifungal, or antiviral compounds; molecules that suppress or modify the effects of the immune system; and chemical defenses that may have physiological effects.

An understanding of ecology is useful even within a species. For instance, populations of plants experiencing high levels of insect attacks may produce more defensive compounds than populations that are not under attack (**Figure 13.19**). Focusing on certain ecological situations can help bioprospectors increase their success in identifying useful compounds from nature.

Reconstructing Evolutionary History

Clues to an organism's chemical traits can also come from knowing the traits found in its closest relatives. This is one reason some scientists argue that a classification system reflecting evolutionary relationships is more useful than one based on more superficial similarities.

The relationship between certain reptiles and birds can help illustrate this point. Komodo dragons and alligators are both large reptiles, and both are ambush predators that lay in wait for their prey. Both animals walk on four legs, are covered with scales, and are cold-blooded. Birds, on the other hand, are warm-blooded, covered with feathers, and have wings (**Figure 13.20**). Not surprisingly, a nonevolutionary classification places crocodiles with Komodo dragons and other cold-blooded, scaly reptiles. However, overwhelming evidence from morphology, physiology, and genetics indicates that crocodiles are more closely related to birds than they are to Komodo dragons.

(a) Komodo dragon

(b) American alligator

(c) Chickadee

Figure 13.20 The challenge of biological classification. The evolutionary relationship between alligators and birds is not evident from their appearance.

An **evolutionary classification** can be quite useful in the study of living organisms. For instance, if scientists wish to know more about the basic biology of Komodo dragons, they should not use what is known about crocodiles—a much more distant relative—as a stepping off point. Instead, they should focus on more closely related lizards. And if bioprospectors want to look for new valuable biological compounds, they could start by screening the close relatives of organisms producing known valuable chemicals.

Developing Evolutionary Classifications. Evolutionary classifications are based on the principle that the descendants of a common ancestor should share any biological trait that first appeared in that ancestor. For example, this principle has been used to uncover the evolutionary relationship among different species of sparrows. A species is a group of organisms whose members can interbreed with each other. All species have a two-part name—the first part of the name indicates the **genus,** or broader group of relatives to which the species belongs; the second part is specific to a particular species within that genus. In effect, a species name is like a full name in which the "family group" is designated by the first name, and the individual in that particular family is identified by the second part of the name. Both parts of the name are always used when identifying a particular species, although if we are discussing a group of species in the same genus, the genus name may be abbreviated to just its first letter.

Figure 13.21 illustrates a hypothesized **phylogeny**—the evolutionary relationship—of sparrow species in the genus *Zonotrichia*. Scientists have used a technique called **cladistic analysis,** an examination of the variation

Figure 13.21 Reconstruction of an evolutionary history. The relationships among these four species of sparrow can be illustrated by their shared physical traits compared to a more distant relative, the dark-eyed junco.
Visualize This: What traits did the ancestral bird likely have?

in these sparrows' traits relative to a closely related species, to determine this phylogeny. For example, if we examine just the heads of the four sparrow species compared to their relative, the dark-eyed junco, we see that three—all but the Harris's sparrow—have dark and light alternating stripes on the crown of their heads. This observation seems to indicate that crown striping evolved early in the radiation of these sparrows. Among the three species with crown stripes, two have seven stripes, while the golden-crowned sparrow has only three stripes. An increase in the number of stripes appears to have evolved after the original striped crown pattern. Finally, of the two species with seven crown stripes, only one has evolved a white throat—the aptly named white-throated sparrow. In the case of these four sparrows, it appears that every step in their radiation involved a visible change in their appearance.

Unfortunately, reconstructing evolutionary relationships is not as simple as the sparrow example suggests. Descendant species may lose a trait that evolved in their ancestor, or unrelated species may acquire identical traits via convergent evolution. You can even see convergent evolution in the phylogeny in Figure 13.22; golden-crowned sparrows, like the white-throated species, have a patch of golden feathers on their heads that appears to have evolved independently. The existence of convergent traits complicates the development of evolutionary classifications.

Stop & Stretch   Amateur bird watchers often consider vultures to be birds of prey, along with hawks, falcons, eagles, and owls. Vultures are scavengers rather than predators, however. What convergent traits do vultures share with these other birds that would cause bird watchers to classify them as birds of prey?

Testing Evolutionary Classifications.   An *evolutionary classification* is a hypothesis of the relationship among organisms. It is difficult to test this hypothesis directly—scientists have no way of observing the actual evolutionary events that gave rise to distinct organisms. However, scientists *can* test their hypotheses by using information from both fossils and living organisms.

By examining the fossils of extinct organisms, scientists can gather clues about the genealogy—the record of descent from ancestors—of various organisms. For example, fossils of birds clearly indicate that these animals evolved from crocodile-like ancestors.

Information from living organisms can provide an even finer level of detail about evolutionary relationships. As illustrated in Figure 13.5, closely related species should have similar DNA. If the pattern of DNA similarity matches a hypothesized evolutionary relationship among species, the phylogeny is strongly supported. This is the case with the relationship between birds and crocodiles; DNA sequence comparisons indicate that the DNA of crocodiles is more similar to the DNA of birds than it is to the DNA of Komodo dragons.

In contrast, DNA sequence comparisons do not support the sparrow phylogeny presented in Figure 13.22. Here, the data suggest that white-crowned birds and golden-crowned birds are closely related, and white-throated sparrows are a more distant relative. In this case, more observations are needed to discern the true evolutionary relationships among the *Zonotrichia*. (In Chapter 10 we described multiple supportive tests of another phylogeny—in that case, the evolutionary relationship among humans and apes.) The examples of phylogeny reconstruction and testing (as described here and in Chapter 10) provide illustrations of the science of systematics, but the species involved in these examples are not likely to contain molecules that are valuable to bioprospectors. The evolutionary reconstructions that interest bioprospectors consist of groups of species that already contain valuable members. For instance, relatives of the curarine-producing curare vine are likely to contain similar secondary biochemical pathways that produce slightly different muscle relaxants (**Figure 13.22**). Once a hypothesis of evolutionary relationship *is* reasonably well supported by additional data, bioprospectors can use the information gathered about one species to predict the characteristics of its relatives.

Although an evolutionary classification of groups that are likely to contain valuable chemicals would increase the speed of discovery, the slow pace of reconstructing and testing phylogenies means that bioprospectors do not always have this information. As an alternative, bioprospectors have turned to the humans who have the most intimate knowledge of useful organisms in their environments—healers in preindustrial cultures.

Learning from the Shamans

To this point, we have been describing the search for biologically active compounds in living organisms as an exploration of the unknown. This is true for organisms in Yellowstone's hot springs; chemicals derived from these organisms were probably truly unknown to humans. For many other species, however, people have known of their usefulness for thousands of years. This knowledge is maintained in the traditions of indigenous people in biologically diverse areas, people who use native organisms as medicines, poisons, and foods.

In many cultures, the repository of this traditional knowledge is the medicine man or woman. A shaman, as these healers are often called, can help direct bioprospectors to useful compounds by teaching about their culture's traditional methods of healing. Shamans employ many remedies that are highly effective against disease. Several bioprospectors have consulted with shamans to increase their chances of finding useful drugs (**Figure 13.23**).

Using the knowledge of native people in developing countries to discover compounds for use in wealthy, developed countries is highly controversial. This process is often referred to as **biopiracy** because organisms and active compounds thus identified can potentially provide enormous financial rewards to the bioprospector with no return to a shaman or the shaman's people.

The United Nations Convention on Biodiversity has sought to reduce biopiracy by asserting that each country owns the biodiversity within its borders. However, because the U.S. government has not signed this legally binding document, companies in the United States are not required to abide by its terms. In addition, even when the government of a country makes a bioprospecting agreement with a pharmaceutical firm, any rewards may not trickle down to the indigenous community that was the source of the knowledge.

The bioprospecting agreement made between Verenium and Yellowstone National Park has not escaped charges of biopiracy. While the company was not planning on relying on information from indigenous people to help locate valuable organisms, critics have charged that the managers of Yellowstone were essentially "selling off" organisms and chemical compounds that belong to the American public. In addition, they argue that the action of bioprospecting itself will damage the very resource that provides these remarkable discoveries. The

Figure 13.22 Curare. Its name translates from the native words *"bird"* and *"kill."* An evolutionary classification would help us identify its closest relatives, which may also have medicinal properties.

Figure 13.23 Indigenous knowledge. This shaman is of the Matses people of the Amazon rain forest. His intimate knowledge of the natural world is the product of the long history of his people in this diverse environment.

federal courts dismissed lawsuits against the U.S. Park Service and Verenium that addressed these points, but the issue remains an ethical dilemma: What is the responsibility of corporations profiting from biological diversity toward the source and survival of that diversity?

Biological diversity represents an enormous resource for humans, but it also comes with an awesome responsibility. Actions of the U.S. Congress protected Yellowstone National Park and perhaps ultimately enabled the discovery of *Thermus aquaticus* and *Taq* polymerase. But thousands of useful organisms are lost every year through the destruction of native habitat, and our ability to use these organisms is diminished by the loss of indigenous cultures and their shamans.

The dramatic rate of biodiversity loss not only denies humans still-undiscovered biological molecules but also diminishes the ability of Earth to sustain our population. At current rates of loss, our generation may be the last to truly enjoy the diverse wonder, beauty, and benefits of wild nature (Chapter 15 discusses the causes, consequences, and possible solutions for the current biodiversity crisis.). Humans can help to reduce the rate at which biodiversity is being lost—but only if we begin to appreciate the value of the diversity that surrounds and sustains us.

SAVVY READER

Is the Government Keeping the Truth from Us?

CNN program: PAULA ZAHN NOW, Aired October 13, 2005

Type "natural cures" into a search engine and it will return thousands of sites promoting different products and services. It can be difficult to determine which are legitimate and based on research and which are unproven or even dangerous. One author, Kevin Trudeau, has made hundreds of millions of dollars selling books and access to his website claiming insider knowledge about natural cures the government and drug companies are hiding from the public. Read the transcript of these excerpts from an interview with Mr. Trudeau broadcast on the CNN program "Paula Zahn Now," and learn to spot some of the techniques used to convince consumers to spend money on unproven products. Note that these techniques are not unique to Mr. Trudeau; they are common in many advertising appeals.

ZAHN: I sat down with Kevin and asked why, despite his book's success, the vast majority of medical professionals seem to view him as a quack.

TRUDEAU: Of course medical doctors are going to say they don't believe in what I'm doing, because the whole book is about exposing the medical business for what it is: fraud. . . .

ZAHN: David Johnson, vice president of the American College of Gastroenterology, he notes that...

TRUDEAU: Drug pusher.

ZAHN: ... he notes that there is no evidence to support many of the book's claims. He says some of Trudeau's suggestions could actually be harmful.

TRUDEAU: Such as?

ZAHN: Digestive enzymes. He says—quote—"These enzymes are very caustic and could burn the esophagus..."

TRUDEAU: OK.

ZAHN: ... and that they are typically only prescribed for people with pancreatic problems, to begin with.

TRUDEAU: OK.

Well, that's a doctor giving his opinion. Now, is it a fact or his opinion?

ZAHN: This is his opinion.

TRUDEAU: Does he say it's his opinion? He's presenting his...

ZAHN: And what you write in your book is your opinion.

TRUDEAU: Correct. And that's the point.

Should people have the option of reviewing opinions? First off, there's over 900 studies in the book that's listed. Under the chapter that says, "Still Not Convinced," I list over three dozen books. And each of those books lists virtually hundreds of studies that back up everything that is said in there.

ZAHN: And one of the things you recommend is getting 15 colonics in 30 days. According to this same Dr. Johnson: "There is no medical reason to have even one of these procedures..."

TRUDEAU: Sure.

ZAHN: "... which typically involve purging the bowels with enemas, let alone 15 in one month."

TRUDEAU: Horror. I know.

ZAHN: He says—quote—"All that purging could lead to dehydration..."

TRUDEAU: Yes.

ZAHN: "... or electrolyte imbalances, which can disturb heart rhythms." True or false?

TRUDEAU: It's false. That's his opinion.

....

TRUDEAU: Let's—let's not mislead the public, Paula. Don't mislead the public. Three million people bought this book. The majority, overwhelming majority, of people that read my book are writing me letters by the tens of the thousands, thanking me.

1. What is the apparent expertise of Mr. Trudeau? How does he use his expertise to establish a sort of credibility with an audience? What other strategies does he use to establish credibility?

2. Based on this interview, to what concerns and opinions of consumers does Mr. Trudeau apparently appeal to get individuals to buy his book or access to his website?

3. Evaluate how both the interviewer and Mr. Trudeau use the terms "fact" and "opinion." Are there cases when a statement of fact is being labeled as an opinion? How about when an opinion is labeled as a fact? Why is it important to distinguish facts and strongly supported hypotheses from unsupported assertions and opinions?

Source: http://transcripts.cnn.com/TRANSCRIPTS/0510/13/pzn.01.html

Chapter Review
Learning Outcomes

Mastering **BIOLOGY**

LO1 Cite the current estimate of the number of species described by science, and explain why this number is only an estimate (Section 13.1).

- The number of known living species is estimated to be between 1.4 and 1.8 million, but the total number of species may be as high as 100 million (pp. 309–310).

LO2 List the three major domains of life, and describe basic characteristics of each domain (Section 13.1).

- Organisms are classified in domains according to evolutionary relationships. Bacteria share a distant common ancestor with Archaea and Eukarya, while the last two groups are more closely related. However, both Bacteria and Archaea are prokaryotic (simple single-celled organisms without a nucleus or other membrane-bound organelles), while organisms in the Eukarya are eukaryotes (pp. 310–314).

LO3 Provide a broad overview of the history of life on Earth (Section 13.2).

- Life on Earth began about 3.6 billion years ago with simple prokaryotes, but it would be 1.5 billion years before eukaryotes evolved (p. 315).
- Multicellular organisms did not appear until approximately 600 million years ago, and this advance in form led to the diversity of species on Earth today (p. 318).

- The history of life on Earth has consisted of successive dynasties of organisms (p. 311).
- Animal groups evolved in a short period of time known as the Cambrian explosion (pp. 318–323).

LO4 **Summarize the endosymbiotic theory for the origin of eukaryotic cells (Section 13.2).**

- Eukaryotes, cells with nuclei and other membrane-bound organelles, probably evolved from symbioses among ancestral eukaryotes and prokaryotes (pp. 316–317).

LO5 **List the major characteristics of the four major kingdoms of Eukarya: Protista, Animalia, Fungi, and Plantae (Section 13.2).**

- The kingdom Protista is a hodgepodge of organisms that are typically unicellular eukaryotes. Organisms in this group may be plant-like, animal-like, or fungus-like in lifestyle (pp. 318–319).
- Animals are motile, multicellular eukaryotes that rely on other organisms for food (pp. 320–323).
- Fungi are immobile, multicellular eukaryotes that rely on other organisms for food and are made up of thin, threadlike hyphae (pp. 323–325).
- Plants are multicellular, photosynthetic eukaryotes that make their own food (pp. 325–328).

LO6 **Define** *adaptive radiation,* **and provide examples of where this has occurred in the history of life (Section 13.2).**

- Adaptive radiation occurs when several new species appear quickly after the evolution of a new "way of life" or the death of a competing group (p. 327).
- The diversity of flowering plants may be due partly to adaptive radiation resulting from their production of defensive chemicals (pp. 327–328).

LO7 **Describe the characteristics of viruses, and explain why they are not considered living organisms (Section 13.2).**

- Viruses are nonliving entities made up of genetic material in a transport container and can reproduce only by hijacking living cells (pp. 328–329).

LO8 **Provide examples of ecological traits in each of the various domains and kingdoms that have led to the evolution of biologically active compounds (Section 13.3).**

- Competition between bacteria has led to the evolution of antibiotic compounds that can kill or disable other bacterial cells (pp. 315–316).
- Algae are protists that store food and thus are targets for predators. They make a variety of defensive chemicals against predators. Because the phyla of algae evolved independently of each other, some produce unique food-storage compounds not found in land plants (pp. 318–319).
- Animal predators may have evolved venom to help them disable large, active prey, while some animal prey species have evolved defensive chemicals (p. 320).
- Fungi compete for food sources with bacteria, so the fungi often produce antibiotics. Some fungi grow on living organisms and have evolved compounds that can suppress their host's immune system response (p. 324).
- Plants, like algae, store food that is tempting to predators, so they have evolved a variety of biologically active anti-predator chemicals (pp. 326–328).

LO9 **Describe how evolutionary classifications of living organisms are created and tested, and explain how these classifications can be useful to scientists (Section 13.3).**

- Phylogenies are created and tested by evaluating the shared traits of different species that indicate they shared a recent ancestor (pp. 330–333).

Roots to Remember

The following roots of words come mainly from Latin and Greek and will help you decipher terms:

anima- means breath or soul. Chapter term: animal

botan- means pertaining to plants. Chapter term: botany

eu- means true. Chapter term: eukaryote

in- means without. Chapter term: invertebrate

-karyo- means kernel. Chapter terms: prokaryote, eukaryote

myco- means fungal. Chapter term: mycologist

zoo- means pertaining to animals. Chapter term: zoology

Learning the Basics

1. **LO1** How many different species have been identified by science? How many are estimated to exist?

2. **LO6** Plants experienced an adaptive radiation after the evolution of the flower. Explain what occurred and list the hypotheses that may explain why it occurred.

3. **LO9** How are hypotheses about the evolutionary relationships among living organisms tested?

4. **LO2 LO5** Which of the following kingdoms or domains is a hodgepodge of different evolutionary lineages?

   A. Bacteria; **B.** Protista; **C.** Archaea; **D.** Plantae; **E.** Animalia

5. **LO2 LO3** Comparisons of ribosomal RNA among many different modern species indicate that _____.

   A. there are two very divergent groups of prokaryotes; **B.** the kingdom Protista represents a conglomeration of very unrelated forms; **C.** fungi are more closely related to animals than to plants; **D.** a and b are correct; **E.** a, b, and c are correct

6. **LO2 LO5** On examining cells under a microscope, you notice that they occur singly and have no evidence of a nucleus. These cells must belong to _____.

   A. Domain Eukarya; **B.** Domain Bacteria; **C.** Domain Archaea; **D.** Kingdom Protista; **E.** more than one of the above could be correct

7. **LO4** The mitochondria in a eukaryotic cell _____.

   A. serve as the cell's power plants; **B.** probably evolved from a prokaryotic ancestor; **C.** can live independently of the eukaryotic cell; **D.** a and b are correct; **E.** a, b, and c are correct

8. **LO5** Fungi feed by _____.

   A. producing their own food with the help of sunlight; **B.** chasing and capturing other living organisms; **C.** growing on their food source and secreting chemicals to break it down; **D.** filtering bacteria out of their surroundings; **E.** producing spores

9. **LO7** Which of the following is/are always true?

   A. Viruses cannot reproduce outside a host cell; **B.** Viruses are not surrounded by a membrane; **C.** Viruses are not made up of cells; **D.** a and c are correct; **E.** a, b, and c are correct.

10. **LO9** Phylogenies are created based on the principle that all species descending from a recent common ancestor _____.

    A. should be identical; **B.** should share characteristics that evolved in that ancestor; **C.** should be found as fossils; **D.** should have identical DNA sequences;

E. should be no more similar than species that are less closely related

Analyzing and Applying the Basics

1. **LO4** Unless handled properly by living systems, oxygen can be quite damaging to cells. Imagine an ancient nucleated cell that ingests an oxygen-using bacterium. In an environment where oxygen levels are increasing, why might natural selection favor a eukaryotic cell that did not digest the bacterium but instead provided a "safe haven" for it?

2. **LO2 LO5** Imagine you have found an organism that has never been described by science. The organism, made up of several hundred cells, feeds by anchoring itself to a submerged rock and straining single-celled algae out of pond water. What kingdom would this organism probably belong to, and why do you think so?

3. **LO8** Vertebrate animals are much less likely to produce antibiotic compounds than bacteria and fungi. What features of vertebrate ecology make antibiotic production unlikely? What features of vertebrate physiology make antibiotic production unlikely?

Connecting the Science

1. Scientists initially rejected the endosymbiotic theory, which is the hypothesis that eukaryotic cells evolved from a set of cooperating independent cells. Most biologists still believe that competition for resources among organisms is the primary force for evolution. Do you think our modern culture, where competition is often valued over cooperation, might make scientists less likely to search for and see the role of cooperation in evolution, or do you think cooperation leading to evolution must be truly rare?

2. Do we have an obligation to future generations to preserve as much biodiversity as possible? Why or why not? Would simply preserving the information contained in an organism's genes (in a zoo or other collection) be good enough, or do we need to preserve organisms in their natural environments?

Answers to **Stop & Stretch, Visualize This, Savvy Reader,** and **Chapter Review** questions can be found in the **Answers** section at the back of the book.

Is the Human Population Too Large?

Population Ecology

From space, Earth doesn't look too crowded.

In its most recent estimate in 2010, the United Nations (UN) reported that the world's population is approximately 6.9 billion—more than double the population in 1965. The UN also predicted that the population would continue to grow for several more decades before stabilizing at as high as 9.1 billion by about 2050. Many observers greeted the report as another piece of bad news. While the UN's population projection is lower than past predictions (previous reports forecast a population of over 12 billion by 2050), many scientists and environmentalists wonder if our planet can support the current population for very long, let alone an additional 2.2 billion people.

Other commentators are skeptical of environmentalists' statements about population growth. They point to predictions made in the best-selling book *The Population Bomb* (1968), in which author Paul Ehrlich forecast worldwide food and water shortages by the year 2000. In fact, most measures of human health have become more upbeat since 1970, including global declines in infant mortality rates, increases in life expectancy, and a 20% increase in per capita income—despite a near doubling in population since the publication of Ehrlich's book. By most measures, the average person is better off today than in 1965, when the population was half as large. Paul Ehrlich was clearly wrong in 1968; why should we believe similar doom-and-gloom predictions about the future today?

Ehrlich and his colleagues counter that while they were wrong about how soon it would happen, there are some indications that the large human population is rapidly reaching a real limit to growth. For example, as of January 2011, according to the UN, 925 million people—including 146 million children

But Earth's human population is 7 billion—and rising.

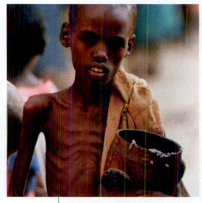

Is this Ethiopian child hungry because the planet is overpopulated?

Or can Earth support everyone at the same level as that of the average North American family?

under the age of 5—do not get enough food regularly for a healthy existence. A staggering 60% of the nearly 11 million deaths each year among children under 5 in the developing world are associated with inadequate nutrition. Despite years of international attention and billions of dollars spent to address this problem, the situation has not improved dramatically—there are only 10% fewer children suffering from malnutrition today than there were in 1980. And rapidly rising food prices around the globe pose a serious threat to future human health and survival.

So what is the truth? Is the human population larger than Earth can support for much longer? Are we headed into a global food crisis and massive famine? Or are we gradually moving toward an era when all people on Earth will be as well fed, long-lived, and affluent as the average North American?

14.1 A Growing Human Population

Ecology is the field of biology that focuses on the interactions among organisms as well as between them and their environment. The relationship between organisms and their environments can be studied at many levels—from the individual, to populations of the same species, to communities of interacting species, and finally to the effects of biological activities on the nonbiological environment, such as the atmosphere.

In ecology, a **population** is defined as all of the individuals of a species within a given area. Populations exhibit a structure, which includes the spacing of individuals (that is, their distribution) and their density (abundance). Ecologists seek to explain the distribution and abundance of the individuals within populations and to understand the factors that lead to the success or failure of a population. The interactions among species make up one set of influences on distribution and abundance (Chapter 15), but another set is the internal dynamics of the population, including the relative numbers of individuals of different sexes and ages and the numbers that are born or die in a given time period.

Population Structure

The first task of a population ecologist is to estimate the size of the population of interest. Certain populations can be counted directly, as in a census of people in the United States or a survey identifying all individuals of a particular tree species in a forest tract.

The size of more mobile or inconspicuous species can be estimated by the **mark-recapture method** (**Figure 14.1**). In this technique, researchers capture many individuals within a defined area, mark them in some way (for instance, with an ear tag), and release them back into the environment. Later, the researchers capture another group of individuals in the same area and calculate the proportion of previously marked individuals in this group. This proportion can be used to estimate the size of the total population.

For an example of the mark-recapture method, imagine that a researcher captured, marked, and released 100 beetles in a small area of forest. One week later, the researcher returns to the same place and captures another group of beetles. If she finds that 10% of the beetles caught on the second round are marked, the researcher can assume that the 100 beetles captured and marked initially represented 10% of the entire beetle population. According to this mark-recapture survey, the total population is approximately 1000 beetles.

Stop & Stretch Imagine the same scenario described in the previous paragraph, but the researcher found that only 1% of the beetles were marked in the second round of captures. How large is the population likely to be in this case?

Another basic aspect of population structure is dispersion—that is, how organisms are distributed in space. Many species show a **clumped distribution**, with high densities of individuals in certain resource-rich areas and low densities elsewhere. Plants that require certain soil conditions and the animals that depend on these plants tend to be clumped (**Figure 14.2a**). On a global

Figure 14.1 The mark-recapture method. 1. In the first step, a researcher captures, marks, and releases animals in a wild population. 2. Some time later, the researcher returns and determines what percentage of animals trapped again were previously marked. 3. In the second capture, the ratio of marked to total animals captured is approximately the ratio of the original number trapped to the whole population.
Visualize This: Animals may become either "trap-happy," that is, attracted to traps, or "trap-shy," after one capture. How might each of these behaviors affect a population estimate?

① Researcher captures 100 beetles in a trap, marks each with a dab of paint.

② After one week, a trap is set again, resulting in a captured group of marked and unmarked individuals.

③ Total population is estimated as equivalent to the percentage of marked individuals in the second trap.

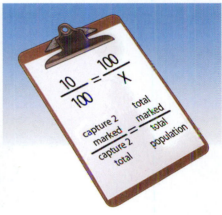

$$\frac{10}{100} = \frac{100}{X}$$

$$\frac{\text{capture 2 marked}}{\text{capture 2 total}} = \frac{\text{total marked}}{\text{total population}}$$

(a) Clumped   **(b) Uniform**   **(c) Random**

Figure 14.2 Patterns of population dispersion. Individuals in a population may be (a) clumped, like these cattails growing only in soil with the correct water content; (b) uniformly distributed, like these territorial nesting penguins; or (c) randomly dispersed, like these seedlings grown from wind-blown seeds in a forest.

scale, humans show a clumped distribution, with high densities found around transportation resources such as rivers and coastlines.

The clumped distribution of humans masks a more **uniform distribution** on a local scale; for instance, the spacing between houses in a subdivision or strangers in a classroom tends to equalize the distances among individual property owners or people. Species that show a uniform distribution are often territorial—they defend their own personal space from intruders. We can observe these same strong reactions among certain species of birds at a breeding site (**Figure 14.2b**).

Nonsocial species with the ability to tolerate a wide range of conditions typically show a **random distribution,** in which no compelling factor is actively bringing individuals together or pushing them apart. The distribution of seedlings of trees with windblown seeds is often random (**Figure 14.2c**).

The distribution and abundance of a population provide a partial snapshot of its current situation. The dispersion of the human population—and recent changes in that pattern—profoundly affects the natural environment (Chapter 16). However, to better understand how a population is responding to its environment, we need to determine how it is changing through time.

An Overview: Exponential Population Growth

Historians have been able to use archaeological evidence and written records to determine the size of the human population at various times during the past 10,000 years. This record, presented in **Figure 14.3**, dramatically illustrates the pattern of population growth.

The graph of human population growth is a striking illustration of **exponential growth**—growth that occurs in proportion to the current total. Populations growing exponentially do not add a fixed number of offspring every year; instead, the quantity of new offspring is a proportion of the size of the previous generation and thus is an ever-growing number. Exponential growth results in the J-shaped growth curve seen in Figure 14.3. *See the next section for **A Closer Look** at exponential population growth.*

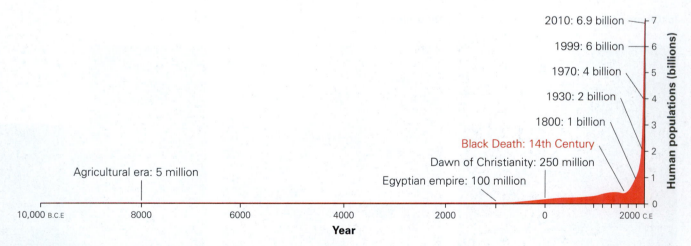

Figure 14.3 Exponential growth. The number of people on Earth grew relatively slowly until the eighteenth century. The rapid growth since then has occurred in proportion to the total, causing a J-shaped curve.

A Closer Look:
Exponential Population Growth

For most of our history, the human population has remained at very low levels. At the beginning of the agricultural era, about 10,000 years ago, there were approximately 5 million humans. There were 100 million people during the Egyptian Empire (7000 years later) and about 250 million at the dawn of the Christian religion in 1 C.E. The population was growing, but at a very slow rate—approximately 0.1% per year.

Beginning around 1750, the rate at which the human population was growing jumped to about 2% per year. The human population reached 1 billion in 1800, had doubled to 2 billion by 1930, and then doubled again to 4 billion by 1970. Although the current growth rate is slower, about 1.1% per year, the rapid increase in population looks quite dramatic on a graph of human population over time.

The larger a population is, the more rapidly it grows because an increase in numbers depends on individuals reproducing in the population. So, while a growth rate of 1.1% per year may seem rather small, the number of individuals added to the 6.9-billion-strong human population every year at this rate of growth is a mind-boggling 76 million (about equal to the combined populations of California, Texas, and Florida). Put another way, three people are added to the world population per second, and about a quarter of a million people are added every day.

What has fueled this enormous increase in human population? The annual **growth rate** of a population is the percentage change in population size over a single year. Growth rate, which is mathematically represented as r, is a function of the birth rate of the population (the number of births as a percentage of the population) minus the death rate (the number of deaths as a percentage of the population). For example, in the entire human population 21 babies are born per year for every 1000 people—that is, the birth rate for the population is 2.1%:

$$\frac{21}{1000} = 0.021 = 2.1\%$$

In addition, each year 10 individuals die of every 1000 people, resulting in a death rate of 1.0%:

$$\frac{10}{1000} = 0.01 = 1\%$$

This results in the current growth rate of 1.1%:

Growth rate = Birth rate − Death rate

1.1% = 2.1% − 1.0%

Today's relatively high growth rate, compared to the historical average of 0.1%, is the result of a large difference between birth and death rates.

It can be easier to think of exponential growth in terms of the amount of time it takes for a population to double in size. At a growth rate of 0.1, it takes 693 years for the population to double. At a rate of 1.1%, it takes only about 64 years (**Figure 14.4**).

(a)

(b)

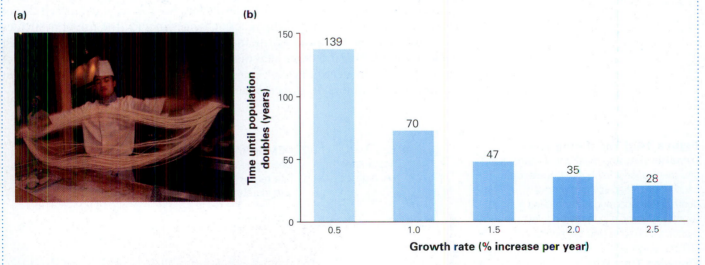

Figure 14.4 Growth rate and doubling time. (a) Hand-pulled noodles are made by folding and pulling dough repeatedly, doubling the number of strands in every fold. A single thick noodle folded only 12 times will produce 4096 fine noodles (seen here), illustrating how rapidly growth occurs when it is exponential. (b) The time it takes a population to double in size depends on its growth rate. A population growing at 2% per year doubles in 35 years, while one growing at 1% takes twice as long to double.

The Demographic Transition

In human populations, the tendency has been for decreases in death rate to be followed by decreases in birth rate. The speed of this adjustment helps to determine population growth in the future. Before the Industrial Revolution, both birth and death rates were high in most human populations. Women gave birth to many children, but relatively few children lived to reach adulthood. Rapid population growth was triggered in the eighteenth century by a dramatic decline in infant mortality (the death rate of infants and children) in industrializing countries. In particular, advances in treating and preventing infectious disease has reduced the number of children who die from these illnesses.

> **Stop & Stretch** Life expectancy is the average age a newborn is expected to reach. Declines in infant mortality dramatically increase life expectancy in a country, even if adults are not living longer. Use your understanding of how averages are calculated to explain why this is the case.

Not long after death rates declined in these countries, birth rates followed suit, lowering growth rates again. Scientists who study human population growth refer to the period when birth rates are dropping toward lowered death rates as the **demographic transition** (**Figure 14.5**). The length of time that a human population remains in the transition has an enormous effect on the size of that population. Countries that pass through the transition swiftly remain small, while those that take longer can become extremely large. Nearly all **developed countries** (those that have industrial economies and high individual incomes) have already passed through the demographic transition and have low population growth rates.

However, global human population growth rates have remained high because the **less developed countries** (countries that are early in the process of industrial development and have low individual incomes) remain in the demographic transition. In addition, several recent changes have decreased infant mortality even more dramatically. These changes include the use of pesticides to reduce rates of mosquito-borne malaria; immunization programs against cholera, diphtheria, and other fatal diseases; and the widespread availability of antibiotics. As the death rate continues to decline but the birth rate remains at historical levels in less developed countries, population growth rates soar.

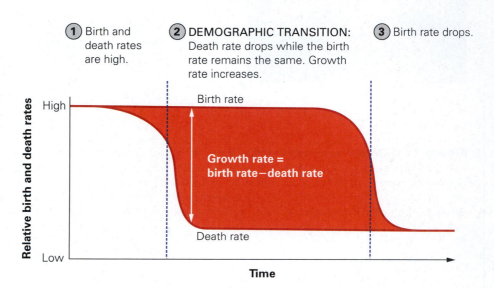

Figure 14.5 The demographic transition. As improvements in sanitation and medical care in human populations cause a decrease in infant mortality, death rates drop, and growth rates soar. Eventually, people in these populations respond by decreasing the number of children they have.
Visualize This: Where on this graph is the growth rate of the population close to zero?

① Birth and death rates are high.

② DEMOGRAPHIC TRANSITION: Death rate drops while the birth rate remains the same. Growth rate increases.

③ Birth rate drops.

Relative birth and death rates

High

Birth rate

Growth rate = birth rate−death rate

Death rate

Low

Time

The vast majority of future population growth will occur within populations in the less developed world, but these countries are also where the vast majority of food crises are occurring. Are the populations in these countries already too large to support themselves? Answering that question requires an understanding of the factors that limit population growth.

14.2 Limits to Population Growth

In their studies of nonhuman species, ecologists see clear limits to the size of populations. They can also observe the sometimes awful fates of individuals in populations that outgrow these limits. For this reason, many professional ecologists are gravely concerned about the rapidly growing human population.

You may know of several instances of nonhuman populations outgrowing their food supplies. The elk population in Yellowstone National Park suffered enormous mortality throughout the 1970s after it grew so large that it degraded its own rangeland. The massive migrations of Norway lemmings that occur every 5 to 7 years and lead to many lemming deaths result from population crowding. While these animals do not commit "mass suicide," as often assumed, the loss of high-quality food in an area as populations increase incites the lemmings to disperse, and they often meet their death in the process. Even yeast in brewing beer grow large populations that eventually use up their food source and as a result die off during the fermenting process. Let us explore what ecology can tell us about the likelihood of human populations suffering the same fate as elk, lemmings, or yeast.

Carrying Capacity and Logistic Growth

The examples of elk in Yellowstone and Norway lemmings illustrate a basic biological principle. While populations have the capacity to grow exponentially, their growth is limited by the resources—food, water, shelter, and space—that individuals need to survive and reproduce. The maximum population that can be supported indefinitely in a given environment is known as the environment's **carrying capacity.**

A simplified graph of population size over time in resource-limited populations is S-shaped (**Figure 14.6**). This model shows the growth rate of a population declining to zero as it approaches the carrying capacity. In other words, birth rate and death rate become equal, and the population stabilizes at its maximum size, which is mathematically represented as K. Not long after ecologists first predicted this pattern of growth, called **logistic growth,** populations of organisms as diverse as flour beetles, water fleas, and single-celled protists were shown in laboratory studies to conform to this projected growth curve.

The declining growth rate near a population's carrying capacity is caused by **density-dependent factors,** which are population-limiting factors that increase in intensity as a population increases in size. Density-dependent factors include limited food supplies, increased risk of infectious disease, and an increase in toxic waste levels. Density-dependent factors cause either declines in birth rate or increases in death rate. In organisms such as fruit flies growing in laboratory culture bottles, high populations lead to increased mortality of the flies as food supplies dwindle and wastes accumulate. Water fleas living in crowded aquariums do not have enough food to support egg production, so birth rates drop. Females of white-tailed deer populations living in crowded natural habitats are less likely

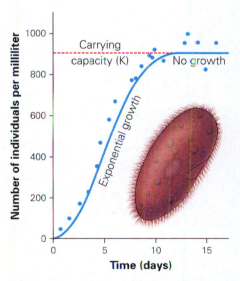

Figure 14.6 The logistic growth curve. The graph illustrates the change in the population of *Paramecium* (a single-celled protist) over time in a laboratory culture. The S-shape of the growth curve is due to a gradual slowing of the population growth rate as it approaches the carrying capacity of the environment.
Visualize This: Where on the graph is the number of individuals added to the population *per unit of time* highest?

Figure 14.7 Limits to growth.
Populations of (a) fruit flies in a laboratory culture, (b) water fleas in an aquarium, and (c) white-tailed deer in the northeastern United States all experience high death rates or low birth rates as their populations approach the carrying capacity of the environment. (d) Do human populations face these same limits?

(a) Fruit flies

(b) Water fleas

(c) White-tailed deer

(d) Humans

to have the energy reserves necessary to carry a pregnancy to term than deer in less crowded environments (**Figure 14.7**). Some populations of humans have experienced density-dependent mortality as well, including the epidemic of bubonic plague that occurred in crowded European cities in the fourteenth century.

Density-dependent factors can be contrasted with **density-independent factors** that influence population growth rates—for instance, severe droughts that increase the death rate in plant populations regardless of their density or increased temperatures that increase the birth rate in insects. The impact of climate change may be to increase the occurence of density-independent factors as Earth warms and dries and storms become more severe (Chapter 5). However, density-independent factors do not occur in a vacuum; they can have more or less severe effects depending on the size of a population. For example, a density-independent factor such as an unusually cold winter can be deadly to individuals in a white-footed mouse population, but the likelihood of survival is also a function of how much food each individual has stored for the winter. How much food is stored by each animal depends on the density of mice competing for that food during the autumn.

Stop & Stretch   Why are infectious diseases considered a density-dependent factor rather than density-independent?

Are density-dependent factors beginning to reduce growth rates in the human population? That is, are humans nearing the carrying capacity of Earth for our population? If we are, will death rates increase as food resources dwindle and more people starve? Or will birth rates decline because fewer women will have enough food to support themselves and a developing baby?

Earth's Carrying Capacity for Humans

One way to determine if the human population is reaching Earth's carrying capacity is to examine whether, and how rapidly, the growth rate is declining. As we saw in Figure 14.6, the S-shaped curve of population size over time results from a gradually declining growth rate as the population approaches carrying capacity.

Human population growth rates were at their highest in the early 1960s, about 2.1% per year, but they have since declined to the current rate of around 1.1%. This steady decline is an indication that the population, though still currently growing, is nearing a stable number. Uncertainty about the future rate of growth has led the UN to produce differing estimates of this number and how soon population stability will be reached (**Figure 14.8**). However, the unique characteristics of humanity make it difficult to determine exactly which population size represents Earth's carrying capacity for humans.

Signs That the Population Is Not Near Carrying Capacity. The rates of population increase of fruit flies and water fleas in the laboratory will slow as these populations near carrying capacity because their growth rates are forced down by density-dependent factors, such as lack of resources, causing increased death rates or decreased birth rates. However, this is not the case in human populations. Even as the human population has rapidly increased, death rates continue to decline—an indication that people are not running out of food. Growth rates are declining because birth rates are falling even faster than death rates. Unlike the water fleas and white-tailed deer, whose females are unable to have offspring when populations are near carrying capacity, birth rates in human populations are falling because women and families, especially those with adequate resources, are *choosing* to have fewer children.

Another way to determine if humans are near Earth's carrying capacity is to estimate the proportion of Earth's resources used by humans now. The amount of food energy available on the planet is referred to as the **net primary productivity (NPP)**. NPP is a measure of plant growth, typically over the course of a single year (**Figure 14.9**).

Several different analyses of the global extent of agriculture, forestry, and animal grazing estimate that humans use roughly one-third of the total land NPP. If we accept these rough estimates, we can approximate that the carrying capacity of Earth is three times the present population, or approximately 21 billion people. This theoretical maximum is the total number of humans that could be supported by all of the photosynthetic production of the planet—leaving no resources for millions of other species. Given the dependence of humans

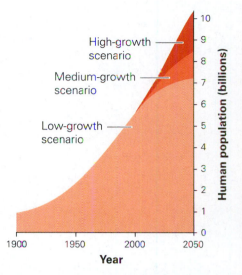

Figure 14.8 Projected human population growth. The United Nation's report is based on a number of uncertainties and returns three projections: a low-growth scenario of 7.2 billion people, medium growth resulting in 8.6 billion, and a high-growth estimate of 10.3 billion.

Figure 14.9 Net primary production. NPP is a measure of the total calories available on Earth's surface to support consumers.

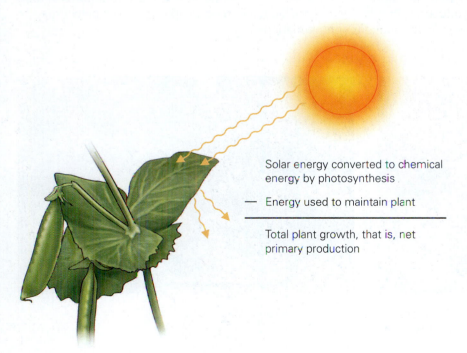

Solar energy converted to chemical energy by photosynthesis

— Energy used to maintain plant

Total plant growth, that is, net primary production

on natural systems (explored in Chapter 15), it is unlikely that our species could survive on a planet where no natural systems remained. However, even the largest population projection by the UN, 10.3 billion, falls well short of this theoretical maximum.

Signs That the Population Is Near Carrying Capacity. Ecologists caution that the resources required to sustain a population include more than simply food, so the carrying capacity deduced from NPP estimates may be much too high. Humans also need a supply of clean water, clean air, and energy for essential tasks such as heating, food production, and food preservation.

The relationship between population size and the supply of these resources is not as straightforward as the relationship between population and food. For instance, every new person added to the population requires an equivalent amount of clean water, but every new person also introduces a certain amount of pollution to the water supply. We cannot simply divide the current supply of clean water by 10.3 billion to determine if enough will be available in the future because increased population leads to increased pollution and therefore less total clean water.

Furthermore, many essential supplies that sustain the current human population are **nonrenewable resources,** meaning that they are a one-time stock and cannot be easily replaced. The most prominent nonrenewable resource is fossil fuel, the buried remains of ancient plants transformed by heat and pressure into coal, oil, and natural gas.

The use of fossil fuel and other nonrenewable resources is a function not only of the number of people but also of average lifestyles, which vary widely around the globe. For example, Americans make up only 5% of Earth's population but are responsible for 24% of global energy consumption. The average American uses as many resources as 2 Japanese or Spaniards, 3 Italians, 6 Mexicans, 13 Chinese, 31 Indians, 128 Bangladeshis, 307 Tanzanians, or 370 Ethiopians (**Figure 14.10**).

Americans also consume a total of 815 billion food calories per day—about 200 billion calories more than is required, or enough to feed an additional 80 million people. Much of modern food production relies on the energy provided by fossil fuel. When these fuels begin to run out, we might find that we need far more of Earth's NPP than we do now to sustain abundant food production. In other words, the actual carrying capacity of our planet may be much lower than our approximations.

American family

Malian family

Figure 14.10 Population plus consumption. Americans consume more than people in the developing world and therefore have a greater impact on resources.

The question posed at the beginning of this section remains unanswered; there is no agreement among scientists concerning the carrying capacity of Earth for the human population. Given that uncertainty, what can ecologists tell us about the risks facing the human population that may result from massive, rapid population growth?

14.3 The Future of the Human Population

Unlike nearly all other species, human populations are not simply at the mercy of environmental conditions. With its ability to transform the natural world, human ingenuity has helped populations circumvent seemingly fixed natural limits. However, ingenuity has a dark side in that it can lull people into believing that nature has an almost infinite capacity to support their ever-growing needs. Managing the growth of human populations before even the most secure of us face environmental and economic disaster requires an understanding of the risks of continued rapid growth and the strategies that help reduce it.

A Possible Population Crash?

The use of nonrenewable resources creates a risk that the human population will overshoot a still-unknown carrying capacity. Ecologists have long known that when populations have high growth rates, they may continue to add new members even as resources dwindle. This causes the population to grow larger than the carrying capacity of the environment. The members of this large population are then competing for far too few resources, and the death rate soars while the birth rate plummets. This results in a **population crash**, a steep decline in number (**Figure 14.11**).

For instance, in some species of water flea, healthy offspring continue to be born for several days after the food supply becomes inadequate. This occurs because females can use their fat stores to produce additional young. The size of the population continues to rise even when there is no food left for grazing; however, when the young water fleas run out of stored fat transferred from their mothers, most individuals die. For many species with high birth rates, rapid growth followed by dramatic crashes produce a **population cycle** of repeated "booms" and "busts" in number.

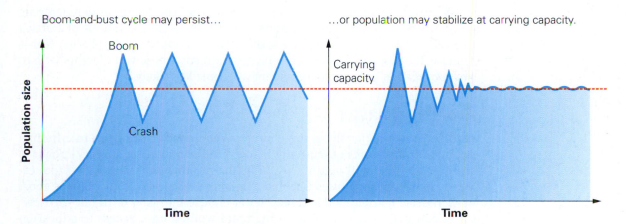

Figure 14.11 Overshooting and crashing. These graphs illustrate rapid population growth followed by a population crash. Over time, the population may stay in a "boom-and-bust" cycle, or it may stabilize at its carrying capacity.

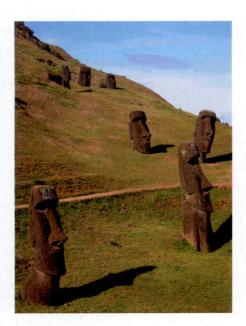

Figure 14.12 The crash of a human population. On Rapa Nui, also known as Easter Island, the human inhabitants created these large statues. Soon after completely deforesting this small island, the large population suffered a severe crash.

A population overshoot and subsequent crash affected the human population on the Pacific island of Rapa Nui (also known as Easter Island) during the eighteenth century (**Figure 14.12**). This 150-square-mile island is separated from other landmasses by thousands of miles of ocean; therefore, its people were limited to using only the resources on or near their island. Archaeological evidence suggests that at one time, the human population on Rapa Nui was at least 7000—apparently a number far greater than the carrying capacity of the island. By 1775, the subsequent overuse and loss of Rapa Nui's formerly lush palm forest had resulted in a rapid decline to fewer than 700 people, a population likely much lower than the initial carrying capacity of the island.

Biological populations may also overshoot carrying capacity when there is a time lag between when the population approaches carrying capacity and when it actually responds to that environmental limit. Scientists who study human populations note a lag between the time when humans reduce birth rates and when population growth actually begins to slow. They call this lag **demographic momentum.** The momentum occurs because while parents may be reducing their family size, their children will begin having children before the parents die, causing the population to continue growing. Even when families have an average of two children, just enough to replace the parents, demographic momentum causes the human population to grow for another 60 to 70 years before reaching a stable level.

Stop & Stretch   How does demographic momentum explain why species with high birth rates are more likely to experience population cycles compared to populations with low birth rates?

The demographic momentum of a human population can be estimated by looking at its **population pyramid,** a summary of the numbers and proportions of individuals of each sex and each age group. As **Figure 14.13** illustrates, the potential for high levels of demographic momentum occurs when the age structure most closely resembles a true pyramid, with a large proportion of young people. As this large group of young people ages, they become potential parents, with the capacity to cause the population to swell. In more stable populations, the proportion who are young is not significantly larger than the proportion who are middle aged, and the pyramid looks more like a column. The number of "parents-in-waiting" is the same as the number of current parents, keeping the population stable.

Whether our reliance on stored resources and the potential demographic momentum in human populations will result in an overshoot of Earth's carrying capacity—followed by a severe crash, as on the island of Rapa Nui—remains to be seen. But human ecologists already know what factors help to slow population growth, so a crash may become less likely.

Avoiding Disaster

As discussed previously in the chapter, when death rates drop in a human population, birth rates eventually follow. Unlike any other species known to science, humans will voluntarily limit the number of babies they produce. When more opportunities become available outside of child rearing, most women delay motherhood and have fewer children. In fact, birth rates are lowest in countries where income is high and women have access to education (**Figure 14.14**).

This information provides a clear direction for public policies attempting to decrease population growth rates: improve conditions for women, including

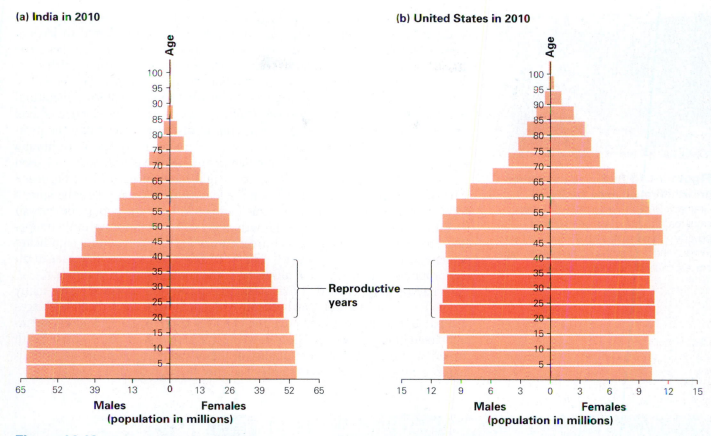

(a) India in 2010

Age

Males   Females
(population in millions)

Reproductive
years

(b) United States in 2010

Age

Males   Females
(population in millions)

Figure 14.13 Demographic momentum. In a rapidly growing human population like that of (a) India in 2010, most of the population is young, and the population will continue to grow as these children reach child-bearing age. In a slower-growing or stable human population, the ages are more evenly distributed, as in (b) the United States in 2010.
Visualize This: According to the graphs, how do the overall population sizes of the United States and India compare?

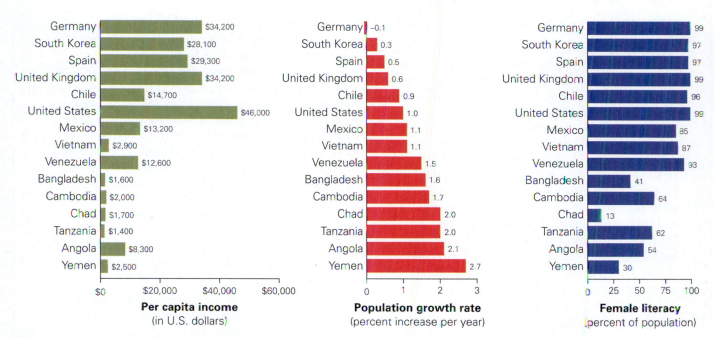

Per capita income
(in U.S. dollars)

Germany	$34,200
South Korea	$28,100
Spain	$29,300
United Kingdom	$34,200
Chile	$14,700
United States	$46,000
Mexico	$13,200
Vietnam	$2,900
Venezuela	$12,600
Bangladesh	$1,600
Cambodia	$2,000
Chad	$1,700
Tanzania	$1,400
Angola	$8,300
Yemen	$2,500

Population growth rate
(percent increase per year)

Germany	–0.1
South Korea	0.3
Spain	0.5
United Kingdom	0.6
Chile	0.9
United States	1.0
Mexico	1.1
Vietnam	1.1
Venezuela	1.5
Bangladesh	1.6
Cambodia	1.7
Chad	2.0
Tanzania	2.0
Angola	2.1
Yemen	2.7

Female literacy
(percent of population)

Germany	99
South Korea	97
Spain	97
United Kingdom	99
Chile	96
United States	99
Mexico	85
Vietnam	87
Venezuela	93
Bangladesh	41
Cambodia	64
Chad	13
Tanzania	62
Angola	54
Yemen	30

Figure 14.14 Income, growth rate, and women's literacy. These three graphs illustrate the relationships among income, population growth, and female literacy. Note that higher income and literacy are correlated with decreased birth rates and thus decreased population growth in most countries.
Visualize This: Why do you suppose the United States has a relatively high population growth rate compared to other wealthy countries?

Figure 14.15 Reducing population growth through opportunity. The Grameen Bank in Bangladesh provides small loans to help raise people from poverty. Women make up 97% of the borrowers, and successful women borrowers are less likely to have large families.

increasing access to education, health care, and the job market, and provide them with the information and tools that allow them to regulate their fertility (**Figure 14.15**). Slowing growth rates before the human population reaches some environmentally imposed limit has additional benefits. Determining Earth's carrying capacity for humans as simply a function of whether food and water will be available also ignores quality-of-life issues, or what some scientists call cultural carrying capacity. An Earth that was wholly given over to the production of food for the human population would lack wild, undisturbed places and the presence of species that nurture our sense of wonder and discovery. With human populations at the limits of growth, much of our creative energy would be used for survival, taking away our ability to make and enjoy music, art, and literature. (Chapter 23 addresses issues of sustainable food production that face the human population now and which will become more pressing as the population grows.)

Limiting human population growth also leaves room for nonhuman species. Human activity is posing a direct threat to the survival of a significant percentage of Earth's biodiversity (Chapter 15)—a threat that increases in direct proportion to the size and affluence of the planet's human population.

What we have learned is that scientists cannot tell us exactly how many people Earth can support, partly because humans make unpredictable choices and partly because humans have the capacity to innovate and adjust seemingly fixed biological limits. Ultimately, the question of how many people Earth should support—and at what quality of life or including support for nonhuman species—is a question not only of science but also of values and ethics.

SAVVY READER

Using Science to Urge Action

What's Happening to Birds We Know and Love?

Audubon's unprecedented analysis of forty years of citizen-science bird population data from our own Christmas Bird Count plus the Breeding Bird Survey reveals the alarming decline of many of our most common and beloved birds.

Since 1967 the average population of the common birds in steepest decline has fallen by 68 percent; some individual species nose-dived as much as 80 percent. All 20 birds on the national Common Birds in Decline list [including northern bobwhite, whip-poor-will, and common grackle] lost at least half their populations in just four decades.

The findings point to serious problems with both local habitats and national environmental trends. Only citizen action can make a difference for the birds and the state of our future.

Which Species? Why?

The wide variety of birds affected is reason for concern. Populations of meadowlarks and other farmland birds are diving because of suburban sprawl, industrial development, and the intensification of farming over the past 50 years.

Greater scaup and other tundra-breeding birds are succumbing to dramatic changes to their breeding habitat as the permafrost melts earlier and more temperate predators move north in a likely response to global warming. Boreal forest birds like the boreal chickadee face deforestation from increased insect outbreaks and fire, as well as excessive logging, drilling, and mining.

The one distinction these common species share is the potential to become uncommon unless we all take action to protect them and their habitat.

1. According to this article, how does the National Audubon Society know that bird species are declining?

2. What information is missing from this article that could influence how you respond to this issue?

From the National Audubon Society "State of the Birds" summer 2007 update http://stateofthebirds.audubon.org/

Chapter Review

Learning Outcomes

Mastering**BIOLOGY**®

Go to the Study Area at www.masteringbiology.com for practice quizzes, myeBook, BioFlix™ 3-D animations, MP3 Tutor sessions, videos, current events, and more

L01 **Define** *population*, **and describe the aspects of populations that are typically measured by ecologists (Section 14.1).**

- A population is defined as a group of individuals of the same species living in a fixed area (p. 340).
- Ecologists measure a population by the number of individuals and their dispersion (p. 340).

L02 **Describe how a mark-recapture estimate of population size is performed, and estimate the size of a population from mark-recapture data (Section 14.1).**

- Population size can be estimated by capturing a sample of the population, marking the individuals in some way, freeing them, and then capturing another sample in the same area. Determining the proportion of marked to unmarked individuals allows calculation of total population size (pp. 340–341).

L03 **Explain how the size of a population that is experiencing exponential growth changes over time (Section 14.1).**

- The human population has grown very rapidly over the last 150 years and exhibits a pattern of exponential growth, which is an increase in numbers as a function of the current population size. On a graph, exponential growth takes the form of a J-shaped curve (pp. 342–343).

L04 **Describe the demographic transition in human populations and how it contributes to population growth (Section 14.1).**

- Human population growth is spurred by decreases in death rate caused by decreased infant mortality. In most populations, this decrease is followed by a decrease in birth rates. The gap between when death rates drop and birth rates follow is called the demographic transition (pp. 344–345).

L05 **Define** *carrying capacity*, **and explain how the existence of a carrying capacity causes a pattern of logistic growth as populations grow (Section 14.2).**

- Nearly all populations eventually reach the carrying capacity of their environment, which is the maximum population that can be supported indefinitely given current conditions. On a graph, logistic growth takes the form of an S-shaped curve (p. 345).

L06 **Compare and contrast the effect of density-dependent and density-independent factors on population growth (Section 14.2).**

- The impact of density-dependent factors depends on the size of the population, whereas density-independent factors affect birth or death rates regardless of population size (pp. 345–346).
- Density-dependent factors include starvation and disease, and density-independent factors include weather (pp. 345–346).

L07 **List the evidence that the human population may not be near carrying capacity and the evidence that it may be near carrying capacity (Section 14.2).**

- The growth rate of the human population is declining, but not as a result of density-dependent factors. Instead, birth rates are dropping because women are choosing to have fewer children (pp. 346–347).
- Rough calculations indicate that the energy received from the sun each year could support a population of 21 billion people, well over the largest population projections (pp. 347–348).
- Humans' reliance on nonrenewable resources may be temporarily inflating the actual carrying capacity of Earth (p. 348).

L08 **Describe the conditions (in both human and other populations) under which a population crash can occur (Section 14.3).**

- Fast-growing populations that overshoot their environment's carrying capacity may experience a crash or go through periodic booms and busts (pp. 349–350).
- It is possible that the human population will overshoot Earth's carrying capacity because of our reliance on nonrenewable resources and due to demographic momentum, the potential for future population growth that comes from a population consisting primarily of young people (pp. 350–352).

Roots to Remember

The following roots of words come mainly from Latin and Greek and will help you decipher terms:

demo- is from the word meaning people. Chapter term: demographic

expo- is from the word meaning to put forth. Chapter term: exponential

logo- means proportion or ratio. Chapter term: logistic

popula- means multitude. Chapter term: population

Learning the Basics

1. **LO4** What factors have led to the explosive increase in the human population over the past 150 years?

2. **LO5** Explain why a decrease in population growth rate is expected as a nonhuman population approaches carrying capacity.

3. **LO1** When individuals in a population are evenly spaced throughout their habitat, their dispersion is termed as _____.

   A. clumped; **B.** uniform; **C.** random; **D.** excessive; **E.** exponential

4. **LO3** A population growing exponentially _____.

   A. is stable in size; **B.** adds a fixed number of individuals every generation; **C.** adds a larger number of individuals in each successive generation; **D.** will likely expand forever; **E.** will not crash.

5. **LO5** According to **Figure 14.16**, the carrying capacity for fruit flies in the environment of the culture bottle is _____.

   A. 0 flies; **B.** 100 flies; **C.** 150 flies; **D.** between 100 and 150 flies; **E.** impossible to determine

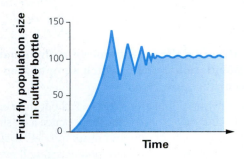

Figure 14.16

6. **LO6** All of the following are density-dependent factors that can influence population size *except* _____.

   A. weather; **B.** food supply; **C.** waste concentration in the environment; **D.** infectious disease; **E.** supply of suitable habitat for survival

7. **LO7** In contrast to nonhuman populations, human population growth rates have begun to decline due to _____.

   A. voluntarily increasing death rates; **B.** voluntarily decreasing birth rates; **C.** involuntary increases in death rates; **D.** involuntary decreases in birth rates; **E.** density-dependent factors

8. **LO8** Populations that rely on stored resources are likely to overshoot the carrying capacity of the environment and consequently experience a _____.

   A. demographic momentum; **B.** cultural carrying capacity; **C.** decrease in death rates; **D.** population crash; **E.** exponential growth

9. **LO8** The current carrying capacity of Earth for the human population may have been inflated by _____.

   A. demographic momentum; **B.** the tendency for women to want to control family size; **C.** an artificially low number of density-independent factors; **D.** our use of fossil fuels; **E.** recent population crashes

10. **LO4 LO8** Demographic momentum refers to the tendency for _____.

    A. low population growth rates to continue to decline; **B.** high population growth rates to continue to increase; **C.** populations to continue to grow in number even when growth rates reach zero; **D.** populations to continue to grow in number even when women are reducing the number of children they bear; **E.** women to continue to have children even though they no longer wish to

Analyzing and Applying the Basics

1. **LO2** A researcher captures 50 penguins, marks them with a spot of paint on their bills, and releases them. One month later, the researcher returns, captures another 50 penguins, and notes that only 1 has a previous mark. What is the likely size of the total penguin population in the researcher's study area?

2. **LO5** Consider a population in which, for every thousand individuals in the population, 25 children are born each year and 27 individuals die. What is the growth rate of this population? Do you expect that it is near carrying capacity? Why or why not?

3. **LO5 LO6** Review Figure 14.16. How would you expect the carrying capacity of the population to change if the flies are supplied with more food? What other factors might influence the carrying capacity in this environment?

Connecting the Science

1. Review your answer to Question 3 in "Analyzing and Applying the Basics." How are the factors that limit fruit fly populations in a culture bottle similar to the factors that limit human populations on Earth? How are they different?

2. Africa is the only continent where increases in food production have not outpaced human population growth. Many of the most severe food crises are in African countries. Should those of us in the more developed world assist African populations? How? What factors influence your thoughts on this question?

Answers to **Stop & Stretch, Visualize This, Savvy Reader**, and **Chapter Review** questions can be found in the **Answers** section at the back of the book.

CHAPTER

15

Conserving Biodiversity

Community and Ecosystem Ecology

The Lost River sucker
faces extinction . . .

They came with chainsaws, pry bars, blowtorches, and American flags to challenge the authorities. On their way to the facility, they passed throngs of cheering supporters holding handmade signs and carrying buckets to fill with plundered resources. Then the farmers from normally sleepy Klamath Falls, Oregon, set upon the fences and gates that were preventing them from working and threatening to destroy their way of life. It was July 2001, and events in the Klamath Basin had come to a head.

The wrath of the people of Klamath Falls and surrounding communities was generated by the federal government's legal requirement to protect species that are recognized as in danger of extinction. In the case of the Klamath crisis, the species at risk are fish—the Lost River sucker and the shortnose sucker, which live in Upper Klamath Lake, and the coho salmon, which lives in the Klamath River. In the midst of a multiyear drought and dangerously low water levels in Upper Klamath Lake, the U.S. Fish and Wildlife Service stopped the pumps providing irrigation water from it. The canals that had provided water to barley, potato, and alfalfa fields in the high desert of the Klamath Basin since the early 1900s suddenly went dry. Without irrigation, thousands of farmers were unable to produce crops and faced the prospect of bankruptcy, foreclosure, and loss of their livelihood.

Lost River and shortnose suckers are dull-colored fish that feed on the mucky bottoms of lakes and streams in the region. These fish have not represented a viable economic resource for humans for several decades. In contrast, the crops produced annually by irrigated fields in the region produce

. . . but saving the fish has angered these farmers.

Who has the right to use Klamath Lake for their survival—farmers or fish?

Why should we care about the fate of such a controversial endangered species—or any endangered species?

millions of dollars in income. Ty Kliewer, a student at Oregon State University whose family farms in the basin, summarized the feelings of many when he told his senator in 2001 that he had learned the importance of balancing mathematical and chemical equations in school. "It appears to me that the people who run the Bureau of Reclamation and the U.S. Fish and Wildlife Service slept through those classes," Kliewer said. "The solution lacks balance, and we've been left out of the equation." The Klamath crisis of 2001 was not unique; thousands of people all over the United States have had their jobs threatened or eliminated by the government's attempts to protect endangered species.

Why should the survival of one or a few species come before the needs of humans? Many biologists and environmentalists say that the Klamath Falls bumper sticker "Fish—or Farmers?" misstates the dilemma. Instead, they argue, humans depend on the web of life that creates and supports natural ecosystems, and they worry that disruptions to this web may become so severe that our own survival as a species will be threatened. In this view, protecting endangered species is not about pitting fish against farmers; it is about protecting fish to ensure the survival of farmers. In this chapter, we explore the causes and consequences of the loss of biological diversity.

15.1 The Sixth Extinction

The government agencies that stopped water delivery to the Klamath Basin farmers were acting under the authority of the **Endangered Species Act (ESA),** a law passed in 1973 with the purpose of protecting and encouraging the population growth of threatened and endangered species. Lost River and shortnose suckers were once among the most abundant fish in Upper Klamath Lake—at one time, they were harvested and canned for human consumption. Now, with populations of fewer than 500 and minimal reproduction, these fish are in danger of **extinction,** defined as the complete loss of a species. Critically imperiled species such as the Lost River and shortnose suckers are exactly the type of organisms that legislators had in mind when they enacted the ESA.

The ESA was passed because of the public's concern about the continuing erosion of **biodiversity,** the entire variety of living organisms. Gray wolves, passenger pigeons, black-footed ferrets (**Figure 15.1**), and spotted skunks— once abundant species—are extinct or highly threatened in the United States as a result of human activity. The ESA was a response to the unprecedented and rapid rate of species loss at our hands.

Critics of the ESA argue that the goal of saving all species from extinction is unrealistic. After all, extinction is a natural process—the approximately 10 million species living today constitute less than 1% of the species that have ever existed. In the next section, we explore the scientific questions posed by ESA critics: How does the rate of extinction today compare to the rates in the past? Is the ESA just attempting to postpone the inevitable, natural process of extinction?

Figure 15.1 A critically endangered species. Black-footed ferrets once numbered in the tens of thousands across the Great Plains of North America but were decimated by human activity. By 1986, they were completely gone from the wild, and only 18 remained alive in captivity.

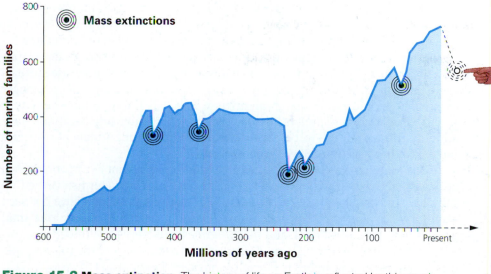

Figure 15.2 Mass extinction. The history of life on Earth is reflected by this graph plotting the change in the number of families of fossil marine organisms over time. The number of families has increased over the past 600 million years. However, this rise has been punctuated by five mass extinctions (marked here with black circles), each resulting in a global decline in biodiversity.

Visualize This: According to the graph, about how long does it take after a mass extinction for diversity to return to preextinction levels?

Measuring Extinction Rates

If ESA critics are correct in stating that the current rate of species extinctions is "natural" and not a result of human activity, then the extinction rate today should be roughly equal to the rate that existed before humans evolved. The rate of extinction in the past can be estimated by examining the fossil record.

Figure 15.2 illustrates what the fossil record tells scientists about the history of biodiversity as measured by the diversity of marine organisms, which have the longest continuous record. Since the rapid evolution of a wide variety of animal groups approximately 580 million years ago, the number of families of organisms has generally increased. However, this increase in biodiversity has not been smooth or steady.

The history of life as pieced together by fossils has been punctuated by five **mass extinctions**—species losses that are global in scale, affect large numbers of species, and are dramatic in impact. Past mass extinctions were probably caused by massive global changes—for instance, climate fluctuations that changed sea levels, continental drift that changed ocean and land forms, or asteroid impact that caused widespread destruction and climate change. Many scientists argue that we are now seeing biodiversity's sixth mass extinction, this one caused by human activity.

Determining whether the current rate of extinction is unusually high requires knowledge of the **background extinction rate,** the rate at which species are lost through the normal evolutionary process. Normal extinctions occur when a species lacks the variability to adapt to environmental change or when a new species arises due to the evolution of its ancestor species. The fossil record can provide clues about the background extinction rate that results from this continual process of species turnover.

The span of rock ages in which fossils of a species are found represents the life span of that species (**Figure 15.3**). Biologists have thus estimated that the "average" life span of a species is around 1 million years (although there is tremendous variation). We would therefore expect that the background rate of extinction is about one species per million (0.0001%) per year. Some scientists

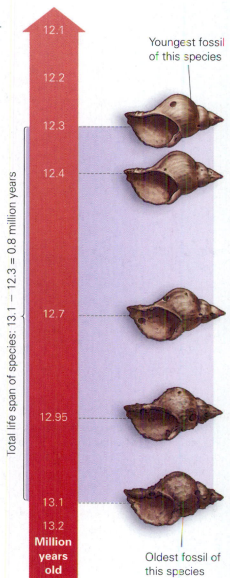

Figure 15.3 Estimating the life span of a species. Fossils of the same species are arranged on a timeline from oldest to youngest. The difference in age between the oldest and youngest fossil of a species is an estimate of the species' life span.

have argued that this estimate is too low because it is based on observations of fossils, a record that may be biased toward long-lived species. However, it remains the best-supported approximation of background extinction rates.

The current rate of extinction is calculated from known species disappearances. Calculating current rates is a challenge because extinctions are surprisingly difficult to document. The only way to conclude that a species no longer exists is to exhaustively search all areas where it is likely to have survived. In the absence of a complete search, most conservation organizations have adopted this standard: To be considered extinct, no individuals of a species must have been seen in the wild for 50 years.

A few searches for specific species give hints to the recent extinction rate. In Malaysia, a 4-year search for 266 known species of freshwater fish turned up only 122. In Africa's Lake Victoria, 200 of 300 native fish species have not been seen for years. On the Hawaiian island of Oahu, half of 41 native tree snail species have not been found, and in the Tennessee River, 44 of the 68 shallow-water mussel species are missing. Despite these results, few of the missing species in any of these searches is officially considered extinct.

The most complete records of documented extinction occur in groups of highly visible organisms, primarily mammals and birds. Since 1600, of an approximate 4500 identified mammal species 83 have become extinct, while 113 of approximately 9000 known bird species have disappeared. The known extinctions of mammals and birds, spread out over the 400 years of these records, correspond to a rate of 0.005% per year. Compared to the background rate of extinctions calculated from the fossil record, the current rate of extinction is 50 times higher. If we examine the past 400 years more closely, we see that the number of mammals and birds lost, and thus the extinction rate, has actually increased over time (**Figure 15.4**). In the years 1850 to 2000, the number of species lost translates to about 0.01% of the total per year, making the current rate 100 times higher than the background rate.

There are reasons to expect that the current elevated rate of extinction will continue into the future. The World Conservation Union (known by its French acronym, IUCN), a highly respected global organization composed of and funded by government agencies and nongovernmental organizations from over 140 countries, collects and coordinates data on threats to biodiversity. According to the IUCN's most recent assessment, 29% of amphibians, 21% of mammals, 12% of all bird species, and 3% of all plant species, including 40% of conifers, are in danger of extinction.

Figure 15.4 Modern extinctions.
This graph illustrates the number of species of mammals and birds known to have become extinct since 1600.

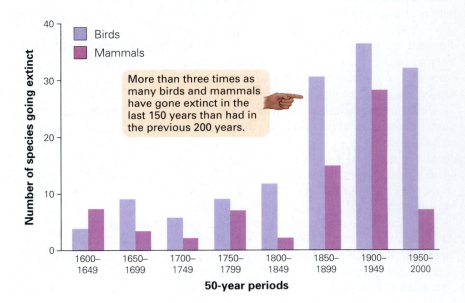

(a) Habitat degradation

(b) Introduced species

(c) Overexploitation of species

(d) Pollution

Figure 15.5 The primary causes of extinction. (a) A road carved through formerly undisturbed rainforest is the first step towards the destruction and fragmentation of habitat for dozens of unique species. (b) The brown tree snake introduced to the Pacific island of Guam from Papua New Guinea is responsible for the extinction of native bird species on the island. (c) These tiger skins were illegally harvested from southeast Asia, primarily for the Chinese market. (d) Polar bears hunt for seals, their primary prey, from sea ice. Some bears contain high levels of persistent pollutants that appear to have disrupted their reproductive organs. Global warming has also reduced the extent of sea ice in the Arctic Ocean over the past 20 years, further threatening the bears' survival.

Causes of Extinction

The IUCN attempts to identify all the threats that put a particular species at risk of extinction. For most, some type of human activity is to blame. The most severe threats belong to one of four general categories: loss or degradation of habitat, introduction of nonnative species, overexploitation, and effects of pollution (**Figure 15.5**). Of these categories, the first poses the most serious threat; the IUCN estimates that 83% of endangered mammals, 89% of endangered birds, and 91% of endangered plants are directly threatened by damage to or destruction of the species' **habitat,** the place where they live and obtain their food, water, shelter, and space.

Habitat Destruction and Fragmentation. The dramatic reduction in numbers of shortnose and Lost River suckers in Upper Klamath Lake is almost entirely due to habitat destruction. Before farmers settled in the Klamath Basin in the nineteenth century, 350,000 acres of wetlands regulated the overall quality and amount of water entering the lake. Most of these wetlands now have been drained and converted to irrigated agricultural fields. The disruption of natural water flows into and out of the lake has interfered with sucker reproduction and has reduced the number of offspring they produce by as much as 95%.

Habitat destruction is not limited to species native to the more developed world (**Figure 15.6**). Rates of habitat destruction caused by agricultural, industrial, and residential development accelerated around the world throughout the twentieth century as the human population swelled from less than 2 billion in 1900 to well over 6 billion today. As the amount of natural landscape declines, the number of species supported by the habitats in these landscapes naturally also decreases.

The relationship between the size of a natural area and the number of species that it can support follows a general pattern called a **species-area curve.** A species-area curve for reptiles and amphibians on a West Indies archipelago

Figure 15.6 Lost habitat = lost species. Lemurs are found only on the island of Madagascar, first settled by humans 1500 years ago. Of the 48 species of lemur present on the island 2000 years ago, 16 have become extinct, and 15 are at risk of extinction.

(a) Species diversity increases with area.

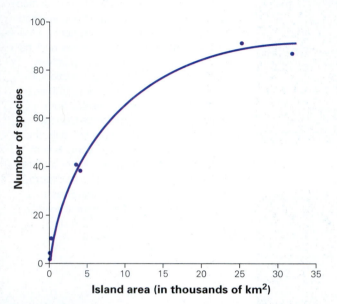

(b) Habitat reduction is predicted to result in loss of species.

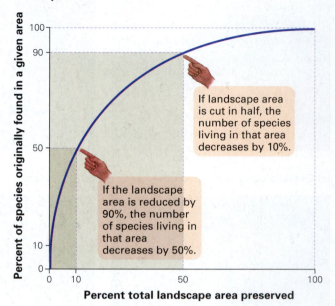

If landscape area is cut in half, the number of species living in that area decreases by 10%.

If the landscape area is reduced by 90%, the number of species living in that area decreases by 50%.

Figure 15.7 Predicting extinction caused by habitat destruction. (a) This curve demonstrates the relationship between the size of an island in the West Indies and the number of reptile and amphibian species living there. (b) We use a generalized species-area curve to roughly predict the number of extinctions in an area experiencing habitat loss.

Visualize This: Using part (a) of this figure, calculate how many species of reptiles and amphibians you would expect to find on an island that is 15,000 square kilometers in area. Imagine that humans colonize this island and dramatically modify 10,000 square kilometers of the natural habitat. Using part (b), calculate the percentage of the species originally found on the island you would expect to become extinct.

is illustrated in **Figure 15.7a**. Similar graphs have been generated in studies of different groups of organisms in a variety of habitats. The general pattern in all these graphs is that the number of species in an area increases rapidly as the size of the area increases, but the rate of increase slows as the area becomes very large. This rule of thumb is shown in **Figure 15.7b**. We can use this curve to estimate rates of extinction in regions that are rapidly being modified by human activity but difficult to survey extensively. For example, from the graph in Figure 15.7b, we can estimate that a 90% decrease in landscape area will cut the number of species living in the remaining area by half.

Using images from satellites, scientists have estimated that approximately 20,000 square kilometers (about 7722 square miles, an area the size of Massachusetts) of rain forest are cut each year in South America's Amazon River basin. At this rate of habitat destruction, tropical rain forests will be reduced to 10% of their original size within about 35 years. According to the species-area curve, this habitat loss translates into the extinction of about 50% of species living there. If this prediction proves accurate, the extinct species in the rain forest would include about 50,000 of the 250,000 known species of plants, 1800 species of birds, and 900 species of mammals.

Of course, habitat destruction is not limited to tropical rain forests. Freshwater lakes and streams, grasslands, and temperate forests are also experiencing high levels of modification. According to the IUCN, if habitat destruction around the world continues at its present rate, nearly one-fourth of *all* living species will be lost within the next 50 years.

Some critics have argued that these estimates of future extinction are too high because not all groups of species are as sensitive to habitat area as the curve in Figure 15.7b suggests. Many species may still survive and even thrive in human-modified landscapes. Other biologists contend that there are other threats to species, including habitat fragmentation, and therefore the rate of species loss is likely to be even higher than these estimates.

Habitat destruction rarely results in the complete loss of a habitat type. Often what results from human activity is habitat fragmentation, in which large areas of intact natural habitat are subdivided. Habitat fragmentation is especially threatening to large predators, such as grizzly bears and tigers, because of their need for large hunting areas.

Large predators require large, intact hunting areas due to a basic rule of biological systems: Energy flows in one direction within an ecological system along a **food chain,** which typically runs from the sun to **producers** (photosynthetic organisms), to the **primary consumers** that feed on them, to **secondary consumers** (predators that feed on the primary consumers), and so on. Along the way, most of the calories taken in at one **trophic level** (that is, a level of the food chain) are used to support the activities of the individuals at that level. In other words, a substantial amount of the energy individuals take in is dissipated as heat. Each level of the food chain thus has significantly less energy available to it compared to the next lower level. You can see this in your own life; an average adult needs to consume between 1600 and 2400 kilocalories per day simply to maintain his or her current weight.

The flow of energy along a food chain leads to the principle of the **trophic pyramid,** the bottom-heavy relationship between the **biomasses** (total weight) of populations at each level of the chain (**Figure 15.8**). Habitat destruction and fragmentation can cause the lower levels of the pyramid to shrink, depriving the top predators of adequate calories for survival.

Habitat fragmentation also exposes wide-ranging predators to additional dangers. For example, grizzly bears need 200 to 2000 square kilometers of habitat to survive a Canadian winter, but the Canadian wilderness is increasingly crisscrossed by roads built for tree harvesting. Each road represents an increased chance of grizzly-human interaction. These encounters are often deadly—to the bears. For example, between 1970 and 1995, of the 136 grizzlies that died in Canada's national parks, 17 died of natural causes, and 119 were killed by humans.

The species that can remain in small fragments of habitat are also made more susceptible to extinction by isolation. Habitat fragmentation often makes it impossible for individuals to follow changes in food sources or available nesting or growing sites. Isolated populations are also subject to a lack of genetic diversity resulting from inbreeding, as we discuss in detail later in the chapter.

Figure 15.8 A trophic pyramid.
Because most of the energy consumed by a trophic level is used within that level for maintenance, biomass decreases as position in the food chain increases.

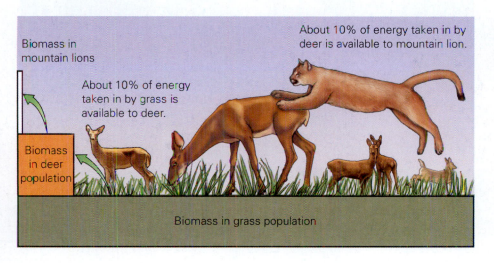

Biomass in mountain lions

About 10% of energy taken in by deer is available to mountain lion.

About 10% of energy taken in by grass is available to deer.

Biomass in deer population

Biomass in grass population

Although habitat destruction and fragmentation are the gravest threats to endangered species, the others are not insignificant. According to the IUCN, activity unrelated to habitat modification plays a role in about 40% of all cases of endangerment. These activities include problems caused by species that actually do well in our presence.

Introduced Species. Introduced species include organisms brought by human activity, either accidentally or purposefully, to new environments. Introduced species are often dangerous to native species because they have not coevolved with them. **Coevolution** occurs when pairs or groups of species adapt to each other via natural selection. For instance, many birds on oceanic islands such as Hawaii and New Zealand evolved in the absence of ground predators and have not evolved behaviors or other strategies to escape or combat these predators—therefore, introduced carnivores can rapidly deplete them via hunting. On Hawaii, the Pacific black rat, accidentally introduced from the holds of visiting ships, became very adept at raiding eggs from nests and contributed to the extinction of dozens of species of honeycreepers, birds found nowhere else on Earth. Even domestic cats, deliberately introduced by people who sought to control rodents around their houses, can take an enormous toll on wildlife. A study in Wisconsin estimated that free-ranging cats in that state kill approximately 39 million birds every year. Multiplied across the entire continent, that is over a billion birds lost to cat predation in North America annually.

Introduced species may also compete with native species for resources, causing populations of the native species to decline. For example, zebra mussels, accidentally introduced to the Great Lakes through the ballast water in European trading ships, crowd out native mussel species as well as other organisms that filter algae from water, and the introduced vine kudzu, deliberately brought to the United States from Japan, grows over native trees and vines in the Southeast (**Figure 15.9**).

Humans continue to move species around the planet. As the global trade in agriculture and other goods continues to expand, the number of species introductions is likely to increase over the next century.

Overexploitation. When the rate of human use of a species outpaces its reproduction, the species experiences overexploitation. Overexploitation may

Figure 15.9 An introduced species. Kudzu is the common name for the vine *Pueraria lobata,* introduced from Japan to the southeast United States as an ornamental plant and promoted as a solution to soil erosion. In the absence of its predators and parasites, kudzu grows extremely rapidly, almost 50 centimeters per day, and effectively smothers other plants.

occur when particular organisms are highly prized by humans, for example, as exotic pets (**Figure 15.10**) or for their medicinal value. For instance, three of the planet's eight tiger species have become extinct, and the remaining species are gravely endangered, in part because their bones are believed to cure arthritis and because their genitals are believed to reverse male impotence. As a result, illegal hunting of these animals is very lucrative. Popular wild-grown herbal medicines, such as echinacea (Chapter 1) and ginseng, are also at risk of overexploitation.

Overexploitation is also likely when an animal competes directly with humans—as in the case of the gray wolf, which was nearly exterminated in the United States by human hunters determined to protect their livestock. Species that cross international boundaries during migration or live in the world's oceans are also susceptible to overexploitation because no single government regulates the total harvest. The near extinction of several whale species in the nineteenth century was the result of the unregulated harvest of these animals by many nations. Stocks of cod, swordfish, and tuna are now similarly threatened.

Pollution. The release of poisons, excess nutrients, and other wastes into the environment—a practice otherwise known as *pollution*—poses an additional threat to biodiversity. For example, the herbicide atrazine poisons frogs and salamanders in agricultural areas of the United States, and nitrogen pollution caused by overuse of fertilizer and car and smokestack exhaust has led to drastic declines of certain plant species within native grasslands in Europe.

Nitrogen and phosphorous fertilizer that pollutes waterways in the Klamath basin has increased the growth of algae in Upper Klamath Lake. When these algae explode in numbers, the bacteria that feed on them flourish and rapidly use up the available oxygen in the water. This process of oxygen-depleting **eutrophication** results in large fish kills. Eutrophication threatens animals in hundreds of waterways all over the United States. In the Gulf of Mexico, for example, fertilizer from farm fields in the midwestern United States creates a low-oxygen "dead zone" the size of New Jersey at the mouth of the Mississippi River.

Perhaps the most serious pollutant released by humans is carbon dioxide, also a principal cause of global climate change (see Chapter 4). Computer models that link predicted changes in climate to known ranges and requirements of over 1000 species of plants indicate that 15% to 37% of these plants face extinction in the next century as the climate changes.

Figure 15.10 Imperiled oddities. Dozens of animal and plant species made vulnerable by habitat loss are further endangered by the exotic wildlife trade. It is not just the beautiful that are endangered in this way; this sawfish, a type of ray, is at risk because of high demand for its unusual "nose."

Stop & Stretch   Many species of amphibians (frogs and toads) around the world are declining, and some are critically endangered. Describe some human activities that might be affecting these animals. What information would help you determine whether these widely dispersed species were put at risk by the same cause?

By nearly any measure, ESA critics who describe modern extinction rates as "natural" are incorrect. Over the past 400 years, humans have caused the extinction of species at a rate that far exceeds past rates. Human activities continue to threaten thousands of additional species around the world. Earth appears to be on the brink of a sixth mass extinction of biodiversity—and the massive global change responsible is human activity.

Many people feel a moral responsibility to minimize the human impact on other species and therefore support conservation. However, there is also a practical, human-centered reason to prevent the sixth extinction from occurring—the loss of biodiversity can hurt us as well.

(a) Rosy periwinkle

(b) Teosinte

(c) Boll weevil wasp

15.2 The Consequences of Extinction

Concern over the loss of biodiversity is not simply a matter of an ethical interest in nonhuman life. Humans have evolved with and among the variety of species that exist on our planet, and the loss of these species often results in negative consequences for us.

Loss of Resources

The Lost River and shortnose suckers were once numerous enough to support fishing and canning industries on the shores of Upper Klamath Lake. Even before the arrival of European settlers, the native people of the area relied heavily on these fish as a mainstay of their diet. They still rely heavily on the salmon community in the Klamath River, of which endangered coho salmon is one species. The loss of the suckers and coho salmon species represents a tremendous impoverishment of wild food sources. And these are not the only wild species humans rely on directly.

The biological resources that are harvested directly from natural areas include wood for fuel and lumber, shellfish for protein, algae for gelatins, and herbs for medicines. The loss of any of these species affects human populations economically. One estimate places the value of wild species in the United States at $87 billion a year or about 4% of the gross domestic product.

Wild species also provide resources for humans in the form of unique biological chemicals. In fact, a number of valuable drugs, food additives, and industrial products derive from wild species (Chapter 13 provides several examples). One dramatic example is the rosy periwinkle (*Catharanthus roseus*), which evolved on the island of Madagascar, one of the regions on Earth where biodiversity is most endangered (**Figure 15.11a**). Two drugs extracted from this plant, vincristine and vinblastine, have dramatically reduced the death rate from leukemia and Hodgkin's disease, another form of cancer. If wild species go extinct before they are well studied, we will never know which ones might have provided compounds that would improve human lives.

Wild relatives of domesticated plants and animals, such as agricultural crops and cattle, are also important resources for humans. Genes and alleles that have been "bred out" of domesticated species are often still found in their wild relatives. These genetic resources are a reservoir of traits that can be reintroduced into agricultural species through breeding or genetic engineering. Agricultural scientists attempting to produce better strains of wheat, rice, and corn look to the wild relatives of these crops for genes conferring pest resistance and improved yields. For example, the Mexican teosinte species *Zea diploperennis* (**Figure 15.11b**), discovered in 1977, is an ancestor of modern corn. This species of teosinte is resistant to several viruses that plague cultivated corn; the genes that provide this resistance have been transferred to our domestic plants via hybridization.

By preserving wild relatives of domesticated crops in their natural habitats, scientists can also find resources that reduce pest damage and disease on the domestic crop. For example, the wasp *Catolaccus grandis* consumes boll weevils and is used to control infestations of these pests in cotton fields (**Figure 15.11c**). *Catolaccus grandis* was discovered in the tropical forest of southern Mexico, where it parasitizes a similar pest in wild cotton populations.

Figure 15.11 Resources from nature. (a) Anticancer drugs vincristine and vinblastine were first isolated from rosy periwinkle (*Catharanthus roseus*), a species of flower native to Madagascar. (b) Teosinte is the ancestor of modern corn, first cultivated in Central America and rediscovered in Mexico in 1968. (c) *Catolaccus grandis*, the boll weevil wasp, is a predator of one of the most damaging pests of cotton. *Catolaccus grandis* was discovered preying on boll weevils on wild cotton in southern Mexico.

Of course, introducing an insect such as *C. grandis* into a new environment carries risk, even if the introduction is meant to reduce environmental damage. We have already noted many examples of environmental disasters caused by introduced species. Often, a less risky approach to reducing pest damage to crops is to preserve nearby habitats and the ecological interactions that persist there.

Predation, Mutualism, and Competition Derailed

Although humans receive direct benefits from thousands of species, most threatened and endangered species are probably of little or no use to people. Even the Lost River and shortnose suckers, as valuable as they once were to the native people of the Klamath Basin, are not especially missed as a food source; no one has starved simply because these fish have become less common.

In reality, most species are beneficial to humans because they are part of a biological **community,** consisting of all the organisms living together in a particular habitat area. Within a community, each species occupies a particular **ecological niche,** which can be thought of as the role or "job" of the species. The complex linkage among organisms inhabiting different niches in a community is often referred to as a **food web** (**Figure 15.12**). As with a spider's

Figure 15.12 The web of life. Species are connected to each other and to their environment in various, complex ways. This drawing shows some of the important relationships among organisms in the Antarctic Ocean. Black arrows represent feeding relationships; for example, penguins eat fish and in turn are eaten by leopard seals. **Visualize This:** Predict what would happen to the other species in the food web if baleen whales were hunted to extinction.

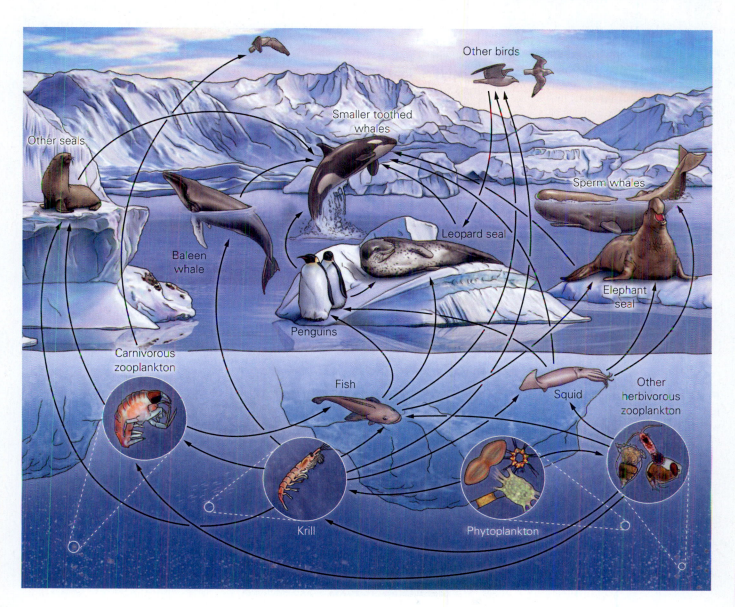

web, any disruption in one strand of the web of life is felt by other portions of the web. Some tugs on the web cause only minor changes to the community, while others can cause the entire web to collapse. Most commonly, losses of strands in the web are felt by a small number of associated species. However, some disruptions caused by the loss of seemingly insignificant species have the potential to be felt by humans.

Mutualism: How Bees Feed the World. An interaction between two species that benefit each other is called **mutualism**. Mutualism can be contrasted with **commensalism**, a relationship in which one species benefits and the other is unaffected—for instance the relationship between cattle egrets, a bird, and domestic cattle. The egrets follow the cattle as they graze, feeding on insects stirred up from the ground by these animals (**Figure 15.13**). The cattle do not appear to benefit or be harmed by the egrets' presence.

We find examples of mutualism in many environments. Cleaner fish that remove and consume parasites from the bodies of larger fish, fungi called mycorrhizae that increase the mineral absorption of plant roots while consuming the plant's sugars, and ants that find homes in the thorns of acacia trees and defend the trees from other insects are all examples of mutualism. The mutualistic interaction between plants and bees is perhaps the most important to us.

Bees occupy a very important ecological niche as the primary pollinators of many species of flowering plants. The role of pollinators is to transfer sperm, in the form of pollen grains, from one flower to the female reproductive structures of another flower. The flowering plant benefits from this relationship because insect pollination increases the number of seeds that the plant produces. The bee benefits by collecting excess pollen and nectar to feed itself and its relatives in the hive (**Figure 15.14**).

Wild bees pollinate at least 80% of all the agricultural crops in the United States, providing a net benefit of about $8 billion. In addition, populations of wild honeybees have a major and direct impact on many more billions of dollars of agricultural production around the globe.

Unfortunately, bees have suffered dramatic declines in recent years. Steady declines in wild and domesticated bee populations occurred from the 1970s through 2005; however, since 2006, the declines have reached crisis proportions, with up to 25% of captive bee colonies perishing as a result of "colony collapse disorder." The exact causes of these dramatic die offs are not known but are believed to result from an increased level of bee **parasites** (infectious organisms that cause disease or drain energy from their hosts), competition with the invading Africanized honeybees ("killer bees"), and habitat destruction. The extinction of populations of either wild or domesticated bees that are mutualists of crop plants would be extremely costly to humans.

Figure 15.13 Commensalism. Cattle egrets are found in close association with domestic cattle. The birds follow the cows, eating the insects stirred up by their activity, but not affecting the cattle at all.

Figure 15.14 Mutualism. Honeybees transfer pollen, allowing a plant to "mate" with another plant some distance away.

Benefit to bee:
It obtains food in the form of nectar and excess pollen.

Benefit to flower:
Its sperm (within the pollen) is carried to the female reproductive structures of another flower, enabling cross-pollination.

Predation: How Songbirds May Save Forests.

A species that survives by eating another species is typically referred to as a **predator.** The word conjures up images of some of the most dramatic animals on Earth: cheetahs, eagles, and killer whales. You might not picture wood warblers, a family of North American bird species characterized by their small size and colorful summer plumage, as predators; however, these beautiful songsters are voracious consumers of insects (**Figure 15.15a**). The hundreds of millions of individual warblers in the forests of North America collectively remove thousands of kilograms of insects from forest trees and shrubs every summer.

Most of the insects that warblers eat prey on plants. By reducing the number of insects in forests, warblers reduce the damage that insects inflict on forest plants. Reducing the amount of damage likely increases the growth rate of the trees. Harvesting trees for paper and lumber production fuels an industry worth over $200 billion in the United States alone. At least some of this wood was produced because warblers were controlling insects in forests (**Figure 15.15b**).

Many species of forest warblers are experiencing declines in abundance. The loss of warbler species has several causes, including habitat destruction in their summer habitats in North America and their winter habitats in Central and South America. Warblers also face increased predation by human-associated animals such as raccoons and housecats. Although other, less vulnerable birds may increase in number when warblers decline, these birds are typically less insect dependent. If smaller warbler populations correspond to lower forest growth rates and higher levels of forest disease, then these tiny, beautiful birds definitely have an important effect on the human economy.

(a) Black-throated blue warbler, predator of insects

(b) Forests suffer when insects are unchecked by predators.

Figure 15.15 Predation. (a) The black-throated blue warbler is one of many warbler species native to North American forests. These birds are active predators of plant-eating insects. (b) Insects can kill trees, as seen in this photo of a spruce budworm infestation. Warblers and other insect-eating birds probably reduce the number and severity of such insect outbreaks.

Competition: How a Deliberately Infected Chicken Could Save a Life.

When two species of organisms both require the same resources for life, they will be in **competition** for the resources within a habitat. In general, competition limits the size of competing populations. To determine whether two species are competing, we remove one from an environment. If the population of the other species increases, then the two species are competitors.

We may imagine lions and hyenas fighting over a freshly killed antelope or weeds growing in our vegetable gardens as typical examples of competition, but most competitive interactions are invisible. The least visible competition occurs among microorganisms. However, microbial competition is often essential to the health of both people and ecological communities.

Salmonella enteritidis is a leading cause of food-borne illness in the United States. Between 2 million and 4 million people in this country are infected by *S. enteritidis* every year, experiencing fever, intestinal cramps, and diarrhea as a result. In about 10% of cases, the infection results in severe illness requiring hospitalization. Four to six hundred Americans die as a result of *S. enteritidis* infection every year.

Most *S. enteritidis* infections result from consuming undercooked poultry products, especially eggs. The U.S. Centers for Disease Control and Prevention estimate that as many as 1 in 50 consumers is exposed to eggs contaminated with *S. enteritidis* every year. Surprisingly, most of these eggs look perfectly normal and intact. These pathogens contaminate the egg when it forms inside the hen. Thus, the only way to prevent *S. enteritidis* from contaminating eggs is to keep it out of hens.

A common way to control *S. enteritidis* is to feed hens antibiotics—chemicals that kill bacteria. However, like most microbes, *S. enteritidis* strains can evolve drug resistance that makes them more difficult to kill off. But there is another way to reduce *S. enteritidis* infection in poultry—make sure another species is occupying its niche.

Most *S. enteritidis* infections originate in an animal's gut. If another bacterial species is already monopolizing the food and available space in a hen's digestive system, then *S. enteritidis* will have trouble colonizing there. Following this principle, some poultry producers now intentionally infect hens' digestive systems with harmless bacteria, via a practice called **competitive exclusion,** to reduce *S. enteritidis* levels in their flocks. This technique involves feeding cultures of benign bacteria to 1-day-old birds. When the harmless bacteria become established in the niche in the chicks' intestines, the chicks will be less likely to host large *S. enteritidis* populations (**Figure 15.16**). There is evidence that this practice is working; *S. enteritidis* infections in chickens have dropped by nearly 50% in the United Kingdom, where competitive exclusion is common practice.

The competitive exclusion of *S. enteritidis* in hens mirrors the role of some human-associated bacteria, such as those that normally live within our intestines and genital tracts. For instance, many women who take antibiotics for a bacterial infection will then develop vaginal yeast infections because the antibiotic kills noninfectious bacteria as well, including species that normally compete with yeast.

Maintaining competitive interactions between larger species can be important for humans as well. For instance, in temporary ponds, the main competitors for the algae food source are mosquitoes, tadpoles, and snails. In the absence of tadpoles and snails, mosquito populations can become quite large—potentially with severe consequences because these insects may carry deadly diseases such as malaria, West Nile virus, and yellow fever. With frogs and their tadpoles increasingly endangered, this risk is a real one.

Figure 15.16 Competition. If poultry producers feed very young chicks non-disease-causing (beneficial) bacteria, the beneficial bacteria take up the space and nutrients in the intestine that would be used by *S. enteritidis.*
Visualize This: Why does the total number of bacteria level off over time in both graphs?

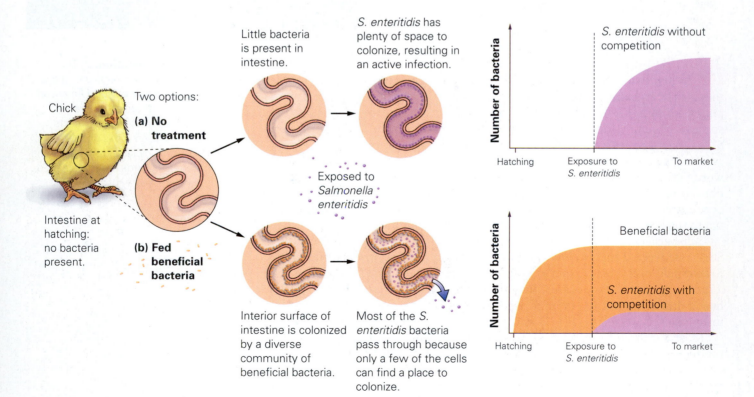

Stop & Stretch Examine the web of relationships among organisms depicted in Figure 15.12 (on page 367). Which of the following species pairs are likely competitors? In each case, describe what the competition involves. A. leopard seals, penguins; B. baleen whales, squid; C. toothed whales, leopard seals; D. sperm whale, elephant seals; E. fish, krill. How could you test your hypothesis that these animals are in competition with each other?

Keystone Species: How Wolves Feed Beavers. **Table 15.1** summarizes the major types of ecological interactions among organisms. However, this table emphasizes the effects of each interaction on the species directly involved; it does not illustrate that many of these interactions may have multiple indirect effects.

TABLE 15.1

Types of species interactions and their direct effects.			
Interaction	**Example**	**Effect on Species 1**	**Effect on Species 2**
Commensalism: Association increases the growth or population size of one species and does not affect the other.	1. Remora 2. Shark	+ As the shark feeds somewhat sloppily, the remora can collect the scraps.	0 The shark seems to suffer no negative effects from its hitchhiker.
Mutualism: Association increases the growth or population size of both species.	1. Ants 2. Acacia tree	+ The swollen thorns of the acacia provide shelter for the ants. The acacia leaves provide "protein bodies" that the ants harvest for food.	+ Ants kill herbivorous insects and destroy competing vegetation, benefiting the acacia.
Predation and Parasitism: Consumption of one organism by another.	1. Brown bear 2. Salmon	+ The brown bear catches the salmon and eats it, obtaining nourishment.	− The salmon does not survive.
Competition: Association causes a decrease or limitation in population size of both species.	1. Dandelion 2. Tomato plant	− The dandelion does not grow as well in the presence of the tomato plant. Dandelion produces fewer seeds and fewer offspring.	− The tomato plant does not grow optimally in the presence of the weed. Tomato plant produces fewer flowers and fruit.

Look again at the food web pictured in Figure 15.12. None of the species in the Antarctic Ocean's biological community is connected to only one other species—they all eat something, and most of them are eaten by something else. You can imagine that penguins, by preying on squid, have a negative effect on elephant seals, which they compete with for these squid, and a more indirect positive effect on other seabirds, which compete with squid for krill. The existence of these **indirect effects** of varying importance has led ecologists to hypothesize that, in at least some communities, the activities of a single species can play a dramatic role in determining the composition of the system's food web. These organisms are called **keystone species** because their role in a community is analogous to the role of a keystone in an archway (**Figure 15.17a**). Remove the keystone, and an archway collapses; remove the keystone species, and the web of life collapses. It is very difficult to predict which species in an intact ecosystem may be a keystone species, but biologists can point to several examples that became apparent after a species disappeared. One example is the population of gray wolves in Yellowstone National Park.

Gray wolves were exterminated within Yellowstone National Park by the mid-1920s because of a systematic campaign to rid the American West of this occasional predator of livestock. However, by the 1980s, thanks to insights gained from the science of ecology and a new interest in environmental health, attitudes about the wolf had changed. A new appreciation of the role of wolves in natural systems led to renewed interest in returning wolves to their historical homeland. In the mid-1990s, 31 wolves originally trapped in Canada were released into Yellowstone National Park, and more were released in surrounding areas. Thanks to protection from hunting and the wolves' own adaptability, by the end of 2010 the number had grown to at least 1500 wolves, living both in the park and in surrounding public lands, and the animal was no longer considered endangered in the area.

During the time that wolves were extinct in Yellowstone Park, biologists noticed dramatic declines in populations of aspen, cottonwood, and willow trees. They attributed this decline to an increase in predation by elk, especially during winter when grasses become unavailable. However, just a few years after wolf reintroduction, aspen, cottonwood, and willow tree growth has rebounded in some areas of the park, even though the wolf population was still too low to make a major dent in elk populations. Besides the regions near active wolf dens, the areas of the park that saw the greatest recovery include places on the landscape where elk have limited ability to see approaching wolves or to escape. Thus, the elk will stay away from these areas to avoid wolf predation. Wolves, primarily by changing elk behavior, appear to be important to maintaining large populations of hardwood trees in Yellowstone Park (**Figure 15.17b**).

Figure 15.17 Keystone species.
(a) The keystone in an archway helps to stabilize and maintain the arch. (b) A keystone species, such as wolves in Yellowstone National Park, helps to stabilize and maintain other species in an ecosystem.

(a)

Keystone

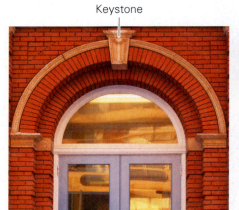

(b)

The rebound of aspen, cottonwood, and willow populations in Yellowstone has effects on other species as well. Beaver rely on these trees for food, and their populations appear to be growing in the park after decades of decline. Warblers, insects, and even fish that depend on shelter, food, and shade from these trees are increasing in abundance as well. Wolves in Yellowstone appear to fit the profile of a classic keystone species, one whose removal had numerous and surprising effects on biodiversity.

Disrupted Energy and Chemical Flows

As the examples in the previous section illustrate, the extinction of a single species can have sometimes surprising effects on other species in a habitat. What may be even less apparent is how the loss of seemingly insignificant species can change the environmental conditions on which the entire community depends.

Ecologists define an **ecosystem** as all of the organisms in a given area, along with their nonbiological environment. The function of an ecosystem is described in terms of the rate at which energy flows through it and the rate at which nutrients are recycled within it. The loss of some species can dramatically affect both of these ecosystem properties.

Energy Flow. In nearly all ecosystems, the primary energy source is the sun. Producers convert sun energy into biomass during the process of photosynthesis. The chemical energy thus captured is passed through trophic levels in the ecosystem. Biomass is partitioned among the trophic levels, with most of it residing at the bottom of the pyramid (review Figure 15.8). As we move up the pyramid, only a portion of the energy available at one level can be converted into the biomass of the next level. The amount of biomass at the producer level effectively determines how large the population of organisms at the highest level can be.

The amount of sunlight reaching the surface of Earth and the availability of water at any given location are the major determiners of both trophic pyramid structure and energy flow through it. (Chapter 16 provides a summary of how variance in sunlight and water availability leads to differences in Earth's ecosystem types.) However, the biodiversity found in an ecosystem can also have strong effects on energy flow within it.

Studies in grasslands throughout the world have provided convincing evidence that loss of species can affect energy flow. By comparing experimental prairie "gardens" planted with the same total number of individual plants but with different numbers of species, scientists at the University of Minnesota and elsewhere have discovered that the overall plant biomass tends to be greater in more diverse gardens. This research indicates that a decline in diversity, even without a decline in habitat, may lead to less energy being made available to organisms higher on the food chain, including people who depend on wild-caught food.

Nutrient Cycling. When essential mineral nutrients for plant growth pass through a food web, they are generally not lost from the environment—hence the term **nutrient cycling. Figure 15.18** (on the next page) illustrates the nitrogen nutrient cycle in a natural prairie. Here, the element moves from inorganic forms in the soil, such as ammonia, into plants, where it is converted to organic forms, and then typically moves from one living organism to the next contained in food. It finally returns to its inorganic form thanks to the activities of decomposers.

Nitrogen is a major component of protein, and abundant protein is essential for the proper growth and functioning of all living organisms. Nitrogen is therefore often the nutrient that places an upper limit on production in most ecosystems—more nitrogen generally leads to greater production, while areas with less available nitrogen can support fewer plants (and therefore animals).

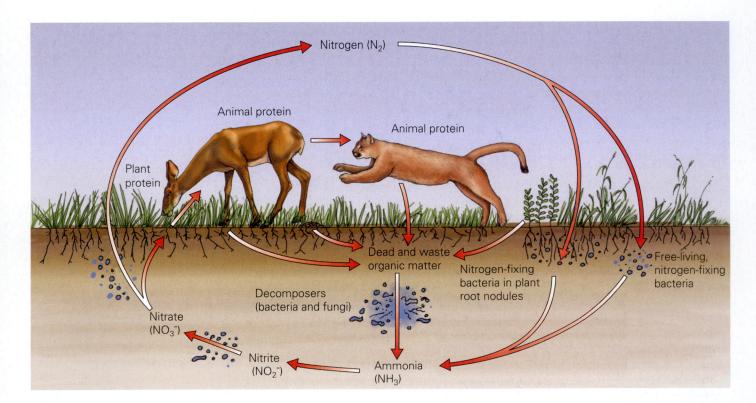

Figure 15.18 Nutrient cycling. Nutrients such as nitrogen, shown here, are recycled in an ecosystem, flowing from soil to producers to consumers and then back into the soil, where complex nutrients are decomposed into simpler forms.

Visualize This: Consider how people obtain nutrition and what happens to our waste after digestion and remains after death. How are nutrient flows in human-made systems different from natural cycles?

(a) Invasive earthworms absent.

(b) Invasive earthworms present.

Stop & Stretch Inorganic nitrogen fertilizer applied to agricultural fields is made from an industrial process using nitrogen gas in the atmosphere. A portion of the nitrogen contained in these agricultural crops is eaten by humans and excreted in urine, which in developed countries is essentially released into nearby waterways (after waste treatment). Use Figure 15.18 as a resource to describe how processing nitrogen gas and treating waste change the nitrogen cycle.

You should note as you review Figure 15.18 that plants absorb simple molecules from the soil and incorporate them into more complex molecules. These complex molecules move through the food web with relatively minor changes until they return to the soil. Here, complex molecules are broken down into simpler ones by the action of **decomposers,** typically bacteria and fungi.

Changes in the soil community can greatly affect nutrient cycling and thus the survival of certain species in ecosystems. Scientists investigating the effects of introduced earthworms, which have invaded forests throughout the northeastern United States, have observed dramatic reductions in the diversity and abundance of plants on the forest floor (**Figure 15.19**). The introduced earthworms have apparently caused changes to the community of native soil organisms. As a result of these changes, the nutrient cycle is disrupted and the native plant community has suffered.

Figure 15.19 Changes in ecosystem function. Notice how barren of living vegetation the worm-infested forest floor appears. One reason for this dramatic change may be a disruption in the native nutrient cycle by this introduced species.

The loss of biodiversity clearly can have profound effects on the health of communities and ecosystems on which humans depend. However, controversy exists over whether the current extinction may negatively affect our own psychological well-being.

Psychological Effects

Some scientists argue that the diversity of living organisms sustains humans by satisfying a deep psychological need. One of the most prominent scientists to promote this idea is Edward O. Wilson, who calls this instinctive desire to commune with nature **biophilia.**

Wilson contends that people seek natural landscapes because our distant ancestors evolved in similar landscapes (**Figure 15.20**). According to this hypothesis, ancient humans who had a genetic predisposition driving them to find diverse natural landscapes were more successful than those without this predisposition because diverse areas provide a wider variety of food, shelter, and tool resources. Wilson claims that we have inherited this genetic imprint of our preagricultural past.

While there is no evidence of a gene for biophilia, there is evidence that our experience with nature has psychological effects. Studies in dental clinics indicate that patients viewing landscape paintings experience a 10- to 15-point decrease in blood pressure. Patients in a Philadelphia hospital who could see trees from their windows recovered from surgery quicker and with less pain than did patients whose views were of brick walls. Individual experiences with pets and houseplants indicate that many people derive great pleasure from the presence of nonhuman organisms. Although not conclusive, these studies and experiences are intriguing since they suggest that a continued loss of biodiversity could make life in human society less pleasant overall.

The consequences of the loss of biodiversity are not confined to our generation. The fossil record illustrated in Figure 15.2 reveals that it takes 5 to 10 million years to recover the biological diversity lost during a mass extinction. The species that replaced those lost in previous mass extinctions were very different. For instance, after the mass extinction of the reptilian dinosaurs, mammals replaced them as the largest animals on Earth. We cannot predict what biodiversity will look like after another mass extinction. The mass extinction we may be witnessing today will have consequences felt by people in thousands of generations to come.

Figure 15.20 Is our appreciation of nature innate? Humans evolved in a landscape much like this one in East Africa. Some scientists argue that we have an instinctive need to immerse ourselves in the natural world.

15.3 Saving Species

So far in this chapter, we have established the possibility of a modern mass extinction occurring, and we have described the potentially serious costs of this loss of biodiversity to human populations. Because the sixth extinction is largely a result of human activity, reversing the trend of species loss requires political and economic, rather than scientific, decisions. But what can science tell us about how to stop the rapid erosion of biodiversity?

Protecting Habitat

Without knowing exactly which species are closest to extinction and where they are located, the most effective way to prevent loss of species is to preserve as many habitats as possible. The same species-area curve that is used to estimate the future rate of extinction also gives us hope for reducing this number. According to the curve in Figure 15.7b, species diversity declines rather slowly as habitat area declines. Thus, in theory we can lose 50% of a habitat but still retain 90% of its species. This estimate is optimistic because habitat destruction is not the only threat to biodiversity, but the species-area curve tells us that if the rate of habitat destruction is slowed or stopped, extinction rates will slow as well.

Protecting the Greatest Number of Species.
Given the growing human population, it is difficult to imagine a complete halt to habitat destruction. However, biologist Norman Myers and his collaborators have concluded that 25 biodiversity "hot spots," making up less than 2% of Earth's surface, contain up to 50% of all mammal, bird, reptile, amphibian, and plant species (**Figure 15.21**). Hot spots occur in areas of the globe where favorable climate conditions lead to high levels of plant production, such as rain forests, and where geological factors have resulted in the isolation of species groups, allowing them to diversify.

Stopping habitat destruction in biodiversity hot spots could greatly reduce the global extinction rate. By focusing conservation efforts on hot spot areas at the greatest risk, humans can very quickly prevent the loss of a large number of species. Of course, even with habitat protection, many species in these hot spots will likely become extinct anyway for other human-mediated reasons.

Figure 15.21 Diversity "hot spots." This map shows the locations of 25 identified biodiversity hot spots around the world. Notice how unevenly these regions of high biodiversity are distributed.

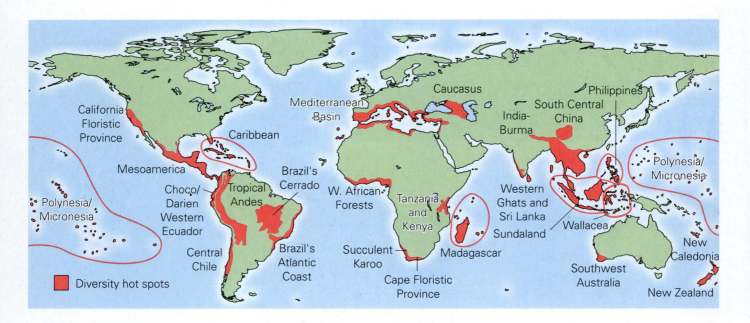

In the long term, we must find ways to preserve biodiversity while including human activity in the landscape. One option is **ecotourism,** which encourages travel to natural areas in ways that conserve the environment and improve the well-being of local people. Some hot spot countries, such as Costa Rica and Kenya, have used ecotourism to preserve natural areas and provide much-needed jobs; other countries have been less successful.

While preserving hot spots may greatly reduce the total number of extinctions, this approach has its critics, who say that by promoting a strategy that focuses intensely on small areas, we risk losing large amounts of biodiversity elsewhere. These critics promote an alternative approach—identifying and protecting a wide range of ecosystem *types*—designed to preserve the greatest range of biodiversity rather than just the largest number of species.

Protecting Habitat for Critically Endangered Species.

Although preserving a variety of habitats ensures fewer extinctions, already-endangered species require a more individualized approach. The ESA requires the U.S. Department of the Interior to designate critical habitats for endangered species—that is, areas in need of protection for the survival of the species. The amount of critical habitat that becomes designated depends on political as well as biological factors.

The biological part of a critical habitat designation includes conducting a study of habitat requirements for the endangered species and setting a population goal for it. The U.S. Department of the Interior's critical habitat designation has to include enough area to support the recovery population. However, federal designation of a critical habitat results in the restriction of human activities that can take place there. The U.S. Department of Interior has the ability to exclude some habitats from protection if there are "sufficient economic benefits" for doing so—a decision that is political in nature.

Decreasing the Rate of Habitat Destruction.

Preserving habitat is not simply the job of national governments that designate protected areas or of private conservation organizations that purchase at-risk habitats. All of us can take actions to reduce habitat destruction and stem the rate of species extinction. Conversion of land to agricultural production is a major cause of habitat destruction, so eating lower on the food chain and reducing your consumption of meat and dairy products from grain-fed animals is one of the most effective actions you can take. Reducing your use of wood and paper products and limiting your consumption of these products to those harvested sustainably (that is, in a manner that preserves the long-term health of the forest) can help slow the loss of forested land.

Other measures to decrease the rate of habitat destruction require group effort. For instance, increased financial aid to developing countries may help slow the rate of habitat destruction. This would allow poor countries to invest money in technologies that decrease their use of natural resources, for example, cooking technologies that reduce the need to harvest large amounts of woody plants for fuel. Strategies that slow the rate of human population growth offer more ways to avoid mass extinction. You can participate in group conservation efforts by joining nonprofit organizations focused on these issues, writing to politicians, and educating others.

Although protecting habitat from destruction can reduce extinction rates for species on the brink of extinction—like the shortnose and Lost River suckers—preserving habitat is not enough. Populations can become so small that they can disappear, even with adequate living space. Recovery plans for both the Lost River and shortnose suckers set a short-term goal of a stable population made up of at least 500 individuals. To understand why at least this many individuals are required to save these species from extinction, we need to review the special problems of small populations.

Small Populations Are Vulnerable to Environmental Disasters

The growth rate of an endangered species influences how rapidly that species can attain a target population size. Shortnose and Lost River suckers have relatively high growth rates and will meet their population goals quickly if the environment is ideal (**Figure 15.22a**). For slower-growing species, such as the California condor (**Figure 15.22b**), populations may take decades to recover.

The rate of recovery is important because the longer a population remains small, the more it is at risk of experiencing a catastrophe that could eliminate it entirely. The story of the heath hen in the United States is a case study on just this point.

The heath hen was a small wild chicken that once ranged from Maine to Virginia. In the eighteenth century, the heath hen population numbered in the hundreds of thousands of individuals. Continued human settlement of the eastern seaboard resulted in the loss of habitat, causing a rapid and dramatic decline of the birds' population. By the end of the nineteenth century, the only remaining heath hens lived on Martha's Vineyard, a 100-square-mile island off the coast of Cape Cod, Massachusetts. Farming and settlement on the island further reduced the habitat for heath hen breeding. By 1907, only 50 birds were present on the island.

In response to this precipitous decline, Massachusetts established a 2.5-square-mile reserve for the remaining birds on Martha's Vineyard in 1908. The response initially seemed effective; by 1915, the population had recovered to nearly 2000 individuals. However, beginning in 1916, a series of disasters struck. First, fire destroyed much of the remaining habitat on the island. The following winter was long and cold, and an invasion of starving predatory goshawks further reduced the heath hen population. Finally, a poultry disease introduced by imported domestic turkeys wiped out much of the remaining population. By 1927, only 14 heath hens remained—almost all males. The last surviving member of the species was spotted for the last time on March 11, 1932.

The final causes of heath hen extinction were natural events—fire, harsh weather, predation, and disease. But it was the population's small size that doomed it in the face of these relatively common challenges. A population of 100,000 individuals can weather a disaster that kills 90% of its members but leaves 10,000 survivors, but a population of 1000 individuals will be nearly eliminated by the

Figure 15.22 The effect of growth rate on species recovery. (a) This graph illustrates the rapid growth of a hypothetical population of quickly reproducing Lost River suckers. (b) The slow growth rate of the California condor has made the recovery of this species a long process. Today, nearly 30 years after recovery efforts began, the population of wild condors is still only in the dozens.

(a) Lost River Sucker

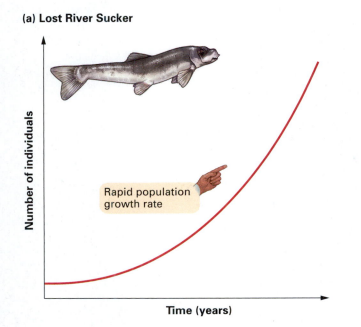

(b) California condor

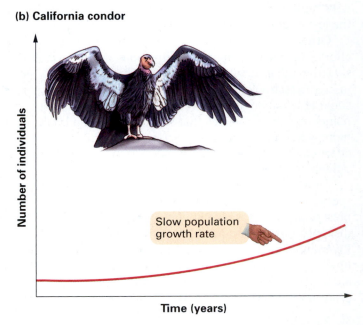

same circumstances. Even when human-caused losses to the heath hen population were halted, the species' survival was still extremely precarious.

Small populations of endangered species can be protected from the fate that befell the heath hen. Having additional populations of the species at sites other than Martha's Vineyard would have nearly eliminated the risk that *all* members of the population would be exposed to the same series of environmental disasters. This is the rationale behind placing captive populations of endangered species at several different sites. For instance, captive whooping cranes are located at the U.S. National Biological Service's Patuxent Wildlife Research Center in Maryland, the International Crane Foundation in Wisconsin, the Calgary Zoo in Canada, and the Audubon Center for Endangered Species Research in New Orleans. Even with multiple habitats, if populations of endangered species remain small in number, they are subject to a more subtle but potentially equally devastating situation—the loss of genetic variability.

An Overview: Conservation Genetics

A species' **genetic variability** is the sum of all of the alleles and their distribution within the species. For example, the gene that determines your ABO blood type comes in three different forms. A population containing all three of these alleles contains more genetic variability than does a population with only two alleles.

The loss of genetic variability in a population is a problem for two reasons. On an individual level, low genetic variability leads to low fitness for two reasons: because homozygotes are more likely to express deleterious, mutant alleles, and because homozygotes generally have a narrower range of environments in which they can survive. On a population level, rapid loss of genetic variability may lead to extinction for two related reasons: due to the combined effects of the low fitness of individuals—a problem known as inbreeding depression—and because the population as a whole does not contain variants that will survive in changed environmental conditions, which are inevitable (**Table 15.2**). *See the next section for A Closer Look at conservation genetics.*

TABLE 15.2

The costs of low genetic diversity.		
Subject of Risk	**Outcomes of Low Diversity**	**Example**
Individuals	**Inbreeding depression:** Deleterious recessive genetic mutations are expressed because low diversity leads to higher numbers of homozygous individuals.	Cheetahs exhibit poor sperm quality and low cub survival, probably because they lack genetic diversity and express more recessive alleles.
	Loss of heterozygote advantage: With only one type of protein produced for a homozygous functional gene, the range of environments in which an individual can be successful may be limited.	In humans, heterozygotes for the sickle-cell allele can have higher fitness than homozygotes for the normal allele because the allele provides protection from malaria.
Populations	**Extinction vortex:** Low population levels resulting in inbreeding depression make the population more vulnerable to the risks of small population size.	The heath hen population may not have grown rapidly enough due to inbreeding depression. As a result, the population could not weather a series of unpredicted environmental challenges.
	Inability to respond to environmental change: Population lacks variants that can survive in the face of environmental challenge.	The potato population in 1840s Ireland did not contain any variants that were resistant to potato blight, leading to the loss of nearly the entire crop.

A Closer Look:
Conservation Genetics

The Importance of Individual Genetic Variability

Fitness refers to an individual's ability to survive and reproduce in a given set of environmental conditions (Chapter 11). Low individual genetic variability decreases fitness for two reasons. As an introduction, we can use an analogy to illustrate them.

First, imagine that you could own only two jackets (**Figure 15.23**). If both are blazers, then you would be well prepared to meet a potential employer. However, if you had to walk across campus to your job interview in a snowstorm, you would be pretty uncomfortable. If you own two parkas, then you will always be protected from the cold, but you would look pretty silly at a job interview. However, if you own one warm jacket and one blazer, you are ready for freezing weather as well as a job interview. In a way, individuals experience the same advantages when they carry two different functional alleles for a gene. Because each allele codes for a functional protein, a heterozygous individual produces two slightly different proteins for the same function.

If the protein produced by each allele works best in different environments, then heterozygous individuals are able to function efficiently over a wider range of conditions. They demonstrate a phenomenon known as **heterozygote advantage** when compared to homozygotes, who only make one version of the protein. The sickle-cell allele in humans (discussed in Chapter 12) is an example of heterozygote advantage. Individuals who carry one copy of the sickle-cell allele have greater fitness than those with no copies of the allele in areas where malaria is common. Their individual diversity in hemoglobin variants allows them to both effectively transport oxygen to their tissues and provides resistance to malaria infection.

The second reason high individual genetic variability increases fitness is that, in many cases, one allele for a gene is **deleterious**—that is, it produces a protein that is not very functional. In our jacket analogy, a nonfunctional, deleterious allele is equivalent to a badly torn jacket. If you have this jacket and an intact one, at least you have one warm covering (**Figure 15.24**). In this case, heterozygosity is valuable because a heterozygote still carries one functional allele. Often, deleterious alleles are recessive, meaning that the activity of the functional allele in a heterozygote masks the fact that a deleterious allele is present (see Chapter 7). An individual who is homozygous (carries two identical copies of a gene) for the deleterious allele will have low fitness—in our analogy, two torn jackets and nothing else. For both of these reasons, when individuals are heterozygous for many genes (or, in our analogy, have two choices for all clothing items), the cumulative effect is greater fitness relative to individuals who are homozygous for many genes.

In a small population, in which mates are more likely to be related to each other simply because there are few mates to choose from, heterozygosity declines. When related individuals mate—known as **inbreeding**—the chance that their offspring will be homozygous for any allele is relatively high. The negative effect of homozygosity on fitness is known as **inbreeding depression.** In cheetahs, high levels of inbreeding have decreased fitness by causing poor sperm quality and low cub survival, both likely due to increased expression of deleterious alleles. The costs of inbreeding are seen in humans as well; the children of first cousins have higher rates of homozygosity and higher mortality rates (thus lower fitness) than children of unrelated parents. In a small population of an endangered species, inbreeding often causes low rates of survival and reproduction and can seriously hamper a species' recovery.

Stop & Stretch What strategies could wildlife managers employ to try to reduce the risk of inbreeding in small animal populations?

Being heterozygous may confer higher fitness for responding to a changing environment.

Homozygote 1: Relatively low fitness (only one type of jacket in wardrobe)

Homozygote 2: Relatively low fitness (only one type of jacket in wardrobe)

Heterozygote: Relatively high fitness (two types of jackets in wardrobe)

Figure 15.23 Heterozygotes can inhabit a wider range of environments. In this analogy, each jacket represents an allele. Just as having two different jackets prepares you for a wider range of situations than having only one type, two different alleles for the same gene may allow for optimal function over a wider range of conditions.

Being heterozygous may confer higher fitness by masking deleterious recessive alleles.

Homozygote 1: Relatively high fitness
(two functional jackets in wardrobe)

Homozygote 2: Relatively low fitness
(two nonfunctional jackets in wardrobe)

Heterozygote: Relatively high fitness
(one functional jacket in wardrobe)

Figure 15.24 Heterozygotes avoid the deleterious effects of recessive mutations.
Again, each jacket represents an allele. If one type of jacket you can receive is nonfunctional, it is better to receive no more than one of them. Heterozygotes are likewise protected against the likelihood of having only nonfunctional alleles, as is present in homozygous, functional recessive individuals.
Visualize This: Explain why homozygotes for the normal allele have higher fitness than homozygotes for the nonfunctional recessive allele.

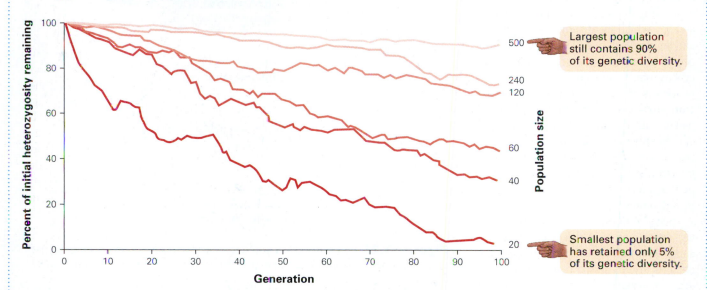

Largest population still contains 90% of its genetic diversity.

Smallest population has retained only 5% of its genetic diversity.

Figure 15.25 Genetic drift affects small populations more than large populations.
In this graph, each line represents the average of 25 computer simulations of genetic drift for a given population size. After 100 generations, a population of 500 individuals still contains 90% of its genetic variability. In contrast, a population of 20 individuals has less than 5% of its original genetic variability.

How Variability Is Lost via Genetic Drift

Small populations lose genetic variability because of **genetic drift**, a change in the frequency of an allele that occurs simply by chance within a population. While genetic drift can be a process for causing evolutionary change (Chapter 12), in a small population, genetic drift can have detrimental consequences also.

Imagine two human populations in which the frequency of blood-type allele A is 1%; that is, only 1 of every 100 blood-type genes in the population is the A form (we use the symbol I^A for this allele). In the first population of 20,000 individuals, there are 40,000 total blood group genes (that is, 2 copies per individual). At 1% frequency, this population contains 400 I^A alleles. If a few of the individuals who carry the I^A allele die accidentally before they reproduce, the number of copies of the allele will drop slightly in the

next generation—let's say to 385. If the population size in this generation is still 20,000, the new allele frequency is 385 divided by 40,000, or 0.96%.

The change in allele frequency from 1% to 0.96% is the result of genetic drift. However, this 0.04% change in allele frequency is relatively minor in this large population; hundreds of individuals will still carry the I^A allele.

Now imagine the situation in a second, much smaller population. In a population of only 200 individuals and with an I^A frequency at 1%, only 4 copies of the allele are present. If 2 individuals carrying a copy of I^A fail to pass it on, its frequency will drop to 0.5%. Another chance occurrence in the following generation could completely eliminate the 2 remaining I^A alleles from the population. Genetic drift is more severe in small populations and is much more likely to result in the complete loss of alleles (**Figure 15.25**).

(continued on the next page)

(A Closer Look continued)

In most populations, alleles that are lost through genetic drift have relatively small effects on fitness at the time. However, many alleles that appear to be nearly neutral with respect to fitness in one environment may have positive fitness in another environment. When this is the case, the loss of these alleles may spell disaster for the entire species.

The Consequences of Low Genetic Variability in a Population

Populations with low levels of genetic variability have an insecure future for two reasons. First, when alleles are lost, the level of inbreeding depression in a population increases, which means lower reproduction and higher death rates, leading to declining populations that are susceptible to all the other problems of small populations. This process is often referred to as the **extinction vortex** (**Figure 15.26**). The heath hen discussed in this chapter is an example of the extinction vortex; once the remnant population was reduced to small numbers by fire, disease, and predation, inbreeding depression prevented it from rebounding, thus dooming it to extinction.

Second, populations with low genetic variability may be at risk of extinction because they cannot evolve in response to changes in the environment. When few alleles are available for any given gene, it is possible that no individuals in a population possess an adaptation allowing them to survive an environmental challenge. For example, there is some evidence that people with type A blood are more resistant to cholera and bubonic plague than are people with type O or B blood. Loss of the I^A allele would make human populations more susceptible to serious declines in the face of these diseases.

Stop & Stretch   Genetic drift occurs rapidly in small populations, but so does natural selection. Use your understanding of natural selection and genetics to explain why loss of recessive deleterious mutations would be faster in small populations compared to large ones.

As is often the case, there are some exceptions to the "rules" just described. For example, widespread hunting of northern elephant seals in the 1890s reduced its population to 20 individuals, thus wiping out much of the species' genetic variation. However, elephant seal populations have rebounded

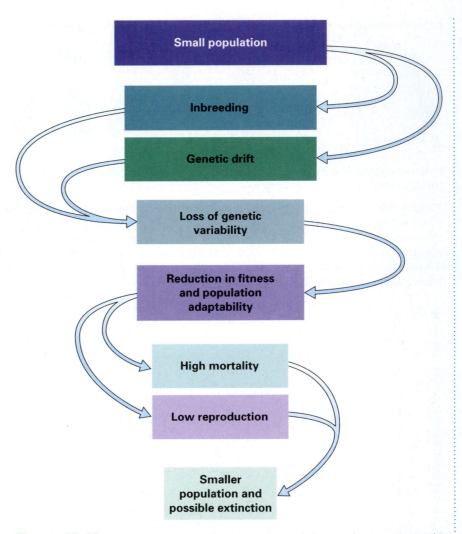

Figure 15.26 The extinction vortex. A small population can become trapped in a positive-feedback process that causes it to continue to shrink in size, eventually leading to extinction.

to include about 150,000 healthy individuals today. The dramatic recovery of elephant seals is apparently unusual. There are many more examples of populations that suffer because of low genetic variability. The Irish potato is perhaps the most dramatic example of this cost of low genetic diversity.

Potatoes were a staple crop of rural Irish populations until the 1850s. Although the population of Irish potatoes was high, it had remarkably low genetic variability. First, potatoes are not native to Ireland (in fact, they originated in South America), so the crop was limited to just a few varieties that were originally imported. Most of the potatoes grown on the island were of a single variety. Second, new potato plants are grown from potatoes harvested from the previous year's plants and thus are genetically identical to their parents. As a result, all of the potatoes in a given plot had identical alleles for every gene, and most potatoes on the whole island were extremely similar.

When the organism that causes potato blight arrived in Ireland in September 1845, nearly all of the planted potatoes became infected and rotted in the fields (**Figure 15.27**). The few potatoes that by chance escaped the initial infection were used to plant the following year's crops. While some varieties of potatoes in South America carry alleles that allow them to resist potato blight, apparently very few or no Irish potatoes carried these alleles. As a result, in 1846 the entire Irish potato crop failed. Because of this failure and another in 1848, along with the ruling British government's harsh policies that inhibited distribution of food relief, nearly 1 million Irish peasants died of starvation and disease, and another 1.5 million peasants emigrated to North America.

Because Irish potatoes lacked genetic diversity, even an enormous population of these plants could not

(a) *Phytophthora infestans*

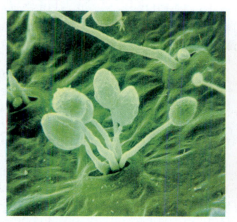

(b) Potatoes with blight

Figure 15.27 Potato blight. (a) The organism, *Phytophthora infestans,* that causes late blight in potatoes. (b) Infected potato tubers.

escape the catastrophe caused by this disease. Similarly, preventing endangered species from declining to very small population levels, thus eroding genetic diversity, is critical to avoiding a similar genetic disaster even once populations recover.

15.4 Protecting Biodiversity versus Meeting Human Needs

Saving the Lost River and shortnose suckers from extinction requires protecting all of the remaining fish and restoring the habitat they need for reproduction. These actions could cause economic and emotional suffering for humans who make their living in the Klamath Basin. In fact, many actions necessary to save endangered species result in immediate problems for people. As pointed out by the Oregon State University student, we need to balance the costs and benefits of preserving endangered species.

A provision of the Endangered Species Act allows members of a committee to weigh the relative costs and benefits of actions taken to protect endangered species. The Endangered Species Committee, which includes the U.S. secretaries of agriculture and the interior as well as the chair of the Council of Economic Advisors, has convened a number of times for this purpose. This so-called God squad decides if it should overrule a federal action protecting an endangered species to protect the livelihoods of people.

Farmers in the Klamath Basin advocated for a God squad ruling on the diversion of water from Upper Klamath Lake but failed to receive one. History suggested that a decision was not likely to be in their favor, anyway. The Endangered Species Committee has convened only four times from 1973 through today, and it has granted two exemptions, one of which was overturned by the subsequent presidential administration.

Instead of convening the God squad in response to the Klamath crisis in 2001, the administration of George W. Bush took a different approach—they asked a committee of scientists from the prestigious National Academy of Science to examine the question. While some critics speculate that the committee was influenced by political concerns, their report indicated that the water-level requirements established in the fish management plan for the Lost River and shortnose suckers were not well supported by science. The committee recommended allowing water releases in subsequent years that could be adequate for the needs of both farmers and fish, at least in the short term. In 2002, water was released for irrigation—with

Figure 15.28 An endangered species success story. The bald eagle was "delisted" from endangered status in 2007, at an event celebrated at this ceremony in Washington, DC. Its dramatic recovery from near extinction was thanks in part to government protections. However, delisting is still a relatively rare event.

apparently no harmful affect on the suckers. However, continued low water flows in the Klamath River have led to the collapse of the coho salmon population and an almost total ban on salmon fishing in the area. The federal government continues to purchase farmland from willing sellers in the Klamath Basin to protect and restore the watershed for all three species.

The ESA has been a successful tool for bringing species such as the peregrine falcon, American alligator, and bald eagle (**Figure 15.28**) back from the brink of extinction, but all of these successes have come with some cost to citizens. If the solution to these and other endangered species controversies is any guide, many Americans are willing to devote tax dollars to efforts that balance the needs of people and wildlife to protect our natural heritage.

As with any challenge that humans face, the best strategy is prevention. **Table 15.3** provides a list of actions that can help reduce the rate at which species become endangered. Meeting the challenge of preserving biodiversity requires some creativity, but it is possible to provide for the needs of people while making space for other organisms. However, it will take all of us to help keep the equation in balance.

TABLE 15.3

Taking action to preserve biodiversity.

Objective		Why Do It?	Actions
Reduce fossil fuel use		• Mining, drilling, and transporting fossil fuels modifies habitat and leads to pollution. • Burning fossil fuels contributes to global climate change, further degrading natural habitats.	• Buy energy-efficient vehicles and appliances. • Walk, bike, carpool, or ride the bus whenever possible. • Choose a home near school, work, or easily accessible public transportation. • Buy "clean energy" from your electric provider, if offered.
Reduce the impact of meat consumption		• The primary cause of habitat destruction and modification is agriculture. • Modern beef, pork, and chicken production relies on grains produced on farms. One pound of beef requires 4.8 pounds of grains or about 25 square meters of agricultural land.	• Eat one more meat-free meal per week. • Make meat a "side dish" instead of the main course. • Purchase grass-fed or free-range meat.
Reduce pollution		• Pollution kills organisms directly or can reduce their ability to survive and reproduce in an environment.	• Do not use pesticides. • Buy products produced without the use of pesticides. • Replace toxic cleaners with biodegradable, less-harmful chemicals. • Consider the materials that make up the goods you purchase and choose the least-polluting option. • Reuse or recycle materials instead of throwing them out.
Educate yourself and others		• Change happens most rapidly when many individuals are working for it.	• Ask manufacturers or store owners about the environmental costs of their goods. • Talk to family and friends about the choices you make. • Write to decision makers to urge action on effective measures to reduce human population growth and curb habitat destruction and species extinction.

SAVVY READER

What Happens When a Species Recovers

Gray Wolf's endangered species status appears close to end

A new federal bill introduced in the U.S. House of Representatives would "de-list" the Gray Wolf from Endangered Species Act protections and allow individual states to design and carry out their own wolf management and recovery plans. The legislation is co-sponsored by Rep. Jim Matheson (D) from Utah, a state that is just beginning to see increases in wolf numbers, and Rep. Mike Ross (D-Arkansas), who co-chairs the Congressional Sportsman's Caucus, made up of advocates of hunting and fishing interests.

According to Rep. Matheson, "Scientists and wolf recovery advocates agree—the gray wolf is back. Since it is no longer endangered, it should be de-listed as a species, managed as others species are—by state wildlife agencies—and time, money, and effort can be focused where it's needed."

Rep. Ross argues that officials within the regions inhabited by particular endangered species should have a greater role in management decisions related to these species. "Excessive wolf populations are having a devastating impact on elk, moose, deer, and other species, and each state has its own unique set of challenges. Both the Obama and Bush administrations have already recommended the de-listing of wolves in many states and turning their management over to state wildlife agencies because it is the right thing to do to keep our nation's sensitive ecosystem in balance."

1. What evidence does the representative from Utah provide to support his statement that "the gray wolf is back"? Is that evidence convincing?

2. Given what you've learned in this chapter, do you think it is likely that the return of the gray wolf is having a "devastating impact" on elk and moose?

Chapter Review

Learning Outcomes

Go to the Study Area at www.masteringbiology.com for practice quizzes, myeBook, BioFlix™ 3-D animations, MP3 Tutor sessions, videos, current events, and more.

L01 **Describe how scientists can estimate current and historical rates of extinction, and provide the evidence that Earth is currently experiencing a mass extinction (Section 15.1).**

- Historical rates of extinction can be estimated by looking at the life span of species in the fossil record, while the rate of modern extinctions can be measured by counting the number of species known to be lost or endangered (pp. 359–360).
- The loss of biodiversity through species extinction is exceeding historical rates by 50 to 100 times (p. 360).

L02 **List the major causes of extinction, and describe how each is related to human activity (Section 15.1).**

- The loss of natural habitat caused by human activities such as agriculture and urban development is the primary cause of the extinctions occurring in the modern era (pp. 361–363).
- Additional threats of habitat fragmentation, the movement of introduced species to regions where they are not native, overexploitation through uncontrolled harvesting, and pollution also contribute to species extinction (pp. 363–365).

LO3 Explain how the species-area curve is used to estimate species number in an area (Section 15.1).

- Species-area curves illustrate the relationship between the number of species and a geographic area. If the actual number of species of a particular type is not known, a species-area curve can provide some estimate of it in a given area (pp. 361–363).

LO4 Explain the principle of the trophic pyramid (Section 15.2).

- The trophic pyramid is the relationship among biomass at different levels of a food chain. In nearly all communities, there is significantly more biomass in the producer level than at higher level consumer levels (p. 363).
- Species at the top of the food chain are more susceptible to extinction because less energy is available for survival at higher trophic levels (p. 363).

LO5 Define *predation*, *mutualism*, and *competition*, and provide examples of each ecological interaction (Section 15.2).

- Species are members of communities; they may interact with one another via mutualism, where both species are benefitted by the relationship; via predation, where one benefits while the other is consumed in the relationship; or via competition, where both are struggling to obtain the same resources (pp. 367–370).

LO6 Define *food web* and *ecosystem*, and explain the role of keystone and nonkeystone species in both (Section 15.2).

- A food web is the connection among species in a community, while an ecosystem includes not only the food web but all of the nonbiological aspects of the environment occupied by a community of organisms (p. 367 and p. 373).
- A keystone species in a food web is one whose absence has a large effect on the entire community (pp. 372–373).
- Species can play a role in ecosystem function, including effects on energy flow and nutrient cycling. (pp. 373–374).

LO7 Provide an example of a nutrient cycle (Section 15.2).

- Nutrients cycle through ecosystems from living organisms to inorganic forms and back again. Nitrogen and carbon dioxide are two examples of nutrients that cycle through the environment (p. 374).

LO8 Explain why small population size and a loss of genetic diversity in a population are risky for both the population and individuals within the population (Section 15.3).

- When species are already endangered, restoring larger populations is critical for preventing extinction (p. 378).

- Small populations are at higher risk for extinction due to environmental catastrophes (pp. 378–379).
- Small populations are at risk when individuals have low fitness due to inbreeding and thus are less able to increase population size (pp. 379–380).
- Genetic variability is lost in small populations because of genetic drift—the loss of alleles from a population due to chance events. Therefore, small populations may be less able to evolve in response to environmental change (pp. 381–383).

LO9 List strategies that will help reduce the number of extinctions going forward (Section 15.3).

- Eating lower on the food chain and using sustainable methods of transportation and energy all help conserve habitat for other species (p. 384).
- The political process enables people to develop plans for helping endangered species recover from the brink of extinction while minimizing the negative effects of these actions on people (pp. 383–384).

Roots to Remember

The following roots of words come mainly from Latin and Greek and will help you decipher terms:

eco-	means home or habitation. Chapter term: ecosystem
eu-	means true or good. Chapter term: eutrophic
-philia	means affection or love. Chapter term: biophilia
-trophic	means food, nourishment. Chapter term: eutrophic

Learning the Basics

1. **LO2** Describe how habitat fragmentation endangers certain species. Which types of species do you think are most threatened by habitat fragmentation?

2. **LO5** Compare and contrast the species interactions of mutualism, predation, and competition.

3. **LO1** A mass extinction _____.

   A. is global in scale; B. affects many different groups of organisms; C. is caused only by human activity; D. A and B are correct; E. A, B, and C are correct

4. **LO1** Current rates of species extinction appear to be approximately _____ historical rates of extinction.

   A. equal to; B. 10 times lower than; C. 10 times higher than; D. 50 to 100 times higher than; E. 1000 to 10,000 times higher than

5. **LO3** According to the generalized species-area curve, when habitat is reduced to 50% of its original size, approximately _____ of the species once present there will be lost.

   A. 10%; B. 25%; C. 50%; D. 90%; E. it is impossible to estimate the percentage

6. **LO2** Which cause of extinction results from humans' direct use of a species?

   A. overexploitation; B. habitat fragmentation; C. pollution; D. introduction of competitors or predators; E. global warming

7. **LO6** The web of life refers to the _____.

   A. evolutionary relationships among living organisms; B. connections between species in an ecosystem; C. complicated nature of genetic variability; D. flow of information from parent to child; E. predatory effect of humans on the rest of the natural world

8. **LO5** Which of the following is an example of a mutualistic relationship?

   A. moles catching and eating earthworms from the moles' underground tunnels; B. cattails and reed canary grass growing together in wetland soils; C. cleaner fish removing and eating parasites from the teeth of sharks; D. Colorado potato beetles consuming potato plant leaves; E. more than one of the above

9. **LO8** The risks faced by small populations include _____.

   A. erosion of genetic variability through genetic drift; B. decreased fitness of individuals as a result of inbreeding; C. increased risk of experiencing natural disasters; D. A and B are correct; E. A, B, and C are correct

10. **LO8** One advantage of preserving more than one population of an endangered species at more than one location is _____.

    A. a lower risk of extinction of the entire species if a catastrophe strikes one location; B. higher levels of inbreeding in each population; C. higher rates of genetic drift in each population; D. lower numbers of heterozygotes in each population; E. higher rates of habitat fragmentation in the different locations

11. **LO4** There are fewer lions in Africa's Serengeti than there are zebras. This is principally because _____.

    A. zebras tend to drive off lions; B. lions compete directly with cheetahs, while zebras do not have any competitors; C. zebras have mutualists that increase their population, while lions do not; D. there is less energy available in zebras to support the lion population than there is in grass to support the zebras; E. zebras are a keystone species, while lions are not.

12. **LO7** Most of the nutrients available for plant growth in an ecosystem are _____.

    A. deposited in rain; B. made available through the recycling of decomposers; C. maintained within that ecosystem over time; D. B and C are correct; E. A, B, and C are correct

Analyzing and Applying the Basics

1. **LO6** Off the coast of the Pacific Northwest, areas dominated by large algae, called kelp, are common. These "kelp forests" provide homes for small plant-eating fishes, clams, and abalone. These animals in turn provide food for crabs and larger fishes. A major predator of kelp in these areas is sea urchins, which are preyed on by sea otters. When sea otters were hunted nearly to extinction in the early twentieth century, the kelp forest collapsed. The kelp were only found in low levels, while sea urchins proliferated on the seafloor. Use this information to construct a simple food web of the kelp forest. Using the food web, explain why sea otters are a keystone species in this system.

2. **LO8** In which of the following situations is genetic drift more likely and why?

   A. A population of 500 in which males compete heavily for harems of females. About 5% of the males father all of the offspring in a given generation. Females produce one offspring per season. B. A population of 250 in which males and females form bonded pairs that last throughout the mating season. Females produce three to four offspring per season.

3. **LO5 LO8 LO9** The piping plover is a small shorebird that nests on beaches in North America. The plover population in the Great Lakes is endangered and consists of only about 30 breeding pairs. Imagine that you are developing a recovery plan for the piping plover in the Great Lakes. What sort of information about the bird and the risks to its survival would help you to determine the population goal for this species as well as how to reach this goal?

Connecting the Science

1. From your perspective, which of the following reasons for preserving biodiversity is most convincing? (A) Nonhuman species have roles in ecosystems and should be preserved to protect the ecosystems that support humans; or (B) nonhuman species have a fundamental right to existence. Explain your choice.

2. If a child asks you the following question 20 or 30 years from now, what will be your answer, and why? "When it became clear that humans were causing a mass extinction, what did you do about it?"

Answers to **Stop & Stretch, Visualize This, Savvy Reader,** and **Chapter Review** questions can be found in the **Answers** section at the back of the book.

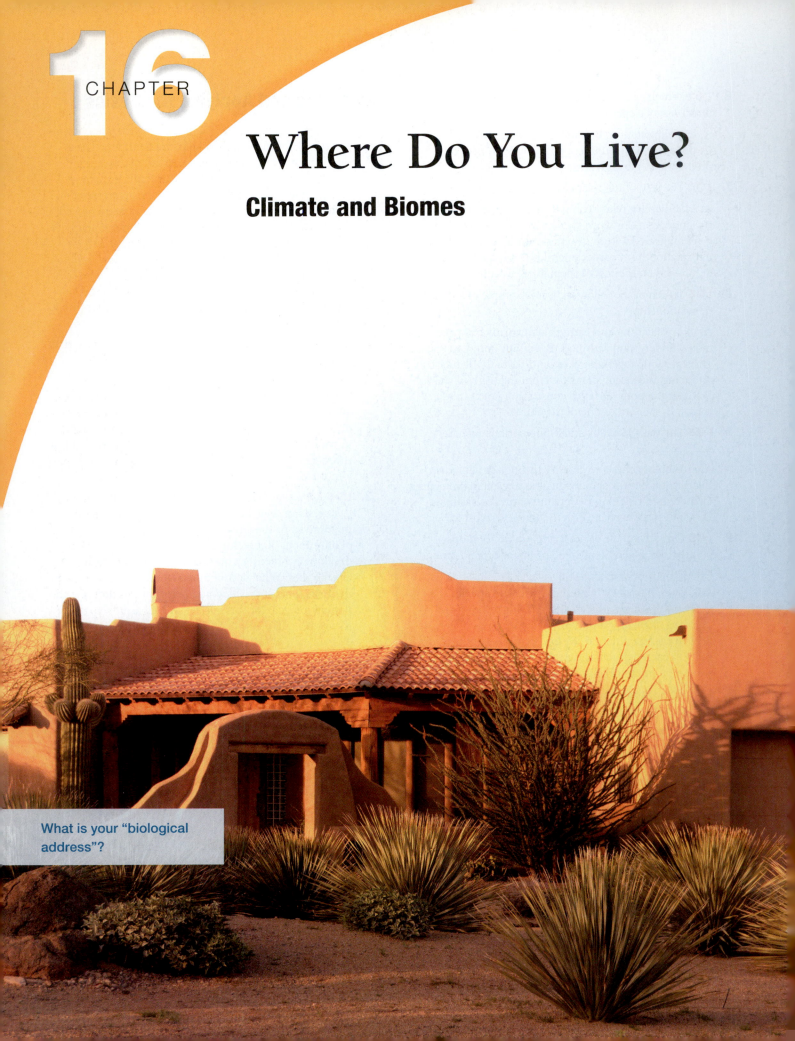

CHAPTER

16

Where Do You Live?

Climate and Biomes

What is your "biological address"?

How do you answer the question, "Where do you live?" Most of us would respond with a neighborhood or street address to someone from our community, a city or town name to someone from elsewhere in our state, or a state or country name to someone who lives far away. Not too many of us would give a reply such as "the Sonoran Desert" or "the boreal forest." However, descriptions of the natural environment in which we live can be thought of as our *bioregional* addresses. Our bioregional addresses include the native vegetation and the resident animals, fungi, and microbes that share, or once shared, our living space.

Who are your neighbors?

In this chapter, we explore factors that help to determine the qualities of a particular bioregional address by guiding you to learn more about your own natural neighborhood. See if you can answer the following questions that test your bioregional awareness:

- Is the native vegetation of the place where you live forest, grassland, or desert?
- What are the seasonal weather changes in your area?
- Can you describe the physical characteristics of three plant species that are native to your area?
- What is the largest mammalian predator that lives, or once lived, in the habitat that is now your neighborhood?
- Can you describe three native bird species that breed in your region?

It can be difficult to identify a biological address in a human-designed landscape.

Many Americans would have a difficult time answering at least some of these questions. But is it important to know the answers to questions like these?

You may have experienced one consequence of poor bioregional awareness— local water shortages occurring when large numbers of homeowners attempt to keep their thirsty lawns green during summertime droughts. Other

And it is sometimes difficult to determine where the resources on which we rely come from.

costs of lack of bioregional awareness may be more severe, including the consequences of home construction in areas where periodic fires or floods are common or on sandy coastlines, which are unstable and prone to storm damage.

Understanding one's own bioregion may allow people to build human settlements that are better for both humans and the natural environment. For example, in the southwestern desert regions of the United States, environmentally sensitive housing developers use xeriscaping, a kind of landscaping that involves native, drought-tolerant plants. Xeriscaping not only prevents the overconsumption of water but also provides a habitat for resident wildlife. In the Midwest, homeowners can support the organisms of their bioregion by replacing their lawn with prairie plants.

Another aspect of bioregional awareness includes understanding how our activities fit into natural cycles. Human populations, like all natural populations, require resources and produce waste. Surprisingly few Americans have an understanding of where our resources come from and where our waste goes. See if you do by considering the following questions:

- What body of water serves as the source of your tap water?
- What primary agricultural crops are produced in your area?
- How is your electricity generated?
- Where does your garbage go?
- How is the waste handled when you flush the toilet or pour liquids down the drain?

Why is knowing the answers to these questions important? Humans remain dependent on the natural world for our resources and waste disposal. If we don't understand how our human communities fit into the surrounding biological community, the results can include air and water pollution that harms our biological neighbors—and ourselves. In this chapter, we explore how the ecology of a bioregion intersects with the biology of human habitats.

16.1 Global and Regional Climate

Why is Buffalo, New York, so much snowier than frigid Winnipeg, Manitoba? Why is Miami, Florida, hot and humid while Tucson, Arizona, is hot and dry? Why does India experience monsoons? Why is the weather on tropical Pacific islands seemingly always beautiful?

The answers to all of these questions require an understanding of **climate,** the average conditions of a place as measured over many years. Climate is different from **weather,** which is the current conditions in terms of temperature, cloud cover, and **precipitation** (rain or snowfall). Simply put, weather information about a region will tell you if you have to shovel snow tomorrow morning, while climate information will tell you if you even need to own a snow shovel.

In general, temperatures are warmer in the tropics than in the Arctic and Antarctic, and at altitudes close to sea level compared to high altitudes. Variation in temperature throughout the year correlates to geographic location as well—the closer one is to the Arctic or Antarctic, the more variation, or seasonality, one experiences.

On a global scale, rain and snowfall patterns are more variable than temperature patterns. There is a wide region of high rainfall in the tropics, while north and south of this point rainfall diminishes considerably and Earth's great deserts can be found. The same pattern is found when comparing the wetter temperate zones to the drier polar regions. **Figure 16.1** summarizes global temperature and precipitation patterns.

The average temperature of any region is determined by the amount of **solar irradiance** it receives—that is, the total solar energy per square meter of land or water surface. Locations that receive large amounts of solar irradiance in a year have a higher average temperature than places receiving small amounts.

Earth's axis is roughly perpendicular to the flow of energy from the sun. The extremes of this axis are called **poles,** while the circle around the planet that is equidistant to both poles is called the **equator.** The amount of solar irradiance varies between these regions because of the planet's spherical shape. **Figure 16.2** illustrates two identical streams of solar energy flowing from the sun.

Latitude and wind patterns affect climate on a very broad scale.

Figure 16.1 Global climate patterns. Average temperature and precipitation in a region are fairly predictable based on the region's distance from the equator.

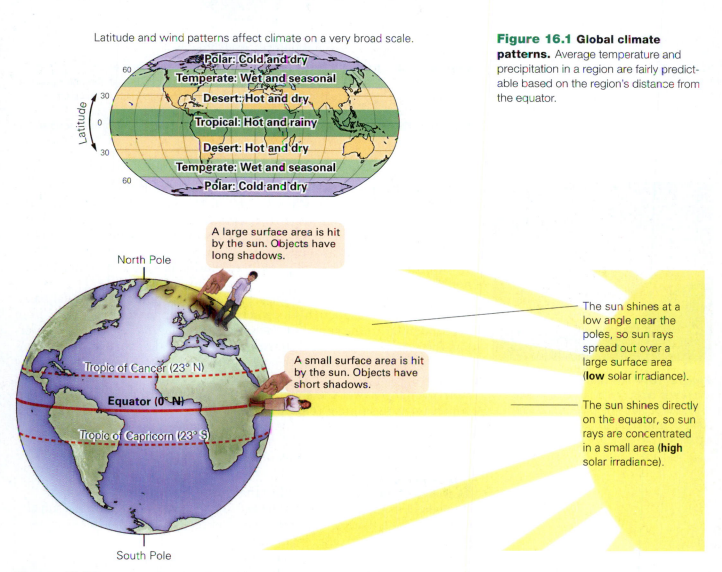

Figure 16.2 Solar irradiance on Earth's surface. The annual average temperature in a location on Earth's surface is most directly determined by its solar irradiance. Areas near the equator receive the greatest amount of solar energy, while areas near the poles receive the least.
Visualize This: How does solar irradiance, and thus climate, differ between Florida and Vermont?

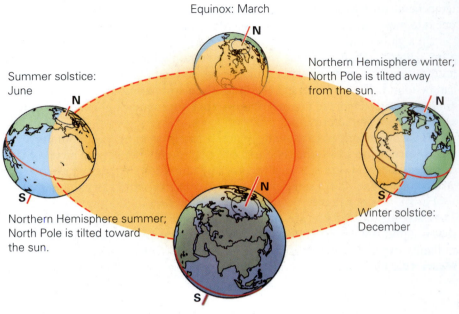

Equinox: March

Summer solstice: June

Northern Hemisphere winter; North Pole is tilted away from the sun.

Northern Hemisphere summer; North Pole is tilted toward the sun.

Winter solstice: December

Equinox: September

Figure 16.3 Earth's tilt leads to seasonality. Because Earth's axis is 23.5° from perpendicular to the rays of the sun, as Earth orbits the sun during a year, the poles appear to move toward and away from the sun. Solar irradiance is increased at and near a pole when it tilts toward the sun and is decreased when that pole tilts away.
Visualize This: Why is it summer in Australia when it is winter in North America?

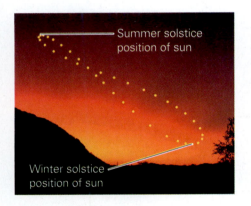

Summer solstice position of sun

Winter solstice position of sun

Figure 16.4 The sun's travels. This image was produced by taking a picture of the sun at the same location and time approximately once each week throughout a year.

One strikes Earth's surface directly at the equator, while the other strikes the surface closer to the pole, where Earth's surface is "curving away" from the sun. The difference in geometry means that the surface area warmed at the equator is much smaller than the surface area warmed at the poles. In other words, the solar irradiance (energy per area) is greatest at the equator and lowest near the poles. This is why southern Florida has a warmer climate than northern Vermont.

Solar irradiance also varies in a particular location on an annual basis. This occurs because Earth's axis actually tilts approximately 23.5° from perpendicular to the sun's rays (**Figure 16.3**). Due to this tilt, as Earth orbits the sun, the Northern Hemisphere (north of the equator) is tilted toward the sun during the northern summer and away from the sun during the northern winter. The tilt also helps explain why the position of the sunrise changes over the course of a year, moving from south to north as winter turns to summer (**Figure 16.4**).

Stop & Stretch There is evidence that Earth has "wobbled" on its axis throughout its history, such that the axis is more perpendicular to the sun at some times and tilted at a greater angle other times. How would a reduction (or increase) in the tilt affect seasonality?

An Overview: Global Temperature and Precipitation Patterns

Energy from the sun is the primary driver of precipitation; that is, rain and snowfall. To understand how sunlight causes rainfall, we must first understand some of the properties of water vapor. For water to remain as vapor, condensation must be less than evaporation: The number of water molecules that clump together to form liquid droplets must be fewer than the number of water molecules that escape from droplets.

The rate of evaporation depends on temperature; at high temperatures, the evaporation rate is high, and the reverse is true at low temperatures. Thus, when air cools and the rate of evaporation decreases, water molecules clump into larger and larger droplets. When the droplets are large enough, concentrations of them can be seen as clouds. As clouds grow even larger, droplets can become heavy enough to fall as rain. If the temperature inside the cloud is cold enough, droplets will freeze into ice crystals, which may fall as snow. The water cycle, which is the relationship between liquid water, water vapor, and rainfall, is illustrated in **Figure 16.5**.

Figure 16.5 The water cycle. Water moves from the oceans and other surface water to the atmosphere and back, with stops in living organisms, underground pools and soil, and ice caps and glaciers on land.

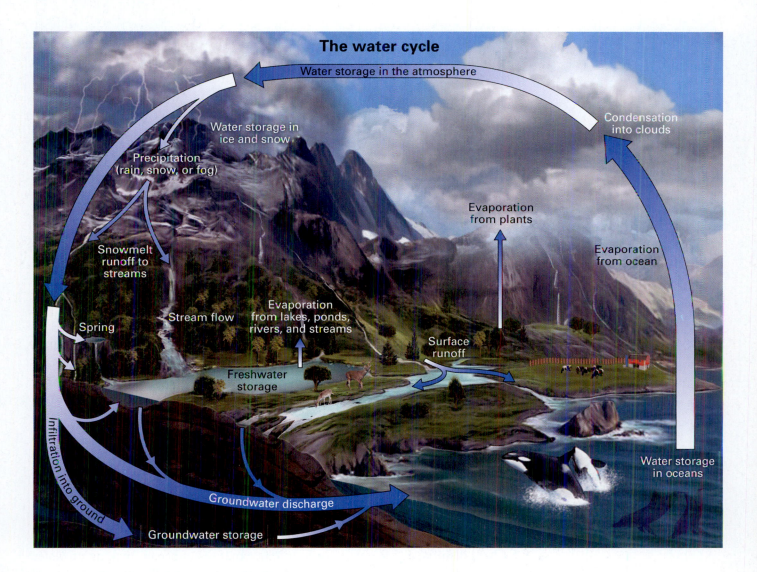

The water cycle

Figure 16.6 Global wind patterns.
High levels of solar irradiance at the solar equator lead to high levels of evaporation and rainfall. This phenomenon drives massive movements of air near the tropics into the temperate zones.

Visualize This: The Doldrums and the Horse Latitudes are regions where there is little wind and thus where sailing ships are likely to stall. Based on these global wind patterns, where do you think these places are on the globe?

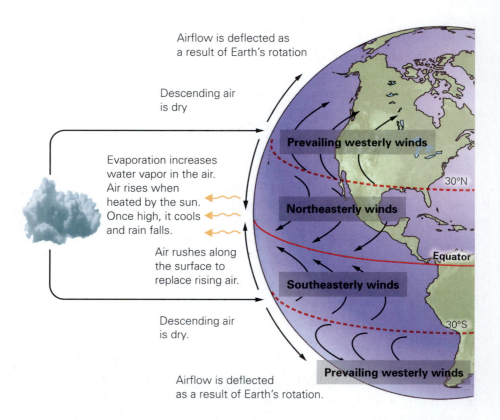

Airflow is deflected as a result of Earth's rotation

Descending air is dry

Evaporation increases water vapor in the air. Air rises when heated by the sun. Once high, it cools and rain falls.

Air rushes along the surface to replace rising air.

Descending air is dry.

Airflow is deflected as a result of Earth's rotation.

Prevailing westerly winds

Northeasterly winds

Southeasterly winds

Prevailing westerly winds

30°N

Equator

30°S

Rainfall patterns result from the air cycling near Earth's surface to high in the atmosphere and back down, as illustrated in **Figure 16.6**. Where solar irradiation is highest, at or near the equator, air temperatures rise quickly during the day. Because hot air is less dense than cold air, air at the equator rises. This leaves an area of low air pressure near Earth's surface, called the intertropical convergence zone, that is filled by breezes blowing from the north and south.

As the air rises, it cools. Water vapor condenses to form clouds and falls to Earth as rain. The now-dry air flows in the upper atmosphere toward the poles, where pressure is lower because of the airflow in the lower atmosphere. The once-warm, wet air mass continues to cool and release water and finally drops back to Earth's surface at about 30° north and south latitudes. This very dry falling air displaces the ground-level air at these latitudes, and that ground-level air, having picked up moisture from the surface, flows toward the poles.

As the air masses drop, they are affected by Earth's rotation and thus deflect to the east or west. The movements of these vast air masses create the prevailing winds in various regions of the globe. The pattern of air movement in the atmosphere helps explain the band of rain forests near the equator, the great deserts at 30° north and south of the equator, and the tendency for weather patterns in North America to come from the west.

Global rainfall patterns exhibit seasonality as well. The area of maximum solar irradiance travels from 23.5° north of the equator to 23.5° south over the course of a year due to the tilt in Earth's axis. Therefore, the intertropical convergence zone also moves, creating distinct rainy seasons wherever it is located. Rainy seasons occur in desert areas when the movement of the intertropical convergence zone results in changes in the prevailing winds. The rainy monsoon seasons in India and southern Arizona are both associated with these wind shifts, as breezes move over long expanses of ocean, picking up water vapor that falls on land as rain. **See the next section for A Closer Look at regional seasons and local climate.**

A Closer Look:
Regional Seasons and Local Climate

Both global and regional factors influence climate locally or in a particular bioregion.

The Solstices and the Seasons

Solar irradiance is at its annual maximum in the Northern Hemisphere during the summer **solstice,** when the sun reaches its northern maximum and the North Pole is tilted closest to the sun. The length of day is also greatest at the summer solstice and least at the winter solstice. For example, in Chicago, the time from sunrise to sunset is approximately 15 hours at the summer solstice and 9 hours at the winter solstice. The closer a region is to a pole, the greater the variance in day length over the course of a year. So, while day length in Chicago varies by 6 hours from winter to summer, in Fairbanks, Alaska, it varies by 18 hours: from less than 4 hours at the winter solstice to nearly 22 hours at the summer maximum.

Although the summer solstice is the day of highest solar irradiance in the Northern Hemisphere, it is not the warmest day of the year. Instead, the warmest days of the year are about 1 month later. This lag occurs because Earth converts the energy from the sun to heat and releases it gradually into the atmosphere in a phenomenon known as thermal momentum. You can think of the solar heat stored in Earth as a bank account. Heat is always dissipating, much like money being withdrawn regularly from a bank account. If money is added to a bank account at the same rate as it is withdrawn, the balance remains the same. Similarly, if heat is added to a location on Earth's surface at the same rate that it is dissipating, the temperature there remains constant. If more money is deposited into the bank than is withdrawn, the account balance increases. Likewise, as day length increases, the amount of heat deposited is greater than the amount withdrawn—that is, the temperature starts to rise. On the longest day of the year, the amount of heat deposited is greatest, but even after that day, the heat deposits are greater than the withdrawals, and the temperature continues to increase. Similarly, the coldest days of the year are about a month after the winter solstice—in late January or early February—because withdrawals of heat from Earth's surface remain larger than deposits for several weeks (**Figure 16.7**).

The same analogy can help explain why the warmest part of the day is in the middle-to-late afternoon, even though the solar irradiance is greatest at noon when the sun is directly overhead. In the first few hours of the afternoon, heat gain remains greater than heat loss, increasing temperature.

Because day length over the course of a year changes more dramatically near the poles than near the equator, the temperature also changes more dramatically closer to the poles. The amount of temperature variance explains why seasonal temperature swings are more pronounced closer to the poles than

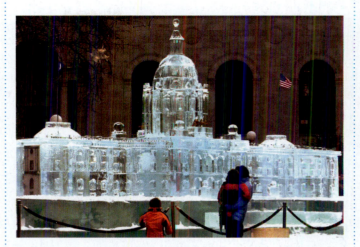

Figure 16.7 Thermal momentum. The winter carnival in St. Paul, Minnesota, is traditionally scheduled for the coldest weeks of the year—at the end of January—both to ensure the survival of the ice sculptures and to celebrate the year's turning point toward the warm summer.

near the equator. The average low temperature for January in Tampa, Florida, is 10°C (50°F), and the average high in July is 32°C (90°F), a difference of 22°C (40°F). On the contrary, the January low in Missoula, Montana, is –10°C (14°F), and the July high is 29°C (84°F), a difference of 39°C (70°F).

Local Influences on Climate

Three characteristics of a location's setting have an effect on its temperatures and precipitation: (1) altitude, (2) the proximity of a large body of water, and (3) characteristics of the land's surface and vegetation (**Figure 16.8** on the next page).

Temperature drops as altitude increases because as gases rise over hundreds of meters, the molecules move away from each other, reducing heat content. Temperature differences due to altitude are dramatic; the summit of Mt. Everest, 8.8 km (29,035 ft) above sea level, averages –27°C (–16°F), while nearby Kathmandu, Nepal, at 1.3 km (4385 ft), averages 18°C (65°F). However, smaller differences in altitude within a region have a converse effect on air temperature. Because cold air masses are denser than warm air, pockets of cold air (for instance, air that has been in shade and not exposed to solar radiation) tend to "drain" to the lowest point on a landscape. Thus, valleys will often be colder than nearby hilltops.

Temperatures in areas near oceans, seas, and large lakes are influenced by the thermal properties of water, including great ability to store heat and a high heat of vaporization (Chapter 5). Water temperature rises and falls slowly in response to solar irradiation when compared to the rate of temperature change

(continued on the next page)

(A Closer Look continued)

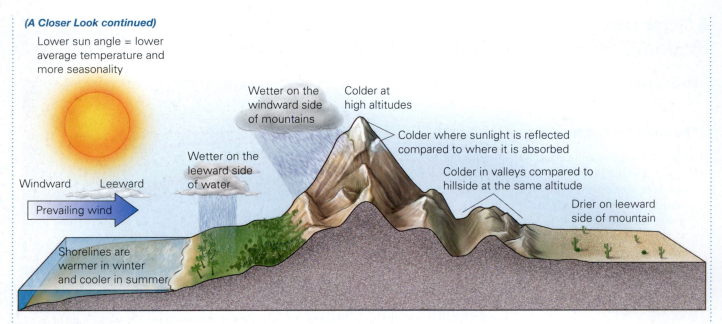

Lower sun angle = lower average temperature and more seasonality

Wetter on the windward side of mountains

Colder at high altitudes

Colder where sunlight is reflected compared to where it is absorbed

Colder in valleys compared to hillside at the same altitude

Wetter on the leeward side of water

Windward   Leeward

Prevailing wind

Drier on leeward side of mountain

Shorelines are warmer in winter and cooler in summer

Figure 16.8 Factors that influence local climate. Local temperature and rainfall patterns result from global climate patterns moderated by several aspects of local geography.

of land surfaces. Thus, air over a large body of water is comparatively cooler in summer and warmer in winter (**Figure 16.9**), and nearby land areas experience more moderate temperatures than do regions further inland. Because of this phenomenon, the growing season—the time at which temperatures are amenable for plant growth—on the shores of the Niagara peninsula in Canada, bordered by Lake Erie and Lake Ontario, is as much as 30 days longer than areas just a few miles inland from the lakes. Oceans also have an effect on local climates because heat is transferred through them via currents. The Gulf Stream is a current that carries water from the tropical Atlantic Ocean to the shores of northern Europe, producing a much milder climate in those regions than in other areas at the same distance from the equator. The warmth of the Gulf Stream makes Dublin, Ireland, as warm as San Francisco, even though Dublin is 1600 km (1000 mi) closer to the North Pole.

The amount of light absorbed or reflected by the surface of the land will also influence surrounding air temperature. A surface that reflects most light energy will have a lower nearby air temperature compared to a surface that absorbs most of that light energy, heats up, and radiates that heat into the air.

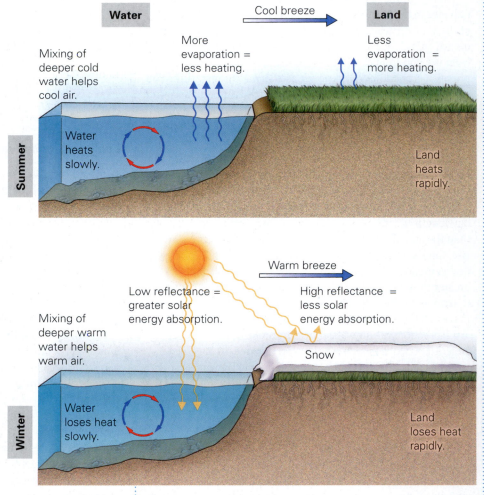

Water   Cool breeze   **Land**

Summer

Mixing of deeper cold water helps cool air.

More evaporation = less heating.

Less evaporation = more heating.

Water heats slowly.

Land heats rapidly.

Winter

Mixing of deeper warm water helps warm air.

Low reflectance = greater solar energy absorption.

High reflectance = less solar energy absorption.

Warm breeze

Snow

Water loses heat slowly.

Land loses heat rapidly.

Figure 16.9 The moderating influence of water. Because water heats slowly, winds blowing across water in spring will cool nearby landmasses. Conversely, because water loses heat more slowly than land, it is a source of warmer breezes in fall and winter.
Visualize This: Consider this figure along with the concept in Figure 16.7. Why is late July through mid-August traditionally the most popular time for people to visit the seashore?

Snow reflects more light than a forest does, so air over a snow-pack remains cold. The low reflectance of asphalt pavement and most building materials contributes to the urban heat island effect, which is the tendency for cities to be from 0.5° to 3°C (1° to 6°F) warmer than the surrounding areas (**Figure 16.10**).

Stop & Stretch One possible consequence of global warming is a decrease in the overall ice cover at the North Pole. Less ice not only is an effect of warming but also is likely to increase the rate of warming. Why?

Cities are also warmer because they contain little vegetation, which tends to reduce air temperature. Much of the solar energy absorbed by plants converts liquid water inside the plant to vapor, which also prevents the energy from being converted to heat.

Less vegetation = less energy used to evaporate water, and more energy converted to heat

Air conditioning = heat vented into outside areas

Dark surfaces = solar energy absorbed rather than reflected

Fossil fuel use = waste heat produced

Figure 16.10 Urban heat island. Urban areas are significantly warmer than nearby rural areas for a variety of reasons.

Local Precipitation Patterns

As with temperature patterns, the amount of rain or snow-fall experienced in any location on Earth is a product of both global and local factors. The amount of precipitation that falls in a given land region is highly dependent on the context of that area—in particular, the proximity of the land to a large body of water. Wind blowing across warm water accumulates water vapor that condenses and falls when it reaches a cooler landmass. Communities surrounding the Great Lakes provide a dramatic example of this effect. Because the prevailing winds are from the northwest, areas immediately to the south and east of these lakes accumulate much more snow than do regions on the northern and western sides. For example, Toronto, on the northwest side of Lake Ontario, averages about 140 cm (55 in) of snow per year, while Syracuse, New York, on the southeast side averages almost twice that—274 cm (108 in).

Precipitation amounts are also affected by the presence of mountains or mountain ranges. When an air mass traveling horizontally approaches a mountain, it is forced upward. Cooling as it moves upward, the water vapor within it condenses to form clouds. Rain or snow then falls on the windward side of the mountain. Warming again as it drops down the other side, the dry air mass causes water to evaporate from land on the sheltered, or leeward, side of the mountain. The dry area that results is often referred to as the mountain's "rain shadow." The Great Basin of North America, encompassing nearly all of Nevada and parts of Utah, Oregon, and California, is a desert because of the rain shadow cast by the Sierra Nevada.

16.2 Terrestrial Biomes

Climate plays the greatest role in determining the characteristics of a **biome,** a geographic area defined by its primary vegetation. Plants (and animals) native to a region are adapted to the water availability and temperatures experienced there. In general, the size of the vegetation is limited by water availability—large trees require large amounts of water. Water availability is a function of total precipitation, but it is also influenced by temperature; frozen water cannot be taken up by plants.

Four basic terrestrial, or land, biomes are typically recognized: forest, grassland, desert, and tundra. Each of these basic biomes may contain variants within them; for instance, a grassland may be prairie, steppe, or savanna. The relationship between climate and biome type is illustrated in **Figure 16.11** (on the next page), and the characteristics of the terrestrial biomes are summarized in **Table 16.1**.

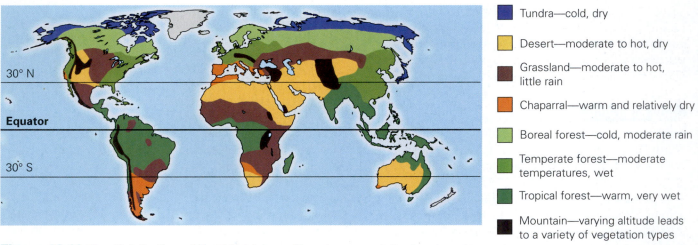

Tundra—cold, dry

Desert—moderate to hot, dry

Grassland—moderate to hot, little rain

Chaparral—warm and relatively dry

Boreal forest—cold, moderate rain

Temperate forest—moderate temperatures, wet

Tropical forest—warm, very wet

Mountain—varying altitude leads to a variety of vegetation types

Figure 16.11 The distribution of Earth's biomes. The primary vegetation type in a given area is determined by the region's climate.
Visualize This: Why is the east side of Australia tropical forest, while the west side is desert?

TABLE 16.1

Terrestrial biomes.

Biome		Characteristics	Location
Forests		**Dominated by woody plants**	
	Tropical	Tall trees, relatively little understory. Great diversity of species.	Around the equator: Central and South America, central Africa, India, southeast Asia, and Indonesia
	Temperate	Three distinct layers: deciduous canopy trees, shrubs, and spring-flowering plants in the understory.	Middle latitudes of the Northern Hemisphere: eastern North America, western and central Europe, and eastern China
	Boreal	Dominated by evergreen coniferous trees with a very short growing season.	Northern latitudes and high altitudes: northern North America, northern Europe and Asia, and mountains in the western United States

TABLE 16.1

Terrestrial biomes.

Biome		Characteristics	Location
Chapparal		Dominated by spiny ever-green shrubs and maintained by frequent fires.	Areas surrounding the Mediterranean Sea and in patches in southern California, South Africa, and southwest Australia
Grasslands		**Dominated by nonwoody grasses with few or no shrubs or trees**	
Savannas		Scattered individual trees, maintained by periodic fires or large-animal grazing.	Tropical: About half of Africa, as well as large areas of India, South America, and Australia
Prairie and Steppe		Tall (prairie) or short (steppe) grasses, no woody plants.	Temperate areas in the middle of large landmasses: central North America, central Asia, parts of Australia, and southern South America
Desert		Dry soils, very sparse vegetation. Plants often covered with spines and adapted to store water.	Near 30° north or south of the equator: northern Africa (Sahara), central Asia (Gobi), the Middle East, central Australia, and the southwestern United States
Tundra		Very short growing season, permafrost. Plants are low to the ground and many animals migrate to warmer climates during the winter season.	Near the poles and at high altitudes

Figure 16.12 Life in the canopy.
This bromeliad, often found more than
60 m from the forest floor, is vase shaped
as an adaptation to acquire water and
nutrients from rainfall.

**Figure 16.13 Loss of tropical
forest.** The amount of tropical forestland
converted to agriculture is greater now than
at any other time in human history.

Forests and Shrublands

Forests occupy approximately one-third of Earth's land surface and, when all
forest-associated organisms are included, contain about 70% of the **biomass,**
the total weight of living organisms, found on land. Forests are generally categorized into three groups based on their distance from the equator: (1) tropical
forests at or near the equator, (2) temperate forests from 23.5° to 50° north and
south of the equator, and (3) boreal forests close to the poles.

Tropical Forests. Extensive areas of tropical forest were once found throughout Earth's equatorial region. Tropical forests contain a large amount of biological diversity; 1 hectare (10,000 m^2) may contain as many as 750 tree species.

One hypothesis regarding why tropical forests are so rich in species is the
high solar irradiance they experience. Because the energy level is high, populations of many species can be supported in a relatively small area. Think of the
available energy as analogous to a pizza—the larger the pizza, the greater the
number of individuals who can be fed to satisfaction.

High energy and water levels also support the growth of very large trees in
tropical forests. As a result, most of the sunlight is absorbed by vegetation before
it hits the ground, and most living organisms are therefore adapted to survive
high in the treetops. Tropical forest animals are able to fly, glide, or move freely
from branch to branch, while small plants are able to obtain nutrients and water
while living on the upper branches of these huge trees (**Figure 16.12**).

With warm temperatures and abundant water, the process of **decomposition,**
the breakdown of waste and dead organisms, is rapid in tropical forests. Dense vegetation quickly reabsorbs the simple nutrients that are produced by decomposition,
resulting in relatively little organic matter stored in the soil. Consequently, when
vegetation is cleared and burned, the ash-fertilized soils can support crop growth
for only 4 to 5 years. Once its soil is depleted of nutrients, a farm field carved out
of tropical forest is abandoned, and a new field is cleared using the same method.

Among human populations in tropical forest areas, this slash-and-burn
(or swidden) agricultural system is common. Tropical forests appear able to
support swidden agriculture for many generations if the human population is
small enough and abandoned plots have several decades to recover. However,
increasing human population levels in tropical countries may be too large to
allow adequate recovery time, and road building into interior forests has exposed untouched wilderness to swidden fields. The resulting **deforestation**—
literally, the removal of forests—endangers thousands of organisms that depend on this biome for survival (**Figure 16.13**).

Temperate Forests. Some areas of tropical forest may demonstrate seasonal
changes between annual wet and dry periods. But most people associate major
seasonal change with forests in temperate areas, where winter temperatures
drop below the freezing point of water.

In temperate forests, water is abundant enough during the growing season
to support the growth of large trees, but cold temperatures during the winter
limit photosynthesis and freeze water in the soil. Trees with broad leaves can
grow faster than narrow-leaved trees, so they have an advantage in the summer. However, broad leaves allow tremendous amounts of water to evaporate
from a plant, leading to potentially fatal dehydration when water supplies are
limited. To balance these two seasonal challenges, most broad-leaved trees in
temperate forests have evolved a **deciduous** habit, meaning that they drop their
leaves every autumn. In preparation for shedding its leaves, a deciduous tree
reabsorbs its chlorophyll, the green pigment essential for photosynthesis. The
colorful fall leaves that grace eastern North America in September and October
result from secondary photosynthetic pigments and sugars that are left behind
in the leaf after the chlorophyll is reabsorbed.

Stop & Stretch Suggest an explanation for why deciduous trees reabsorb chlorophyll, but not all pigments, from their leaves before shedding them. Hint: chlorophyll contains significantly more nitrogen than most other pigments.

The short time lag between the onset of warm temperatures and the releafing of deciduous trees provides an opportunity for plants on the temperate forest floor to receive full sunlight. Spring in these forests triggers the blooming of wildflowers that flower, fruit, and produce seeds quickly before losing light and water to their towering companions (**Figure 16.14**). The thinner leaves of deciduous trees allow more sunlight through than do their tropical forest equivalents, permitting a shrub layer typically missing in the tropics. The forest floor and shrub layer are collectively known as the forest **understory**. In addition, animals in temperate forests are more evenly distributed throughout the forest rather than concentrated in the treetops.

Nearly all of the forested lands in the eastern United States were converted to farmland within 100 years of the American Revolution. However, in the late nineteenth and early twentieth centuries, farms in the eastern United States were abandoned as production switched to the south and west. The regrowth of these forests provided a unique opportunity for ecologists to examine **succession**, the progressive replacement of different suites of species over time. Succession in most forests follows a predictable pattern. The first colonizers of vacant habitat are fast-growing plants that produce abundant, easily dispersed seeds. These pioneers are replaced by plants that grow more slowly but are better competitors for light and nutrients. Eventually, a habitat patch is dominated by a set of species—called the climax community—that cannot be displaced by other species without another environmental disturbance (**Figure 16.15**).

Today, these resurgent forests are once again threatened—this time by expanding urban development. The World Wildlife Fund estimates that worldwide, only 5% of temperate deciduous forests remain relatively untouched by humans.

Figure 16.14 Springtime in a temperate forest. Forest wildflowers are adapted to take advantage of a short period between snowmelt and tree leafing by emerging from the soil, flowering, and producing fruit rapidly.

(a)

(b)

(c)

Figure 16.15 Forest succession in the eastern United States.
(a) Abandoned farms are quickly colonized by fast-growing herbs. (b) The second stage of succession consists of a dense landscape of shrubs and quickly growing trees. (c) The climax community in the northeastern United States is made up of beech and maple trees.

Figure 16.16 Coniferous trees. Coniferous trees in the boreal forest are evergreen, an adaptation that allows them to dominate this biome. These plants produce sperm and eggs on cones instead of in flowers, which may be an advantage in cold climates because they do not rely on animal pollinators.

Boreal Forests. Coniferous plants that produce seed cones instead of flowers and fruits dominate boreal forests (**Figure 16.16**). In fact, boreal forests are the only land areas where flowering plants are not the dominant vegetation type.

Climate conditions for boreal forests include very cold, long, and often snowy winters and short, moist summers. The dominant conifers in these forests are evergreens, meaning that they maintain their leaves throughout the year. Coniferous tree leaves are needle shaped and coated with a thick, waxy coat—both adaptations to reduce water loss during winter. The evergreen habit likely explains conifers' dominance over flowering trees in boreal forests—the growing season is so short that the ability to begin photosynthesizing immediately after thawing gives conifers an advantage over faster-growing but slower-to-start deciduous trees.

Because they occupy regions with such cold winters and short summers, boreal forests tend to be lightly populated by humans. These areas represent some of the "wildest" landscapes on Earth—home to moose, wolves, bobcat, beaver, and a surprising diversity of summer-resident birds. Trees in these landscapes are valuable for both building materials and paper products, and logging in the boreal forest is extensive. There are increasing concerns that the boreal forest in North America is being cut at rates faster than it can be replaced.

Chaparral. One major biome dominated by woody plants is not a forest. This landscape is known as chaparral, and its vegetation consists mostly of spiny evergreen shrubs.

Long, dry summers and frequent fires maintain the shrubby nature of chaparral. In fact, chaparral vegetation is uniquely adapted to fire. Several species have seeds that will germinate only after experiencing high temperatures. Many chaparral plants have extensive root systems that quickly resprout after aboveground parts are damaged. Chaparral will grow into temperate forest when fire is suppressed. As a result, natural selection has favored chaparral shrubs that actually encourage fire—such as rosemary, oregano, and thyme, which contain fragrant oils that are also highly flammable.

Fire can be an important contributor to the structure and function of many biomes, including chapparal. In Southern California, the flammability of chaparral vegetation has come directly into conflict with rapid urbanization. In recent years, this area has experienced extensive and expensive fire seasons—for example, in 2009, fires near Los Angeles burned nearly 400,000 acres and cost almost $100 million. In response to these events, support has increased for a policy of suppressing fires immediately rather than letting them burn. In fact, such a strategy may contribute to the buildup of fuel and more intense fires that are destructive to both people and the wildlands. Fire policy in Southern California along with the long-term human modification of chaparral around the Mediterranean makes this biome one of the most threatened on the planet.

Grasslands

Grasslands are regions dominated by nonwoody grasses and containing few or no shrubs or trees. These biomes occupy geographic regions where precipitation is too limited to support woody plants. Grasslands can be further categorized into tropical and temperate categories.

Tropical grasslands are known as **savannas** and are characterized by the presence of scattered individual trees. Savannas are maintained by periodic fires or clearing. In regions where wet and dry seasons are distinct, yearly grass fires during the dry season kill off woody plants; in regions where elephants and other large grazing animals are present, the damage these animals do to trees helps to maintain the grass expanse (**Figure 16.17**).

Figure 16.17 Savanna maintenance. Elephants severely prune acacia trees when feeding on them, keeping the size of the trees relatively small; they also actively destroy thorny myrrh bushes that they do not eat, which provides additional habitat for grass to grow.
Visualize This: Why aren't domestic cattle as effective as elephants at maintaining savanna?

Because grazing animals eat the tops of plants, natural selection in these environments has favored plants, like grasses, that keep their growing tip at or below ground level. Grass grows from its base, so grazing (or mowing) is equivalent to a haircut—trimming back but not destroying. (Woody plants grow from the tips, so intense grazing can destroy these plants.)

Grass leaves are very fibrous and thus difficult to digest. The most common adaptation of animals that use grass as food is a mutualism with fiber-digesting bacteria. The bacteria live in a specialized chamber of the grazer's digestive system and partially digest the grass so that more nutrients are available to the animal. Although grazing mammals are characteristic of savannas, the introduction of large numbers of cattle and other grazers in these habitats has transformed the landscape from grassland to bare, sandy soil in a process known as **desertification**. In the Sahel region of central Africa, overgrazing and resulting desertification has destroyed thousands of square kilometers of savanna.

Temperate grassland biomes include tallgrass **prairies** and shortgrass **steppes.** Generally, the height of the vegetation corresponds to the precipitation—greater precipitation can support taller grasses. These landscapes are generally flat or slightly rolling and contain no trees.

In the cooler temperate regions, decomposition is relatively slow, and the soil of prairies and steppes is rich with the remains of partially decayed plants. These soils provide an excellent base for agriculture. As a result, most native prairies and steppes have been plowed and replanted with crops. In North America, less than 1% of native prairie remains.

Desert

Where rainfall is less than 50 cm (20 in) per year, the biome is called **desert.** Most of these deserts are close to 30° north or 30° south of the equator, where the air masses that were "wrung" of water in the rain-forest regions around the equator fall back to Earth's surface as hot and dry winds.

Although the image of a desert is often hot and sandy, some deserts can be quite cold; most deserts have vegetation, although it can be sparse. Plants and animals in desert regions have evolutionary adaptations to retain and conserve water. For example, plants are often thickly coated with waxes to reduce evaporation, contain photosynthetic adaptations that reduce water loss through leaf pores, and may be protected from predators of their stored water by spines and poisonous compounds (**Figure 16.18**).

Deserts are also home to many flowering plants that complete their entire life cycle from seed to seed in a single season. In deserts, the wet season is quite short; many desert-adapted plants can germinate, flower, and produce seeds in a matter of 2 or 3 weeks. The seeds they produce are hardy and adapted to survive in the hot, dry soil for many years until the correct rain conditions return.

Some animals in these dry environments have physiological adaptations that allow them to survive with little water intake. The most amazing of these animals are the various species of kangaroo rat, which apparently never consume water directly. Instead, they conserve water produced during the chemical reactions of metabolism and have kidneys that produce urine four times more concentrated than our own.

The appeal of the sunny, warm, and dry climate of the deserts of the southwestern United States has made this region one of the fastest-growing areas in the country. The increasing population of the desert Southwest is putting stress on water supplies, causing conflicts, and depleting water sources for native animals. The Colorado River is so extensively used that often it cannot sustain a flow through its historic outlet at the Gulf of California in Mexico. As a result, up to 75% of the species that lived in the Colorado River delta are now gone.

Figure 16.18 Desert plant adaptations to constant drought. Thick, whitish, and waxy coatings reflect light and reduce evaporation. Column-like forms reduce exposure to the high-intensity midday sun. Stems store water. The production of spines discourages predators from accessing their water or damaging their protective surface.

Tundra

The biome type where temperatures are coldest—close to Earth's poles and at high altitudes—is known as **tundra**. Here, plant growth can be sustained for only 50 to 60 days during the year, when temperatures are high enough to melt ice in the soil. High temperatures are not sustained long enough to melt all of the ice stored there. Therefore, places like the arctic tundra near the North Pole are underlain by **permafrost**, icy blocks of gravel and finer soil material. The permafrost layer impedes water drainage, and soils above permafrost are often boggy and saturated.

Plants in tundra regions are adapted to windswept expanses and freezing temperatures, often growing in mutualistic multispecies "cushions" where all individuals are the same height and shelter each other. This low vegetation supports a remarkably large and diverse community of grazing mammals, such as caribou and musk oxen, and their predators, such as wolves.

Animals in tundra regions have evolved to survive long winters with structural adaptations, such as storing fat and producing extra fur or feathers. Other animals, such as ground squirrels, have adapted by evolving hibernation; they enter a sleep-like state and reduce their metabolism to maximize energy conservation. Grizzly bears and female polar bears also spend many of the coldest months in deep sleep; although this is not true hibernation, these bears are so lethargic that females give birth in this state without fully awakening. Other animals survive by migrating south to avoid the hardships of a long, frigid winter. Hundreds of millions of birds migrate from the arctic tundra or other polar regions toward warmer climates and back again and thus take part in one of the great annual dramas of life on Earth (**Figure 16.19**).

Tundra is very lightly settled by humans, but it is threatened by our dependence on fossil fuels—oil, natural gas, and coal that formed from the remains of ancient plants. Some of the largest remaining untapped oil deposits are found in tundra regions, including northern Alaska, Canada, and Siberia, and the proposed development of these resources threatens thousands of square kilometers of tundra with destruction caused by road building, oil spills, and water

Figure 16.19 Migration. Birds that breed in the Arctic, such as these red knots massed on the beach at Delaware Bay, undertake a fatiguing and perilous journey to take advantage of the abundant daylight and insect life available during summer on the tundra at the tip of South America.

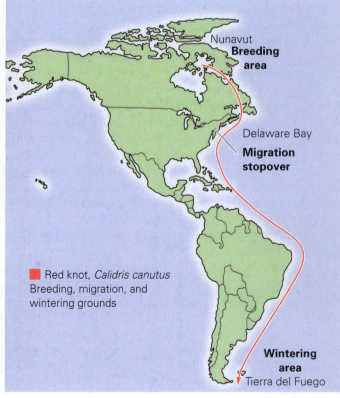

pollution. Fossil fuel use also contributes to global warming, which has had the greatest impact at the poles, where tundra is predominant. Winters in Alaska have warmed by 2° to 3°C (4°–6°F), whereas elsewhere they have warmed by about 1°C. As climate conditions change, areas that were once tundra have begun to support shrub and tree growth and are changing into boreal forest.

16.3 Aquatic Biomes

Nearly all human beings live on Earth's land surface, but most people also live near a major body of water and are both influenced by and have an influence on these **aquatic** systems. Aquatic biomes are typically classified as either freshwater or saltwater, and their characteristics are summarized in **Table 16.2.**

TABLE 16.2

Aquatic biomes.	
Biome	**Characteristics**
Freshwater	**Less than 0.1% total volume of salts**
Lakes and Ponds	Inland, some dry up seasonally. Yearly seasonal turnovers in larger lakes cause mixing of nutrients that would settle to the bottom.
Rivers and Streams	Flowing water moving in one direction. Different habitats along the way, from the fast-flowing, oxygen-rich regions at the headwaters to the sluggish, sediment-filled waters near the mouth.
Wetlands	Standing water with emergent plants. High species diversity due to high nutrient levels at water-land interface.

(continued)

TABLE 16.2 *(continued)*

Aquatic biomes.

Biome		Characteristics
Marine (Saltwater)		
Ocean		Contains multiple habitats, including intertidal zones, where organisms are exposed to fluctuating water levels; deep abyssal zones where no light penetrates; and open ocean.
Coral Reef		Tropical areas in relatively shallow water. Reefs are made up of the remains of coral animals and provide habitat for a diverse array of species.
Estuary		Where freshwater rivers flow into intertidal areas; mix of salt- and freshwater. Highly productive due to nutrients from the rivers and warmer temperatures.

Freshwater

Freshwater is characterized as having a low concentration of salts—typically less than 0.1% total volume. Scientists usually describe three types of freshwater biomes: lakes and ponds, rivers and streams, and wetlands.

Lakes and Ponds. Bodies of water that are inland, meaning surrounded by land surface, are known as lakes or ponds. Typically, ponds are smaller than lakes, although there is no set guideline for the amount of surface area required for a body of water to reach lake status. Some ponds, however, are small enough that they dry up seasonally. These vernal (springtime) ponds are often crucial to the reproductive success of frogs, salamanders, and a variety of insects that spend part of their lives in water.

(a) Lake in summer

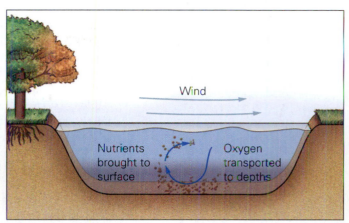

Adequate oxygen

Warm, less dense water

Low oxygen

Cold, denser water

Nutrients settle

(b) Lake in fall and spring

Wind

Nutrients brought to surface

Oxygen transported to depths

Figure 16.20 Nutrients in lakes. (a) A lake is stable during the summer, and oxygen levels in deeper regions diminish as nutrient levels near the surface decline. (b) Nutrients and oxygen are mixed throughout a lake during fall and spring "turnover," providing the raw materials for algae growth.

The aquatic environment of lakes and ponds can be divided into different zones: the surface and shore areas, which are typically warmer, brighter, and thus full of living organisms; and the deepwater areas, which are dark, low in oxygen, cold, and home to mostly decomposers. The biological productivity of lakes in temperate areas is increased by seasonal turnovers—times of the year when changes in air temperature and steady winds lead to water mixing, which redistributes nutrients from the bottom of the lake to the surface and carries fresh oxygen from the surface to deeper water (**Figure 16.20**).

Fertilizers applied to agricultural lands and residential lawns near lakes and ponds can increase their algae populations as these nutrients leach into the water. Ironically, too many nutrients, a process known as eutrophication, can lead to the "death" of these water bodies. Their degradation occurs because large populations of algae lead to large populations of microbial decomposers, which consume a large fraction of oxygen in the water, thereby suffocating the native fish (**Figure 16.21**).

The life in freshwater lakes may also be threatened by **acid rain**, precipitation that has a high acid content resulting from high levels of air pollution. In lakes with little ability to buffer pH changes, acid rain lowers the pH and causes sensitive species to die off. Antipollution legislation in the United States has reduced the impact of acid rain in the northeastern states, but acid precipitation remains a significant problem in countries with less strict pollution regulations.

Rivers and Streams. Rivers and streams are flowing water moving in one direction. These waterways can be divided into zones along their lengths: the headwaters near the source, the middle reaches, and the mouth where they flow into another body of water.

At the headwaters of a river—often by the source lake, an underground spring, or a melting snowpack—the water is clear, cold as a result of the temperature of the source, and generally flowing quickly because its volume is small; it is thus high in oxygen, providing an ideal habitat for cold-water fish such as trout (**Figure 16.22**). Near the middle reaches of a river, the width typically increases; thus, the flow rate slows. Also increasing is the diversity of fish, reptile, amphibian, and insect species that the river supports because the warming water provides a better habitat for their food source of photosynthetic plants and algae. At the mouth of a river, the speed of water flow is even

Figure 16.21 Eutrophication. High nutrient levels from fertilizer runoff lead to "blooms" of algae growth, severe oxygen depletion, and a risk of fish kills.

Figure 16.22 A river headwater. Near its source, for example, a cold mountain lake, a river is cold, fast flowing, and rich in oxygen.

slower, and the amount of sediment—soil and other particulates carried in the water—is high. High levels of sediment reduce the amount of light in the water and therefore the diversity of photosynthesizers that survive there. Oxygen levels are also typically lower near the mouth because the activities of decomposers increase relative to photosynthesis. Many of the fish found at the mouth of a river are bottom-feeders, such as carp and catfish, which eat the dead organic matter that flowing water has picked up along its way.

Rivers and streams are threatened by the same pollutants that damage lakes. But their habitats also face wholesale destruction with the development of dams and channels—dams that provide hydropower or reservoirs for cooling fossil-fuel-powered, electricity-generating plants, and channels that simplify and expedite boat traffic.

Wetlands. Areas of standing water that support emergent, or above-water, aquatic plants are called wetlands. In number of species supported, wetlands are comparable to tropical rain forests. The high biological productivity of wetlands results from the high nutrient levels found at the interface between aquatic and land environments.

Besides their importance as biological factories, wetlands provide health and safety benefits by slowing the flow of water. Slower water flows reduce the likelihood of flooding and allow sediments and pollutants to settle before the water flows into lakes or rivers.

Stop & Stretch How might the loss of wetlands contribute to the death of lakes and rivers due to eutrophication?

Since the European settlement of the continental United States, over 50% of wetlands have been filled, drained, or otherwise degraded. Although legislation passed in the last few decades has greatly slowed the rate of wetland loss in the United States, about 58,000 acres of this biome type are still destroyed around the country every year.

Saltwater

About 75% of Earth's surface is covered with saltwater, or **marine**, biomes. Marine biomes can be categorized into three types: oceans, coral reefs, and estuaries.

Oceans. The open ocean covers about two-thirds of Earth's surface but is the least well-known biome of all. In truth, it can be thought of as a shifting mosaic of different environments that vary according to temperature, nutrient availability, and depth of the ocean floor. About 50% of the oxygen in Earth's atmosphere is generated by single-celled photosynthetic plankton in the open ocean (**Figure 16.23**). The open ocean also generates most of Earth's freshwater because water molecules evaporating from its surface condense and fall on adjacent landmasses as rain and snow, which eventually flow into lakes, ponds, rivers, and streams.

Photosynthetic plankton serve as the base of the ocean's food chain, providing energy to microscopic animals called zooplankton, which in turn feed fish, sea turtles, and even large marine mammals such as blue whales. These predators provide a source of food for yet another group of predators, including sharks and other predatory fish, as well as ocean-dwelling birds such as albatross.

Figure 16.23 The ocean's oxygen factories. These phytoplankton are single-celled algae that are abundant in the ocean's surface waters. Their photosynthesis provides 70% of Earth's oxygen, as well as food to support an enormous diversity of life in the ocean.

Unlike lakes, oceans experience tides—regular fluctuations in water level caused by the gravitational pull of the moon. As a result of tides, oceans contain unique habitats known as **intertidal zones**, which are underwater during high tide and exposed to air during low tide. Organisms in intertidal zones must be able to survive the daily fluctuations and rough wave action they experience. Adaptations such as strong anchoring structures, burrow building, and water-retaining, gelatinous outer coatings allow animals and seaweeds to take advantage of the high-nutrient environment found along the shore.

Oceans also contain a habitat known as the abyssal plain, the deep ocean floor. In these areas, sunlight never penetrates, temperatures can be quite cold, and the weight of the water above creates enormous pressure. Once thought to be lifeless because of these conditions, the abyssal plain is surprisingly rich in life, supported primarily by the nutrients that rain down from the upper layers of the ocean. In the 1970s, researchers studying the deep ocean discovered a previously unknown ecosystem supported by bacteria that use hydrogen sulfide escaping from underwater volcanic vents as an energy source. Animals in this ecosystem either use the bacteria as a food source directly or have evolved a mutualistic relationship with them, providing living space while benefiting from the metabolism of the bacteria. The abyssal plain represents the last major unexplored frontier on Earth.

The open ocean is heavily exploited by human fishing fleets—recent estimates found that species diversity in the open ocean has declined by 50% over the last 50 years as a result of fishing pressure.

Coral Reefs. Coral reefs are unusual in that the structure of the habitat is not determined by a geological feature but composed of the skeletons of the dominant organism in the habitat: coral animals. Coral animals are very simple in structure but have a unique lifestyle—they not only filter dead organic material from the water but also feed on the photosynthetic algae they harbor inside their bodies. Up to 90% of the nutrition of a coral is provided by the algae, which receive in return a protected site for photosynthesis and easy access to the coral's nutrients and carbon dioxide. Reef-building coral live in large colonies, and each individual coral animal secretes a limestone skeleton that protects it from other animals and from wave action.

Coral reefs are found throughout the tropics in warm and well-lit water, providing ample resources for abundant plankton and algae growth. Their complex structure and high biological productivity make them the most diverse aquatic habitats, rivaling terrestrial tropical rain forests in species per area.

Coral reefs are sensitive to environmental conditions and prone to "bleaching," which occurs when host coral animals expel their algae companions. Without their photosynthetic mutualists, the animals may starve. Bleaching can occur for various reasons, but recent episodes seem to be associated with high ocean temperatures. Although coral can recover from bleaching, increased global temperatures due to global warming may lead to more frequent bleaching and the death of especially sensitive reef systems.

Estuaries. The zone where freshwater rivers drain into salty oceans is known as an **estuary.** The mixing of fresh and saltwater combined with water-level fluctuations produced by tides creates an extremely productive habitat. Some familiar and economically important estuaries are Tampa Bay, Puget Sound, and Chesapeake Bay.

Estuaries provide a habitat for up to 75% of commercial fish populations and 80% to 90% of recreational fish populations; they are sometimes called the "nurseries of the sea." Estuaries are also rich sources of shellfish—crabs, lobsters, and clams. Vegetation surrounding estuaries, including extensive salt marshes consisting of wetland plants that can withstand the elevated saltiness compared to freshwater, provides a buffer zone that stabilizes a shoreline and prevents erosion. Unfortunately, estuaries are threatened by human activity as well, including eutrophication from increasing fertilizer pollution and outright loss as a result of housing and resort development.

16.4 Human Habitats

As is clear from our discussion of both aquatic and terrestrial biomes, no habitat on Earth has escaped the effects of humans. Consequently, to truly know our bioregion, we must learn how human populations use the environment. Preserving our biological neighborhoods requires being able to meet human needs while respecting the needs of other species with which we coexist.

According to the United Nations, humans have modified 50% of Earth's land surface (**Figure 16.24**). Most of this modification is for agriculture and forestry, but a surprisingly large amount—2% to 3% of Earth's land surface, or 3.0 to 4.5 million square kilometers, larger than the area of India—has been modified for human habitation. And this is only direct conversion; human activities have environmental effects far beyond their geographic boundaries, including changes to Earth's atmosphere, which receives some of our waste. In fact, it is fair to say that no natural habitat has escaped the human fingerprint.

Today, half of the human population—approximately 3.5 billion people—live in cities. If current migrations from rural to urban areas continue, by the year 2050 two-thirds of the then population, or about 6 billion people, will live in urban areas. Clearly, cities are going to become increasingly important in determining the human impact on native biomes.

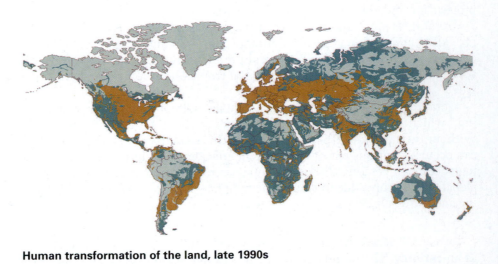

Human transformation of the land, late 1990s

☐ Almost pristine   ☐ Partially transformed   ☐ Almost fully transformed

Figure 16.24 Human modification of Earth's land surface. Most of the land that is relatively untouched by humans lies in the vast boreal forest covering much of Canada and Siberia, the great deserts of Africa and Australia, and the dense Amazonian rain forest in north and central South America.

Energy and Natural Resources

Cities are different from natural landscapes in a variety of ways. Consider the requirements of a forest. The only energy required for its growth is the solar irradiance striking the area, and most of the nutrients available to support this growth are already present in the soil. In addition, nearly all of the waste produced by the organisms in a forest is processed on site; that is, it becomes part of the soil or air and is recycled in the system.

In contrast, cities rely extensively on energy imported from elsewhere and tend to be the central link in a linear flow of materials—from natural or agricultural landscapes and into waste disposed elsewhere.

Energy Use. In more developed countries, much of the energy required to power the activities within cities—from running buses and heating and cooling buildings to lighting traffic signals and energizing wireless phones—is derived from fossil fuels. These fuels are extracted from wells and mines around the globe, including the region surrounding the Persian Gulf, source of much of the world's oil, and the Appalachian Mountains, a rich source of coal. For most people in developed countries, the energy used is not associated with the bioregion in which they live. The environmental impacts of acquiring and transporting fossil fuel can be substantial, ranging from the degradation of oceans and estuaries resulting from oil spills to the wholesale dismantling of forested mountains during coal mining (**Figure 16.25**).

In less developed countries, fossil fuel use is still relatively low, and energy sources are more directly tied to the surrounding natural environment. A primary source of energy for heating and cooking in these countries is wood and other plant-based materials (including animal dung). As urban areas in less developed countries grow, surrounding forests are stripped of trees, often at a faster rate than they can regrow.

In both less developed and more developed countries, an awareness of cleaner sources of energy within a given bioregion—such as abundant solar or wind resources—could help to reduce the environmental costs of energy consumption.

Figure 16.25 The cost of fossil fuel extraction. In the 1970s, mining companies developed a technique known as "mountaintop removal"—using explosives to blast off the rock above a coal deposit and dumping the waste rock into nearby valleys—to access coal. Since 1981, more than 500 square miles of West Virginia have been destroyed by this method of mining.

> **Stop & Stretch** Which historical and present factors do you think cause fossil fuels, rather than solar or wind power, to be the most common source of energy in the developed world?

Natural Resources. In addition to energy, human settlements require materials for survival—food from agricultural production and harvesting of natural resources, metals and salts from mines, freshwater for human and industrial consumption, petroleum for asphalt and manufactured goods, and trees for processing into paper and packaging. Developing and extracting all these raw materials changes biomes far from a city's center.

The amount of material needed to support a city can be tremendous. A year 2000 estimate calculated that the city of London's **ecological footprint**—that is, the amount of land needed to support the human activity there—was 293 times the actual size of the city, equal to twice the entire land surface of the United Kingdom. Clearly, the resources to support London must come from other countries around the world—from the boreal forests of Russia and Scandinavia that supply wood for building and paper production to the tropical forests of India and China that supply Londoners' daily spot of tea.

London is not exceptional in its importation of natural resources or its ecological footprint. In fact, most cities in more developed countries have similar size footprints. A city's footprint can be reduced by its citizens, especially if they reduce the amount of energy used for transportation and make sensitive consumer choices, including buying locally produced foods and less meat, which requires significant additional land to raise.

Figure 16.26 Treatment of human waste. In more developed countries, human waste is treated in expensive and technologically advanced sewage treatment plants. This is a great improvement over the past, when untreated waste was dumped directly in rivers, lakes, and oceans.

Waste Production

The waste produced by a city presents a special problem. The density of the population is so great in cities that specialized systems are required to handle the enormous amounts of human waste generated in a small area. Human wastes can be liquid, in the form of wastewater; solid, in the form of garbage; or gaseous, in the form of air pollution.

Wastewater. In developed countries, cities have sewage treatment systems to handle the wastewater that drains from sinks, tubs, toilets, and industrial plants (**Figure 16.26**). These systems typically treat water by removing semi-solid wastes through settling, using chemical treatment to kill any disease-causing microorganisms, and eventually discharging the treated water to lakes, streams, or oceans. Unfortunately, older treatment systems can be overwhelmed by storm water, causing the discharge of large volumes of wastewater directly into waterways. Untreated wastewater can cause nutrient levels to spike and algae and bacteria growth to increase, thus leading to beaches being closed to swimming and other water-based recreation.

Disposal of semisolid wastes, called sludge, is a significant challenge. Sludge is often composted (allowed to decompose) and applied to land as fertilizer or trucked to landfills. Land application of composted sludge has its own problems; this material contains not only human waste—a valuable fertilizer if properly composted—but also industrial wastewater, which can contain a wide variety of toxic chemicals.

In less developed countries, safe disposal of treated sewage is often a distant dream. In many rapidly industrialized regions, the emigration of people to urban areas has overwhelmed antiquated and inadequate sewer systems. Large numbers of people in these cities live in slums where running water is scarce and untreated human waste flows in open gutters. The consequences of inadequate disposal of human waste can be severe—intestinal diseases due to contact with waste-contaminated drinking water cause the preventable deaths of over 2 million children under 5 years old every year.

Garbage and Recycling. In addition to dealing with wastewater, people in urban areas must find ways to control their solid waste—that is, their garbage. In more developed countries, most of the solid waste finds its way into sanitary landfills, which are pits lined with resistant material such as plastic. Landfills have systems for collecting liquid that drains through the waste and exhaust pipes to vent dangerous gases that result from decomposition (**Figure 16.27**).

Figure 16.27 Disposal of garbage. Solid waste contained in a sanitary landfill is locked within a thick liner. Garbage is compacted and "capped" with soil or other material.

In most areas, landfill space is becoming more and more limited and farther removed from cities. To stave off a looming "garbage crisis," states and communities mandate recycling of paper, glass, metal, and plastic, and many have or are considering composting programs to reduce the amount of food and yard waste. Unfortunately, even where recycling rates are relatively high, household garbage production continues to increase.

In less developed countries, the problem of solid waste disposal is more severe. Many large cities in these areas have large, open dumps in which unstable, fire-prone piles of garbage provide living space for desperately poor immigrants in the city.

Air Pollution.   Because of human dependence on fossil fuels, urban areas produce large amounts of gaseous waste. These air emissions include carbon dioxide as well as byproducts of the combustion of oil, gasoline, and coal (from power plants), including nitrogen and sulfur oxides, small airborne particulates, and fuel contaminants such as mercury. Exposure to sunlight and high temperatures can cause some of these byproducts to react with oxygen in the air to form ground-level ozone or smog (**Figure 16.28**). For individuals with asthma, heart disease, or reduced lung function, smog exposure can lead to severe illness, even death.

When gaseous pollution enters the upper reaches of Earth's atmosphere, it can be carried on air currents to less settled areas throughout the globe. Air emissions from coal-fired power plants throughout the Midwest cause severe acid rain in lightly settled regions of the northeastern United States. Airborne toxins such as benzene and PCBs (polychlorinated biphenyls), byproducts of manufacturing processes, have been found in high levels in animals in the Arctic as a result of **bioaccumulation,** the process that causes relatively small amounts of pollutants to become concentrated to dangerous levels in animals that are high on the food chain.

Air pollution is a problem in both developed and less developed countries. In less developed countries, pollution control is weak or lacking altogether; in more developed countries, the sheer volume of fossil fuel use contributes to poor air quality. In the United States, the number of miles driven by car per household has increased by 50% since the 1970s partially due to an increase in the distance that individuals live from their workplaces.

Figure 16.28 Smog. This brownish haze over downtown Los Angeles, California, results from automobile emissions and other air pollutants causing the conversion of oxygen to ozone in sunny, still weather conditions.

Figure 16.29 Fitting human needs into the bioregion. Lake Erie was once so polluted that it was considered "dead" in the 1960s. Thanks to government and citizen efforts, the lake has now recovered and provides clean swimming beaches and boasts a healthy fish population.

Development of suburban settlements outside the geographical limits of cities has been termed **urban sprawl.** Urban sprawl not only contributes to an increase in fossil fuel consumption but also negatively affects wildlife through habitat destruction and fragmentation. Urban sprawl also impairs water quality via the destruction of wetlands and the increasing amount of paved surfaces, which funnel pollutants and warmed water into lakes, streams, and rivers.

Reducing Human Impacts. The effect of human settlements on surrounding natural biomes can be significant and severe, as we have discovered in this chapter. However, many of these impacts can be mitigated with thoughtful planning and the use of improved technology. In the United States, laws such as the Clean Air Act and the Clean Water Act—passed in 1970 and 1972, respectively—have greatly reduced air and water pollution and have contributed to the recovery of once severely impaired habitats (**Figure 16.29**).

Cities throughout the more developed world are supporting projects aimed at creating sustainable communities that are both economically vital and environmentally intelligent. **Table 16.3** poses a set of questions that will help you become more knowledgeable about your own bioregion. Perhaps by getting to know our biological neighbors and understanding how our choices affect these organisms, we can be inspired to help create communities that are safe and healthy for humans and other species.

TABLE 16.3

Know your bioregion.	
Environmental Factor	**Your Bioregion**
Climate	Describe the climate of the place where you live, including the annual average temperature, average precipitation, average number of sunny days, and names of seasons. Which global and local factors influence the climate where you live?
Vegetation	Describe the native vegetation of the place where you live. How much native vegetation remains, and what has taken its place? Which ecological factors (climate, fire, etc.) have influenced the native vegetation type?
Aquatic habitats	What are the aquatic habitats nearest to you? Can you name some of the dominant species in these habitats? What threats do these habitats face in your area?
Energy resources	What is the primary source for the electricity you use in your home? Is your bioregion rich in any less environmentally-damaging energy resources?
Food resources	What agricultural products are produced in your bioregion? How easy is it to buy locally produced food?
Natural resources	What natural resources does your bioregion supply?
Wastewater and sewage	Where does your wastewater go? Does your community have any problems handling wastewater and sewage?
Solid waste	Where does your garbage go? What is the rate of recycling in your community, and can it be improved?
Air quality	How is the air quality in your community? What are the major air pollutants and their sources?

SAVVY READER

Greenwashing

A clean and healthy environment is a high priority for many consumers—which is why many businesses tout their environmental bonafides in their advertisements and products. While many companies have business practices that represent a true choice of placing conservation above profits, some companies have made a practice of using public relations techniques to claim that they are ecologically friendly when they have not made the same choice. This technique is called *greenwashing*.

One way to spot greenwashing, or indeed any deceptive advertising practice, is to use the guidelines from the Federal Trade Commission (FTC). In short, the FTC makes clear that any claim made in advertising should be supported by objective evidence. Consider the following statements posted on a variety of company websites:

From two global oil and gas companies:

"Many of the locations in which we operate present challenging environmental sensitivities, so managing our impact in these areas is always at the core of our activities."

"We consider biodiversity early in new projects, develop biodiversity action plans, and collaborate with biodiversity experts to help protect areas with rich and delicate ecosystems."

From two multinational computer technology corporations:

"Today's energy- and climate-related issues are at the top of our strategic agenda."

"We know that the most important thing we can do to reduce our impact on the environment is to improve our products' environmental performance. That's why we design them to use less material, ship with smaller packaging, be free of toxic substances used by others, and be as energy efficient and recyclable as possible."

From two luxury automobile industries:

"Every . . . manufacturing site has achieved Zero Landfill Status, meaning that any waste resulting from manufacture is recycled, repurposed, or used to generate additional energy."

"Our measurement of success must go beyond the traditional bottom line and extend to recognition of the environment and the needs of individuals both inside and outside the company."

1. Which if any of these statements by itself contains objective information to support the claims?

2. Where would it be reasonable to expect these companies to report tests, analyses, research, studies, or other evidence to support these claims?

3. Choose one of these statements and list the evidence that you would want to see in order to determine if the claim has merit.

Chapter Review

Learning Outcomes

Go to the Study Area at www.masteringbiology.com for practice quizzes, myeBook, BioFlix™ 3-D animations, MP3Tutor sessions, videos, current events, and more.

LO1 Describe the factors that determine temperature and precipitation patterns on a global scale (Section 16.1).

- Climate is the weather of a place as measured over many years. Climate in an area is determined by global temperature patterns, which are driven by solar irradiance and local factors such as proximity to large bodies of water (pp. 390–392).

- Temperatures are warmer at the equator than at the poles because solar irradiance is greater at the equator (pp. 391–392).

- Seasonal changes are caused by the tilt of Earth's axis. For example, during summer in the Northern Hemisphere, total solar irradiance during the day is much greater than it is during winter (pp. 391–392).

- Global precipitation patterns are driven by solar energy, with areas of heavy rainfall near the solar equator and dry regions at latitudes of 30° north and south of the equator (pp. 393–394).

LO2 List factors that influence climate on a local scale, and explain how they affect local temperature and precipitation patterns (Section 16.1).

- Local temperatures are influenced by altitude, proximity to a large body of water, nearby ocean currents, the light reflectance of the land surface, and the amount of surrounding vegetation (pp. 395–397).
- Local precipitation patterns are influenced by the presence of a large body of water, which tends to increase rainfall, and the presence of a mountain range, which tends to reduce rainfall on its leeward or sheltered side (p. 397).

LO3 Briefly describe the climate and vegetation characteristics of the terrestrial biomes, including tropical, temperate, and boreal forests; chaparral; tropical and temperate grasslands; deserts; and tundra, and give examples of animal and plant adaptations in these environments. Describe some of the threats to these biomes imposed by human activity (Section 16.2).

- Forests are dominated by trees and categorized by distance from the equator into tropical, temperate, and boreal types. Tropical forests are dominated by large trees with little understory, while temperate forests have a diverse understory that flourishes when the trees drop their leaves. Boreal forest is dominated by evergreen conifers that conserve water during the long, cold winter (pp. 398 and 400–402).
- The chaparral biome is found in slightly hotter and drier areas than forests and is dominated by highly flammable woody shrubs that regenerate quickly after fire (pp. 399 and 402).
- Grasslands are found where precipitation is relatively low. The major vegetation type on grasslands is nonwoody grasses. These biomes are categorized by distance from the equator into tropical and temperate types. Both tropical and temperate grasslands are maintained by periodic fires and adapted to regrow quickly (pp. 399 and 402–403).
- Deserts are found where precipitation is 50 cm (20 in.) per year or less and contain mostly drought-resistant plants and animals adapted to the low-water environment (pp. 399 and 403).
- Tundra occurs in areas where the growing season is 60 days or less, both near the poles and at high altitudes, and is dominated by short plants and grazing animals (pp. 399 and 404–405).
- Forests are threatened by logging and agricultural development, while chaparral is primarily threatened by human fire-suppression activities. Grasslands are threatened by agricultural development and overgrazing, and deserts by human settlement and water depletion. Tundra is threatened by climate change and oil and gas exploration (pp. 399 and 400–402).

LO4 Briefly describe the characteristics of the aquatic biomes, including freshwater lakes, rivers, and wetlands, and saltwater oceans, coral reefs, and estuaries, and provide examples of organisms found

in these biomes. Describe some of the threats to these biomes imposed by human activity (Section 16.3).

- Freshwater biomes include lakes, areas of standing water dominated by aquatic plants and algae and fish; rivers, made up of flowing water with relatively few plants; and wetlands, which consist of waterlogged soil and emergent vegetation (pp. 405–408).
- Marine biomes include oceans, which can be divided into multiple zones depending on water temperature, light, and nutrient availability. The dominant organisms in the open ocean are photosynthetic plankton, while the great diversity of marine life is found in coral reefs and estuaries (pp. 406 and 408–410).
- Aquatic habitats are subject to pollution via acid rain and nutrient runoff from agricultural lands. Organisms in these biomes are often threatened by overharvesting (pp. 406 and 408–410).

LO5 List the ecological requirements of human communities, including food, energy, and waste disposal, and describe how these needs are met in various human societies (Section 16.4).

- Humans have modified over 50% of Earth's land surface. Much of this modification now supports the activities of cities (p. 411).
- Cities must rely on imported energy and other resources to survive; extraction of these resources carries an environmental cost felt by other bioregions (pp. 410–411).
- Waste disposal is a significant challenge for large urban areas. Sewage must be treated to avoid contaminating drinking water; garbage must be disposed of effectively, and air emissions must be controlled. More developed countries have more resources to consume and handle the vast amounts of waste that they produce, while less developed countries struggle with inadequate systems (pp. 412–413).
- Increasing the bioregional awareness of citizens and communities can help them devise methods for supporting human activities in a more sustainable manner (p. 414).

Roots to Remember

The following roots of words come mainly from Latin and Greek and will help you decipher terms:

    aqua- means water. Chapter term: aquatic

    equ- means equal. Chapter term: equator

    sol- means sun. Chapter term: solstice

    -stice means to stand still. Chapter term: solstice

Learning the Basics

1. **LO1** Explain why the northern United States experiences a cold season in winter and a warm season in summer.

2. **LO2** Why does proximity to a large body of water moderate climate on a nearby landmass?

3. **LO1** Areas of low solar radiation are _____.

   A. closer to the equator than to the poles; B. closer to the poles than the equator; C. at high altitudes; D. close to large bodies of water; E. more than one of the above is correct.

4. **LO1** The solar equator, the region of Earth where the sun is directly overhead, moves from 23.5° N to 23.5° S latitudes and back over the course of a year. Why?

   A. Earth wobbles on its axis during the year; B. The position of the poles changes by this amount annually; C. Earth's axis is 23.5° from perpendicular to the rays of the sun; D. Earth moves 23.5° toward the sun in summer and 23.5° away from the sun in winter; E. Ocean currents carry heat from the tropical ocean north in summer and south in winter.

5. **LO3** Which of the following biomes is most common on Earth's terrestrial surface?

   A. chaparral; B. desert; C. temperate forest; D. tundra; E. boreal forest

6. **LO3** Tundra is found _____.

   A. where average temperatures are low and growing seasons are short; B. near the poles; C. at high altitudes; D. A and B are correct; E. A, B, and C are correct

7. **LO3** Which statement best describes the desert biome?

   A. It is found wherever temperatures are high; B. It contains a larger amount of biomass per unit area than any other biome; C. Its dominant vegetation is adapted to conserve water; D. Most are located at the equator; E. It is not suitable for human habitation.

8. **LO4** Which of the following biomes has a structure made up primarily of the remains of its dominant organisms?

   A. coral reefs; B. freshwater lakes; C. rivers; D. estuaries; E. oceans

9. **LO5** An ecological footprint _____.

   A. is the position an individual holds in the ecological food chain; B. estimates the total land area required to support a particular person or human population; C. is equal to the size of a human population; D. helps determine the most appropriate wastewater treatment plan for a community; E. is often smaller than the actual land footprint of residences in a city.

10. **LO5** In more developed countries, wastewater containing human waste _____.

    A. is dumped directly into waterways; B. is applied directly to farm fields; C. is piped into sanitary landfills; D. is treated to remove sludge and disease-causing organisms and released into nearby waterways; E. is a major source of infectious disease in human populations

Analyzing and Applying the Basics

1. **LO1 LO2 LO3** Consider the following geographic factors and predict both the climate and biome type found in the location described. Explain the reasoning that you used to determine your answer. This small city is

   - On the coast of the Pacific Ocean
   - 20° north of the equator
   - 20 m above sea level
   - At the base of a mountain range

2. **LO1 LO2** One prediction of global climate change models is that significant amounts of melting ice will change the salt content of the ocean, causing the Gulf Stream current in the Atlantic Ocean to stop altogether. How will this change likely affect Europe?

3. **LO1 LO2** What can you infer about the geographical relationship among the cities in the following table and the Cascades, the primary mountain range that influences their climate?

City	Approximate Average Annual Rainfall (inches)
Bend, OR	12
Eugene, OR	43
Portland, OR	36
Tacoma, WA	39
Walla Walla, WA	21
Yakima, WA	8

Connecting the Science

1. How many biomes do you rely on to supply your food? Many grocery stores label the origin of their produce. The next time you go to the grocery store, try to determine the number of different countries from which your groceries come. Could you easily change your diet and shopping habits to rely on locally produced food? Why or why not?

2. Consider the setting of your home. What kind of changes to the environment of your home would help it fit better into the bioregion? Are these changes feasible? Are they desirable?

Answers to **Stop & Stretch, Visualize This, Savvy Reader,** and **Chapter Review** questions can be found in the **Answers** section at the back of the book

Organ Donation

Tissues, Organs, and Organ Systems

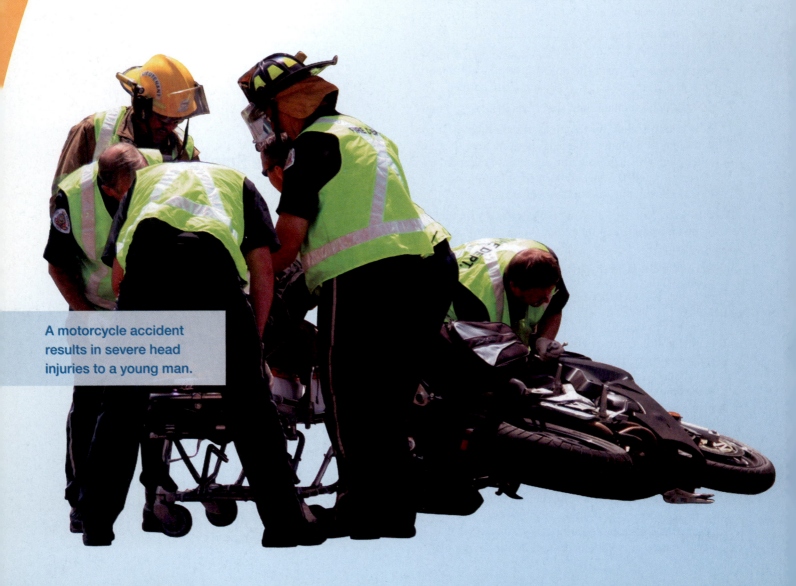

A motorcycle accident results in severe head injuries to a young man.

The drama plays out in emergency rooms around the country thousands of times every year. A formerly healthy young man is rushed to the hospital in an ambulance after having been involved in a motorcycle accident. After many failed attempts to find brain activity, emergency room physicians decide that the young man is brain dead. In the meantime, his family has assembled in the waiting room, overcome by the loss. A physician enters the waiting room and asks the family whether they would be willing to donate their loved one's tissues and organs.

The family is understandably confused and has many questions. Is there no chance that their loved one will ever be okay—even many years from now? If they agree, will the medical staff not work as hard to keep their loved one alive? What will happen to the tissues and organs if they do decide to donate them? The family knows that their loved one would have liked the idea of helping others, but they have never really discussed organ donation. Even though his driver's license lists him as a donor, the family is not sure what to do.

This chapter will help you understand the types of tissues and organs that can be donated and why properly functioning tissues and organs are necessary to maintain life.

LEARNING OUTCOMES

LO1 Describe the structure and function of epithelial tissues.

LO2 Compare the structures and functions of the six connective tissues.

LO3 Compare the structures and functions of the three types of muscle tissues.

LO4 Describe the structure and function of nervous tissues.

LO5 Explain what an organ is and how organs interact in an organ system.

LO6 List the structures and functions of the liver.

LO7 List the organs of the digestive system, and outline their functions.

LO8 List the accessory organs of the digestive system, and outline their roles in digestive processes.

LO9 Explain how positive and negative feedback mechanisms help in the maintenance of homeostasis.

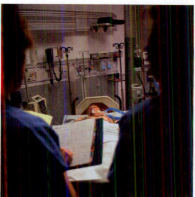

He is rushed to the hospital, where it is determined that his brain has stopped functioning.

Medical personnel search his wallet, looking for information on relatives to contact and seeing if he carried a signed organ donor card.

His family is asked whether they will donate his organs.

17.1 Tissues

Groups of similar cell types that perform a common function comprise a **tissue.** Damage to tissues can sometimes prevent them from recovering their function, which may be restored by a tissue transplant. We will first learn about the structure and function of different tissue types and then see some situations in which transplants can help replace damaged tissues. The human body has four main types of tissue: epithelial, connective, muscle, and nervous.

Epithelial Tissue

Epithelial tissue, or **epithelium,** is tightly packed sheets of cells that cover organs and outer surfaces and line hollow organs, vessels, and body cavities (**Figure 17.1**).

The sheets of epithelial tissue, unlike other tissue types, are usually not attached on all surfaces. Instead, epithelial tissues are anchored on one face of the tissue but free on the other tissue face. The unattached tissue can be exposed to body fluids or the external environment. For example, the epithelial tissue that lines blood vessels is attached to the wall of the blood vessel, but its outer surface is exposed to the bloodstream. The epithelium that comprises the outer layer of skin, called the epidermis, is anchored to the underlying tissue but has a surface that is exposed to air. This is also true of the epithelia that line other parts of the body, including respiratory surfaces and the the digestive, urinary, and genital (or urogenital) tracts. Epithelia can be a single layer or many layers thick.

Epithelial cells function in protection, secretion, and absorption. Epithelial cells in the skin help protect the body from injury and ultraviolet light. Some glands in the skin are formed by outgrowths of epithelial tissue. These glands secrete substances such as mucus, oils, and sweat. Epithelial

Figure 17.1 Epithelial tissues. (a) Epithelia line many organs. (b) Epithelial tissues are tightly packed cells that usually have one unattached surface, as seen in the epidermis of the skin. (c) Epithelia can be multilayered or a single layer, as is the epithelium that lines the small intestine.

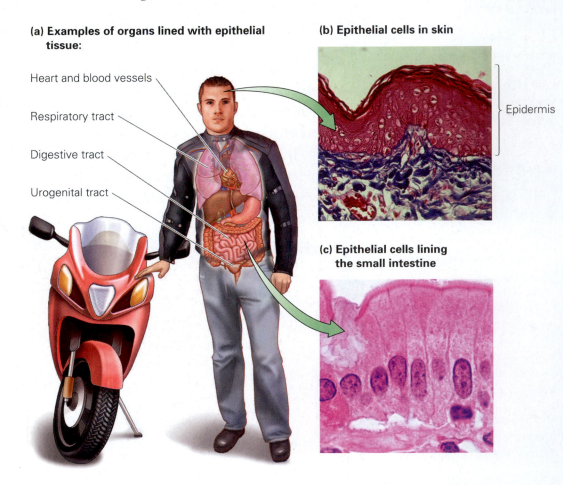

(a) Examples of organs lined with epithelial tissue:

Heart and blood vessels

Respiratory tract

Digestive tract

Urogenital tract

(b) Epithelial cells in skin

Epidermis

(c) Epithelial cells lining the small intestine

tissues can also function to protect the body from water loss and pathogens. Epithelial cells that line blood vessels and intestines function in absorbing nutrients.

Epithelial tissues constantly slough off cells, which are replaced by cell division. Skin cells are rubbed off by clothing, and cells that line the mouth and intestines are scraped off by food. These cells are replaced, resulting in a replacement of the epidermis every 4–6 weeks.

Some epithelial tissues can be transplanted, as we will see at the end of this section.

Connective Tissue

Connective tissues function, as their name implies, to form connections. They usually bind organs and tissues to each other. Connective tissue is loosely organized, not tightly packed like epithelium. Our bodies contain six different types of connective tissue, all of them composed of cells embedded in a **matrix** made of protein fibers and an amorphous, gel-like ground substance. The matrix can be likened to Jell-O® with carrot shavings in it: The Jell-O is analogous to the ground substance, and the carrots are analogous to the protein fibers. The proportions of these components vary from one part of the body to another, depending on structural requirements. In some areas, the connective tissue is highly cellular; in others, the predominant feature is the protein fibers or ground substance of the matrix. The consistency of the matrix is a function of the types of fibers and ground substances present. Blood, for example, is a connective tissue with a liquid matrix. Adipose or fat tissue has a thick fluid, or viscous, matrix, and bone has a solid matrix. Specific compositions and functions of the six types of connective tissue follow.

Loose Connective Tissue.   Loose connective tissue (**Figure 17.2a**) is the most widespread connective tissue in animals. It connects epithelia to

Figure 17.2 Connective tissues. The six different types of connective tissue are each composed of cells and a specialized matrix.

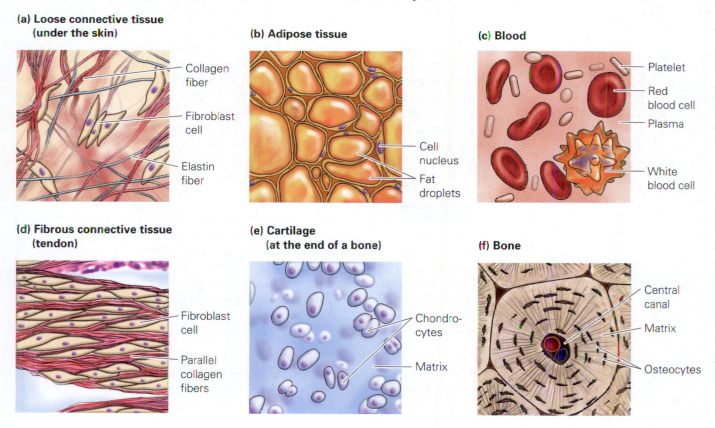

(a) Loose connective tissue (under the skin)
- Collagen fiber
- Fibroblast cell
- Elastin fiber

(b) Adipose tissue
- Cell nucleus
- Fat droplets

(c) Blood
- Platelet
- Red blood cell
- Plasma
- White blood cell

(d) Fibrous connective tissue (tendon)
- Fibroblast cell
- Parallel collagen fibers

(e) Cartilage (at the end of a bone)
- Chondrocytes
- Matrix

(f) Bone
- Central canal
- Matrix
- Osteocytes

underlying tissues, holds organs in place, and acts as padding under the skin and elsewhere. The tissue is called loose connective tissue due to the loose weave of its constituent fibers. The cells of loose connective tissue, called fibroblasts, secrete the proteins comprising the matrix. The matrix of loose connective tissue is its chief structural feature. Two proteins are important constituents of this matrix—collagen and elastin. Collagen fibers give connective tissue tremendous strength, and the elastin fibers allow connective tissue to stretch without breaking. The degradation of collagen and elastin in skin that occurs with age, exposure to sunlight, and cigarette smoke causes weakening of the connection between the skin and the underlying muscle, leading to wrinkles.

Adipose Tissue. Adipose (fat) tissue (**Figure 17.2b**) connects the skin to underlying structures and insulates and protects organs. Adipose cells are specialized for the synthesis and storage of energy-rich reserves of fat (lipids). A fat droplet fills the cytoplasm of these cells and shrinks when the fat is used for energy. Cells are the predominant constituent of this connective tissue; only a small amount of matrix is associated with adipose tissue.

Blood. Blood (**Figure 17.2c**) is a type of connective tissue that circulates throughout the body through arteries, veins, and capillaries and transports oxygen and nutrients to cells. The cellular component of blood tissue includes red cells, which carry oxygen; white cells, which help fight infection; and cell fragments called platelets, which function in clotting. Blood cells are suspended in a liquid- and protein-rich matrix called plasma.

Fibrous Connective Tissue. Fibrous connective tissue forms the tendons that connect muscles to bones and the ligaments that connect bones to each other at joints. As in loose connective tissue, the cells of this tissue are called fibroblasts. The densely packed collagen fibers of the matrix, arranged in parallel, are the most conspicuous feature of fibrous connective tissue (**Figure 17.2d**).

Cartilage. Cartilage is a type of connective tissue composed of cells, called chondrocytes, that secrete substances to form the dense matrix surrounding them (**Figure 17.2e**). The matrix is rich in collagen and other structural proteins. Cartilage connects muscles with bones, provides flexible support for the ears and nose, and allows for shock absorption. At joints, the ends of bones are covered in cartilage, which helps permit smooth gliding where joint surfaces contact each other. Cartilage has no blood vessels, so injuries heal very slowly if at all.

Stop & Stretch Some athletes experience a loss of cartilage in their knee joints due to injury or wear and tear. When the cartilage that covers the bones wears down enough to allow direct contact between the heads of the bones comprising this joint, a painful condition called osteoarthritis results. Scientists can grow cartilage cells in the lab to provide cartilage replacements. What characteristic of cartilaginous tissues needs to be overcome for this type of transplant to succeed?

Bone. Bones that make up the skeleton are connected to each other at the joints. In addition to being a framework for the body, bones support and protect other tissues and organs. Marrow, inside the bone, produces blood cells.

Bone is a rigid type of connective tissue whose branched cells, called **osteocytes,** secrete substances that harden into a solid matrix of collagen and calcium and other minerals (**Figure 17.2f**). Bone serves as a reservoir of calcium and minerals that the body can use if dietary levels of these substances are low. Central canals inside bones house nerves and blood vessels.

Many connective tissues can be transplanted from tissue donors. Some of these tissues, including blood and bone (marrow), can be transplanted from living donors.

Muscle Tissue

While connective tissue serves to support and connect, another important tissue is muscle tissue. In fact, muscle is the most abundant type of tissue in animals. Anything we think of as "meat" is actually muscle tissue.

Muscle is a highly specialized tissue capable of contracting. Most muscle tissue is composed of bundles of long, thin, cylindrical cells called muscle fibers. Muscle fibers contain specialized proteins, **actin** and **myosin,** that cause the cell to contract when signaled by nerve cells.

Muscle tissues are differentiated according to whether their function requires conscious thought—such as walking—and is **voluntary** or whether their function requires no conscious thought, such as the beating of the heart, and is **involuntary.** Muscle tissues are also differentiated by the presence or absence of bands that look like stripes under a microscope. Muscle that is banded is called **striated muscle.** Unbanded muscle is called **smooth muscle.** The striated appearance is due to the banding pattern formed by actin and myosin deposits in the cells. Muscle tissue that lacks these striations also contains actin and myosin deposits, just not in a banded pattern. Humans and other vertebrates have three types of muscle tissues. These contracting tissues are skeletal, cardiac, and smooth muscle (**Figure 17.3**).

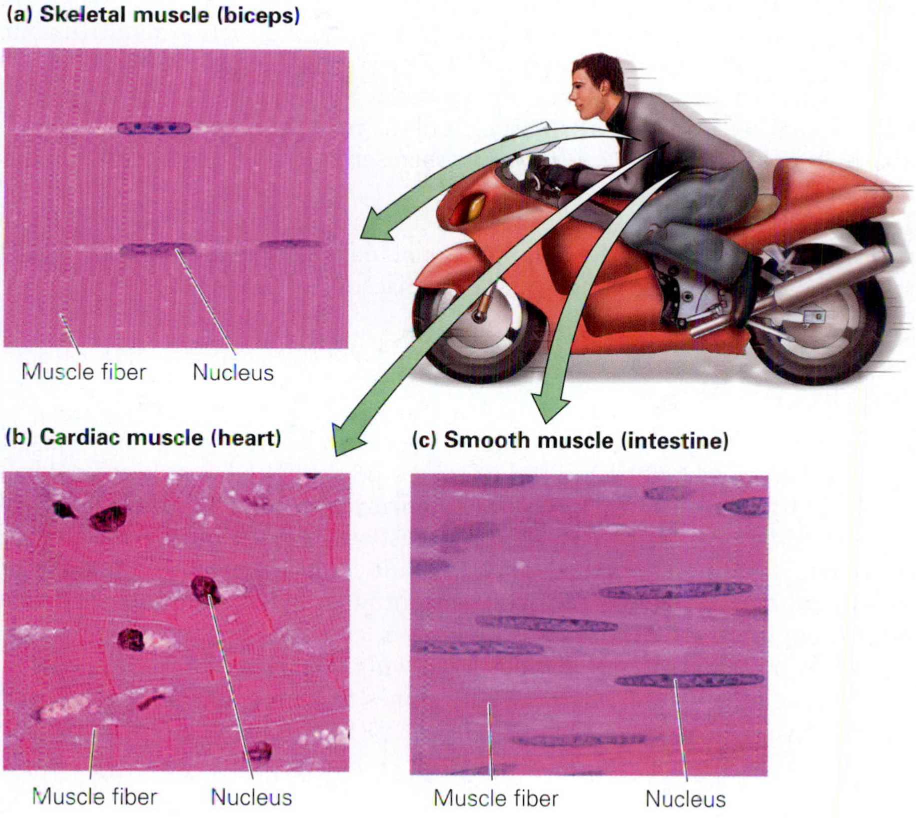

(a) Skeletal muscle (biceps)

Muscle fiber   Nucleus

(b) Cardiac muscle (heart)

Muscle fiber   Nucleus

(c) Smooth muscle (intestine)

Muscle fiber   Nucleus

Figure 17.3 Muscle tissue. Skeletal (a) and cardiac (b) muscles are striated, while smooth muscle (c) is not.
Visualize This: What causes the striped pattern seen in cardiac and skeletal muscle?

Skeletal Muscle. **Skeletal muscle** is usually attached to bone and produces all the movements of body parts in relation to each other. Skeletal muscle is responsible for voluntary movements such as walking. This striated muscle tissue is shown in **Figure 17.3a**. Exercise causes an increase in the size of skeletal muscle cells, not an increase in the number of cells.

Cardiac Muscle. **Cardiac muscle** is found only in heart tissue. This involuntary striated tissue undergoes rhythmic contractions to produce the heartbeat. Cardiac muscle cells are highly branched and interwoven (**Figure 17.3b**); this allows a contraction signal to be propagated.

Smooth Muscle. **Smooth muscle,** as its name implies, is not striated (**Figure 17.3c**). This involuntary muscle is composed of spindle-shaped cells that comprise the musculature of internal organs, blood vessels, and the digestive system. Smooth muscle contracts more slowly than skeletal muscle, but it can remain contracted for a long time. Consider the difference between the contractions of the smooth muscle lining the digestive tract and helping to move food along as opposed to a biceps contraction. Contractions of the smooth muscles of the digestive tract are involuntary and are held for long periods of time. Contractions of the biceps are voluntary and occur more quickly.

Cardiac muscle tissue can be transplanted as long as the heart has not been damaged. The last of the four types of tissues, and the least likely to be involved in a transplant, is nervous tissue.

Nervous Tissue

Nervous tissue is composed mainly of cells called neurons (**Figure 17.4**) that can conduct and transmit electrical impulses. The main function of nervous tissue is to help the body sense stimuli, process the stimuli, and transmit signals from the brain out to the rest of the body and back. For example, if a car door suddenly opens in front of you while you are biking, you react quickly by trying to swerve and maneuvering out of the way. This action requires your body to recognize the threat and coordinate a response to it very quickly. It is the nervous tissue that makes this possible. Nervous tissue is found in the brain, spinal cord, and nerves. Most cells of the nervous system do not undergo cell division and therefore cannot repair themselves. This is why injuries to the brain and spinal cord are so devastating.

From this overview, you can now see the importance of the various tissue types. For an organism to be fully functional, damage to these tissues must be repaired or the damaged tissue must be replaced. (In Chapter 22, nervous tissue is covered in greater detail.)

Tissue Donation

If a family consents to tissue donation, tissues can be removed from a person who has been declared brain dead or from a person who has suffered cardiac death. Brain death occurs when the cerebrum and brain stem of the brain cease to function. The cerebrum, the largest and most sophisticated part of the brain, is composed of two hemispheres. It controls movement and higher mental functions, such as thought, reasoning, emotion, and memory. The brain stem, located at the base of the brain, governs such automatic functions as heartbeat, respiration, and swallowing. When the brain stem ceases to function, a person has to rely on machines to perform these functions to remain alive.

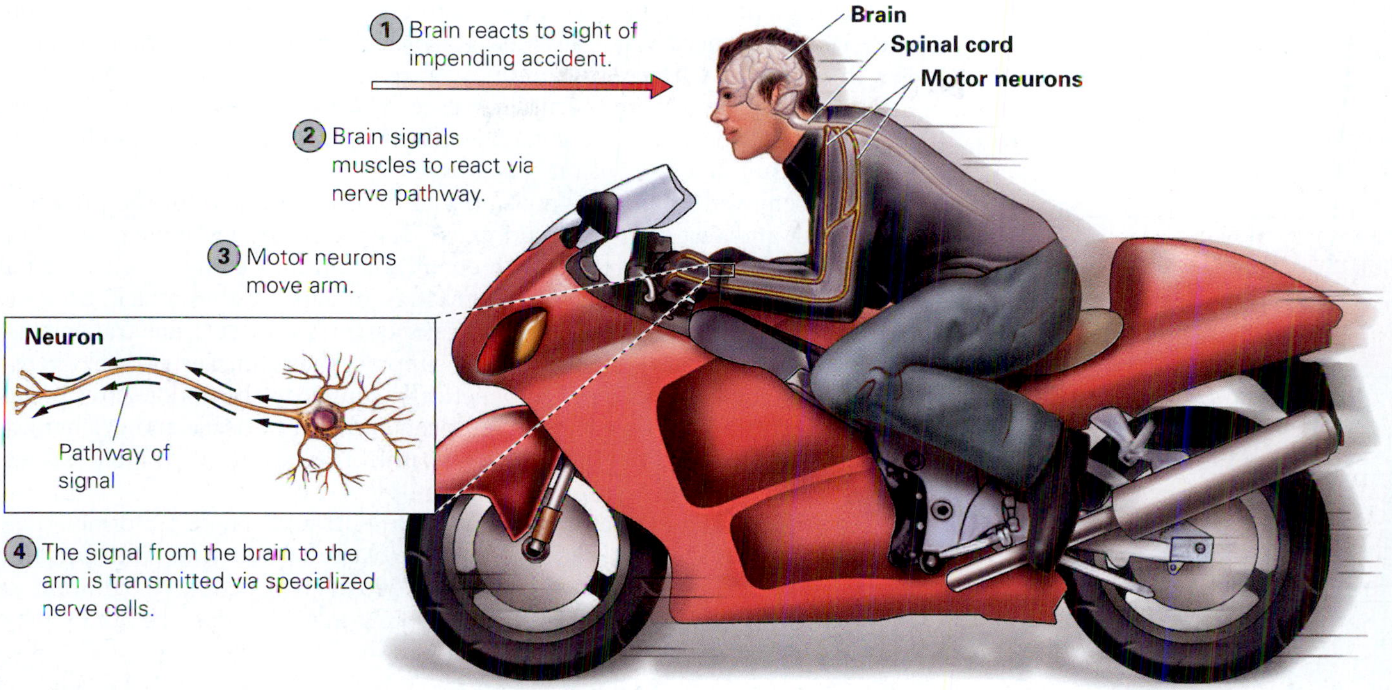

1. Brain reacts to sight of impending accident.

Brain
Spinal cord
Motor neurons

2. Brain signals muscles to react via nerve pathway.

3. Motor neurons move arm.

Neuron

Pathway of signal

4. The signal from the brain to the arm is transmitted via specialized nerve cells.

Figure 17.4 Nervous tissue. The brain and spinal cord are composed of nervous tissues. Nerves are composed of neurons that transmit signals to and from the brain.

Head injuries caused by motor vehicle and cycling accidents, ruptured blood vessels, and drowning are all common causes of brain death. When a person suffers this type of injury, the person's brain cells die because swelling and damage to the tissues prevent them from being supplied with oxygen and nutrients. Once dead, brain cells cannot recover. A person who is brain dead has a different prognosis from that of someone who is in a coma or vegetative state. A person in a coma or vegetative state might recover some functions, while a brain-dead person will not.

Organ donations can come from individuals who have suffered a cardiac death as well. Cardiac death occurs when the heart and lungs stop functioning.

In the case of the young motorcyclist discussed at the beginning of the chapter, the family was asked whether they would donate his tissues and organs because testing showed that all activity had ceased in his cerebrum and brain stem. If the family grants permission for tissue donation, a specially trained team of surgeons will carefully remove any usable tissues that the family agrees to donate. These tissues can include bones, tendons, ligaments, cartilage, veins, and skin.

The motorcyclist's bones may be donated to someone who has had bone cancer and would require an amputation without the transplant. His tendons, ligaments, and cartilage may be used to reconstruct and stabilize the damaged joints of many different people, helping them resume normal activities after accidents and injuries. Veins from the young man's legs can be used to help recipients with circulatory problems and as bypasses for people undergoing heart bypass surgery. Donated skin can be used for burn patients as a type of biological dressing, not only to help prevent infections but also to decrease blood loss and pain due to exposed nerve endings until the patients' own skin can grow back.

The transparent cornea of the eye can be transplanted to a living person who has experienced vision loss due to corneal damage. The cornea consists of a thin layer of surface epithelium overlaying a layer of fibrous connective tissue. In a healthy eye, the cornea serves as a barrier that shields the eye from dust and germs. Damage to the cornea caused by injury, infection, or chemical burn can result in loss of vision.

Once removed, the young man's tissues will be checked for the presence of bacteria and viruses that could cause disease in the recipients. Many of his tissues will then be treated with chemicals to remove any proteins that a recipient's immune system might recognize as foreign and thus attack. Because of these treatments, tissue recipients do not normally reject tissue transplants and usually do not need to take drugs to suppress their immune system. Therefore, tissue recipients need not be genetically similar to the donor. After being tested and treated, the tissues will be frozen and sent to tissue and eye banks, where they will be catalogued and stored until they are needed. Most tissues can be stored for up to 5 years.

Tissue removal and treatment takes a few hours and can be performed during the time that the family begins to make plans for a funeral. One person's tissues can improve the lives of as many as 50 different people. In addition to improving people's lives by donating the young man's tissues, the family may save several lives by donating his organs.

17.2 Organs and Organ Systems

Many structures in the body are actually assemblages of different tissue types. **Organs** are structures composed of two or more tissues packaged together, working in concert to produce the organ's specific function. When many organs interact to perform a common function, these organs are said to be part of an **organ system.** All of the organ systems in a body work together to help an organism function.

This activity can be likened to the manner in which the engine (or organ) of a motor vehicle (organism) functions due to the combined actions of its different component parts, such as valves and pistons (tissues). Many different systems in the vehicle, such as the electrical system (organ system) or the fuel injection system (organ system), work together to keep the vehicle (organism) functioning.

Living things display increasingly complex levels of organization, from cells, to tissues, to organs, to organ systems, to an entire organism (**Figure 17.5**). For example, the regulation of the motorcyclist's body temperature while riding through different climates requires several levels of organization. The brain (an organ) sends signals through the cells of the nervous system (an organ system) to the epidermis (tissue) that result in changes in individual units (cells) that comprise sweat glands and blood vessels, resulting in an overall change in the amount of sweat secreted and in the dilation of blood vessels.

There are many different organs in humans and other animals, such as the heart, lungs, kidney, and brain. (Some of these organs are covered in more

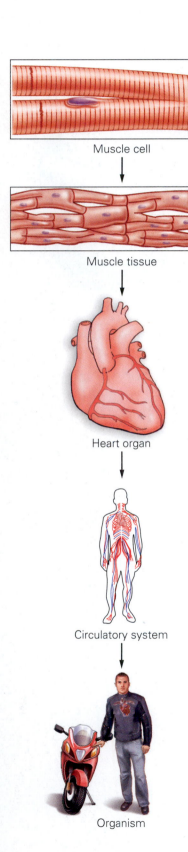

Muscle cell

Muscle tissue

Heart organ

Circulatory system

Organism

Figure 17.5 Levels of organization. Cells give rise to tissues. Tissues with a common function comprise organs. Organs working together make up an organ system, and an entire organism often has many organ systems that communicate with each other. **Visualize This:** What other organ systems might communicate with the circulatory system?

detail in subsequent chapters of this unit.) In the following section, we use the liver as a model organ to illustrate how an organ's ability to perform its required function is made possible by the actions of different tissue types. We also look at how an organ can function as part of an organ system.

Organs: The Liver as a Model Organ

The **liver** is a large, reddish-brown organ found on the right side of the abdominal cavity below the diaphragm. The human liver is divided into four lobes. A greenish sac called the gallbladder is held in close contact with the liver (**Figure 17.6**).

Structurally, the epithelial tissue comprising the liver is covered with connective tissue that branches and extends throughout the body of the liver. This connective tissue acts as a sort of scaffolding to support the tissue and subdivides the tissue into hexagonal structures called **lobules.** In the middle of each lobule is a central vein that allows blood to reach all parts of the liver. Blood flows through the liver and is filtered by liver cells called **hepatocytes.** Filtering results in the removal of toxic materials, dead cells, pathogens, drugs, and alcohol from the bloodstream.

As described, an organ (in this case, the liver) is composed of different tissues (epithelial tissue and connective tissue). These tissues work together to help the organ perform a function (such as filtering).

The liver is an important part of both the circulatory and the digestive systems. As a part of the circulatory system, the liver synthesizes blood-clotting factors, detoxifies harmful substances in the blood, regulates blood volume, and destroys old blood cells. As an accessory part of the digestive system, the liver produces bile, which helps metabolize fats. Bile is secreted into the small intestine and is composed of water, pigments, and salts that help dissolve fats. Bile salts cause emulsification, which is a process that prevents fats from clumping. Fats that have been emulsified are smaller and have more surface area exposed than do large clumps of fat. Thus, secretion of bile into the small intestine facilitates digestion because smaller, emulsified fats are easier for enzymes to act on.

In addition to helping the body digest fats by secreting bile into the small intestine, the liver helps to metabolize and store nutrients. Portal veins carry nutrients from the intestines to the liver, where they can be metabolized. The liver is able to convert excess carbohydrates and proteins into fat, which is then exported and stored in adipose tissue. The liver also serves as a storage site for some vitamins and excess glucose. Glucose monomers are joined to each other to produce a polymer of glucose called glycogen. The liver removes glucose from the blood when glucose levels are high, such as after a meal, and stores them as glycogen. When a meal has not been eaten recently, blood glucose levels fall, and the liver breaks the bonds between glucose molecules in glycogen and releases glucose monomers into the bloodstream.

Liver transplants are performed when the liver is failing due to a scarring of the liver called cirrhosis. Cirrhosis can be caused by infection with the hepatitis C virus and by chronic alcohol abuse. Liver donors can be either living or brain-dead individuals. Living donors can be used for liver transplants because the liver has a remarkable characteristic that other adult organs do not share—it can regenerate itself. If a piece of the liver is removed from a living donor and placed in a compatible recipient, then both the donor and the recipient's liver will grow to full size. Close to 7000 liver transplantations occur every year in the United States.

Unlike the liver, most organs of the digestive system are not suitable for transplantation. After you have gained an understanding of the digestive system as a whole, we will discuss the few digestive system organs that can be transplanted.

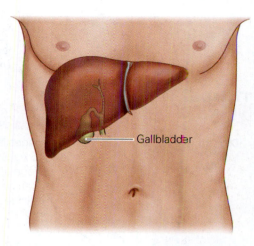

Figure 17.6 The liver. The human liver is located on the right side of the abdomen below the diaphragm. A duct connects the gallbladder to the liver.

Organ Systems: The Digestive System

The digestive system is 1 of 12 organ systems in the human body. **Table 17.1** briefly outlines the functions and component organs involved in each system.

TABLE 17.1

Organ systems and functions in humans.

Digestive
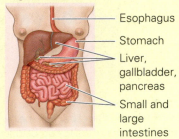
- Esophagus
- Stomach
- Liver, gallbladder, pancreas
- Small and large intestines

- Ingests and breaks down food so that it can be absorbed by the body

Cardiovascular
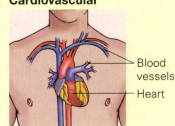
- Blood vessels
- Heart

- Enables the transport of nutrients, gases, hormones, and wastes to and from cells of the body

Excretory
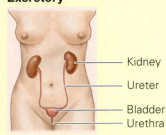
- Kidney
- Ureter
- Bladder
- Urethra

- Eliminates liquid wastes; regulates water balance

Endocrine
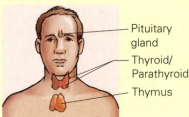
- Pituitary gland
- Thyroid/ Parathyroid
- Thymus

- Secretes hormones into bloodstream for regulation of body activities

Respiratory
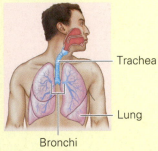
- Trachea
- Lung
- Bronchi

- Enables gas exchange, supplying blood with oxygen and removing carbon dioxide

Nervous
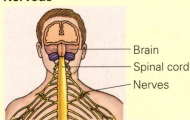
- Brain
- Spinal cord
- Nerves

- Senses environment, communicates with and activates other parts of the body

Skeletal
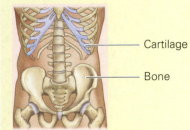
- Cartilage
- Bone

- Provides mechanical support for the body; stores minerals and produces red blood cells

Lymphatic and Immune
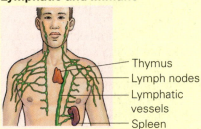
- Thymus
- Lymph nodes
- Lymphatic vessels
- Spleen

- Protects against infections

Muscular
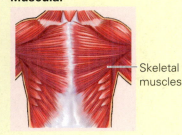
- Skeletal muscles

- Enables movement, posture, and balance via contraction and extension of muscles

Reproductive—female
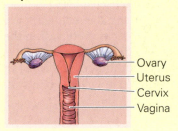
- Ovary
- Uterus
- Cervix
- Vagina

- Produces eggs and supports the development of offspring

TABLE 17.1

Organ systems and functions in humans.

Integumentary		Reproductive—male	
Hair Nails Skin	• Protects body from environment, injury, and infection; stores fat	Prostate Testicle Penis	• Produces and delivers sperm and associated fluids

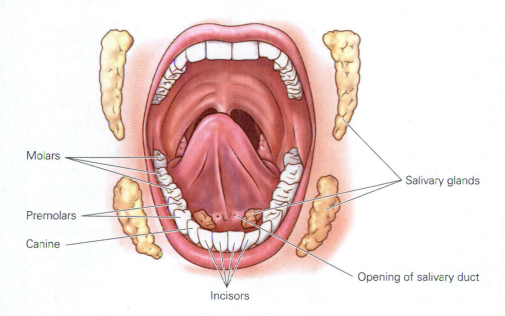

Molars

Premolars

Canine

Incisors

Salivary glands

Opening of salivary duct

Figure 17.7 The human oral cavity.
Food is ground into smaller pieces by the 32 adult teeth. Incisors are specialized to enable biting off pieces of food. Canine teeth are sharp enough to help rip food apart. Molars have broad surfaces that enhance grinding.

The job of the digestive system is to break down or metabolize food so that the body can utilize the energy stored in its chemical bonds. Food is broken down into successively smaller units as it is passed through a long tube with openings at the mouth and anus. Called the alimentary canal or digestive tract, this tube allows food to be exposed to the actions and secretions of various organs and glands. *Focus on Evolution* outlines the evolutionary trends resulting in development of the digestive tract in both humans and other animals.

The breakdown of food begins in the mouth, or oral cavity, where chewing fractures food into smaller pieces (**Figure 17.7**). Mechanical digestion of food occurs as teeth grind down food, which increases the surface area exposed to enzymes within the mouth. Chemical digestion also begins in the mouth with the secretion of saliva from salivary glands. Saliva contains the enzyme **salivary amylase,** which breaks down sugars.

The carbohydrate breakdown that begins in the mouth is the first step in the conversion of polymers that we consume (carbohydrates, proteins, and fats) into their component subunits (monosaccharides, amino acids, and glycerol and fatty acids, respectively). This conversion is necessary because the body cannot absorb the polymers you ingest or utilize them to build cellular structures directly. Instead, the polymers are broken down and rebuilt according to your body's specifications.

The **tongue** has taste buds that help you taste food. This muscular structure shapes food into a ball and pushes the ball of food, called the **bolus,** to the

The Digestive System

The human digestive system is complex. It consists of numerous glands and organs, each of which performs a specific role, and a variety of tissue types. If we compare the digestive structures of very simple organisms to evolutionarily more complex organisms, we can see the adaptations acquired during the evolution of the vertebrate digestive system.

Digestive systems have evolved from digestion inside structures called food vacuoles, to the development of a saclike digestive cavity, to the development of a one-way tube with a mouth and anus to help maximize nutrient gathering, to the development of specialized organs and glands for digestion.

The simplest eukaryotes, the protozoans, do not have a true digestive system made up of specialized cell types. However, these single-celled organisms do contain specialized organelles called food vacuoles, which contain enzymes capable of breaking down biological molecules (**Figure E17.1**). Paramecia use projections of their membranes, called *cilia*, to sweep food into the "mouth." Once inside the cytoplasm, the food is packaged into vacuoles. The engulfed food is digested by enzymes in the vacuole. Wastes are released to the outside via an anal pore.

Hydra are simple animals that have a digestive system consisting of a saclike structure with a single opening that serves as both the mouth and anus (**Figure E17.2**). Because it does not have two openings, this type of digestive system is called an **incomplete digestive system.** Hydra feed on entire organisms, which they first paralyze by stinging them with their tentacles. The paralyzed organism is then brought into the hydra's cavity, where cells secrete

enzymes to aid in digestion. This digestive sac has its limitations; all cells of the body must be in close proximity to this cavity to receive nutrients. Animals with this type of digestive system, by necessity, have very simple body plans.

While an incomplete digestive system is adequate for organisms with very simple body structures, natural selection has resulted in the production of a more complex mechanism for delivering food energy to tissues of animals with more elaborate internal anatomy. The evolution of a **complete digestive system,** consisting of two openings (a mouth and an anus), paired with a circulatory system or some other mechanism for transporting nutrients, permitted the other internal organs of animals to become more elaborate. The one-way flow of food through a complete digestive tract permits more efficient and thorough digestion, and close proximity of the digestive tract and the circulatory system allows food molecules to be carried throughout the body. The earthworm has a simple one-way digestive tract with specialized organs able to carry out digestion (**Figure E17.3**). Food enters the mouth and moves through the digestive system. Nutrients are absorbed through the intestine, and undigested foods are released through the anus.

An even greater degree of organ specialization can be seen in organisms having accessory organs that help support the function of the main digestive tract. Digestion and absorption of nutrients from food is extremely efficient in organisms with such specialized digestive systems, but at a cost—up to 30% of the energy obtained from the diets of animals with these complex systems is required simply to support the activities of digestion.

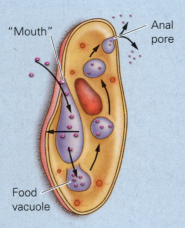

Figure E17.1 Food vacuoles in a paramecium.

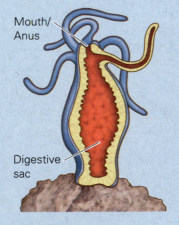

Figure E17.2 Incomplete digestive system of the hydra.

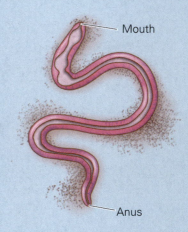

Figure E17.3 Complete digestive system in an earthworm.

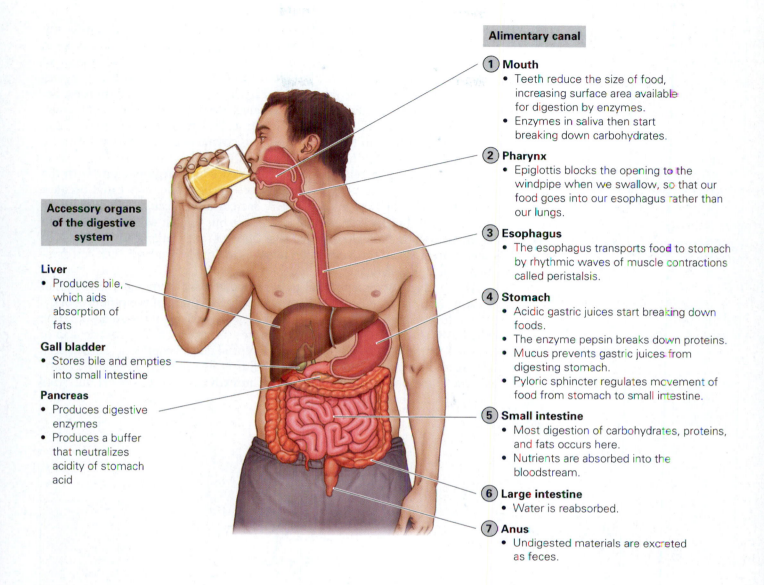

Alimentary canal

① Mouth
- Teeth reduce the size of food, increasing surface area available for digestion by enzymes.
- Enzymes in saliva then start breaking down carbohydrates.

② Pharynx
- Epiglottis blocks the opening to the windpipe when we swallow, so that our food goes into our esophagus rather than our lungs.

③ Esophagus
- The esophagus transports food to stomach by rhythmic waves of muscle contractions called peristalsis.

④ Stomach
- Acidic gastric juices start breaking down foods.
- The enzyme pepsin breaks down proteins.
- Mucus prevents gastric juices from digesting stomach.
- Pyloric sphincter regulates movement of food from stomach to small intestine.

⑤ Small intestine
- Most digestion of carbohydrates, proteins, and fats occurs here.
- Nutrients are absorbed into the bloodstream.

⑥ Large intestine
- Water is reabsorbed.

⑦ Anus
- Undigested materials are excreted as feces.

Accessory organs of the digestive system

Liver
- Produces bile, which aids absorption of fats

Gall bladder
- Stores bile and empties into small intestine

Pancreas
- Produces digestive enzymes
- Produces a buffer that neutralizes acidity of stomach acid

Figure 17.8 The digestive system.
The digestive system consists of accessory organs and the alimentary canal.

back of the mouth. From there it is swallowed and enters the digestive system (**Figure 17.8**).

When you swallow chewed food, it moves from your mouth to your **pharynx.** The pharynx, forming the back of your throat, branches to feed into both the trachea (which leads to the lungs) and the esophagus (which leads to the stomach). A flap of tissue called the **epiglottis** keeps swallowed food from entering the trachea. The **esophagus** brings food to a large digestive organ called the **stomach.** The movement of the bolus of food is hastened by rhythmic waves of smooth muscle contraction down the esophagus, through a process called **peristalsis.**

Further breakdown of food occurs in the fluid of the stomach, which has a very low pH. This acidic pH hastens the metabolism of proteins. In addition, many digestive enzymes are secreted by stomach epithelia. Stomach epithelial cells also secrete mucus to prevent the digestive enzymes from damaging the stomach.

Digested food mixes into a slurry with digestive enzymes, becoming a substance called **chyme.** Chyme is slowly allowed to enter the **small intestine,** where it is neutralized by other secretions so that the small intestine is not damaged by low pH. The small intestine of an adult is about 6 m, or close to 20 ft, long and is the major organ for chemical digestion and the absorption of nutrients into the bloodstream via its epithelium. Many of the digestive enzymes used in the small intestine are produced by an organ called the **pancreas.** Secretions from the pancreas neutralize stomach acids that enter the small intestine and contain enzymes that break down carbohydrates, fats, proteins, and nucleic acids. While in the small intestine, chyme is exposed to a substance called bile, which is synthesized by the liver.

The liver and pancreas are considered **accessory organs** along with the gallbladder. In a sense, they are accessories to the alimentary canal because they are outside the tube but produce or secrete substances required for digestion. The **gallbladder** stores and concentrates bile, which will be released into the small intestine to help dissolve fats. Concentrating bile involves removing water from it. When the gallbladder removes too much water, the bile can crystallize, causing gallstones.

The products of digestion can be absorbed across the epithelia of the small intestine into the bloodstream (**Figure 17.9**). Absorption of the products of digestion in the small intestine is facilitated by projections of the small intestine called **villi** that increase the small intestine's surface area. The surface area of the small intestine is larger than that of a typical dorm room floor! Surface area is increased even further because each villus is also covered with smaller projections of the plasma membrane, called **microvilli,** that transport nutrients into the blood vessels inside each villus.

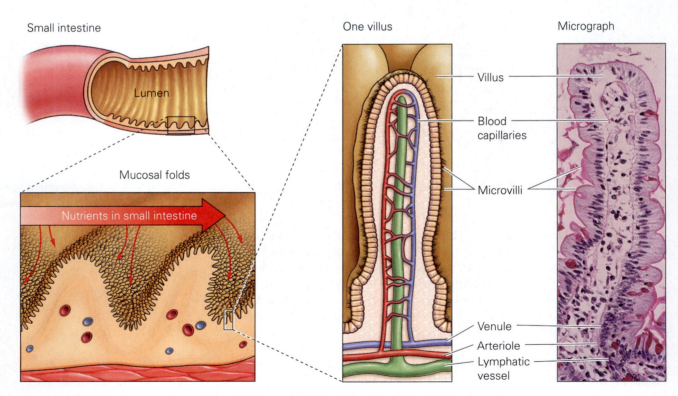

Figure 17.9 Absorption of nutrients in the small intestine. The foldings of the small intestine to produce finger-like projections called villi and microvilli increase the surface area across which nutrients can enter the bloodstream. Once in the bloodstream, nutrients can be transported to cells throughout the body.

Regulation of Digestive Secretions. The secretion of digestive juices is regulated hormonally (**Figure 17.10**). After a meal, the stomach produces the hormone *gastrin,* which stimulates the upper part of the stomach to produce more gastric juices, thus facilitating digestion.

The duodenum of the small intestine secretes the hormones *secretin* and *cholecystokinin* (CCK). Hydrochloric acid (HCl) from the stomach moves to the small intestine, where it stimulates the release of secretin. Partially digested proteins and fats stimulate the release of CCK. Once in the bloodstream, secretin and CCK cause the pancreas and gallbladder to increase their output of digestive juices.

> **Stop & Stretch**   Gastrin release is inhibited by HCl. What effect would this inhibition have on the process of digestion?

After traversing the small intestine, most polymers have been converted into monomers and can move across the small intestine into the bloodstream, where they are transported to individual cells. Materials that are not absorbed are passed through the **large intestine** (also known as the **colon**) to the anus and will exit the body as feces. Fecal matter consists largely of indigestible plant fibers.

We have now seen how a body system, in this case the digestive system, consists of many organs working together to perform a common function. The failure of some of these organs can compromise the system's ability to function properly.

Several hundred intestinal transplants are performed each year to replace intestines that have been damaged by infection or cancer. Pancreatic transplants are becoming most common for people with diabetes. When a pancreatic transplant takes place, the diabetic's original pancreas is left in place. That way, if the transplant fails, the diabetic's original pancreas can still perform its digestive functions. The gallbladder is not usually transplanted because it is possible to live a normal life without this organ. This is because the gallbladder does not synthesize bile; it only secretes it. Likewise, stomach transplantation is not commonly performed because a person can survive without a stomach. In that case, the person obtains nutrition from digestion and absorption in the small intestine only.

Failure of an individual organ can affect an entire organ system or more than one organ system. Liver failure, for instance, would have a large impact not only on the digestive system but also on the circulatory system. In addition to affecting organ systems, organ failure compromises the body's overall ability to maintain a steady state. The ability to maintain a consistent internal environment is imperative for proper health.

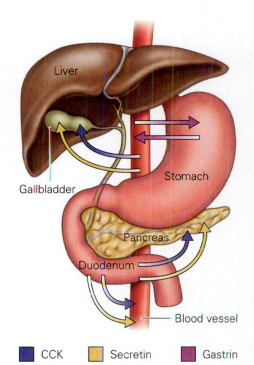

■ CCK   ■ Secretin   ■ Gastrin

Figure 17.10 Hormonal control of the digestive system. After eating, the hormone gastrin enters the bloodstream and stimulates the upper part of the stomach to produce gastric juice. Secretin and CCK from the small intestine cause the pancreas and gallbladder to increase their output of digestive substances.

17.3 Regulating the Internal Environment

An organism's ability to maintain relatively constant internal conditions under a wide array of environmental conditions is called **homeostasis** (Chapter 2). Maintaining this steady state requires the coordinated efforts of cells, tissues, organs, and organ systems. Each cell engages in basic metabolic activities that will ensure its own survival. Cells of a tissue perform activities that contribute to the proper functioning of organs; organs perform required functions, alone or in conjunction with organ systems, to contribute to the maintenance of a stable internal environment.

Animals must keep their heart rate, blood pressure, water and mineral balance, and temperature and blood glucose levels within a narrow range for survival. Regulation of temperature, called **thermoregulation,** is an example of a bodily function that in many organisms is tightly regulated. Humans and all other mammals and birds are **endotherms**—that is, organisms with the ability to maintain a body temperature that is warmer than the surrounding environment. Endotherms derive most of their body heat from metabolism. Most invertebrates, fish, amphibians, and reptiles are **ectotherms,** or organisms that obtain their body heat primarily by absorbing it from their surroundings.

Keeping physical and chemical properties within tolerable ranges utilizes feedback from one system to another. Feedback is information that is sent to a control center, which in turn directs a cell, tissue, or organ to respond by turning up or down a given process. Feedback can be negative or positive. In the following examples, we illustrate that negative feedback negates change, and positive feedback promotes change.

Negative Feedback

Negative feedback occurs when the product of the process inhibits the process. Thermoregulation in endotherms occurs by negative feedback in a manner very similar to the way that temperature is controlled in a house. Temperature in a house is regulated by a thermostat that functions as a sort of control center to turn the heater on when the room becomes too cold. When the room becomes too hot, the heater is turned off. Therefore, the negative feedback of the temperature change regulates heating systems.

Body temperature is controlled in a similar manner. When the body temperature increases, a control center in the brain sends signals to the body to dilate blood vessels near the skin and activate sweat glands, allowing heat to escape. When the body temperature cools, blood vessels near the surface of the skin constrict, muscles shiver, and body temperature increases. These negative-feedback mechanisms keep the body temperatures of endotherms within a very narrow range.

Regulation of blood glucose levels by the liver provides another example of negative feedback (**Figure 17.11**). When food is digested, the amount of glucose in the blood increases. This stimulates the pancreas to produce insulin, a hormone that stimulates the liver (along with fat and muscle tissues) to remove glucose from the blood and store it as glycogen.

When most of the glucose has been removed from the blood, the pancreas secretes a hormone called glucagon, which stimulates the liver to break down glycogen and secrete glucose units into the bloodstream. Consequently, negative regulation of blood glucose levels reverses the direction of diet-induced changes in glucose levels, keeping glucose levels from swinging wildly after a meal and confining glucose levels to within a very narrow range—thus promoting homeostasis.

Positive Feedback

Positive feedback occurs when the product of the process intensifies the process. For example, during labor and childbirth, hormones cause the muscles of the uterus to contract, pushing the baby out of its mother's body. The uterine contractions stimulate the release of more hormones, which cause the contractions to intensify in rate and duration. This in turn causes even more hormones to be released and results in even more and longer contractions. In animals, most homeostatic regulation is via negative feedback because positive feedback amplifies a response and can actually drive the process away from homeostasis.

Damaged and diseased organs do not respond properly to signals that help promote homeostasis. Transplanting healthy organs to replace defective ones can restore, at least temporarily, homeostasis to a diseased individual.

(a) If blood glucose level rises...

Liver converts glucose to glycogen.

(b) If blood glucose level falls...

Liver breaks down glycogen into glucose and releases glucose into bloodstream.

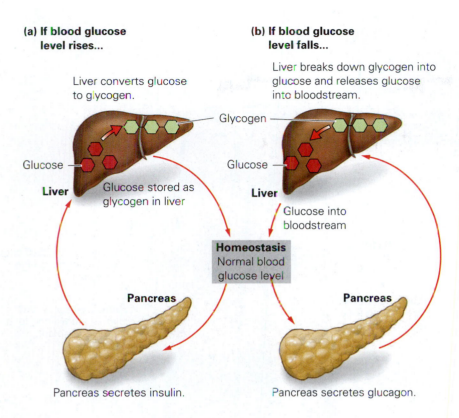

Glycogen

Glucose

Liver   Glucose stored as glycogen in liver

Glucose

Liver   Glucose into bloodstream

Homeostasis Normal blood glucose level

Pancreas

Pancreas

Pancreas secretes insulin.

Pancreas secretes glucagon.

Figure 17.11 Feedback in the regulation of blood glucose levels.
(a) When intake of glucose is high, such as after a meal, the amount of glucose in the blood increases. The excess is stored as glycogen. (b) When blood glucose is low, glucose is secreted into the bloodstream.
Visualize This: Blood glucose homeostasis can be disrupted by factors other than food you ingest. What might also cause a disruption of glucose homeostasis?

Organ Donation

The young motorcyclist was a good candidate for organ donation because he died from a brain injury. People who die from cardiac death can be tissue donors, but the lack of oxygen experienced during cardiac death causes organs to deteriorate, making such people less suitable organ donors. Since many more people will die of cardiac death than brain injury, tissues can be banked for future use, but there is a shortage of organs for use in transplant operations.

If the family agrees, the young man will be kept on a ventilator so that his organs continue to be nourished with oxygen and blood until the organs can be surgically removed. While the organs are being removed, medical personnel will attempt to find the best recipient. To select the recipient, a search of a computerized database is performed. Starting at the top of the list, medical staff will narrow down possible recipients, based in part on how long the recipient has been waiting. People at the top of the waiting list for a given organ tend to be very ill and will die if a transplant does not become available in a matter of days or weeks. These people are often hospitalized or waiting at home or in hotels near the hospital, hoping to receive a call telling them that an organ has become available. Recipients must be located close to the hospital since donor organs cannot be preserved indefinitely; they must be transplanted as soon as possible. Lungs, for instance, can be preserved for around 6 hours before the transplant operation must begin.

Organ donors and recipients must have the same blood type so that the recipient's immune system does not react against the organ. Likewise, donors and recipients are matched for the presence of certain markers on the surface of their tissues. When it has been determined that blood and marker types of donor and recipient are a close enough match, a transplant may occur. The most commonly transplanted organs are the liver, kidney, heart, and lungs. Around 10% of people on waiting lists for replacement organs will die before one becomes available (**Figure 17.12** on the next page).

If you would like to be an organ donor, discuss your plans with your family. Even if you carry a signed organ donor card and indicate your preference to be

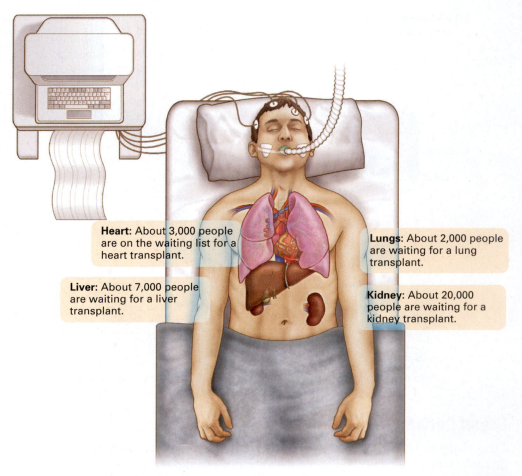

Heart: About 3,000 people are on the waiting list for a heart transplant.

Lungs: About 2,000 people are waiting for a lung transplant.

Liver: About 7,000 people are waiting for a liver transplant.

Kidney: About 20,000 people are waiting for a kidney transplant.

Figure 17.12 Organ donation saves lives. Thousands of lives can be saved each year through organ donation.

a donor on your driver's license, your family will need to agree to donate your tissues and organs. If you should suffer an accidental death, like the young motorcyclist, your family will be forced to make many difficult decisions quickly. Having a discussion about organ donation now will relieve them of the burden of trying to make this decision for you.

SAVVY READER

Trafficking in Kidneys

Individuals who sell a kidney do not receive a long-term economic benefit from the sale and may have a worsening of their health, according to a study in the October 2 issue of *The Journal of the American Medical Association* (JAMA).

Study Reveals Consequences for People Who Sell Kidneys

The investigation, led by Madhav Goyal from Geisinger Health System in State College, Pa., and Ashwini Sehgal from the CWRU School of Medicine, involved 305 individuals in southern India who had sold a kidney an average of six years earlier for about $1,000 apiece (about twice their annual family income).

The study compared the sellers' current economic and health status with their economic and health status before they sold a kidney. The income of the sellers declined by one-third after selling a kidney, and 86 percent had a worsening of their health status.

Because nearly every country has a shortage of organs for transplantation, providing financial incentives to donate is often proposed or justified as a way to benefit recipients by increasing the supply of organs and to benefit donors by improving their economic status.

In the United States, most organs for transplantation come from brain-dead individuals whose families have consented to donate. Providing financial incentives such as payment for funeral expenses has been proposed as a way to increase the supply of organs.

In many developing countries such as India, most organs come from living donors, and the sale of kidneys is widespread. This practice is justified as a way to both increase the supply of kidneys for transplantation and to benefit the seller economically. However, critics argue that purchasing kidneys amounts to exploitation of the poor and that the poor do not overcome poverty as a result of the sale.

The researchers also sought to determine whether individuals who sell a kidney actually benefit from the sale.

"We were surprised by our findings," Goyal said. "The sellers' economic and health status appeared to worsen after selling a kidney. Most of them would not recommend that others sell a kidney."

"These findings undercut arguments in favor of financial incentives," Sehgal said. "Even modest financial incentives such as those proposed in the United States may be perceived by the general public as taking advantage of poor families. If this happens, such incentives may actually lead to fewer total organ donations."

1. What is the affiliation of the scientists who performed this study?

2. The scientists responsible for this study claimed that they were surprised by their findings. What did they do when they found that data did not turn out the way they had expected?

3. Would you find a report by an organ-trafficking agency outlining the benefits of an organ sale more or less credible than this article? Why?

Campus News Case Western Reserve University, October 17, 2002.

Chapter Review

Learning Outcomes

Mastering BIOLOGY®

Go to the Study Area at www.masteringbiology.com for practice quizzes, myeBook, BioFlix™ 3-D animations, MP3Tutor sessions, videos, current events, and more.

LO1 Describe the structure and function of epithelial tissues (Section 17.1).

- Epithelia line and cover organs, vessels, and body cavities. They are tightly packed tissues with one free surface. Outgrowths of epithelia form some glands. These tissues function in protection, secretion, and absorption (pp. 420–421).

LO2 Compare the structures and functions of the six connective tissues (Section 17.1).

- Loosely packed connective tissues bind tissues and organs to each other. The six types of connective tissues are each composed of a characteristic cell type, embedded in a characteristic matrix (pp. 421–423).
- Loose connective tissue connects epithelia to underlying tissues and holds organs in place. The cellular component is fibroblasts, and the matrix is rich in the proteins collagen

and elastin, which provide this tissue with tensile strength and elasticity (pp. 421–422).
- Adipose tissue connects the skin to underlying structures and insulates and protects organs. Cells of this tissue synthesize and store fat and have very little extracellular matrix (p. 422).
- Blood is a type of connective tissue that transports oxygen and nutrients to body cells. Blood cells have a liquid matrix called plasma (p. 422).
- Fibrous connective tissue forms the tendons and ligaments. The cells of this tissue are fibroblasts, and the matrix is rich in collagen (p. 422).
- Cartilage is a flexible, shock-absorbing tissue composed of cells called chondrocytes that secrete a dense, collagenous matrix (p. 422).
- Bone tissue provides support for the body. The cells of this tissue are called osteocytes, and the matrix is rich in collagen and minerals (pp. 422–423).

LO3 Compare the structures and functions of the three types of muscle tissues (Section 17.1).

- Muscle tissues are composed of fibers that contract and conduct electrical impulses (p. 423).
- The three types of muscle tissue can be either voluntary or involuntary and smooth or striated (p. 423).
- Skeletal muscle is attached to bones, responsible for voluntary movements, and striated (pp. 423–424).
- Cardiac muscle comprises the heart; it is involuntary and striated (pp. 423–424).
- Smooth muscle comprises the musculature of internal organs and blood vessels. Smooth muscle is involuntary and not striated (pp. 423–424).

LO4 Describe the structure and function of nervous tissues (Section 17.1).

- Nervous tissue, found in the brain and spinal cord, is composed of neurons; it senses stimuli and transmits signals throughout the body (pp. 424–425).

LO5 Explain what an organ is and how organs interact in an organ system (Section 17.2).

- Organs are groups of tissues working in concert (pp. 426–427).
- Organ systems are suites of organs working together to perform a function or functions (pp. 426–427).

LO6 List the structures and functions of the liver (Section 17.2).

- The liver, like all organs, is composed of different tissues whose combined functions give rise to the organ's overall function (p. 427).
- The epithelial tissue of the liver is divided into lobes. Lobes are divided by connective tissue into lobules. Lobules contain a central vein that allows blood to reach the hepatocytes, which function as filters to remove toxins and pathogens (p. 427).
- Other functions of the liver include the production of proteins and cholesterol, secretion of bile, and storage of vitamins and glycogen (p. 427).

LO7 List the organs of the digestive system, and outline their functions (Section 17.2).

- The digestive system is a group of organs and glands working together to break foods into their component parts for reassembly into forms that the body can use or for use in generating energy (pp. 428–429).
- Food moves from the mouth to the pharynx and through the esophagus to the stomach to the small intestine. Digested nutrients are absorbed into the bloodstream across the small intestine and brought to cells (pp. 429 and 431).

LO8 List the accessory organs of the digestive system, and outline their roles in digestive processes (Section 17.3).

- The pancreas, liver, and gallbladder are accessory organs that secrete substances that aid in digestion (pp. 431–432).
- Digestive enzymes produced by the pancreas help break down most food molecules. Bile produced by the liver facilitates the breakdown of fats, and the gallbladder stores and concentrates bile before its release into the small intestine (pp. 431–432).

LO9 Explain how positive and negative feedback mechanisms help in the maintenance of homeostasis (Section 17.3).

- Homeostasis is the ability to maintain a constant internal environment even when the external environment changes (pp. 433–434).
- Negative feedback is one mechanism that helps facilitate homeostasis. When negative feedback occurs, the product of the process acts to turn down the process. Negative feedback negates change (p. 434).
- Positive feedback is a less commonly used homeostatic mechanism. When positive feedback occurs, the product of the process intensifies the process. Positive feedback promotes change (pp. 434–435).

Roots to Remember

The following roots of words come mainly from Latin and Greek and will help you decipher terms:

adipo- means fat. Chapter term: adipose tissue

dia- and **trans-** mean through. Chapter terms: diaphragm and transplant

epi- means on or upon. Chapter term: epithelium

homeo- means the same. Chapter term: homeostasis

lig- means to join or connect. Chapter term: ligament

neuro- means nerve. Chapter term: neuron

osteo- relates to bone. Chapter term: osteoblast

thermo- relates to warmth and heat. Chapter term: thermoregulation

Learning the Basics

1. **LO1 LO2 LO3 LO4** Describe the four tissue types found in animals.

2. **LO2** Name the six types of connective tissue.

3. **LO3** What are the differences between the three separate types of muscle tissue?

4. LO1 Epithelia _____.

A. are loosely packed tissues that hold organs in place;
B. can be free-floating tissues such as blood; C. line
organs and cavities and have one surface exposed;
D. include the tissues that hold joints together; E. are
tissues that conduct nerve impulses

5. LO2 Connective tissues include _____.

A. cartilage; B. blood; C. fat; D. bone; E. all of the above

6. LO3 Muscle tissues _____.

A. contain actin and myosin, whether or not they are
striated; B. that facilitate movements requiring conscious
thought are involuntary; C. increase in cell number with
exercise; D. that comprise the heart are called smooth;
E. all of the above

7. LO6 Which of the following is not a function of the liver?

A. storing bile; B. filtering toxins; C. producing
cholesterol; D. storing glycogen

8. LO7 The pharynx _____.

A. forms the connection between the small and large
intestine; B. keeps swallowed food from entering the
epiglottis; C. connects the esophagus to the stomach;
D. branches to feed into the trachea and esophagus

9. LO9 When a woman is breastfeeding, the more her
infant drinks, the more milk she produces. This is an
example of _____.

A. negative feedback; B. positive feedback;
C. thermoregulation; D. independent regulation

10. LO8 Bile _____.

A. is stored in the pancreas; B. helps break down
glycogen; C. emulsifies fats; D. removes water from
indigestible materials in the large intestine; E. all of
the above

11. LO4 Which of the following cell types is found in
nervous tissue?

A. osteocyte; B. melanocyte; C. leukocyte; D. neuron;
E. brain and spinal cord

12. LO5 **True or False:** An organ system is a group of
tissues, composed of similar cell types, with a common
function.

Analyzing and Applying the Basics

1. LO7 Describe what happens to food as it moves through
the alimentary canal.

2. LO8 List the accessory organs in the digestive system.

3. LO9 Describe a nonbiological example of both negative
and positive feedback.

Connecting the Science

1. Why might families refuse to donate the tissues and organs of a loved one?

2. Do you think private citizens should be able to sell their
own organs? Why or why not?

Answers to **Stop & Stretch, Visualize This, Savvy
Reader,** and **Chapter Review** questions can be found in
the **Answers** section at the back of the book.

CHAPTER 18

Clearing the Air

Respiratory, Cardiovascular, and Urinary Systems

Many communities have banned smoking in all public places, including restaurants and bars.

Supporters of these bans say that secondhand tobacco smoke is a public health threat.

Like many trends, it started in California. In 1977, Berkeley became the first city in the United States to limit smoking in public places, including restaurants. In 1983 the city of San Francisco passed the first ban on smoking in private workplaces. In 1995 the state banned smoking in all indoor spaces but gave bars and restaurants a few years to come up with strategies for effectively protecting employees and nonsmoking customers from the smoke generated by their cigarette-using patrons. By 1998, a total ban went into effect.

States and cities across the country—even Louisville, Kentucky, once home to the world's third-largest tobacco company—have followed California's lead and enacted their own bans on indoor smoking. Today, 22 states have complete indoor smoking bans, and over 50% of the U.S. population is subject to smoking restrictions of some kind. The trend has even carried overseas; Ireland became the first smoke-free country, and Bhutan became the first country to ban the sale of all tobacco-containing products in 2004. England, Norway, and several eastern European countries, such as Estonia, Albania, and Hungary, have also joined the smoke-free club since then.

Opponents include smokers, who feel discriminated against while engaging in a lawful activity . . .

While many nonsmokers celebrate the enforcement of smoking bans because they remove what they see as annoying odor and haze from their favorite restaurants, bars, and bowling alleys, ban supporters have also justified these restrictions as essential to protecting public health. Over the last dozen years, an emerging medical and scientific consensus about the dangers of "secondhand smoke" has linked the airborne particles and gases that are emitted by the lit end of cigarettes, cigars, and pipes to illnesses from lung cancer to heart disease.

. . . as well as bar owners, who worry that smoking bans will cost them business.

LEARNING OUTCOMES

LO1 List the components of the mammalian respiratory system, and describe the path of air into the body.

LO2 Describe the muscles involved in breathing, including the diaphragm, and explain how their movements facilitate air movement into and out of lungs.

LO3 Explain the role of hemoglobin in gas exchange.

LO4 List the organs and tissues of the mammalian circulatory system, and describe the function of each.

LO5 Describe how blood moves through the double circulation system of the heart.

LO6 Explain the steps of the cardiac cycle, including the role of the heart's electrical system.

LO7 List the structures of the mammalian urinary system.

LO8 Describe the steps in the process of urine excretion.

LO9 Describe the effects of smoking on the respiratory, cardiovascular, and urinary systems.

The passage of these bans has never been without controversy, however. While nonsmokers demand a right to smoke-free air, smokers claim a right to engage in a lawful activity where and when they wish. While restaurant employees advocate for work environments that are healthful, bar owners worry about the effects of a ban on their bottom line. Citizens in every community that is considering a ban try to find the right balance between protecting public health and respecting the rights of private individuals and property owners.

Is there a public place for smoking, or do the risks demand that we protect all nonsmokers from the consequences of secondhand smoke? In this chapter, we examine the pathway of secondhand smoke into a nonsmoker's lungs and body and investigate the evidence that links exposure to this smoke to preventable diseases that affect the lungs, heart, and other organs.

18.1 Effects of Smoke on the Respiratory System

The "secondhand" smoke emitted by the lit end of a cigarette, combined with the smoke exhaled by active smokers, is more accurately referred to as **environmental tobacco smoke,** or **ETS.** This smoke affects not only the **active smoker,** who is holding and inhaling from the cigarette, but also nonsmoking individuals in the environment. A nonsmoker in an environment high in ETS is called a **passive smoker.** ETS is similar to the smoke inhaled by an active smoker, but it has some key differences.

According to the National Cancer Institute, more than 4500 chemicals have been identified in tobacco smoke, and there may be as many as 100,000 chemicals in this mix of gases and particles. Passive smokers are exposed to the same mix of chemicals as active smokers, although typically in much lower concentrations. Surprisingly, certain chemical compounds are actually higher in concentration in ETS than in actively inhaled smoke. These compounds are produced when the tobacco burns incompletely on the end of a cigarette or in the bowl of a pipe in between puffs. Just like when someone blows on a smoldering fire, when a smoker takes a drag on a cigarette, oxygen is provided to the smoldering tobacco leaves, causing them to burn more completely.

Carbon monoxide is a by-product of incomplete combustion. It is also the most abundant gas in ETS. In fact, carbon monoxide is approximately five times more abundant in ETS than it is in the smoke inhaled by active smokers. You may have heard of carbon monoxide as a poisonous air pollutant that occurs inside homes with faulty furnaces or other gas-burning appliances. While the levels of carbon monoxide in ETS alone are not deadly to passive smokers, effects of exposure to elevated levels of carbon monoxide can lead to serious consequences.

Also as a result of incomplete combustion, ETS contains high concentrations of airborne **particulates**, particles with a diameter less than half the width of

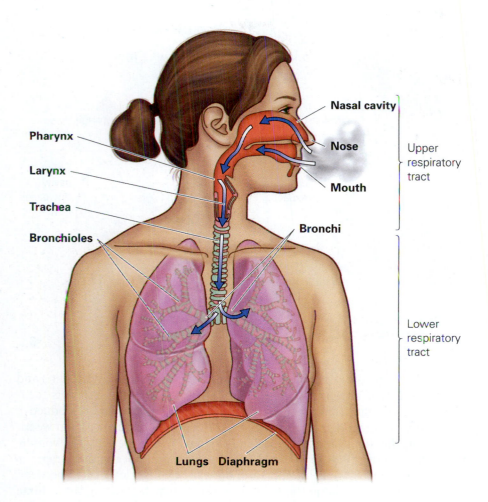

Figure 18.1 The human respiratory system. Smoke enters the upper respiratory system through the nasal cavity or mouth, then passes through the pharynx and larynx into the lower respiratory system. Here, it flows into a series of smaller and smaller tubes within the lungs until it comes into close contact with circulating blood.

a human hair. These particles make up the visible portion of ETS and are commonly known to cigarette smokers as tar. Because most active smokers inhale cigarette smoke through a filter tip, the amount of particles acquired by primary smoking is reduced. However, because neither active nor passive smokers breathe through a filter otherwise, they are both exposed to the full concentration of particles present in ETS. Chronic exposure to airborne particulates can lead to a number of negative health effects, including emphysema and cancer.

Airborne particulates enter the human body via the mouth and nose; pass through the **pharynx** (throat) and **larynx** (voicebox) to the distribution network of the **trachea** (windpipe), **bronchi,** and **bronchioles;** and ultimately settle in the **lungs,** the site of gas exchange (**Figure 18.1**), where they cause some of their most severe consequences. However, some of the chemicals in these particles, as well as carbon monoxide gas, have the ability to cross from the lungs into the bloodstream and thus can be carried throughout the body. To understand how chemicals pass from the air into our bodies, first we must understand the structure and function of lungs.

What Happens When You Take a Breath?

You take a breath approximately 12 times per minute, or over 6 million times per year, without thinking much about it. In each normal breath at rest, you take in about 500 milliliters—approximately 2 cupfuls—of air and release the same amount. Over the course of a single minute, about 6 liters of air come into direct contact with the surfaces of your lungs.

Figure 18.2 How the diaphragm works. (a) Contraction of the dome-shaped diaphragm flattens it out, increasing the volume of the chest cavity and lowering air pressure. (b) Relaxation of the diaphragm causes the chest cavity to lose volume, increasing air pressure. **Visualize This:** How does a hiccup, an involuntary contraction of the diaphragm, affect normal breathing?

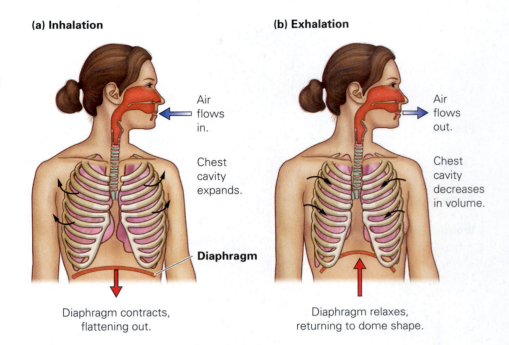

(a) Inhalation

Air flows in.

Chest cavity expands.

Diaphragm

Diaphragm contracts, flattening out.

(b) Exhalation

Air flows out.

Chest cavity decreases in volume.

Diaphragm relaxes, returning to dome shape.

Diaphragm. In contrast to many other animals, respiration in humans and other mammals is an active process. The human respiratory system is separated from the organs that make up the bulk of the digestive and reproductive systems by a strong, dome-shaped muscle called the diaphragm. When we inhale at rest, the diaphragm contracts and flattens out, increasing the volume of our chest cavity (**Figure 18.2a**). At the same time that the diaphragm is contracting, the rib cage is lifted up and outward by muscles located between the ribs, further increasing the volume of the cavity. Increasing the volume decreases the air pressure inside the chest, causing air to flow into the lungs.

The change in pressure in our chest cavity during inhalation is similar to the change in the air pressure inside a syringe when the plunger is pulled back. If the syringe is in a liquid solution, we can see the liquid flow toward the lower pressure inside the syringe, just as air flows to the lower-pressure environment inside the lungs.

When we are resting, exhalation requires no muscle contraction. The diaphragm relaxes, and the volume of the chest cavity decreases (**Figure 18.2b**). The air that had rushed into the lungs is squeezed into a smaller volume, increasing the pressure and causing air to flow back out of the trachea and the nasal cavity. Passive exhalation is similar to the release of air from an expanded balloon. Because the elastic walls of the balloon exert pressure on the contents, air rushes out of an open nozzle even without actively squeezing the balloon.

Exhalation becomes an active process during intense aerobic exercise, when the body must actively rid itself of carbon dioxide. In this case, muscles in the abdomen contract, forcing the diaphragm upward, and muscles in the chest wall tighten, squeezing the chest cavity further. Forcing liquid out of a syringe by depressing the plunger and reducing the volume is much like the process of active exhalation.

Stop & Stretch The Heimlich maneuver is a strategy for forcing an obstruction out of the trachea of a person who is choking by forcibly squeezing the abdomen directly under the rib cage. Use your understanding of inhalation and exhalation to explain why this maneuver is effective.

Our rate of breathing is regulated by control centers in the most ancient part of the brain, called the brain stem. The brain stem signals the diaphragm to contract in response to carbon dioxide levels in the blood. The breathing rate increases when carbon dioxide levels in the body increase, such as during exercise. Recall that the activities of cells require energy in the form of ATP, which is produced when carbon-containing food molecules are broken down, resulting in the production of carbon dioxide (Chapter 3). Rapid breathing moves this carbon dioxide out of the body more quickly while also providing oxygen for optimal ATP production.

Contraction of the diaphragm can also be controlled voluntarily. Holding your breath requires consciously overriding the signals from the brain stem—at least for a time. Speaking or making any vocal noises requires active exhalation as air forced out past the vocal cords in the larynx causes these structures to vibrate and produce sound. Taking a drag on a cigarette requires stronger contraction of the diaphragm than normal breathing does to pull air through the burning tobacco and into the mouth and lungs.

The amount of air forced past the vocal cords determines the volume of our speech, while muscles that control the length of the vocal cords help to determine the pitch of our speech. The shape of our mouth, lips, and tongue and the position of our teeth determine the actual sound that is produced. Sustained exposure to tobacco smoke can cause parts of the larynx to become covered with scar tissue, often making long-time smokers sound quite hoarse. Non-smokers regularly exposed to environmental tobacco smoke demonstrate some changes to their vocal cords relative to other nonsmokers, although it is not clear that these changes affect the voice.

The diaphragm also may contract involuntarily for other reasons, for instance, when we inhale deeply before a cough. Coughing is a reflex that helps remove irritants from the trachea. Keeping air passageways into the lungs clear is crucial for maintaining adequate oxygen uptake.

Lungs. Healthy lungs are pink, rounded, and very spongy in texture (**Figure 18.3a**). Even at rest, these organs almost completely fill the chest cavity. The external surface of the lungs is linked to the lining of the chest cavity by two thin membranes, one covering the lungs and one lining the chest wall. These membranes stick together because they are moist, much like two wet pieces of plastic wrap stick together because of the cohesive attraction among water molecules. This ensures that when the chest cavity enlarges, the lungs expand. If the two membranes become separated—for instance, as a result of injury to the chest wall—then air will enter the space between them, sometimes causing the underlying lung to collapse.

(a) Human lungs

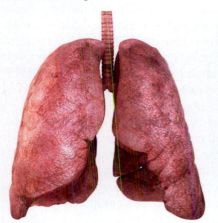

(b) Tennis court

Figure 18.3 The respiratory surface of the human body. (a) Lungs contain the surfaces over which gases are exchanged between the body and the environment. (b) The total respiratory surface area contained within a pair of healthy human lungs is approximately the size of a tennis court.

Figure 18.4 Alveoli. (a) Alveoli occur in clusters at the end of bronchi. The clusters are enmeshed in a dense network of thin-walled capillaries, which bring blood into close proximity with the gas in the alveoli. (b) An electron micrograph of alveoli in lungs.

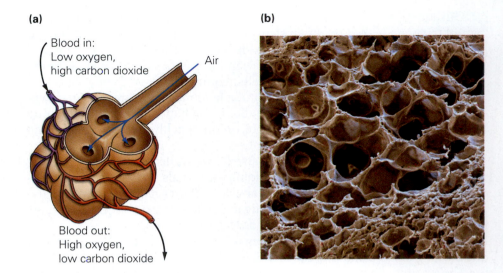

(a)

Blood in:
Low oxygen,
high carbon dioxide

Air

Blood out:
High oxygen,
low carbon dioxide

(b)

Once air enters the lungs, it flows through the bronchi and bronchioles, each of which dead-ends at structures called **alveoli.** The alveoli are 300 million tiny sacs within our lungs that contain the **respiratory surface** of our bodies, the tissue across which gases from inside the body are exchanged with gases in the air. The area of this respiratory surface in humans is approximately 160 square meters (1725 square feet), about the size of a tennis court (**Figure 18.3b**). A large surface area is required to supply oxygen and remove carbon dioxide from large-bodied, active mammals. Each grape-like cluster of alveoli is surrounded by a net of tiny blood vessels, called **capillaries,** connecting the gases exchanged in these structures with the entire body (**Figure 18.4**).

Exposure to smoke can damage and eventually destroy the respiratory surfaces within alveoli. A reduction in the total amount of respiratory surface results in shortness of breath, wheezing, and an inability to participate in even moderately vigorous activity. These symptoms are all caused by the reduced capacity for gas exchange in the lung.

Gas Exchange

The primary function of the lungs is **gas exchange,** the process that allows us to acquire oxygen for cellular respiration and to expel carbon dioxide, one of the waste products of respiration. Exchange of gases occurs between the air spaces of the alveoli and the blood contained in neighboring capillaries. The exchange works by simple diffusion across the thin membranes surrounding the blood vessels and the alveoli walls. Recall that diffusion is the passive movement of substances from where they are in high concentration to areas where they are in low concentration. Carbon dioxide in high concentration within the capillaries passes via diffusion from these structures into the alveoli, where it is maintained in low concentration by exhalation. Oxygen in high concentrations, maintained by inhalation, passes by diffusion from the alveoli to the deoxygenated blood in the capillaries (**Figure 18.5**).

Carbon dioxide and oxygen gas dissolve in the fluid that coats the respiratory surface prior to diffusing across the cell membrane. Because the size of the alveoli increases and decreases over the course of a single breath, this layer of fluid must be somewhat slippery to maintain constant coverage of the surface and to prevent the alveoli walls from sticking together. A soap-like substance called surfactant provides this slippery consistency. The composition of surfactant can be negatively affected by tobacco smoke.

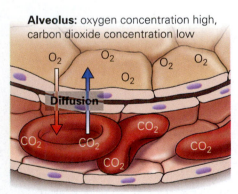

Alveolus: oxygen concentration high, carbon dioxide concentration low

O_2 O_2 O_2 O_2

O_2

Diffusion

CO_2 CO_2 CO_2 CO_2

CO_2

Capillary: oxygen concentration low, carbon dioxide concentration high

Figure 18.5 Gas exchange in the lungs. The movement of oxygen and carbon dioxide gas at the respiratory surface occurs via simple diffusion.
Visualize This: Oxygen levels in the atmosphere are low at high altitudes. How do you think this affects gas exchange in the lungs?

Stop & Stretch Respiratory distress syndrome (RDS) is the most common cause of death for premature infants (born earlier than 7 months of gestation). Babies with RDS typically have low levels of surfactant in the lung. Describe how a reduction in surfactant can lead to difficulty inhaling and exhaling.

The Role of Hemoglobin in Gas Exchange. Carbon dioxide is in relatively low concentrations in the atmosphere, so it readily moves from the blood into the alveoli and out into the atmosphere. However, oxygen does not dissolve as readily into the blood from the air; thus, all animals (and even some plants) produce specialized proteins called respiratory pigments that can bind oxygen. Respiratory pigments contain a metal atom that produces a color when oxygen binds to it. For instance, the pigment in insects contains copper, which produces the greenish "blood" you can see when a large insect is crushed. When these respiratory pigments are located in a circulatory system, they quickly whisk oxygen away from the respiratory surface. This process helps to maintain the maximum concentration difference between the outside air and circulatory fluid, increasing the rate of oxygen diffusion.

Hemoglobin is the respiratory pigment in humans that acquires and transports oxygen. Hemoglobin is actually made up of four separate protein chains, each containing a single iron atom that can bind to a single oxygen molecule (**Figure 18.6**). In its deoxygenated form, hemoglobin is a brownish color. When oxygen attaches to the iron atom in the protein chain, the hemoglobin becomes bright red. Each of our red blood cells contains approximately 250 million hemoglobin molecules; therefore, each cell can carry about 1 billion oxygen atoms.

In areas of the body where oxygen levels are low and carbon dioxide production is high, hemoglobin releases oxygen molecules. In this way, red blood cells are loaded with oxygen in the lungs and will drop their load in areas of the body where high levels of cellular respiration are occurring. Hemoglobin in the veins still carries quite a bit of oxygen, giving the blood in these vessels the familiar deep "blood-red" color. Although in this book and nearly all anatomy drawings the deoxygenated blood in the body is colored blue, it is not this color and is only drawn so by convention. Blood in veins near the skin in fair individuals appears blue only because of the way the skin scatters light—essentially the same effect that causes the sky and ocean to appear blue as well.

Hemoglobin picks up oxygen effectively, but surprisingly, it binds to the smoke by-product carbon monoxide about 200 times more strongly. Carbon monoxide inhaled by breathing tobacco smoke is preferentially loaded into red blood cells at the respiratory surface. Because the binding of carbon monoxide is so strong, hemoglobin is slow to release it. As a result, fewer hemoglobin molecules are available to transport oxygen. Even small amounts of carbon monoxide can tie up large amounts of hemoglobin in the body, causing severe oxygen shortages in body tissues. While the amounts of carbon monoxide in ETS are not high enough to cause death from lack of oxygen, chronic low levels of oxygen deprivation can damage tissues and organs.

Carbon monoxide is especially damaging to developing embryos and fetuses because they must acquire the oxygen they need through exchange with their mother's blood supply. The lower-than-average birth weights of babies born to smoking mothers may be due to their relative oxygen deprivation. There is some evidence from animal studies that long-term exposure to the carbon monoxide levels found in ETS may contribute to diminished brain function in infants and children for the same reason—oxygen deprivation.

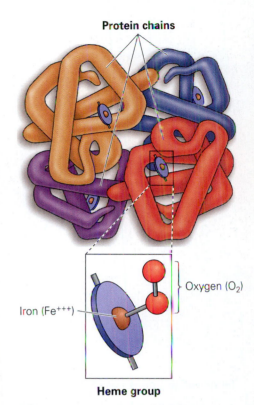

Protein chains

Oxygen (O_2)

Iron (Fe^{+++})

Heme group

Figure 18.6 Hemoglobin, the blood's oxygen shuttle. The four protein chains in a single hemoglobin molecule are colored differently here, but function identically. The heme group in each chain is the oxygen-binding site.

The biggest threat ETS poses to respiratory function does not stem from the interaction between hemoglobin and carbon monoxide. Instead, the threat comes from damage caused to lungs by microscopic particles of smoke.

Smoke Particles and Lung Function

The particulates in ETS are a significant hazard to infants and children as well as to anyone with impaired lung function, including individuals who have been chronically exposed to particulate-laden air.

Bronchitis and Asthma. Our lungs have various mechanisms for expelling and removing foreign objects. Coughing is our first response to larger particles or droplets that enter the trachea. Particles too small to trigger a cough can settle in the upper respiratory tract, trachea, or larger bronchi. These particles become trapped in mucus produced by the tissue lining these structures. The mucus and particles are swept upward toward the mouth and nose by the actions of tiny hairs called cilia on the tissue surface. When a glob of mucus reaches the tip of the larynx, it is coughed up, expelled out of the nose or mouth, or swallowed.

Airborne particles in tobacco smoke not only increase the production of mucus but also damage the cilia lining the bronchi, making it more difficult to expel these particles. This damage in turn can lead to inflammation of the bronchi, known as **bronchitis.** Particulates are also known to exacerbate **asthma,** an allergic response that results in the constriction of bronchial walls and an overproduction of mucus. According to the U.S. Environmental Protection Agency (EPA), the number of additional cases of asthma caused by ETS among children in the United States is likely to be 26,000 per year.

Emphysema. The physical consequences of chronic bronchitis and asthma can lead to the formation of scar tissue in the lungs, permanently blocking bronchi. The inflammation also causes damage to alveoli walls, causing small alveoli to merge into fewer, larger sacs (**Figure 18.7**). This merging of alveoli reduces the overall area of the respiratory surface, interfering with gas exchange. The sacs themselves become surrounded by thick scar tissue, which interferes with the passage of oxygen into the capillaries. The buildup of scar tissue also makes the lungs less elastic, meaning that the passive process of exhalation is ineffective. As more "dead air" remains in the lungs, they become overinflated, gradually increasing the size of the chest. A barrel-shaped chest is a characteristic physical sign of the underlying disease, called emphysema.

Because alveoli cannot be regenerated, the lung damage that results in emphysema is permanent and irreversible. Individuals with emphysema are chronically short of breath and unable to participate in vigorous activity. The most severely affected individuals require supplemental oxygen simply to engage in the activities of daily living. The EPA estimates that adult nonsmokers exposed to ETS have a 30% to 60% higher occurrence of emphysema, asthma, and bronchitis when compared to unexposed nonsmokers.

(a) Alveoli in a nonsmoker's lung

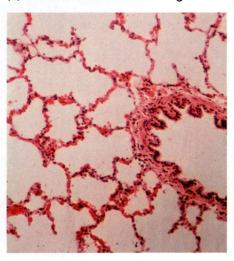

(b) Alveoli in a smoker's lung

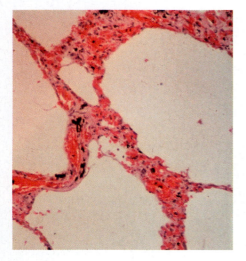

Figure 18.7 The effect of chronic smoke exposure on alveoli. (a) In this cross section of a normal lung, the alveoli are tiny. The walls surrounding each air pocket are areas where gas exchange occurs. (b) In a smoker's lung, many of the alveoli walls have been destroyed, leading to larger air pockets and less surface for gas exchange.
Visualize This: Smaller and more numerous alveoli would increase the respiratory surface area even more. What do you think are disadvantages to more, smaller alveoli?

Stop & Stretch If damage to alveoli is irreversible, is there any potential cure for emphysema?

Lung Cancer. The tiniest particulates in ETS can be drawn deeply into the lungs, even into the alveoli. Because alveoli lack cilia, the movement of foreign materials out of these structures is much more limited. Small particles can remain in the alveoli for long periods. We can see the accumulation of particles in the lungs of a long-time smoker, which are often black with tar trapped inside the alveoli (**Figure 18.8**). These accumulated particles promote inflammation and fluid accumulation and generally reduce gas exchange. Even more seriously, because many components of tobacco smoke are carcinogens, chemicals that are known to cause mutations, lung cells are susceptible to transforming into cancer cells long after smoke inhalation has stopped.

About 170,000 new cases of lung cancer are diagnosed in the United States each year, 90% in current or former smokers. However, for the 17,000 cases of lung cancer not due to active smoking, epidemiologists at the National Institutes of Health (NIH) have estimated that 3000 cases are directly attributable to exposure to ETS, and that passive smokers have a 20% to 30% higher risk of lung cancer than unexposed nonsmokers.

Epidemiological studies, such as the one that linked ETS exposure to lung cancer risk, have the same problem as all other correlations—it is impossible to eliminate the chance that some other difference between individuals in the study is, in reality, causing a difference between groups (Chapter 1). In the case of ETS and lung cancer, the hypothesized link is also supported by the presence of known carcinogens in secondhand smoke, by experiments using animals that showed higher rates of cancer in ETS-exposed individuals, and by a link established by hundreds of studies between active smoking and lung cancer.

The relationship between exposure to secondhand smoke and diseases in other organs and organ systems is less clear from the data collected. However, given the connection between the lungs and the bloodstream, one would expect that passive smokers have a higher risk than unexposed individuals of all the diseases that are more common in active smokers. To understand how tobacco smoke breathed into the lungs can affect the entire body, we must understand first how blood and its components are moved around via the bloodstream.

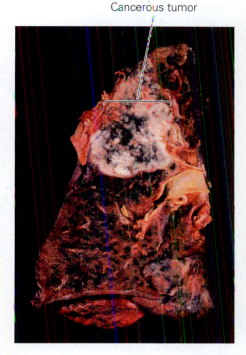

Cancerous tumor

Figure 18.8 Tobacco tar trapped in lungs. These lungs show the dark staining caused by the accumulation of particulates. The lighter-colored mass is a cancerous tumor.

18.2 Spreading the Effects of Smoke: The Cardiovascular System

The risks of active smoking, in addition to lung and airway damage, include increased rates of throat, bladder, and pancreatic cancer; higher rates of heart attack, stroke, and high blood pressure; and even premature aging of skin. All of these effects occur because many of the components of tobacco smoke can cross the thin walls of alveoli into the bloodstream and move throughout the body.

Structure of the Cardiovascular System

Gases and other materials are distributed around the body of most animals by a **cardiovascular system**, which consists of three major components: a circulating fluid (often called blood), a pump (in most cases, this is called a heart), and a vascular system (blood vessels and capillaries).

Blood. The 5 liters (or 11 pints) of blood in the vascular system of an adult human are made up of both liquid and solid (that is, cellular) portions. The liquid portion, called plasma, consists of water and dissolved proteins, salts,

(a) Plasma   **(b) Red blood cells**   **(c) White blood cells**   **(d) Platelets**

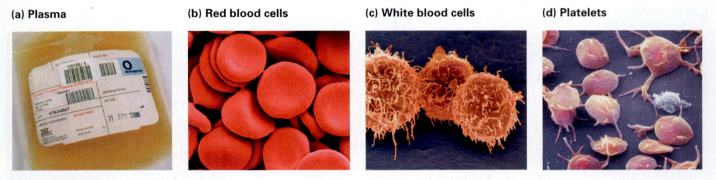

Figure 18.9 Components of blood. Blood consists of the fluid plasma as well as large amounts of red blood cells and lesser amounts of white blood cells and platelets.

Figure 18.10 Blood stem cells.
Blood cells are produced by stem cells found in the marrow of larger bones. Notice all cellular elements in the blood derive from these stem cells.
Visualize This: Leukemia is a form of cancer caused by the uncontrolled division of blood stem cells. How do you think leukemia manifests itself in the blood?

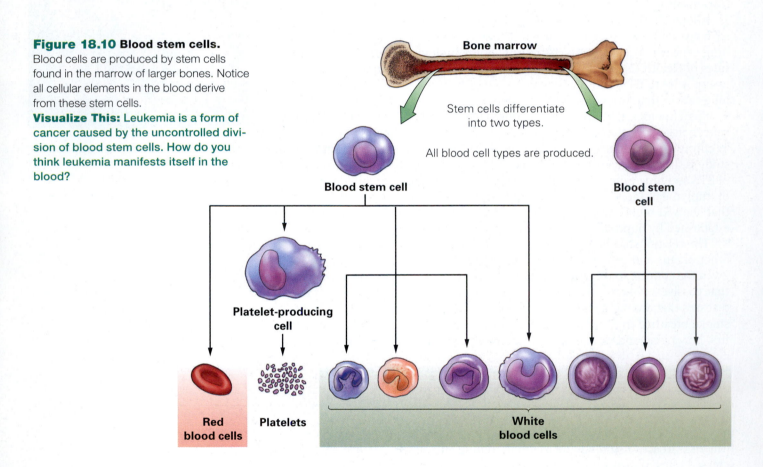

and gases. The cellular, or solid, portion is primarily made up of red blood cells (giving blood its reddish color), with much smaller numbers of white blood cells and platelets (**Figure 18.9**). In adults, all of the cellular components of blood are produced by stem cells in the bone marrow, tissue found in the cavities of certain bones. Stem cells are unique in adults—instead of being completely differentiated into a single cell type, such as nerve or skin, they have the capacity to produce descendants of a variety of cell types. Blood stem cells give rise to two classes of more specialized stem cells that together can generate nine different blood cell types (**Figure 18.10**).

Red blood cells are uniquely adapted to their primary task—shuttling oxygen from the lungs to the rest of the body. These cells are packed with hemoglobin molecules and lack a nucleus and other organelles. Red blood cells are also small and pinched in the middle, which makes their surface area large relative to their volume, ensuring rapid diffusion of oxygen into and out of the cells. **White blood cells** come in many different varieties and are the essential component of immune system function, attacking invading organisms as well as removing toxins, wastes, and damaged cells throughout the body. The principal function of **platelets** is to prevent blood loss. These cell-like structures are actually membrane-bound fragments of larger cells.

Platelets work with proteins in the blood in **blood clotting**, the process that stems the flow of blood out of damaged blood vessels. Immediately after injury occurs, muscles in the walls of damaged blood vessels contract to restrict the flow of blood. Within seconds, sticky platelets become attached to any breaches in the vessels, forming a temporary plug to help restrict blood flow. At the same time, chemical signals released by the damaged tissue initiate the complex clotting process. A blood clot consists of a net made up of the protein fibrin, which forms a patch over the damaged area (**Figure 18.11**). Cell division in the walls of the damaged vessel and surrounding tissues eventually seals the cut, and the patch dissolves.

A substance in tobacco smoke, perhaps nicotine, increases the stickiness of platelets and promotes production of fibrinogen, the precursor to fibrin, making blood clots form more readily, even when a clot is not needed to stop blood loss. When a clot forms within a blood vessel, it can block blood flow where it forms (a condition called a thrombosis) or break free and travel throughout the bloodstream until it becomes lodged in another blood vessel (in which case it is called an embolism). When a thrombus or embolus becomes lodged within the blood vessels of the heart or any other organ, it can restrict the flow of oxygen to nearby cells, killing them and causing severe damage to the organ.

Heart. We often think of the heart in poetic terms as the seat of our emotions, intuition, and spirituality, but in physiological terms, the heart is a fairly simple and very utilitarian organ. The fist-sized heart in a human consists of two muscular pumps that are coordinated but also somewhat independent. One pump on the right side of the heart receives oxygen-poor blood from the body and sends it to the lungs, while the left pump receives oxygen-rich blood from the lungs and sends it into general circulation within the body. The two pumps are each divided into two chambers, a relatively thin-walled atrium and a thick-walled ventricle. Each pump also contains two valves to help control blood movement: an **AV valve** between the atrium and ventricle and a **semilunar valve** between the ventricle and the major artery that it supplies (**Figure 18.12**).

The heart muscle contracts with an intrinsic rhythm determined by a small patch of muscle tissue in the wall of the right atrium. This patch, called the **sinoatrial (SA) node,** sends out electrical signals that first cause both left and right atria to contract and then, one-tenth of a second later, cause both ventricles to contract. The complete sequence within the heart of filling with blood and then pumping is called the

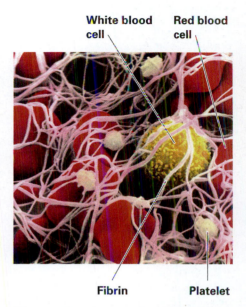

Figure 18.11 A blood clot. A blood clot consists of a net of protein fibers that trap red blood cells, forming a temporary patch over a damaged blood vessel.

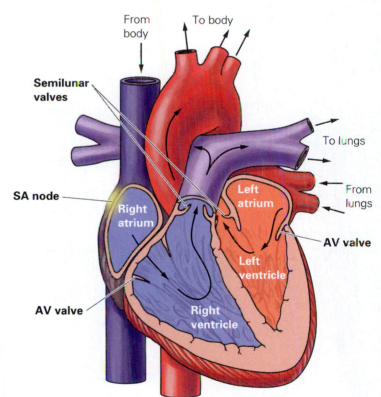

Figure 18.12 The human heart. The heart consists of four chambers making up two mostly independent pumps. One pump sends deoxygenated blood to the lungs, and the other sends blood returning from the lungs to the rest of the body.

Visualize This: The left and right ventricles move the same amount of blood per stroke, but the left ventricle walls are thicker and more powerful. Why is this necessary?

cardiac cycle, alternating between a relaxed period (called **diastole**) and a contraction phase (called **systole**).

During systole, the AV valves close as the ventricle contracts (making the first heart sound, "lubb"). At the beginning of diastole, the semilunar valves close as the ventricle relaxes ("dupp"). This cycle of contraction and relaxation leads to that distinctive heartbeat sound ("lubb-dupp, lubb-dupp").

The speed of the cardiac cycle, also known as the heart rate, is initially controlled by the SA node, which functions as an internal pacemaker (although its function can be replaced by an implanted artificial pacemaker). The speed of the pacemaker can also be affected by signals from the brain stem and spinal cord at rest, and from other parts of the brain in response to experiencing danger or intense emotion. Certain drugs, such as nicotine, may also affect heart rate.

Compared to muscles elsewhere in the body, the heart is much more dependent on oxygen to generate energy for its activities. A thrombus or embolus that restricts blood flow to the heart muscle can cause portions of the muscle to die because of oxygen starvation. Formation of these blockages depends not only on clotting factors in the blood but also on the condition of the blood vessels.

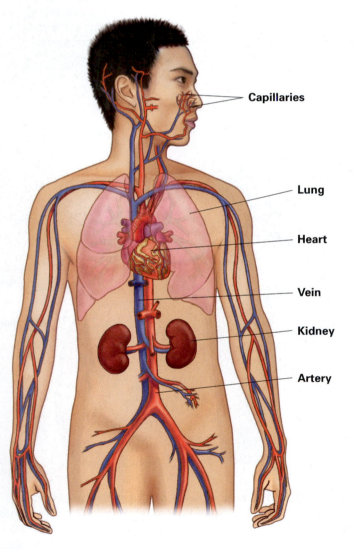

Figure 18.13 The vascular system. Arteries carry blood away from the heart, while veins carry blood to the heart. The two types of blood vessels are connected by nets of capillaries.

Stop & Stretch In what organ besides the heart would you expect a blood clot, which restricts oxygen delivery, to have the most severe effect on health and survival?

Blood Vessels. The system of tubes that carries blood to and from the heart is generally referred to as the vascular system. Components of the vascular system, broadly termed the blood vessels, include arteries, veins, and capillaries (**Figure 18.13**). Arteries are the branching blood vessels that carry blood from the heart, and veins are the converging vessels that bring it back. As in the lungs, capillaries throughout the body are tiny, thin-walled blood vessels that create a net of channels between the smallest arteries (arterioles) and the smallest veins (venules).

Arteries have thick, elastic walls that balloon out as contraction of the ventricles causes a mass of blood to flow into the system and that snap back to resting size once the blood passes by. The wave of blood is called a **pulse;** you can measure your heart rate by feeling the pulse as it passes through an artery close to the surface of the skin.

Exchange of gases and other materials between the circulating fluid and the body's tissues and organs occurs across the porous walls of the capillaries. Capillaries are extremely diffuse and abundant; an adult human body contains an estimated 100,000 kilometers (60,000 miles) of capillaries, and few living cells are more than 0.1 mm (about the thickness of a sheet of paper) away from a capillary.

Liquid and materials in these vessels are forced out of capillaries due to higher blood pressure near the arterial end of a net of capillaries, called a **capillary bed.** Materials and liquid from the body tissues then flow back into the capillaries as a result of concentration differences at the venous (vein) end of the capillary bed (**Figure 18.14**). Muscles surrounding the arterial ends of capillary beds can contract to cut off blood flow to less needy organs to effectively deliver blood and nutrients to more essential regions. Capillaries that nourish the skin often become constricted in response to a crisis in another organ, which is why a common symptom of illness is paler skin (most apparent in the tissues lining the eyes and inside the mouth). Skin capillaries also play an essential role in the regulation of body temperature, opening when temperature rises to allow body heat to dissipate and closing to conserve heat in cold temperatures.

Veins have much thinner, less elastic walls than arteries, and the pressure of the blood is much lower once it reaches these vessels. Blood tends to pool within veins, a fact you can easily see by lowering your hand to your side. Blood pooling in the veins on the back of the hand will make these veins become distended and stand out. Movement of blood from the veins back to the heart is facilitated by the contraction of skeletal muscles, which compress the veins and squeeze the blood through them. The blood flows in only one direction, toward the heart, due to the presence of one-way valves within the veins (**Figure 18.15**).

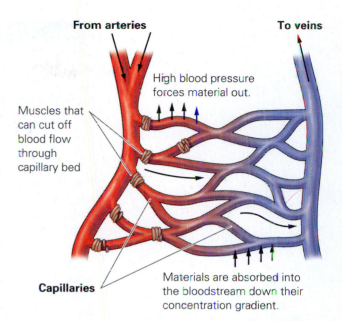

Figure 18.14 Capillary function. Liquid forced out of the upstream end of a capillary bed carries oxygen and nutrients to the tissues. Wastes and other materials are absorbed into the blood at the downstream end. Muscles on the upstream side can contract to restrict flow through the bed.

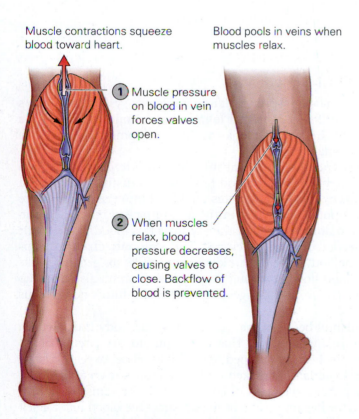

Figure 18.15 Flow of blood in veins. Blood in the veins is under low pressure and returns to the heart due to the contraction of skeletal muscle. The blood flows in one direction thanks to the presence of one-way valves.

Stop & Stretch Varicose veins are distended veins in which blood pools and can return only slowly to the heart. They typically form in the legs when the force of gravity on slow-moving blood causes disruption of the internal anatomy of the veins. What part of the vein must be disrupted to cause this condition?

The efficient flow of oxygen and materials to the body's tissues requires that the blood is maintained under enough pressure to move it throughout the vascular system. **Blood pressure,** the force of the blood against blood vessel walls, is created in part by the pulse of blood from the contracting heart and in part by the diameter of the arteries. Blood pressure rises when arteries become constricted by the action of muscles in the vessel walls. For instance, this happens when release of the stress hormones epinephrine and norepinephrine indicates that muscles need oxygen delivered quickly. In a healthy individual, blood pressure will rise in response to vigorous activity or excitement but return to normal under restful conditions.

Chronic high blood pressure, called **hypertension,** may be caused by arteries that are narrowed by constant psychological stress (causing stress hormone release) or by the accumulation of fatty material within the walls of the arteries, a condition called **atherosclerosis** that occurs as a result of aging, a high-fat diet, or exposure to tobacco smoke. Hypertension can lead to damage in the walls of blood vessels, and it causes the heart to work much harder to push blood through the arteries.

The components of tobacco smoke can exert effects on all organ systems because materials crossing the respiratory surface enter the bloodstream, which then circulates these materials throughout the body.

Movement of Materials through the Cardiovascular System

The relationship between the respiratory and circulatory system in humans and other mammals is different from that found in other animals. The Focus on Evolution feature in this chapter looks at the development of these two systems through the animal kingdom.

In humans, lungs and the cardiovascular system are connected by a double circulation system in which blood flows in two distinct but related circuits. The **pulmonary circuit** circulates the blood into the lungs, where oxygen and other components inhaled in the air are picked up and carbon dioxide released. The pulmonary circuit then returns this blood to the heart, where it enters the second or systemic circuit. This **systemic circuit** pumps blood to the rest of the body, where the blood drops off its oxygen and picks up carbon dioxide. Blood flowing through the systemic circuit also picks up and distributes materials from the digestive system and filters nongaseous wastes via the kidneys.

The path of circulation between the heart, lungs, and body is detailed in **Figure 18.16**. From this, you can see that in the pulmonary circuit, blood from the right side of the heart is pumped to the lungs. Here, oxygen, carbon monoxide, and other materials are picked up by the bloodstream at the lungs and then return to the heart via a pair of large veins. These veins empty into the left atrium, where the beginning of a heartbeat forces the blood into the left ventricle. There is a slight lag between when the atrium contracts and when the contraction signal reaches the ventricles. This lag allows the ventricles to fill completely with blood, increasing the efficiency of the heartbeat.

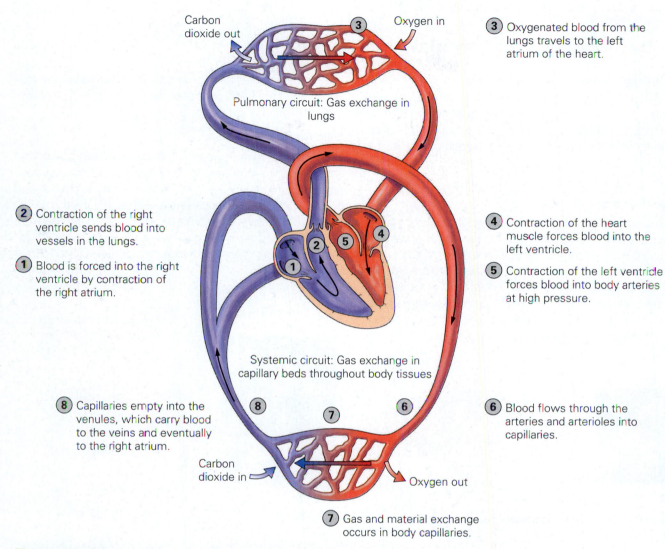

Carbon dioxide out

Oxygen in

3 Oxygenated blood from the lungs travels to the left atrium of the heart.

Pulmonary circuit: Gas exchange in lungs

2 Contraction of the right ventricle sends blood into vessels in the lungs.

1 Blood is forced into the right ventricle by contraction of the right atrium.

4 Contraction of the heart muscle forces blood into the left ventricle.

5 Contraction of the left ventricle forces blood into body arteries at high pressure.

Systemic circuit: Gas exchange in capillary beds throughout body tissues

8 Capillaries empty into the venules, which carry blood to the veins and eventually to the right atrium.

6 Blood flows through the arteries and arterioles into capillaries.

Carbon dioxide in

Oxygen out

7 Gas and material exchange occurs in body capillaries.

Figure 18.16 Connecting the respiratory and circulatory system. The lungs are connected to the other organ systems by the heart and blood vessels.

Once the left ventricle contracts, the blood is forced at high pressure into the systemic circulation, first to the arteries, which repeatedly branch out and become smaller in diameter, and then into capillary beds. In the capillaries, oxygen and other components in high concentration within the blood diffuse out, and carbon dioxide and wastes diffuse in. The deoxygenated blood then travels to the systemic veins, which empty into the right atrium. Contraction of the right atrium forces blood into the right ventricle. Contraction of the right ventricle sends blood into the arteries and capillary beds of the pulmonary circuit.

A single red blood cell carrying a load of oxygen—or carbon monoxide—can travel through the entire circulatory system in approximately 1 minute. Thus, the chemicals present in ETS have almost immediate effects on the body. These quickly occurring effects can cause more severe long-term damage as well.

Smoke and Cardiovascular Disease

Most people believe that lung cancer and other lung diseases are the primary risk associated with exposure to tobacco smoke. In reality, most of the deaths due to smoking result from heart and blood vessel damage, or **cardiovascular disease**.

The Respiratory and Circulatory Systems

Most animals have active lifestyles and extremely high energy needs compared to more sessile (stationary) organisms. As a result, organ systems have evolved that allow for gas exchange and for the distribution of materials throughout their bodies. The following paragraphs detail the evolution of respiratory and circulatory systems in the animal kingdom.

Simple Diffusion

Small animals have a large surface-to-volume ratio—that is, their surface area is great compared to their internal volume. As a result, most of their body cells are close to the external environment. Flatworms such as planaria are an example of animals small enough to allow diffusion of gases across body walls (**Figure E18.1**). Therefore, these organisms have no need for a specialized respiratory system or a circulatory system to distribute gases throughout the body.

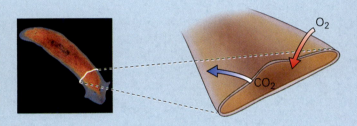

Figure E18.1 Flatworm (planarium). Simple diffusion.

Increased Surface Area and Open Circulation

Insects and spiders have pores called spiracles that allow gases to flow in and out of the body through a complex branching system of tracheae (**Figure E18.2**). These structures essentially increase the surface of the animal and bring outside air into close contact with every cell of the body. Food molecules are distributed via an **open circulatory system,** in which vessels empty fluid directly into the body's tissues. The fluid carried by these vessels, called *hemolymph,* is pumped into the body by a simple,

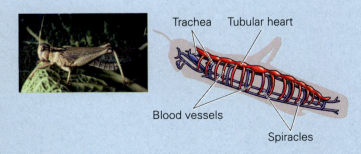

Figure E18.2 Insect (grasshopper). Open circulation.

tube-shaped heart and returns to the heart via pores that close when the heart contracts, allowing a one-way flow of material.

Gills for Gas Exchange

Larger water-dwelling animals like clams, sea stars, and fish have gills—thin sheets of tissue that rely on the constant movement of water past them to bring in oxygen and diffuse out carbon dioxide (**Figure E18.3**). Like the alveoli in lungs, gills have an enormous surface area and network of capillaries. Clams pair gills with an open circulatory system and tubular heart to deliver oxygenated blood to their tissues.

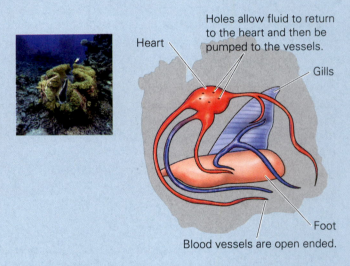

Figure E18.3 Mussel (giant clam). Open circulation with a pump.

Countercurrent Gas Exchange in Water and Closed Circulatory Systems

Fish are especially efficient at employing gills to remove oxygen from relatively oxygen-poor water via a process called countercurrent exchange (**Figure E18.4**). In this process, water flows in a single direction over the surface of the gills, while blood flows in the opposite direction. As a result, oxygen is able to diffuse out of water over the entire surface of the gills because the blood always contains a lower concentration of oxygen than the water.

Fish, as well as reptiles, mammals, birds, and even squid, have **closed circulatory systems** in which the circulating fluid, blood, is maintained in vessels. Among closed systems, fish have the simplest circulation, with a single loop from heart to gills, to the rest of the body, and then back to the heart. Because blood loses pressure as it passes through capillaries, circulation of oxygenated blood

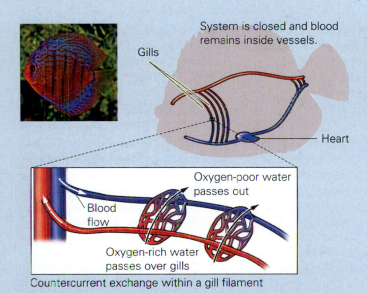

System is closed and blood remains inside vessels.

Gills

Heart

Oxygen-poor water passes out

Blood flow

Oxygen-rich water passes over gills

Countercurrent exchange within a gill filament

Figure E18.4 Fish (discus fish). Closed circulation with a single pump.

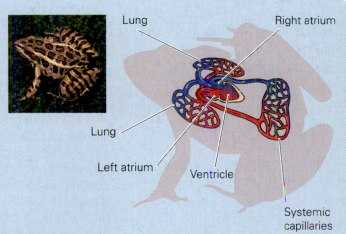

Lung

Right atrium

Lung

Left atrium

Ventricle

Systemic capillaries

Figure E18.5 Amphibian (leopard frog). Closed circulation with a partial double pump.

is relatively slow in fish, and oxygen delivery to tissues is limited. Squid and their relatives have evolved a multiple-pump system in which two hearts force blood into each set of gills, and a third systemic heart pumps this oxygenated blood to the rest of the body.

Lungs and Double Circulation

Larger animals on land evolved internal lungs that allow the respiratory surface to remain moist for gas exchange. The advent of lungs permitted the **double circulation** system as seen in humans, with a compact heart that pumps blood to the lungs for gas exchange and then back to the heart, where it can be pumped at high pressure to the rest of the body. In amphibians, the heart has only one ventricle, allowing mixing between oxygenated and deoxygenated blood (**Figure E18.5**). This system is less efficient than the one for mammal and bird hearts but sufficient for ectotherms—animals that do not generate much internal body heat and thus have lower metabolic rates and lower oxygen needs. Reptiles have a single ventricle, like amphibians, but a muscular half-wall between the two sides of the ventricle provides more separation between oxygenated and deoxygenated blood, providing more efficient delivery of oxygen to tissues.

Four-Chambered Hearts and Countercurrent Gas Exchange in Air

Birds and mammals are endothermic animals, which means they maintain a regular body temperature and must use significant energy to do so. They are able to maintain this lifestyle because of adaptations that increased the efficiency of respiratory and circulatory systems. Birds have the most efficient respiratory system of all, using countercurrent exchange in a one-way flow of air, allowing for maximum oxygen "capture" by the blood (**Figure E18.6**). The four-chambered hearts of birds and mammals ensure that blood with high oxygen content can be pumped directly to the body's tissues, providing ample oxygen for maintaining regular body temperature as well as for engaging in physical activity.

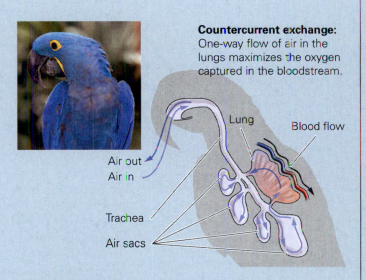

Countercurrent exchange: One-way flow of air in the lungs maximizes the oxygen captured in the bloodstream.

Lung

Blood flow

Air out

Air in

Trachea

Air sacs

Figure E18.6 Bird (macaw). Closed circulation with a double pump and countercurrent exchange of oxygen with the bloodstream.

(a) Normal artery

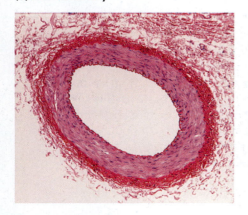

(b) Atherosclerotic artery

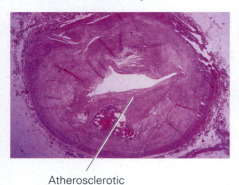

Atherosclerotic
plaque

Figure 18.17 Atherosclerosis. Fat and cholesterol accumulate in the walls of arteries, reducing their diameter and their ability to carry blood.

Visualize This: Angioplasty is a procedure that threads a tiny balloon into a narrowed artery and inflates it near a plaque. The balloon is then removed. How must this improve the condition of the artery?

Most of the cardiovascular damage that results from tobacco smoke is caused by **nicotine**, the primary active ingredient in tobacco. Nicotine is a compound produced by tobacco plants as a natural pesticide—in fact, nicotine sulfate spray is sold commercially for use by organic gardeners to combat aphids and other insects. The nicotine in tobacco smoke is readily absorbed across the alveoli walls and enters the bloodstream almost immediately after smoke is drawn into the lungs.

In high doses, nicotine is toxic to humans and other mammals. In low doses, nicotine interacts with cells in the brain to stimulate the release of epinephrine, which in turn increases heart rate and blood pressure. Some of the brain cells activated by nicotine include those in the brain's "reward pathways," regions with activity that is associated with feelings of well-being and happiness. The reward pathways of the brain are an adaptation that reinforces behaviors important for our survival and reproduction; for instance, these pathways are activated in response to tasting delicious food or engaging in sexual activity.

By hijacking this pathway, nicotine and other drugs such as cocaine and methamphetamine become addictive, driving the user to seek out these drugs again and again to get the same response. Nicotine addiction is part of what keeps smokers smoking, despite the extensive financial and physical costs associated with tobacco use. Nicotine-containing patches, chewing gum, and nasal spray provide low doses of nicotine that can help dull the drug cravings that occur when a smoker quits.

Nicotine has effects on other parts of the body besides the brain. It appears to increase production of LDL (low-density lipoprotein), "bad cholesterol," and reduce production of HDL (high-density lipoprotein), or "good cholesterol" (Chapter 3). As a result, individuals exposed to tobacco smoke have higher levels of circulating lipids and thus a higher risk of atherosclerosis, the accumulation of fats and other debris on the interior walls of arteries (**Figure 18.17**). Recall that nicotine stimulates blood clot formation. A clot can cut off blood flow in arteries narrowed by both atherosclerosis and nicotine-induced hypertension. When a blockage occurs in an artery that brings blood to the brain, it results in a **stroke**, which causes the death of brain tissue. When a blockage stops blood flow to part of the heart muscle, it can result in the death of that muscle, or a **heart attack**. Stroke and heart attack are the primary cardiovascular diseases that cause the death of affected individuals.

Combining the artery-damaging and clot-inducing results of nicotine consumption with the oxygen-robbing increase in inhaled carbon monoxide, exposure to tobacco smoke represents a serious risk to heart function. According to the U.S. Centers for Disease Control and Prevention (CDC), heart disease accounts for about 147,000 deaths per year among active smokers. A smoker has a 200% to 300% greater chance of dying from heart disease than does a nonsmoker. However, the CDC also estimates that 35,000 deaths per year from heart disease among nonsmokers result from exposure to carbon monoxide in environmental tobacco smoke. There is good reason to suspect that the artery-damaging effects of smoke exposure also occur in passive smokers, but the epidemiological studies have not convincingly demonstrated this link.

Unlike the permanent lung changes caused by tobacco smoke, the effects of nicotine and carbon monoxide on cardiovascular health are much more reversible. According to the World Health Organization (WHO), the risk of heart disease decreases by 50% in ex-smokers within 1 year of quitting, and by 15 years after quitting, the risk to ex-smokers is the same as that for people who have never smoked. The effects of nicotine are reversed over time because this drug is metabolized, that is, changed into simpler forms and excreted. To understand how the body rids itself of nicotine and other toxins, we must first understand the structure and function of the urinary system.

18.3 Removing Toxins from the Body: The Urinary System

The metabolism of nicotine results in waste products that circulate through the bloodstream until they are expelled from the body. The **urinary system** has the task of efficiently removing these and other wastes while retaining valuable materials that can be reused and recycled. In humans, the major organs of the urinary system are the **kidneys,** which filter and cleanse circulating blood before sending the waste through **ureters** to the **bladder,** where it is stored until it is expelled via the **urethra** (**Figure 18.18**). The urinary system is sometimes referred to as the renal system, a term derived from the Latin word for kidney.

Kidney Structure and Function

The kidneys are paired, approximately fist-sized organs that sit behind the liver and stomach in the upper abdominal cavity. These compact organs are packed with a dense network of looped tubules, called **nephrons.** Each kidney contains approximately 1,250,000 nephrons, with a combined length in an adult human of about 145 kilometers (85 miles). Networks of capillaries surround the nephrons, allowing wastes to diffuse out of the blood and into these tubules for excretion. The total volume of blood passes through the kidneys hundreds of times per day, such that each kidney filters about 1000 liters of blood every 24 hours.

Blood is brought to the kidneys by large vessels called the **renal arteries.** The processing of waste in the kidneys has four distinct phases, which we can follow on a diagram of a single, loop-shaped nephron embedded in the body of a kidney (**Figure 18.19** on the next page). The first step, **filtration,** occurs within the Bowman's capsule, a structure at the head of a nephron. The Bowman's capsule encloses the glomerulus, a compact ball of blood vessels. The capillaries in the glomerulus contain tiny pores, and blood pressure forces the plasma portion of the blood through these pores and into the upstream end of the nephron. When blood reaches the glomerulus, this filter allows water and small molecules through, but retains large proteins in the plasma. The fluid that enters the interior of the nephron via this process is called filtrate.

Because filtrate contains both wastes and valuable substances, such as sugars, amino acids, and water, the next step of waste processing is **reabsorption** of these materials across the walls of each nephron. Water moves by osmosis out of the nephron and into the kidney interior as it descends into this salty environment. On the ascending limb, the nephron walls are impermeable to water but actively transport salt into the kidney. The structure of the nephron allows for a gradient of salt concentration to be maintained, from relatively low at the top of the loop to relatively high at the base. What remains in the filtrate after the reabsorption phase is water containing a high concentration of urea, the waste product of the breakdown of amino acids.

The last segment of the nephron permits the **secretion** into the filtrate, via active transport, of certain wastes that are in low concentration in the plasma, including toxins such as the products of nicotine metabolism. The filtrate then moves from the nephron into a **collecting duct** that leads to the center of the kidney, called the renal pelvis. As the collecting duct crosses the salty kidney interior, more water moves by osmosis from the filtrate into the kidney tissue. Water and other materials removed from the filtrate are drawn into the capillaries, returned to the bloodstream, and eventually leave the kidney via the **renal vein.** The fluid that collects in the renal pelvis is called **urine** and is made up of water, organic wastes including urea, and various ions.

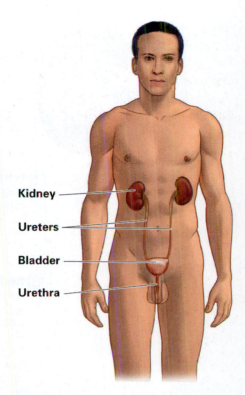

Kidney

Ureters

Bladder

Urethra

Figure 18.18 The human urinary system. Kidneys filter the blood and transfer waste urine to the bladder, where it is held until voluntarily released through the urethra.

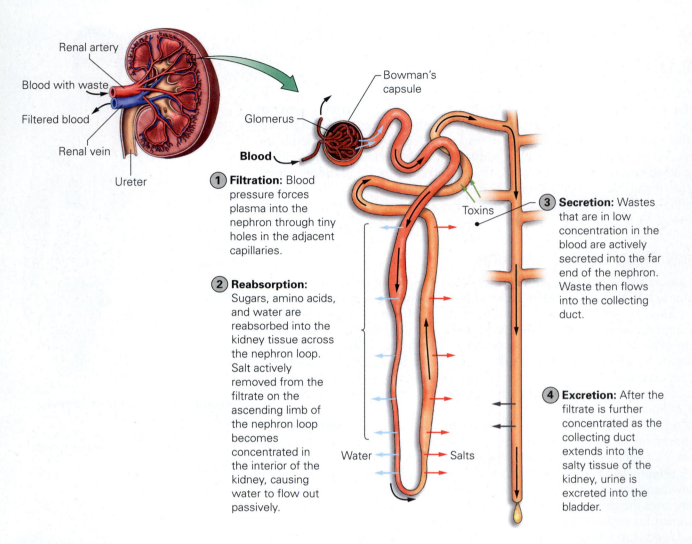

Renal artery

Blood with waste

Filtered blood

Renal vein

Ureter

Bowman's capsule

Glomerus

Blood

① **Filtration:** Blood pressure forces plasma into the nephron through tiny holes in the adjacent capillaries.

② **Reabsorption:** Sugars, amino acids, and water are reabsorbed into the kidney tissue across the nephron loop. Salt actively removed from the filtrate on the ascending limb of the nephron loop becomes concentrated in the interior of the kidney, causing water to flow out passively.

Toxins

③ **Secretion:** Wastes that are in low concentration in the blood are actively secreted into the far end of the nephron. Waste then flows into the collecting duct.

④ **Excretion:** After the filtrate is further concentrated as the collecting duct extends into the salty tissue of the kidney, urine is excreted into the bladder.

Water

Salts

Figure 18.19 Nephron function. The nephron, the functional unit of a kidney, controls the excretion of nongaseous waste from the blood.

Because the kidneys can exert fine control over the makeup of urine, the urinary system also plays an important role in regulating blood volume, acidity, and salt balance. The kidneys regulate acidity and salt balance by actively excreting acids and reabsorbing salts during the secretion and reabsorption steps of urine formation. The kidneys help maintain blood pressure by regulating water excretion. When blood pressure is low, which can happen because water intake is low or water loss is high, antidiuretic hormone (ADH) released by the pituitary gland increases the permeability of the collecting duct to water, allowing more water from the filtrate to return to the bloodstream. When blood pressure is high, ADH release is curtailed, and more water is excreted. Nicotine consumption results in the inhibition of ADH, so individuals exposed to smoke retain more water in the blood, further increasing blood pressure.

During urine **excretion,** urine leaves the kidneys and flows to the bladder. Release of urine from the bladder, called **micturition,** is primarily under our conscious control.

Stop & Stretch Chronic high blood pressure damages cells in the filtering structure of the nephron, making the pores between cells much larger. How is the urine of someone with damage to these structures likely to be different from someone with an intact filter?

Smoking and the Urinary System

Smoking appears to damage the filtration capacity of kidneys, probably due to increased blood pressure and blood vessel wall damage. Constant exposure to nicotine and its breakdown products may also explain why studies showed that the risk of kidney cancer is at least 38% greater among smokers than among those who have never smoked.

The toxins in tobacco smoke are even more dangerous to the bladder, where they are retained for long periods of time. Bladder cancer is nearly three times more common in smokers than in nonsmokers. As with many of the health consequences of ETS, it is difficult to determine from epidemiological studies whether kidney and bladder cancers are more common in passive smokers than in nonsmokers. However, given our understanding of kidney function, there is certainly reason to believe that exposure to the toxins in ETS causes damage to these important organs.

The links between our respiratory, cardiovascular, and urinary systems provide a mechanism by which what we inhale affects all the organs of our bodies (**Figure 18.20**). While many of the links between environmental tobacco smoke

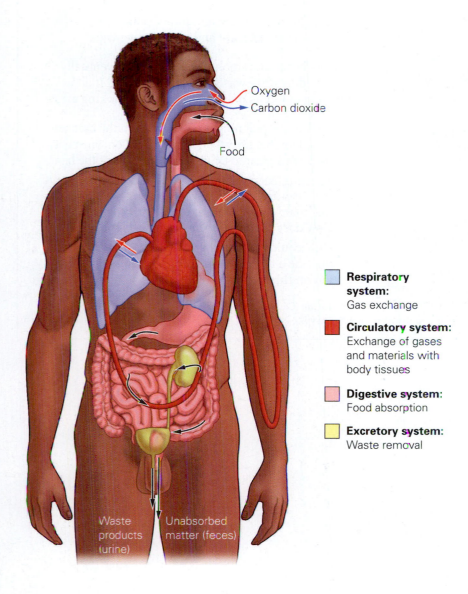

Oxygen
Carbon dioxide
Food

■ **Respiratory system:**
Gas exchange

■ **Circulatory system:**
Exchange of gases and materials with body tissues

■ **Digestive system:**
Food absorption

■ **Excretory system:**
Waste removal

Waste products (urine)

Unabsorbed matter (feces)

Figure 18.20 Digestion, respiration, circulation, and excretion. This figure illustrates the relationship between these four functions by simplifying the actual structure of their associated organs. These organ systems provide the primary routes of materials into, out of, and throughout the body.

and disease have not been "proven beyond reasonable doubt," there is accumulating evidence from both epidemiology and the known risks of active smoking that ETS is dangerous.

As citizens, we are often called on to make decisions in the absence of perfect knowledge. In most of the communities that have considered imposing smoking bans, voters have decided that the risk of ETS is large enough to justify restricting smokers. Given that nonsmokers outnumber smokers by almost 4 to 1, clearing the air in formerly smoky eating establishments and entertainment venues has proven to be mostly good for business. In this case at least, resolution of the controversy has been a win for both the business owners and the public's health.

SAVVY READER

Opinion Polling

Students Speak Out in Smoking Ban Survey

By Sabrina Porter

Thirty-eight percent of students voted in favor of the university's proposed smoking ban, while 52 percent voted against it, according to survey results released Tuesday night.

The survey, administered by Student Senate, solicited students' views on a September 2006 proposal to ban smoking everywhere on campus by 2010.

. . .

The survey results represent the views of 1837 Carnegie Mellon students, whose participation was voluntary. . . .

"With about 30 percent of the undergraduate student body participating in the referendum, I think we got a fairly balanced view of how the campus feels about the ban," said Lauren Hudock, communications chair on Student Senate and junior public policy and management and philosophy major."

Editor's Note: The smoking policy CMU eventually instituted in 2009 was not a complete ban. It allows smoking outdoors only in "designated smoking areas.

1. Look at the poll numbers described in the first paragraph of this story. Do Carnegie Mellon University students appear to be largely in support of the ban?

2. The third paragraph of the story indicates that participation in the survey was voluntary. How does this change your interpretation of the survey results? How does voluntary participation affect the sample?

3. How could this survey have been improved to provide a better understanding of student opinion on this topic?

The Tartan (Carnegie Mellon University Student Newspaper), February 26, 2007.

Chapter Review

Learning Outcomes

Mastering BIOLOGY

Go to the Study Area at www.masteringbiology.com for practice quizzes, myeBook, BioFlix™ 3-D animations, MP3Tutor sessions, videos, current events, and more.

L01 **List the components of the mammalian respiratory system, and describe the path of air into the body (Section 18.1).**

- Air flows into the body via the mouth and nose and enters the respiratory system via the pharynx and trachea. Once in the lungs, air flows through bronchi and into alveoli, where gas exchange occurs (pp. 443, 445–446).

L02 **Describe the muscles involved in breathing, including the diaphragm, and explain how their movements facilitate air movement into and out of the lungs (Section 18.1).**

- The movement of air into and out of the lungs depends largely on the action of the diaphragm, a dome of muscle that sits directly below the lungs (p. 444).
- Contraction of the diaphragm increases the volume of the chest cavity, decreasing air pressure and allowing air to flow in. Relaxation of the diaphragm causes the opposite to occur. Muscles surrounding the rib cage and in the abdomen can also contract or relax to cause changes in the chest cavity volume (pp. 444–445).

L03 **Explain the role of hemoglobin in gas exchange (Section 18.1).**

- As blood flows through the lungs, hemoglobin reversibly binds to oxygen molecules in high concentrations there. In the body tissues, the hemoglobin releases some of its oxygen load to supply active tissues (pp. 447–448).

L04 **List the organs and tissues of the circulatory system, and describe the function of each (Section 18.2).**

- Circulatory systems in animals consist of a fluid for gas and material exchange (e.g., blood), "tubes" for carrying the fluid throughout the body (e.g., veins and arteries), and a pump to facilitate fluid flow (the heart) (pp. 449–455).

L05 **Describe how blood moves through the double circulation system of the heart (Section 18.2).**

- Blood is made up of a liquid portion, called plasma, and a cellular portion, consisting of red blood cells, white blood cells, and platelets (pp. 449–451).
- The heart is a double pump consisting of four chambers—two atria and two ventricles. The right-side pump sends oxygen-poor blood to the lungs; the left-side pump sends oxygen-rich blood to the body (pp. 451–452).

- Blood from the lungs flows to the heart, where it is pumped into the systemic circulation. After dropping off a load of oxygen and picking up carbon dioxide in the body tissues, the blood returns to the heart and is pumped back to the lungs (pp. 454–455).

L06 **Explain the steps of the cardiac cycle, including the role of the heart's electrical system (Section 18.2).**

- The heart can generate its own beat, via the electrical activities of the SA node. Signals from the SA node are transmitted to the atria, causing these chambers to contract and forcing blood into the ventricles. The signal is carried by conductive fibers to the ventricles, which then contract to force blood out of the heart and into circulation (pp. 450–455).

L07 **List the structures of the mammalian urinary system (Section 18.3).**

- The urinary system consists of the kidneys, bladder, ureters, and urethra (p. 459).

L08 **Describe the steps in the process of urine excretion (Section 18.3).**

- In the nephrons of the kidney, the process of filtration forces most liquid, but not cells or larger molecules, from the plasma into the kidney tubules. As the filtrate travels through the nephron, water, glucose, and other valuable molecules are reabsorbed through both active and passive mechanisms. Secretion occurs when waste materials that did not leave with the filtrate are actively brought into the nephron tubules from surrounding capillaries (pp. 459–460).
- Hormones that regulate blood pressure control the concentration of water in urine. Urine collects in the renal pelvis, then travels down the ureters to be stored in the bladder. During urination, urine is released from the bladder through the urethra and exits the body (p. 460).

L09 **Describe the effects of smoking on the respiratory, cardiovascular, and urinary systems (Sections 18.1, 18.2, and 18.3).**

- Small particles in tobacco smoke enter the lungs, causing cell damage that eventually leads to chronic bronchitis. Chronic bronchitis in turn can lead to emphysema. The tiniest smoke particles are drawn into the alveoli, where they may remain for long periods, exposing the alveolar cells to carcinogens within the smoke particles (pp. 442 and 448–449).

- The nicotine in smoke increases heart rate and blood pressure, putting strain on the heart muscle. Nicotine also increases the production of LDL ("bad cholesterol"), causing atherosclerosis, and increases the likelihood of blood clot formation, causing blockages in blood flow (p. 458).
- Toxins from tobacco smoke are stored in urine within the bladder, leading to a greatly increased risk of bladder cancer among smokers (p. 461).

Roots to Remember

The following roots of words come mainly from Latin and Greek and will help you decipher terms:

card- and **cardio-**   relate to the heart. Chapter term: cardiovascular

-hale   means to breathe. Chapter terms: inhale, exhale

-itis   is used to describe inflammation (of an organ). Chapter term: bronchitis

vascul-   means vessel. Chapter term: vascular

Learning the Basics

1. **LO2** Describe how contraction of the diaphragm allows the lungs to fill and how oxygen from inhaled air enters the lungs.

2. **LO8** List the three processes that occur within the nephrons of kidneys.

3. **LO1** Alveoli are _____.

   A. small air sacs at the ends of bronchi in the lungs; B. the respiratory surface in mammals; C. surrounded by a net of capillaries; D. subject to damage because of exposure to tobacco smoke; E. more than one of the above is correct

4. **LO3** All of the following statements about hemoglobin are true, except _____.

   A. it is a protein that can bind oxygen; B. it is carried by red blood cells; C. it can bind carbon monoxide; D. it picks up oxygen in the body tissues and releases it in the lungs; E. it contains iron, which is responsible for its red color.

5. **LO4** The blood vessels that carry blood from the heart are called _____.

   A. arteries; B. veins; C. atherosclerosis; D. capillaries; E. nephron tubules

6. **LO6** The sound of a heartbeat as heard through a stethoscope is produced by _____.

   A. electrical signals from the SA node; B. rush of blood into the ventricles and out of the atria; C. the ribs expanding as the diaphragm contracts; D. the closing of valves between heart chambers and vessels; E. the pulse of blood flowing through the arteries

7. **LO5** Blood flowing from body tissues to the heart _____.

   A. is pumped via veins to the rest of the body; B. is pumped via arteries to the lungs; C. is under relatively high pressure compared to blood leaving the heart; D. returns via millions of capillaries tied directly to the heart's two atria; E. has used up all of its hemoglobin

8. **LO5** Deoxygenated blood from the body first enters the _____ of the heart, and oxygenated blood from the lungs is pumped to the body by the _____.

   A. right atrium, right ventricle; B. right atrium, left atrium; C. left atrium, right ventricle; D. right atrium, left ventricle; E right ventricle, left atrium

9. **LO7** After blood is filtered in the nephrons of the kidneys, _____.

   A. water and nutrients are reabsorbed; B. toxins are actively secreted into the urine; C. carbon dioxide is actively loaded into unfiltered blood cells; D. A and B are correct; E. A, B, and C are correct

10. **LO9** Which of the following diseases is associated with exposure to tobacco smoke?

    A. heart disease; B. lung cancer; C. bladder cancer; D. asthma; E. all of the above

Analyzing and Applying the Basics

1. **LO3** Some endurance athletes, such as cyclists, engage in the practice of "blood doping," in which packed red blood cells are transfused into their bloodstream immediately before an event. How might increasing the volume of red blood cells provide an advantage to an athlete? What do you think could be the risks of this practice?

2. **LO5** About 17,000 babies born in the United States each year have a heart defect that allows blood from the right side of the heart to mix with blood in the left side of the heart. How do you think a "hole in the heart," as this

condition is commonly termed, affects the function of the cardiovascular system?

3. **LO9** As tobacco plants grow, radioactive minerals found in soil stick to the plants' leaves. Minerals found in phosphate fertilizer, such as radium, lead-210, and polonium-210, can also accumulate on the tobacco plant. Radioactive substances on tobacco are not removed as the tobacco is processed to make cigarettes. Therefore, each cigarette delivers a dose of radiation along with its dose of nicotine. Over the course of a year, someone who smokes about 1.5 packs per day is exposed to a radiation dose equivalent to 300 chest X-rays. How could radioactive minerals in cigarettes produce disease throughout the human body?

Connecting the Science

1. Exposure to tobacco smoke causes early death, but it also causes years or even decades of disabling, chronic health problems such as emphysema and lung cancer.

Some people have argued that the best way to reduce the number of smokers is to make smoking, and the health care costs associated with it, more expensive. This strategy entails placing high taxes on cigarettes and other tobacco products and restricting current and former smokers' access to subsidized medical care. What do you think of these strategies? Do you think they would be successful? Do you think they might be unfair?

2. Although tobacco causes more cases of long-term disability, alcohol is more deadly to young men and women than cigarettes because of automobile and other accidents. Do you think the success of tobacco bans means that alcohol will become more strongly restricted? Do you think this is a good policy? Why or why not?

Answers to **Stop & Stretch, Visualize This, Savvy Reader,** and **Chapter Review** questions can be found in the **Answers** section at the back of the book.

CHAPTER 19

Vaccinations: Protection and Prevention or Peril?

Immune System, Bacteria, Viruses, and Other Pathogens

Can vaccines cause autism as some celebrities claim?

LEARNING OUTCOMES

LO1 Compare the structure of bacteria to that of viruses.

LO2 Contrast bacterial replication with viral reproduction.

LO3 List the structures involved in the first line of defense against infection, and describe how they function in protecting the body against pathogens.

LO4 List the participants in the second line of defense against infection, and describe their functions.

LO5 List the two different cell types involved in the third line of defense against infection, and compare how they recognize and eliminate antigens.

LO6 Explain allergic reactions.

LO7 Explain the process immune cells undergo to be able to respond to millions of different antigens.

LO8 Describe the process used to determine whether a cell is foreign or native to the body and what happens when this process fails.

LO9 Explain the role of memory cells in helping to protect against infection.

LO10 Explain the role of helper T cells in allowing infection by the AIDS virus.

LO11 Describe the mechanism by which vaccines help confer immunity.

Should you get a flu shot? For many college students, this is one of the first health care decisions that will be made without much parental input. University health services personnel want you to, but you hear conflicting information from fellow students. The flu shot, like all vaccines, contains substances that resemble the disease-causing organisms. When you are vaccinated, your body is tricked into preparing for an attack of the disease in case one does occur.

It's not just college students who are vaccinated. Newborns and children under 6 years of age in the United States are vaccinated against 15 different diseases. The first vaccinations happen before a baby even leaves the hospital to go home and continue every few months until the child is a year and a half old. Another series of vaccinations occurs when the child is school aged, and a few additional vaccinations are given during adolescence. Virtually all parents would prefer not to subject their child to these vaccinations, which are often administered via uncomfortable injections, not to mention the inconvenience of taking children to the doctor for so many visits. But for most parents, the promise of protection from diseases makes the choice an easy one.

There are, however, a growing number of parents who are questioning whether to vaccinate their children. Fueled by Internet obtained information and championed by celebrities who believe vaccines have hurt their children, many parents are skipping some or all vaccinations. Many college students also choose not to get the flu shot, believing the shot might make them ill or that it's not such a big deal to get the flu. To understand the risks and benefits of vaccinations, it is first necessary to learn about the causes of diseases that vaccines attempt to prevent.

Should you get a flu shot?

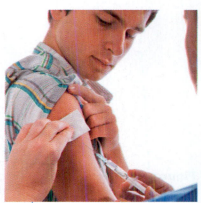
Why would a male be vaccinated against cervical cancer?

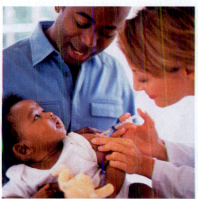
Are parents justified in not allowing their babies to be vaccinated?

467

19.1 Infectious Agents

Diseases that can be transmitted from one individual to another result from the activity of infectious agents. Such communicable or transmissible diseases differ from genetic diseases in that they are caused by organisms rather than by malfunctioning genes—although malfunctioning genes can make an organism more susceptible to infection.

Disease-causing organisms are called *pathogens*. There are many different types of pathogens, including viruses, bacteria, and eukaryotic pathogens. When a pathogen can be spread from one organism to another, it is said to be *contagious*. Pathogens that find a tissue inside the body that will support its growth are said to be *infectious*. Organisms that obtain nutrients and shelter required for growth and development from a different organism while contributing nothing to the survival of the host are *parasites*. Most infectious agents are parasites. Some organisms can cause an infection in an individual, but they are not contagious if the infected individual cannot pass the infection to other individuals.

Stop & Stretch Lyme disease comes from the bite of an infected tick but cannot be passed from one person to another. Is this disease infectious, contagious, or both?

Organisms that can be seen only when viewed through a microscope are called *microscopic organisms*, or *microbes*. The most common infectious microbes are bacteria and viruses.

Bacteria

Bacteria are a diverse group of tiny, single-celled prokaryotic organisms. There are more bacteria in your mouth than there are humans on Earth. Bacteria are commonly rod shaped (bacilli), spherical (cocci), or spiral (spirochetes) (**Figure 19.1**). While we often think of bacteria only in their roles as disease agents, many different types of bacteria are beneficial to living organisms and the environment.

Figure 19.1 Bacteria. The drawing shows the structure of a typical bacterium. The photos show the three characteristic bacterial shapes.
Visualize This: Note that the bacterial cell does not contain any membrane-bound organelles such as a Golgi apparatus or an endoplasmic reticulum. Would you expect that this prokaryotic cell could have ribosomes? Why or why not?

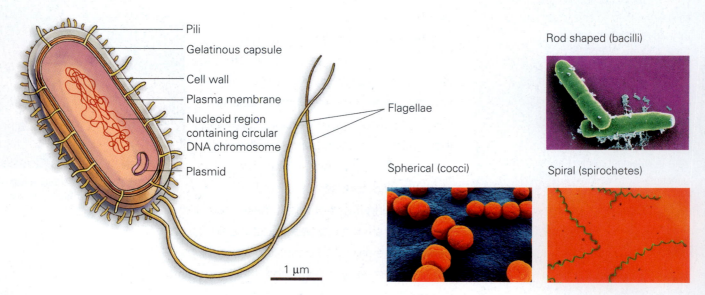

Pili
Gelatinous capsule
Cell wall
Plasma membrane
Nucleoid region containing circular DNA chromosome
Plasmid
Flagellae
1 μm

Rod shaped (bacilli)
Spherical (cocci)
Spiral (spirochetes)

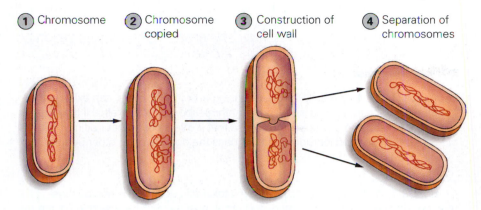

① Chromosome ② Chromosome copied ③ Construction of cell wall ④ Separation of chromosomes

Figure 19.2 Binary fission. (1) The bacterial cell begins with one copy of the circular DNA chromosome that is wound around itself. (2) The chromosome is copied, and each copy is attached to the plasma membrane. (3) Continued growth separates the two chromosomes. The plasma membrane pulls inward in the middle, and a new cell wall is constructed. (4) Daughter cells separate.

Like other prokaryotes, bacteria lack a nucleus. Instead, they have a nucleoid region containing a double-stranded, circular DNA chromosome. In addition to the large DNA chromosome, bacteria may also contain small, circular extrachromosomal DNA (DNA that is separate from the chromosome) called plasmids. Bacteria have evolved to carry genes on these plasmids that provide protection against chemicals, called antibiotics, designed to kill them.

Most bacterial cells are surrounded by a cell wall that provides rigidity and protection; it is composed of carbohydrate and protein molecules. The cell walls of many bacteria are surrounded by a gelatinous capsule, which helps the bacteria attach to cells within tissues they will infect. This capsule also allows bacteria to escape destruction by cells of the immune system. Bacteria also may have one or more external flagellae to aid in movement, and pili, which help some bacterial cells attach to each other and pass genes.

Bacteria reproduce by a process called binary fission (**Figure 19.2**), in which a single bacterial parent cell gives rise to two genetically identical daughter cells. When a bacterial cell divides by binary fission, the single circular chromosome that is attached to the plasma membrane inside the cell wall is copied. This copy is then attached to another site on the plasma membrane, and the membrane between the attachment sites grows and separates the two copies of the original chromosome, eventually producing two separate, genetically identical daughter cells.

Bacteria can reproduce rapidly under favorable conditions, doubling their population approximately every 20 minutes or so. In theory, after 8 hours at room temperature, a single *Salmonella* bacterium in a chicken salad sandwich could give rise to millions of bacterial cells. This exponential growth occurs because 1 cell gives rise to 2, and those 2 yield 4, the 4 divide to become 8, then 16, and so on (**Figure 19.3**). *Salmonella* reproduces quickly in food because the bacteria have access to all the nutrients they need. Fortunately, refrigeration slows the rate at which bacteria divide.

Bacterial Infections. When a bacterial infection occurs in your body, the rapidly increasing number of bacteria support their growth by using your cells' nutrients, effectively preventing your cells from functioning properly. However, most symptoms of bacterial infection result from more than just the large numbers of bacteria in your body. The symptoms arise due to the effects of molecules called *toxins* that are secreted by the bacterial cells, which can result in cell death. Many examples of diseases caused by bacteria are listed in **Table 19.1** (on the next page).

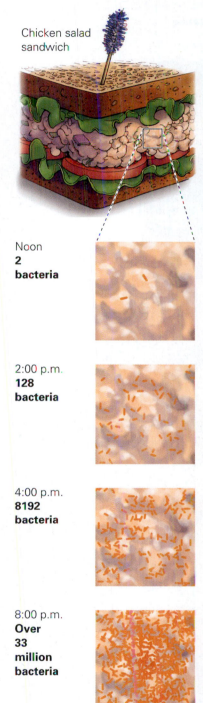

Chicken salad sandwich

Noon
2 bacteria

2:00 p.m.
128 bacteria

4:00 p.m.
8192 bacteria

8:00 p.m.
Over 33 million bacteria

Figure 19.3 Exponential growth of bacteria. One kind of bacterium that can cause food poisoning reproduces every 20 minutes. If a sandwich with 2 bacteria is left out for 20 minutes, each bacterial cell will make a copy of itself, yielding 4 bacteria. After 40 minutes, there will be 8 bacteria. Within 8 hours, there will be 33 million bacteria.

TABLE 19.1

Examples of diseases caused by bacteria.

Disease	Disease Information
Anthrax 	*Bacillus anthracis* lives mainly in soil and, as a part of its life cycle, can exist as a tough, resistant structure called a spore. Anthrax spores can be lethal if they are inhaled. Anthrax is an example of a disease that is infectious without being contagious because infected individuals do not pass the disease on to other people.
Botulism 	*Clostridium botulinum* produces a powerful toxin that acts on the nervous system, causing paralysis of respiratory and facial muscles. Botox—injections of the toxin produced by this bacterium—temporarily paralyzes facial muscles to prevent drooping and the appearance of wrinkles. This soil-dwelling bacterium can adhere to vegetables. Improper sterilization allows the bacteria to proliferate and produce a gas that causes telltale bulges in canned foods.
Escherichia coli infection 	*E. coli* bacteria live in the intestines of humans and other organisms. Disease can occur when human sewage contaminates water supplies. *E. coli* infection can also occur from eating undercooked, contaminated beef or drinking contaminated, unpasteurized milk.
Staphylococcus (commonly known as staph infection) 	*Staphylococcus* bacteria that reside on human skin and mucous membranes can cause infections; they invade and destroy tissue through an opening in the skin. *S. aureus* can cause toxic shock syndrome when tampons are left in place long enough to allow bacterial growth. Methicillin-resistant *S. aureus* (MRSA) infection is deadly because these bacteria have evolved resistances to antibiotics. Recent cases of healthy young athletes dying from exposures to this bacterium were likely caused by exposure to resistant bacteria that enter the skin through cuts or scrapes and are present on locker room floors or shared equipment.
Tuberculosis 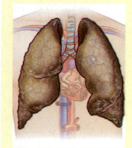	Tuberculosis is caused by the bacterium *Mycobacterium tuberculosis*. This infectious agent is spread through infected droplets when a person with the infection coughs or sneezes. Once inside the body, the bacteria take up residence in the lungs. The immune system responds by walling off the bacteria in a manner much like the formation of a scab. The bacteria can live within the walls for years, during which time the infection is not considered to be active, and it is not transmissible. If the bacteria escape the immune system's attempts at control, the infection becomes active. Renewed activity is most common in people with compromised immune systems. Approximately one-third of the human population is thought to be infected with TB. The disease is extremely common in areas where impoverished people have minimal health care.

Many bacterial infections are routinely treated with medications called *antibiotics*. Unfortunately, the days of easily treatable bacterial infections may soon be over because the bacteria that cause many diseases—including tuberculosis, ear infections, and gonorrhea—have become resistant to the antibiotics produced to cure them.

Antibiotic resistance develops as a result of natural selection on a bacterial population. Variation exists within any population, including within a population of bacterial cells. Some bacteria, even before exposure to an antibiotic, carry genes that enable them to resist the antibiotic. You can develop a drug-resistant infection when your body selects for the resistant bacteria or when you contract the resistant bacteria from someone else.

Vaccines against Bacterial Diseases. Immunizations against many bacterial diseases are recommended by the U.S. Department of Health and Human Services Centers for Disease Control and Prevention (CDC). In fact, children are supposed to have proof of vaccination to attend school. Young children receive a series of injections containing vaccines against three bacterial diseases commonly known as the DPT vaccine. The first letter in the name of this vaccine is for diphtheria, a respiratory infection that has been eradicated in the United States but has a 10% death rate in countries that don't vaccinate. Continued vaccinations are necessary in the United States, in spite of the eradication of this disease, because travelers can bring the disease back into the country. The DPT vaccine also protects against pertussis, another respiratory tract infection. This potentially deadly infection results in violent coughing that makes breathing so difficult that infected people must struggle to get a breath in. The resulting intake of air causes a whooping sound, leading to the disease's more commonly known name, whooping cough. The T in DPT stands for tetanus, a disease caused by a toxin produced by the *Clostridium tetani* bacterium. This bacterium can persist as a resistant spore in the soil. Puncture wounds produced by objects carrying contaminated soil allow the bacteria into tissues where it can grow, producing a toxin that has devastating, often fatal, effects on the nervous and muscular systems.

There is also a vaccine against bacterial meningitis, a disease caused by a bacterium that can cause inflammation of the membranes surrounding the brain. Around 10% of people who have meningitis will die, and many more will suffer permanent disability. Because this disease can be spread when an infected person sneezes or coughs, outbreaks of meningitis are most common in places where many people live in close quarters, like university dormitories. For this reason, the CDC recommends that the vaccine against meningitis be given to adolescents.

Viruses

Many of the CDC-recommended vaccines protect against viral diseases. Viruses (**Figure 19.4**) are little more than packets of nucleic acid (DNA or RNA) surrounded by a protein coat. Viruses are not considered to be living organisms for two reasons: (1) They cannot replicate themselves without the aid of a host cell, and (2) they are not themselves composed of cells. Viruses lack the enzymes for metabolism and contain no ribosomes. Therefore, they cannot make their own proteins. They also lack cytoplasm and membrane-bound organelles. They cannot produce toxins like bacteria can.

The genetic material, or genome, of a virus can be DNA or RNA; it can be double stranded or single stranded, and it can be linear or circular. For example, the herpes virus has a double-stranded DNA genome, and the poliovirus has a single-stranded RNA genome. The genes of a virus can code for the production of all the proteins required to produce more viruses inside a host cell. RNA viruses called retroviruses are packaged with the enzyme reverse transcriptase, which synthesizes DNA from the RNA genome rather than the reverse in "normal" genetic transcription, where RNA is copied from DNA.

The protein coat surrounding a virus is called its *capsid*. Many of the viruses that infect animals have an additional structure outside the capsid called the viral envelope. The envelope is derived from the cell membrane of the host cell and may contain additional proteins encoded by the viral genome. Viruses that are surrounded by an envelope are called *enveloped viruses*.

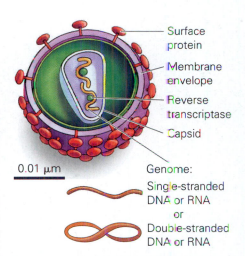

0.01 μm

Surface protein

Membrane envelope

Reverse transcriptase

Capsid

Genome:

Single-stranded DNA or RNA

or

Double-stranded DNA or RNA

Figure 19.4 Viral structure. Viruses are composed of genetic material surrounded by a protein coat. Some viruses, such as this one, are also surrounded by an envelope.

Visualize This: Which structures on the virus might help it attach to a cell?

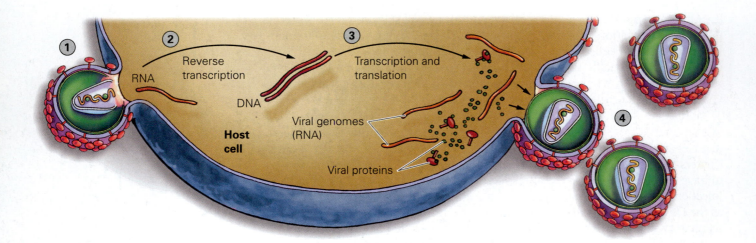

Figure 19.5 Viral replication.
(1) Viral membrane fuses with the host cell; the capsid is removed, and the genome enters. (2) In the case of an RNA virus, such as the AIDS virus, the viral genome is used to synthesize many copies of DNA by the reverse transcriptase enzyme. (3) The DNA is transcribed and translated by the host cell. More viral proteins and copies of the viral genome are produced. (4) Once assembled, new viruses leave the host cell and infect other cells.

Viral Infection. Infection by an enveloped virus occurs when the virus gains access to the cell by fusing its envelope with the host's cell membrane (**Figure 19.5**). An unenveloped virus uses its capsid proteins to bind to receptor proteins in the plasma membrane of a host cell. Some capsid proteins function as enzymes that digest holes in the plasma membrane, thereby allowing the viral genome to enter cells. Once inside the host cell, the capsid is removed.

Regardless of how it enters, after the genome enters the host cell, the infection continues when the virus makes copies of itself. First, the genome is copied, then the virus uses the host-cell ribosomes and amino acids to make viral proteins for building new capsids and synthesizing some of the envelope proteins. Once assembled, the daughter viruses exit the cell, leaving behind some viral proteins in the host's cell membrane. The daughter viruses then move to other cells, spreading the infection.

Some viruses, called latent viruses, enter a state of dormancy in the body. During that time, the virus is not replicating. Herpes simplex viruses, for example, which cause outbreaks on the mouth and genitals, can undergo long periods of dormancy, during which time they are present in the body but not causing symptoms such as painful sores and blisters. Herpes simplex is sexually transmitted, as are many viruses (which are discussed in detail in Chapter 21). Examples of nonsexually transmitted diseases caused by viruses are listed in **Table 19.2**.

Vaccines against Viral Diseases. Newborns are vaccinated against hepatitis B, a sexually transmitted virus. Newborns are vaccinated, in part, because this disease is easily passed from mothers, who do not realize they have the virus, to their newborns during delivery. The potentially fatal hepatitis viruses attack the liver and prevent it from functioning properly. Infants are also vaccinated against rotavirus, which causes severe diarrhea, fever, abdominal pain, and dehydration, resulting in half a million deaths worldwide in children under the age of 5. Young children are vaccinated against polio, a disease that can cause paralysis and death. They are also given the MMR triple shot, a vaccine against measles, a severe and sometimes deadly rash and fever disease still common in many parts of the world; mumps, a swelling of the salivary glands that can, though rarely, cause deafness or reproductive difficulties; and rubella, a disease that causes fever and rash and can cause birth defects if a pregnant woman acquires the infection.

The MMR triple shot is given at around 15 months of age, which happens to coincide with the age when the first symptoms of autism appear. This correlation has led many parents of autistic children to assume that the vaccine can cause autism. However, many scientific studies have been performed to test this hypothesis, and no evidence has emerged to support it.

TABLE 19.2

Examples of diseases caused by viruses.	
Disease	**Disease Information**
Common Cold 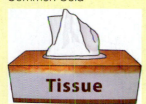	As many as 200 different viruses cause colds, with rhinovirus the most common cause. Cold viruses spread easily in droplets from sneezes or coughs or on objects such as doorknobs and utensils that an infected person contaminated.
Mononucleosis 	Mononucleosis is caused by the Epstein-Barr virus. Most people will have mononucleosis before their mid-30s. If infected as a child, the symptoms are less severe and may go unnoticed. In older adolescents and young adults, the symptoms—fatigue, weakness, fever, headache, and sore throat—can be severe and may last a month or two. The virus can be transmitted in the saliva (kissing) and by coughing and sneezing or by touching an object contaminated by an infected person.
Rabies 	Rabies is a viral disease that typically spreads to humans through a bite from an infected animal. The initial signs of infection include flu-like symptoms but progress to more serious symptoms, including paralysis, convulsions, and hallucinations. Most domesticated animals are vaccinated against rabies, so transmission is most common from bites by wild animals.
West Nile 	The West Nile virus is transmitted to humans via the bite of an infected mosquito. The virus first emerged in the United States in New York and quickly spread west and south throughout the country. In infected persons, the brain and spinal cord can become inflamed, resulting in paralysis and death.

More recently, the CDC began recommending childhood vaccination against chicken pox, a skin rash of blisterlike lesions that usually resolves itself but can cause encephalitis (swelling of the brain) and pneumonia, typically when older children come down with the disease. Annual vaccination against the influenza (flu) virus is now recommended for children as well as adults. The flu shot is composed of parts of three or four viruses and is designed to stimulate your body's defenses, giving it a head start against your next exposure to influenza.

Of particular interest to college students is a new vaccine called Gardasil that provides protection against some strains of human papillomavirus (HPV), known to cause cancer in women. The lower third of the uterus, called the cervix, is very susceptible to infection with HPV. Even with the best medical care, 37% of women with cervical cancer will die from the disease. In poor countries, the toll is even more dramatic—cervical cancer is the leading cause of cancer deaths in the developing world, killing over 300,000 women every year. If a larger percentage of the population were vaccinated against HPV, most of these cervical cancer deaths could be prevented.

The success of vaccination programs lies not with providing immunity to every susceptible individual, but with reducing the total number of susceptible

Figure 19.6 Disease eradication.
Not all members of a population need to be immunized for a disease to die out. As long as enough individuals are protected, a person with the infection cannot transmit it effectively to someone else, so the pathogen cannot reproduce.
Visualize This: If large numbers of people refuse vaccinations, what effect may that have on the likelihood of an epidemic occurring?

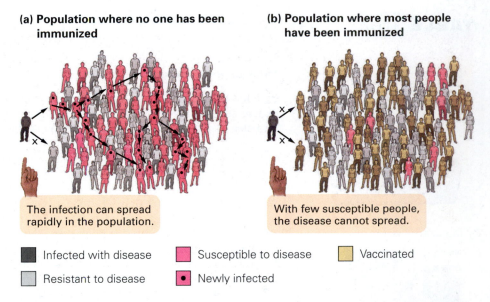

(a) Population where no one has been immunized

The infection can spread rapidly in the population.

(b) Population where most people have been immunized

With few susceptible people, the disease cannot spread.

| ⬛ Infected with disease | 🟥 Susceptible to disease | 🟨 Vaccinated |
| ⬜ Resistant to disease | ⬛ Newly infected | |

individuals so that the disease cannot easily spread (**Figure 19.6**). Effectively reducing the toll of cervical cancer requires that as many potential carriers of HPV be immunized as possible. In other words, men should receive the vaccine as well to lower the population's infection rate.

To improve its appeal to men, the makers of Gardasil developed it so that it would also prevent infection by two strains of HPV that are not associated with cervical cancer but do cause genital warts. Including these strains in the vaccine will presumably make it more attractive to men interested in reducing the relatively high risk of developing genital warts.

Stop & Stretch Many people cannot get vaccinated. For instance, newborns are not vaccinated against most diseases, and people undergoing cancer treatments are not vaccinated against the flu. Children undergoing treatment for cancer are also not vaccinated against childhood diseases. In addition, no vaccine is 100% effective. What is likely to happen to infants and people with compromised health if they are exposed to someone who has refused to be vaccinated and is sick with disease? Likewise, many people feel that it is unnecessary to get vaccinated against diseases that are mostly eradicated in the United States. What would happen if a person traveled to a foreign country where the disease still exists and brought it back to the United States?

Eukaryotic Pathogens

While many infectious diseases are caused by bacteria or viruses, some are also caused by eukaryotic organisms. Eukaryotic pathogens such as the single-celled protozoans, from the kingdom Protista, along with worms and some fungi, cause tremendous human suffering and death worldwide. Protozoans are often spread by water and food contaminated with animal feces. Most worms gain access to the body because of poor sanitation or by consumption of contaminated meat or fish. Worms that take up residence in intestines or other organs can cause extensive damage to internal organs and tissues. Fungi cause diseases of the skin and internal organs. They damage tissue by secreting digestive enzymes into it and absorbing the products of this digestion. Examples of diseases caused by eukaryotic pathogens are found in **Table 19.3**. There are no vaccines routinely given in the United States against diseases caused by these pathogens.

TABLE 19.3

Examples of diseases caused by eukaryotic pathogens.	
Disease	**Disease Information**
Giardiasis 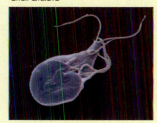	Giardiasis is caused by the waterborne *Giardia lamblia*. During one part of its life cycle, *Giardia* exists as a tough, resistant structure called a cyst. *Giardia* cysts enter bodies of water or food supplies in contaminated feces. The organism completes its life cycle in the intestines of an infected person, causing severe diarrhea and gas.
Malaria 	Malaria is caused by the protozoans from the genus *Plasmodium*. The protozoans are transmitted when infected mosquitoes bite humans. This disease kills close to 2 million people annually, mostly in Africa.
Schistosomiasis 	Freshwater snails carry this parasitic disease, caused by flatworms. Humans affected by this disease live around contaminated water, commonly found in Asia, Africa, and South America.
Tapeworm 	Tapeworm infections are usually caused by ingesting infected raw or undercooked pork or beef. This worm lives in the intestines and can cause abdominal pain and diarrhea.
Athlete's foot and jock itch 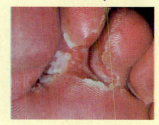	Athlete's foot and jock itch are both types of fungal infections that affect the top layer of skin. Athlete's foot typically grows between the toes. Jock itch affects the skin of the genitals and inner thighs. This fungus spreads easily in public places such as locker rooms and the floors of communal showers. Excessive perspiration can wash away naturally occurring fungus-killing oils present on skin. Due to their routine exposure to locker room floors and communal showers, and their tendency to sweat during workouts, athletes are more prone to these infections.

19.2 The Body's Response to Infection: The Immune System

Organisms have evolved defenses against pathogens. In humans and most vertebrates, it is the job of the immune system to protect against infection. The immune system consists of three different lines of defense (**Figure 19.7**). The first line of defense—the skin and mucous membranes—helps prevent access to the body, while the remaining two lines of defense—white blood cells and inflammation, among others—operate if the pathogen gets past the first line.

First Line of Defense: Skin and Mucous Membranes

The skin and mucous membrane secretions comprise the first line of defense against pathogens. These external physical and chemical barriers are nonspecific defenses; that is, they do not distinguish one pathogen from another.

Organisms living on the body's surface are kept out by the skin. In addition to being a physical barrier, the skin sheds, taking pathogens with it, and has a low pH that can help to repel microorganisms. Likewise, glands in the skin secrete chemicals that slow the growth of bacteria. For example, tears and saliva contain enzymes that break down bacterial cells, and earwax traps microorganisms.

Mucous membranes lining the respiratory, digestive, urinary, and reproductive tracts also secrete mucus that traps pathogens, which are coughed or sneezed away from the body, destroyed in the stomach, or excreted in the urine

Figure 19.7 Three lines of defense.
(1) The skin and mucous membranes serve as a nonspecific barrier to infection.
(2) Macrophages attack pathogens that gain access to the body and cause an immune response. (3) Lymphocytes initiate a specific response by targeting specific pathogens and attempting to contain them.

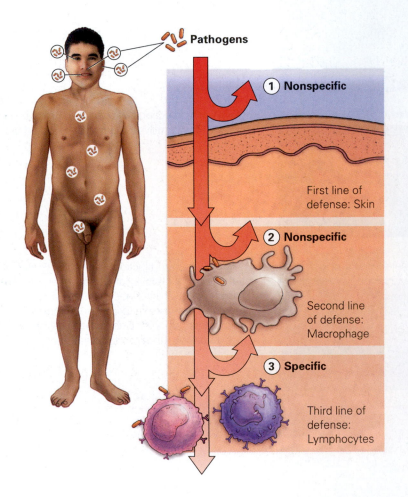

or feces. Digestive secretions, including acids, kill many microorganisms that gain access to the stomach. Vomiting can also rid the body of toxins or infectious agents. Pathogens that are able to evade the first line of defense next encounter an internal, second line of defense.

Second Line of Defense: Phagocytes and Macrophages, Inflammation, Defensive Proteins, and Fever

If a pathogen is able to get past the first line of defense, the internal second line of defense, which is also nonspecific, can often stop the infection. Like the first line of defense, participants in the second line of defense do not target specific pathogens.

Phagocytes and Macrophages. Phagocytes are white blood cells that indiscriminately attack and ingest invaders by engulfing and digesting the invader. Macrophages are phagocytic white blood cells that move throughout the lymphatic fluid, cleaning up dead and damaged cells. To destroy the cells, macrophages extend their long pseudopodia (cellular extensions used for eating and moving), grab the invading organism, and engulf it (**Figure 19.8**). Enzymes inside the macrophage help break the invader apart. Macrophages can also clean up old blood cells, dead tissue fragments, and other cellular debris. They also release chemicals that stimulate the production of more white blood cells. Much of the destruction of the offending cells occurs in the lymph nodes.

When an invader is too big to be engulfed by phagocytosis, other white blood cells cluster around the invader and secrete digestive enzymes that irritate or may even destroy the organism. Invaders that get this treatment include larger organisms such as protozoans and worms.

Additional white blood cells that circulate through the blood and lymph destroying invaders are natural killer cells. These nonspecific cells attack tumor cells and virus-invaded body cells on first exposure. These cells release chemicals that break apart plasma membranes of their target cells, causing them to burst. When white blood cells are actively fighting an infection, the body increases production of white blood cells, and white cell counts can increase dramatically. In addition, tissue fluid, dead cells, and microorganisms accumulate at the site of the infection, producing *pus*. If the pus cannot drain, the body may wall it off with connective tissue, producing an *abscess*.

Another component of the second line of defense is the inflammatory response.

Inflammation.   The inflammatory response is a reaction producing redness, warmth, swelling, and pain.

Whenever a tissue injury occurs, the inflammatory response begins. Damaged cells release chemicals that stimulate specialized cells to release histamine. Histamine promotes increased size of blood vessels, or *vasodilation*, near the injury. As more blood and cells arrive at the site of infection to speed cleanup and repair, redness, warmth, and swelling occur. Swollen tissues cause pain by pressing against nearby nerves, causing one to want to rest the affected area. The extra blood flow also brings oxygen and nutrients required for tissue healing.

Defensive Proteins.   In addition to white blood cells, some proteins act as nonspecific defenders. Interferons are proteins produced by virus-infected body cells to help uninfected cells resist infection. When infected cells die, they release interferons that

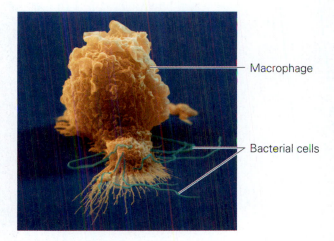

Macrophage

Bacterial cells

Figure 19.8 Phagocytosis. A macrophage attacks bacterial cells.

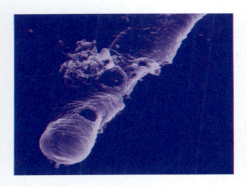

Figure 19.9 A bacterial cell after complement exposure. Complement proteins cause holes to form in the bacterial cell, penetrating the cell wall and cell membrane and causing the bacterium to fill with fluids and burst.

bind to receptors on uninfected cells and stimulate the healthy cells to produce proteins that inhibit viral reproduction.

Another class of defensive proteins, called complement proteins, includes at least 20 different proteins that circulate in the blood and help, or complement, other defense mechanisms. These proteins can coat the surfaces of microbes, making them easier for macrophages to engulf. Complement proteins can also poke holes in the membranes surrounding microbes, causing them to break apart (**Figure 19.9**). These proteins also increase the inflammatory response.

Fever. A body temperature above the normal range of 97–99°F is called a fever. Macrophages release chemicals called pyrogens as a weapon in their assault on invaders. Pyrogens cause body temperature to increase. A slightly higher-than-normal temperature decreases bacterial growth and increases the metabolic rate of healthy body cells. Both of these processes help to fight infection by slowing pathogen reproduction and allowing repair of tissue to occur more quickly. When the infection is controlled, macrophages stop releasing pyrogens, and body temperature returns to normal.

Third Line of Defense: Lymphocytes

If a pathogen makes it past the first two nonspecific defense systems, the immune system presents a third line of defense. Cells of the immune system identify and attack specific microorganisms that are recognized as foreign. This specific defense system consists of millions of white blood cells, called lymphocytes. Lymphocytes travel throughout the body by moving through spaces between cells and tissues or being transported via the blood and lymphatic systems (**Figure 19.10**).

Figure 19.10 The lymphatic system. Pathogens trigger a response by the lymphatic system. The various organs of this system work to eliminate infections.

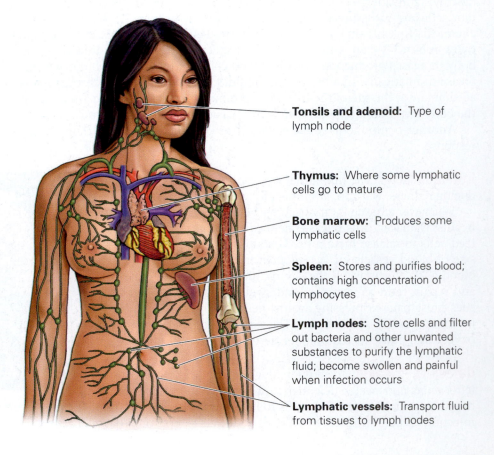

Tonsils and adenoid: Type of lymph node

Thymus: Where some lymphatic cells go to mature

Bone marrow: Produces some lymphatic cells

Spleen: Stores and purifies blood; contains high concentration of lymphocytes

Lymph nodes: Store cells and filter out bacteria and other unwanted substances to purify the lymphatic fluid; become swollen and painful when infection occurs

Lymphatic vessels: Transport fluid from tissues to lymph nodes

The specific response generated by the immune system is triggered by proteins and carbohydrates on the surface of pathogens or on cells that have been infected by pathogens. Molecules that are foreign to the host and stimulate the immune system to react are called *antigens*. Examples of antigens include molecules found on invading viruses, bacteria, fungi, protozoans, and worms, as well as molecules on the surface of foreign substances such as dust, pollen, or transplanted tissues.

Regardless of the source, when an antigen is present in the body, the production of two types of lymphocytes is enhanced: the B lymphocytes (B cells) and T lymphocytes (T cells). Like macrophages, these lymphocytes move throughout the circulatory and lymphatic systems and are concentrated in the spleen and lymph nodes. Lymphocytes display specificity because they recognize specific antigens.

The specificity of lymphocytes is based on the presence of proteins, called *antigen receptors*, whose shape fits perfectly to a portion of the foreign molecule. The receptor is able to bind to the antigen much the way that a key fits into a lock. These receptors are either attached to the surface of the lymphocyte or secreted by the lymphocyte. B and T cells recognize different types of antigens: B cells recognize and react to small, free-living microorganisms such as bacteria and the toxins they produce (**Figure 19.11a**). T cells recognize and respond to body cells that have gone awry, such as cancer cells or cells that have been invaded by viruses. T cells also respond to transplanted tissues and larger organisms such as fungi and parasitic worms (**Figure 19.11b**).

B and T cells both recognize and help eliminate antigens, but they do so in different ways. B cells secrete proteins called antibodies that bind to and inactivate antigens. T cells, in contrast, do not produce antibodies but instead directly attack invaders.

Antibodies are found in the lymph, intestines, and tissue fluids. They are also found in breast milk. When a baby breast-feeds, some antibodies are passed from the mother to the baby, conferring short-term passive immunity to the child. Because the child did not make the antibodies by himself or herself, this immunity will last only as long as the antibody lasts in his or her blood. An infant's passive immunity is different from the immunity conferred by exposure to an

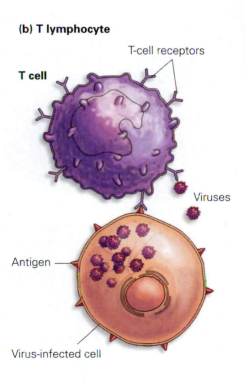

(a) B lymphocyte

B-cell receptors   Antibodies

B cell

Bacteria

Antigen

(b) T lymphocyte

T-cell receptors

T cell

Viruses

Antigen

Virus-infected cell

Figure 19.11 B cells and T cells.
(a) The receptors produced by B cells function as cell-surface antigen receptors or as secreted antibodies that are present in bodily fluids. (b) T cells have surface antigen receptors only. They do not secrete antibodies. B cells and T cells differ with respect to the types of cells and organisms they respond to.

antigen. Exposures to antigens cause the production of antibodies to combat the infection for the individual's lifetime, called active immunity.

Allergy. An allergy is an immune response that occurs even though no pathogen is present. The body simply reacts to a nonharmful substance as though a pathogen were present. For example, some people have allergic reactions to peanuts or ragweed pollen. Asthma may also be the result of an allergic reaction.

The fact that infants today receive around two times as many immunizations as they did 30 years ago has led many people to speculate that the increased rate of immunization may be the cause of increasing rates of allergies and asthma. Well-controlled scientific studies, many of which were performed by independent nonprofit organizations such as the Institute of Medicine (IOM), show no connection between the rates of allergy and asthma and vaccination.

Anticipating Infection

The ability of the body's lymphocytes—B and T cells—to respond to specific antigens begins before humans are born. Thus, we are able to respond to infectious agents the very first time we are exposed. This ability continues into adulthood because these cells are manufactured, at the rate of about 100 million per day, throughout our lives. The ability to respond to an infection, the immune response, actually results from the increased production of B and T cells.

Lymphocytes are produced from special cells called stem cells, immature cells that can differentiate into adult cell types. Many parts of the body, including the bone marrow, retain a supply of stem cells that can develop into more specialized cells. Bone marrow stem cells enable the bone marrow to produce blood cells throughout a person's lifetime. Lymphocytes are produced from the stem cells of bone marrow and released into the bloodstream.

Some lymphocytes continue their development in the bone marrow and become B cells. Others take up residence in the thymus gland. The thymus gland, located behind the top of the sternum, stimulates T cells to develop. When immune cells finish their development in the bone marrow, they are called B cells; when they mature in the thymus, they are called T cells (**Figure 19.12**).

Figure 19.12 Lymphocyte development. Lymphocytes develop from cells in the bone marrow. Those that continue their development in the bone become B cells, while those that move to the thymus to continue their development become T cells.

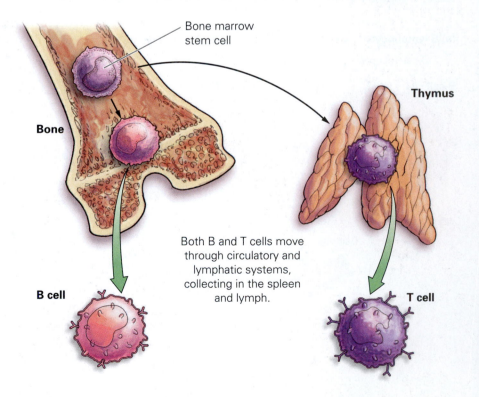

Bone marrow stem cell

Thymus

Bone

Both B and T cells move through circulatory and lymphatic systems, collecting in the spleen and lymph.

B cell

T cell

Lymphocytes can recognize trillions of different antigens. This extraordinary ability results from the trillions of different antigen receptors that are produced by B and T cells. How is it that such an astonishing diversity of receptors can be produced?

Antigen Diversity. If every antigen receptor were encoded by its own gene, trillions of genes would have to code immune system proteins alone. Yet the entire human genome contains approximately 20,000 genes, so this cannot be the case. Instead, B and T cells have evolved a mechanism to generate an enormous variety of receptors from a limited number of genes.

As B and T cells develop, each cell's DNA segments that code for the production of antibodies do something very unusual: They rearrange themselves. Some portions are cut out, and the remaining DNA segments are joined together (**Figure 19.13**). Each unique arrangement of DNA produced by a given cell encodes a different receptor protein. Once synthesized, the proteins move to the surface of the B or T cell and act as antigen receptors. Thus, from a pool of only a few hundred genes, a virtually unlimited variety of diverse antigen receptors can be created.

Stop & Stretch   Based on what you know about antigenic diversity, explain why a particular individual might become ill from an infection that most other members of the population are able to fight off.

Another important facet of the immune system is its ability to determine whether a molecule is part of the host cell or foreign.

Self versus Nonself. While B and T cells are maturing, their antigen receptors are tested for potential self-reactivity. The cells of a given individual have characteristic proteins on their surfaces, and developing lymphocytes are tested to determine whether they will bind to self-proteins. Any developing lymphocyte with antigen receptors that bind to self-proteins is eliminated, making an immune response against one's own body less likely. Lymphocytes with receptors

Figure 19.13 Genetic rearrangements allow for the production of millions of different antigen receptors. The genetic information encoding antibody structures is stored in bits and pieces. These bits and pieces are arranged in various orders during development of the antibody-producing cells. Removal of various DNA segments produces a wide variety of antigen receptors and increases the number of antigens to which each of us can respond.

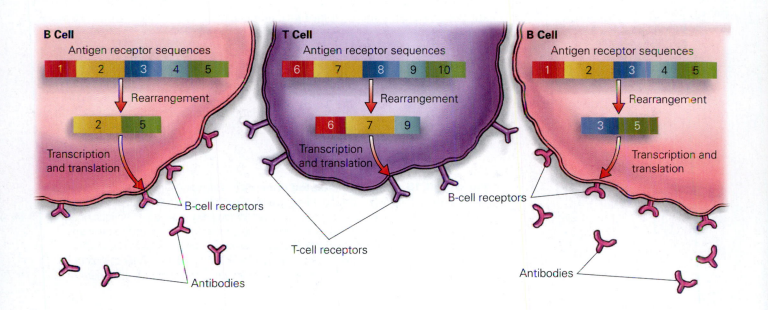

Figure 19.14 Testing developing lymphocytes for self-proteins. Cells have characteristic proteins and other molecules on their surfaces that function as antigens. Developing lymphocytes are tested in the thymus to determine whether they are self or nonself. Lymphocytes that bind to antigens on testing cells are destroyed; those that do not bind are allowed to develop.

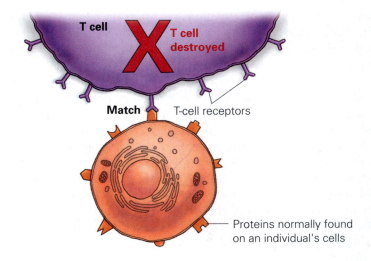

that do not bind are then allowed to develop to maturity (**Figure 19.14**). Thus, the body normally has no mature lymphocytes that react against self-proteins, and the immune system exhibits self-tolerance.

When this testing fails, the cells of the immune system attack normal body cells. Diseases that result when a person's immune system is attacking the body are called *autoimmune diseases.*

Multiple sclerosis is an autoimmune disease that occurs when T cells specific for a protein on nerve cells attack these cells in the brain. In insulin-dependent diabetes, T and B cells attack cells that produce the hormone insulin in the pancreas.

Even though we have a single immune system, it is diversified into two subsystems so that we can combat the multitude of infectious agents encountered in our lifetime. This diversification is a result of the differing approaches of B and T cells to ridding the body of infectious agents once they are found.

Antibody-Mediated and Cell-Mediated Immunity

The protection afforded by B cells is called *antibody-mediated immunity.* T cells provide immunity that depends on the involvement of cells rather than antibodies and is called *cell-mediated immunity.*

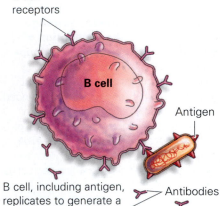

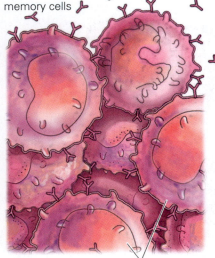

Clonal population

Antibody-Mediated Immunity. When a B cell responds to a specific antigen, the B cell immediately makes copies of itself, resulting in a population of identical cells, called plasma cells, able to help fight the infection (**Figure 19.15**). This population of cells is called a *clonal population.* The sheer number of cells in the clonal population strengthens the immune system's ability to rid the body of the infectious agent.

Figure 19.15 Antibody-mediated immunity. B lymphocytes do not directly kill cells bearing antigens. Instead, they make and secrete antibodies that aid in the process of ridding the body of the antigenic cell. When a B cell binds an antigen, the B cell makes copies of itself, with each copy, called a memory cell, carrying the same antibody. A population of such identical cells is called a clonal population. Now, all the cells of the clonal population can aid in overcoming the infection. After the infection, some memory cells remain and are able to recognize the antigen in the future.

The entire clonal population has the same DNA arrangement. Therefore, all the cells in a clonal population carry copies of the same antigen receptor on their membrane and secrete the same antibody. Some cells of the clonal population, called memory cells, will help the body respond more quickly if the infectious agent is encountered again. Should subsequent infection occur, the presence of memory cells facilitates a quicker immune response.

Inactivation of the infectious agent occurs when antibodies encounter a pathogen that matches the variable region. The antibody then binds to the antigen, forming the antibody-antigen complex. Formation of this complex marks the pathogen for phagocytosis or degradation by complement proteins. Complement proteins cause destruction of the foreign and infected cells. The complement proteins circulate in the blood and lymph in an inactive state until they come in contact with antibodies bound to the surface of microorganisms, at which time they are activated. Some antibodies cause the pathogen and attached antibodies to clump together or *agglutinate*, rendering them unable to infect other cells.

Vaccinations take advantage of the long-term protection provided by antibody-producing memory cells. Usually consisting of components of the disease-causing organisms, such as proteins from the plasma membrane of a bacterial cell, parts of a virus, or a whole virus that has been inactivated, the immune system responds by producing the clonal population of memory cells that will be prepared for a real infection should it happen.

Some vaccines require multiple doses or boosters before a sufficient response is generated. Others such as flu vaccines must be given every year because the flu virus rearranges its genetic sequences swiftly. This shifting genetic sequence results in different proteins being encoded and placed on the surface of the virus, thereby preventing the body's existing memory cells from recognizing a virus that they have already encountered. Because the proteins on the surface of a flu virus change so quickly, a vaccine that was prepared to protect you from last year's flu virus will not likely work against this year's flu virus. Unfortunately, scientists have not been able to make vaccines for every pathogen—some pathogens are easier to work with than others. For instance, intensive efforts have been made to produce a vaccine against human immunodeficiency virus (HIV), to little avail.

Cell-Mediated Immunity. T cells also respond to infection by undergoing rapid cell division to produce memory cells, which then become specialized cells (**Figure 19.16** on the next page). However, unlike B cells, T cells do not secrete antibodies. Instead, they directly attack other cells. Two of these attacking cell types are the cytotoxic T cells and helper T cells. Cytotoxic T cells attack and kill body cells that have become infected with a virus. When a virus infects a body cell, viral proteins are placed on the surface of the host cell. Cytotoxic T cells recognize these proteins as foreign, bind to them, and destroy the entire cell. By releasing a chemical that causes the plasma membranes of the target cell to leak, they break down the cell before the virus has had time to replicate.

Helper T cells, also called T4 cells, can be thought of as boosters of the immune response. These cells detect invaders and alert both the B and T cells that infection is occurring. Without helper T cells, there can be almost no immune response. Helper T cells also secrete a substance called interleukin 2 that greatly increases the level of cytotoxic T-cell response. HIV infects and destroys helper T cells, thus crippling the body's ability to respond to any infection.

As you can see, the various cells (and proteins) of the immune system work in concert to strengthen the overall response of the immune system. However,

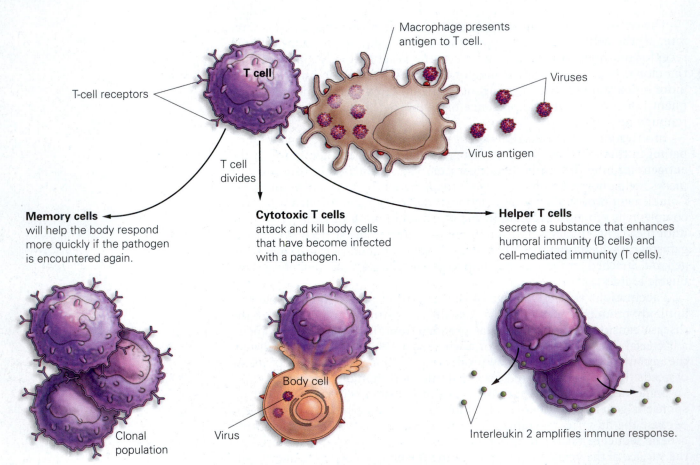

Macrophage presents antigen to T cell.

T cell

T-cell receptors

Viruses

Virus antigen

T cell divides

Memory cells
will help the body respond more quickly if the pathogen is encountered again.

Cytotoxic T cells
attack and kill body cells that have become infected with a pathogen.

Helper T cells
secrete a substance that enhances humoral immunity (B cells) and cell-mediated immunity (T cells).

Clonal population

Virus

Body cell

Interleukin 2 amplifies immune response.

Figure 19.16 Cell-mediated immunity. T lymphocytes divide to produce different populations of cells: Memory cells carry the specific antigen receptor. Cytotoxic T cells attack and dismantle infected bodily cells. Helper T cells boost the immune response.

even with all the complexity of the immune system, some pathogens, such as the virus that causes AIDS, are able to disable the immune system (see the Focus on Evolution feature).

AIDS Weakens the Immune System

Acquired immunodeficiency syndrome, or AIDS, is caused by the human immunodeficiency virus or HIV. HIV primarily kills or disables helper T, or T4, cells. The loss of helper T cells causes immune deficiency—that is, affected individuals experience diseases that are normally controlled by healthy immune systems. These include infections by organisms commonly found on our bodies in low levels, such as *Pneumocystis jiroveci*, a fungus that is found in nearly everyone's lungs by age 30. In healthy people, *P. jiroveci* is held in check by the immune system, but in AIDS patients, this organism often causes pneumonia and extensive lung damage. Diseases like *P. jiroveci* pneumonia are called *opportunistic infections* because they occur only when the opportunity arises due to a weakened immune system.

focus on EVOLUTION

HIV, Evolution, and the Immune System

Infection with HIV that goes untreated typically shows some early signs of infection, followed by a long period of time during which the infected person is symptomless, followed by the symptom-filled period called full-blown AIDS. The reason the infection usually follows this pattern has to do with the evolution of the virus.

Early symptoms of HIV infection typically resemble the flu. This generalized feeling of fatigue and illness is caused by the initial nonspecific immune response to any viral invader. Because HIV is killing off immune system cells that protect people from a variety of infectious organisms, the rundown feeling may last for a number of weeks. After about 6 to 12 weeks, the immune system of most infected people has had enough time to mount an immune response that can control the virus, and people recover from these flu-like symptoms.

At this point, the person is asymptomatic. The asymptomatic phase of HIV infection may last for as long as 10 years. During this time, the number of viruses in the bloodstream is relatively low, but the virus is never totally eliminated. Instead, the virus persists in lymph nodes.

In nearly all HIV-infected individuals who are not receiving drug treatment, the immune system eventually loses control over the virus (**Figure E19.1**). Why does HIV eventually win its battle with the immune system? Primarily, this is due to the evolution of HIV within its host.

The population of HIV is not unchanging during the asymptomatic period; instead, it evolves in response to the environment created by the infected individual's immune system. HIV particles are constantly reproduced as they continue to infect cells in the lymph nodes; any time there is reproduction, mutation can occur.

The reverse transcriptase enzyme of HIV is highly error prone. Some scientists estimate that every single HIV particle produced has at least one difference from the HIV from which it arose. As a result, during the asymptomatic period, new variants of HIV continually arise, and antibodies produced against earlier variants do not recognize the new variants.

The frequent mutation and rapid reproduction of HIV result in a population of the virus within an asymptomatic host that contains about 1 billion distinct variants. With this many variants, it is almost ensured that one or more has antigens that are not immediately recognized by the host's immune system. Because these variants avoid the host's antibodies, they can grow to large populations very quickly. Evolution of the HIV population within an infected individual can be extremely fast.

HIV's rapid evolution appears to be the cause of the eventual end of the asymptomatic period in an infected person. The immune system is able to produce antibodies to many different HIV antigen variants, but eventually the

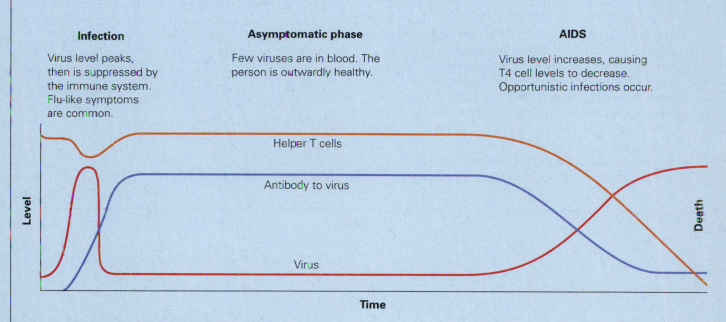

Infection

Virus level peaks, then is suppressed by the immune system. Flu-like symptoms are common.

Asymptomatic phase

Few viruses are in blood. The person is outwardly healthy.

AIDS

Virus level increases, causing T4 cell levels to decrease. Opportunistic infections occur.

Helper T cells

Antibody to virus

Virus

Level

Time

Death

Figure E19.1 The course of HIV infection.

(continued on the next page)

sheer number of different HIV variants that the immune system must respond to becomes overwhelming. Finally, one variant arises that escapes immune system control for a long period; large numbers of T helper cells become infected with this variant and are killed or disabled, and the infected individual becomes increasingly immune deficient. This change initiates the onset of AIDS. Without expensive drug therapies, the relentless evolution of HIV within an infected person's body eventually exhausts his or her ability to control this deadly virus. The rapidly changing HIV genome has also hampered the development of a vaccine against HIV.

HIV is a fragile virus that is transmitted only through direct contact with bodily fluids—primarily blood, semen, vaginal fluid, or occasionally via breast milk to newborns. There is no evidence that the virus is spread by tears, sweat, coughing, or sneezing. It is not spread by contact with an infected person's clothes, phone, or toilet seat. It is not transmitted by an insect bite. And it is unlikely to be transmitted by kissing, unless both individuals have lesions or cuts. HIV is frequently spread through needle sharing among injection drug users, but the primary mode of HIV transmission is via unprotected sex, including oral sex, with an infected partner.

SAVVY READER

Denying AIDS

Thabo Mbeki was the president of South Africa from 1999 to 2008. During his tenure, he became convinced that AIDS was not caused by a virus and that antiviral medicines should not be used to treat this disease. He even turned down free anti-retroviral medications offered by health organizations. Hundreds of thousands of South Africans have died of AIDS, even though lifesaving medications are available. Many of the dead are babies who were infected by their mothers during childbirth, a form of transmission that can be prevented with the proper medications.

1. Is it possible that a comprehensive review of the scientific literature can lead to different interpretations of the science being presented? Is this likely what informed Mbeki's opinion about HIV not causing AIDS?

2. Do you think government officials have an obligation to accept prevailing scientific opinions when making decisions that affect the health of their constituents?

3. Other scientific claims are also denied by governments. For example, climate change is denied by many U.S. politicians. When these differing views occur, should sound scientific evidence trump the opinions of nonscientists? Why or why not?

Chapter Review

Learning Outcomes

Mastering**BIOLOGY**®

Go to the Study Area at www.masteringbiology.com for practice quizzes, myeBook, BioFlix™ 3-D animations, MP3 Tutor secessions, videos, current events, and more.

LO1 Compare the structure of bacteria to that of viruses (Section 19.1).

- Bacteria are single-celled organisms with no nucleus or membrane-bound organelles. They are surrounded by a cell wall (pp. 468–469).
- Viruses are composed of nucleic acids encased in a capsid and sometimes an envelope (p. 469).

LO2 Contrast bacterial replication with viral reproduction (Section 19.1).

- Bacteria replicate by binary fission (p. 471).
- Viruses replicate their genome inside host cells. The viral genome then directs the synthesis of viral components (p. 472).

LO3 List the structures involved in the first line of defense against infection, and describe how they function in protecting the body against pathogens (Section 19.2).

- The skin and mucous membrane secretions are nonspecific defenses that comprise the first line of defense against infection (pp. 476–477).

LO4 List the participants in the second line of defense against infection, and describe their functions (Section 19.2).

- The second line of defense consists of nonspecific internal defenses, including white blood cells such as phagocytic macrophages, which engulf and digest foreign cells, and natural killer cells, which release chemicals that disintegrate cell membranes of tumor cells and virus-infected cells (p. 477).
- Inflammation attracts phagocytes and promotes tissue healing (p. 477).
- Defensive proteins, including interferon, help protect uninfected cells from becoming infected. Complement proteins coat microbes and make them easier for macrophages to ingest (pp. 477–478).

LO5 List the two different cell types involved in the third line of defense against infection, and compare how they recognize and eliminate antigens (Section 19.2).

- Lymphocytes help comprise the third specific line of defense in response to antigens on the surface of pathogens. Exposure to antigens causes increased production of B and T lymphocytes (pp. 478–479).
- B cells secrete antibodies against pathogens, and T cells attack invaders (p. 479).

LO6 Explain allergic reactions (Section 19.2).

- Allergic reactions are the immune system responding to antigenic substances that are not normally pathogens (p. 480).

LO7 Explain the process immune cells undergo to be able to respond to millions of different antigens (Section 19.2).

- Lymphocytes are produced during fetal development and allow the immune system to respond to trillions of different antigens (p. 481).
- Antigen receptors on the surface of immune cells are produced by rearrangements of gene segments (p. 481).

LO8 Describe the process used to determine whether a cell is foreign or native to the body and what happens when this process fails (Section 19.2).

- The antigen receptors of B and T cells are tested for self-reactivity, and those that react against self are eliminated (pp. 481–482).

LO9 Explain the role of memory cells in helping to protect against infection (Section 19.2).

- Memory cells carry copies of the antigen receptor and secrete antibody against a previously encountered pathogen (pp. 482–484).

LO10 **Explain the role of helper T cells in allowing infection by the AIDS virus (Section 19.2).**

- AIDS is an immunodeficiency caused when HIV attacks helper T cells (p. 484).
- Helper T cells increase immune response (p. 485).

LO11 **Describe the mechanism by which vaccines help confer immunity (Section 19.2).**

- Vaccines contain parts of disease-causing organisms or entire organisms that have been inactivated. The immune system produces memory cells in response to vaccination, thereby affording protection in the case of actual infection (p. 483).

Roots to Remember

The following roots of words come mainly from Latin and Greek and will help you decipher terms:

auto- means self. Chapter term: autoimmune

enceph- and **encephalo-** mean brain. Chapter term: encephalitis

-path- or **patho-** relates to disease. Chapter term: pathogen

pseudo- means false. Chapter term: pseudopodia

-podia or **-pod** and **-ped** are from words for the foot. Chapter term: pseudopodia

Learning the Basics

1. **LO1** Describe the structure of a typical bacterium and a typical virus.

2. **LO3** How do bodily secretions help protect against infection?

3. **LO5** What roles do B cells and T cells play in the immune response?

4. **LO2** Binary fission _____.

   A. allows bacterial cells to produce more plasmids; B. is a type of sexual cell division that viruses undergo; C. allows immune cells to replicate; D. is the method by which bacteria replicate

5. **LO2** Viral replication _____.

   A. occurs only inside living cells; B. requires that copies of the viral genome be produced; C. requires that

copies of viral proteins be synthesized; D. all of the above

6. **LO7** The immune system _____.

   A. has a gene for each antigen; B. undergoes genetic re-arrangement in response to different antigens; C. is able to make many antigen receptors by rearranging the DNA of immune cells; D. can always devise an antibody that will bind to an antigen

7. **LO8** Autoimmune diseases result when _____.

   A. a person's endocrine system malfunctions; B. liver enzymes malfunction; C. B cells attack T cells; D. the immune system fails to differentiate between self and nonself cells

8. **LO5** Which of the following cell types divides to produce cells that make antibodies?

   A. helper T cells; B. B cells; C. cytotoxic T cells; D. all of the above

9. **LO10** Helper T cells secrete substances that _____.

   A. help prevent leukemia; B. prevent bacteria from entering cells; C. boost B-cell and cytotoxic T-cell response; D. inhibit reverse transcriptase; E. stimulate the thymus to make more B cells

10. **LO9** The immune system can recognize a virus you have been exposed to once because _____.

    A. you harbor the virus for many years; B. we have genes to combat every type of virus; C. a cell that makes receptors to a virus multiplies on exposure to it and produces memory cells; D. a copy of the viral genome is inserted into a memory cell.

11. **LO4** Which of the following is not involved in the second line of defense?

    A. natural killer cells; B. inflammation; C. complement proteins; D. macrophages; E. T lymphocytes

12. **LO10** **True or false:** The AIDS virus attacks antibodies against unrelated pathogens, weakening the immune system's ability to fight off many diseases.

13. **LO6** **True or false:** Allergic reactions occur when the body can't make antibodies against an antigen.

Analyzing and Applying the Basics

1. **LO1 LO2** Why are viruses not considered to be living organisms?

2. **LO11** Why do you need a flu shot every year but only one inoculation against some other diseases?

3. **LO9** Why is your immune system usually more effective at fighting off infection the second time you are exposed to a pathogen?

Connecting the Science

1. Many states allow parents to opt out of the vaccinations required to start school. Should parents be allowed to opt out of vaccinating their children? Defend your answer.

2. What would you say to a friend who refuses the flu vaccine because he did not get vaccinated last year and did not get sick?

Answers to **Stop & Stretch, Visualize This, Savvy Reader,** and **Chapter Review** questions can be found in the **Answers** section at the back of the book.

Sex Differences and Athleticism

Endocrine, Skeletal, and Muscular Systems

Many young athletes dream of playing professional sports.

The scene on the baseball field is one that many budding athletes dream about. For one young baseball pitcher, the dream has come true. On a warm Friday evening late in July, the sun is beginning to set over left field in a beautiful, open-air, brick baseball stadium. On the mound, a 23-year-old left-hander stands motionless, squinting to better see the catcher's signals.

Ila Borders was the first female pitcher to win a professional baseball game.

The scoreboard above the center field wall shows that it is the top of the sixth inning, and the Duluth-Superior Dukes are leading the Sioux Falls Canaries by a score of 2 to 0. The Dukes' pitcher has given up only three hits this game; but the Canaries may be putting together a rally, with one runner on base and their number three hitter at bat.

The pitcher glances at the runner on second base, turns to face the batter, and in one strong, fluid motion, rears back and hurls the ball across the inside corner of home plate. "Strike three. The batter is out," yells the home-plate umpire.

The stadium erupts as the fans jump to their feet cheering wildly, and the 5 ft. 10 in. pitcher walks off the mound toward the dugout to a standing ovation. For the next half-inning, the hometown fans chant the pitcher's name in unison, until the southpaw finally comes out of the dugout to salute the cheering fans. When the pitcher raises the black-and-purple Dukes baseball cap toward the fans behind home plate, her long hair spills across the back of her uniform, obscuring her name and number.

Michelle Wie has had success golfing in men's tournaments.

The pitcher is Ila Borders, the first woman in the starting lineup of a men's professional baseball game and the first female pitcher to record a win in a men's professional baseball game. This accomplishment was so impressive that *Sports Illustrated* honored Borders as one of the top 100 female athletes in history; the uniform and glove that served her in this historic victory adorn the prestigious Baseball Hall of Fame in Cooperstown, New York, and the national television stations ESPN and CNN broadcasted footage of her pitching that night.

Danica Patrick has had success in auto racing against men.

Will Ila Borders be the first of many women to play major league base-ball? Will women infiltrate professional football and hockey leagues next? Will the successes of women golfers and auto racers be replicated in other sports as opportunities increase for women to compete? Or are the bodies of women and men simply constructed so differently that they will be able to compete against each other only rarely?

To answer these questions, we need to delve into the study of sex differences—average differences between males and females. They can be biological, such as differences in reproductive anatomy, or cultural, such as differences in opportunities to practice and access to top-quality coaching.

20.1 The Endocrine System

Hormones act as chemical messengers. They are carried in the blood to all parts of the body and help to regulate many processes, including metabolism, reproduction, and development. One of the main ways that sex differences originate and are maintained has to do with the sexes secreting different amounts of sex-specific hormones. Hormones are secreted by specialized organs and glands that comprise the endocrine system.

The **endocrine system** is an internal system of regulation and communication involving hormones, the glands that secrete them, and the particular cells that respond to the hormones. A given hormone will elicit a response only in some cell types. The cells that respond to a particular hormone are called its **target cells.**

Hormones

Hormones are chemicals that travel through the circulatory system and act as signals to elicit a response from target cells. There are two general mechanisms by which hormones elicit a specific response: (1) Hormones can bind to receptors on the surface of the target cell and trigger a change inside the cell, or (2) hormones can diffuse across the cell membrane and bind to receptors inside the cell to trigger the response.

Protein hormones (**Figure 20.1a**) cannot cross cell membranes. Typically, they bind to a receptor on the surface of the target cell, and that receptor stimulates a series of other proteins to become involved in relaying the message to the inside of the cell, eliciting a specific cellular response. The chain reaction that relays, or transduces, a signal from the outside of the cell to the inside is called **signal transduction.**

Steroid hormones are fat-soluble hormones that can cross cell membranes and bind to receptors inside the target cell, causing cells to turn specific genes on or off (**Figure 20.1b**). The steroid hormones most involved in producing anatomical sex differences are called **sex hormones.** These include the male hormone **testosterone** and the female hormone **estrogen.** Once a sex hormone passes into the cell and binds to a receptor, the hormone and receptor bind to the DNA in the nucleus of the cell and either enhance or inhibit the expression of different genes (**Figure 20.2**).

The role of a particular hormone, whether it be acting as a messenger or a regulator, begins when it is synthesized and secreted by a particular endocrine gland.

Endocrine Glands

Endocrine glands are groups of cells or organs that secrete hormones. Many glands are involved in the endocrine system (**Figure 20.3a**). For example, the

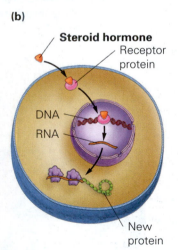

(a)

Protein hormone
Hormone receptor
Relay molecules
Cellular response

(b)

Steroid hormone
Receptor protein
DNA
RNA
New protein

Figure 20.1 Hormone actions. (a) A protein hormone binds to a receptor on the cell membrane of its target cell. This binding can induce a shape change in the receptor molecule, serving as a way of transmitting a signal to the inside of the cell that some response is required. (b) Steroid hormones diffuse across the membrane and bind to receptors, which in turn regulate gene expression.

(a) Breast tissue can respond to estrogen.

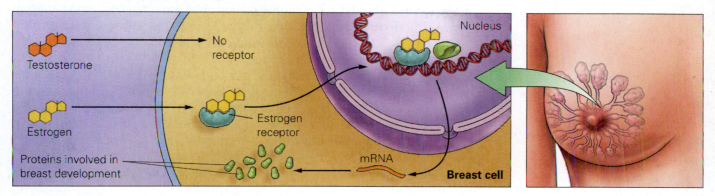

(b) The larynx responds to testosterone.

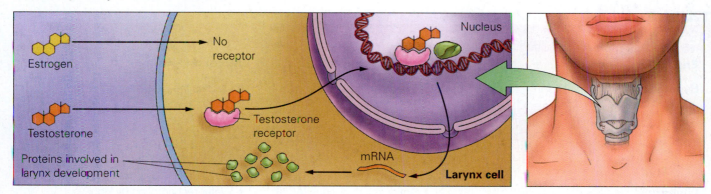

Figure 20.2 Sex hormones. (a) Estrogen produced by females enters the cells that compose the breast tissue, where it binds to its receptor. The estrogen and its receptor together bind to DNA to increase transcription of genes required for the growth and development of the breast at puberty. (b) In males, testosterone enters the cells of the larynx and binds receptors that increase the production of proteins that change the structure of the larynx, altering the sounds it produces and resulting in voice deepening.
Visualize This: Why is the breast tissue less developed in males than in females?

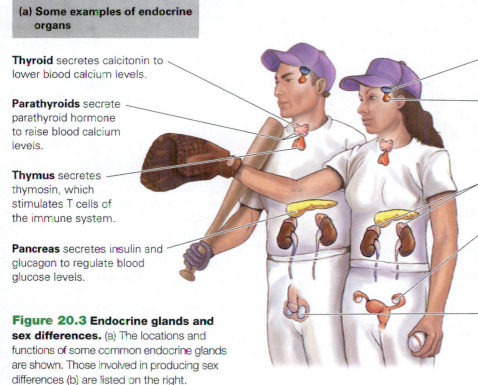

(a) Some examples of endocrine organs

Thyroid secretes calcitonin to lower blood calcium levels.

Parathyroids secrete parathyroid hormone to raise blood calcium levels.

Thymus secretes thymosin, which stimulates T cells of the immune system.

Pancreas secretes insulin and glucagon to regulate blood glucose levels.

Figure 20.3 Endocrine glands and sex differences. (a) The locations and functions of some common endocrine glands are shown. Those involved in producing sex differences (b) are listed on the right.

(b) Endocrine organs involved in producing sex differences

Hypothalamus secretes gonadotropin-releasing hormone (GnRH).

Pituitary gland responds to GnRH by secreting the pituitary gonodatropins—follicle-stimulating hormone (FSH) and luteinizing hormone (LH).

Adrenal glands secrete adrenaline, testosterone (masculinizing hormone), and estrogen (feminizing hormone).

Ovaries respond to FSH and LH by secreting **estrogen**, which regulates menstruation, maturation of egg cells, breast development, pregnancy, and menopause.

Testes respond to FSH and LH by secreting **testosterone**, which aids in sperm production, increased muscle mass, and voice deepening.

thyroid gland secretes hormones that stimulate metabolism. Along with the **parathyroid glands,** the thyroid gland also helps to regulate blood calcium levels. The **thymus** produces T cells active in immune response, and the **pancreas** helps regulate blood glucose levels. **Table 20.1** lists all the endocrine glands and the hormones they produce.

Some cells and organs that are not typically considered to be endocrine organs also produce hormones. The kidneys produce erythropoietin, which stimulates production of red blood cells, and the digestive tract produces the hormone gastrin, which stimulates the stomach to release hydrochloric acid during digestion.

Five endocrine organs are involved in the production of biological sex differences and are investigated more thoroughly in this chapter: the hypothalamus, the pituitary gland, the adrenal glands, the ovaries, and the testes (**Figure 20.3b**).

TABLE 20.1

Endocrine glands, their hormones, and targets and actions.

Endocrine Gland	Hormones Secreted	Target and Action
Hypothalamus	• Releasing and inhibiting hormones	• Causes the pituitary to store or secrete hormones
Pituitary	• Antidiuretic hormone (ADH) • Oxytocin • Thyroid stimulating hormone (TSH) • Adrenocorticotropic hormone (ACTH or corticotropin) • Follicle-stimulating hormone (FSH) • Luteinizing hormone (LH) • Prolactin • Growth hormone (GH)	• Water reabsorption by kidneys • Breast milk release • Uterine contractions during delivery • Causes thyroid to secrete T_3 and T_4, which increase metabolic activities of many cells • Causes adrenals to secrete glucocorticoids • Egg and sperm production • Ovulation and testosterone production • Breast milk production • Bone growth, protein synthesis, and cell division
Pancreas	• Insulin • Glucagon	• Lowers blood glucose levels • Raises blood glucose levels
Adrenals	• Glucocorticoids • Mineralocorticoids • Epinephrine and norepinephrine • Small amounts of sex hormones	• Raises blood glucose levels, promotes fat breakdown, and suppresses inflammation • Regulates level of sodium and potassium in blood • Acts on liver, muscle, and fat to raise blood sugar, increase heart rate, regulate blood vessel diameter, and increase respiration • Promotes development and maintenance of sex-specific characteristics

TABLE 20.1

Endocrine glands, their hormones, and targets and actions.		
Endocrine Gland	**Hormones Secreted**	**Target and Action**
Gonads	• Testes produce androgens • Ovaries produce estrogen and progesterone	• Promotes development of sperm, male reproductive structures, and male-specific sex characteristics • Promotes development of egg cells and regulates menstruation, pregnancy, and other female-specific sex characteristics
Thyroid	• Triiodothyronine T_3 and thyroxine T_4 • Calcitonin	• T_3 and T_4 act on most body cells to increase metabolic activities • Acts on the bone to help lower blood calcium levels
Parathyroids	• Parathyroid hormone (PTH)	• Acts on the bones, digestive tract, and kidneys to raise blood calcium levels
Pineal	• Melatonin	• Helps regulate sleep
Thymus	• Thymosin • Thymopoietin	• Helps T cells mature, especially in children

Hypothalamus and Pituitary Gland. The hypothalamus, located deep inside the brain, regulates body temperature and affects behaviors such as hunger, thirst, and reproduction. Growth hormone, secreted from the pituitary, acts on many organs, including bones that help children reach adult height. In the reproductive system, the hypothalamus secretes a hormone, called gonadotropin-releasing hormone (GnRH), that stimulates the activities of the gonads (testes or ovaries). GnRH moves through a web of veins downward from the hypothalamus to the pituitary gland. The pituitary gland secretes many different hormones; two in particular are involved with producing sex differences: follicle-stimulating hormones (FSHs) and luteinizing hormones (LHs). In males,

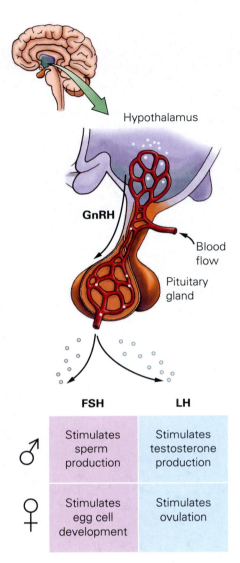

Figure 20.4 **FSH and LH secretion.** The hypothalamus secretes GnRH, causing the pituitary gland to secrete FSH and LH. FSH and LH are secreted by the brain of both males and females but stimulate a different response in each sex.

FSH stimulates sperm production, and LH stimulates testosterone production. In females, FSH stimulates egg cell development and LH stimulates the release of an egg cell during ovulation (**Figure 20.4**).

Adrenal Glands. An adrenal gland sits atop each kidney. These glands secrete the hormone adrenaline (epinephrine) in response to stress or excitement. They also secrete androgens, which include masculinizing hormones such as testosterone and feminizing hormones such as estrogen.

The cells of the outer covering of each adrenal gland synthesize sex hormones from cholesterol. While cholesterol can be synthesized in many body tissues, it is converted into sex hormones only within the adrenal glands, testes, and ovaries. Thus, the adrenal glands of both males and females secrete small amounts of both testosterone and estrogen. Most sex hormone production and secretion takes place in the testes and ovaries.

Testes. The paired testes are oval organs suspended in the scrotum of human and other mammalian males. Testes secrete testosterone, the hormone that aids in sperm production, hair thickness and distribution, increased muscle mass, and voice deepening. Beginning at puberty, sperm are produced by the cells of the testes. Sperm production is most efficient at temperatures that are lower than body temperature. Therefore, the testes are situated outside the body cavity, in the scrotum.

Some athletes use synthetic muscle-building steroids that mimic the activity of male sex hormones. These drugs are available legally with a prescription but are often sold illegally to athletes to enhance performance. Abuse of such steroids can lead to cancer, high blood pressure, depression, irritability, insomnia, and mood swings in both sexes. In males, the use of steroids may cause shrunken testes and decreased sperm production, which can lead to infertility. Hair loss and breast development may also occur. In females, steroids can disrupt the menstrual cycle, increase facial hair growth, and deepen the voice.

Stop & Stretch Why might the use of illegal steroids result in shrunken testes?

Ovaries. The paired ovaries are about the size and shape of almonds in the shell. They produce and secrete estrogen. Estrogen regulates many functions in the female body, including menstruation, the maturation of egg cells, breast development, pregnancy, and the cessation of menstruation after reproductive age, called menopause. Inside the ovaries are all of the cells that can mature into the egg cells for ovulation. Production of egg cells begins while a female is developing inside her mother's uterus, pauses at birth, resumes at puberty, and continues until menopause, at around 50 years of age.

The hormones of the female endocrine system regulate pregnancy and menstruation (Chapter 21). Many people have questioned whether a woman's athletic performance or capabilities are altered during any portion of the menstrual cycle. The answer is an unequivocal no. Researchers who measured speed and strength among sprint swimmers and weight lifters found that the women's performances remained consistent throughout their menstrual cycles. Other research has shown no differences in maximum bicycling and jumping power at different stages of the menstrual cycle. In fact, study after study has shown no loss of physical function during or prior to menstruation.

Puberty. The endocrine systems of both males and females increase the secretion and synthesis of hormones around the time of puberty. Puberty marks the beginning of sperm production in males and the beginning of egg cell maturation and menstruation in females. Boys typically begin puberty at the age of 9 to 14 years, with the average being 13 years. Puberty in males includes enlargement of the penis and testes; an overall growth spurt; growth of muscles and the skeleton, resulting in wide shoulders and narrow hips; and changes in hair growth, including the production of pubic, underarm, chest, and facial hair. In addition, the larynx enlarges and vocal cords lengthen to produce a deeper voice.

For girls, the first signs of puberty occur between ages 8 and 13, most commonly around age 11. These signs include breast development, increased fat deposition, a growth spurt, pubic and underarm hair growth, and the commencement of menstruation.

It is at puberty that biologically induced sex differences in athleticism begin to emerge, largely because the endocrine system begins to affect skeletal and muscle development.

20.2 The Skeletal System

The skeletal system provides support for the body, protects internal organs, aids in movement, and stores minerals. The Focus on Evolution feature describes the development of the bony skeleton in many different animals. In humans, the skeleton is an internal framework composed of bones and cartilage. Bones are rigid body tissues that function to provide support for the actions of muscles and protection for soft parts of the body—for example, what the skull does for the brain. Bones also help produce blood cells within the bone marrow and serve as a reservoir for minerals like calcium that circulate in the body. Cartilage protects the ends of bones from degradation.

The Human Skeleton

The human skeleton is composed of 206 bones and is organized into two basic units, the axial and the appendicular skeleton (**Figure 20.5**). The **axial skeleton** supports the trunk of the body and consists largely of the bones making

Figure 20.5 The human skeleton. The human skeleton can be divided into those bones that support the trunk, called the axial skeleton, and those that compose the limbs, called the appendicular skeleton. Cartilage is found between the joints.

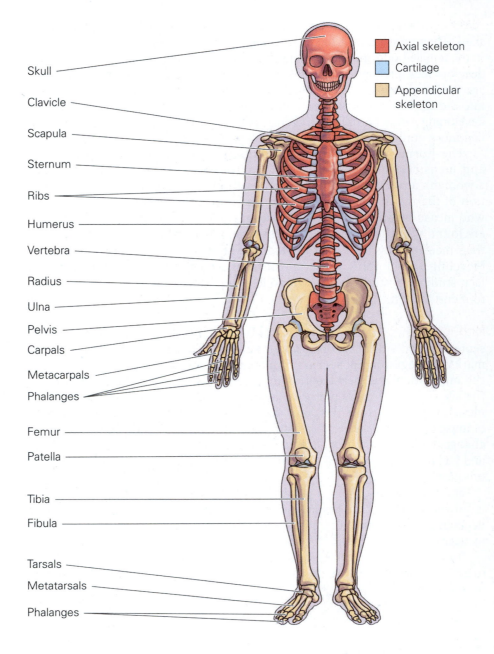

Axial skeleton

Cartilage

Appendicular skeleton

Skull

Clavicle

Scapula

Sternum

Ribs

Humerus

Vertebra

Radius

Ulna

Pelvis

Carpals

Metacarpals

Phalanges

Femur

Patella

Tibia

Fibula

Tarsals

Metatarsals

Phalanges

(a) Ball and socket joint (hip)

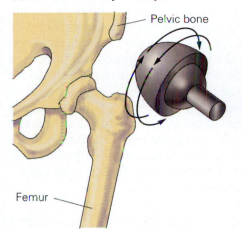

(b) Hinge joint (knee)

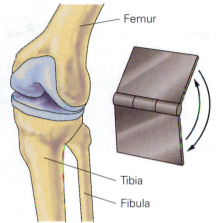

(c) Pivot joint (neck)

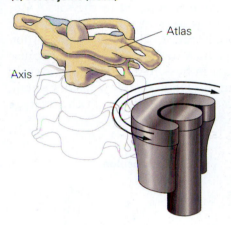

Figure 20.6 Joints in the human skeleton. (a) The hip is an example of a ball-and-socket joint. The ball-shaped end of the femur fits into a cup-shaped socket, allowing a wide range of motion. (b) The hinge joint of the knee allows movement in a single plane. (c) The pivot joint of the neck allows for freedom of movement of the head.

up the vertebral column or spine and the skull. The **appendicular skeleton** is composed of the bones of the hip, shoulder, and limbs.

To allow for a variety of movements, humans have different types of movable joints (**Figure 20.6**). **Ball-and-socket joints** in the hips and shoulders enable arms and legs to move in three dimensions. **Hinge joints** such as those in the knee and elbow allow back-and-forth movement similar to opening and closing a door. The **pivot joint** in the neck allows the head to turn from side to side. To facilitate movement, joints are lined with cartilage and lubricated by fluids. The bones are kept in place by ligaments and moved by muscles.

Bone Structure and Remodeling

Bones are living tissues composed of cells that actively remodel the skeleton. Calcium, the most abundant mineral of the body, is absorbed by cells as bone is built and released from cells when bone breaks down. Bone tissue can be tightly or loosely packed. **Compact bone** forms the hard outer shell of bones, and **spongy bone** is the porous, honeycomb-like inner bone (**Figure 20.7**). Because bone is a living tissue, it has blood vessels throughout. The interior of bone contains **bone marrow,** which helps to produce blood cells.

Osteoblasts, Osteoclasts, and Calcium. The cells that make up bone are called osteoblasts and osteoclasts. Osteoblasts help bone tissue regenerate itself, by a process called bone deposition. Osteoclasts are involved in breaking down or reabsorbing bone tissue. The delicate balance between deposition and reabsorption is integral to the maintenance of bones.

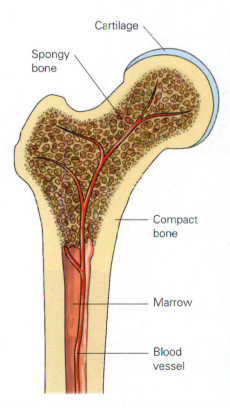

Figure 20.7 Bone structure. Bone is composed of compact and spongy bone. There are blood vessels running through bone, and the marrow inside bones helps produce blood cells.

Calcium is delivered to the bloodstream in response to the effects of the parathyroid hormone, which is produced by the parathyroid glands (**Figure 20.8**). Release of this hormone signals the kidneys to decrease the amount of calcium secreted in the urine and stimulates the breakdown of bone so that stored calcium is released into the blood. The thyroid gland secretes another hormone, **calcitonin,** which decreases the concentration of calcium within the blood when the level rises too high.

When calcium levels in the blood are high, osteoblasts remove calcium from the blood and use it to make new bone. When calcium levels in the bloodstream are low, osteoclasts break down bone and release calcium into the bloodstream. The maintenance of bloodstream calcium levels is important in many physiological processes, including blood clotting, muscle contraction, production of nerve impulses, and the activity of many enzymes.

Sex Differences in Bone Structure. Average differences in male and female skeletons arise due in part to the timing of puberty. From infancy to adulthood, humans grow in overall size, but a larger proportion of growth occurs in the extremities (legs and arms) as compared to the torso—for example, consider how short toddlers' legs look in relationship to their bodies. A growth spurt occurs during adolescence, with a larger proportion of growth occurring in the arms and legs than in the torso. In part because puberty both occurs later and lasts longer in boys than in girls, the average adult man has longer arms and legs than the average adult woman. Overall, the average man is 15 centimeters (5.9 inches) taller than the average woman.

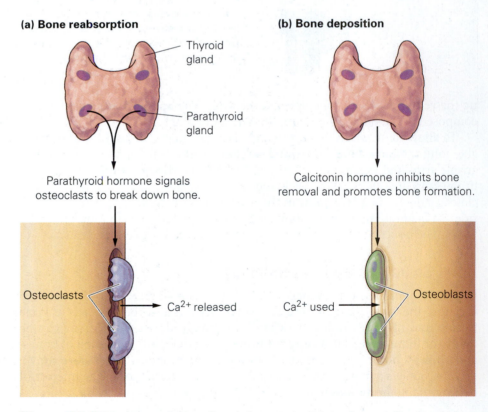

(a) Bone reabsorption

Thyroid gland

Parathyroid gland

Parathyroid hormone signals osteoclasts to break down bone.

Osteoclasts

Ca^{2+} released

(b) Bone deposition

Calcitonin hormone inhibits bone removal and promotes bone formation.

Ca^{2+} used

Osteoblasts

Figure 20.8 Bone remodeling. Bone is in a constant state of remodeling, by which old bone is removed by osteoclasts and new bone is laid down by osteoblasts. (a) Bone breakdown is stimulated by the parathyroid hormone, resulting in the release of calcium. (b) Bone formation is stimulated by the hormone calcitonin and results in the uptake of calcium. **Visualize This:** Is bone formation homeostasis maintained by negative or positive feedback?

The Skeletal System

Natural selection has provided a variety of means by which organisms move. In animals that actively travel through the environment, the skeleton provides a surface on which muscles function and a protective cage for internal organs. Three types of skeletal systems have evolved in animals: hydrostatic skeletons, endoskeletons, and exoskeletons. Each type of skeletal system has advantages for organisms in certain ecological roles but may be ill-suited for others.

The simplest skeleton consists of fluid held under pressure in a closed body compartment. These hydrostatic skeletons are found in jellyfish, flatworms, and earthworms, among other creatures. For these organisms, movement is controlled by contraction of muscles against the fluid-filled compartment. Because water, the typical fluid found in a hydrostatic skeleton, cannot compress much, muscles can "push off" against the compressed cylinder, forcing the body forward. For instance, some worms hold fluid in their body cavity at high pressure and contract longitudinal muscles against it to produce a whiplike movement. In other worms, different segments of the body can change shape individually by using circular and longitudinal muscles to produce rhythmic contractions of the hydrostatic skeleton. This produces waves of head-to-tail movement (**Figure E20.1**). Hydrostatic skeletons are adapted for life in aquatic environments or in soil. However, these fluid-filled cylinders cannot maintain an animal's shape above ground, as anyone who has ever picked up an earthworm knows. Animals that move through the air by walking, running, or flying require a stiffer type of skeleton.

One solution that has evolved for animals that require a stiffer form is an external skeleton. An exoskeleton is a hard encasement, on the surface of the animal to which muscles attach. Clams and many other mollusks produce a hard shell that serves primarily as protection but also as a point of attachment for muscles that open and close their shells (**Figure E20.2**). These shells grow throughout the life of the animal. Arthropods such as lobsters and many insects have evolved a more flexible exoskeleton that helps them move more efficiently and provides more limited protection than a shell. Because an arthropod's exoskeleton is like a form-fitting coat, it must be periodically shed as the animal grows.

Endoskeletons, which consist of hard supporting elements such as bone, are located inside the soft body tissues and can grow with the animal. However, they provide very little protection from predators. Endoskeletons may be made primarily of a mixture of protein and inorganic material, as in sponges and sea stars; cartilage, as in sharks and rays; or bones, as in humans. Harder endoskeletons must be made flexible by joints, which are gaps between bones held together by ligaments.

Figure E20.1 Earthworms move by contracting their hydrostatic skeletons.

Figure E20.2 The clam's shell is an exoskeleton.

The differences between women's and men's skeletons contribute to average differences in success at performing certain physical activities. Because women generally have long torsos relative to their leg length, their center of gravity (the point on the body where the weight above equals the weight below) is lower than in men. Individuals with a lower center of gravity are better able to maintain balance. Competitive gymnastics recognizes this difference in center of gravity—women, but not men, compete on the balance beam apparatus. Ice skating, ballet, and other forms of dance also require a lot of balance, and the lower center of gravity may enhance a woman's ability to stay on her feet.

Men, with their generally longer legs and arms, have more difficulty balancing, but they bring more power to activities that rely on the lever action of their extremities. A longer arm or leg transmits more force; therefore, men have faster and stronger slap shots in hockey, stronger kicks in soccer, and faster swings and throws in baseball. Longer legs also have longer stride lengths, which means that men will usually be faster runners than women are.

Aside from differences in leg and arm length, there are few differences in the human skeleton that affect athleticism (**Figure 20.9a**). In fact, there are only six bones that differ between same-sized males and females. Four of the bones that differ are found on the head. These are the **mandible,** or jawbone, which is larger

Figure 20.9 Sex differences in the skeleton. Human male and female skeletons are fairly hard to distinguish. (a) Consistent differences are found in the frontal bone, temporal bones, mandible, and the two ossa coxae. The female pelvic inlet is larger and rounder than the male pelvic inlet. (b) One side of the Q angle goes up along the femur. The other side is a line extending up from the patellar tendon, which connects the kneecap to the tibia. Q angles in males are typically smaller than Q angles in females.

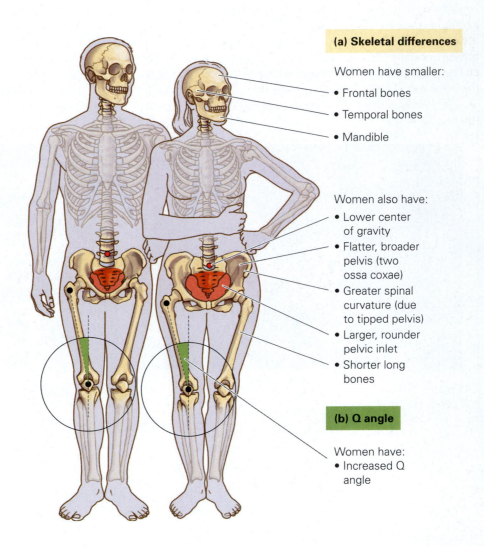

(a) Skeletal differences

Women have smaller:

• Frontal bones

• Temporal bones

• Mandible

Women also have:

• Lower center of gravity

• Flatter, broader pelvis (two ossa coxae)

• Greater spinal curvature (due to tipped pelvis)

• Larger, rounder pelvic inlet

• Shorter long bones

(b) Q angle

Women have:
• Increased Q angle

in males; the paired **temporal bones** found near the temple, which have a larger opening in males to allow for the connection of thicker muscles to support the large jaw; and the **frontal bone** of the cranium, or forehead, which is generally more rounded in females and has a less-pronounced ridge above the eyes. The other two bones that show general differences between men and women are the two **ossa coxae**, which form the bony pelvis.

The differences between the pelvises of men and women evolved in response to different selective pressures. The evolution of upright walking in early humans required that these changes take place. However, the changes to the ossa coxae, combined with our large head size, results in one of the most difficult births of any mammal. Early human females with rounder pelvic inlets may have had a selective advantage because they were less likely to die during childbirth since rounder inlets generally permit the easier passage of a child's head through the birth canal.

The round pelvic inlet that is typical of most women is produced by a bony pelvis that is flatter and broader than a man's pelvis, which has an inlet that resembles an elongated oval. A flatter pelvis also requires that the bony pelvis be tipped forward to bring the hip bones to the front of the body; this tipping forward is maintained by the curvature of the lower spine. The spinal curvature and greater pelvic tilt in individuals with broad pelvises elevate the buttocks and gives a curvy appearance to the profile of women. Conversely, the male pelvis lowers the buttocks and gives a flat appearance to a man's profile.

When you bring your feet together, your femurs, or thighbones, extend diagonally to your knees from where they attach at the hips. A broader pelvis means that the femurs are farther away from each other at the point of attachment than are femurs attached to a narrower pelvis. This leads to a sex difference in the **Q angle**, or the angle formed between the kneecap, the femur, and the line of the tendon from the kneecap to the shinbone (**Figure 20.9b**).

An increased Q angle has been thought to be a sex-influenced risk factor for knee injury. For instance, knee injuries among basketball players are much more common in women, who generally have greater Q angles than men do.

Stop & Stretch Devise an experiment that would help determine whether differences in rates of knee injuries in males and females are due to differences in Q angles or differences in conditioning levels.

The male body has one clear advantage in sports requiring greater strength. A taller, heavier skeletal frame supports more muscle, and muscle mass does affect athleticism.

20.3 The Muscular System

Skeletal muscle is one of three types of muscle in humans; the other two types are cardiac and smooth muscle. The major function of the skeletal muscular system is movement. Skeletal muscle attaches to bones at tendons and produces voluntary body movements by interacting with the skeleton. Skeletal muscle is nourished by blood vessels that supply oxygen and nutrients and controlled by nerves that help signal the process of muscle contraction.

Muscle Structure and Contraction

The structure of a muscle involves smaller and smaller parallel arrays of filaments (**Figure 20.10a**). A muscle contains a sheathed bundle of parallel **muscle fibers.** Each muscle fiber is a single cell. Within each muscle fiber is a parallel array of threadlike filaments called **myofibrils.** Bands of all myofibrils in a cell are aligned together, like pipes of the same length stacked neatly with their ends aligned (**Figure 20.10b**). This gives the skeletal muscle its characteristic striated or striped appearance. Each myofibril is a linear arrangement of **sarcomeres,** which are the unit of contraction of a muscle fiber. A sarcomere is composed of thin filaments of the protein **actin** and thick filaments of the protein **myosin.** The sarcomere is the region between two dark lines, called **Z discs,** in the myofibril.

When a shortstop moves to field a ball, his bones move because the skeletal muscles attached to them shorten. To contract the muscles, the sarcomeres shorten. This shortening takes place through a series of steps outlined in the next section.

Sliding-Filament Model of Muscle Contraction.
Sarcomere shortening involves coordinated sliding and pulling motions within the sarcomere (**Figure 20.11**). Actin filaments are anchored at the Z discs at the ends of a sarcomere. They overlap a parallel, stationary set of myosin molecules that are not attached to Z discs. During contraction, actin filaments slide over the fixed myosin filaments, pulling the ends of the sarcomere with them. The sarcomere ends are brought toward the center when the head of the myosin molecule attaches to binding sites on the actin molecules and pulls toward the center.

The sliding-filament mechanism proceeds as the sarcomere shortens in a stepwise manner. First, the high-energy compound ATP binds to the myosin

Figure 20.10 Skeletal muscle structure. (a) Muscle cells (fibers) are arranged in parallel bundles inside the muscle's outer sheath. (b) A skeletal muscle cell, with its myofibrils in register, produces the characteristic banded or striated pattern. The myofibrils inside a muscle cell are linear arrangements of sarcomeres. Sarcomeres are bounded by Z discs.

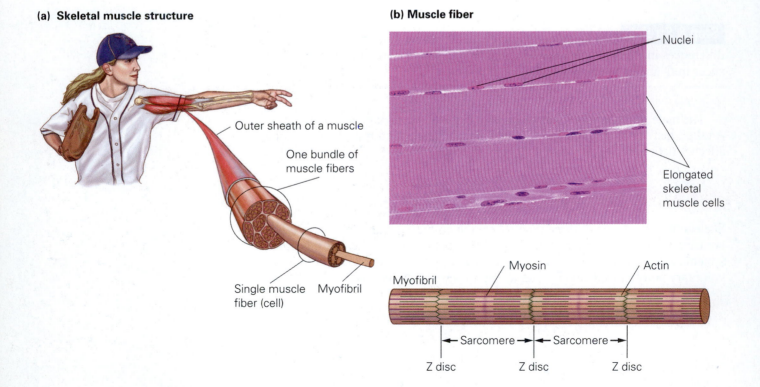

(a) Skeletal muscle structure

Outer sheath of a muscle

One bundle of muscle fibers

Single muscle fiber (cell)   Myofibril

(b) Muscle fiber

Nuclei

Elongated skeletal muscle cells

Myofibril   Myosin   Actin

Sarcomere   Sarcomere

Z disc   Z disc   Z disc

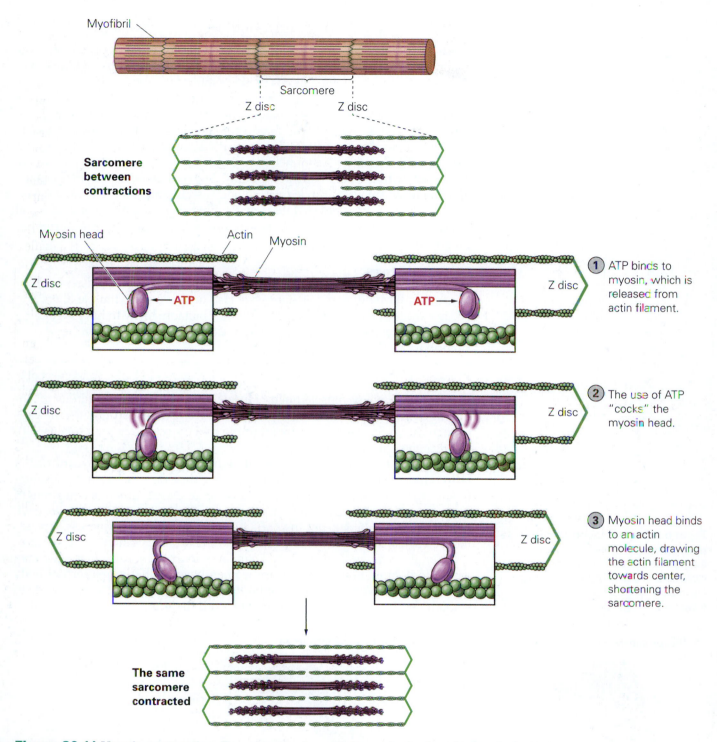

Figure 20.11 Muscle contraction. The sarcomere of a myofibril consists of actin molecules attached to the Z disc and myosin molecules. Using energy from ATP, the myosin head binds to actin and pulls it toward the center. Because this is happening at both ends of the sarcomere, the sarcomere shortens, allowing contraction of the muscle cell and movement of the muscle.

head, causing the myosin head to detach from the binding site on the actin filament. Second, the breakdown of adenosine triphosphate (ATP) provides energy to the myosin head, which changes shape or "cocks." Third, the myosin head binds to another actin-binding site; pulls in short, powerful strokes; releases the actin; and reattaches further along. This process serves to pull the Z discs inward, shortening the sarcomere.

Muscle Interaction with Bones

Figure 20.12 illustrates how skeletal muscles and tendons interact with bone. The ability to move the arm to pitch requires that two muscle groups perform opposite tasks. Contraction of the biceps muscle shortens the muscle while the triceps muscle is relaxed. Likewise, when the biceps muscle is relaxed, the triceps muscle is contracted. Such **antagonistic pairs** of muscles, in which each muscle is paired with a muscle of the opposite effect, are a common feature of the muscular system.

Skeletal muscles can be of different varieties based on how quickly they use energy, ATP, to produce a contraction. Slow-twitch fibers contract more slowly, while fast-twitch fibers contract quickly. Slow-twitch fibers contain many mitochondria, are well supplied with blood vessels, and can therefore use oxygen as quickly as it is delivered. Fast-twitch fibers have fewer mitochondria and blood vessels. Contractions by fast-twitch muscle are rapid and powerful but cannot be sustained for long.

Stop & Stretch Do you think athletes who are successful at different sports may have different percentages of fast- or slow-twitch muscle fibers? For instance, would world-class marathoners be more likely to have higher-than-average percentages of fast-twitch or slow-twitch fibers in their legs? How about world-class weight lifters?

Sex Differences in Muscle. Testosterone released by the testes at puberty increases the size of muscle fibers. Muscle fibers contain testosterone receptors in their cytoplasm, and the presence of testosterone stimulates the cells to increase in mass—in part by retaining dietary nitrogen, which is used to produce more muscle proteins.

Although the ovaries and adrenal glands do release small quantities of testosterone, males have more of this hormone than do females, and the subsequent difference in muscle mass accounts for an overall difference in strength between males and females. In sports that require brute strength, such as football, men have an advantage, but in sports like baseball, in which success depends more on agility and hand-eye coordination, the effect of sex differences is reduced.

Even if a man and a woman had similar amounts of muscle mass, the man would look more muscular because women have a different pattern of body fat deposition. Women have more body fat under their skin, which tends to smooth out the appearance of muscles beneath the skin.

Figure 20.12 Antagonistic muscle pairs. Movements are often produced through the actions of two opposing muscles. (a) Contraction of the biceps raises the forearm, and (b) contraction of the triceps lowers the forearm.

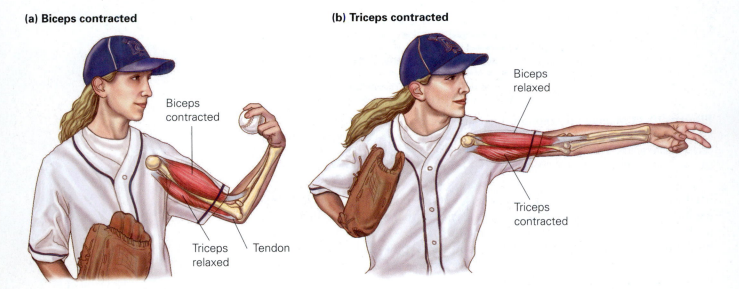

(a) Biceps contracted

Biceps contracted

Triceps relaxed   Tendon

(b) Triceps contracted

Biceps relaxed

Triceps contracted

We have seen that sex differences in the endocrine system can lead to differences in skeletal structure and muscle mass, all of which can affect athleticism. In addition to these differences, a few other sex differences can influence a person's athletic abilities.

20.4 Other Sex Differences that Influence Athleticism

As we have seen, some of the endocrine, skeletal, and musculature differences between the sexes may affect athleticism. In this section, we investigate a few other sex differences and then turn our attention to what these differences really mean.

Body Fat Differences

Differences in body fat percentages and locations may affect athleticism. These differences arise at puberty, when a young woman begins to store fat in her torso, abdomen, hips, thighs, and buttocks. This body fat, in addition to skeletal differences, gives women's bodies their curvy shape. Men usually carry most of their fat in the abdomen. Overall, women have about 10% more body fat than men do; this body fat is necessary to maintain fertility. When a female does not have enough body fat, the secretion of hormones that regulate menstruation is altered and menstruation ceases. Lack of menstruation can be permanent, resulting in sterility, and it can result in bone damage because estrogen helps bones stay strong. Excessive exercise or starvation that leads to the cessation of menstruation, called **amenorrhea,** causes permanent bone damage when the estrogen that normally increases prior to ovulation is not produced.

In addition, women may metabolize fat differently than men do. Females seem to utilize more fat to produce energy than do males, who tend to use the body's stored sugars more readily. In women, this has the effect of slowing glucose metabolism, meaning that more sugars are available for prolonged exercise. This difference may result in a greater tolerance for endurance events in women. In the 30 years that women have been allowed to compete in marathons, their finishing times are edging closer and closer to males. Whether this trend continues, ultimately making women as fast as—or faster than—men at endurance events, remains to be seen.

The extra body fat that gives women increased endurance may also be a physical advantage in long-distance swimming events—fat increases buoyancy and enables women to maintain the most energy-conserving streamlined position with less effort (**Figure 20.13**). In addition, fat not only provides increased insulation, which slows the rate of body heat loss, but also stores energy that can be converted to ATP during endurance events.

While women's physiology may be favored in endurance events, men have the advantage when it comes to events requiring a high level of intensity over a shorter period of time; this advantage is partially due to sex differences within the cardiovascular system.

Cardiovascular Differences

Women have smaller hearts and less blood volume relative to body size, and they have fewer blood cells per unit volume than similarly sized men do. This means that men are better able to supply oxygen and nutrients to their greater muscle mass. Women also have smaller lungs that work harder to transport

Figure 20.13 Swimmer in stream-line position. Fat makes a woman's body smoother and more buoyant.

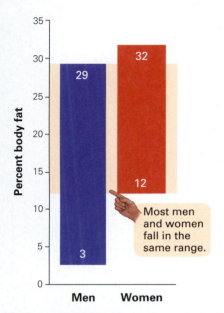

Figure 20.14 Male and female body fat percentages. This graph shows that there is a lot of overlap in body fat percentages between the sexes.

Visualize This: Is it possible to determine a person's gender based on their body fat percentage?

oxygen to the cells and tissues that need it. If a man and woman of similar size are running side by side at the same speed, the woman's respiratory and cardiovascular systems are working harder to bring oxygen to her tissues.

Group Differences and Individual Differences

While we can determine average differences between males and females in terms of skeletal size, muscle mass, body fat, and cardiovascular activity, these differences are often of little value in the real world. Average differences hide the fact that there is a wide range of values for each of these characteristics influencing athleticism, and the ranges for males and females typically overlap a great deal. Consider the example of body fat: The average healthy woman has 22% body fat, and the average healthy man has 14% body fat. This seems like a large difference, but most males and females fall in the same range (**Figure 20.14**). Healthy women have from 12% to 32% body fat, and the range is between 3% and 29% for men. Most males and females fall within the 12% to 29% range of body fat.

The range of body fat percentages for normal women shows a 20% difference from lowest to highest; for men, the range varies by 26%. However, difference between a man of average body fat (14%) and a woman of average body fat (22%) is only 8%. As is often the case when we try to study differences between human groups, the differences within a group are greater than the differences between two groups.

To understand the whole story of athleticism and gender, we must evaluate the impact that our culture has on intensifying average differences in athletic performance. Strong cultural forces often determine the amount of time athletes spend practicing as well as the amount of support they receive for doing so.

As a young girl, Ila Borders was prevented from registering for Little League baseball when she first tried to play at age 10. In college, where she was the first woman to be given a baseball scholarship and the only woman on the team, she was taunted by her own teammates as well as by members of the opposing team. In a television interview with Mike Wallace of CBS's *60 Minutes,* Borders spoke of how her own teammates would throw baseballs at her when she had her back turned and how players from opposing teams yelled obscenities at her.

Today, more than 100,000 women participate in intercollegiate athletics—a fourfold increase since 1971. This is due in part to Title IX federal legislation, which mandates equal athletic opportunities for schools receiving federal monies. Financial support for female athletes, and their social acceptance, have grown tremendously.

Physical activity is important to overall health, regardless of gender. Studies have shown that children of both sexes who participate in sports are less likely to drop out of high school, smoke, or drink. Participation in sports teaches values such as teamwork, discipline, pride in accomplishment, drive, and dedication.

Children who exercise are also more likely to continue exercising in adulthood than children who do not exercise. In a country where most adults do not get the recommended 30 to 60 minutes of exercise most days of the week, it makes sense to encourage everyone to become more athletic. When good exercise habits are carried into adulthood, there is a decreased risk of heart disease, obesity, diabetes, and many cancers. Additional benefits include lowered cholesterol, and studies suggest that exercise may decrease anxiety and depression. Sex differences that lead to differences in strength, speed, balance, and endurance are meaningless for recreational athletes. What really matters to most athletes is finding an activity they enjoy so that they can keep healthy for as long as possible.

SAVVY READER

Anabolic Steroids Use

Drug and Alcohol Dependence
Volume 90, Issues 2–3, 8 October 2007, pages 243–251

by S. McCabe et al.

The peer reviewed article cited above reports the results of a study on the non-medical use of anabolic steroids among U.S. college students. Non-medical use means the illegal use of these drugs to gain muscle mass, instead of physician prescribed use for various medical conditions.

Surveys were sent to over ten thousand randomly chosen college students in four different years between the late 1990s and early 2000s. The students queried were all from the same suite of colleges. The scientists performing this study found that non-medical use of steroids occurred in around 1% of those surveyed. They found also that steroid use was increasing among male college students. Statistical analysis showed that male student athletes were the most likely users. There were also correlations between non-medical steroid use and the use of other illegal drugs, alcohol and cigarette use, as well as with drinking and driving.

1. List two controls used in this study.
2. The scientists who performed this study also found that users of anabolic steroids were more likely to use drugs and alcohol. How might the symptoms of steroid abuse lead to increased drug use?
3. Propose a hypothesis to explain why nonmedical use of anabolic steroids increased among men during the course of this study.

Chapter Review

Learning Outcomes

Go to the Study Area at www.masteringbiology.com for practice quizzes, myeBook, BioFlix™ 3-D animations, MP3 Tutor sessions, videos, current events, and more.

LO1 **Differentiate between protein and steroid hormones (Section 20.1).**

- Protein hormones trigger a response by binding to receptors on the surface of the target cell, which relay the signal to the inside of the cell (p. 492).
- Steroid hormones diffuse into the cell and act on receptors there (p. 492).

LO2 **List the major endocrine glands, the hormones they secrete, and the effects these hormones have on their target organs (Section 20.1).**

- The thyroid gland secretes hormones that stimulate metabolism; along with the parathyroid gland, it regulates calcium levels (pp. 493–495).
- The thymus helps the immune system form, and the pancreas helps to regulate blood glucose levels (pp. 493–495).
- The hypothalamus regulates body temperature, hunger, thirst, and reproduction. The hypothalamus secretes GnRH, which stimulates the pituitary gland to release FSH and LH (pp. 493–496).
- Adrenal glands secrete adrenaline, androgens, and estrogens (pp. 494–496).

- Testes secrete testosterone, which affects the development of male characteristics. Ovaries secrete estrogen, which fosters the development of female characteristics (pp. 495–497).

LO3 Describe the main structural components of the skeleton and their functions (Section 20.2).

- The skeletal system is composed of bones and cartilage. It provides support, protection, movement, and mineral storage (p. 498).
- Joints are regions where bones come together. Different types of joints allow for different kinds of movements (p. 499).
- Ligaments hold bones together at joints (p. 499).

LO4 Discuss the process of bone remodeling (Section 20.2).

- Bone remodeling is under the control of the endocrine system. Hormones stimulate the osteoblasts to use calcium to build bone, and the osteoclasts to break down bone and release calcium (pp. 499–500).

LO5 Describe the main structural components of the muscular system and their functions (Section 20.3).

- The muscular system functions in movement, which is coordinated by the combined actions of antagonistic pairs of muscles (pp. 504–506).
- Each muscle in the antagonistic pair has an opposite effect (p. 506).

LO6 Outline the structure of a muscle fiber, and describe how muscles contract (Section 20.3).

- Muscles are bundles of parallel fibers. Muscle fibers are bundles of parallel myofibrils. Myofibrils are linear arrays of sarcomeres, which are the unit of muscle contraction (pp. 504–505).
- Sarcomeres are composed of actin and myosin. Actin filaments attached to the ends of a sarcomere are pulled toward the middle of the sarcomere by the movement of myosin heads. These myosin heads then attach to the filaments, pull them toward the sarcomere, release the filaments, and reattach further down the filament. This process is called the sliding-filament model of muscle contraction (pp. 504–505).

LO7 List several sex differences in body composition (Section 20.4).

- Women, on average, have more body fat than do men (p. 507).
- Women store fat on their hips and thighs; men store fat in their abdomen (p. 507).
- Men have larger hearts, more blood volume, and more blood cells than do women (p. 508).

Roots to Remember MB

The following roots of words come mainly from Latin and Greek and will help you decipher terms:

-blast	refers to an early formation or to a place where something forms. Chapter term: osteoblast
-clast	is from a Greek verb to break or to break down. Chapter term: osteoclast
my- or **myo-**	means muscle. Chapter terms: myosin, myofibril
osteo-	relates to bones (the Latin form is os- or ossi-). Chapter terms: osteoblasts and osteoclasts
sarc- or **sarco-**	means body. Chapter term: sarcomere
-ule	is an ending for something small. Chapter terms: tubule, molecule

Learning the Basics

1. **LO2** List the endocrine organs involved in producing sex differences and describe their functions.

2. **LO3** Describe bone structure.

3. **LO6** Describe muscle structure.

4. **LO1** Steroid hormones act by _____.

   A. causing neurons to fire; B. increasing muscle contraction; C. diffusing across membranes and regulating the expression of genes; D. binding to receptors on the cell surface and transducing a response without entering the cell

5. **LO2** The endocrine organ that sits atop a kidney is _____.

   A. the pituitary gland; B. the hypothalamus; C. the ovary; D. the adrenal gland; E. the testicle

6. **LO3** Bones of the limbs form part of the _____.

   A. appendicular skeleton; B. axial skeleton; C. spongy bone; D. compact bone

7. **LO4** Osteoblasts _____.

   A. are found in the marrow of bone; B. regulate bone deposition; C. regulate bone reabsorption; D. are more active when calcium is low

8. **LO5** The antagonistic pairs of arm and leg muscles _____.

   A. allow healthy muscles to compensate for injured ones; B. allow muscles to produce opposing movements; C. allow different types of rotations of joints; D. allow myofibrils to fire in response to different stressors

9. **LO6** During muscle contraction, _____.

A. the actin heads pull the sarcomere closed;
B. myosin attaches to actin, pulls it toward the center of the sarcomere, releases it, and reattaches further along;
C. myofibrils shrink due to the actions of testosterone;
D. muscle fibers contract under the actions of testosterone

10. **LO6** Which of the following is a true statement regarding muscle contraction?

A. Actin filaments are stationary, and myosin heads can move; B. Z discs are pulled toward the center of the sarcomere; C. Actin is used up during this process; D. No ATP is required for movement to occur.

Analyzing and Applying the Basics

1. **LO7** When it comes to sex differences, the overlap between genders, even in terms of anatomical differences, can be quite significant. For instance, the female and male pelvises have different shapes, which can affect how the legs are attached. Yet according to the classic study on pelvises, only 40–50% of actual women have a female pelvis as defined in anatomy books; 33% have an "android" or male-type pelvis. How is our ability to make generalizations about the sexes impacted by such differences?

2. **LO2** How would a chemical that blocks testosterone production affect muscle development?

3. **LO7** Why are most men taller than most women?

Connecting the Science

1. Provide several examples of cultural influences that sway our views of biological differences between males and females.

2. The use of anabolic steroids to build muscle is illegal, but the practice has become widespread in certain professional sports. Do you agree that using steroids is a form of cheating? How is it different from other forms of physical enhancement (for instance, special diets, vision correction, even drugs that reduce asthma symptoms)?

Answers to **Stop & Stretch, Visualize This, Savvy Reader,** and **Chapter Review** questions can be found in the **Answers** section at the back of the book.

Is There Something in the Water?

Reproductive and Developmental Biology

Frogs with extra limbs . . .

Is something wrong with the water? People in many communities are concerned about the health of organisms that live in their local lakes, rivers, and streams. All over the Midwest, scientists are finding frogs with extra limbs. A pesticide spill in Lake Apopka, in central Florida, killed most of the resident alligators. The surviving alligators then produced male alligators with abnormally small penises. Fish near a waste treatment plant in Colorado's Boulder Creek have both male and female reproductive organs. Populations of chinook salmon in the Pacific Northwest are almost all female, and most of these female fish have chromosomes typical of males of this species. The salamander population in Tecumseh, Missouri, has decreased by 75%, and their eggs and larvae are no longer found in local rivers.

Although there is, as yet, no conclusive evidence linking reproductive changes in wildlife to similar effects in humans, the following points are worth considering. According to some scientific studies and the popular press, human males have ever-decreasing sperm counts and reproductive organ abnormalities. Human females are reported to be experiencing more fertility problems and menstrual irregularities than in the past.

If several women working in the same office or living in close proximity become pregnant at the same time, people will often joke that "it must be something in the water." Current fertility problems have turned that notion on its head, and now scientists and laypersons alike are wondering if there *is* something in the water that is causing a decrease in fertility among both humans and animals.

Alligators with altered reproductive structures . . .

Fish that are all female . . .

Is there something in the water altering how organisms develop?

LEARNING OUTCOMES

LO1 Contrast asexual and sexual reproduction.

LO2 List the male reproductive structures and their functions.

LO3 List the female reproductive structures and their functions.

LO4 Compare the processes of spermatogenesis and oogenesis.

LO5 Describe the events that occur during the menstrual cycle and how they are hormonally regulated.

LO6 Outline the paths sperm and egg must follow for fertilization to occur.

LO7 Describe the stages of early development from the rapid, early divisions through early tissue differentiation.

LO8 Compare the events leading to differentiation of male and female reproductive organs.

LO9 Outline the hormonal controls regulating pregnancy and birthing.

Studies of aquatic animals with certain reproductive problems seem to implicate a class of chemicals called **endocrine disruptors.** Endocrine disruptors disrupt the actions of the hormone-producing endocrine system. Evidence of endocrine disruption in aquatic animals has led to concern about the water humans are drinking. Are these chemicals present in the water supply, and if so, are they affecting human reproductive health?

To answer these questions, a more thorough understanding of reproduction and development is required. We begin by looking at basic reproductive processes.

(a) Sponge

(b) Hydra

Figure 21.1 Asexual reproduction in animals. (a) Sponges can reproduce by fission. The white cloud at the top of this photograph consists of cells that are breaking away from the parent. (b) Hydra can reproduce by budding. The newly formed hydra on the left will break away from the parent.

21.1 Principles of Animal Reproduction

Reproduction is the process by which organisms produce offspring, allowing for the continuation of their species. Organisms compete with other individuals for the resources required to survive and reproduce. During sexual reproduction, cells such as sperm and eggs, which are nonessential to the animal's own survival but indispensable to the perpetuation of their species, produce and then join to develop offspring. Because reproduction begins with the production of single cells, any environmental factor that affects these cells can dramatically influence the offspring produced. This is why an early clue to the presence of a biologically active pollutant in an area is often a decline in the survival and health of offspring produced in that environment.

Animals employ a wide variety of strategies to reproduce offspring. However, the two basic schemes are asexual and sexual reproduction.

Asexual Reproduction

If you have ever propagated plants from cuttings, you know that these organisms can reproduce without the aid of a partner. This process also occurs in some animals. Called **asexual reproduction**, this method of reproduction occurs when a single organism reproduces by itself, without mating with another organism. The offspring of asexual reproduction are genetically identical to the parent and to each other. In effect, the parent clones itself.

Prokaryotic organisms such as bacteria produce two genetically identical cells, called daughter cells, by first copying their single chromosome and then splitting the single cell into two cells. This process is called **binary fission.** Some single-celled eukaryotic organisms, such as amoebas, divide by the asexual process of mitosis to produce two genetically identical offspring. Sometimes, multiple fissions can occur, fragmenting one individual into many individual daughter cells that can develop into adults. Sponges are multicellular animals that can reproduce by fission, during which masses of cells break away from the parent to form offspring (**Figure 21.1a**).

Budding is a form of asexual reproduction that is similar to fission. A yeast cell budding in bread dough undergoes cell division to produce a daughter cell that remains attached to the parent until it achieves sufficient growth to break away. Hydra are simple animals that can also reproduce by budding (**Figure 21.1b**).

Asexual reproduction has many advantages as well as disadvantages. There is no need to find a partner to reproduce, which is particularly advantageous for organisms that do not move or are isolated from other organisms. When conditions are favorable, asexual reproduction also allows for the rapid production of large numbers of offspring. However, because all offspring are identical to the parent, there is not much ability to adapt to environmental change. When a population consists of identical individuals, an environmental change could decimate the entire population.

Sexual Reproduction

Environments with variable conditions favor **sexual reproduction.** In sexual reproduction, the offspring differ genetically from their parents and from each other. Populations of individuals that are not identical will likely have a few members with a genetic makeup that will allow them to survive and perpetuate the species when conditions change.

Creation of offspring by the fusion of sex cells from two different individuals— for example, a mating between one male and one female parent—is required for sexual reproduction. Individuals that produce large sex cells or **gametes** are classified as females, and individuals that produce small gametes are classified as males. Focus on Evolution discusses why natural selection has resulted in two classes of different-sized gametes.

Offspring are produced after the gametes from two different individuals combine genetic information at fertilization. The gamete-producing structures are called **gonads.** In males, the gametes are sperm, and the gonads are **testes,** which produce the sperm. In females, eggs are the gametes, and the gonads are **ovaries,** which produce the eggs.

Because two different individuals contribute genetic information to their gametes, the offspring will be a genetic mixture of both parents. Each gamete is haploid (n) and contains half the genetic information of the individual that produced it. Fusion of two haploid gametes produces a diploid ($2n$) **zygote,** or fertilized egg cell. Because a mixing of parental genetic information occurs in the offspring, sexual reproduction produces an almost infinite variety of genetically unique individuals. Most commonly, an organism will produce either male or female gametes but not both. However, there are exceptions. **Hermaphrodites** are animals that have both male and female reproductive systems. Earthworms, for example, can both donate and receive sperm during mating. (Because the production of male and female gametes by one individual is very common in the plant kingdom, the term *hermaphrodite* applies only to animals.)

The mechanism of fertilization can differ among sexually reproducing animals. Fertilization can occur inside or outside the body. Depositing sperm in or near the female reproductive tract through **copulation,** or sexual intercourse, allows **internal fertilization** to occur. Sharks, reptiles, birds, and mammals reproduce via internal fertilization. Many aquatic invertebrates and most fish and some amphibians use **external fertilization,** during which the parents need not have physical contact with each other. The female lays eggs in the water, and the male releases sperm over the eggs. Because these eggs are fertilized and develop outside the body, in water, they are very susceptible to aquatic contaminants.

Figure 21.2 Dr. Tyrone Hayes.
Dr. Hayes at work collecting frogs from
an atrazine-contaminated site.

Environmental Contaminants and Sexual Reproduction

Frogs are sexually reproducing organisms that use external fertilization. The fertilized eggs of frogs give rise to tadpoles that develop into adults in the water. Because so much of a frog's development occurs in the water and because many chemicals in the water can be absorbed directly, frogs in early developmental stages are highly sensitive to environmental contaminants. Whether these frogs with extra limbs or other obvious deformities arose as a result of exposures to human-made chemicals that have found their way into the frog's habitat is the object of intensive scientific study. In a sense, frogs can be used as environmental early-warning signals in the same way that canaries were once used to warn of high concentrations of carbon monoxide in coal mines. When canaries that were brought into mines showed distress or died, the miners knew to get out of the mine quickly before the carbon monoxide levels were high enough to kill them also.

Because problems in frog reproduction and development can foreshadow problems for other organisms, the results of work done by Dr. Tyrone Hayes (**Figure 21.2**) concern many scientists. Hayes's research has shown that a widely used pesticide called atrazine, which makes its way from farmers' fields into groundwater, marshes, and ponds, is causing male frogs to grow eggs inside their testes.

Why would male frogs produce egg cells inside their testes? Because these "feminized" frogs were collected from atrazine-contaminated ponds and marshes, Hayes hypothesized that this pesticide might be an endocrine disruptor that affects the development of the frog gonad.

To test the hypothesis that atrazine was responsible for the conversion of sperm-producing tissue into egg-producing tissue, Hayes moved his research from ponds and marshes into his laboratory. He injected normal male tadpoles with atrazine. He found that this did indeed disrupt the development of testicular tissue, resulting in gonads that could also produce egg cells.

Further research by Hayes also suggested that atrazine does this by activating the enzyme **aromatase,** which converts testosterone, a hormone normally produced in higher quantities within males, into estrogen, a hormone normally produced in higher quantities within females. This resulted in the "feminization" of the male testicle. Alarmingly, Hayes found that this endocrine disruption will occur with doses of atrazine that are lower than those found in our drinking water.

Stop & Stretch In an experiment designed to determine whether estrogen secreted in the urine of women using birth control pills is making its way into the waterways and affecting fish, Canadian scientists placed estrogens in a remote lake. When scientists compared fish exposed to estrogen with those from the same lake before estrogen treatment, they found significant increases in the numbers of male fish that were producing eggs. They found no increase in such fish in two nearby untreated lakes. List the controls used in this study.

The work of these and other researchers shows that organisms living in waters contaminated by endocrine-disrupting chemicals can develop altered reproductive structures. While humans do not grow and develop in contaminated waters, they might be drinking contaminated water. Are human reproductive structures affected by exposures to endocrine disruptors?

The Development of Two Different-Sized Gametes

Males produce small gametes, and females produce large gametes. A female bird produces 3 to 5 eggs that each weigh 15% to 20% of her body weight. In the same breeding season, a male bird may produce trillions of sperm that in aggregate weigh less than 5% of his body weight. In humans, women produce eggs that are 1 million times larger than a single sperm.

Biologists agree on one fairly simple explanation for why there are sex differences in male and female gamete size. Once sexuality evolved as a method of reproduction, the formation of two different sexual types (i.e., female and male) was probably inevitable. This diversification occurred through the process of natural selection and is diagrammed in **Figure E21.1**.

When a range of gamete sizes is produced in a population, individuals producing gametes near either the small or large end of the size distribution will be more successful than individuals producing medium-sized gametes.

Consider three different ways of distributing the same amount of energy into gametes: one could produce many small gametes, few large gametes, or a moderate number of medium-sized gametes. Individuals that produce few, but large, gametes provide a large supply of nutrients in each cell. These individuals will therefore have larger, healthier offspring at birth. Their offspring, because of their size and hardiness, will be better able to compete for food

than will offspring produced from smaller gametes. By the process of natural selection, individuals with large gametes should therefore become prevalent in the population.

On the other hand, individuals who produce many small gametes can potentially parent more offspring than can those who produce fewer, larger gametes because they can distribute these abundant gametes more widely. In addition, small gametes may be more likely than larger gametes to win the race to fertilization sites because they can move faster. In this scenario, natural selection would favor individuals who produce many small gametes over individuals who produce fewer, larger ones.

The two successful gamete production strategies coexist in populations because each strategy alone is unstable. In a population in which almost all gametes were large, individuals who produced smaller gametes would be successful at parenting offspring because they could fertilize many other gametes, thereby creating many more offspring. In a population in which almost all gametes were small, individuals that appeared with larger gametes would be more successful at generating healthy offspring because their larger gametes would provide more nutrition to the zygote. The most stable situation is a population in which 50% of the individuals produce a few large gametes and 50% produce many small gametes—in other words, the population is 50% female and 50% male.

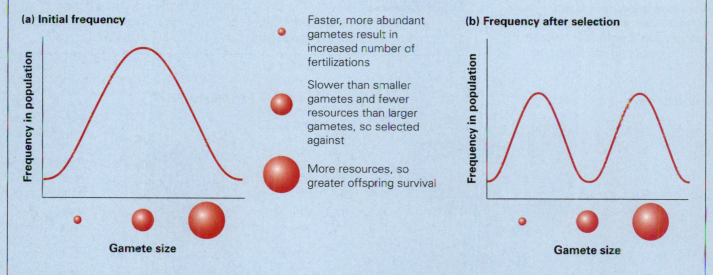

Figure E21.1 The evolution of gamete size. Natural selection resulting in two successful gamete production strategies.

21.2 Human Reproduction

To understand how human reproduction might be affected by environmental contaminants, we start by considering the anatomical structures that comprise the human male and female reproductive systems.

Human Reproductive Systems

The reproductive systems of males and females consist of external and internal structures. These structures are designed to allow the production and maturation of gametes, signal the synthesis and secretion of substances required for reproductive function, and provide a route to deliver the gametes through ducts.

Male Reproductive Anatomy. The external and internal components of the male reproductive system are shown in **Figure 21.3**.

The **penis** delivers sperm to the female reproductive tract during sexual intercourse. The penis is composed of spongy erectile tissue. During sexual arousal, this tissue fills with blood. Pressure from the increased volume of blood in the penis seals off the veins that drain blood from it. This causes the penis to become engorged with blood, keeping it erect. The erection is essential for insertion of the penis into the vagina, which facilitates delivery of sperm to the egg. A tube inside the penis, called the **urethra,** provides passage out of the body for both sperm and urine. The head, or **glans penis,** consists of highly sensitive skin that is covered by a fold of skin called the foreskin. In circumcised males, the foreskin is removed.

The **scrotum** is a pouch below the penis that contains the testes, or **testicles.** The skin of the scrotum is thin and devoid of fatty tissue. It tends to be folded or wrinkled with little hair. Below the skin is a layer of involuntary smooth muscle that regulates the position of the testes relative to the body to keep them at a temperature that maximizes sperm production. The scrotum contracts in the cold to keep the sperm-producing testes closer to the body.

Stop & Stretch In some men, one or both of the testes fail to descend from their location in the abdomen to the scrotum during fetal development. This condition results in decreased fertility. Why might this be the case?

Figure 21.3 Male reproductive anatomy. (a) Side view and (b) front view of the male reproductive system. The male external genitalia consist of the penis and scrotum. The scrotum houses the testicles. The testicles of a male produce sperm, which is stored in the epididymis and carried by the vas deferens through the urethra to the outside of the body. Secretions from the seminal vesicle, prostate, and bulbourethral glands help sperm mature and gain motility.

Visualize This: Use the front view to trace the path a sperm cell would take from its production in the testes to its exit via the urethra at ejaculation.

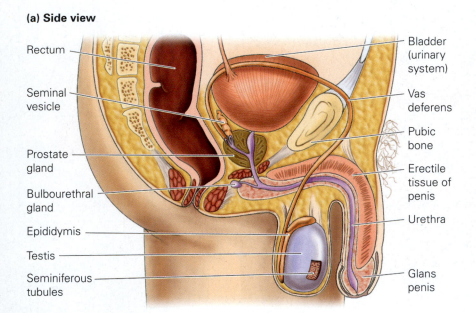

(a) Side view

Rectum
Seminal vesicle
Prostate gland
Bulbourethral gland
Epididymis
Testis
Seminiferous tubules

Bladder (urinary system)
Vas deferens
Pubic bone
Erectile tissue of penis
Urethra
Glans penis

(b) Frontal view

Seminal vesicle (behind bladder)
Bladder (urinary system)
Prostate gland
Urethra
Erectile tissue
Vas deferens
Epididymus
Testis
Glans penis
Scrotum

In addition to producing sperm, testes also produce male hormones called **androgens.** The testes consist of many highly coiled tubes, called **seminiferous tubules,** where sperm form. The seminiferous tubules are held in place by connective tissues that are rich in hormone-producing cells called **Leydig cells.** The process of sperm formation takes about 60 days and begins in the seminiferous tubules. From the seminiferous tubules, sperm pass through the coiled **epididymis,** which is an approximately 6-meter (20-foot) tube that rests atop each testis. During the 20 or so days that sperm pass through the epididymis, the sperm become motile.

During ejaculation, sperm are propelled from the epididymis through sperm-carrying ducts called the **vas deferens.** Each duct is surrounded by smooth muscle and is capable of peristaltic contractions (contractions in waves) that move sperm through the duct.

Several glands add secretions to the developing sperm as they make their way through the ducts of the reproductive system. **Seminal vesicles** are glands that secrete mucus and fructose (a sugar) for the sperm to use as energy. The **prostate** gland secretes a thin, milky white fluid into the urethra. These secretions contain nutrients for sperm. The **bulbourethral glands** are a pair of small glands that lie below the urethra between the prostate and the penis. Before ejaculation takes place, these glands secrete clear mucus that helps neutralize any acidic urine present in the urethra. Sperm and these secretions make up **semen.**

Female Reproductive Anatomy. **Figure 21.4** shows the female reproductive anatomy. The most obvious feature of the female external genitalia is a structure called the vulva. The vulva consists of two sets of lips, or labia: the outer labia majora, which are fatty and have a hairy external surface, and the inner labia minora, which contain neither fat nor hair. At the front of the vulva, the labia minora divide around the clitoris, an important organ for female sexual arousal.

Between the folds of the labia, there is an opening for the urethra, which in females serves as a passageway for urine from the bladder. The urethra in women is short (4 centimeters on average) compared to the urethra in men (18 centimeters on average). Partially due to this difference in length, women are more susceptible to bladder infections because bacteria have a shorter distance to travel from outside the body. A few centimeters below the urinary opening is the vaginal opening.

The organs that make up the internal genitalia are the ovaries, vagina, uterus, and oviducts. As mentioned, the ovaries are the female gonads. They produce the

Figure 21.4 Female reproductive anatomy. (a) Side view and (b) front view of the female reproductive system. The female external genitalia consist of the vulva and the clitoris. Openings to the outside allow for the expulsion of urine (via the urethra) and menstrual tissue (via the vagina). The ovaries of the female produce egg cells and hormones required for reproduction. Egg cells move through the oviducts to the uterus. The uterus can house a developing fetus. The cervix is the bottom portion of the uterus. The cervix widens during birthing. The vagina is the passageway into and out of the uterus.

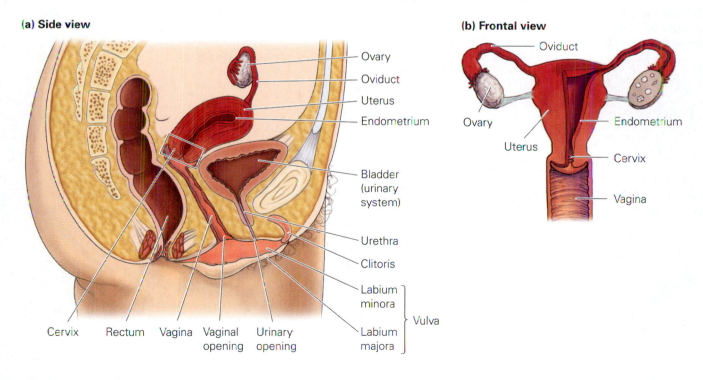

(a) Side view — Ovary, Oviduct, Uterus, Endometrium, Bladder (urinary system), Urethra, Clitoris, Labium minora, Labium majora, Vulva, Cervix, Rectum, Vagina, Vaginal opening, Urinary opening

(b) Frontal view — Oviduct, Ovary, Uterus, Endometrium, Cervix, Vagina gametes and sex hormones that circulate in a woman's body during her reproductive years. Ovaries are about the size of a walnut, but at birth they contain about 2 million ova (eggs) each. During the menstrual cycle, one or sometimes more eggs mature and are released from a structure called a **follicle.** The follicle, a fluid-filled sac containing the developing egg, secretes the ovarian hormone estrogen. During ovulation the follicle ruptures, much like a blister popping, to release the egg. The remnant of the follicle, called the **corpus luteum,** secretes both estrogen and progesterone.

The **vagina** is a muscular organ that acts primarily as a passageway into and out of the uterus. The **uterus** is about the size of a fist. The wall of the uterus is thick (about 1 centimeter) and is composed of some of the most powerful muscles in the human body. These muscular walls contract rhythmically during labor, childbirth, and orgasm. The internal surface of the uterine wall is called the **endometrium,** which changes in thickness during the course of the menstrual cycle. The lower third of the uterus becomes narrower than the upper portion and is called the **cervix.**

The **oviducts** are actually an extension of the top surface of the uterus. These tubes extend from the body of the uterus toward the ovaries, which are suspended within the abdominal cavity. The oviducts are not attached to the ovaries directly. Instead, they end in brushy structures that move over the surface of the ovary. These movements, along with suction inside the oviducts, direct the eggs released by the ovary into the oviducts.

The pathway from the vagina to the uterus to the oviduct represents the only natural pathway in humans directly from outside the body to inside the abdominal cavity. This means that women can experience bacterial infections inside their abdominal cavity without experiencing an injury that punctures the cavity wall. These infections are often sexually transmitted. For a description of common sexually transmitted diseases (STDs) in both men and women, see **Table 21.1.**

TABLE 21.1

Sexually transmitted diseases. The causes, symptoms, prevention, and prevalence of common STDs.

Disease and the Bacterial Pathogen That Causes It	Mode of Infection	Symptoms	Treatment and Prevention	Incidence
Chlamydia is caused by *Chlamydia trachomatis.*	• Infects the urethra, cervix, and oviducts of women and the urethra of men	• Pelvic pain, fluid discharge • Many cases undiagnosed because symptoms may not appear until years after infection; if untreated, can lead to pelvic inflammatory disease and infertility	• Treat with antibiotics • Prevent transmission with condom use	• Most common bacterial STD • Approximately 3 million cases each year in the United States
Gonorrhea is caused by *Neisseria gonorrhoeae.*	• Known as "the clap"; can be sexually transmitted or spread from infected mothers to babies during birth	• 80% asymptomatic • Thick discharge from penis or vagina • If untreated and bacteria spread to the oviducts, can cause infertility and pelvic inflammatory disease in women	• Treat with antibiotics • Prevent transmission with condom use	• Around 700,000 cases diagnosed in the United States each year

TABLE 21.1

Sexually transmitted diseases. The causes, symptoms, prevention, and prevalence of common STDs.

Disease and the Bacterial Pathogen That Causes It	Mode of Infection	Symptoms	Treatment and Prevention	Incidence
Pelvic inflammatory disease (PID) is caused by gonorrhea or chlamydia.	• Infection of the female reproductive tract; occurs when sexually transmitted bacteria ascend from the vagina into the uterus and oviducts	• Pelvic pain • Difficulty becoming pregnant due to scarring and blockage of reproductive organs	• Treat with antibiotics; no cure for damaged reproductive organs • Prevent transmission with condom use	• About 1 million U.S. women diagnosed each year • Approximately 100,000 infertile • Thousands of pregnancy complications
Syphilis is caused by *Treponema pallidum*.	• Sexually transmitted	• A small sore, or chancre, at the site of infection • If untreated, possible neurological problems, paralysis, and death	• Antibiotics (if caught early)	• Around 40,000 annually in the United States

Viral Pathogens	Mode of Infection	Symptoms	Treatment and Prevention	Incidence
AIDS is caused by human immunodeficiency virus (HIV).	• Oral, anal, and vaginal sex, as well as via blood transfusion	• Over time weakens the immune system; eventually severe tissue and organ damage	• Combination-drug therapies to halt progression; no cure • Reduced transmission with condom use	• Worldwide, about 40 million with AIDS/HIV
Genital warts are caused by human papilloma virus (HPV).	• Sexual contact (intercourse or oral-genital contact) with an infected partner	• Growths or bumps on the pubic area, penis, vulva, or vagina • Genital warts (from some types of HPV); cervical cancer (from certain types of HPV)	• Vaccination to prevent infection by many strains • Reduced transmission with condom use	• At least 1 or more of the over 70 different types of HPV in the sexually active
Hepatitis B is caused by hepatitis B virus (HBV).	• Blood and other bodily fluids • Possible for newborns to obtain infections from mother during delivery	• Inflammation and scarring of the liver (cirrhosis), which may be fatal	• Vaccination for prevention • Reduced transmission with condom use	• Approximately 200,000 U.S. cases diagnosed annually • About 4000–5000 deaths annually

(continued)

TABLE 21.1 *(continued)*

Sexually transmitted diseases. The causes, symptoms, prevention, and prevalence of common STDs.				
Viral Pathogens	**Mode of Infection**	**Symptoms**	**Treatment and Prevention**	**Incidence**
Herpes is caused by two types of herpes simplex viruses.	• Direct skin-to-skin contact, usually by kissing or by oral, vaginal, or anal intercourse	• Cold sores or fever blisters on mouth or face • Sores in the genital area, known as genital herpes	• Antiviral medications to lessen symptoms; no cure • Reduced transmission with condom use	• For genital herpes nearly 20% of adults in the United States
Insect, Protozoan, and Fungal Pathogens	**Mode of Infection**	**Symptoms**	**Treatment and Prevention**	**Incidence**
Pubic lice are caused by *Pediculus pubis* (insect).	• Also known as "crabs"; transmitted via skin-to-skin contact or contact with an infected towel or clothing	• After insect bite an allergic reaction; subsequent itching of pubic area around 5 days after initial infection	• Cured by washing the affected area with a delousing agent	• Approximately 3 million cases in the United States each year
Trichomoniasis is caused by *Trichomonas vaginalis* (protozoan).	• Also known as "trich"; transmitted by sexual intercourse	• In women, vaginal itching with a frothy yellow-green vaginal discharge • In some men irritation in urethra after urination or ejaculation	• Treated with antibiotics • Reduced transmission with condom use	• An extremely common STD, infecting up to 15% of sexually active women in the United States
Yeast infections are caused by *Candida* (fungi).	• Normal inhabitants of the female reproductive tract • Antibiotics (which kill vaginal bacteria and allow the yeast to grow) • Passed from person to person via sexual intercourse	• In women, thick whitish discharge from vagina and vaginal itching	• Cured by antifungal medicines	• At least one case in nearly 75% of all adult U.S. women

Sexually transmitted diseases can cause **infertility,** defined as the failure to achieve pregnancy after 1 year of unprotected intercourse. Infertility can result from physical blockages of ductal structures caused by scarring in response to infection.

For pregnancy to occur, the oviducts must be open, allowing the sperm to reach the egg and the fertilized egg to travel back down the oviduct to the uterus.

Likewise, the sperm must be able to move from the testes through the vas deferens and urethra to be delivered to the egg. Physical blockages or obstructions to the ducts can occur in males also, but this is less often a cause of infertility in males than it is in females.

Although STDs are one cause of infertility, they are not the only cause. Problems with sperm development, number, and motility can lead to infertility, as can problems with the maturation and ovulation of egg cells. In addition, endocrine-disrupting chemicals can cause changes to reproductive structures that lead to infertility.

Endocrine Disruptors, Reproductive Anatomy, and Infertility. A well-studied example of an endocrine disruptor that affects human fertility is provided by a chemical called diethylstilbestrol, or DES. Between 1938 and 1971, many pregnant women were given this synthetic drug to help prevent miscarriage. However, the physicians who prescribed this drug did not know that it behaved as a mimic of the female hormone estrogen and altered the development of the uterus in the developing baby, resulting in a misshapen uterus less able to sustain a pregnancy. Daughters of women who took this drug have a higher-than-average rate of infertility. Approximately 24% of these women are infertile.

The most common cause of male infertility is the inability to produce enough healthy sperm. Many scientific studies have shown that sperm counts are declining in developed countries. Although no one knows for sure why this is happening, some scientists are studying whether endocrine disruptors are interfering with human males' ability to produce sperm, an effect these chemicals have been shown to have on other animals. The production of gametes is a multistep process allowing many different opportunities for endocrine disruptors to act.

Gametogenesis

The production of sex cells, or gametes, is called **gametogenesis**. In human males and females, gametogenesis involves the process of meiosis (discussed in Chapter 6), which is a type of cell division that reduces the chromosome number by one-half. Since human body cells contain 46 chromosomes, meiosis in humans produces gametes that contain 23 chromosomes. Human body cells with 46 chromosomes (23 pairs) are diploid ($2n$), and the gametes produced after meiosis are haploid (n).

Meiosis alone is not enough to produce a functional gamete. Other changes to sex cells occur as they go through gametogenesis to enable them to mature into gametes capable of being involved in fertilization. For instance, sperm cells gain motility, and egg cells increase in size and nutrient content to allow for the development of the early embryo.

Spermatogenesis. The production of sperm, called spermatogenesis (**Figure 21.5a** on the next page), begins to occur in the testes at puberty. Cells that will undergo gametogenesis line the walls of the seminiferous tubules. During spermatogenesis, each parent cell first duplicates by mitosis, and then one of the two daughter cells undergoes meiosis. The other daughter cell maintains the function of the parent cell. Since each round of mitosis produces two cells, only one of which goes on to complete meiosis, a man will never exhaust the supply of cells that can be used to make sperm cells.

The diploid cell that begins meiosis is called a **primary spermatocyte**. After the first meiotic division, which separates the 23 homologous pairs of chromosomes, the cell is called a **secondary spermatocyte**. Secondary spermatocytes, with their 23 duplicated but unpaired chromosomes, are haploid. These cells then undergo meiosis II to produce haploid cells that no longer have duplicated chromosomes; these haploid cells are called **spermatids**.

Cells that aid the developing sperm, called **Sertoli cells**, are also located in the seminiferous tubules. These cells secrete the substances required for further

(a) Spermatogenesis

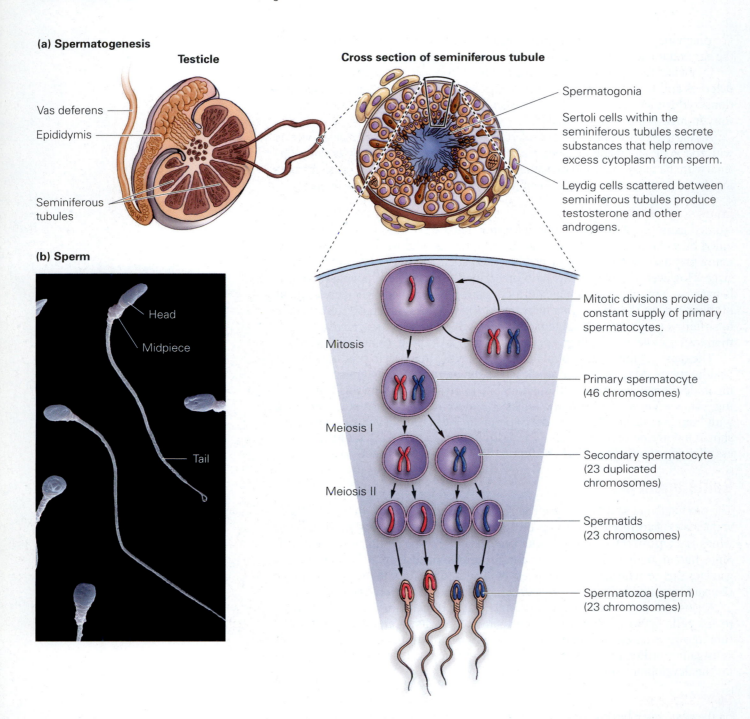

Testicle

Vas deferens

Epididymis

Seminiferous tubules

Cross section of seminiferous tubule

Spermatogonia

Sertoli cells within the seminiferous tubules secrete substances that help remove excess cytoplasm from sperm.

Leydig cells scattered between seminiferous tubules produce testosterone and other androgens.

Mitotic divisions provide a constant supply of primary spermatocytes.

Mitosis

Primary spermatocyte (46 chromosomes)

Meiosis I

Secondary spermatocyte (23 duplicated chromosomes)

Meiosis II

Spermatids (23 chromosomes)

Spermatozoa (sperm) (23 chromosomes)

(b) Sperm

Head

Midpiece

Tail

Figure 21.5 Spermatogenesis.
(a) Spermatozoa (sperm) are produced in the seminiferous tubules of each testicle. Cells in the testes first undergo a round of mitosis, producing two identical daughter cells. The primary spermatocyte begins meiosis with 46 chromosomes. After meiosis I, the two result-ant daughter cells are the secondary sper-matocytes. Meiosis II results in the production of four haploid daughter cells containing 23 chromosomes, called spermatids. Sperma-tids are then modified to become mature sperm. (b) The head of the sperm contains the DNA. The midpiece contains mitochon-dria to provide energy for the flagellum or tail.

sperm development and convert the spermatids into **spermatozoa** (sperm) by re-moving some of their excess cytoplasm. The mature sperm is composed of a small head containing the DNA, a midpiece that has mitochondria to provide energy for the journey to the oviduct, and a tail (or flagellum) to propel the sperm (**Figure 21.5b**). At the tip of the sperm's head lies the **acrosome**. This structure contains digestive enzymes that help a sperm cell gain access to the egg cell.

Oogenesis. The formation and development of female gametes, called oogen-esis, occurs in the ovaries and results in the production of egg cells. A small percentage of these egg cells will be released over a woman's reproductive life span, and an even smaller percentage will be fertilized.

While spermatogenesis begins at puberty, oogenesis actually begins while the female is still in her mother's uterus and then pauses until puberty. A fe-male produces around 2 million potential egg cells prior to her birth, but some of these cells degenerate, from about 700,000 at birth to around 350,000 at puberty. At puberty, development of preexisting eggs continues each month until the cessation of menstruation, or **menopause**. Amazingly, the egg cell that helped produce you started developing inside your mother while she was developing inside your grandmother's uterus!

The ovary contains many **follicles**, with each follicle containing an immature egg called an *oocyte*. The ovarian cycle includes all the events that occur as a *primary follicle* develops into a *secondary follicle* and then into the mature *Graafian follicle* (**Figure 21.6a**). The stepwise development of the follicle occurs as the primary follicle (which houses the primary oocyte) secretes estrogen. The secondary follicle includes the secondary oocyte and the cells that surround it. The secondary oocyte is surrounded by pools of fluid and follicle cells that secrete estrogen. In the next step in the ovarian cycle, the mature Graafian follicle develops from the secondary follicle. The Graafian follicle contains a fluid-filled cavity that increases in volume, causing the ovary wall to balloon until it bursts, expelling the secondary oocyte from the ovary, a process called **ovulation**, which typically occurs around the 14th day of the menstrual cycle.

The remnant of Graafian follicle (minus the oocyte that was ovulated) is called the **corpus luteum**, which secretes reproductive hormones but degenerates after about 10 days if fertilization does not occur.

After ovulation, the secondary oocyte moves into the oviduct where, if there are sperm present within 12 hours of ovulation, fertilization is likely.

An ovulated egg cell is large enough to be viewed without a microscope. This is because the egg cell undergoes an unequal meiosis that produces both very small structures that will not be involved in fertilization, called **polar bodies**, and a much larger cell that will give rise to the female gamete (**Figure 21.6b**).

Figure 21.6 Oogenesis. (a) One follicle per month goes through all of these steps to produce an egg cell that is ovulated. **Visualize This:** Would twins produced when ovulation occurs from both ovaries in 1 month be genetically identical? Why or why not? (b) Meiosis in human females produces only one fertilizable cell because of the production of nutrient-poor polar bodies. The larger egg cell, if fertilized, can provide all the nutrients and organelles needed to serve as the progenitor of all the millions of cells of the human body.

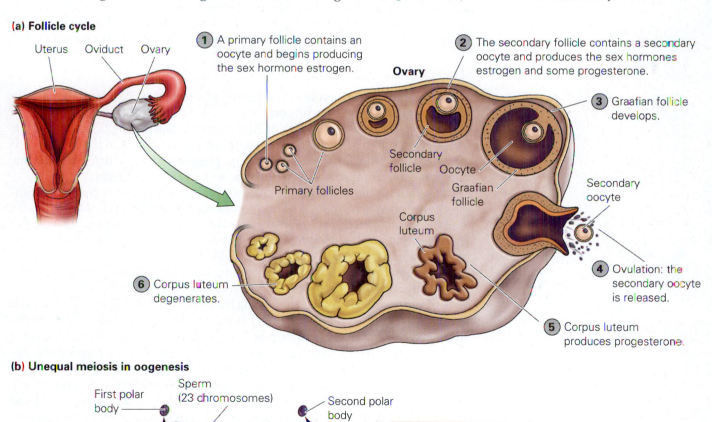

(a) Follicle cycle

Uterus   Oviduct   Ovary

1 A primary follicle contains an oocyte and begins producing the sex hormone estrogen.

2 The secondary follicle contains a secondary oocyte and produces the sex hormones estrogen and some progesterone.

Ovary

3 Graafian follicle develops.

Secondary follicle   Oocyte

Primary follicles

Graafian follicle

Secondary oocyte

Corpus luteum

6 Corpus luteum degenerates.

4 Ovulation: the secondary oocyte is released.

5 Corpus luteum produces progesterone.

(b) Unequal meiosis in oogenesis

First polar body

Sperm (23 chromosomes)

Second polar body

Sperm nucleus and egg nucleus fuse: zygote with 46 chromosomes results.

Primary oocyte (46 chromosomes)   Meiosis I   Secondary oocyte (23 chromosomes)   Meiosis II   Zygote

This larger cell contains enough nutrients to help nourish the embryo through its very early development.

If not fertilized, it takes about 3 days for the secondary oocyte—an object about the size of a pinhead—to move from the ovary through the oviduct, uterus, and cervix to be shed with menstrual fluid.

Women are fertile, or able to produce secondary oocytes, only a few days per month from puberty until menopause. Men are fertile throughout their lifetimes, with a slow but progressive decrease in sperm count as they age. Many different chemicals, potentially including endocrine disruptors, can hasten this decline in sperm counts.

Endocrine Disruptors, Gametogenesis, and Infertility. The vast majority (around 90%) of male infertility cases are due to problems with sperm production and formation, including low numbers of sperm, improperly shaped sperm, or sperm with impaired motility. The likelihood of male infertility increases with alcohol and drug use, including cigarette smoking. In addition to these factors, many scientists are investigating whether exposures to endocrine disruptors can cause altered sperm production in humans.

One study that compared reproductive health of men from around the United States showed that sperm counts of men from rural Missouri were lower than they were for men from many large cities across the country. Scientists who performed this study found higher levels of three commonly used agricultural pesticides (including atrazine) in the bodies of men with lower sperm counts than were present in the bodies of men with higher sperm counts. Most of the men with decreased sperm counts did not work directly with these chemicals, leading researchers to question whether these men may have been exposed to these chemicals in their drinking water.

Another important reproductive function in humans may be affected by the water we drink. That function is the menstrual cycle, which we discuss next.

The Menstrual Cycle

The term **menstrual cycle** refers specifically to cyclic changes that occur in the uterus. This cycle depends on intricate interrelationships among the brain, ovaries, and lining of the uterus or endometrium. During the course of a single menstrual cycle, a woman's body prepares an egg for potential fertilization and her uterus for a potential pregnancy. If no pregnancy has occurred, the uterine lining is excreted during **menstruation,** and a new cycle begins.

Figure 21.7 illustrates the changes in hormone levels, ovaries, and condition of the endometrium that occur throughout the 28-day cycle. Most women's bodies do not adhere precisely to a 28-day cycle—some women have longer cycles, and others have shorter cycles. The first day of a menstrual cycle is considered to be the first day of actual bleeding.

The hypothalamus, located in the brain, secretes gonadotropin-releasing hormone (GnRH), which stimulates another endocrine structure in the brain—the pituitary gland—to release follicle-stimulating hormone (FSH) and luteinizing hormone (LH) (Chapter 20). In males, FSH and LH are involved in the production of sperm; in females, these hormones help regulate the development of egg cells and ovulation.

Like many other biological processes, the menstrual cycle is self-regulating, operating by responding to feedback. High levels of estrogen provide positive feedback to the hypothalamus. The hypothalamus secretes GnRH, which acts on the pituitary gland to increase the secretion of FSH and LH. In this manner, FSH and LH progressively increase in concentration. Conversely, high levels of the hormone **progesterone** have a negative-feedback effect on the hypothalamus; GnRH secretion is decreased, and FSH and LH levels decline (**Figure 21.8** on page 528).

When a woman menstruates, the endometrial tissue is shed and released through the cervix and vagina. Inside the ovary, the primary follicle begins to grow under the influence of FSH. As the follicle grows, it produces estrogen.

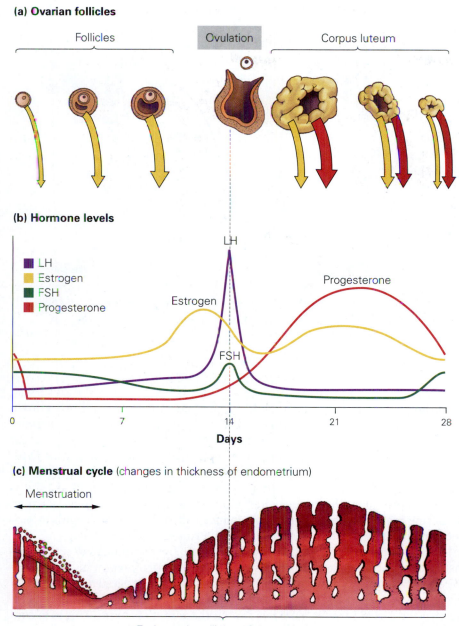

(a) Ovarian follicles

Follicles Ovulation Corpus luteum

(b) Hormone levels

- LH
- Estrogen
- FSH
- Progesterone

LH

Estrogen

Progesterone

FSH

0   7   14   21   28

Days

(c) Menstrual cycle (changes in thickness of endometrium)

Menstruation

Endometrium (lining of the uterus)

Figure 21.7 The menstrual cycle.
Changes in hormone levels are linked to (a) the state of ovarian follicles, (b) hormone levels, and (c) uterine condition over the course of the menstrual cycle.

When the primary follicle is large enough, it produces enough estrogen to stimulate GnRH release. This leads to a spike in both FSH and LH levels, which lasts for about 24 hours. Ovulation occurs 10 to 12 hours after the LH peak, around 14 days before menstruation in a typical 28-day cycle. The follicle then bursts open, releasing the egg, which is drawn into the oviducts (**Figure 21.9** on page 528).

After ovulation, the corpus luteum produces estrogen and progesterone. Progesterone helps maintain the blood flow to the uterine lining to support early fetal development. In addition, progesterone inhibits continued LH production, so the LH surge is inhibited by the ovarian hormones that it stimulates—an example of negative feedback.

If fertilization does not occur, the corpus luteum degenerates in about 10 days after ovulation. Because the corpus luteum is no longer secreting them, progesterone and estrogen levels fall, triggering the arteries that supply the uterus to spasm. This allows the lining of the uterus to be shed and causes menstrual cramps. Decreasing levels of progesterone also serve to release the hypothalamus from inhibitory control. Therefore, LH and FSH levels rise, and the cycle starts over.

Figure 21.8 Feedback during the menstrual cycle. Gonadotropin-releasing hormone (GnRH) stimulates the pituitary to release follicle-stimulating hormone (FSH) and luteinizing hormone (LH). FSH stimulates the follicle to produce estrogen. LH stimulates the corpus luteum to secrete progesterone. Estrogen (a) exerts positive-feedback control over the hypothalamus. The positive-feedback loop stimulated by estrogen only occurs a few days before ovulation. Progesterone (b) exerts negative-feedback control over the hypothalamus.

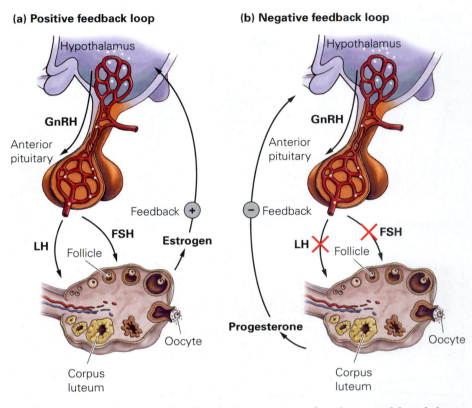

(a) Positive feedback loop

(b) Negative feedback loop

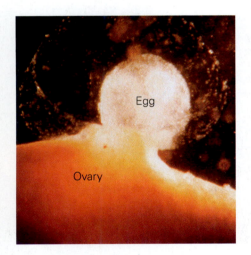

Figure 21.9 Ovulation. The egg cell is released from the ovary.

If, however, the egg is fertilized, the process of endometrial breakdown does not occur. The early embryo produces a hormone called *human chorionic gonadotropin* (HCG) that extends the life of the corpus luteum. Women test for the presence of this hormone in most over-the-counter pregnancy tests. With the corpus luteum intact, progesterone and estrogen levels remain high, and the endometrium is maintained. In a pregnant woman, the corpus luteum discontinues producing hormones after about 12 weeks of pregnancy, when an endocrine organ that forms during pregnancy called the *placenta* begins to produce enough progesterone to maintain the uterine lining on its own.

The birth control pill supplies the body with low continuous doses of synthetic versions of estrogen and progesterone. The estrogen present in the pill is at a level low enough to prevent follicle development. The progesterone contained in the pill serves as a backup (in case ovulation does occur) by making the mucus that the cervix secretes more resistant to the ascent of sperm. This is in contrast to the cervical mucus secreted during ovulation, which is thin and stringy and has the consistency of raw egg white. This type of cervical mucus actually facilitates fertilization by providing channels through the cervix for the sperm to swim through and gain access to the egg. The low hormone levels present in the birth control pill also prevent the uterine lining from developing enough to support a pregnancy. See **Table 21.2** for a list of other methods of birth control.

Stop & Stretch Women who live together often report that their menstrual cycles are synchronized. In other words, they start menstruating at around the same time each month. Although scientists have tried to determine whether this is actually happening and if so why, results have been inconclusive. Assume that four fertile women move in together. Design an experiment, with appropriate controls, that would measure whether their menstrual cycles were synchronizing over time.

TABLE 21.2

Birth control methods. The merits and efficacy of various birth control methods.

	Method	Mode of Action	Risks
Abstinence	Abstinence	• Sperm and egg never have contact.	• No associated risks • When used correctly, 100% effective
Hormonal methods	Combination birth control pill	• Synthetic estrogen and progesterone given at continuous doses (vs. the cyclic fluctuations of a menstrual cycle) prevent ovulation.	• Increased risk of heart disease and fatal blood clots for women over age 35 who smoke; slightly increased risk of breast cancer; does not protect against STDs • 5% of women become pregnant every year
	Minipill	• The minipill contains progesterone only. Since there is no estrogen to combat the effects of progesterone, cervical mucus is thickened, sperm ascent is impeded, and the uterine lining is not prepared to support a pregnancy.	• Increased risk of ovarian cysts • Does not protect against STDs • Pregnancy in 5% of women using this method every year
	Emergency birth control pills	• High doses of progesterone or estrogen prevent sperm from reaching egg or prevent fertilized egg from attaching to wall of uterus. Can also prevent ovulation for one cycle if it has not occurred. These so-called morning after pills are not the same as the mifepristone abortion pill. Emergency birth control pills prevent pregnancy, while mifepristone terminates an established pregnancy.	• Does not protect against STDs • Nausea, vomiting, abdominal pain, fatigue, headache • Pregnancy in 20% of women using this method
	Patch	• When applied to skin, the patch delivers progesterone and estrogen to prevent ovulation and thicken cervical mucus.	• Same as combination pill • Does not protect against STDs • Pregnancy in 1% of women using this method every year
	Vaginal ring	• When inserted in vagina, the vaginal ring delivers progesterone and estrogen to prevent ovulation and thicken cervical mucus.	• Same as combination pill • Does not protect against STDs • 99% effective
	Injectables	• This method requires the user to have progesterone injections every 3 months, and the mode of action is the same as the minipill.	• Irregular vaginal bleeding • Does not protect against STDs • Pregnancy in 5% of women using this method every year
	Implantables	• Implantables, such as Implanon, secrete progesterone to prevent ovulation and to thicken cervical mucus.	• Under investigation • Does not protect against STDs • Effectiveness unknown

(continued)

TABLE 21.2 *(continued)*

Birth control methods. The merits and efficacy of various birth control methods.

	Method	Mode of Action	Risks
Sterilization	Female sterilization 	• Female sterilization permanently blocks oviducts with a tightly coiled tube.	• Does not protect against STDs • Pregnancy in 0.5% of sexually active women using this method every year
	Male sterilization (vasectomy) 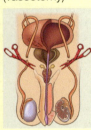	• Male sterilization cuts each vas deferens to prevents sperm from being ejaculated in the semen.	• Does not protect against STDs • Fails to prevent pregnancy in 0.1% of sexually active men using this method every year
Barrier methods	Cervical cap 	• When inserted against cervix before intercourse, the cervical cap prevents sperm and egg contact.	• No known risks • Does not protect against STDs • Pregnancy in 20% of sexually active women using this method (combined with spermicide) every year
	Diaphragm 	• When inserted into vagina before intercourse, the diaphragm prevents sperm and egg contact.	• No known risks • Does not protect against STDs • Pregnancy in 20% of sexually active women using this method (combined with spermicide) every year
	Sponge 	• The spermicide-soaked sponge fits against the cervix to prevent sperm and egg contact.	• Pregnancy in 10% of sexually active women using this method (combined with spermicide) every year
	Female condom 	• The female condom is held against the cervix by a flexible ring to prevent sperm and egg contact.	• No known risks • Pregnancy in 14% of sexually active women using this method alone every year
	Male condom 	• The male condom fits over the penis to prevent sperm and egg contact.	• No known risks • Every year, pregnancy in 15% of women when the couple used this method alone

TABLE 21.2

Birth control methods. The merits and efficacy of various birth control methods.

Method	Mode of Action	Risks
Other methods Spermicides	• When inserted into vagina 1 hour before intercourse, spermicide kills sperm.	• No known risks • Does not protect against STDs • Every year, pregnancy in 26% of women when the couple used this method alone
Fertility awareness	• Fertility awareness requires abstinence for the 4 days before and 4 days after predicted time of ovulation.	• None • Does not protect against STDs • Pregnancy in 85% of women when the couple used no birth control
Intrauterine device	• This small plastic device inserted into the uterus by a physician prevents fertilization and prevents the uterus from supporting a pregnancy.	• May increase risk of pelvic inflammatory disease

Endocrine Disruptors, Menstruation, and Infertility. Some cases of female infertility can be related to a condition called endometriosis, which occurs when menstrual tissues that normally line the uterus begin to grow in other areas, such as the oviducts and ovaries. These tissues continue to respond to the hormonal cycle, growing and shedding in sync with the lining of the uterus. In addition to being very painful, endometriosis can lead to scarring and inflammation of the oviducts and disruption of ovulation. An endocrine disruptor, the chemical di-(2-ethylhexyl)-phthalate (DEHP), is under investigation to determine whether it plays a role in causing endometriosis. DEHP, a type of chemical called a plasticizer, is added to plastics to keep them soft and pliable. Some scientists believe that drinking water from plastic bottles or heating food in plastic containers might increase an individual's exposure to this and other plasticizers, which leach out of the plastic and into the water or food. Researchers at the University of Siena in Italy found that women with endometriosis had higher levels of DEHP in their blood than did women who did not have endometriosis. These studies provided no direct evidence that chemicals present in plastics are causing reproductive problems. These women might also have high levels of other, untested-for chemicals that are actually causing the problem. The increased level of environmental chemicals also may be due to some other factor. Research on the hormone-disrupting effects of plasticizers is still in its infancy; before drawing any definitive conclusions, scientists must wait until many studies that provide direct evidence of human reproductive problems caused by plasticizers have been completed.

Reproductive problems can arise because of exposures to toxins that a person experiences during his or her lifetime. Such problems also can result from exposures experienced in the womb during development, when reproductive and other structures are forming.

21.3 Human Development

Development is the series of events that take place after fertilization. It leads to the formation of a new multicellular organism. Fertilization sets the stage for development to occur.

Fertilization

Of the roughly 300 million sperm ejaculated, only about 200 reach the site of fertilization in the oviduct. Initially, sperm must survive the acidic (low pH) conditions of the upper vagina. The natural acidity of the upper vagina works to help prevent bacterial infections in women but also can kill sperm. Some sperm get stuck in folds of the cervix and in the uterine wall, and only half of the sperm that make it to the uterus enter the oviduct that contains the ovulated egg.

The oviducts themselves undergo muscular contractions during the time of ovulation, and these contractions allow the sperm to move toward the upper portion of the oviduct to wait for an egg cell to be ovulated. It takes about half an hour to move a sperm from the upper vagina to the oviduct.

How Fertilization Takes Place. When both sperm and egg are present in the oviduct, fertilization can occur (**Figure 21.10**). First, the sperm cells must move through the follicle cells that surround and nourish the egg. Then they must pass through a translucent covering on the egg cell called the zona pellucida. This protective layer shields the egg from mechanical damage and acts as a species-specific barrier. Only sperm produced by a male of the same species as the female that produced the egg can traverse the zona pellucida. This is because getting through the zona pellucida requires binding of a specific receptor on the sperm head. This binding of the sperm head to the zona pellucida triggers the release of enzymes present in the acrosome of a sperm. The acrosomal enzymes interact with the egg cell's zona pellucida to allow the formation of a tunnel through the zona pellucida toward the egg cell's plasma membrane. Once this

Figure 21.10 Fertilization. The sperm cell must traverse follicle cells and the zona pellucida before fusing with the plasma membrane and depositing its chromosomes.
Visualize This: At what stage would the process be halted if the sperm and egg were from different species?

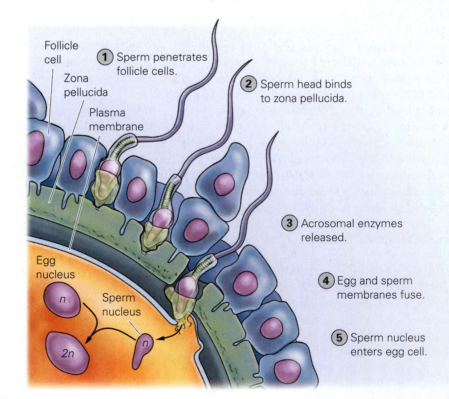

Follicle cell
Zona pellucida
Plasma membrane
Egg nucleus
Sperm nucleus
n
2n
n

① Sperm penetrates follicle cells.
② Sperm head binds to zona pellucida.
③ Acrosomal enzymes released.
④ Egg and sperm membranes fuse.
⑤ Sperm nucleus enters egg cell.

sperm cell makes it through the zona pellucida, its plasma membrane and the egg's plasma membrane fuse, and the egg cell draws the sperm cell's nucleus inward. The fertilized egg is now ready to begin the divisions of early development.

Chemicals in the environment can affect the ability of sperm to bind to the egg during fertilization. Some scientists are concerned that this could be a cause of infertility.

Endocrine Disruptors and Fertilization. Trichloroethylene (TCE) is an industrial solvent that is used to remove grease from metal and is found in many household products. Large amounts of this chemical can get into the water supply via air emissions from metal-degreasing plants and via wastewater from factories that produce metals, paints, cleaning products, electronics, and rubber products. Studies have shown that males exposed to TCE while working as mechanics or dry cleaners have a larger number of abnormally shaped sperm than men not exposed to TCE. Abnormally shaped sperm are less likely to be able to fertilize normal egg cells.

Studies showing an effect of endocrine-disrupting chemicals on reproductive health, as we described in this and the previous sections, seem to support the hypothesis that exposures during adulthood can decrease fertility, but these studies provide no definitive evidence. In the next section, we describe something scientists are surer of—that minute chemical exposures during critical windows of development can have tremendous impacts on a developing baby.

Human Embryonic Development

After fertilization, the fertilized egg cell, or zygote, undergoes a series of changes to produce a multicellular structure capable of producing adult tissues (**Figure 21.11**). In humans, the term **embryo** is used to describe the stage of development from the first divisions of the zygote until body structures begin to appear at about the ninth week. After that time, the developing human is referred to as a **fetus**. Early development of the embryo involves a series of rapid cell divisions, called **cleavage**, that begin while the zygote is still in the oviduct. These divisions continue until the now-multicellular embryo is propelled down the oviduct toward the uterus.

When the embryo reaches the uterus, a few days after fertilization, it is in the form of a hollow ball of cells known as the **blastocyst**. Inside the blastocyst, a cluster of cells called the **inner cell mass** forms. Some cells of the inner cell mass give rise to the embryo. The outer cells of the blastocyst form part of the placenta. The inner cell mass undergoes a dramatic reshuffling of cells to produce a structure with three layers of cells, called the **gastrula**. The cells comprising the three layers of the gastrula will begin to specialize, or **differentiate**, into the adult cells and tissues with their specific functions.

These three layers of tissue include an outer, middle, and inner layer. The outer layer, or **ectoderm**, gives rise to the skin, nervous system, and sense organs. The middle layer, or **mesoderm**, gives rise to muscles, excretory organs, circulatory organs, gonads, and the skeleton. The innermost layer, or **endoderm**, lines the digestive and respiratory organs.

Figure 21.11 Development of the early embryo. The human zygote undergoes cleavage divisions to produce the blastocyst, which implants in the uterus. Continued cell divisions along with specialization result in the formation of the three-layered gastrula.

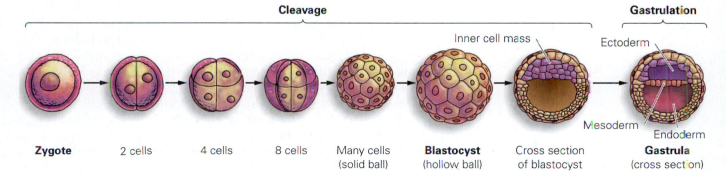

| Cleavage | | | | | | | Gastrulation |

Inner cell mass — Ectoderm — Mesoderm — Endoderm

| **Zygote** | 2 cells | 4 cells | 8 cells | Many cells (solid ball) | **Blastocyst** (hollow ball) | Cross section of blastocyst | **Gastrula** (cross section) |

This differentiation occurs in response to chemical signals that pass between the different cell layers and adjacent cells, turning on some genes in a given cell and leaving others off. In this manner, the specialized cells that comprise the adult body all arise from the single fertilized egg cell that undergoes mitotic cell division and cell differentiation.

Exposures to toxins of any kind very early in development tend to have an all-or-nothing impact. Exposures can be so damaging that they kill many cells, and the early embryo is spontaneously aborted—often without a woman ever realizing that she was pregnant. Conversely, exposures can have very little effect since cells are dividing so rapidly that if a few cells die due to exposure, they are quickly replaced. Exposures of greatest concern occur while organs are developing. The development of reproductive organs in particular can be derailed by exposures to endocrine-disrupting chemicals.

Development of Human Reproductive Organs.

Prior to week 7, the reproductive organs of male and female embryos are indistinguishable except at the level of their chromosomes.

In the early embryo, the structures that will become the gonads are located in the abdomen. The same starting material can be used to produce either the paired ovaries of a female or the paired testes of a male. Differentiation of the embryonic gonads into ovaries or testes is determined by the expression of sex-specific genes. Once differentiated, the male gonads produce androgens; but the female gonads do not start to produce estrogen until puberty.

The cells of the embryonic testes make the androgen testosterone, which directs the development of internal, gamete-carrying, ductal structures. Prior to differentiation, two different ductal structures exist side by side in both male and female embryos. In each sex, one embryonic duct system persists and the other regresses, based in part on the presence or absence of androgens. In males, the ductal structure that persists develops into the sperm-carrying vas deferens, epididymis, and urethra. In female embryos, the ductal structure that persists develops into the oviducts, uterus, cervix, and vagina.

Androgens also affect the development of external genitalia. In male and female embryos, one embryonic structure can be molded into either the male or female external genitalia. In fact, the penis and clitoris arise from the same starting tissue, as do the scrotum and vulva. **Table 21.3** summarizes the developmental pathways followed by embryonic males and females.

In this section, we highlighted the important role that androgens play in driving the development of male reproductive structures. Endocrine disruptors that interfere with the action of androgens can alter this developmental pathway.

Endocrine Disruptors and the Development of Reproductive Organs.

Earlier in the chapter, we related the story of DES—the synthetic estrogen taken by women in the 1960s and 1970s that caused some of their daughters to have deformed uteruses. The development of male reproductive structures also appears to be vulnerable to the effects of environmental contamination.

According to a report issued by the Tufts University School of Medicine, mice exposed to commonly used fungicides and herbicides have a higher-than-average rate of undescended testes, a condition called **cryptorchidism**. This anomaly decreases reproductive success because the sperm of males with this condition are held inside the abdomen, where the temperature is higher than optimum for the development of sperm. It is thought that these fungicides and herbicides function as endocrine disruptors by blocking androgen receptors.

Cryptorchidism occurs in 2% to 4% of human male newborns and is usually surgically corrected. Endocrine disruptors have been proposed to be one of several causes of cryptorchidism, although no particular chemical exposure is known to cause the condition.

TABLE 21.3

Development of reproductive structures. Development is color coded, using the same color for adult structures that develop from a common embryonic structure.

Embryonic Structure	Structure in Adult Female	Structure in Adult Male
Gonads 	Ovaries 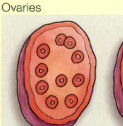 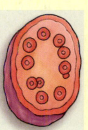	Testes 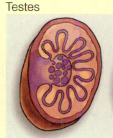 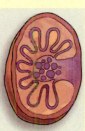
Gamete-carrying ducts 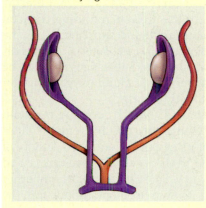	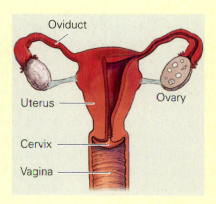 Oviduct, Uterus, Ovary, Cervix, Vagina	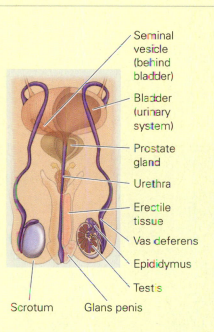 Seminal vesicle (behind bladder), Bladder (urinary system), Prostate gland, Urethra, Erectile tissue, Vas deferens, Epididymus, Testis, Scrotum, Glans penis
External genitalia 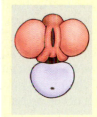	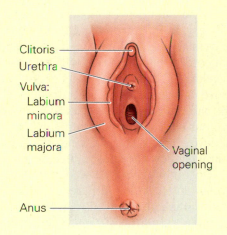 Clitoris, Urethra, Vulva: Labium minora, Labium majora, Vaginal opening, Anus	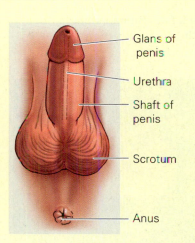 Glans of penis, Urethra, Shaft of penis, Scrotum, Anus

Deformities in reproductive or other structures in humans occur because certain chemicals have the ability to cross from the mother's bloodstream to the developing offspring during pregnancy. This is why pregnant women must be very careful to limit their exposure to toxins.

Pregnancy

Pregnancy or gestation involves carrying a developing baby within the female reproductive tract. Gestation encompasses the period of time between fertilization and childbirth; in humans, this lasts about 38 weeks.

To sustain a pregnancy, the placenta must form inside the uterus. The placenta is produced by the cooperative actions of the embryo and the mother's uterus. By the time the embryo has made its way to the uterus, it is composed of the inner cell mass, which will become the fetus, and an outer ring of cells called the **trophoblast**, which becomes part of the placenta. Cells of the trophoblast secrete enzymes that enable the embryo to implant in the wall of the uterus.

Eventually, the implanted trophoblast begins to infiltrate the uterine lining, forming fingerlike projections that are able to carry blood. Maternal uterine blood vessels erode to form pools of blood in the area surrounding the fetal blood. This close positioning of fetal and maternal blood supplies allows the exchange of nutrients and wastes to occur. Blood cells and bacteria do not normally pass between fetal and maternal blood supplies. Substances that can be freely exchanged between the fetus and the mother include oxygen, carbon dioxide, water, salts, hormones, viruses, and many drugs.

The cells of the early blastocyst synthesize HCG, which serves to maintain the corpus luteum. Recall that the corpus luteum is the ruptured, empty follicle remaining in the ovary after the ovulation of its egg cell. Continued activity of the corpus luteum results in a steady increase in the levels of progesterone and estrogen during the first 5 weeks of pregnancy. In humans and many other mammals, once the placenta has established its own hormone production, the corpus luteum is no longer necessary. Consequently, HCG is secreted only during the first months of pregnancy.

An additional hormone synthesized by the placenta is called **somatomammotropin**, which works with estrogen, progesterone, and another hormone, **prolactin**, to stimulate development of the mother's mammary glands. Near term, the mother's estrogen level drops, and prolactin is produced by the pituitary gland of the brain. Prolactin continues to stimulate the production and secretion of breast milk during breastfeeding or **lactation**.

Figure 21.12 Human embryo at 1 month.

From the time the embryo implants itself in the lining of the uterus until delivery, it undergoes remarkable growth and development. A month-old embryo is about 7 millimeters long and is beginning to develop a brain and spinal cord. It has buds that will give rise to limbs (**Figure 21.12**). A 9-week-old developing fetus has all of its organs and limbs in place (**Figure 21.13**). It can move its arms and legs and turn its head. For the remainder of the pregnancy, the baby increases in size; its features become more refined, with developments such as eyebrows, eyelashes, fingernails, and toenails. During the last stages of development, the fetal circulatory and respiratory systems undergo changes that will allow the fetus to breathe air.

During pregnancy a woman must be very careful about exposing a developing fetus to many different chemicals. It has been clearly established that environmental chemicals can have negative effects on fetal development. For example, exposure to secondhand cigarette smoke is known to cause decreased birth weight, exposure to mercury can affect brain development, and exposure to alcohol can cause severe developmental problems. Exposure to endocrine disruptors may also have negative effects.

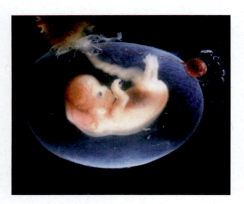

Figure 21.13 Fetus at 9 weeks.

Endocrine Disruptors and Pregnancy. Polychlorinated biphenyls (PCBs) are pollutants that once were used in hundreds of products, including electrical and hydraulic equipment, plastic and rubber products, and paints and dyes. Concern over the toxicity and persistence of PCBs in the environment led Congress in 1976 to prohibit their manufacture and distribution. Prior to this ban, billions of pounds of PCBs were manufactured in the United States.

Fetal exposure to PCBs is correlated with low birth weight and premature birth. Low-birth-weight babies weigh less than 2500 grams (5 pounds, 8 ounces) at birth and are at increased risk of respiratory and heart problems at birth as well as of cerebral palsy and learning problems as they grow. Low birth weights are more likely when childbirth occurs prematurely—before development of the fetus is complete.

Childbirth

Giving birth involves both labor and delivery. **Labor** involves strong, rhythmic contractions of the uterus that result in the birth of the baby. Induction of labor is hormonally controlled (**Figure 21.14**). Maternal estrogen levels increase toward the end of pregnancy, triggering the formation of receptors on the uterus for a hormone called **oxytocin** that serves to stimulate the smooth muscles lining the wall of the uterus to contract. Oxytocin is produced by fetal cells and by the mother's pituitary gland. During the induction of labor, oxytocin and estrogen are regulated by positive feedback. Oxytocin causes uterine contractions. Uterine contractions stimulate the release of more oxytocin, resulting in progressively intensifying muscle contractions that ultimately expel the baby from the uterus. Oxytocin also acts on the brain to increase feelings of pleasure, generosity, and bonding.

Uterine contractions normally signal the beginning of labor. Passage of the thick mucus plug blocking the cervix during pregnancy is another sign that labor is imminent. When the cervix begins to dilate, the plug loosens and is passed through the vagina. Beginning of labor can also be signaled by the rupture of a fluid-filled sac, called the **amnion**, that surrounds and protects the fetus. This is followed by the loss of fluid, occurring as either a gush of fluid or a slow trickle. It is often referred to as a woman's "water breaking" prior to delivery.

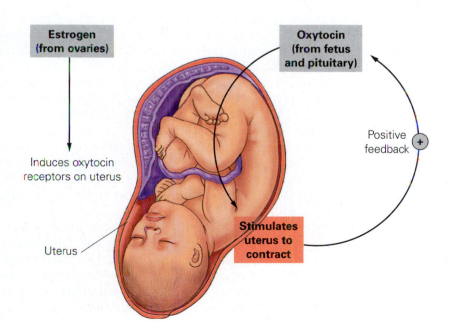

Estrogen
(from ovaries)

Oxytocin
(from fetus
and pituitary)

Induces oxytocin
receptors on uterus

Uterus

Stimulates
uterus to
contract

Positive
feedback   +

Figure 21.14 Induction of labor. Estrogen and oxytocin are involved in a feedback loop that stimulates labor. **Visualize This:** To artificially induce labor in their pregnant patients, physicians will provide an oxytocin-like drug. Where does this fit into in the feedback loop illustrated here?

Stage 1: Dilation

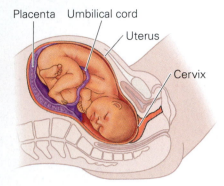

Stage 2: Expulsion

Stage 3: Delivery of placenta

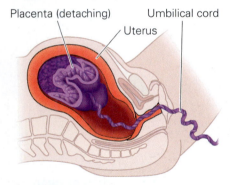

Figure 21.15 The three stages of labor. During labor, (1) the cervix dilates, (2) the baby is expelled, and (3) the placenta is delivered.

Stages of Labor. Labor can be thought of as occurring in three stages (**Figure 21.15**). During the first stage of labor, the cervical opening enlarges or dilates, going from a tiny, pencil-lead-sized opening to 10 centimeters. Contractions of the uterine muscle tissues help to dilate the cervix because they apply pressure to the baby. The baby is forced against the cervical opening, which forces the opening to enlarge. The second stage of labor involves delivery of the baby through the narrow pelvic opening and birth canal, often aided by the mother's active "pushing" (contractions of the abdominal muscles); the third stage of labor is delivery of the placenta.

As any new parent knows, concern about exposure to toxic chemicals does not end when the baby is born. Newborns must be protected against toxic exposures as well.

Endocrine Disruptors and the Newborn. Concern about exposure to endocrine disruptors in newborns has again focused on plasticizers. Unease about materials leaching out of plastics has led the Consumer Product Safety Commission to recommend that parents dispose of any soft vinyl baby products such as pacifiers and teething rings that contain the plasticizer DEHP. Even though DEHP has not definitively been shown to act as an endocrine disruptor in humans, negative effects of DEHP exposure have not been ruled out. Parents have also been cautioned about feeding babies from plastic bottles and feeding cups containing the chemical bisphenol A (BPA). It has been shown that this chemical can leach out of hard plastic bottles and be ingested. While toxicological studies on humans ingesting low levels of BPA have so far shown the chemical to be safe, newer studies, able to test for more subtle effects, have led to concern about the effects of BPA on the brains and behavioral development of infants and young children.

Babies fed infant formula are also exposed to significant quantities of tap water. This brings us back to our original questions about endocrine disruptors in the water supply.

21.4 Is the Water Safe to Drink?

At first glance, the issue of endocrine-disrupting chemicals in the water supply seems quite disturbing, but how concerned do we really need to be? What is being done to monitor our tap water?

The U.S. Environmental Protection Agency (EPA) is the governmental organization charged with regulating the amount of various chemicals, including known endocrine disruptors, present in drinking water. City water suppliers collect water samples every 3 months and analyze contaminant levels. If the level of harmful chemicals is too high, steps are taken to lower the contaminant level, and the water supplier must immediately notify the public. Since there are regulations in place to protect the water supply, the more pertinent question becomes whether the EPA-allowed levels are low enough to prevent reproductive problems in humans. This we do not yet know.

What we do know is that wildlife are affected by endocrine disruptors found in aquatic environments. We also know from laboratory studies of animals that exposure to endocrine-disrupting chemicals can have negative consequences for reproductive health, and that animal studies can be predictive of what might happen in humans.

As the DES and TCE data make clear, the human body can mistake a synthetic compound for a naturally occurring hormone, and this is harmful to both female and male reproductive health. However, no study to date has been able to show that endocrine disruptors in the water supply have caused human reproductive problems. All studies to date have been correlative. Abnormal sperm, decreased ability of sperm to fertilize eggs, and cryptorchidism are correlated with endocrine disrupter exposure, as is endometriosis. Correlation does not equal causation (Chapter 1). For instance, there is a high correlation between milk consumption and the number of teachers living in a particular geographic area. If we relied on correlation only, we would assume that teachers drink more milk than do people with other occupations. In reality, another factor is causing the increase in milk consumption. Where there are many teachers, there are also many students and consequently high levels of milk consumption. The correlation between endocrine-disrupting chemicals and reproductive health could also result from a shared factor rather than demonstrating a cause-and-effect relationship. For instance, men in agricultural areas may be exposed to more atrazine than urban men are, but they may also have jobs with a greater component of manual labor. It may be that the consistently higher body temperatures experienced by manual laborers interfere with sperm production, and that atrazine exposure has little effect.

While there is no conclusive evidence that endocrine disruptors in the water supply are harming human reproductive health, neither is there any reason to disregard this hypothesis. Further research is necessary before the role of endocrine disruptors in human reproductive health can be firmly established.

SAVVY READER

Endocrine Disruption and Early Puberty

Much Too Soon; Disorder Seems to be Striking More Preschoolers

by Jim Ritter

Before she turned 4, Xaviera Stanciel-Ohora already had begun to develop breasts. Her mother, Tarissa Stanciel, recalls thinking: "This is really weird." Xaviera was experiencing "precocious puberty," defined as puberty that starts before age 8 in girls and age 9 in boys. It's more common in girls. The signs are the same as in normal puberty: increased growth and pubic and underarm hair in boys and girls, breast growth and menstruation in girls, and deepening voice and genital growth in boys. Precocious puberty sometimes is caused by hormone disorders, tumors, brain abnormalities, etc. But in many cases, including Xaviera's, there's no known cause.

Between 1 in 5,000 and 1 in 10,000 kids experience precocious puberty, according to Oak Park-based Magic Foundation, a growth disorders support group. The incidence might be increasing because of the childhood obesity epidemic and chemical pollution. A 2001 study in the journal *Pediatrics* found that girls who had precocious puberty were significantly more overweight than other girls. This link was stronger for white girls than black girls.

Pesticides and other chemicals in the environment also might be triggering precocious puberty, said Xaviera's endocrinologist, Dr. Richard Levy of Rush University Medical Center.

1. The title of this article implies an increased rate of precocious puberty in preschoolers. Is this assertion supported by data in the article?

2. Is there any evidence presented that correlates increases in pesticide exposure to increased rates of precocious puberty?

3. If you were interested in finding out more about the link between endocrine disruptors and precocious puberty, what could you do?

4. Does the absence of data showing a direct link between pesticide exposure and precocious puberty mean that scientists should reject this hypothesis?

Chicago-Sun-Times, October 29, 2007.

Chapter Review
Learning Outcomes

Go to the Study Area at www.masteringbiology.com for practice quizzes, myeBook, BioFlix™ 3-D animations, MP3Tutor sessions, videos, current events, and more.

L01 Contrast asexual and sexual reproduction (Section 21.1).

- Asexual reproduction is a method of reproduction by which one parent produces offspring that are genetically identical to the parent. Large numbers of offspring can be reproduced quickly during asexual reproduction, but the absence of any genetic diversity makes the offspring less able to adapt to changes in the environment (pp. 514–515).
- Sexual reproduction requires a mating between two parents. Males and females produce gametes in structures called gonads. Gametes unite at fertilization to produce genetically distinct offspring (p. 515).

L02 List the male reproductive structures and their functions (Section 21.2).

- The penis of the male reproductive system is involved in sperm delivery; the urethra delivers both sperm and urine, the scrotum houses the testes, and the testes produce sperm and androgens. The seminal vesicles, prostate, and bulbourethral glands add secretions to sperm that help them develop and provide a source of energy. Semen is composed of the secretions from these glands combined with sperm (pp. 518–519).

L03 List the female reproductive structures and their functions (Section 21.2).

- The female reproductive system consists of the external vulva and clitoris, the internal vaginal passageway, the cervix at the base of the uterus that opens during childbirth, the uterus that houses the developing baby, oviducts for the passage of gametes, and ovaries that produce egg cells and hormones (pp. 519–520).

L04 Compare the processes of spermatogenesis and oogenesis (Section 21.2).

- Spermatogenesis occurs in the seminiferous tubules of the testes and begins at puberty. Mature sperm are composed of a small head containing the DNA, a midpiece that has mitochondria to provide energy for the journey to the oviduct, and a flagellum. An acrosome at the sperm's head contains enzymes that help the sperm gain access to the egg (pp. 523–524).
- Oogenesis occurs in the ovaries and results in the production of egg cells. This process begins while the female is still in her mother's uterus, then pauses until puberty. At puberty, development of preexisting eggs continues each month until menopause. The primary follicle develops into the secondary follicle and then the Graafian follicle. The secondary oocyte

is released from the Graafian follicle at ovulation. The leftover corpus luteum secretes progesterone for a while and then regresses if a pregnancy has not occurred (pp. 524–526).

L05 Describe the events that occur during the menstrual cycle and how they are hormonally regulated (Section 21.2).

- During the menstrual cycle, levels of estrogen, produced by the follicle, begin to rise during menstruation, eventually stimulating GnRH release and causing a spike in FSH and LH levels. LH causes ovulation to occur (pp. 526–528).
- The remnant of the follicle that stays in the ovary after ovulation is the corpus luteum, which secretes progesterone and estrogen. If fertilization does not occur, the corpus luteum degenerates, and progesterone and estrogen levels fall. Menstruation occurs. Then LH and FSH levels rise, and the cycle starts over. If fertilization has occurred, the embryo implants itself within the uterus and secretes HCG, which extends the life of the corpus luteum. Progesterone and estrogen levels remain high. Eventually, the placenta begins to produce these hormones (pp. 527–528).

L06 Outline the paths sperm and egg must follow for fertilization to occur (Section 21.3).

- To fertilize an egg cell, sperm slide through the follicular cells that surround the egg to reach the zona pellucida. Binding of the sperm head to the zona pellucida triggers the release of acrosomal enzymes, which break down the zona pellucida and allow the sperm to reach the egg cell's plasma membrane. After fusion of the egg and sperm plasma membranes, the sperm nucleus enters the egg cell, and the zygote can begin to develop (pp. 532–533).

L07 Describe the stages of early development from the rapid, early divisions through early tissue differentiation (Section 21.3).

- The zygote undergoes rapid cleavage divisions to produce the blastocyst. The inner cell mass of the blastocyst becomes the embryo and eventually the fetus. The blastocyst undergoes many divisions to produce a three-layered gastrula. Each layer of the gastrula (ectoderm, mesoderm, and endoderm) differentiates into specific tissues (pp. 533–534).

L08 Compare the events leading to differentiation of male and female reproductive organs (Section 21.3).

- Embryonic gonads become either testes or ovaries. Ductal structures exist side by side in male and female embryos. In each sex, one structure regresses. Male and female external genitalia are fashioned from the same embryonic material. The expression of sex-specific genes and hormones determines whether a particular embryo will become male or female (p. 534).

LO9 Outline the hormonal controls regulating pregnancy and birthing (Section 21.3).

- Pregnancy hormones are produced by the placenta, which forms in the uterus from the trophoblast of the blastocyst. These hormones help maintain the uterus, prevent ovulation, and prepare the breasts for lactation. Childbirth involves both labor and delivery. Labor involves contractions that help expel the baby from the uterus. During labor, the cervix dilates to allow the baby's passage out of the uterus. After her child's birth, the mother delivers the placenta (pp. 536–538).

Roots to Remember (MB)

The following roots of words come mainly from Latin and Greek and will help you decipher terms:

-genesis	means generation of or birth of. Chapter terms: oogenesis and spermatogenesis
gon-, gono-, and **gonado-**	mean seed or generation. Chapter term: gonad
oo-	is from the Greek word for egg. Chapter term: oocyte

Learning the Basics

1. **LO1** In what ways do asexual and sexual reproduction differ?

2. **LO2 LO3** Describe the male and female reproductive structures.

3. **LO4** Describe gametogenesis in males and females.

4. **LO6** A sperm cell follows which path?

   A. seminiferous tubules, epididymis, vas deferens, urethra; B. urethra, vas deferens, seminiferous tubules, epididymis; C. seminiferous tubules, vas deferens, epididymis, urethra; D. epididymis, seminiferous tubules, vas deferens, urethra; E. epididymis, vas deferens, seminiferous tubules, urethra

5. **LO6** An egg cell that is not fertilized follows which path?

   A. ovary, oviduct, uterus, cervix, vagina; B. ovary, uterus, oviduct, cervix, vagina; C. oviduct, ovary, cervix, uterus, vagina; D. oviduct, ovary, uterus, cervix, vagina; E. ovary, oviduct, cervix, uterus, vagina

6. **LO2** Which of the following is mismatched?

   A. urethra: sperm passage; B. Leydig cells: androgen production; C. vas deferens: semen production; D. seminiferous tubules: sperm production

7. **LO4** Gametogenesis _____.

   A. begins at puberty in males and females; B. requires that the Leydig cells of males produce semen; C. results in the production of diploid cells from haploid cells; D. begins at puberty in females; E. requires that meiosis halve the chromosome number so that sperm and eggs carry half the number of chromosomes

8. **LO5** The release of the oocyte from the follicle is caused by _____.

   A. a decrease in estrogen; B. an increase in FSH; C. an increase in LH; D. an increase in progesterone

9. **LO5** Menstruation is regulated such that _____.

   A. increasing estrogen levels have a positive feedback effect on FSH and LH; B. increasing FSH levels lead to ovulation; C. as progesterone levels increase, so do FSH and LH levels; D. ovulation occurs on the fifth day of the cycle; E. the placenta produces FSH, which stimulates ovulation

10. **LO6** The path of a sperm during fertilization is _____.

    A. follicle cells, egg plasma membrane, zona pellucida; B. egg plasma membrane, zona pellucida, follicle cells; C. egg plasma membrane, follicle cells, zona pellucida; D. zona pellucida, follicle cells, egg plasma membrane; E. follicle cells, zona pellucida, egg plasma membrane

11. **LO8** Development of female reproductive structures _____.

    A. requires the SRY gene; B. requires that estrogen is secreted by the developing fetus; C. will occur only when no testosterone is present; D. occurs when the ovaries secrete hormones that stop testicular development

Analyzing and Applying the Basics

1. **LO5** Why does the continuous dose of estrogen provided by the birth control pill prevent ovulation?

2. **LO7** Why must women planning to become pregnant be careful about exposures to toxins?

3. **LO9** Is childbirth regulated by positive or negative feedback? Justify your answer.

Connecting the Science

1. Most birth control methods are designed for and marketed to women. Why do you think this is the case? What could be the advantages of creating more birth control methods for men?

2. Why should pregnant women be very careful about limiting their exposures to toxins?

Answers to **Stop & Stretch, Visualize This, Savvy Reader,** and **Chapter Review** questions can be found in the **Answers** section at the back of the book.

Attention Deficit Disorder

Brain Structure and Function

Many college students are taking the prescription drugs Ritalin™, Dexedrine™, and Adderall™ to help control attention deficit disorder.

Drugs that affect the nervous system are some of the most common legal—and illegal—drugs used by college students. Drugs like Ritalin™, Dexedrine™, and Adderall™ are legally used by students diagnosed with attention deficit disorder (ADD) and its counterpart attention deficient/hyperactivity disorder. The ADD diagnosis is typically given to people who consistently display a suite of behaviors including forgetfulness, impulsivity, distractibility, fidgeting, impatience, and restlessness. It is estimated that around 5% of college students have been diagnosed with ADD.

Some college students who have not been prescribed the drug use it illegally to pull all-night study sessions . . .

Sometimes these prescription drugs find their way into the hands of students who do not have ADD. Such illegal use by nondiagnosed students includes using someone else's medication as a study aid, weight loss aid, or a way to get high. The drugs used to treat ADD are stimulants, and all stimulants, whether prescription or street drugs like cocaine and speed, have similar effects on the body. On their way to the brain, stimulants affect the heart, blood vessels, and lungs by increasing heart rate, elevating blood pressure, and expanding airways. Once in the brain, these drugs can cause euphoria, increased energy and endurance, and a feeling of mental sharpness. Studies indicated that close to 10% of U.S. college students have used someone else's ADD prescription stimulant.

. . . or to help them stay up later.

Many students think that these prescription drugs can't hurt them, but addiction to and overdose of prescription drugs is possible. Research indicated that people with ADD are not likely to become addicted to these medications when taken in the form and dosage prescribed. However, when misused, these drugs can be addictive, and lethal overdose can occur.

How do these drugs exert their effects? To understand this, we must first learn about the biology of the human nervous system.

Abuse of prescription drugs can lead to overdose, which seems to have been the cause of Heath Ledger's death in 2008.

22.1 The Nervous System

Every second, millions of signals make their way to your brain and inform it about what your body is doing and feeling. Your **nervous system** receives and interprets these messages and decides how to respond. Receiving and responding to stimuli require the actions of specialized cells called **neurons.** Neurons, or nerve cells, carry electrical and chemical messages back and forth between your brain and other parts of the body. These electrochemical message-carrying cells can be bundled together, producing structures called **nerves.**

Neurons form an extensive network throughout your body. They are highly concentrated in your brain and spinal cord, which together form the **central nervous system (CNS).** Other neurons in sense organs such as your eyes, ears, or skin transmit information from the world around you to your CNS. There the information is processed, and the appropriate response is relayed back to your body by yet other neurons out of the CNS (**Figure 22.1**).

To carry information between parts of the body, the cells of the nervous system pass electrical and chemical signals to each other. Signals are transmitted from one end of a neuron to the other end, between neurons, and from neurons to the cells of tissues, organs, or glands that respond to nerve signals. These responsive tissues are called **effectors.** Information is typically carried along nerves by electrical changes called **nerve impulses** and transmitted between neurons and from neurons to the cells they act on by changes in ion concentration or by chemicals, called **neurotransmitters,** released from nerve cells.

Neurons can be grouped into three general categories (**Figure 22.2**): (1) **sensory neurons,** which carry information toward the CNS; (2) **motor neurons,** which carry information away from the CNS toward effector tissues; and (3) **interneurons,** which are located between sensory and motor neurons within the brain or spinal cord. Most actions involve input from all three sources; sensory activity is followed by integration and motor output.

Sensory input is detected by **sensory receptors.** Receptors are usually neurons or other cells that communicate with sensory neurons. They detect changes in conditions inside or outside the body. When receptors are stimulated, signals are generated and carried to the brain.

The **general senses** are temperature, pain, touch, pressure, and body position, also called proprioception. The sensory receptors for the general senses are scattered throughout the body. The **special senses** are smell, taste, equilibrium, hearing, and vision. The sensory receptors for these five special senses are found in complex sense organs in the head (**Table 22.1**).

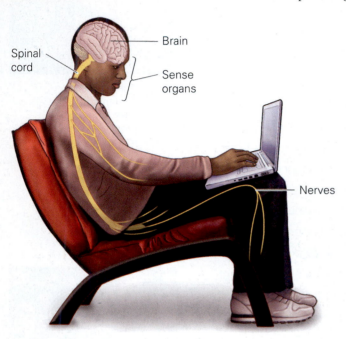

Brain
Spinal cord
Sense organs
Nerves

Figure 22.1 The nervous system. The nervous system includes the brain, spinal cord, sense organs, and the nerves that interconnect those organs.

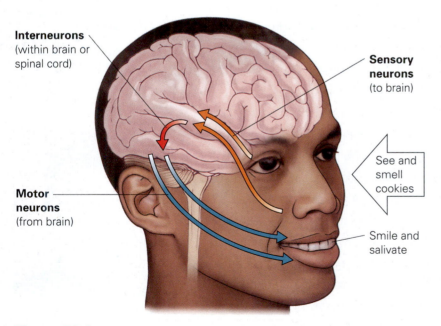

Interneurons (within brain or spinal cord)
Sensory neurons (to brain)
See and smell cookies
Motor neurons (from brain)
Smile and salivate

Figure 22.2 Neurons. The nervous system's three main functions are sensory input, integration, and motor output. Most actions, such as being tempted to eat cookies, involve input from sensory neurons, followed by integration via interneurons and motor output from motor neurons.

TABLE 22.1

The senses. There are two groups of senses: general and special. Both groups are illustrated here, along with their locations and functions.

Sense and Its Location	Sensory Receptors	Description
General senses—throughout body		
Temperature, pain, pressure, touch 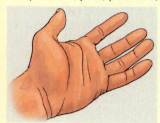	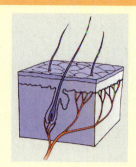	Sensory nerve endings in the skin react to temperature, pain, pressure, and touch. These receptors respond to different levels of sensation (e.g., slight cold vs. freezing cold).
Proprioception 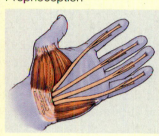	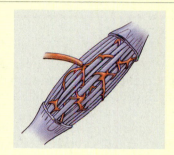	Specialized neurons sense joint position, tension in joints and ligaments, and muscle contractions. These neurons relay messages to the brain, along with balance information from the inner ear.
Special senses—found in complex sense organs		
Smell 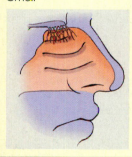	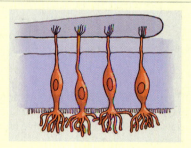	Olfactory (smell) organs with specialized receptors are located within the nasal cavity. Chemicals stimulate the receptors on hair-like projections called cilia. The receptors then send nerve impulses to the brain.
Taste 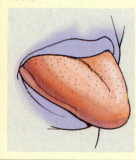	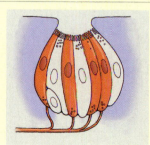	Taste buds on the surface of the tongue are specialized to respond to one of five primary taste sensations: sweet, salty, sour, bitter, or umami (savory or meaty).
Vision 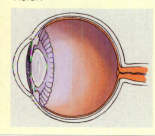	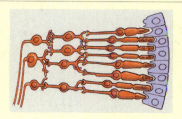	Neuron receptors in the retina of the eye allow for sight and relay visual information to the brain.

(continued)

TABLE 22.1 *(continued)*

The senses. There are two groups of senses: general and special. Both groups are illustrated here, along with their locations and functions.

	Sense and Its Location	Sensory Receptors	Description
Special senses—found in complex sense organs	Equilibrium and hearing		Equilibrium and hearing receptors are located on cilia in the inner ear. Movement of cilia help determine the position of the body in space, monitoring gravity, acceleration, and rotation. Cilia also detect and interpret sound waves.

As sensory information is passed to your brain, it travels through the main nerve pathway, the **spinal cord.** Your spine, which protects your spinal cord from injury, is made up of bones called **vertebrae.** Spinal nerves branch out between the vertebrae and go to every part of the body (**Figure 22.3**).

In addition to transmitting messages to and from the brain, the spinal cord serves as a reflex center. **Reflexes** are automatic responses to a stimulus. They are prewired in a circuit of neurons called a **reflex arc,** which often consists of a sensory neuron that receives information from a sensory receptor, an interneuron that passes the information along, and a motor neuron that sends a message to the muscle that needs to respond.

Figure 22.3 Spinal nerves and vertebrae. Spinal nerves branch out between the vertebrae and go to all parts of the body.

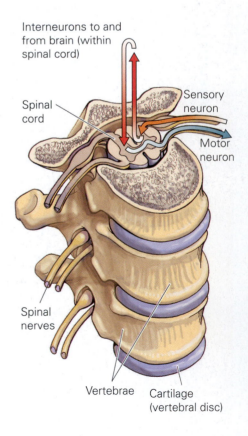

Interneurons to and from brain (within spinal cord)

Spinal cord

Sensory neuron

Motor neuron

Spinal nerves

Vertebrae

Cartilage (vertebral disc)

Reflexes allow a person to react quickly to dangerous stimuli; for instance, the withdrawal reflex occurs when you encounter a dangerous stimulus such as touching something hot (**Figure 22.4**). When you touch something hot, sensory neurons from touch receptors send the message to your spinal cord. Within the spinal cord, interneurons send the message to motor neurons to withdraw your hand from the hot surface. While the spinal reflexes are removing your hand from the source of the heat, pain messages are also being sent through your spinal cord to your brain.

Stop & Stretch When you touch something hot, you will often remove your hand before actually feeling the pain. Why might this be?

Even though these parts of the nervous system work together, they are separated into two different subdivisions. You have already learned that the CNS consists of your brain and spinal cord. The CNS is responsible for interpreting and acting on information received by the senses and is the seat of functions such as intelligence, learning, memory, and emotion. The second subdivision of the nervous system, the **peripheral nervous system (PNS),** includes the network of nerves that radiates out from your brain and spinal cord. Not all organisms have such complex nervous systems. Focus on Evolution outlines some evolutionary trends in the development of nervous systems from more primitive to more complex organisms.

Many scientists who study ADD believe that the brains of those affected differ from the brains of those who are not affected. If this is true, ADD medications may have different effects on people who do not have the disorder, something to consider for those who choose to take such drugs not prescribed for them. In the next section, we take a look at the structure and function of the brain and learn what science tells us about brain differences between people with and without ADD.

22.2 The Human Brain

The brain is the region of the body where decisions are reached and where bodily activities are directed and coordinated. The human brain is about the size of a small cantaloupe. Brains of the evolutionary ancestors to humans tend to be smaller and less complex (**Figure 22.5**). The brain is housed inside the skull, where it sits in a liquid bath, called **cerebrospinal fluid,** which protects and cushions it.

The functions of the brain are divided between the left and right hemispheres. Because many nerve fibers cross over each other to the opposite side, the brain's left hemisphere controls the right half of the body and vice versa. The areas that control speech, reading, and the ability to solve mathematical problems are located in the left hemisphere, while areas that govern spatial perceptions (the ability to understand shape and form) and the centers of musical and artistic creation reside in the right hemisphere.

In addition to housing 100–200 billion neurons, the brain is composed of other cells called **glial cells.** There are 10 times as many glial cells in the brain as there are neurons. In contrast to neurons, glial cells do not carry messages. Instead, they support the neurons by supplying nutrients, helping to repair the brain after injury, and attacking invading bacteria. Structurally, the brain is

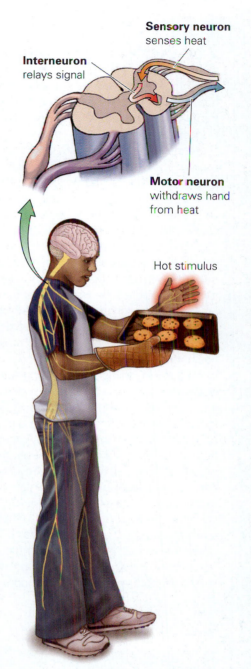

Figure 22.4 A reflex arc. A reflex arc can consist of a sensory receptor, a sensory neuron, an interneuron, a motor neuron, and an effector. Touching a hot baking sheet evokes the withdrawal reflex.

Sensory neuron senses heat

Interneuron relays signal

Motor neuron withdraws hand from heat

Hot stimulus

Figure 22.5 Evolution of larger brains. These skulls of early human ancestors show that the volume of the skull taken up by the brain has increased over the last 3–5 million years from about 400 to 1400 milliliters.

The Nervous System

A defining feature of the animal kingdom is the ability to move about the environment and to sense living prey. Animals perform these activities by using two tissues unique to their kingdom: muscles and nerves. Like muscle tissue, nervous tissue in all animals is remarkably similar in basic structure, but the arrangement of these tissues into systems differs in the various animal phyla due to their different lifestyles.

The beginnings of a nervous system are found in invertebrate organisms such as jellyfish, sea anemones, and hydras, all of which have nerve cells that are distributed throughout their bodies in nets that allow limited response to stimuli (**Figure E22.1**). The cells of the nerve net can communicate with each other. The ends of nerve nets form structures that secrete substances similar to neurotransmitters.

Communication among nerve cells can allow more complex movements to evolve. In some invertebrates, such as insects, the development of more complex nervous systems has permitted a wider range of behaviors. Concentration of nerves in the head region allows for greater complexity; insect brains, like human brains, are subdivided into different regions that control different functions. One region of the insect brain receives information from the eyes and other sense organs and controls complex behaviors. A second region controls antennae. The third region controls the mouthparts and digestive canal. In many of these invertebrates, longitudinal nerve cords provide a conduit for information that is transmitted to and from the extremities (**Figure E22.2**).

All vertebrates, at some point in their life cycle, develop a rod-shaped bar called a notochord. The notochord runs down the animal's back between the digestive tube and nerve cord. The notochord develops into the backbone of higher vertebrates, and the nerve cord becomes the brain and spinal cord. The notochord in higher vertebrates fuses with the bottom of the skull (**Figure E22.3**). The skull protects the sensitive nerve tissue of the brain, allowing for much larger brains to develop.

Grasshopper

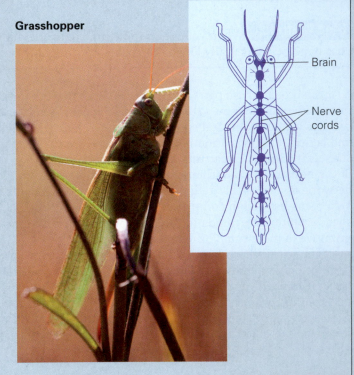

Brain

Nerve cords

Figure E22.2 The nervous systems of insects show a concentration of nerves in the brain.

Goose

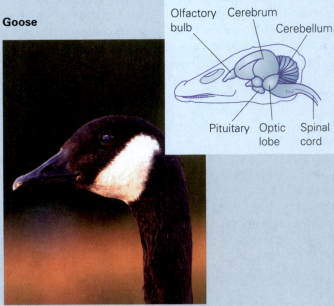

Olfactory bulb Cerebrum Cerebellum

Pituitary Optic lobe Spinal cord

Figure E22.3 Vertebrates have skulls that protect their brains and the highly specialized organs therein.

Hydra

Nerve net

Figure E22.1 The cells comprising the nerve net of hydra can communicate with each other.

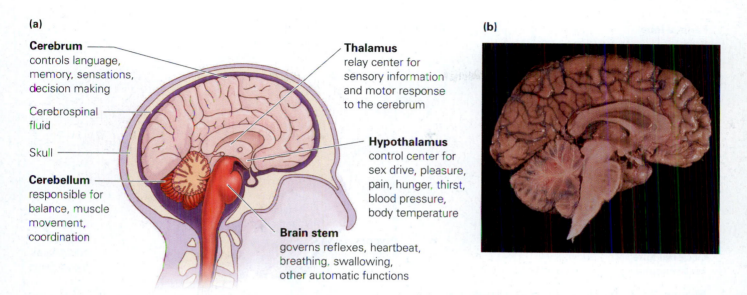

(a)

Cerebrum
controls language,
memory, sensations,
decision making

**Cerebrospinal
fluid**

Skull

Cerebellum
responsible for
balance, muscle
movement,
coordination

Thalamus
relay center for
sensory information
and motor response
to the cerebrum

Hypothalamus
control center for
sex drive, pleasure,
pain, hunger, thirst,
blood pressure,
body temperature

Brain stem
governs reflexes, heartbeat,
breathing, swallowing,
other automatic functions

(b)

Figure 22.6 Anatomy of the human brain. (a) The location and function of the cerebrum,
thalamus, hypothalamus, brain stem, and cerebellum are shown. (b) Photo of a human brain.
Visualize This: Find the locations of the cerebrum, thalamus, hypothalamus, brain stem,
and cerebellum in this photo.

subdivided into many important anatomical regions, including the cerebrum,
thalamus, hypothalamus, cerebellum, and brain stem (**Figure 22.6**).

Cerebrum

The **cerebrum** fills the whole upper part of the skull. This part of the brain
controls language, memory, sensations, and decision making. The cerebrum has
two hemispheres, each divided into four lobes (**Figure 22.7** on the next page):

1. The **parietal lobe** processes information about touch and is involved in
   self-awareness.
2. The **frontal lobe** processes voluntary muscle movements and is involved in
   planning and organizing future expressive behavior.
3. The **temporal lobe** is involved in processing auditory and olfactory information and is important in memory and emotion.
4. The **occipital lobe** processes visual information from the eyes.

The deeply wrinkled outer surface of the cerebrum is called the **cerebral
cortex.** In humans, the cerebral cortex, if unfolded, would be the size of a
16-inch pizza. The folding of the cortex increases the surface area and allows
this structure to fit inside the skull. The cortex contains areas for understanding and generating speech, areas that receive input from the eyes, and areas
that receive other sensory information from the body. It also contains areas that
allow planning.

The cerebrum and its cortex are divided from front to back into two
halves—the right and left cerebral hemispheres—by a deep groove or
fissure. At the base of this fissure lies a thick bundle of nerve fibers, the
corpus callosum, which functions as a communication link between the
hemispheres. The **caudate nuclei** are paired structures found deep within
each cerebral hemisphere. These structures function as part of the pathway
that coordinates movement.

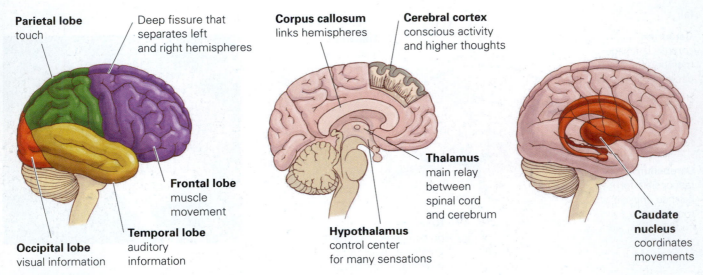

Parietal lobe
touch

Deep fissure that
separates left
and right hemispheres

Corpus callosum
links hemispheres

Cerebral cortex
conscious activity
and higher thoughts

Frontal lobe
muscle
movement

Thalamus
main relay
between
spinal cord
and cerebrum

**Caudate
nucleus**
coordinates
movements

Occipital lobe
visual information

Temporal lobe
auditory
information

Hypothalamus
control center
for many sensations

Figure 22.7 Structure of the cerebrum. The highly convoluted cerebral cortex is divided down the midline into left and right hemispheres, and each hemisphere is divided into four lobes: parietal, frontal, temporal, and occipital. The hemispheres are connected by the corpus callosum. Deep inside and between the two cerebral hemispheres are the thalamus and the hypothalamus. A caudate nucleus is located within each cerebral hemisphere.
Visualize This: Consider the functions of the parts of the cerebrum as depicted. Structural abnormalities of one of these structures is correlated with ADD. Which structure is the most likely candidate?

Thalamus and Hypothalamus

Deep inside the brain, lying between the two cerebral hemispheres, are the thalamus and the hypothalamus. The **thalamus** relays information between the spinal cord and the cerebrum. The thalamus is the first region of the brain to receive messages signaling such sensations as pain, pressure, and temperature. The thalamus suppresses some signals and enhances others, which are then relayed to the cerebrum. The cerebrum processes these messages and sends signals to the spinal cord and to neurons in muscles when action is necessary. The **hypothalamus,** located just under the thalamus and about the size of a kidney bean, is the control center for sex drive, pleasure, pain, hunger, thirst, blood pressure, and body temperature. The hypothalamus also releases hormones, including those that regulate the production of sperm and egg cells as well as the menstrual cycle.

Cerebellum

The **cerebellum** controls balance, muscle movement, and coordination (*cerebellum* means "little brain" in Latin). Since this brain region ensures that muscles contract and relax smoothly, damage to the cerebellum can result in rigidity and, in severe cases, jerky motions. The cerebellum looks like a smaller version of the cerebrum. It is tucked beneath the cerebral hemispheres and, like the cerebrum, has two hemispheres connected to each other by a thick band of nerves. Additional nerves connect the cerebellum to the rest of your brain.

Brain Stem

The **brain stem** lies below the thalamus and hypothalamus (**Figure 22.8**). It governs reflexes and some spontaneous functions, such as heartbeat, respiration, swallowing, and coughing.

Figure 22.8 The cerebellum and brain stem. The brain stem consists of the midbrain, pons, and medulla oblongata.

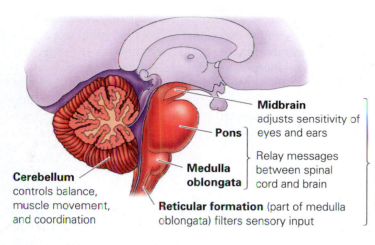

Figure 22.8 The cerebellum and brain stem. The brain stem consists of the midbrain, pons, and medulla oblongata.

Midbrain adjusts sensitivity of eyes and ears

Pons

Relay messages between spinal cord and brain

Medulla oblongata

Brain stem controls reflexes and spontaneous functions

Cerebellum controls balance, muscle movement, and coordination

Reticular formation (part of medulla oblongata) filters sensory input

The brain stem is composed of the midbrain, pons, and medulla oblongata. Highest on the brain stem is the **midbrain,** which adjusts the sensitivity of your eyes to light and of your ears to sound. Below the midbrain is the **pons** (*pons* means "bridge"). The pons functions as a bridge, allowing messages to travel between the brain and the spinal cord. The **medulla oblongata** is the lower part of the brain stem. It helps control heart rate and conveys information between the spinal cord and other parts of the brain.

The **reticular formation,** found in the medulla oblongata (the lower part of the brain stem), is an intricate network of neurons that radiates toward the cerebral cortex. The reticular formation functions as a filter for sensory input; it analyzes the constant onslaught of sensory information and filters out stimuli that require no response. This prevents the brain from having to react to repetitive, familiar stimuli such as the sound of automobile traffic outside your dorm room or the sound of your roommate talking while you are trying to sleep.

ADD and Brain Structure and Function

In the same way that X-rays help physicians check for broken bones, scientists have access to brain-scanning technologies that enable them to view the brain. Some **neurobiologists**—biologists who study the nervous system—use these technologies to find physical evidence in the brain structures of people diagnosed with ADD. Some neurobiologists have reported that they have found subtle differences in the structure and function of the brains of people with ADD.

In some studies, ADD was associated with decreased volume of the cerebral cortex as well as decreased surface area and folding throughout the cerebral cortex. In other studies, researchers found that the corpus callosum (the band of nerve fibers connecting the two hemispheres of the brain) was slightly smaller in people with ADD. A similar study discovered size differences in the caudate nuclei (involved in coordinating movement). The caudate nucleus in the right hemisphere of people with ADD was slightly smaller than it was in people who did not have the disorder. Other studies have shown that the cerebellum of affected persons tends to be smaller than those in unaffected people.

Stop & Stretch Would it be important to know whether persons diagnosed with ADD who participate in brain differences studies had been treated with medicines? Why or why not?

Because some scientific studies have shown differences in brain structure between those affected with ADD and those without it, you might come to the conclusion that people are simply born with ADD. However, brain differences

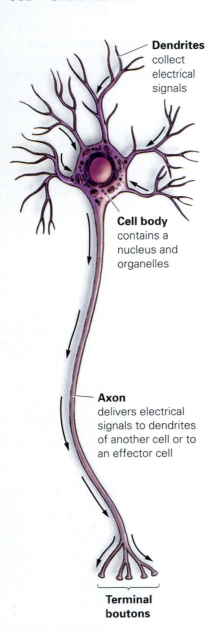

Dendrites
collect electrical signals

Cell body
contains a nucleus and organelles

Axon
delivers electrical signals to dendrites of another cell or to an effector cell

Terminal boutons

Figure 22.9 The structure of a generalized neuron. A neuron consists of the branching dendrites, the cell body, the axon, and terminal boutons. Nerve impulses are propagated in the direction of the arrows.

between people with ADD and people who do not have the diagnosis may have arisen because of each person's unique experiences. Likewise, brain differences may arise from some combination of these two factors, in much the same way that people increase strength in different muscles in response to exercise.

Differences in brain structure and function, whether present at birth or developed later in life, are closely related to how neurons work. By investigating the ways in which messages are transmitted by neurons, researchers have been able to propose additional hypotheses about the causes and effects of ADD.

22.3 Neurons

Neurons are highly specialized cells that usually do not divide. Therefore, damage to a neuron cannot be repaired by cell division and often results in permanent impairment. For example, damage to spinal motor neurons results in lifelong paralysis because messages can no longer be transmitted from the CNS to muscles. Likewise, injury to the brain can result in permanent brain damage if neurons in the brain are harmed.

Neuron Structure

Neurons are composed of branching **dendrites** that radiate from a bulging **cell body,** which houses the nucleus and organelles, and a long, wirelike **axon,** terminating in knobby structures called the **terminal boutons** (**Figure 22.9**). A nerve impulse usually travels down the axon of one neuron and is transmitted to the dendrites of another neuron.

How Neurons Work. The axons of many neurons are coated with a protective layer called the **myelin sheath** (**Figure 22.10a**). The myelin sheath is formed by supporting cells such as **Schwann cells.** The myelin sheath is composed mostly of lipids and has the same appearance as animal fat. Due to its white color, nervous tissue composed of myelinated cells is called **white matter.** The myelin sheath functions like insulation on a wire by preventing sideways message transmission, thus increasing the speed at which the electrochemical impulse travels down the axon.

Long stretches of the myelin sheath are separated by small interruptions in the Schwann cells, leaving tiny, unmyelinated patches called the **nodes of Ranvier.** Nerve impulses "jump" successively from one node of Ranvier to the next (**Figure 22.10b**). This jumping transmission of nerve impulses is up to 100 times faster than signal conduction would be on a completely unmyelinated axon—nerve impulses can travel along myelinated axons at a rate of 100 meters per second (more than 200 miles per hour).

Some neurons are not myelinated and thus transmit nerve impulses more slowly. Unmyelinated axons, combined with dendrites and the cell bodies of other neurons, look gray in cross sections and thus are called **gray matter.**

Neuron Function

Neurons transmit impulses from one part of the body to another. Many kinds of stimuli, including touch, sound, light, taste, temperature, and smell, cause neurons to respond. When you touch something, signals from touch sensors travel along sensory nerves from your skin, through your spinal cord, and into your brain. Your brain then sends out messages through your spinal cord to the motor neurons, telling your muscles how to respond. To evoke this response, nerve cells must transmit signals along their length and from one cell to the next.

(a) Axons covered in myelin sheaths   **(b) Nerve impulses travel more quickly on myelinated than on unmyelinated neurons.**

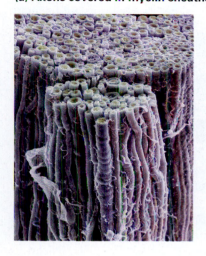

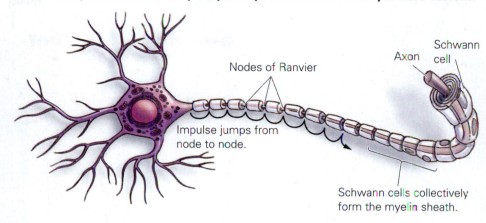

Nodes of Ranvier

Axon

Schwann cell

Impulse jumps from node to node.

Schwann cells collectively form the myelin sheath.

Action Potential.
Another name for a nerve impulse is an action potential. An action potential transmits a signal along the length of a neuron. An action potential is a brief reversal of the electrical charge across the membrane of a nerve cell. The action potential is propagated as a wave of electrical current down the length of the neuron.

In the resting neuron, a charge difference exists between the inside and outside of the cell. The resting neuron is negatively charged on the inside compared to the outside of the cell. Because opposite charges tend to move toward each other, the membrane serves as a source of stored energy, like a battery, by keeping these opposite charges apart. This difference in the charge across the membrane is called **polarization.** This polarization exists because the there is a high concentration of negatively charged ions inside the cell.

The membranes of neurons house a specialized protein pump, called the **sodium-potassium pump,** which uses the energy of ATP to move three sodium molecules out of the cell for every two potassium molecules returned back into the cell, thus setting up a concentration gradient (**Figure 22.11a** on the next page).

Stimulation of a neuron opens the gates of a few sodium channels in a small part of the membrane, shifting that particular region inside the cell toward a less negative state of polarization. If this local change in polarization reaches a critical level, then it stimulates nearby sodium channels to open as well. This causes sodium to flood into the cell, quickly eliminating the charge difference across the cell and causing an action potential. This decrease of charge difference, or **depolarization,** moves in a wave down the cell, activating sodium channels in adjacent parts of the membrane and thereby moving the impulse along the length of the neuron (**Figure 22.11b**).

The entry of sodium into the neuron must be balanced by the exit of another positively charged ion, or the action potential could not decline and return the neuron to its resting state. The action potential declines as sodium channels close and potassium channels open, allowing potassium out of the cell. **Repolarization** occurs when potassium ions diffuse out of the cell and the inside of the cell again becomes more negative than the outside. This occurs because potassium channels are "leaky" and allow potassium to leak out of the cell down its own concentration gradient, leaving the negatively charged ions behind (**Figure 22.11c**).

Synaptic Transmission.
Once the signal has traveled along the length of the axon, it must be passed to the next neuron or to the effector organ, muscle, or gland. Because most neurons are not directly connected to each other, the signal must be transmitted to the next neuron across a gap between the two neurons,

Figure 22.10 Myelination. (a) This photo shows a cut bundle of nerve fibers. Each separate nerve consists of an internal nerve cell axon (greenish) covered in an insulating layer of myelin (whitish). (b) By jumping from one node of Ranvier to the next, nerve impulses travel more quickly on myelinated nerves than on unmyelinated nerves.

(a) Resting nerve cell and polarization

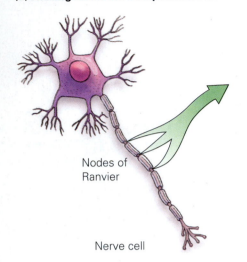

Nodes of Ranvier

Nerve cell

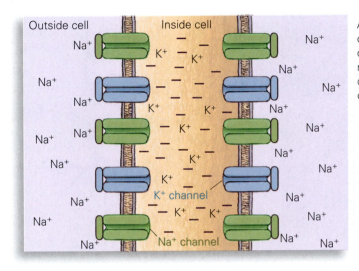

Outside cell   Inside cell

K+ channel

Na+ channel

All channels are closed. The inside of the cell has a more negative charge than the outside of the cell.

(b) Propagation of an action potential, or nerve impulse, for depolarization

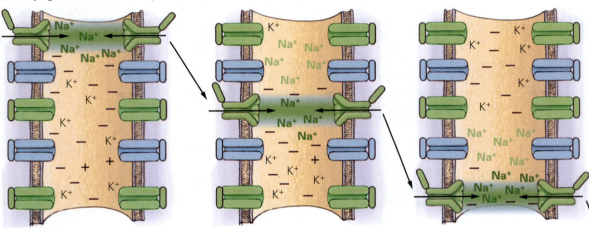

Sodium channels open sequentially, and sodium comes into the cell. This depolarizes the membrane. As the positively charged sodium is spread, the next sodium channel opens to let in more sodium.

(c) Decline of an action potential, or nerve impulse, for repolarization

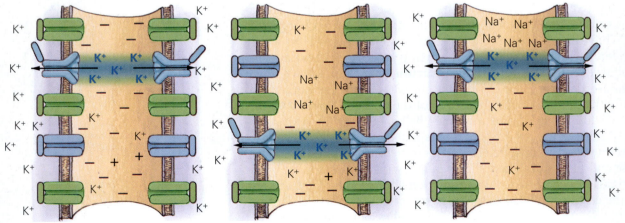

The diffusion of potassium ions allows repolarization of the cell.

Figure 22.11 Generating and propagating a nerve impulse. (a) In a resting nerve cell, the inside of the neuron is more negatively charged than the outside; the cell is polarized. Closed channels stop these ions from moving back across the membrane and help prevent the ions from diffusing back toward their chemical equilibrium. (b) A nerve impulse starts with an influx of positively charged sodium ions, which depolarizes the nerve cell. A domino effect occurs as the sodium ions spread their charges toward each successive voltage-sensitive channel, causing each to open and let in more sodium, thus propagating the depolarization along the length of the neuron. (c) Repolarization results from the diffusion of potassium ions out of the cell, which causes the inside of the cell to again become more negative than the outside.

Visualize This: Predict the effects of a mutation to a gene that prevents the sodium potassium pump from binding to sodium.

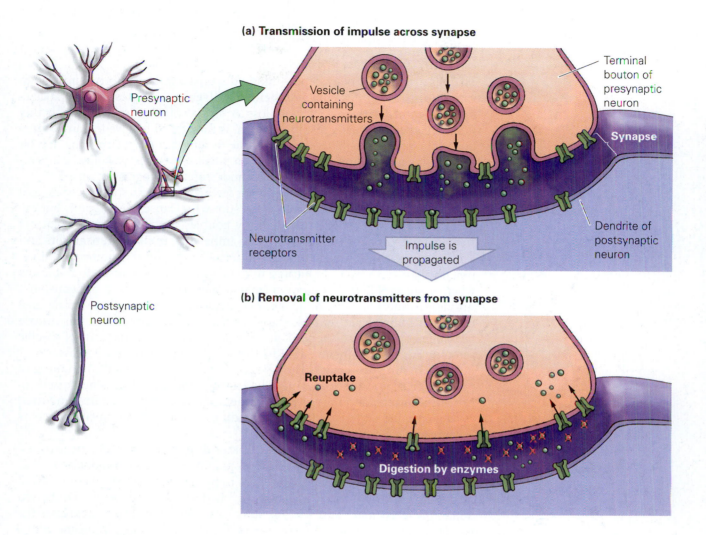

(a) Transmission of impulse across synapse

Presynaptic neuron

Vesicle containing neurotransmitters

Terminal bouton of presynaptic neuron

Synapse

Neurotransmitter receptors

Impulse is propagated

Dendrite of postsynaptic neuron

(b) Removal of neurotransmitters from synapse

Reuptake

Digestion by enzymes

Postsynaptic neuron

Figure 22.12 Propagating the nerve impulse between neurons.
(a) The nerve impulse is transmitted from the terminal bouton of one neuron to the dendrite of the next. Neurotransmitters are released from the presynaptic neuron, travel across the synapse to the postsynaptic neuron, and bind to receptors on dendrites of the postsynaptic neuron. (b) After the neurotransmitter evokes a response, it is removed from the synapse by enzymes present in the synapse that break the neurotransmitter apart or by reuptake via the presynaptic cell.

called the synapse (**Figure 22.12a**). The synapse consists of the terminal boutons of the presynaptic neuron, the space between the two adjacent neurons, and the cell body or dendrite of the postsynaptic neuron. Saclike structures called vesicles are found in the terminal boutons of the presynaptic neuron. Each vesicle in a particular neuron is filled with a specific chemical neurotransmitter. When an electrical impulse arrives at the terminal bouton of a nerve cell, neurotransmitters are released. Once released, they diffuse across the synapse and bind to specific receptors on the cell membrane of the postsynaptic neuron. The binding of a neurotransmitter to the postsynaptic cell membrane once again stimulates a rapid change in the uptake of sodium ions by the postsynaptic neuron. The sodium channels open, and sodium ions flow inward, causing another depolarization and generating another action potential. In this manner, the nerve impulse is propagated from one neuron to the next until the signal reaches the tissue that it will affect.

At least 30 known chemicals function as neurotransmitters. Examples of neurotransmitters include glutamate, which is an important CNS neurotransmitter, and acetylcholine, which is an important neurotransmitter in the muscle. Each type of neuron secretes one type of neurotransmitter, and each cell responding to it has receptors specific to that neurotransmitter.

After the neurotransmitter evokes a response, it is removed from the synapse. Some neurotransmitters are broken down by enzymes in the synapse; others are reabsorbed by the neuron that secreted them via a process called **reuptake** (**Figure 22.12b**). Reuptake occurs when neurotransmitter receptors on the presynaptic cell permit the neurotransmitter to reenter the cell. This allows the neuron that released the neurotransmitter to use it again. Both breakdown by enzymes and reuptake by receptors prevent continued stimulation of the postsynaptic cell.

Neurotransmission, Alzheimer's, Depression, Parkinson's, and ADD

There is a great deal of evidence linking disturbances in neurotransmission to various diseases. For example, **Alzheimer's disease,** a progressive mental deterioration in which there is memory loss along with the loss of control of bodily functions that eventually results in death, is thought to involve impaired function of the neurotransmitter **acetylcholine** in some neurons. Drugs that inhibit the enzyme **acetylcholinesterase,** which breaks down acetylcholine, can temporarily improve mental function but cannot stop this progressive illness.

Depression is a disease that involves feelings of helplessness, despair, and thoughts of suicide. No one knows for sure what causes depression, but experts believe that a genetic susceptibility, coupled with environmental factors such as stress or illness, may trigger an imbalance in neurotransmitters and lead to depression. Three neurotransmitters—serotonin, dopamine, and norepinephrine—have been implicated in depression, but scientists do not know whether changes in these neurotransmitters are a cause or result of depression. Antidepressants blocking the actions of enzymes that degrade these neurotransmitters or inhibit their reuptake help alleviate many symptoms of this disease in some people. Prozac® and Zoloft®, for instance, fall into the class of specific serotonin reuptake inhibitors (SSRIs) that are prescribed to treat depression.

Parkinson's disease is thought to be caused by the malfunctioning of neurons that produce dopamine. The loss of dopamine causes nerve cells to fire without regulation; as a result, patients have difficulty controlling their movements and experience tremors, rigidity, and slowed movements. There is no cure for this progressive disease.

Abnormal levels of the neurotransmitter **dopamine** may also be involved in producing the symptoms of ADD. Dopamine controls emotions as well as complex movements. Some researchers hypothesized that people with ADD symptoms may have lower dopamine levels than other people do. Decreased dopamine levels may cause ADD symptoms because dopamine suppresses the responsiveness of neurons to new inputs or stimuli. Therefore, someone with a low concentration of dopamine may respond impulsively to situations in which pausing to process the input would be more effective.

The cause of decreased dopamine levels in some people with ADD may actually involve an overabundance of dopamine receptors on the presynaptic cell. During reuptake, dopamine receptors on the presynaptic cell remove dopamine from the synapse to prevent continued stimulation of the postsynaptic neuron. Some studies have shown elevated numbers of these receptors in people with ADD.

Methylphenidate, or **Ritalin,** is one of the drugs prescribed to treat symptoms of ADD. Ritalin may work to decrease the impact of these extra reuptake receptors. Ritalin is thought to increase dopamine's ability to stimulate the postsynaptic cell by blocking the actions of the dopamine reuptake receptor (**Figure 22.13a**). This leaves more dopamine around for a longer time, resulting in a decrease in the symptoms of ADD.

Adderall and Dexedrine cause their effects in a slightly different manner (**Figure 22.13b**). These drugs belong to a class of stimulants called amphetamines and work by increasing the levels of dopamine and other neurotransmitters in the synapse. Amphetamines either diffuse or are transported into the neuron and cause the neuron to release any stored dopamine into the synapse. They can also bind to dopamine transporters to prevent reuptake.

Stop & Stretch A neurotoxin carried by pufferfish binds to sodium channels and prevents passage of ions. What effect would this neurotoxin have on the nervous system of someone who ingested it?

(a) The mechanism of Ritalin action

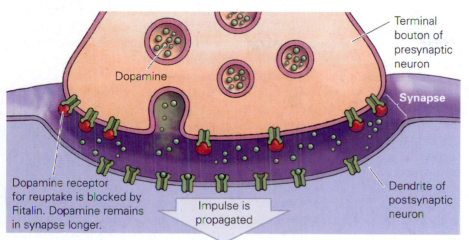

Terminal bouton of presynaptic neuron

Dopamine

Synapse

Dopamine receptor for reuptake is blocked by Ritalin. Dopamine remains in synapse longer.

Impulse is propagated

Dendrite of postsynaptic neuron

(b) Amphetamines like Adderall and Dexedrine cause dopamine to be released into the synapse.

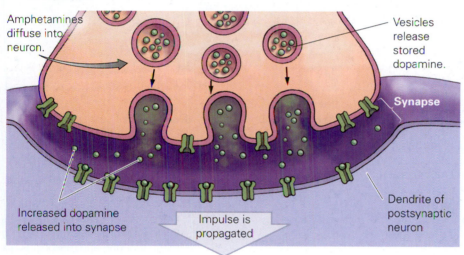

Amphetamines diffuse into neuron.

Vesicles release stored dopamine.

Synapse

Increased dopamine released into synapse

Impulse is propagated

Dendrite of postsynaptic neuron

Figure 22.13 Mechanism of ADD prescription drug action. Some ADD drugs like (a) Ritalin and (b) Adderall and Dexedrine work to increase the amount of dopamine in the synapse.

The dopamine idea is not the only hypothesis about the biological cause of ADD symptoms, and it does not explain all cases of ADD. Many times, changes in environmental conditions can help improve the symptoms of ADD.

The nervous system is a dynamic, changing system, and many different parts of the brain are responsive to environmental influences. There is some evidence of biological differences in the brains of people diagnosed with ADD, but there is also evidence that environmental factors may exacerbate or even cause ADD. In the last decade, the number of people diagnosed with ADD has risen from about 1 million to 6–11 million. Such a dramatic increase in a very short time is not likely due to biology alone. It would take thousands of years for evolution to affect this magnitude of biological change within a population.

As is often the case in science, the data on increased diagnosis of ADD are difficult to interpret, and there are many different ways to explain the same findings. Doctors may simply be getting better at diagnosing ADD cases, or they may be misdiagnosing overactivity as ADD. It also may be that societal changes are having a negative impact on children. This is an area of active research and debate involving a wide variety of people, including scientists, medical doctors, psychologists, teachers, and parents.

What about the problem of nonmedical use of these drugs? Due to their availability and status as prescription drugs—not street drugs, such as speed or other amphetamines—students may see these stimulants as less serious drugs and be unaware of the dangers of abusing them. However, like all drugs, stimulants have many unwanted affects. For these stimulants, side effects include heightened fatigue, insomnia, irritability, tearfulness, depression, hostility, paranoia, skin rashes, fever, nausea, and headaches. **Table 22.2** lists the ways in which other recreational drugs affect the nervous system.

Nonmedical users who snort or inhale these drugs are particularly at risk because they tend to use more of the drug than those who use it orally. Like most drugs, increased consumption not only increases the good feelings but also increases the crash. This is also more likely to lead to dependence, as users need to increase the amount ingested to obtain the same response. Abuse of stimulants can lead to psychotic episodes, delusions, seizures, hallucinations, and sudden death.

If you have not been prescribed these drugs but use them to help focus or study for an exam, there is no evidence that this drug use will actually help you learn. This is because you probably do not have low dopamine levels. For nonmedical users, the euphoria they experience, coupled with the fact that they are more likely to sit in a chair and focus on their schoolwork, may make them believe that these drugs are helping them learn.

TABLE 22.2

Recreational drugs and the nervous system.			
Drug	**Mechanism of Action**	**Desired Mental Effect**	**Side Effect**
Alcohol is a by-product of fermentation.	• A depressant that diffuses easily across cell membranes and does not require specific receptors • Inhibits neurotransmission and interferes with the activity of neurons throughout the brain	• Reduced anxiety and a sense of well-being, loss of concern for social constraints	• Impaired judgment, slurred speech, unsteady gait, slower reaction times, uncontrollable emotions • Chronic alcohol abuse damages nerve cells in frontal lobes, leading to impaired intellect, judgment, thought, and reasoning. Liver damage is also likely.
Amphetamines are used legally to treat obesity, asthma, and narcolepsy. Methamphetamine and ice are illegal amphetamines.	• Increase dopamine release • Block reuptake and inhibit the enzyme that breaks down norepinephrine, a hormone produced in response to stress	• Small doses make a person feel more energetic, alert, and confident.	• Effects wear off quickly, causing sudden "crashes" from depleted neurotransmitter stores and resulting in depression and fatigue. • Prolonged use results in aggressiveness, delusions, hallucinations, violent behaviors, and death.
Caffeine is a naturally occurring chemical found in plants such as coffee, tea, and cocoa.	• A general stimulant that affects all cells, not just those of the central nervous system • Increases cellular metabolic activities	• Mental alertness, increased energy	• Insomnia, anxiety, irritability, and increased heart rate

TABLE 22.2

Recreational drugs and the nervous system.

Drug	Mechanism of Action	Desired Mental Effect	Side Effect
Cocaine is extracted from the leaves of the coca plant of South America. 	• A stimulant that increases levels of dopamine and norepinephrine by decreasing reuptake	• A rush of intense pleasure, increased self-confidence, and increased physical vigor	• Increased heart rate and blood pressure, dilation of pupils, increased temperature • Crash involves deep depression, anxiety, and fatigue. • Abuse may change neurons to prevent experience of positive feelings without the drug.
Ecstasy or MDMA is a white crystalline powder that is primarily ingested in pill or capsule form. 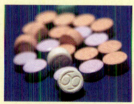	• A stimulant and hallucinogenic that acts to prevent serotonin reuptake • Also floods neurons with several other neurotransmitters	• Euphoria, enhanced emotional and mental clarity, increased energy, heightened sensitivity to touch, and enhanced sexual response	• Confusion, anxiety, paranoia, depression, and sleeplessness lasting several days • Permanent memory damage • When combined with physical exercise, can lead to severe dehydration and the inability to regulate body temperature
Lysergic acid diethylamide (LSD) and mushrooms: LSD is a derivative of the fungus *Claviceps purpurea*, which grows on rye. 	• Hallucinogenic that binds to serotonin receptors in the brain, increasing the normal response to serotonin	• Heightened sensory perception and bizarre changes in thought and emotion, hallucinations	• Hallucinations can lead users to dangerous actions. • Heavy use leads to permanent brain damage, including losses of memory and attention span, and can lead to psychosis.
Marijuana is derived from the *Cannabis sativa* plant and contains delta-9-tetrahydrocannabinol (THC). 	• Receptors for THC are in the areas of the brain that influence mood, pleasure, memory, pain, and appetite. • THC is thought to work by increasing dopamine release.	• Altered sense of time, enhanced feeling of closeness to others, increased sensitivity to stimuli • Large doses can cause hallucinations.	• Slowed reaction time and decreased coordination • Impaired short-term memory, slowed learning, decreased attention span, and lessened ability to store and acquire information • Damage to THC receptors on the hypothalamus can decrease testosterone production and disrupt menstruation.
Nicotine is found in tobacco plants. 	• Stimulant that triggers neurons of cerebral cortex to produce acetylcholine, epinephrine, and norepinephrine	• Increased alertness and awareness, appetite suppression, relaxation	• Increased odds of most cancers • Also causes increased heart rate and blood pressure

(continued)

TABLE 22.2 *(continued)*

Recreational drugs and the nervous system.

Drug	Mechanism of Action	Desired Mental Effect	Side Effect
Opiates—including heroin, morphine, and codeine—are derived from the opium poppy. 	• Bind to opiate receptors in neurons that control feelings of pleasure • Endorphins produced during exercise bind to opiate receptors also	• A quick, intense feeling of pleasure, followed by a sense of well-being and drowsiness	• Poor motor coordination, depression • High doses can cause coma and death. • Thought to change the ability of the brain to respond to normal pleasures

SAVVY READER

Prescription Drug Abuse

Non-medical Use of Prescription Stimulants Among US College Students: Prevalence and Correlates from a National Survey

The peer reviewed article cited above reports the results of a study on the illegal use of prescription attention deficit medications, such as Ritalin™ and Adderall™, by U.S. college students who have not been prescribed the medication. Surveys were mailed to over ten thousand randomly selected students at over 100 four year colleges in the United States. The scientists who performed this study report that around 7% of the college students surveyed had used such stimulants in the past. Statistical analysis showed that use was highest among white male fraternity members and that schools with more rigorous admissions standards had more students who used these drugs.

The authors of this article also present evidence that students who used these stimulants are more likely to use other legal and illegal drugs.

1. List several controls used by the authors of this peer-reviewed study.
2. Did the authors use statistics to analyze their data?
3. Did the authors speculate about anything other than the data they present?

McCabe SE, Knight JR, Teter CJ, Wechsler H. *Addiction 2005*, Jan; 100(1):96–106.

Chapter Review

Learning Outcomes

Masting**BIOLOGY**®

Go to the Study Area at www.masteringbiology.com for practice quizzes, myeBook, BioFlix™ 3-D animations, MP3Tutor sessions, videos, current events, and more.

L01 Describe the roles that sensory, motor, and interneurons play in the nervous system (Section 22.1).

- Neurons are specialized cells of the nervous system that carry chemical and electrical messages (p. 544).
- Sensory neurons carry information toward the brain; motor neurons carry information away from the brain (p. 544).
- Interneurons are located between sensory and motor neurons (p. 544).

L02 Describe the manner in which general senses relay information (Section 22.1).

- Sensory receptors for general senses, found throughout the body, relay information to the brain. Receptors for the special senses are found in more complex sense organs (pp. 544–546).

L03 Explain how reflexes work (Section 22.1).

- Reflexes are prewired responses that carry messages from sensory receptors to interneurons to a motor neuron for a response (pp. 546–547).

L04 List the components of the nervous system, and differentiate between the CNS and PNS (Section 22.1).

- The nervous system consists of the brain and spinal cord of the central nervous system as well as the nerves of the peripheral nervous system that carry information to and from the brain (pp. 544–547).

L05 Outline the structure and function of the cerebrum, thalamus, hypothalamus, cerebellum, and brain stem (Section 22.2).

- The cerebrum of the brain is where most thinking occurs. The two hemispheres of the cerebrum consist of four lobes—temporal, occipital, parietal, and frontal. The outer surface of the cerebrum is the cortex (pp. 549–550).
- The thalamus and hypothalamus are located between the two cerebral hemispheres. The thalamus relays information between the spinal cord and the cerebrum, and the hypothalamus regulates many vital bodily functions (p. 550).
- The cerebellum is located at the base of the brain, beneath the cerebral hemispheres. It regulates balance and coordination (pp. 550–551).
- The brain stem is located below the thalamus and hypothalamus. It controls many unconscious functions (pp. 550–551).

L06 Describe the structure of a neuron (Section 22.3).

- Neurons are specialized cells with a structure that consists of the branching dendrites, the cell body, the axon, and terminal boutons (p. 551).

L07 Differentiate between neurons comprising white and gray matter (Section 22.3).

- Axons that are covered in myelin conduct impulses faster and comprise the white matter (p. 552).
- Unmyelinated axons conduct impulses more slowly and comprise the gray matter (p. 552).

L08 Describe the events involved in generating an action potential (Section 22.3).

- Nerve impulses, or action potentials, are generated when depolarization of the cell membrane occurs (pp. 552–555).
- A resting neuron is polarized in the sense that it is negatively charged in comparison to the outside of the cell. This polarization is maintained by sodium-potassium pumps in the membrane, which pump three sodium molecules out of the cell for every two potassium molecules they pump into the cell. Stimulation of a neuron partially opens the gates on the sodium channel proteins in the membrane, allowing sodium to diffuse into the cell and depolarizing the cell (pp. 553–554).
- The depolarization moves in a wave down the cell, activating sodium channels in adjacent parts of the membrane and thereby moving the impulse along the length of the neuron (pp. 553–554).

LO9 Describe how an action potential is relayed across a synapse (Section 22.3).

- The electrical impulse is propagated to the ends of the axon, which house chemical neurotransmitters (pp. 553, 555).
- Nerve impulses cause neurotransmitters to be released into the synapse, from which they can bind to receptors on the next neuron (pp. 553, 555).

LO10 List several neurological disorders, including a description of the biology underlying the disorder (Section 22.3).

- Neurological disorders can be caused by structural abnormalities of the brain or by defects in neurotransmission (pp. 556–558).

Roots to Remember MB

The following roots of words come mainly from Latin and Greek and will help you decipher terms:

cereb- is from the Latin word for the brain. Chapter term: cerebrum

cortex is the Latin word for the bark of a tree. Chapter term: cerebral cortex

hypo- means under or below. Chapter term: hypothalamus

ret- or **reti-** means a net or network. Chapter term: reticular formation

Learning the Basics

1. **LO4** How do the CNS and PNS differ?
2. **LO1** How do sensory, motor, and interneurons differ?
3. **LO2** What are the general senses?
4. **LO5** Which brain structure controls language, memory, sensations, and decision making?

   A. cerebrum; **B.** thalamus; **C.** hypothalamus; **D.** cerebellum; **E.** brain stem

5. **LO5** Where in the brain is the thalamus located?

   A. in the pons; **B.** in the medulla; **C.** in the cerebral cortex; **D.** between the cerebral hemispheres; **E.** in the occipital lobe

6. **LO7** Myelin _____.

   A. is a protective layer that coats some neurons; **B.** is formed by cells called Schwann cells; **C.** gives neurons a white appearance; **D.** prevents sideways transmission of nerve impulses, thereby increasing the rate of transmission; **E.** all of the above

7. **LO8** Neurons maintain a negative charge in their cytoplasm relative to outside the cell via the _____.

   A. Krebs cycle; **B.** electron transport chain; **C.** dopamine receptor; **D.** sodium-potassium pump

8. **LO8** An action potential _____.

   A. is a brief reversal of temperature in the neuronal membrane; **B.** is propagated from the terminal bouton toward the cell body; **C.** begins when sodium channels close in response to stimulation; **D.** is propagated as a wave of depolarization moves down the length of the neuron

9. **LO9** Neurotransmitters _____.

   A. are electrical charges that move down myelinated axons; **B.** are released across the nodes of Ranvier to hasten nerve impulse transmission; **C.** are released when an electrical impulse arrives at the terminal bouton of the postsynaptic neuron; **D.** diffuse across a synapse and bind to receptors on the cell membrane of the postsynaptic neuron; **E.** are steroid hormones released from the thalamus

10. **LO9** The effects of a neurotransmitter could be increased by _____.

    A. increasing the number of receptors on the postsynaptic cell; **B.** preventing reuptake; **C.** providing more enzymes involved in synthesizing the neurotransmitter; **D.** inhibiting enzymes involved in breakdown of the neurotransmitter from the synapse; **E.** all of the above

11. **LO3** True or false: A reflex arc is generated after multiple exposures to a negative stimulus such as heat.

Analyzing and Applying the Basics

1. **LO10** Develop a neurobiological hypothesis for why ADD might be more severe in some people than others.

2. **LO9** How does blocking neurotransmitter receptors affect neuronal function?

3. **LO3** How do reflexes differ from other nervous system responses?

Connecting the Science

1. Some forms of depression are thought to be caused by decreased amounts of the neurotransmitter serotonin. Does this mean that all depression has a biological cause? Why or why not?

2. Medicines are used to treat many health problems that are partially the result of an individual's lifestyle. For example, antacids are taken to treat stomach disorders that in some cases would be effectively treated through dietary changes. Likewise, heart disease and high cholesterol are medicated when, for many people, diet and exercise would work as well or better. When a health problem can be treated with medicine or via changes in behavior, which approach do you think is preferable? Why?

Answers to **Stop & Stretch, Visualize This, Savvy Reader,** and **Chapter Review** questions can be found in the **Answers** section at the back of the book.

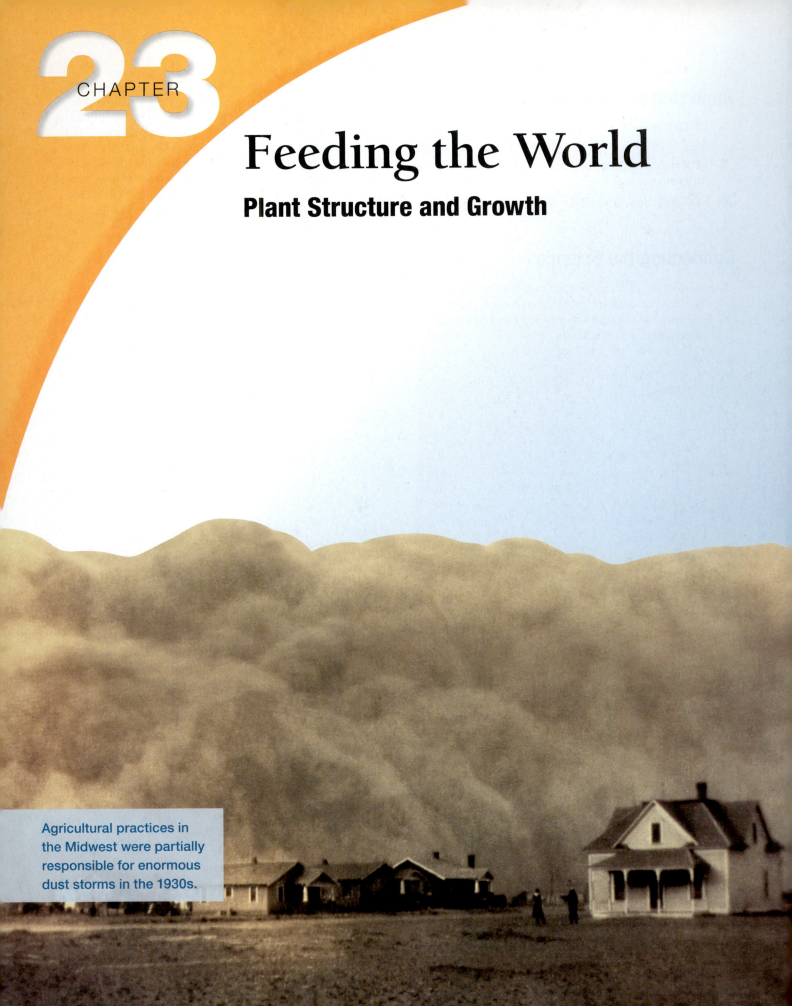

Feeding the World

Plant Structure and Growth

Agricultural practices in the Midwest were partially responsible for enormous dust storms in the 1930s.

The damage to soil that occurred during this period caused many families to lose their farms and become homeless.

A nd then the dispossessed were drawn west. . . . Car-loads, caravans, homeless and hungry; twenty thousand and fifty thousand and a hundred thousand and two hundred thousand. They streamed over the mountains, hungry and restless—restless as ants, scurrying to find work to do—to lift, to push, to pull, to pick, to cut—anything, any burden to bear, for food. The kids are hungry. We got no place to live.

This passage from John Steinbeck's classic novel *The Grapes of Wrath* describes the movement of people from the American Midwest to California during the era of the Dust Bowl. From 1931 to 1936, a combination of intense drought, severe storms, and inappropriate farming practices in the short-grass prairies of the Great Plains led to crop failures and widespread soil loss. Impoverished residents of the panhandles of Oklahoma and Texas faced a desperate choice between waiting out the bad times or pulling up stakes and leaving their homes. Nearly 25% of the population in these regions left over a period of just 3 or 4 years.

Similar dust storms have occurred more recently in China, carrying soil across the Pacific Ocean.

The farming practices that helped create the Dust Bowl were initially heralded as great improvements. Slow horse-drawn plows and harvesters were replaced with gasoline-powered tractors and combines, allowing more and more of the native prairie to be turned into wheat fields— but plowing prairie grasses destroyed the root systems that held the soil in place. Once these grasses were gone, soil could be picked up by fierce midwestern winds and blown thousands of miles in tremendous dust storms.

The environmental and human disaster during the Dust Bowl years was not a unique event. Today, thousands of acres of formerly productive farmland in China are turning to shifting sand. As a result, hundreds of thousands of

The Dust Bowl is now one of the most productive regions on Earth thanks to modern technology. But is this production sustainable?

people are moving away from this Asian Dust Bowl to cities along China's Pacific coast. Similar migrations have occurred in west-central Africa, parts of the Middle East, southern Russia, and elsewhere around the world as agriculture expands to ever-larger areas of Earth's surface. The mistakes of the Dust Bowl era are being repeated again and again.

Aggressive efforts to reverse soil erosion, combined with the return of normal rainfall levels, finally ended the Dust Bowl era in the United States. Thanks to the advances of modern agriculture, including the use of chemical fertilizers and irrigation from underground water stores, the Great Plains is now part of the most productive corn- and wheat-producing region on Earth. Can newly destroyed regions recover in the way that areas depleted in the American Dust Bowl did? Or is the Dust Bowl recovery an illusion maintained by huge inputs of resources that will eventually run out? To answer these questions, we must first understand the objectives and methods of agricultural practice—beginning by understanding the source of our crops and what these plants need to produce abundant food.

23.1 Plants as Food

For the first 250,000 years of humans' existence, our ancestors survived by hunting game and gathering fruits, seeds, roots, and other edible plant parts. Beginning about 11,000 years ago, human groups in what is now modern Iraq, Mexico, Peru, China, New Guinea, and elsewhere independently began to practice **agriculture**—that is, to plant seeds, cultivate the soil, and harvest crops. What triggered the development of this radically different way of life?

The Evolution of Agriculture: Food Plant Diversity

Nearly all agricultural plants belong to a single phylum—the **angiosperms,** known as flowering plants, which produce energy-storing seeds within fruit. Around the time agriculture developed, certain angiosperms that were suitable for cultivation may have become more common due to climate change and increased human activity.

Clues about past climate conditions can be gathered from a variety of sources, including plant remains trapped in sediments, the size of annual growth rings in ancient trees and fossil wood, and air bubbles trapped in ice sheets. According to this record, Earth's climate changed dramatically beginning about 11,000 years ago from a cool and wet period to a warmer, drier condition. The new climate also included more dramatic seasonality—wet seasons alternating with dry seasons.

Annual plants, angiosperms that complete their life cycle from seed to adult plant to seed in the course of a single season, were successful in these new climate conditions. Annuals are also quite suitable for agriculture, providing a rapidly produced, abundant, and easily transported food source. Nearly all of the major crops are annual plants.

The annuals that provide the majority of calories to most societies evolved from weedy ancestors. These ancestors were well suited to growing on open and disturbed land, which often surrounds human settlements. The hardiest weeds in these environments were usually from the same two plant families: Poaceae, or grasses, in the class known as **monocots,** and Fabaceae, or peas and beans, from the class commonly known as **dicots.** Monocots and dicots get their names from the number of **cotyledons,** or embryonic leaves, they produce in their seeds (**Table 23.1**). The seeds of grasses (known in agriculture as **cereal grains**) provide a rich source of carbohydrates and 80% of the calories in a typical diet. On the other hand, the seeds of peas and beans (commonly called **legumes**) are rich in proteins, especially in certain amino acids that are uncommon in cereal grains. Thus, these two groups of plants together provide a reasonably balanced diet. In the Middle East, the grain-legume combinations are wheat or barley with lentils or peas; in China, it is rice and soybeans; and in Mexico, corn and kidney beans.

Humans who realized that nutritious and highly productive wild grasses and peas were colonizing "waste areas" probably deliberately cleared additional land and scattered seed to encourage the plants' growth. Over time, the wild weeds became **domesticated** so that their growth and reproduction came under human control (**Figure 23.1**). The process of artificial selection, by which farmers chose which plants would produce the next generation based on their growth rates and ease of harvesting, produced crop plants that cannot survive without human intervention—and human societies that cannot survive without crop plants.

Grains and peas are not the only domesticated plants, of course, but they are among the most important. The **staple crops,** or major sources of calories, in Europe, Asia, and Central America are wheat, rice, and corn, respectively. But in other places, people get most of their calories from other crops—for example, sweet potatoes in tropical Africa and Indonesia, white potatoes in South America, and manioc (also known as cassava or tapioca) in Central and South America and Western Africa. While potatoes and manioc are also angiosperms, their

(a) Teosinte, ancestor of modern corn

(b) Modern corn

Figure 23.1 The effect of domestication. (a) Teosinte, the ancestor of corn, produces a few small, loosely held kernels in a fragile "ear" that disintegrates when the kernels are ripe. (b) In contrast, corn's kernels are large, numerous, and do not detach from the cob.
Visualize This: Describe the actions of early human farmers that would have caused teosinte to evolve so dramatically into the familiar ear of corn.

TABLE 23.1

Monocot and dicot seeds. The two major classes of angiosperms are named for the number of cotyledons contained in the seed.		
Name	**Number of Cotyledons (Embryonic Seed Leaves)**	**Example**
Monocot	1 cotyledon	Corn Wheat Rice Lily
Dicot	2 cotyledons	Pea Tomato Maple Dandelion

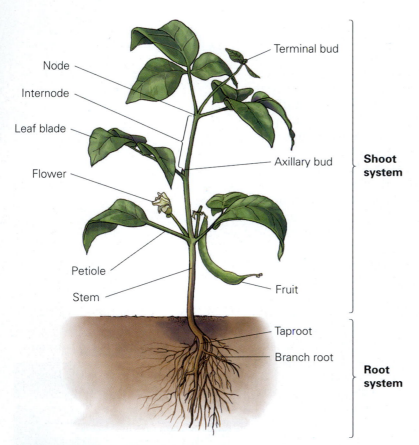

Figure 23.2 Plant structure. Plants consist of five organs: the vegetative structures—roots, stems, and leaves—and the reproductive structures—flowers and fruit.

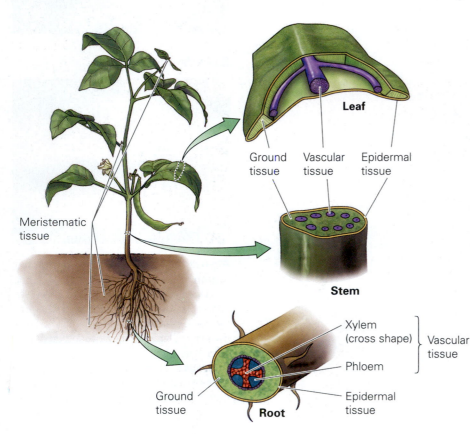

Figure 23.3 Plant tissue systems. Plants possess four tissue types—epidermal, ground, vascular, and meristematic.

food value comes not from their fruit or seeds but from other plant organs.

Plant Structure

Angiosperm plants have a relatively simple structure, consisting of three nonreproductive **vegetative organs** and two **reproductive organs.** The three vegetative organs are the roots, stems, and leaves; the reproductive organs are flowers and fruit (**Figure 23.2**).

Plant organs are made up of only four types of tissue (**Figure 23.3**). **Meristematic tissue** has cell division as its primary function and is most commonly found at the tips of roots and stems. Cells arising from meristematic tissue are undifferentiated, and their adult form depends on their location in a growing plant; thus, unlike animals, a plant does not have a completely specified final form. **Epidermal tissue** serves as an outer covering of the plant body, functioning in water absorption below ground and water retention above ground. **Vascular tissue** is made up of two subtypes of liquid-conducting tissue: **xylem,** which carries water and minerals from the roots to the leaves, and **phloem,** which carries food, in the form of sap, throughout the plant. **Ground tissue** makes up the remainder of the plant body and may consist of cells that perform photosynthesis or store starch, among other roles. The relative amounts of these types of tissue in a vegetative organ often determine its nutritional value; more vascular tissue means more fiber, but more ground tissue means a larger percentage of usable nutrients.

Roots. The primary function of a plant's root system is to absorb water and minerals from the soil. The roots of most monocots are fibrous and spread out broadly, giving these plants the ability to exploit water as soon as it becomes available in the soil. However, the large central taproot of most dicots permits these plants to access minerals and water found much deeper in the soil (**Figure 23.4**). Most of the water and minerals obtained by a plant are absorbed across the surface of root hairs, tiny

(a) Typical monocot root system **(b) Typical dicot root system**

Taproot

Fibrous roots

Figure 23.4 Monocot and dicot root systems. (a) Most monocots produce a fibrous root system that competes well for water but is not a strong support structure. (b) Most dicots produce a much sturdier, deeper taproot.

projections of the epidermal cells (**Figure 23.5**). The meristematic tissue at the tips of roots allows the root system to extend through the soil via growth.

Stop & Stretch If a gardener removes the stem and leaf of a weedy plant without pulling out its roots, the weed will very often regrow. What does this imply about the capabilities of plant cells?

Roots may also serve as the primary storage organ of a plant, and some that store carbohydrates in their ground tissue, such as carrots, parsnips, and sugar beets, have been modified into agricultural crops. Many plants with storage roots are not annuals but **perennials** that live for many years. Perennials survive difficult environmental conditions by using food they have stored during favorable environmental conditions. Perennials grown as crops, such as sweet potatoes and manioc, are harvested after 1 year of growth to take advantage of this stored carbohydrate (**Figure 23.6**).

Figure 23.5 Root hairs. The growing ends of a plant's roots are covered with tiny, hairlike extensions of the epidermal cells.
Visualize This: How do root hairs improve a root's capacity to absorb water?

Figure 23.6 Manioc, or cassava. The storage roots of this plant are a staple in the diet of people throughout Brazil and elsewhere in the tropics. The roots contain large amounts of a cyanide-producing acid and must be extensively processed to make them edible.

Figure 23.7 A modified stem. A tuber is a modified stem, as the presence of sprouting buds on these potatoes indicates. These tubers are produced on another stem modified for underground life, a rhizome.

Figure 23.8 Plants as food. Our diet includes a diversity of plant parts, including fruits, seeds, roots, and flowers.

Stems. The principal role of the stem is to serve as a support structure for the other organs of the shoot system (see Figure 23.2). At the apex, or top, of the stem is the terminal bud, which encloses the growing tip, or **apical meristem,** of the plant. The nodes of the stem, the point where leaves are attached, are separated from each other by sections of stem called internodes. At the junction between the leaf and stem at each node is an axillary bud, an embryonic shoot containing meristematic tissue. Most axillary buds have the ability to grow into a new lateral node-internode structure (that is, a vegetative branch), although some will become flowering branches that cannot branch further.

Most stems contain significant vascular tissue, which is fibrous and of limited food value. However, like roots, stems can be modified for food storage, and some storage stems have become major crops. White potatoes and yams are **tubers,** enlarged structures found on underground stems called **rhizomes.** The "eyes" on a potato are axillary buds, each with the ability to sprout into a stem containing nodes and internodes (**Figure 23.7**).

Leaves. The "food factories" of a plant are its leaves, the primary site for photosynthesis. While no staple crops are leaves, leafy greens, including lettuce, spinach, chard, kale, and bok choy, are rich in essential vitamins and minerals and are important components of diets in many parts of the world. The leaf part that is more commonly consumed is the flat, photosynthetic blade, rather than the stiff and fibrous petiole, which consists primarily of vascular tissue. Humans have also domesticated plants that store food in modified leaves. Onions and other similar plants produce bulbs, underground storage structures that consist of short stems and fleshy, carbohydrate-filled leaves.

Humans have domesticated crops based on all imaginable plant parts (**Figure 23.8**). However, the vast majority of our foods are based on reproductive organs, primarily fruits and seeds.

Seed   Fruit   Leaf   Tuber (modified stem)

Flower   Root   Endosperm (of seed)   Bulb (modified stem with leaves)

Plant Reproduction

We may not think of our breakfast toast and cornflakes as belonging to the same category of foods as our orange juice or bananas, but all of these foods are products of **fruit,** defined by botanists as mature flower ovaries containing seeds. Flowers and fruit are structures for sexual reproduction that are unique to angiosperms.

Flowers are the sexual organs of angiosperms (**Figure 23.9**). The typical flower produces both male and female structures. Male gametes are contained in pollen. The pollen is produced in structures called **anthers** that are held up by **filaments** in the male reproductive organ, called the **stamen.** Female gametes are contained in structures called **ovules** and produced in a chamber, called the **ovary,** of a flower's female reproductive organ, the **carpel.** The transfer of pollen to the **stigma,** a sticky pad on the top of a carpel, is called **pollination.** Pollination may occur via wind, as in corn and wheat, or via insects or other animals, as in soybeans. A flower's **petals** are adapted to facilitate pollination. After landing on the stigma, pollen grains germinate into tubes that extend down the length of the carpel's **style** to reach the ovules. A developing flower is enclosed in protective, modified leaves called **sepals** that in some plants also facilitate pollination.

In the case of wind-pollinated grass, the petals are much reduced, whereas in insect-pollinated legumes, they are colorful and contain nectar to attract and

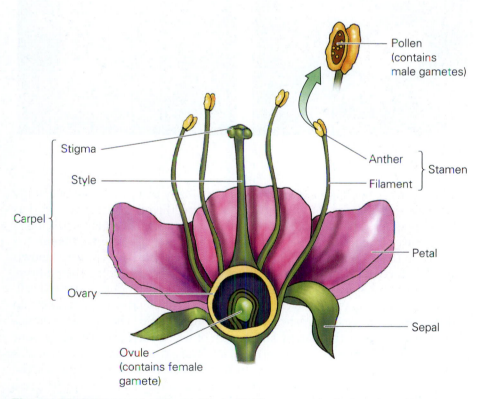

Figure 23.9 Flower structure. The typical flower contains both male and female reproductive organs, although many plant species produce two types of flowers: male only and female only.

Visualize This: What are the advantages and disadvantages of producing both male and female parts in the same flower?

(a) Wind-pollinated flower

(b) Bee-pollinated flower

(c) Bird-pollinated flower

(d) Bat-pollinated flower

(e) Fly-pollinated flower

Figure 23.10 Petals and pollination. A flower's petals are adapted to aid in the process of pollination. (a) The most successful wind-pollinated plants have reduced petals that do not interfere with pollen transfer. (b) Bee-pollinated flowers often provide a nectar reward at the base of a petal tube, while (c) bird-pollinated flowers typically have petals fused into a single deep, highly visible floral tube. (d) Flowers that are pollinated by bats often open their petals only at night. (e) The petals of fly-pollinated flowers may have the appearance—and odor—of rotting flesh.

reward their bee pollinators (**Figure 23.10a–b**). Many plants are pollinated only by a single species or group of species. These specific relationships evolved as pollinators selected plants that gave them the best rewards, and plants developed traits that were highly attractive to particular pollinators (**Figure 23.10c–e**).

Once in the ovules, the pollen tube releases two sperm. In a process known as **double fertilization,** one sperm fertilizes the egg to become the embryo, and the other fuses with two nuclei within the ovule to produce a triploid cell, containing three copies of each chromosome. This cell divides and develops into **endosperm,** tissue that nourishes the developing embryo. The primary component of wheat flour is the starchy endosperm of wheat seeds; whole wheat flour also contains the embryo (the germ) and accessory tissues (the bran), whereas white flour is strictly the endosperm. In some species, such as in legumes, the embryo consumes the endosperm as it develops and the seed matures. The two halves of a bean seed or peanut are enlarged cotyledons produced by the digestion of the endosperm.

The embryo, endosperm, and remainder of the ovule tissues make up the **seed.** Triggered by successful fertilization, other parts of the flower—in nearly all cases, the ovary, but sometimes including petals, sepals, or the flower stalk—develop into a fruit. Most fruits are adapted to be vehicles for dispersing seeds far from the parent plant. Some fruits help seeds disperse on wind or water; other fruits help seeds disperse by using animals to carry the fruit away and drop its seeds elsewhere (**Figure 23.11**). Because humans have similar sensory systems and nutritional needs to those of other mammals, many of the crop plants that we rely on were selected from fruits adapted to attract mammal dispersers. Even grains are adapted to attract mammals, primarily rodents. **Figure 23.12** illustrates the relationship among flowers, seeds, and fruit in angiosperms.

Stop & Stretch If a parent plant is already successful, the conditions where it is growing must be ideal for that species. What is the advantage of dispersing seeds rather than keeping them near an obviously successful parent plant?

In Focus on Evolution, we explore how plants that do not produce flowers reproduce.

Figure 23.11 Seed dispersal units. Fruits are adapted to disperse seeds away from the parent plant. Many mechanisms are possible, including (a) wind, (b) water, (c) mammal fur, and (d) through an animal's digestive tract.

(a) Wind-dispersed fruit

(b) Water-dispersed fruit

(c) Animal-dispersed hitchhiker fruit

(d) Animal-dispersed edible fruit

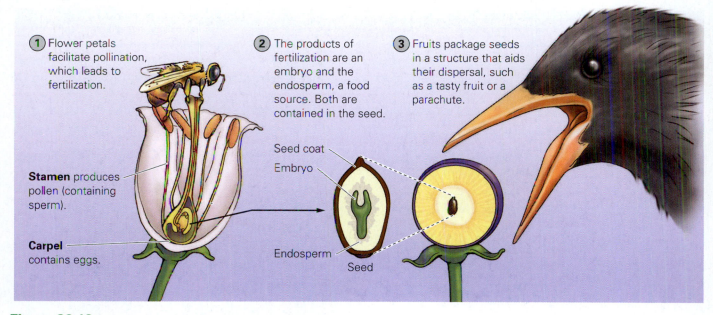

1. Flower petals facilitate pollination, which leads to fertilization.

2. The products of fertilization are an embryo and the endosperm, a food source. Both are contained in the seed.

3. Fruits package seeds in a structure that aids their dispersal, such as a tasty fruit or a parachute.

Stamen produces pollen (containing sperm).

Carpel contains eggs.

Seed coat
Embryo
Endosperm
Seed

Figure 23.12 A summary of angiosperm reproduction.

Plant Reproduction

Land plants have a unique life cycle. Instead of one multicellular "adult" form as is found in animals, plants have two, which alternate with each other. In fact, the life cycle of plants is commonly referred to as **alternation of generations.** The two life-forms result because the product of meiosis in plants is not a sperm or egg (gamete); it is a haploid structure (containing half the typical number of chromosomes) called a **spore,** which undergoes mitosis to produce a multicellular plant called the **gametophyte** (or gamete-producing plant). The gametophyte produces sperm or eggs (or both) by mitosis. When gametes fuse, they produce a diploid embryo that divides and eventually produces spores by meiosis. This stage in the life cycle is called the **sporophyte.** The generalized life cycle is illustrated in **Figure E23.1.** Because the algal ancestors of land plants were haploid for most of their lives, but sexual reproduction requires a short period of diploidy, the alternation between two forms is a vestige of evolutionary history. In fact, the trend in land plant evolution has been that the haploid gametophyte generation has become increasingly reduced.

In mosses and other nonvascular plants, the gametophyte generation is the most prominent form, reflecting their close relationship with the green algae. Sperm and eggs are produced in structures at the tips of the plant, and sperm swim in a film of water, such as a drop of rain, to reach the eggs. The embryo develops within the female reproductive organ, eventually producing a sporophyte that emerges from the tip of the gametophyte and is dependent on it for survival. Tiny spores produced in the sporophyte are dispersed on the wind and germinate into new gametophytes if they land in moist conditions.

On land, a large haploid body is a disadvantage because DNA-damaging UV light, which is scattered by water molecules, is much more of a threat in dry air. Because diploid cells have two copies of each chromosome, UV damage to DNA is not as detrimental in a diploid compared to a haploid cell. This may be why ferns and other seedless vascular plants, which arose from the nonvascular plants, have a dominant sporophyte. Spores are produced in structures on leaves and are dispersed by wind or water to germinate in warm, wet conditions. The small independent gametophyte that emerges from a spore may produce both sperm and eggs. Sperm swim on a film of water to reach the egg, and the diploid embryo grows out of the female reproductive structure, although it quickly becomes much larger and independent of the gametophyte.

The seed plants that arose from the ferns represent the next step in the evolution of reproductive cycles—another reduction in gametophyte size and a change in the dispersal structure. Pine trees and other nonflowering seed plants, called gymnosperms, produce two types of gametophyte, both of which are dependent on the sporophyte for survival. Pollen, released from small cones produced by the adult sporophyte, is actually tiny, not-yet-mature male gametophyte plants. Pollen is carried by the wind (or occasionally insects) to the female reproductive structures, which are typically produced in woody cones on the sporophyte. Here, the pollen completes maturation and releases sperm to fertilize the egg. The female gametophyte, which contains the egg, is a small structure on the surface of a cone scale. The embryo begins development inside

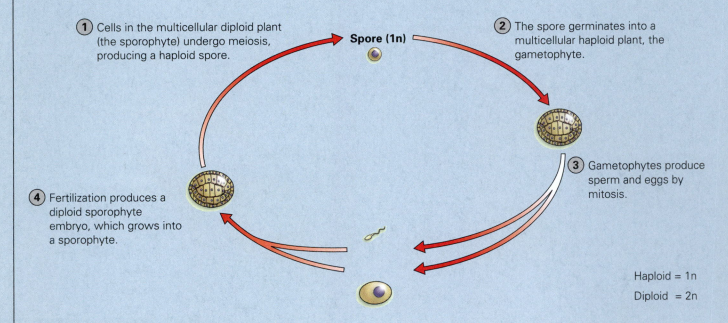

① Cells in the multicellular diploid plant (the sporophyte) undergo meiosis, producing a haploid spore.

Spore (1n)

② The spore germinates into a multicellular haploid plant, the gametophyte.

③ Gametophytes produce sperm and eggs by mitosis.

④ Fertilization produces a diploid sporophyte embryo, which grows into a sporophyte.

Haploid = 1n

Diploid = 2n

Figure E23.1 Alternation of generations.

the female gametophyte, and the embryo, female gametophyte, and surrounding sporophyte tissue become the seed. The seed, rather than the spores, is the primary dispersal unit in gymnosperms. The seed provides an additional advantage on land—an internal food supply and protection from dry conditions—compared to spores. The adult sporophyte develops when the seed has landed in an appropriate site for growth.

Like gymnosperms, angiosperms also produce male and female gametophytes that are mostly retained on the parent sporophyte plant. The male gametophyte is pollen, and female gametophytes are even smaller than the ones in gymnosperms, consisting of only seven cells tucked inside the female parts of a flower. **Figure E23.2** provides a comparison of the gametophyte and sporophyte generations in the four major land plant groups.

Figure E23.2 Gametophtyes and sporophytes in the groups of land plants.

Producing abundant fruit for human consumption requires more than simply selecting plants for optimal fruit production. The agricultural practices of people around the world are designed to maximize the growth of these plants.

23.2 Plant Growth Requirements

Agricultural practices such as turning over soil, irrigating soil with water, and adding nutrients to soil developed independently everywhere that plants were domesticated by people. To understand why these techniques enhance agricultural production, we must understand first how plants grow.

How Plants Grow

The first stage of plant growth is the emergence of the seedling, supported by its internal food stores. Later plant growth can be divided into two types—growth in length, called **primary growth**, and growth in girth or circumference, called **secondary growth**.

Germination. The emergence of an embryo from a seed is called **germination.** The process begins when the seed begins to take up water, expands, and ruptures its confining seed coat (**Figure 23.13a**). As long as the seed coat is impermeable to water, the seed remains dormant, which means that it will not germinate. Dormancy allows seeds to act as "time travelers," passing through poor environmental conditions in a state of suspended animation. Many seeds require exposure to specific environmental conditions before they can escape dormancy.

Stop & Stretch In temperate regions with cold winters, some seeds need to be exposed to low temperatures for a certain amount of time before they can germinate. What is the advantage of this strategy in these environments?

Figure 23.13 Germination. (a) Seed germination begins when the seed takes up water. (b) The first organ to emerge is the root. (c) Water obtained by the root makes the cells in the embryonic shoot enlarge, allowing the shoot to emerge from the soil. (d) Germination is complete when the seedling is functioning independently of the stored food in the seed.

The first structure to emerge from a germinating seed is the root, which enables the developing young plant, or **seedling**, to become anchored in the soil and absorb water (**Figure 23.13b**). Growth of the shoot soon follows as the seed's stored food is depleted, and photosynthesis becomes necessary (**Figure 23.13c**). The time between germination and when the seedling is established as an independent organism is the most vulnerable period in the life of a plant. Damage to the plant during this period is usually fatal. Once a seedling becomes established, however, damage may be overcome by continued growth (**Figure 23.13d**).

(a) **(b)** **(c)** **(d)**

Primary Growth. In plant stems, primary growth takes place through the production of additional nodes and internodes by growth at the shoot tips and by the growth of axillary buds. Growth in length occurs when cells at a meristem divide in a single plane, creating columns of daughter cells below them and pushing the apical meristem farther up into the air (**Figure 23.14**). A hormone produced by the terminal bud inhibits cell division in the axillary buds in many plants. Once the terminal apical meristem has grown away from a particular axillary bud, the bud's meristem begins to divide and produces a lateral branch, also containing axillary buds.

The node-internode system of the main stem and branches supports the **indeterminate growth** of a plant's shoot system—its ability to grow throughout its lifetime. This is in contrast to the **determinate growth** of most animals, which grow to a set size that they typically reach early in life.

In roots, primary growth at the tip occurs due to cell division in the apical meristem. Cell division in the root occurs in two opposing directions, resulting in columns of cells that elongate and push the root tip forward through the soil, as well as the production of a loosely organized **root cap** that covers the tip of the root. As the root tip pushes through the soil, exterior cells of the root cap are worn off and replaced by younger cells more recently produced by the apical meristem. The primary meristematic tissue producing lateral roots is found near the center of the root, surrounding the vascular tissue. As a result, lateral roots can form anywhere along the length of a root (**Figure 23.15**).

Primary growth results in plants that are taller and bushier and have more extensive root systems. In longer-lived plants, continued primary growth presents a dilemma—a tall plant cannot support itself with a skinny stem. Many plants can increase their stability by making wider stems and roots via secondary growth.

Secondary Growth. All of the flowering plants that can undergo true secondary growth are dicots. In these plants, a ring of cells running through the bundles of vascular tissue in the stem becomes a meristem called the vascular cambium (**Figure 23.16**). This cambium produces phloem cells to the outside of the stem and xylem cells to the inside.

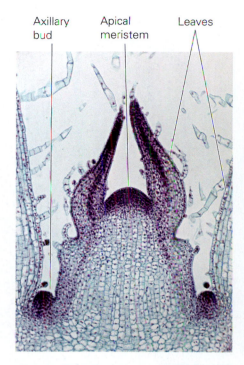

Figure 23.14 Primary growth in stems. This microscopic image of an apical meristem clearly shows the repeating node-internode system.
Visualize This: Identify the nodes and internode visible in this micrograph.

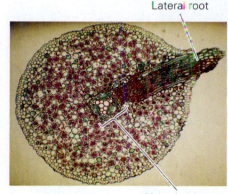

Figure 23.15 Lateral root formation. In this cross section, meristematic tissue in the center of the root has produced a lateral root.
Visualize This: Why must the lateral root originate from the vascular cylinder in the center of the primary root?

(a) Monocot stem
Vascular tissue scattered

(b) Dicot stem
Vascular tissue in a ring

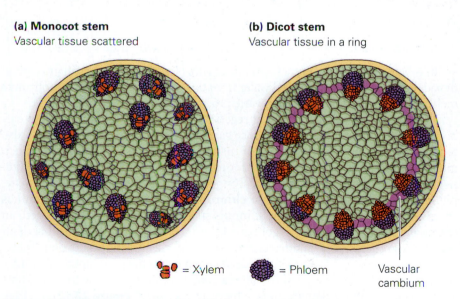

= Xylem   = Phloem   Vascular cambium

Figure 23.16 Monocot and dicot stems. Unlike monocot stems, which have scattered bundles of vascular tissue, dicot stems have vascular bundles arranged in a ring. This organization allows a single ring of meristematic tissue, the vascular cambium, to form between the xylem and phloem.

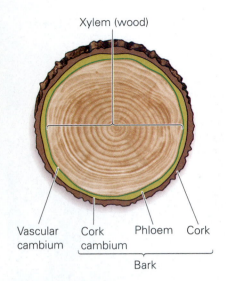

Xylem (wood)

Vascular cambium   Cork cambium   Phloem   Cork

Bark

Figure 23.17 Dicot wood. The wood of a dicot stem is made up of xylem cells, while the bark consists of phloem and cork.

(a) Palm tree trunk cross section

(b) Palm trees bending in hurricane-force winds

Figure 23.18 Monocot "wood." Secondary growth in monocots, such as this palm, produces much more fibrous wood (a), which is not nearly as strong as dicot wood but is very resilient (b).

The additional xylem cells produced by the cambium make up the bulk of the material in these stems, and their stiffened cell walls produce the stem's **wood** (**Figure 23.17**). Epidermis on the surface of the stem breaks apart and is shed as the stem's surface area increases. Cells immediately under the epidermis become another meristem (called the **cork cambium**). Cell division in the cork cambium results in the production of **cork**, a water-resistant outer covering. Phloem, cork, and cork cambium make up the **bark** of a woody dicot stem. All the components of bark are continually shed and replenished by secondary phloem cells, which are produced by the vascular cambium as the tree increases in diameter.

Stop & Stretch Removing a small strip of bark around the complete circumference of a tree, a technique called girdling, always eventually kills the tree. In contrast, removing a large patch of bark from one side of a tree often does not kill the tree. What function of bark is totally disrupted by girdling but not totally disrupted by loss of a large patch?

Interestingly, although dicot trees are much more common than monocot trees, the only tree-based staple in human societies is the monocot coconut, which is important to communities in Indochina and Central America. The thick stem of coconut palms consists of additional strands of vascular tissue, produced by a **thickening meristem** that surrounds the apical meristem. The structure of this stem makes palm trees incredibly flexible (**Figure 23.18**).

Primary and secondary growth of plants can be maintained only by abundant energy. Plants obtain this energy by transforming light into chemical energy in the process of photosynthesis. Our agricultural practices are intended to maximize the photosynthesis of crop plants so that they will produce a surplus of carbohydrates that we can harvest.

Maximizing Plant Growth: Water, Nutrients, and Pest Control

The basic equation of photosynthesis is as follows:

Carbon dioxide + Water + Light energy $\longrightarrow$ Carbohydrate + Oxygen gas

In other words, photosynthesis in plants converts energy from sunlight, carbon dioxide from the atmosphere, and water from the soil into energy-rich carbohydrate molecules. Oxygen is essentially a "waste product" of this reaction (Chapter 5).

If we look only at this chemical reaction, it is clear that plants require carbon dioxide, water, and light to produce carbohydrates. However, the process also requires a few more components. For instance, to survive, plants require much more water than that needed simply for the chemical reactions of photosynthesis. Plants also require other chemical inputs in smaller amounts. For example, photosynthesis requires the pigment **chlorophyll** to proceed, and magnesium is an essential element in chlorophyll. Inputs such as magnesium and nitrogen, an essential element in proteins, are obtained from the soil. Pest organisms that damage photosynthetic tissue and fruits and seeds also affect plant production. Therefore, to increase crop production, modern farming practices attempt to maximize the amount of light, water, carbon dioxide, and nutrients available to crop plants as well as to minimize damage caused by plant predators (**Figure 23.19**).

Water and Sunlight. To hold its leaves perpendicular to the sun's rays, a plant's cells must be full of water; that is, they must be turgid (**Figure 23.20**). Recall

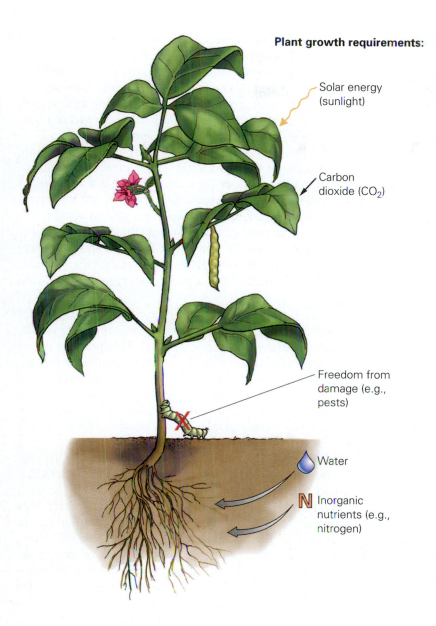

Plant growth requirements:

Solar energy (sunlight)

Carbon dioxide (CO$_2$)

Freedom from damage (e.g., pests)

Water

N Inorganic nutrients (e.g., nitrogen)

Figure 23.19 Five requirements for plant growth. Plants require the basic ingredients for photosynthesis—water, carbon dioxide, and light—as well as nutrients from the soil and freedom from damage to produce excess carbohydrates for human consumption.

Figure 23.20 Water in plant cells. (a) When a plant is fully hydrated, the cell's vacuoles balloon with water, and the cells press against each other so that the plant remains "crisp." (b) As the plant dries out, the vacuoles lose water, and the cells no longer support each other, causing the plant to wilt.

Visualize This: What do you think would happen to the cells of a plant if it was watered with very salty water, and why?

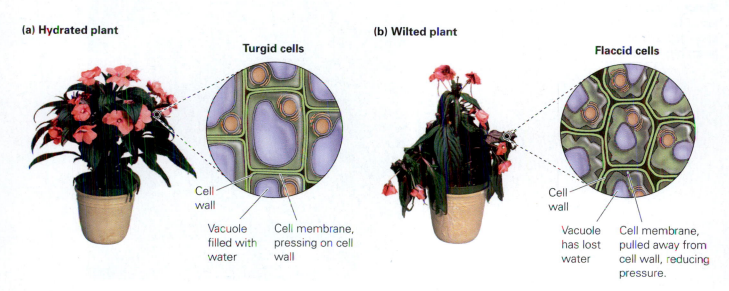

(a) Hydrated plant

Turgid cells

Cell wall

Vacuole filled with water

Cell membrane, pressing on cell wall

(b) Wilted plant

Flaccid cells

Cell wall

Vacuole has lost water

Cell membrane, pulled away from cell wall, reducing pressure.

that the membrane of a plant cell is surrounded by a stiff cell wall. The cell wall is tough, but it is also elastic, meaning that it can stretch and still hold its basic shape. As a plant cell's central vacuole fills with water, the cell wall balloons slightly, becoming turgid. It is the pressure of adjoining turgid cells that holds the leaves upright. When cells lose water, the pressure of cells pushing against each other decreases; the cells become flaccid, and the plant wilts. Thus, part of maximizing a plant's exposure to light is ensuring that it has plenty of water.

In many regions, it is necessary to prevent wilting of crop plants by adding water to the soil through **irrigation** (**Figure 23.21**). The enormous production of corn, wheat, and soybeans in the former Dust Bowl region is a result of an equally enormous amount of irrigation. The water for irrigation in this dry region comes from a huge underground reservoir called the Ogallala aquifer.

Another way to increase the amount of water that reaches preferred plants is to reduce the amount of competition for that water. Farmers do this by trying to minimize the number of nonpreferred plants, commonly called **weeds,** in agricultural fields. Controlling weeds has the additional benefit of reducing competition for sunlight.

Controlling Weeds. Farmers have several techniques for controlling weeds. One is to remove competitors before the crop is planted. This process, called tilling, involves turning over the soil to kill weeds that have sprouted from seed. After a crop begins growing, it is possible to prevent the growth of competitors and reduce water loss from the soil by mulching the crop. To mulch, farmers spread a thick layer of straw, other dead plant material, sheets of newspaper, or dark plastic around the base of preferred plants. These materials block sunlight from reaching the soil, preventing weed seeds from germinating.

Tilling does not kill all weeds, and some spring back to life quickly. In these cases, farmers may use a **herbicide,** a chemical that kills plants. Herbicides that exploit chemical differences between dicot and monocot plants are the most commonly used sprays in agriculture. These herbicides are used to control dicot weeds in monocot crops such as corn and wheat. Broad-spectrum herbicides, such as glyphosate (Roundup®), kill all plants; until recently, these herbicides have been useful only for controlling weeds before planting. Development of genetically modified organisms (GMOs) has increased the applicability of broad-spectrum herbicides to many more crop-and-weed combinations in recent years (Chapter 9).

Figure 23.21 Irrigation. The circle pattern of lush plant growth derives from center-pivot irrigation sprayers, which spray water in an arc from the center point.

(a) When water is abundant:

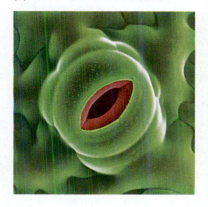

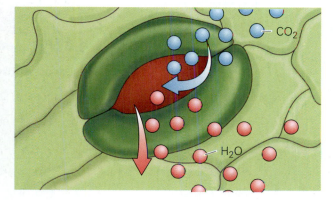

- Guard cells swell
- Stoma (pore) is large
- High carbon dioxide uptake
- High water loss

(b) When water is scarce:

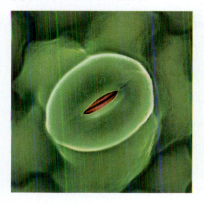

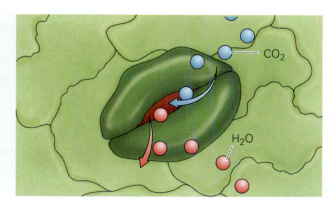

- Guard cells shrink
- Stoma (pore) is small
- Low carbon dioxide uptake
- Low water loss

Even when plants are fully hydrated and not wilting, their efforts to prevent water loss can interfere with growth. This is because plant strategies for conserving water can impose a cost to photosynthetic production.

Figure 23.22 Guard cells. Guard cells help to regulate the size of stomata on a plant's surface, thus minimizing water loss and maximizing carbon dioxide intake.

Water and Carbon Dioxide. The primary adaptation that plants have to reduce water loss is a waxy covering, called the cuticle, on the epidermis of the shoot system. The cuticle also restricts the diffusion of carbon dioxide into the plant. To enable plant cells to obtain adequate carbon dioxide for photosynthesis despite the cuticle, the shoot epidermis is pocked with tiny pores called **stomata** (singular: stoma). However, these pores come with a price—stomata also provide portals through which water can escape.

In most plants, each stoma is encircled by a pair of **guard cells** that regulate the size of the pore (**Figure 23.22**). When the guard cells have abundant water, they increase in size, and their unique shape causes the pore to open. When the guard cells are water deprived, they shrink and relax, causing the pore to be small. Thus, to maximize carbon dioxide uptake, plants need enough water to keep their stomata open.

Where precipitation or irrigation is sufficient to keep the stomata open full time, the most important factor affecting plant growth is nutrient availability.

Nutrients and Soil. Elements needed by plants in large amounts, such as nitrogen and magnesium, are called macronutrients (**Table 23.2** on the next page). Nutrients needed in much smaller amounts—such as iron, which is required in the processes of chlorophyll synthesis and aerobic respiration—are known as micronutrients. Both categories of nutrients are available in the soil.

Soil is a unique material composed of both living and nonliving components. Healthy soil consists of a base layer of recently eroded rock; a top layer

TABLE 23.2

Plant nutrients and their functions.	
Primary Functions	
Macronutrient	
Nitrogen	Component of proteins, DNA, and RNA
Potassium	Involved in opening and closing of stomata
Calcium	Component of cell walls; involved in cell membrane permeability
Magnesium	Component of chlorophyll
Phosphorus	Component of ATP, DNA, RNA, and cell membranes
Sulfur	Component of proteins
Micronutrient	
Chlorine	Involved in moving water into and out of cells
Iron	Required for chlorophyll synthesis and aerobic respiration
Boron	Influences calcium use; required for DNA and RNA synthesis
Manganese	Required for integrity of chloroplast membrane
Zinc	Component of many enzymes
Copper	Component of some enzymes
Nickel	Required for nitrogen metabolism
Molybdenum	Required for nitrogen metabolism

Figure 23.23 Soil structure and development. Soil forms from the breakdown of rock and the accumulation of organic matter from dead and decaying plants, fungi, animals, and animal wastes.

made up of dead and decaying remains of plants and animals, along with living organisms that eat them, and waste; and a middle layer where components of the base and top layers are mixed by earthworms and other soil organisms (**Figure 23.23**). The primary way to maximize a crop plant's access to soil nutrients is to reduce competition with weeds—using the same techniques used to reduce competition for water.

In natural systems, nutrients become available to plants through nutrient cycling (Chapter 15). In agricultural systems, the nutrient cycle is often broken. Instead of plants being eaten by animals, which are then eaten by other animals and decompose all in the same location, the plants from agricultural fields are removed and trucked to other locations where they will be consumed and their nutrients released. Most of the nutrients that are consumed end up in human or animal waste, which often flows into lakes or oceans. In other words, nutrients on croplands are not recycled back into the soil but mined from it (**Figure 23.24**).

Because nutrient cycles are broken in agricultural fields, farmers must add nutrients to soils in the form of **fertilizer.** Fertilizer that is applied to the soil can be **organic**—that is, carbon-containing molecules made up of the partially decomposed waste products of plants and animals. Fertilizer can also be **inorganic**—meaning simple molecules that lack carbon, such as ammonia and nitrate, produced by an industrial process. Because plants require nutrients in inorganic form, soil organisms must first decompose organic fertilizer before its nutrients are available for plant growth.

Most farmers prefer inorganic fertilizer for two reasons: because the nutrients in it are more concentrated and much easier to transport, store, and apply than most organic fertilizers; and because plants can utilize these nutrients immediately. Most farmers in the United States fertilize with inorganic nitrogen, phosphorus, and potassium and occasionally will add other macronutrients such as calcium.

Stop & Stretch Most people do not harvest their houseplants, yet many "feed them" inorganic fertilizer every few months. Why do houseplants need fertilizer?

Farmers can also replace some nutrients in the soil through **crop rotation**—that is, by varying the crops that are planted on any one field over time. Surprisingly, some plants actually increase the levels of nutrients in the soil. Some plants have a mutualistic relationship with **nitrogen-fixing bacteria** (**Figure 23.25**), which are microbes with the ability, unique among living organisms, to convert nitrogen gas from the atmosphere into a solid form—namely, ammonia.

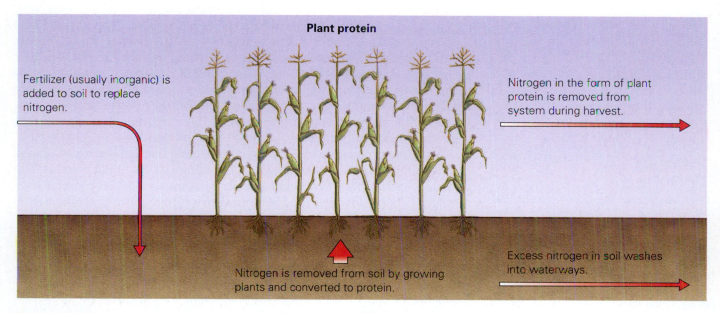

Plant protein

Fertilizer (usually inorganic) is added to soil to replace nitrogen.

Nitrogen in the form of plant protein is removed from system during harvest.

Nitrogen is removed from soil by growing plants and converted to protein.

Excess nitrogen in soil washes into waterways.

Figure 23.24 Nutrient mining. The movement of nitrogen in an agricultural system is illustrated here.
Visualize This: How would this image differ if the system were a natural environment?

(a) Alfalfa

(b) Nodules on alfalfa roots

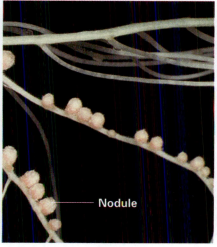

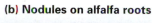

Nodule

Figure 23.25 Nitrogen-fixing crop.
(a) Alfalfa can increase the nitrogen content of soil. (b) The nodules on the roots of alfalfa contain nitrogen-fixing bacteria.

Legumes develop structures called **nodules** on their roots in response to chemical signals from certain species of nitrogen-fixing bacteria. The nodules house the bacteria and supply them with carbohydrates. In return, the plants receive the excess nitrogen that the bacteria fix. When a legume is growing well, the bacteria fix so much nitrogen that the excess is released into the surrounding soil. If legumes are planted in alternating years with a crop that does not fix nitrogen (such as corn or wheat), the need for nitrogen fertilizer can be greatly reduced because of the release of this excess nitrogen and because the nitrogen-fixing crop remains can be plowed back into the soil. The benefits of crop rotation may be one reason that so many early agricultural societies relied on staples of grains and legumes.

Once a farmer has maximized the growth of his or her plants by providing abundant water and nutrients, these plants become attractive targets to other consumers—including other mammals, birds, insects, fungi, and bacteria. To ensure that most of the food produced by an agricultural crop goes to feed people, a farmer must try to reduce the damage these competitors do to the crop.

Freedom from Pest Damage. Living organisms that cause damage to agricultural products are generally referred to as pests. Up to 40% of the potential food production in major crops is lost to pests every year. The primary tools farmers use to reduce pest impact are pesticides, chemicals designed to kill or reduce the growth of a target pest. (Herbicides are often considered a class of pesticides. In this book, however, we use the word *pesticide* only when referring to chemicals that control insect, fungal, and bacterial pests.) Pesticide use became widespread in the years following World War II. Today, millions of tons of these chemicals are applied annually to crops all over the world.

Surprisingly, pesticide use has not considerably reduced the overall loss of crops to pests. Instead, these chemicals have dramatically influenced farming methods—primarily by allowing farmers to plant a single crop over a wide acreage. This practice is called **monoculture.** Monocultural production of crops is very efficient. Farmers have one planting, fertilizing, and pest control schedule, and they require only the planting and harvesting equipment specific to their crop (**Figure 23.26a**). Monoculture and the accompanying mechanization of agriculture have greatly reduced the amount of labor required to produce crops. The percentage of the population working on farms in the United States has dropped from 25% to less than 2% in the past 60 years due to this change in farming practice.

On the other hand, monocultures are very susceptible to outbreaks of pests. If a small population of insects or disease infestation consumes or kills a host plant, another appropriate host for the pest is right next door. The easy availability

Figure 23.26 Monoculture and polyculture. (a) Planting the same crop over many hundreds of acres (monoculture) allows for the efficient use of massive planting and harvesting equipment. (b) Planting many different crops together (polyculture) can reduce pest problems and buffer farmers from crop failures.

(a) Monoculture

(b) Polyculture

of resources to pests in monocultural crops allows their populations to grow exponentially and can lead to the loss of an entire crop. Thus, while monoculture increases the efficiency of agricultural production, it results in an increased need for pesticides just to keep pest damage at historical levels.

Farmers can also use **cultural control** to minimize pest populations. Because most pests have evolved to attack a single crop, they will not cause damage to other, unrelated crops. Crop rotation, in addition to conserving soil nutrients, is a method of cultural control that moves plants away from their pests. After the growing season, a pest population spends the winter in the soil or on plant waste left in the field. If the same crop is planted the following year, the pest population has easy access to more resources and can continue to increase in population. However, if a different crop is planted, the pests must disperse in the spring in an attempt to find their required crop. Only some of these dispersers will find their host, and the population of pests in the new site will be relatively small.

Another form of cultural control is **polyculture**—planting many different crop plants over a single farm's acreage (**Figure 23.26b**). Unlike monocultures, polycultural planting keeps pest populations relatively small because, after the pests have consumed a patch of plants, they must disperse long distances to find another source of food. Polyculture also minimizes risk to farmers by ensuring that a pest outbreak on a single crop does not destroy all of the farm's production for that year.

Over the last 60 years, the use of pesticides, fertilizer, and large farm machinery has dramatically changed the way food is produced around the world. The revolution in biology—particularly the new science of genetics—has also played a role in this change.

Designing Better Plants: Hybrids and Genetic Engineering

Traditionally, farmers used artificial selection to modify the characteristics of their crop plants and animals; that is, they selected for next year's seed stock or breeding animals those that performed best under the conditions on their farm. However, this process limits farmers to only the variation that is currently available. Two developments in plant science during the 20th century reduced and finally eliminated this limitation.

Beginning in the 1920s, agricultural scientists began using plant breeding to produce better crops. Plant breeding generally consists of creating hybrids, the offspring of two different varieties of an agricultural crop. One of the first hybrids produced on a large scale was dwarf spring wheat. One parent variety was a nitrogen-loving, high-producing wheat that was difficult to harvest because it tended to fall over when the wheat grains were ripe. The other parent variety was a lower-producing dwarf wheat plant. The hybrid of these two varieties was high producing but short, so it responded well to fertilizer application and was easy to harvest with large machines. Because hybrids are the first-generation offspring of two distinct varieties, new hybrid seeds have to be produced every year from the parent populations. This is necessary because independent assortment occurs during meiosis; thus, seeds produced by the hybrid plant will not contain the mix of chromosomes found in the first-generation plants.

In addition to the difficulty of producing hybrid seeds, plant breeding limits agricultural scientists to the genetic variation within a single species. More recently, DNA technology has permitted the creation of novel agricultural organisms containing genes from other species. These genetically modified organisms, or GMOs, include herbicide-resistant crop plants. In the United States, over 40% of farmland planted with corn, more than 70% of the cotton crop, and nearly 80% of soybean fields are planted with genetically modified varieties. However, many questions remain about GMOs, including their safety as human food, the likelihood that their widespread use will cause the evolution of resistance in pests, their effects on nontarget organisms, and the consequences of the "escape" of engineered genes into noncrop plants (see Chapter 9).

The concerns related to GMOs should not be downplayed. However, in several ways, these concerns pale in comparison to those created by conventional agricultural practices, including environmental degradation and loss of soil.

23.3 The Future of Agriculture

The use of pesticides, herbicides, fertilizer, and irrigation over the past 60 years—a phenomenon known as the "Green Revolution"—has increased food production dramatically. However, many biologists argue that current high levels of production are **unsustainable**—they compromise the ability of future generations to grow enough food to feed the human population. The future of agricultural production depends on finding ways to maximize food production while minimizing environmental damage.

Modern Agriculture Causes Environmental Damage

What evidence supports the hypothesis that current agricultural practices are unsustainable? First, the production and application of agricultural chemicals such as pesticides and fertilizers requires large amounts of nonrenewable fossil fuels; second, the water sources used to provide irrigation water are limited and under increasing demand; third, the environmental costs of fertilizer and pesticide use may be greater than the human population is able to bear.

Precious Liquids: Oil and Water. Petroleum is required not only to run large pieces of farm equipment, but also for use as a raw material for the production of fertilizer and pesticides. While scientists and oil industry experts disagree about if and when the world's oil supply will run out, it is clear that demand for oil continues to increase, causing prices to climb as a result. Dependence on the finite and ever-more-costly resource of oil makes current agricultural practices unsustainable.

Some environmental scientists argue that freshwater is the "new oil." Several statistics support this view. In the United States, 65% of total water withdrawals not used in power generation are used to irrigate crops. In western states, the portion of water use devoted to irrigation exceeds 90%. During the 1990s, it became clear to geologists studying the Ogallala aquifer (**Figure 23.27**)

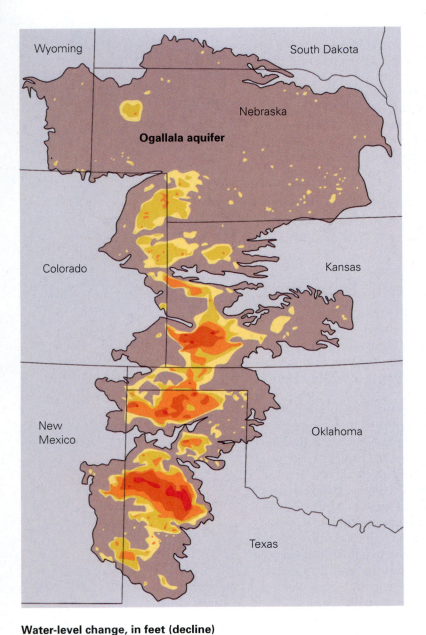

Water-level change, in feet (decline)

☐ 10–25   ☐ 26–50   ☐ 51–100   ☐ 101–150   ☐ More than 150

Figure 23.27 Unsustainable water use. The Ogallala aquifer is the largest single pool of freshwater in the world; it underlies and supplies the "bread basket" region of the United States. However, water is being removed from the aquifer much more rapidly than it refills, leading to dramatic declines in some areas. This figure illustrates the drop in water level in areas of the aquifer in the time period 1980–1995.

that irrigation in many of the regions above the aquifer was using water much more quickly than it was being replaced by precipitation. Analysis of the rate of water loss indicates that some areas of the aquifer will dry up within decades.

In addition to its reliance on nonrenewable or limited resources, such as oil and freshwater, many of the processes and inputs of modern agriculture can cause serious environmental damage.

Fertilizer Pollution.   The minerals in inorganic fertilizer that are not taken up quickly by growing plants are carried away by water moving through the soil. This runoff accumulates in water reservoirs in lakes, streams, and oceans or deep underground in aquifers. When the nutrients from inorganic fertilizer reach high levels in the water, it becomes unsafe for human consumption. High levels of nitrate fertilizer in water wells in the American Midwest correlate with greater risks of miscarriage and bladder cancer in women who use this water. Some forms of nitrogen in drinking water can cause brain damage or death among infants who drink it in infant formula.

Fertilizer runoff from farms can also cause eutrophication (described in Chapter 15) and result in extensive fish kills (**Figure 23.28**). If the use of fertilizer increases as human populations grow, the loss of drinking water sources and fishing grounds will probably also increase.

The Problem of Pesticides.   Over the past 50 years, the amount and toxicity of pesticides applied to crops in the United States has increased 10-fold. Pesticides directed toward one crop pest can cause another, secondary pest to thrive if the second pest is a competitor with the original target. In addition, pest populations can evolve resistance to pesticides that target them. To control populations of secondary pests or now-resistant target pests, higher levels of pesticide or new, more toxic pesticides must be applied. This situation is called the pesticide treadmill because farmers must continually expend en-

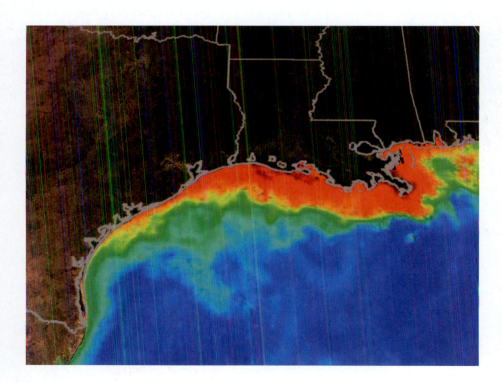

Figure 23.28 Dead zone.   This satellite image shows the dispersal of fertilizer-laden sediment—the orange colors in the photo—flowing out of the mouth of the Mississippi River. Thousands of square miles of the Gulf of Mexico are deadly to fish for much of the year as a result of the eutrophication this runoff causes.

Figure 23.29 Rachel Carson. Silent Spring, Rachel Carson's most influential book, made well known the hypothesis that pesticides like DDT bioaccumulate and can kill birds and negatively affect people.

Stop & Stretch Explain how natural selection causes a pest population to become resistant to a pesticide.

ergy and resources simply to "stay in place" and keep pest populations under reasonable control.

One side effect of the increase in pesticide use has been an increase in the number of incidents of wildlife poisonings, especially of birds. Although trends can only be inferred from sketchy reports, it appears that the number of wildlife poisonings has tripled since 1980. Even at low doses, many pesticides may have effects that do not cause death but may weaken individuals and make them more susceptible to predation or less able to compete—or may even affect their endocrine system (Chapter 21).

An additional problem associated with some types of pesticides is **biomagnification,** the concentration of persistent toxic chemicals at higher levels of a food web. Biomagnification is a product of the trophic pyramid, the bottom-heavy relationship between the biomass (total weight) of populations at each level of a food chain (Chapter 15). Thanks to the activism of ecologist Rachel Carson (**Figure 23.29**), the problem of biomagnification was recognized in the late 1960s—particularly the effects of magnified concentrations of the pesticide DDT in fish-eating birds such as bald eagles. This recognition led to a ban on DDT application. However, many persistent pesticides are still used in the United States and elsewhere, and these can accumulate in any organism that is high on a food chain—including humans.

Long-term health effects of low-level exposure to pesticides are still mostly unknown, but some evidence suggests that these effects include increased risks of cancer, birth defects, and permanent nerve damage. The pesticide treadmill demands ever-more-powerful and toxic chemicals, and high pesticide levels that result from biomagnification are likely to negatively affect more and more of the human population.

Wise use of fertilizers and pesticides may help minimize the environmental effects of modern agriculture; these negative consequences fade when their use is reduced. However, one consequence of modern agriculture results in environmental damage that is not easily recovered—the loss of soil via erosion.

Soil Erosion. Soil is held together by the organisms living within it and by the roots of plants growing within it. Modern agricultural techniques destroy soil in two ways: by regularly removing the plants (through tilling or harvest) and by the application of inorganic fertilizer, which is in the form of salts and can be deadly to soil organisms. When soil loses its structure, it is easily picked up by strong winds and blown away—the Dust Bowl and the dust storms in China are dramatic examples of this type of soil erosion caused by agricultural practices.

Irrigation water running over the surface of disturbed soil can also result in erosion. In addition, irrigation can cause **salinization**—the "salting" of soil. Water from underground aquifers contains tiny amounts of mineral salts. When the water is applied to the soil, some evaporates, leaving the salts behind. Eventually, so much salt accumulates that the soil becomes completely infertile. The production of soil from rock and organic material takes time, and soil lost through erosion and salinization is not easily replaced. Agricultural practices that rapidly degrade soil are unsustainable.

If we also consider the effects of rapid climate change on agriculture, it seems clear that current agricultural practices will not be sufficient to feed a growing human population indefinitely. Fortunately, some alternatives to these practices may help provide enough food more sustainably.

How to Reduce the Damage

A more **sustainable** agricultural system is one that meets the needs of the current generation of humans without compromising the ability of future generations to meet their needs. One path to sustainability focuses on requiring changes in agricultural practices. However, another path is to encourage consumers to change their eating and food-buying habits.

Reducing Harmful Inputs. We have already reviewed many of the farming techniques that can reduce the environmental costs of agriculture. Minimizing monocultural plantings may be the single most effective strategy. Not only does planting polycultures reduce the need for pesticides, it also helps minimize fertilizer inputs and preserves soil by increasing crop rotation. Additionally, planting polycultures helps to insulate farmers from environmental factors that destroy a single crop. Moving away from monocultural production will not be an easy task; it must be preceded by changes in government policies and agricultural economics. For instance, government subsidies that pay farmers to produce a few commodities, such as wheat, corn, cotton, and rice, encourage monocultural production.

Even without a switch to polyculture, farmers have good reason to reduce their use of fertilizer and pesticides—after all, these inputs cost money. One technique employs technology to closely monitor the production of agricultural fields so that fertilizer and pesticide applications are targeted more effectively.

Another strategy to reduce inputs has been adoption of GMOs that produce pesticide directly. In non-GMO crops, farmers can employ **integrated pest management (IPM),** which utilizes releases of predator insects, introduction of pest competitors, and changes in planting techniques to reduce pesticide use. Unfortunately, neither GMO nor IPM use has been a silver bullet. Since the introduction of both strategies in conventional agriculture, pesticide and fertilizer use has still continued to rise in the United States.

The Role of the Consumer. Farmers are only one-half of the food production equation. Consumers also determine how and what food is produced. If consumers consistently select certain types of foods, farmers will try to supply them. In many cases, consumer demand encourages some of the more damaging environmental consequences of modern agriculture.

Most of the pesticides sprayed on vegetable and fruit crops are applied to reduce crop "losses" that are caused when consumers refuse to purchase fruits and vegetables with superficial signs of pest damage. The only way to produce pristine products is to apply hundreds of pounds of pesticides to completely eliminate pest insects. If consumers were willing to accept a visible but small amount of pest damage to fresh vegetables, then the amount of pesticides used on these crops could drop tremendously (**Figure 23.30**).

Consumer demand for meat also fuels unsustainable farming practices. Most of the beef, pork, and poultry produced in the United States comes from animals that

(a) Conventional apple

(b) Organic apple

Figure 23.30 Cosmetic pesticide use. (a) This apple was produced on a conventional farm, with high levels of pesticides to control superficial damage. (b) This apple was produced on an organic farm, with no pesticides.

Figure 23.31 Modern meat production. Most of the meat and dairy products available in modern supermarkets are produced in high-density "factory farms," such as the one pictured here.

are fed field-grown grain in enormous feed lots, called factory farms (**Figure 23.31**). The same production process holds for hogs, chickens, and turkeys. In fact, 66% of the cereal grain consumed in the United States, including 80% of the corn and 95% of the oats, is used to feed these animals. Because much of the energy consumed by an animal is used for maintenance rather than growth, the grain used as animal feed yields much fewer calories in the form of meat. Put another way, a 10-acre field of corn can support 10 people if they eat the corn directly, but only 2 people if they feed the corn to cattle and then eat the beef.

When individuals consume many of their calories as conventionally produced meat and dairy products, they are forcing the high-intensity farming that causes the environmental problems discussed in this chapter. Reducing meat consumption would also reduce the number of acres farmed; as a result, some soils that are highly susceptible to degradation through erosion or salinization could be left undisturbed. Less land needed for agricultural production also leaves more natural habitat for other species to survive.

Factory farming of cattle, hogs, and poultry carries environmental costs in addition to its prodigious use of crops. For example, the U.S. Department of Agriculture estimated that the meat industry produced 2 billion tons of animal waste in 2005; this is 100 times the amount of waste produced by the human population of the United States. Failures of manure storage ponds and inadequate disposal have led to spills, water contamination, and fish kills. In addition, the high-density populations of animals in factory farms provide prime conditions for the spread of disease. Many meat producers in the United States handle this risk by feeding large amounts of antibiotics to all of their animals, sick and healthy alike. The use of antibiotics in agricultural settings has led to the development of antibiotic resistance in several dangerous human pathogens, including E. coli O157:H7, a bacterium found in beef that sickens thousands of people—and kills over 100—every year.

Figure 23.32 A consumer movement. Sales of organic foods have grown at a rate of 20% per year over the past 15 years thanks to changes in the buying choices of millions of individuals.

Organic Farming. There is little doubt that consumer demand can change agricultural practices. In recent years, many consumers with concerns about the costs of modern agriculture have changed their buying habits (**Figure 23.32**). As a result, some farms have switched to organic farming, a practice that shuns the use of chemical fertilizers and pesticides as well as GMOs. The U.S. Department of Agriculture reported 20% growth in the retail sales of organically grown food in every year from 1990 to 2007. Organic farm acreage in the United States increased by 400% between 1992 and 2005. Organic farms are still very much the exception in American agriculture, representing less than 1% of the total land in agriculture, but consumer interest in organic food is clearly having an effect.

The focus of organic farming is the natural maintenance of soil fertility (using crop rotation and organic fertilizers) and the development of biological diversity on the farm (using polycultures) as a method to control pests and provide income stability. Organically grown food has lower levels of pesticide residues and is produced in a way that minimizes the environmental costs of agriculture (**Table 23.3**). Organic food does often cost more for consumers; grocery bills average 60% higher according to several analyses, although the cost differences decrease when consumers purchase seasonal foods and buy directly from the grower at farm markets. And because food is relatively inexpensive in the United States, a 60% increase in food cost for a family of four totals about $3500 a year—about what an average family spends annually on meals at restaurants. Organic food also does not avoid other costs seen in the conventional agricultural system, including the transport of food over long distances, although purchasing from local organic producers can minimize that environmental impact.

Feeding the human population in a way that preserves environmental quality and leaves room for biodiversity will require thoughtful decision making. These decisions include those we make as individuals, such as how much meat to consume and what sort of produce we buy, and those we make collectively, such as how federal subsidies for crops are distributed. Better knowledge of plant and agricultural science and an understanding of the power and limitations of scientific knowledge can help you become an effective participant in this decision-making process.

TABLE 23.3

Reducing the environmental impact of modern agriculture. This table summarizes the environmental costs of modern agriculture along with the individual and collective actions we can employ to reduce these costs.

Environmental Problem	Solution	Actions We Can Take
Bioaccumulation of persistent pesticides 	• Minimize monocultural plantings. • Reduce use of pesticides. • Increase use of integrated pest management.	• Accept produce that has superficial pest damage. • Purchase organic foods. • Add diversity of foods to our diets to promote polyculture. • Reduce meat consumption and thus overall crop production. • Support policies that reduce use of monocultural production (for example, elimination of subsidies for crops). • Support integrated pest management research.

(continued)

TABLE 23.3 (*continued*)

Reducing the environmental impact of modern agriculture. This table summarizes the environmental costs of modern agriculture along with the individual and collective actions we can employ to reduce these costs.

Environmental Problem	Solution	Actions We Can Take
Eutrophication of surface water and nitrate pollution of groundwater 	• Reduce use of inorganic fertilizer.	• Purchase organic foods. • Purchase foods grown in polyculture, where crop rotation is employed. • Reduce meat consumption.
Salinization of soil 	• Reduce number of acres irrigated. • Increase efficiency of irrigation.	• Reduce meat consumption. • Support research and national agricultural policies to reduce agricultural water use.
Soil erosion and desertification 	• Reduce use of tilling on marginal lands. • Reduce amount of land used for crop production.	• Reduce meat consumption. • Support research on effective nonchemical means of weed control. • See also suggestions regarding population and aid to developing countries.
Loss of habitat for nonhuman species 	• Minimize the amount of land converted to agricultural production.	• Reduce meat consumption. • Support policies that effectively lead to decreased human population growth rates. • Support policies that provide developing countries with the tools to increase crop yield sustainably.

SAVVY READER

The Benefits of Organic Food

Is Going Organic Worth the Price?

Organic foods are taking on new popularity, despite their higher costs. But is it really worth the extra expense to buy organic? …

We also know that studies show that by eating organic food you can greatly reduce your exposure to chemicals found in conventionally produced food. But critics argue that there is no evidence that regular foods pose health risks, and there is no evidence that organic foods reduce your health risk.

Some people choose organic foods because they appreciate the effort by farmers to be environmentally friendly.

Dr. Francisco Diez-Gonzalez teaches food science and nutrition classes at the University of Minnesota. He has done extensive research on organic food and says while they are a great option for consumers, they are not "superfoods," as some people seem to believe.

"In my opinion, the only really substantiated claim [is] that if you are buying organic vegetables you are definitely reducing your pesticide intake," he said. …

Diez-Gonzalez said conventionally produced fruits and vegetables have pesticide levels lower than the tolerance level regulated by the government. Those foods are quite safe to eat. …

He also says people shouldn't [feel] badly if they choose not to buy organic.

"You have to remember that the conventionally produced veggies or crops have pesticide levels below the tolerance level regulated by government. They are there, but way below the tolerance level," he said.

1. What is the hypothesis the reporter is attempting to "test" in her story? Is it a scientific hypothesis?
2. What evidence does she report that sheds light on this hypothesis?
3. What evidence (of the costs or benefits of organic food) is missing from this story? What additional information would you need to determine if organic food is "worth" the cost?

WCCO Television News, Minneapolis, MN. Reporter: Angela Davis, October 8, 2007.

Chapter Review

Learning Outcomes

LO1 **List the tissues and structures found on angiosperm plants, and describe their functions (Section 23.1).**

- Nearly all major crops belong to the angiosperm phylum, which are flower- and fruit-producing plants (p. 566).
- Roots absorb water and minerals from the soil but also may be modified to store carbohydrates (pp. 568–569).
- Stems serve as support for leaves and reproductive organs but may also be modified for carbohydrate storage (p. 570).

MasteringBIOLOGY®

Go to the Study Area at www.masteringbiology.com for practice quizzes, myeBook, BioFlix™ 3-D animations, MP3Tutor sessions, videos, current events, and more.

- Leaves are the photosynthetic organs of the plant and contain some essential nutrients for humans (p. 570).
- Flowers contain both male and female reproductive organs (p. 571).

LO2 **Define *alternation of generations*, and describe how the importance of the two plant generations has changed over evolutionary time (Section 23.1).**

- Land plants produce two multicellular generations: a haploid gametophyte and a diploid sporophyte. The sporophyte produces a haploid spore, which develops into the gametophyte, and the gametophyte produces sperm and eggs by mitosis (p. 574).

- The earliest-evolving land plants—such as the mosses—look like their green algae ancestors, in that the gametophyte stage is the most prominent stage. Seedless vascular plants arose from these ancestors and have a dominant sporophyte, with spores for dispersal. The more modern seed plants have a much reduced gametophyte stage that is fully dependent on the sporophyte and produce a seed for dispersal (pp. 574–575).

L03 Describe the process of reproduction in angiosperms, including the role of double fertilization (Section 23.2).

- When pollen is transferred from the male to the female angiosperm, two sperm are released. One sperm fertilizes the egg, and the other fuses two other nuclei inside the ovule, a process called double fertilization (pp. 571–572).
- The resulting seed remains inside the flower structure, which becomes modified into a fruit that functions in seed dispersal (pp. 572–573).
- Seeds may survive difficult environmental conditions by becoming dormant. The emergence of a plant embryo from a seed is called germination (pp. 571–573).

L04 Compare and contrast primary and secondary growth in dicot plants (Section 23.2).

- Dicot plants undergo primary growth, in the length of stems and roots, and secondary growth, in the width of these organs (pp. 573–578).
- Primary growth occurs because of cell division in the apical meristems. This meristem can produce additional meristems, leading to indeterminate growth patterns in plants (p. 577).
- Secondary growth in dicot stems occurs when the vascular cambium produces additional xylem, which becomes wood, and phloem, which becomes the inner part of the bark (pp. 577–578).

L05 Describe the role of water availability in maximizing a plant's exposure to sunlight and carbon dioxide (Section 23.2).

- Plants require abundant water for photosynthesis, for effective interception of light, and to maximize the uptake of carbon dioxide. Farmers not only provide water through irrigation but also remove noncrop plants that compete for water (pp. 578–580).
- Plants regulate water loss but also cut off the uptake of carbon dioxide by closing the stomata on their leaf and stem surfaces (p. 581).

L06 Provide examples of important nutrients for plant growth, and distinguish between how these nutrients occur in natural systems and how they are made available in agricultural systems (Section 23.3).

- Soil is a matrix composed of minerals from eroded rock, dead and decaying material, and the living organisms that decompose and mix the components. Soil is the major source of nutrients for plant growth (pp. 581–582).
- In agricultural systems, nutrients are mined from the soil. Farmers make up for the loss of nutrients by adding fertilizer and growing nitrogen-fixing crops (pp. 582–583).

L07 Describe the costs associated with modern agriculture, and list some possible ways these costs could be reduced (Section 23.3).

- Widespread use of pesticides has led to monocultural production of crops. Before pesticides, polycultural plantings kept pest populations down and insulated farmers from pest outbreaks (pp. 584–585).
- Petroleum and water are essential to modern agriculture and are in limited or finite supply (p. 586).
- Fertilizer runs off into waterways, causing human health problems and damage to fish and other animal populations (p. 587).
- Farmers continually increase pesticide use as pests become resistant (pp. 587–588).
- Pesticides can be persistent and accumulate in organisms at higher levels of the food web, including humans (p. 588).
- In many areas, soil is being lost through erosion and salinization faster than it is being replaced (pp. 588–589).
- Fertilizer and pesticide use can be minimized using polyculture, by crop rotation, and by careful monitoring of the agricultural environment (pp. 587–588).
- Consumers can help reduce the environmental costs of modern agriculture by accepting a small amount of pest damage on produce, eating less meat, and purchasing organically grown products (pp. 589–592).
- Organic agriculture may provide a sustainable model for future food production (pp. 590–591).

Roots to Remember

The following roots of words come mainly from Latin and Greek and will help you decipher terms:

angio- means vessel. Chapter term: angiosperm

enni- pertains to a year. Chapter term: perennial

-phyte means plant. Chapter terms: gameotophyte, sporophyte

rhizo- means root. Chapter term: rhizome

sperm- means seed. Chapter terms: angiosperm, gymnosperm

Learning the Basics

1. **LO1** Sketch and label the parts of a plant and flower.

2. **LO2** Describe the generalized life cycle of plants—alternation of generations—and explain how this life cycle has been modified over evolutionary time.

3. **LO1** All of the following plant organs have been modified in certain plants to store large amounts of carbohydrate, except _____.

   A. fruit; **B.** stems; **C.** roots; **D.** leaves; **E.** apical meristems

4. **LO1** One difference between roots and leaves is _____.

   A. roots have vascular tissue, but leaves do not; **B.** roots can store carbohydrates, but leaves cannot; **C.** leaves have only phloem, roots only xylem; **D.** root epidermis absorbs water, leaf epidermis conserves it; **E.** leaves are only produced by flowering plants, roots by all plants

5. **LO3** As a result of fertilization in flowering plants, _____.

   A. pollen is picked up at the stigma by an animal for transfer to an anther; **B.** a single sperm is released by a pollen grain; **C.** a seed begins to absorb water and break free of its coat; **D.** a triploid tissue called endosperm is formed; **E.** a carpel is produced that later becomes a seed

6. **LO4** In dicot woody plants, secondary growth results in _____.

   A. production of new xylem and phloem; **B.** increase in the girth or width of the stem; **C.** the production of additional nodes and internodes; **D.** A and B are correct; **E.** A, B, and C are correct

7. **LO5** The function of stomata is _____.

   A. to prevent pests from damaging the photosynthetic surfaces of the plant; **B.** to allow carbon dioxide to enter the plant body; **C.** to sense water availability in the air; **D.** to provide stability for a growing plant; **E.** to increase nutrient availability in soil

8. **LO4** The first organ to emerge from a germinating seed is the _____.

   A. axillary bud; **B.** root; **C.** cotyledon; **D.** endosperm; **E.** stem

9. **LO6** Soils in agricultural systems require fertilizer because _____.

   A. farming mines nutrients from the soil; **B.** weeds compete with crop plants for soil nutrients; **C.** most crop plants have nitrogen-fixing bacteria in their roots and need lots of nitrogen; **D.** most nutrients run off soils into waterways, causing eutrophication; **E.** pests remove most nutrients from crop plants

10. **LO7** When inorganic fertilizer ends up in waterways, _____.

    A. the waterways become eutrophic; **B.** algae populations explode in surface waters; **C.** fish kills result; **D.** drinking water may become contaminated; **E.** all of the above

Analyzing and Applying the Basics

1. **LO3** Agricultural scientists have been successful at creating only a few types of seedless fruits through traditional breeding—notably bananas, apples, grapes, pineapples, navel oranges, and a few melons. Why might it be difficult to produce seedless fruits?

2. **LO4** One way to make a houseplant look fuller is to cut off its primary growing tip. Based on what you've learned about primary growth, how does this action promote a bushy form?

3. **LO6 LO7** Your friend has been planting the same variety of tomatoes in the same spot in his garden for the past 8 years. In recent years, he has noticed that his plants are less healthy, and that pests damage more of the tomato fruits. Apply what you have learned in this chapter to explain why his tomato crop is declining and suggest several ways that he can improve his tomato yield.

Connecting the Science

1. The human population relies on approximately 6 staple crops and about 24 additional crops for most of our calories. In the United States, nearly 80% of the cropland is planted with three species: corn, wheat, and soybeans. What are the risks and benefits of relying on so few agricultural species? Are you concerned about the lack of diversity in agricultural production? Why or why not?

2. Have you ever considered the environmental impact of your diet? Are you willing to change your diet to reduce its environmental impact? Why or why not?

Answers to **Stop & Stretch, Visualize This, Savvy Reader,** and **Chapter Review** questions can be found in the **Answers** section at the back of the book.

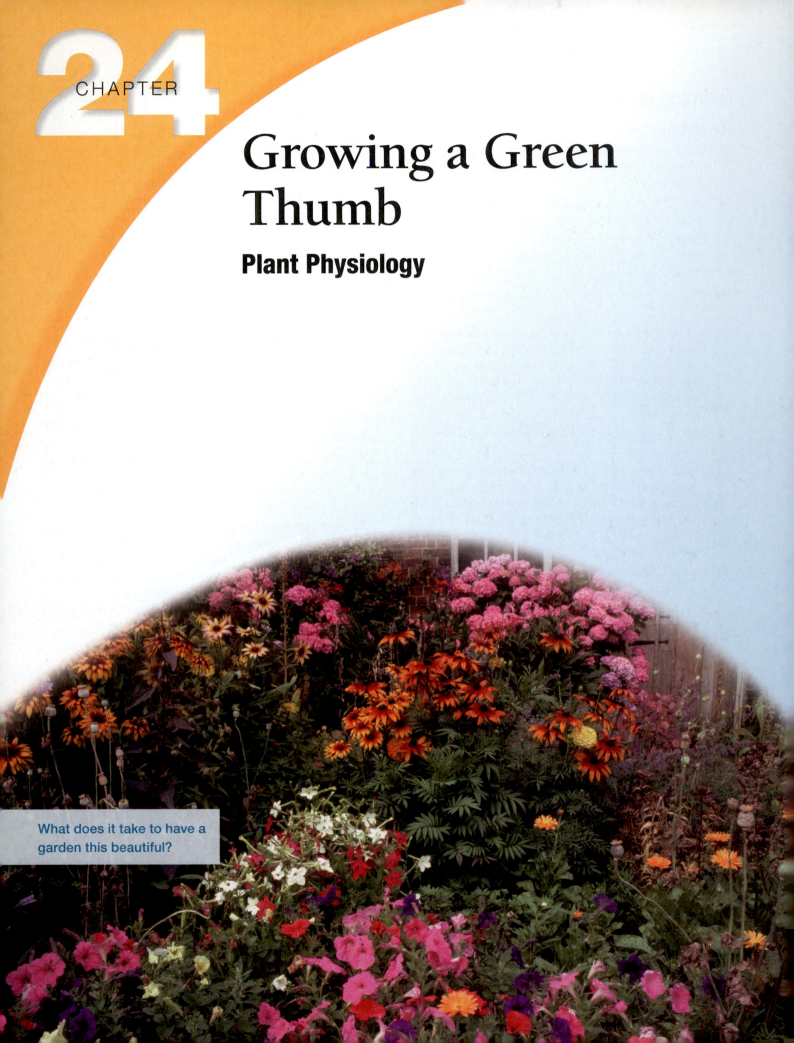

CHAPTER

24

Growing a Green Thumb

Plant Physiology

What does it take to have a garden this beautiful?

Many of us know people who are described as "having a green thumb"—men and women whose homes are a jumble of thriving houseplants, who produce enough vegetables in their backyard gardens to saturate friends with excess tomatoes and zucchini, or who have flower gardens that make the neighbors, well, green with envy. How do they do it? Is it simply a gift, a green thumb that some are born with and some without?

A significant contributor to a gardener's success is time. Years of trial and error, weeks of preparation for a growing season, and hours of work tending to gardens and indoor plants will create an experienced grower who can get the most out of his or her plants. Part of what a gardener gains through experience is a more thorough understanding of how plants work. This knowledge helps a plant lover choose the right plants for the environment, design a landscape that is always in bloom, and even shape the form of plants to fit his or her needs.

Fortunately for those of us who are eager but inexperienced gardeners, it is possible to start developing our green thumb by learning the basics of plant function. Our discussion of plant physiology fits within the context of planning a plant-filled space, such as a garden or houseplant collection. Let's get growing!

How can you make a backyard vegetable plot yield abundant produce . . .

. . . or fill your home with thriving houseplants?

It helps to have knowledge of plants, gained through experience and understanding.

LEARNING OUTCOMES

LO1 Describe how water moves up a plant stem.

LO2 List the modifications of plant physiology and anatomy that reduce water loss or make plants more drought tolerant.

LO3 Describe the pressure flow mechanism of phloem transport.

LO4 Define *photoperiodism*, and explain the mechanism by which plants respond to day length.

LO5 List the environmental factors that cause tropism in plants, and describe how differences in cell expansion lead to directional growth.

LO6 Define "plant hormone," and describe a number of ways in which one hormone, auxin, affects plant growth.

24.1 The Right Plant for the Place: Water Relations

One of the first considerations of gardening is choosing plants that can thrive in the available environment. Many factors influence plant growth in a particular space, including the soil's chemical nature, light availability, and the length of the **growing season,** which is the time from the last frost in the spring to the first frost in the fall. However, one of the most important factors influencing plant production is the amount of water available to the plant. How do plants obtain and manage water, and how can we use our understanding of this process to help plants thrive?

Transpiration

Water and dissolved minerals obtained from the soil, together called **xylem sap,** travel up the xylem, the water-conducting cells inside a plant. This occurs primarily as a result of the loss of water vapor from the leaves through evaporation. Through this process, called **transpiration,** the plant essentially pulls a column of sap from the soil, through the roots, up tubes made of dead xylem cells in the stem, and into the leaves. Xylem sap can be pulled via transpiration from the roots to the leaves of even the tallest trees (over 100 meters, or 330 feet) because of the hydrogen bonds formed by water molecules (Chapter 2). Transpiration evolved in plants after they colonized land 400 million years ago, and it has allowed plants to colonize nearly all areas of Earth's surface.

The tendency for identical molecules to stick together, such as water molecules linked by hydrogen bonding, is referred to as **cohesion;** the tendency for unlike molecules to stick together, such as when water is hydrogen bonded to other polar molecules, is called **adhesion.** Water adheres to cellulose in plant cell walls. Cohesion and adhesion together maintain the continuity of a sap column during transpiration.

The pulling force on a column of xylem sap is generated by the evaporation of water from the leaves (**Figure 24.1**). Stomata (singular stoma), the pores on leaf surfaces, control the rate of gas exchange and water loss in land plants. When water evaporates out of stomata, the forces of cohesion and adhesion on the water molecules remaining inside the leaves create **tension,** or negative water pressure. Tension is the force that allows a column of water to be pulled into a syringe when the plunger is pulled up. In plants, tension causes water molecules from the xylem to be drawn out into the leaf to replace the water that has evaporated. This in turn increases tension on the water molecules immediately below them in the xylem, causing them to move toward the leaves, and so on, all the way down to the roots and soil.

When a stem is injured, the continuity of the stream of water is disrupted and the tension is released. As a result, water pulls away from the cut surface. This leaves an air gap that remains even when the cut stem is placed in a water-filled vase. Because the stream of water in the xylem is no longer continuous, water from the vase cannot replace water evaporating from the xylem, and the stem will rapidly dry up and wilt. Gardeners who understand this know either to cut flowers in the early morning, when the plants are full of water and tension is lowest—producing little or no air gap—or to reestablish a continuous stream of water by cutting the flower stem a second time, once it is placed in a water-filled vase.

While transpiration is a process that requires only the energy of the sun, causing evaporation, and the strong hydrogen bonds in liquid water, plants can actively modify the rate of transpiration by regulating the size of stomata (Chapter 23). In addition to these active movements, certain plants also possess

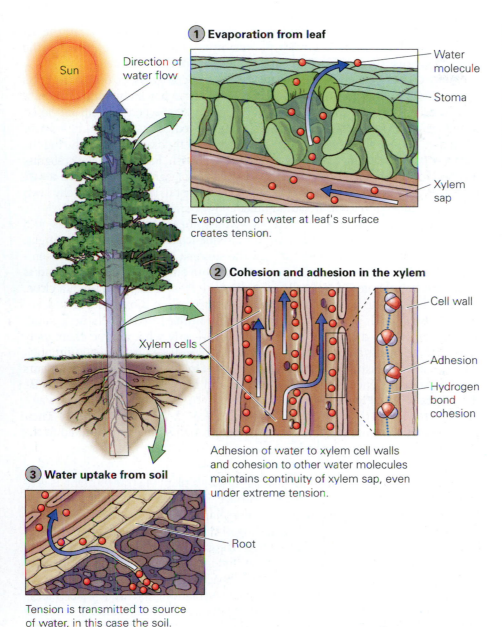

① **Evaporation from leaf**

Water molecule

Stoma

Xylem sap

Evaporation of water at leaf's surface creates tension.

② **Cohesion and adhesion in the xylem**

Xylem cells

Cell wall

Adhesion

Hydrogen bond cohesion

Adhesion of water to xylem cell walls and cohesion to other water molecules maintains continuity of xylem sap, even under extreme tension.

③ **Water uptake from soil**

Root

Tension is transmitted to source of water, in this case the soil.

Sun

Direction of water flow

Figure 24.1 Transpiration. Evaporation of water from the leaves creates a tension that is transmitted to a column of water in the xylem. Water flows upward against the pull of gravity and maintains its continuity thanks to the forces of cohesion and adhesion.
Visualize This: How are the conditions in the tree different at night? Does transpiration occur then?

adaptations to photosynthesis or leaf shape that can affect the rate of transpiration. The presence or absence of these adaptations helps determine which plants are best suited for particular climates and environments.

Adaptations That Affect Transpiration

Plants have evolved several strategies for reducing transpiration water loss. These strategies fit into two general categories: (1) modifications to photosynthesis that make carbon dioxide acquisition more efficient and (2) modifications to leaf shape and stomata placement that increase water conservation.

Photosynthetic Adaptations. Two adaptations to a plant's photosynthetic pathways help reduce water loss: C_4 and CAM photosynthesis (Chapter 5). Plants with C_4 photosynthesis have an additional chemical process that precedes the Calvin cycle, the step that converts carbon dioxide into sugar in all plants. The Calvin cycle is also known as C_3 photosynthesis. The additional

Figure 24.2 A C$_4$ plant. Crabgrass has the C$_4$ photosynthetic pathway, which gives it an advantage over Kentucky bluegrass in hot, dry midsummer and late summer.

Figure 24.3 Jade, a CAM house-plant. Plants with CAM photosynthesis are adapted to very dry conditions and thus are ideal houseplants to survive dry indoor air.

Figure 24.4 Small leaves limit transpiration loss. In desert plants, small leaves help reduce the surface area for water evaporation. Some leaves may be so reduced that they are modified to spines.

chemical cycle in a C$_4$ plant essentially pumps carbon dioxide from the air toward the chloroplasts, allowing photosynthesis to proceed even when the stomata are just barely open and carbon dioxide uptake is limited.

Many C$_4$ plants, including corn, sugarcane, and crabgrass, belong to the grass family. These plants have an advantage over C$_3$ plants in hot, sunny, and dry conditions because they will continue photosynthesizing even when stomata are closed to reduce water loss. Homeowners across the United States can see the relative advantage of C$_4$ crabgrass over lawn grasses like Kentucky bluegrass during the warmest parts of summer. At these times, the yellowish, broad leaves of crabgrass will become dominant as growth of the less tolerant bluegrass slows (**Figure 24.2**). If the crabgrass spreads, patches of bluegrass may die off, leading to a patchy lawn when the temperature cools in late summer and the crabgrass dies off.

CAM plants, including all cacti, as well as pineapples, many orchids, and aloe and jade plants, accumulate carbon dioxide from the air during the night, when less water will evaporate through the stomata because of the cooler temperatures. The carbon dioxide is converted to a simple carbohydrate and is stored in the vacuoles of leaf cells as malic acid. At sunrise, the stomata close, and the accumulated acid is degraded, releasing carbon dioxide for photosynthesis. The growth of a CAM plant is greatly limited by the relatively small amount of carbon dioxide it can use in a day—only what can be chemically stored. These plants are most common in extremely dry environments, although they can be successful in other environments. One of the driest environments a plant can experience is a house in winter, and many popular houseplants use the CAM process (**Figure 24.3**).

So, what about the plants without either C4 or CAM photosynthesis? Their strategies for reducing the rate of water loss through stomata are limited to the structural adaptations of their leaves.

Leaf Adaptations. The amount of water lost by a plant via transpiration is generally correlated to its photosynthetic surface. In particular, a plant with many large leaves tends to transpire more water than does a plant with fewer small leaves. Gardeners use their understanding of this relationship between leaf surface area and transpiration to help transplanted plants survive. Because removing plants from soil (no matter how carefully) inevitably results in root damage, newly transplanted plants have a reduced capacity to acquire water. To compensate for this, gardeners remove leaves when transplanting to reduce water loss.

Of course, decreased leaf area also means less photosynthesis, so plants with smaller leaves tend to grow more slowly than plants with broader leaves in the same environment. Plants that thrive in shady sites often have broad leaves because evaporation is lower and intercepting light is more crucial, while narrow-leaved plants tend to be more successful in bright sun. Leaf size is reduced to its extreme in cacti, which have photosynthetic stems and leaves modified into spines (**Figure 24.4**). Cacti grow quite slowly, but they make for another ideal houseplant that can survive even the most inattentive gardener.

Some broad-leaved plants have other adaptations to reduce water loss, such as the placement of stomata on the cooler, shaded undersurface of leaves or in hair-lined pits, which trap water molecules and thus reduce water loss even when the stomata are open (**Figure 24.5**). Where periods of drought occur reliably on an annual basis, many broad-leaved plants have evolved to become **deciduous,** that is, to actively drop their leaves to reduce the rate of transpiration. In Section 24.2, we describe this process in greater detail.

Stop & Stretch Even if they are watered at the same rate, houseplants are less likely to dry out when they are displayed in groups rather than individually. Why would grouping pots of plants improve their water retention?

In environments where water is abundant, plants compete with each other for access to light. In these environments, faster-growing plants often have an advantage. Plants that have evolved strategies for maximizing transpiration, including changes in water-conducting tissues, can grow rapidly and quickly tower over their neighbors.

Xylem Adaptations. When a column of xylem is under extreme tension inside a plant, the forces of cohesion and adhesion may not be strong enough to prevent the water column from breaking. Breakage of the water column results in the formation of a bubble, called an embolism, that inhibits water flow within the xylem tubes. In general, larger-diameter xylem tubes are more likely to suffer embolisms than are smaller-diameter tubes. Many drought-adapted species have only very narrow or tapering xylem tube cells, called **tracheids** (**Figure 24.6a** on the next page).

However, the narrow passageway produced by a column of tracheids leads to increased friction, reducing the rate of water flow—a disadvantage in moist environments. In moist conditions, plants with wider, perforated **vessel elements** are favored because they move water to their leaves up to 10 times faster, resulting in a proportionally faster growth rate (**Figure 24.6b**).

The diameter of the xylem tubes in any given plant thus represents a trade-off between the risk of embolism and maximum growth rate. For example, in cold climates, where liquid water may be unavailable for a large part of the year, trees are at high risk of embolism and have much less porous-looking wood than do trees in warmer, wetter climates where embolism is rare but competition for light is intense. Focus on Evolution illustrates how the distribution of divisions of plants on Earth reflects the trend of adaptation to increasingly drier environments.

The growth of a plant in a natural landscape is strongly influenced by the balance between water availability and transpiration rate. However, in gardens and houses, water is rarely limited as long as the gardener pays attention to the plant's needs. In fact, a common problem of houseplants in the hands of inexperienced gardeners is not excessive dryness but too much water.

Epidermal hairs   Stomata

Figure 24.5 Shaded stomata help conserve leaf water. The stomata of this leaf are found on the shaded underside, where temperatures are lower and evaporation slower. The hair-lined pits trap evaporating water molecules and keep humidity high around them.

Overwatering

Excess water, by filling the air spaces in soil, prevents root cells from obtaining oxygen. As a result, the water-absorbing root hairs die, decreasing the transpiration rate and causing the plant to wilt. Inexperienced gardeners respond to an overwatered, wilting plant by adding more water, further destroying the roots and eventually killing it. As a rule of thumb, potting soil should be allowed to become completely dry down to 2 centimeters below the surface before more water is added.

Overwatering is a typical problem for houseplants, but garden and agricultural plants can also suffer from excess moisture. Soils that are high in clay, the tiniest mineral particles, are prone to becoming waterlogged because each clay particle can trap a shell of water around it, eventually filling up all of the soil's air spaces and leading to the same symptoms as those in overwatered houseplants. Gardeners with clay soils often need to amend the soil before planting to avoid this problem. The most common soil amendment is organic matter, such as compost or peat moss. Additions of organic matter help separate clay particles, provide air holes in the soil, and have the additional benefit of adding nutrients when the organic matter decomposes. Sandy soil, which has large particles and cannot hold as much water, can be amended by adding water-holding organic matter as well.

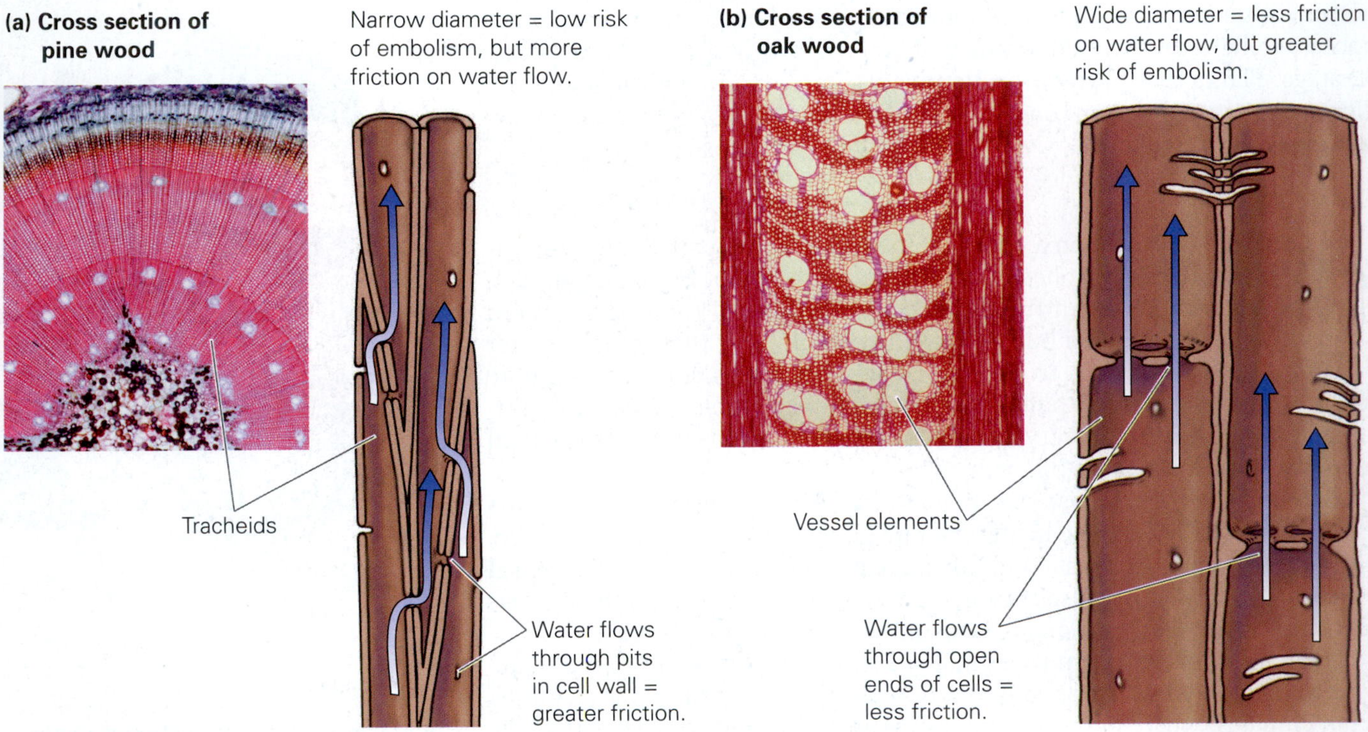

(a) Cross section of pine wood

Narrow diameter = low risk of embolism, but more friction on water flow.

(b) Cross section of oak wood

Wide diameter = less friction on water flow, but greater risk of embolism.

Tracheids

Vessel elements

Water flows through pits in cell wall = greater friction.

Water flows through open ends of cells = less friction.

Figure 24.6 Xylem cells. Water flows through xylem tubes made up of one or two cell types. (a) Plants in dry environments have a high proportion of long, narrow tracheids. (b) In moist environments, short, wide vessel elements are the most dominant xylem cell type.
Visualize This: Wood grain is the pattern seen when a tree is sliced lengthwise, parallel to the xylem cells. How do you think grain differs between these two species?

Even in the best soils, overwatering can become a problem when water remaining on leaves and the soil surface encourages the growth of bacteria and fungi. In addition, because plants produce roots where water is available, consistently wet soils promote shallow rooting and thus less-stable plants. In most areas of the country, plants require about 2.5 centimeters (1 inch) of water per week, through rainfall or supplemental watering, for optimal growth. Experienced gardeners know to look for signs of water stress (limp or wilting plants, a yellow or bluish cast to leaves) and to check how dry the soil is a few centimeters below the surface before watering their plants.

Stop & Stretch If you have a plant that is suffering from overwatering, one solution is to remove a large number of leaves while waiting for the soil to dry out (a process that could take weeks if the soil is really soaked). Why does removing leaves help the plant survive this period?

A gardener's ability to supplement natural rainfall allows the growth of water-needy plants even in dry landscapes. However, it is far more difficult for gardeners to control the temperatures to which their landscape plants are exposed, so most garden plant choices are based on temperature requirements. Ability to tolerate freezing is one factor that determines whether a plant can survive in a particular temperature range. Cold tolerance is primarily a function of a plant's ability to control the freezing of water within its cells.

Plants on Land

The ancestors of plants were green algae, photosynthetic organisms that spent their lives immersed in water. For plants to thrive on land, adaptations that allowed them to move water from the ground to the leaves were required.

The earliest land plants were small and close to the ground, with no water- or food-conducting tissues. Some groups of mosses evolved primitive water-conducting tissues, but true xylem (and phloem, the nutrient-conducting tissue) evolved later, in the ancestors of modern ferns and seed plants. The most modern group, flowering plants, possesses additional adaptations that greatly increase the speed of both water and food transport. These adaptations have allowed flowering plants to dominate Earth's surface.

Thalloid Liverworts

The simplest land plants, the hornworts and liverworts, consist primarily of thin, platelike (thalloid) structures with no roots or stems (**Figure E24.1**). These plants obtain water via diffusion from wet soil, and carbon dioxide enters the plant through stomata-like pores on the plant surface. Hornworts and liverworts have no internal water-conducting tissues; therefore, they must remain small and are limited to land areas that are consistently damp.

Figure E24.2 Moss.

Ferns

Ferns belong to the oldest group of plants with true xylem and phloem—the seedless vascular plants (**Figure E24.3**). Tracheids first appeared in the ancestors of modern ferns. Because tracheids have stiff secondary walls that provide a skeleton of support for a stem, they allow true vascular plants to grow much taller than nonvascular plants. More efficient water transport also allows vascular plants to grow in much drier places. The phloem cells of ferns and other seedless vascular plants are also identical to those in all seed plants and probably evolved at the same time. The vascular tissue in ferns forms a solid ring of mixed xylem and phloem tubes just under the stem's surface, forming a sturdy scaffold to support the stem.

Figure E24.1 Thalloid liverwort.

Figure E24.3 Fern.

Mosses

The center of the "stem" of many mosses consists of water-conducting cells called *hydroids* (**Figure E24.2**). Like xylem, hydroids are dead cells that persist in the plant only as hollow tubes of cell walls, but unlike true xylem, hydroids are thin and do not function as support structures. Some groups of moss possess food-conducting cells called *leptoids*, which are similar in structure to true phloem.

Conifers

Gymnosperms, like the pine tree in **Figure E24.4** (on the next page), are plants that produce seeds but no flowers. Gymnosperms organize their vascular tissue in discrete bundles scattered within the stem. Production of new vascular tissue during secondary growth (growth in girth) occurs at

(continued on the next page)

Figure E24.4 Conifer.

a ring of tissue called the vascular cambium. The vascular cambium produces new phloem to the outside of the stem, which will become bark that flakes off as the stem expands, and new xylem to the inside, making up the wood (Chapter 23). The rings of wood produced by yearly vascular cambium growth provide a solid core to support the increasing height of a tree.

Flowering Plants

The most successful group of plants on Earth is the angiosperms, or flowering plants (**Figure E24.5**). A key adaptation that appears to have contributed to their success is more efficient vascular tissue. Angiosperm vessel elements produce large-diameter xylem tubes and thus can move water from soil to plant tissues more rapidly than can narrower tracheids. Angiosperm phloem tubes are also wider and more efficient, possibly allowing for the more rapid growth of organs in response to environmental resources.

Figure E24.5 Flowering plants.

Water Inside Plant Cells

Ice formation in plants is damaging in three ways: (1) When water outside plant cell walls turns into ice, liquid water from inside the cells will begin flowing out by osmosis—that is, down a concentration gradient across the cell membrane (**Figure 24.7**). As the cell loses increasing amounts of water to the ice crystal, normal cell processes become disrupted and may cease to function. (2) When water freezes inside the central vacuole of a plant cell, it produces an ice crystal (Figure 24.7). The jagged edges of this crystal can pierce the membrane, allowing cytosol (the fluid inside the cell) to leak into the vacuole, thereby killing the cell. (3) When ice freezes within xylem tubes, air comes out of the water solution, forming embolisms and blocking future water uptake. All three of these processes can severely damage or kill plants.

A plant's tolerance of cold temperatures, known as its **hardiness,** depends on its ability to either prevent or manage freezing inside cells. Water freezes at temperatures lower than 0°C (32°F) when it has solutes dissolved in it, in the same way that added antifreeze prevents water in a car's radiator from freezing. Tomato plants can survive

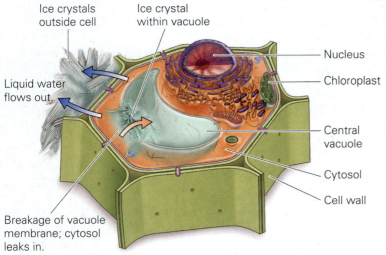

Ice crystals outside cell

Ice crystal within vacuole

Nucleus

Chloroplast

Liquid water flows out.

Central vacuole

Cytosol

Cell wall

Breakage of vacuole membrane; cytosol leaks in.

Figure 24.7 Freezing inside plant cells. Ice outside plant cells can cause the cell to dehydrate, interfering with normal functions. An ice crystal inside the central vacuole can pierce its membrane, allowing the cytosol to leak into the vacuole.

temperatures down to −1°C (31°F), apple fruits −4°C (24°F), and cabbage as low as −9°C (16°F) due to their sugar-based cell antifreeze. At lower temperatures, certain adaptations can improve freezing tolerance. These adaptations include controlling ice formation within cells to prevent sharp edges and internal structures that prevent xylem embolisms.

Many cold-tolerant plants require both a gradual cooling period to become maximally hardy and uninterrupted cold temperatures to remain hardy. Cold snaps that occur in late summer and early fall may damage plants that can normally thrive in even colder temperatures because they have not become **hardened,** or fully able to resist the cold. Woody plants can become damaged by cold even after they are fully hardened when bright winter sunshine warms their stems, allowing water to transpire that cannot be replaced by frozen water in the soil. This is a common occurrence in evergreens in landscaping, which can suffer from winter "burn." Freezing temperatures are most likely to cause damage during the spring, however, when intolerant leaves and flowers are beginning to emerge from formerly cold-tolerant buds.

Because of the extreme damage that freezing temperatures can cause, gardeners must know two things before planning an outdoor garden: (1) the hardiness zone of their landscape, that is, how cold it gets (**Figure 24.8**), and (2) the rated hardiness—that is, the cold tolerance—of the plants they wish to grow. Experienced gardeners know, however, that with special care, plants that are less cold tolerant can survive freezing conditions. Plants that are grown outside their hardiness zone can be placed on the north sides of buildings, where they are shaded from the sun (preventing winter burn) and where cooler local temperatures retard budding and leafing until the threat of frost has decreased. Plants can also be covered with mulch such as straw or snow, and they can be protected from unusually early or late frosts by covering with plastic tents.

Less tolerant plants are also better able to survive cold winters if they have sufficient water and sugar reserves. These reserves allow the plants to produce adequate antifreeze at the appropriate times during the hardening process. Gardeners can ensure that their sensitive plants are ready for winter by giving them plenty of water during the fall and by preventing them from using up their stored starch and sugar reserves.

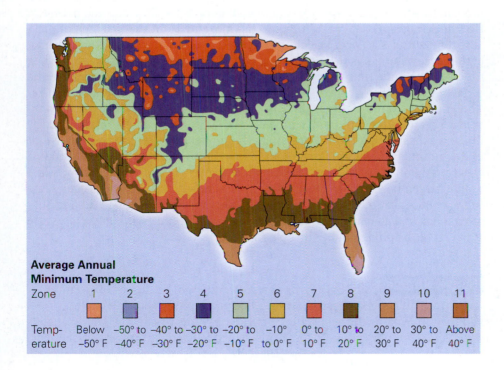

Figure 24.8 Hardiness zones. The U.S. Department of Agriculture generated this hardiness zone map based on average lowest temperatures experienced in thousands of sites across the country.

Average Annual Minimum Temperature

Zone	1	2	3	4	5	6	7	8	9	10	11
Temperature	Below −50° F	−50° to −40° F	−40° to −30° F	−30° to −20° F	−20° to −10° F	−10° to 0° F	0° to 10° F	10° to 20° F	20° to 30° F	30° to 40° F	Above 40° F

To best manage a plant's sugar supply, a gardener needs to understand how plants store and move sugar. This knowledge helps gardeners not only ensure that their favorite plants survive harsh conditions but also maximizes the beauty and productivity of these plants.

24.2 A Beautiful Garden: Sap Translocation, Photoperiodism, and Flower and Fruit Production

Each gardener has personal goals for a garden—from a summer-long supply of fresh vegetables, to a continually changing palette of showy flowers, to an aesthetically pleasing but low-maintenance decor. Because plants grow throughout their lives, each plant can allocate its resources in many possible directions at any time—to greater leaf production, flower production, fruit size, and so on. By understanding how plants allocate resources, gardeners can encourage plants to transfer resources to their different plant organs at appropriate times.

Translocation of Sugars and Nutrients

Sugars and other dissolved nutrients are moved around a plant's body in phloem tissue. The solution in the phloem, or **phloem sap,** always flows from a **source,** where sugar is in high concentration, to a **sink,** where nutrients are being used or accumulated. As a result, the movement of phloem sap, called **translocation,** can be in any direction within a plant. Different organs can act as sources and sinks throughout the year; in fact, separate phloem tubes in the same branch can be translocating sugars in opposite directions at the same time. For instance, some tubes in an apple tree branch may be sending sugar to the roots for storage, while others are sending sugar to the branch tips for fruit production.

Translocation of phloem sap results from pressure differences within a phloem tube between a source and a sink. **Figure 24.9** illustrates the **pressure flow mechanism** that causes phloem sap movement. In the first step of pressure flow, sugar is actively loaded into a phloem tube at the source. As sugar accumulates, water flows in passively via osmosis from nearby tissues and xylem tubes. At the same time, respiring cells at the sink are acquiring sugar from the phloem tube, either through active transport or passively, down a concentration gradient from the sugary phloem sap into a sugar-consuming cell. As a result of this movement into the sink cells, the sugar concentration in the nearby phloem tubes declines and water exits the phloem tube passively. As a result of the movements of sugar at the source and the sink and subsequent osmosis, water pressure in the phloem tube is high at the source and low at the sink. Just as in a garden hose, in which high water pressure at the faucet and low pressure in the open air cause water to flow out, this pressure differential causes the bulk movement of water from the high-pressure source end to the lower-pressure sink end of the phloem tubes in a plant.

Stop & Stretch Maple syrup producers tap the phloem tubes of sugar maple trees in early spring to draw off the sugary phloem sap. Before they add the tap, what is the sugar source in the tree, and what is the sink? When a tap is inserted, how does the identity of the source or sink change?

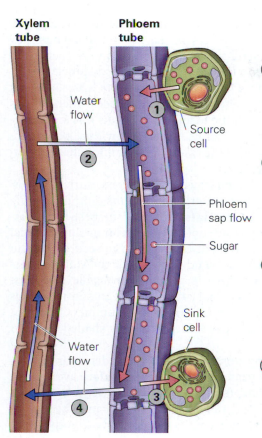

Xylem tube **Phloem tube**

Water flow

Source cell

Phloem sap flow

Sugar

Water flow

Sink cell

① Sugar is actively loaded into phloem tubes at the source.

② Water flows into phloem passively, down its concentration gradient, increasing the water pressure.

③ Sugar moves into sink cell (actively if it is a storage cell, passively if it is a growing or dividing cell).

④ Water flows out of the phloem tube passively, down its concentration gradient, decreasing the water pressure.

Figure 24.9 The pressure flow mechanism of phloem transloca-tion. Phloem sap moves from source to sink because of pressure differentials in the phloem tube.

(a) Annual plants

(b) Biennial plants

Managing Translocation

The key feature of translocation in plants is that materials always move from source to sink. Gardeners who recognize which plant organs are function-ing as sources or sinks at any given time can modify a plant's structure to promote flowering, increase the quantity and quality of produce, and prepare plants for winter.

Promoting Flowering. How a gardener manages translocation to increase flower production varies depending on the plant type. Plants that complete their life cycle within a single year and do not survive past that year are called annuals. Many common garden flowers, including petunia, pansy, morning glory, and cosmos, are annuals (**Figure 24.10a**).

The evolutionary strategy of annual plants is to produce as many seeds as possible in their single year of life; to do this, annuals produce many flow-ers to increase their chances of pollination and fertilization. The larger the sugar source available for flower production, the greater the number of flowers

(c) Perennial plants

Figure 24.10 Plant seasonal growth cycles. (a) Annual plants live only a single season or year, growing from a seed and producing a new seed crop at the end of their lives. Their evolutionary strategy is to put a significant amount of energy into flower produc-tion, making them ideal for garden displays. (b) Biennial plants typically make a low-profile set of leaves in the first year and send sugar into a storage root. In the second year, sugar from storage supplies a large floral display, after which the plant dies. (c) Perennials live for many years, putting energy into flower production and storage roots every year. Keeping competition to a minimum helps maintain garden perennials.

(sinks) that can be produced. To manage flower production in annuals, gardeners remove, or **prune,** many of the flowers on the young transplants they purchase to encourage the plant to send sugars to different sinks—meristems that are developing more leaves. Once the annual has reached sufficient size, the plant can be encouraged to produce flowers instead of more leaves; this can often be accomplished by restricting both water and fertilizer, which signals the plant that the growing season is coming to an end and thus flower production is a priority. If the plant is grown just for its flowers, individual blooms are removed once they fade (a process called *deadheading*) to eliminate the competitive sink of developing fruits and seeds.

Plants that live for 2 years—producing leaves the first year, surviving dormancy by storing nutrients in roots or underground stems, and producing flowers the second year—are called **biennials.** Some of the best-loved garden flowers, including hollyhock, forget-me-not, and delphinium, are biennial plants (**Figure 24.10b**). Gardeners who plant biennials from seed know that the role of the plant in its first year is to send sugars into the storage roots so that the floral display the following year is as large as possible. The most effective way to do this is to encourage leaf growth early in summer, thus maximizing the amount of sugar sent into storage in late summer. Typically, this means ensuring that the biennial leaves are not overly shaded by taller neighbors.

Plants that live for many years by storing nutrients in underground structures and producing flowers every year are called **perennials.** Favorite garden perennials include phlox, peony, daylilies, and tulips (**Figure 24.10c**). Perennials with stiffened aboveground stems, such as trees and shrubs, are called **woody plants.** Most garden perennials do not have woody stems and instead die back every year. Because of their long lives, the evolutionary strategy for perennials is more balanced—sinks are evenly divided between flower and fruit production and storage organ production. As with annuals, if perennials are grown for flowers, they are deadheaded to minimize the fruit and seed sink.

Many gardeners will remove perennial flowering stalks well before the last flower has bloomed so that adequate sugars go into the storage sink before **dormancy,** the plant's yearly rest period. In addition, because the underground storage organs continue to enlarge and become crowded over time, the stems and roots of older perennials may have difficulty accessing water and light. To deal with this problem, gardeners must regularly divide perennials by lifting the storage organs from the soil, removing old tissue that no longer produces buds, and replanting smaller, more vigorous portions.

Increasing Produce Quality and Quantity. When annuals, biennials, and perennials are grown for fruit or vegetable production, the goal is to increase the number and quality of these sinks. Many of the same techniques for flower production apply. For example, for annual crops, making sure that the plant puts energy into producing leaves and stems early ensures better production of fruit later. Other gardening strategies focus on different sinks; for instance, plants grown for leaves or stems instead of fruit (such as lettuce and asparagus) may be actively discouraged from flowering or setting seed because these activities will rob their edible parts of quality.

In many crop plants grown for fruit, gardeners must strike a balance between the number of fruits produced and their quality. If plants are allowed to continue making new fruit sinks, each fruit will be smaller and less flavorful because the sugar has to be divided among more sinks. Experienced gardeners know to prune additional tomato blossoms after a plant has a significant load of fruit and to pick off a proportion of the developing apples on a tree to produce larger, more flavorful fruit (**Figure 24.11**).

Although most fruits do not contain vascular tissue, the same source-sink principle can be applied when harvesting fruits to maximize flavor. Once a fruit is harvested, living cells within the fruit continue to act as a sink; however, because the fruit no longer has an external source, it begins to use its own stored sugars as an energy source. As these sugars are used up, the flavor of the fruit suffers. Plant cells use sugar at higher rates when they are warm, so keeping fruit as cool as possible during harvesting helps retain maximum sweetness. To keep vegetable quality high, gardeners try to harvest during the coolest parts of the day and load their vegetables in baskets that allow for air movement.

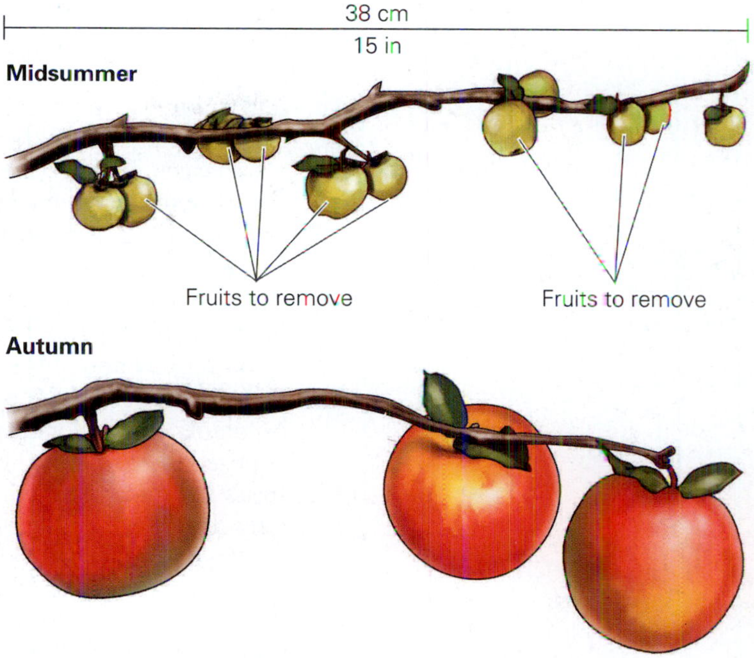

Preparing Plants for Winter. Many plants produce carbohydrate antifreeze throughout the late fall. Perennials with inadequate carbohydrate stores are susceptible to freezing damage, so gardeners must discourage new leaf growth, which uses up stored carbohydrates, late in the summer. Avoiding pruning and fertilization in late summer and fall keeps growth down. Managing fruit production so that moderate numbers are produced and harvesting fruit early also increase the amount of sugar that can be stored over the winter.

Figure 24.11 Maximizing fruit quality over quantity. A flavorful apple requires about 40 leaves to support it. A good rule of thumb is to thin fruit until they are spaced about the width of your hand apart.

Stop & Stretch The best time to prune woody plants in seasonal areas is at the end of their dormant period, right before they open their leaf buds. According to your understanding of source-sink relationships, why is this the best time?

Modifications of plant sources and sinks work for gardeners because plants develop through indeterminate growth, in which their ultimate size and shape are subject to environmental conditions rather than a genetic program. However, other responses of a plant to its environment result from innate traits. In particular, the timing of plant flowering and the beginnings of dormancy are often controlled by genetic factors.

Photoperiodism

In the early part of the 20th century, scientists from the U.S. Department of Agriculture began studying a unique tobacco plant that had appeared in a Maryland farmer's field. This plant grew to twice the height of normal tobacco plants and continued producing leaves until the end of the growing season. Obviously, farmers were thrilled with this variety, except for one small problem—it rarely flowered in the field and thus produced little seed with which to sow an entire crop.

The scientists who worked on the strange tobacco suspected that the "Maryland mammoth" was not flowering because it was not responding to day length in the same way that other tobacco plants did. Through a process of trial and error, researchers discovered that these tobacco plants could be triggered to flower if the day length (that is, total sunlight) they experienced was shorter than a critical number of hours. Later investigations of other plants illustrated that many, like tobacco, exhibit **photoperiodism**—a biological response that is tied to a change in the proportion of light and dark within a 24-hour cycle.

Tobacco, like chrysanthemums, strawberries, primroses, and poinsettias, is considered a short-day plant because it will not flower if day length surpasses a critical value. Short-day plants tend to bloom in spring or fall. Other plants will flower only if day length is longer than a critical value. These long-day plants, like irises, spinach, and lettuce, are typically summer flowers. Not all plants exhibit flowering photoperiodism. Those that do not are called day-*neutral* plants.

Long-day and short-day plants are actually responding to the concentrations of two forms of a protein called **phytochrome.** Phytochrome is produced by plants in one form that changes chemical shape, or conformation, when exposed to red light, a wavelength that is abundant in sunlight (**Figure 24.12**). In its light-activated conformation, phytochrome acts to trigger flowering in long-day plants but delays flowering in short-day plants.

Phytochrome is produced continuously within a photoperiodic plant. The switch between the inactive and active forms of phytochrome is nearly instantaneous on exposure to light, but because proteins are regularly recycled within cells, the active form disappears after a time when no light is present. The amount of time that active phytochrome is present provides the measure of day length. Because a few seconds of light will convert phytochrome to its active form, scientists more accurately measure a photoperiodic plant's **critical night length,** the minimum amount of darkness required to inhibit the effects of phytochrome. If night is shorter than this critical length, the active form of phytochrome can exert its effect (**Figure 24.13**). Indoor gardeners can manipulate a photoperiodic plant's exposure to artificial light during the night, or shade it during the day, to promote or prevent flowering.

Flowering is not the only activity that is triggered by light cues. The opening of leaf buds in spring and the dropping of leaves in fall are also photoperiodic in many woody plants. **Abscission,** or leaf drop, is a complex process that is first triggered by increasing night length as winter approaches. In preparation for abscission, leaves cease to produce chlorophyll for photosynthesis. As chlorophyll levels decline, other pigments in the leaf become visible. As the leaf turns, cells at the boundary between leaf and stem break down, allowing the

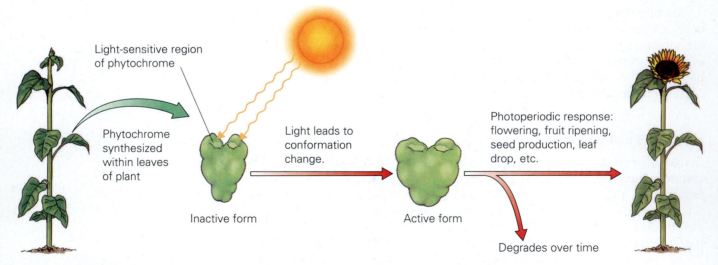

Figure 24.12 How plants tell time. The ratio between the active and inactive form of phytochrome allows plants to "sense" day length. A plant synthesizes phytochrome in an inactive form, while exposure to sunlight turns it into its active form. In the active form, phytochrome can have a number of effects, but it is slowly degraded over time.

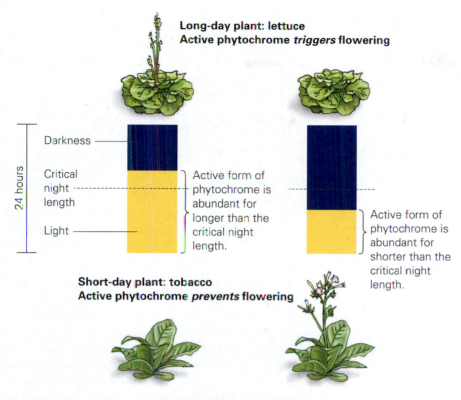

Long-day plant: lettuce
Active phytochrome *triggers* flowering

24 hours

Darkness

Critical
night
length

Light

Active form of
phytochrome is
abundant for
longer than the
critical night
length.

Active form of
phytochrome is
abundant for
shorter than the
critical night
length.

Short-day plant: tobacco
Active phytochrome *prevents* flowering

Figure 24.13 Phytochrome, short-day plants, and long-day plants. The presence of the active form of phytochrome has the opposite effect on short-day and long-day plants.
Visualize This: Lettuce is typically planted so that it reaches edible size in early spring (April and May) and in fall (August to October). Why is it less common to plant lettuce in late spring so that it reaches large size in midsummer?

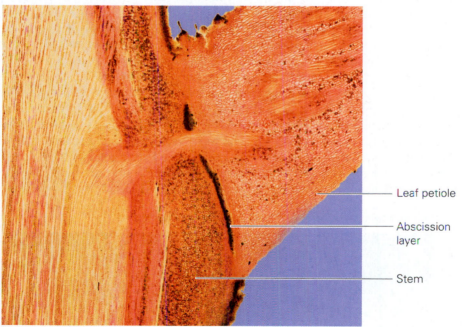

Leaf petiole

Abscission
layer

Stem

Figure 24.14 Abscission. Leaf drop occurs because cells at the junction between a leaf and a stem break down, forming a weak layer that separates and leaves a layer of scar tissue.

leaf to separate from the stem and leaving a protective cap over the junction (**Figure 24.14**).

Experienced gardeners know that they should match the photoperiod of a plant's native environment to their own environment. The same species of plant growing at different latitudes may have different critical night lengths and will thus respond differently to light conditions. Spinach, strawberries, and onions are all especially susceptible to being planted "out of place" and thus flower, fruit, and produce bulbs too early or too late in the growing season. Southern varieties of widespread deciduous tree species can suffer from damage if planted so far north that they freeze before dropping their leaves.

In addition to finding the right plant for the site and ensuring that their plants produce when and how they wish them to, gardeners are also concerned with the aesthetic form of their gardens. For many, this means actively managing the way that plants fill and share space with each other. To modify form, we must understand how plants grow in response to environmental cues.

24.3 Pleasing Forms: Tropisms and Hormones

On the surface, plants seem to be captives to their environment. They cannot run away when something tries to eat them. They cannot seek shade on hot days or find a sheltered spot from a freezing wind. However, plants are not as inert as they appear. We learned in the preceding section that many plants are able to measure day length. Here, we will see that plants can respond to many different environmental signals, sometimes in surprising ways.

Tropisms

Although a rooted plant cannot physically move itself from one environment to another, plants can grow toward the best possible environment. Directional growth in plants is called **tropism.** Tropic responses are visible in various plant organs in response to different environmental cues. Roots in germinating seeds show positive **gravitropism** (growing downward toward the source of gravity), while stems show negative gravitropism (**Figure 24.15a**). This pattern ensures that emerging plant roots maximize exposure to water in the soil and that young shoots maximize exposure to light. Houseplants often demonstrate **phototropism,** directional growth in response to light, when placed on a windowsill (**Figure 24.15b**). Other plants grow in response to touch, exhibiting **thigmotropism.** Vining plants show thigmotropism by twining around support stakes (**Figure 24.15c**).

Indoor gardeners can use their understanding of plant tropisms to maintain the beauty of their plants. Rotating houseplants regularly ensures that the phototropic growth of a plant is balanced around the whole pot. Handling plants occasionally by touching their leaves can promote a thigmotropic response of shorter, sturdier stems. It may be that the widespread belief that talking to plants is good for them comes from this response—they grow fuller not because of the kind words, but because of the increased amount of touch and movement that a well-tended plant experiences.

Figure 24.15 Plant tropisms.
(a) This plant stem is exhibiting negative gravitropism, orienting itself opposite the pull of gravity. The plant was allowed to reorient in the dark, indicating that the response is not to light. (b) This houseplant is growing toward a light source, exhibiting phototropism. (c) The tendril on this vine is curled around the support as a result of thigmotropism.

(a) Gravitropism

(b) Phototropism

(c) Thigmotropism

Plants bend toward light, or away from gravity, when cells on one side of a stem increase in length and cells on the other side do not, forcing the top of the plant to turn away from the expanding cells. This differential cell expansion is caused by different amounts of growth hormone in these cells.

Hormones

Plant hormones regulate a plant's internal environment and control its responses to environmental conditions. Plant hormones are similar to animal hormones in several ways: (1) They are produced in tiny amounts but have large effects, (2) they are mobile so that they can be produced in one organ and have an effect on another, and (3) they have multiple and complex effects and interactions. **Table 24.1** lists the five major plant hormones, their effects, and their commercial applications. We will focus on the first plant hormone identified, and the one most commonly manipulated, auxin.

Auxin is the hormone that triggers the expansion of cells and is thus responsible for causing plants to increase in size. In tropic responses, auxin is destroyed in some cells in response to an environmental signal. As the cells that retain auxin grow in length while those without it remain small, the organ itself bends away from the auxin-containing side (**Figure 24.16**).

Like many hormones, auxin can have opposing effects on different plant organs. While auxin causes cells to expand in a main stem, it inhibits the growth of axillary buds along the sides of the stem. Thus, auxin is responsible for **apical dominance**—the tendency for the plant apex, or uppermost bud, to grow more rapidly than the lower buds. Auxin produced at the shoot tip strongly inhibits branching right below the tip. Further from the shoot tip, auxin concentrations are lower, and its inhibitory effects are less dramatic. Auxin-mediated apical dominance is responsible, for instance, for the "Christmas tree" shape of young conifers.

Apical dominance increases the height and leaf area of a plant but suppresses bushiness as well as flower and fruit production. For this reason, gardeners are often interested in reducing the amount of apical dominance that

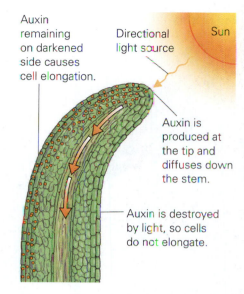

Auxin remaining on darkened side causes cell elongation.

Directional light source

Sun

Auxin is produced at the tip and diffuses down the stem.

Auxin is destroyed by light, so cells do not elongate.

Figure 24.16 Directional growth.
Tropisms are caused when one side of a plant organ (in this case, a phototropic stem) grows more rapidly than the opposite side, causing the tip to bend.

TABLE 24.1

Plant hormones.		
Hormone Type	**Effects**	**Commercial Applications**
Auxin	Promotes cell expansion, apical dominance, tropic responses, differentiation of vascular tissue; promotes roots on plant cuttings; inhibits fruit and leaf abscission; stimulates fruit development; stimulates ethylene production (in high doses)	Rooting powder for plant cuttings; prevents fruit and flower drop; promotes seedless fruit production; used as weed killer by triggering too-rapid growth
Gibberellin	Stimulates leaf and stem cell elongation and division; hastens seed germination, breaks seed dormancy; stimulates fruit development	Used to cause uniform germination of barley used in brewing; prevents flower initiation; promotes seedless fruit production
Cytokinin	Promotes cell division; triggers lateral bud growth even in presence of auxin; with auxin, regulates root and shoot production; delays leaf aging	Used to trigger growth of whole plants from genetically engineered plant cells
Ethylene	Inhibits cell expansion; promotes fruit ripening; promotes abscission	Hastens ripening of tomatoes, walnuts, and grapes after harvest; used as fruit-thinning agent in commercial orchards; used to defoliate plants
Abscisic acid	Prevents seed germination; promotes dormancy	Inhibits growth and promotes flowering of potted plants for florists

Early spring

Pruning cut

Late summer

Axillary buds, released from apical dominance, produce new branches.

Figure 24.17 Overcoming apical dominance. Apical dominance can be countered by light pruning, removing the terminal bud. Loss of the terminal bud reduces auxin concentration, lifting growth suppression on the lateral buds, which sprout.

a plant exhibits. The primary method for reducing apical dominance is light pruning, also called pinching back (**Figure 24.17**). When the terminal bud is pruned, the plant's auxin levels are reduced, inhibition on lateral buds is lifted, and branches form. Eventually, these branches begin to exert apical dominance and must also be pinched back to maintain a plant's bushy form.

Auxin also interacts with other growth regulators. Abscission is promoted by the hormone **ethylene.** Recall that during abscission, cells at the boundary between the stem and a leaf, flower, or fruit break down so that the abscised organ can separate from the plant. Auxin reduces the sensitivity of boundary layer cells to the effects of ethylene and thus delays abscission. Levels of auxin production are positively correlated to a plant's overall health; as a result, stressed plants tend to abscise fruits and leaves earlier than do healthy plants. The anti-abscission role of auxin makes it useful in commercial operations; for instance, citrus growers spray their orchards with auxin to prevent early fruit drop.

Stop & Stretch Roses are among the most sensitive cut flowers. They drop their petals more rapidly if they are stored in a cooler with ripening fruit. Why might this be the case?

All of the plant hormones identified by scientists have been adopted for some commercial uses. Knowledge of the effects of these hormones, along with knowledge of transpiration, translocation, and photoperiodism, can contribute to the development of a gardener's "green thumb." But this knowledge has a deeper meaning as well. The discovery and human use of plant hormones reflects the way that evolution has created both highly complex biological processes and highly complex human brains. These brains have enabled us to modify organisms to meet human needs, and as a result, we have the opportunity to improve lives for ourselves, our descendants, and other species.

As this chapter has illustrated, an experienced gardener understands that he or she gets the best results by working with natural processes, not by ignoring them. And as many of the stories in this book have demonstrated, this basic gardening principle can be a guide for all of our interactions with the biological world (**Figure 24.18**).

Figure 24.18 Earth as a garden. The future of humanity depends on the wise use of our biological heritage. Learning about the natural world ideally inspires people to live within natural processes rather than trying to dominate them.

SAVVY READER

The Benefits of Gardening

Houseplants Offer Health Benefits

by Dennis Douda

If you get a good feeling from having plants around the house, research shows you may be on to something.

Some scientists say plants can go beyond a mental boost and actually help people breathe more easily.

Former NASA scientist Dr. Bill Wolverton has researched plants for decades and insists they can actually help filter the air.

"In our studies with plants, we found that plants can absorb and destroy common toxic chemicals found in homes, such as formaldehyde (and) benzene," Wolverton said.

Wolverton says the plants can take low-level toxic chemicals often found in things such as carpets, furniture, appliances, computers and common cleaners and make them useful.

"They take what is a toxic chemical to us, and they convert it into a source of energy and food for them," Wolverton said.

. . .

The Environmental Protection Agency isn't completely sold, saying it would take too many plants to really make a difference.

. . .

Government officials say indoor contaminants, even at low levels, can cause allergic or respiratory reactions in some people.

Plant lovers also need to protect against mold. Wolverton says it's crucial not to over-water plants and suggests laying aquarium gravel over the topsoil.

1. How could you find the research this story is based on?

2. According to the story, how might houseplants actually reduce air quality inside a home?

3. Does this story provide any way to evaluate the relative costs and benefits of houseplants on indoor air quality?

WCCO-TV, Minneapolis, MN, May 21, 2006.

Chapter Review

Learning Outcomes

Go to the Study Area at www.masteringbiology.com for practice quizzes, myeBook, BioFlix™ 3-D animations, MP3Tutor sessions, videos, current events, and more.

LO1 **Describe how water moves up a plant stem (Section 24.1).**

- Water is drawn from the soil and into the stems and leaves of a plant because of evaporation at the leaf's surface (pp. 598–599).
- Loss of water results in tension, which pulls on the water column. The column maintains its continuity via cohesion among water molecules and adhesion between water and the xylem walls (pp. 598–599).

LO2 **List the modifications of plant physiology and anatomy that reduce water loss or make plants more drought tolerant (Section 24.1).**

- Certain plants have modifications to photosynthesis that reduce water loss. These include C_4 photosynthesis, which pumps carbon dioxide to photosynthetic cells, and CAM photosynthesis, which allows carbon dioxide to be harvested from the air at night (pp. 599–600).

- Smaller size, shaded stomata, and deciduousness are leaf adaptations that reduce transpiration (pp. 600–601).
- Narrow tracheids in xylem prevent the occurrence of embolisms, which destroy a plant's capacity to take up water. However, narrow vessels deliver water more slowly, making this trait a disadvantage in moist environments (p. 601).
- Overwatered plants wilt because their roots die from lack of oxygen (pp. 601–602).
- Cold-resistant plants control the rate of freezing inside cells to prevent cell damage and prevent or manage the ice-induced formation of embolisms in the xylem (pp. 604–605).

LO3 Describe the pressure flow mechanism of phloem transport (Section 24.2).

- Sap moves in phloem from source to sink. Sugar is loaded into phloem tubes at the source, causing water to flow in as well. Sugar is unloaded at the sink, and water flows out. The difference in water pressure between the tube at the source and the tube at the sink causes sap to travel through the phloem tubes (pp. 604–607).
- Gardeners can modify relationships between sources and sinks to ensure that energy is directed to the gardener's preferred plant organ (pp. 607–609).

LO4 Define *photoperiodism*, and explain the mechanism by which plants respond to day length (Section 24.2).

- Many plants exhibit photoperiodism—sensing the length of daylight and responding to it. One system for sensing light relies on a protein called phytochrome (pp. 609–611).
- Light-mediated responses in plants include the initiation of flower or fruit production, the emergence of leaves, and leaf abscission in deciduous trees (pp. 610–611).

LO5 List the environmental factors that cause tropism in plants, and describe how differences in cell expansion lead to directional growth (Section 24.3).

- Plants grow in response to environmental cues, such as the direction of a light source, gravity, and contact with other objects (pp. 612–613).
- Directional growth is caused by differential cell expansion on opposite sides of an organ (p. 613).

LO6 Define "plant hormone," and describe a number of ways in which one hormone, auxin, affects plant growth (Section 24.3).

- Plant growth regulators, called hormones, tune a plant to its environment and like animal hormones are produced in one organ but have effects on different organs (p. 613).
- Auxin causes cell expansion in stem cells but inhibits the growth of lateral buds, leading to apical dominance. Light pruning can reduce apical dominance (pp. 613–614).
- Auxin interacts with the hormone ethylene to control abscission (p. 614).

Roots to Remember MB

The following roots of words come mainly from Latin and Greek and will help you decipher terms.

ad-	means to. Chapter term: adhesion
co-	means with. Chapter term: cohesion
-chrome	means color. Chapter term: phytochrome
photo-	means light. Chapter terms: photoperiodism, phototropism
phyto-	means plant. Chapter term: phytochrome
trans-	means across. Chapter terms: transpiration, translocation
trop-	means to turn. Chapter term: tropism

Learning the Basics

1. **LO1** Why does a column of water in a xylem tube remain intact despite being under tremendous tension?

2. **LO2** How do the C_4 and CAM adaptations to photosynthesis reduce water loss from plants?

3. **LO5** How does light shining on a phototropic stem change conditions in the stem, and how does this change affect the stem's growth?

4. **LO1** Water moves up a plant's stem as a result of _____.

   A. the xylem pump; B. diffusion of water into roots; C. translocation of phloem sap; D. evaporation of water from the leaves; E. photosynthesis in xylem cells

5. **LO1 LO2** When water is under extremely high tension in a xylem tube, _____.

   A. transpiration slows; B. water cannot be removed from the soil; C. the water column can "break," blocking further water flow; D. stomata open wide to permit greater water absorption; E. the plant is not photosynthesizing

6. **LO2** Which of the following adaptations provide an advantage to plants in warm, moist environments?

   A. closing stomata in response to decreased water availability; B. large-diameter xylem vessels, permitting rapid water uptake; C. deciduousness, reducing water loss during dry periods; D. small leaf size, reducing sun exposure; E. vertically oriented photosynthetic stems, shading the stomata

7. **LO3** Sap travels in phloem as a result of _____.

   A. evaporation of sugar water from the leaves; B. the phloem pump; C. a countercurrent to xylem flow; D. differences in water pressure in a phloem tube between a sugar source and a sugar sink; E. cohesion and adhesion of water to the phloem walls

8. **LO3** Each of the following pairs represents a likely sugar source and its sink *except* _____.

A. an apple seed and the fruit in which it is contained; B. mature leaves of a tomato plant and a developing tomato; C. newly emerged leaves on a bean seedling and the seedling's developing roots; D. a sugar beet root in spring and its developing flower stalk; E. the roots of an oak tree and its leaf buds in spring

9. **LO4** A short-day plant can be prevented from flowering by _____.

A. keeping it constantly in the light; B. preventing phytochrome from converting into the active form; C. ensuring that day length does not exceed the plant's critical length; D. growing it in the same environment as a long-day plant; E. more than one of the above is correct

10. **LO6** Plant hormones are unlike animal hormones because _____.

A. they are produced only in small amounts; B. they have different effects on different organs; C. they are not produced in specialized glands; D. they are mobile and can move throughout the plant body; E. interactions among different plant hormones can determine their ultimate effect on an organ

Analyzing and Applying the Basics

1. **LO1 LO2** You have decided to divide an overgrown daylily and move a portion of it to a new spot in your yard. Thinking about water relations in plants, how should you prepare the site and the lily for transplanting?

2. **LO4** Poinsettias are short-day plants with a critical night length of about 14 hours. Describe how you could induce a poinsettia to flower during the long days of summer.

3. **LO6** Christmas tree farmers want to encourage apical dominance so that their plants have the "correct" form. They encourage this when plants are a few years old by pruning. What should they prune to maintain a single tip to the tree? What should they prune to ensure bushy lower branches?

Connecting the Science

1. In even the smallest apartments, there is usually enough window space and light to grow a few herbs or small lettuce plants, and people with larger yards can grow a significant amount of the fresh produce they consume during the growing season. Do you grow food to eat? Why or why not? Do you see a value in growing some of your own food?

2. In the last paragraph of the chapter, we suggest taking a gardener's approach—working with natural processes and having a desire to understand—in all of our interactions with the natural world. Describe how this approach could be applied to a current issue concerning human modification of nature. Do you agree that it might be helpful to "think like a gardener," or do you see disadvantages of this attitude as well?

Answers to **Stop & Stretch, Visualize This, Savvy Reader,** and **Chapter Review** questions can be found in the **Answers** section at the back of the book.

Appendix

Metric System Conversions

To Convert Metric Units:	Multiply by:	To Get English Equivalent:
Length		
Centimeters (cm)	0.3937	Inches (in)
Meters (m)	3.2808	Feet (ft)
Meters (m)	1.0936	Yards (yd)
Kilometers (km)	0.6214	Miles (mi)
Area		
Square centimeters (cm^2)	0.155	Square inches (in^2)
Square meters (m^2)	10.7639	Square feet (ft^2)
Square meters (m^2)	1.196	Square yards (yd^2)
Square kilometers (km^2)	0.3831	Square miles (mi^2)
Hectare (ha) (10,000 m^2)	2.471	Acres (a)
Volume		
Cubic centimeters (cm^3)	0.06	Cubic inches (in^3)
Cubic meters (m^3)	35.30	Cubic feet (ft^3)
Cubic meters (m^3)	1.3079	Cubic yards (yd^3)
Cubic kilometers (km^3)	0.24	Cubic miles (mi^3)
Liters (L)	1.0567	Quarts (qt), U.S.
Liters (L)	0.26	Gallons (gal), U.S.
Mass		
Grams (g)	0.03527	Ounces (oz)
Kilograms (kg)	2.2046	Pounds (lb)
Metric ton (tonne) (t)	1.10	Ton (tn), U.S.
Speed		
Meters/second (mps)	2.24	Miles/hour (mph)
Kilometers/hour (kmph)	0.62	Miles/hour (mph)

To Convert English Units:	Multiply by:	To Get Metric Equivalent:
Length		
Inches (in)	2.54	Centimeters (cm)
Feet (ft)	0.3048	Meters (m)
Yards (yd)	0.9144	Meters (m)
Miles (mi)	1.6094	Kilometers (km)
Area		
Square inches (in^2)	6.45	Square centimeters (cm^2)
Square feet (ft^2)	0.0929	Square meters (m^2)
Square yards (yd^2)	0.8361	Square meters (m^2)
Square miles (mi^2)	2.5900	Square kilometers (km^2)
Acres (a)	0.4047	Hectare (ha) (10,000 m^2)
Volume		
Cubic inches (in^3)	16.39	Cubic centimeters (cm^3)
Cubic feet (ft^3)	0.028	Cubic meters (m^3)
Cubic yards (yd^3)	0.765	Cubic meters (m^3)
Cubic miles (mi^3)	4.17	Cubic kilometers (km^3)
Quarts (qt), U.S.	0.9463	Liters (L)
Gallons (gal), U.S.	3.8	Liters (L)
Mass		
Ounces (oz)	28.3495	Grams (g)
Pounds (lb)	0.4536	Kilograms (kg)
Ton (tn), U.S.	0.91	Metric ton (tonne) (t)
Speed		
Miles/hour (mph)	0.448	Meters/second (mps)
Miles/hour (mph)	1.6094	Kilometers/hour (kmph)

Metric Prefixes		
Prefix		**Meaning**
giga-	G	10^9 = 1,000,000,000
mega-	M	10^6 = 1,000,000
kilo-	k	10^3 = 1,000
hecto-	h	10^2 = 100
deka-	da	10^1 = 10
		10^0 = 1
deci-	d	10^{-1} = 0.1
centi-	c	10^{-2} = 0.01
milli-	m	10^{-3} = 0.001
micro-	μ	10^{-6} = 0.000001

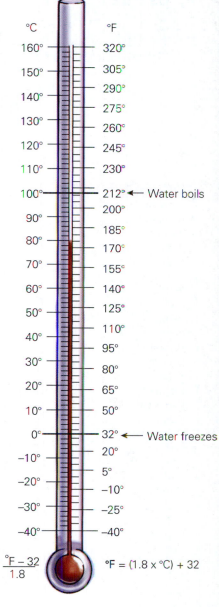

Water boils

Water freezes

$$°C = \frac{°F - 32}{1.8}$$

$$°F = (1.8 \times °C) + 32$$

Periodic Table of the Elements

Metals | Metalloids | Nonmetals

Group number
Atomic number
Transition elements

1																	8

Period 1
- 1 **H** 1.008
- 2 **He** 4.003

Period 2
- 3 **Li** 6.941
- 4 **Be** 9.012
- 5 **B** 10.81
- 6 **C** 12.01
- 7 **N** 14.01
- 8 **O** 16.00
- 9 **F** 19.00
- 10 **Ne** 20.18

Period 3
- 11 **Na** 22.99
- 12 **Mg** 24.31
- 13 **Al** 26.98
- 14 **Si** 28.09
- 15 **P** 30.97
- 16 **S** 32.07
- 17 **Cl** 35.45
- 18 **Ar** 39.95

Period 4
- 19 **K** 39.10
- 20 **Ca** 40.08
- 21 **Sc** 44.96
- 22 **Ti** 47.87
- 23 **V** 50.94
- 24 **Cr** 52.00
- 25 **Mn** 54.94
- 26 **Fe** 55.85
- 27 **Co** 58.93
- 28 **Ni** 58.69
- 29 **Cu** 63.55
- 30 **Zn** 65.41
- 31 **Ga** 69.72
- 32 **Ge** 72.64
- 33 **As** 74.92
- 34 **Se** 78.96
- 35 **Br** 79.90
- 36 **Kr** 83.80

Period 5
- 37 **Rb** 85.47
- 38 **Sr** 87.62
- 39 **Y** 88.91
- 40 **Zr** 91.22
- 41 **Nb** 92.91
- 42 **Mo** 95.94
- 43 **Tc** (98)
- 44 **Ru** 101.1
- 45 **Rh** 102.9
- 46 **Pd** 106.4
- 47 **Ag** 107.9
- 48 **Cd** 112.4
- 49 **In** 114.8
- 50 **Sn** 118.7
- 51 **Sb** 121.8
- 52 **Te** 127.6
- 53 **I** 126.9
- 54 **Xe** 131.3

Period 6
- 55 **Cs** 132.9
- 56 **Ba** 137.3
- 57 **La** 138.9
- 72 **Hf** 178.5
- 73 **Ta** 180.9
- 74 **W** 183.8
- 75 **Re** 186.2
- 76 **Os** 190.2
- 77 **Ir** 192.2
- 78 **Pt** 195.1
- 79 **Au** 197.0
- 80 **Hg** 200.6
- 81 **Tl** 204.4
- 82 **Pb** 207.2
- 83 **Bi** 209.0
- 84 **Po** (209)
- 85 **At** (210)
- 86 **Rn** (222)

Period 7
- 87 **Fr** (223)
- 88 **Ra** (226)
- 89 **Ac** (227)
- 104 **Rf** (261)
- 105 **Db** (262)
- 106 **Sg** (266)
- 107 **Bh** (264)
- 108 **Hs** (269)
- 109 **Mt** (268)
- 110 **Ds** (271)
- 111 — (272)
- 112 — (285)
- 113 — (284)
- 114 — (289)
- 115 — (288)

Lanthanides
- 58 **Ce** 140.1
- 59 **Pr** 140.9
- 60 **Nd** 144.2
- 61 **Pm** (145)
- 62 **Sm** 150.4
- 63 **Eu** 152.0
- 64 **Gd** 157.3
- 65 **Tb** 158.9
- 66 **Dy** 162.5
- 67 **Ho** 164.9
- 68 **Er** 167.3
- 69 **Tm** 168.9
- 70 **Yb** 173.0
- 71 **Lu** 175.0

Actinides
- 90 **Th** 232.0
- 91 **Pa** 231.0
- 92 **U** 238.0
- 93 **Np** (237)
- 94 **Pu** (244)
- 95 **Am** (243)
- 96 **Cm** (247)
- 97 **Bk** (247)
- 98 **Cf** (251)
- 99 **Es** 252
- 100 **Fm** 257
- 101 **Md** 258
- 102 **No** 259
- 103 **Lr** 260

Name	Symbol	Name	Symbol	Name	Symbol	Name	Symbol
Actinium	Ac	Einsteinium	Es	Mendelevium	Md	Samarium	Sm
Aluminum	Al	Erbium	Er	Mercury	Hg	Scandium	Sc
Americium	Am	Europium	Eu	Molybdenum	Mo	Seaborgium	Sg
Antimony	Sb	Fermium	Fm	Neodymium	Nd	Selenium	Se
Argon	Ar	Fluorine	F	Neon	Ne	Silicon	Si
Arsenic	As	Francium	Fr	Neptunium	Np	Silver	Ag
Astatine	At	Gadolinium	Gd	Nickel	Ni	Sodium	Na
Barium	Ba	Gallium	Ga	Niobium	Nb	Strontium	Sr
Berkelium	Bk	Germanium	Ge	Nitrogen	N	Sulfur	S
Beryllium	Be	Gold	Au	Nobelium	No	Tantalum	Ta
Bismuth	Bi	Hafnium	Hf	Osmium	Os	Technetium	Tc
Bohrium	Bh	Hassium	Hs	Oxygen	O	Tellurium	Te
Boron	B	Helium	He	Palladium	Pd	Terbium	Tb
Bromine	Br	Holmium	Ho	Phosphorus	P	Thallium	Tl
Cadmium	Cd	Hydrogen	H	Platinum	Pt	Thorium	Th
Calcium	Ca	Indium	In	Plutonium	Pu	Thulium	Tm
Californium	Cf	Iodine	I	Polonium	Po	Tin	Sn
Carbon	C	Iridium	Ir	Potassium	K	Titanium	Ti
Cerium	Ce	Iron	Fe	Praseodymium	Pr	Tungsten	W
Cesium	Cs	Krypton	Kr	Promethium	Pm	Uranium	U
Chlorine	Cl	Lanthanum	La	Protactinium	Pa	Vanadium	V
Chromium	Cr	Lawrencium	Lr	Radium	Ra	Xenon	Xe
Cobalt	Co	Lead	Pb	Radon	Rn	Ytterbium	Yb
Copper	Cu	Lithium	Li	Rhenium	Re	Yttrium	Y
Curium	Cm	Lutetium	Lu	Rhodium	Rh	Zinc	Zn
Darmstadtium	Ds	Magnesium	Mg	Rubidium	Rb	Zirconium	Zr
Dubnium	Db	Manganese	Mn	Ruthenium	Ru		
Dysprosium	Dy	Meitnerium	Mt	Rutherfordium	Rf		

Answers

Chapter 1

STOP & STRETCH

p. 7: Given the information here, it is most likely that someone brought a box of doughnuts to share, and that Homer ate a large number of them. This hypothesis is based on inductive reasoning.

p. 7: No, because even though Homer is a likely culprit, there are probably other individuals in your workplace who may have eaten some or all of the doughnuts.

p. 10: The independent variable is the presence of *H. pylori* infection, the dependent variable is number of ulcers per animal.

p. 11: Participants could have been exposed to different cold viruses; early arrivals may have been infected by a severe virus that was spreading in the hospital, while later arrivals may have been infected by a milder virus.

p. 14: The independent variable is stress, and the dependent variable is susceptibility to developing a cold when exposed to a virus.

p. 19: According to the poll, the actual percentage of the population who favor candidate A has a 95% probability of being between 44% and 50%, while the percentage who favor B is likely between 48% and 54%. Since these ranges overlap, it is currently unclear whether most of population favor A or B.

p. 21: It is much easier to relate to individual experiences than to a more abstract piece of information. Anecdotes are the way humans have learned from each other throughout history; polling or statistical data are much more recent.

VISUALIZE THIS

Figure 1.3: Even with a well-designed experiment, it is possible that an alternative factor can explain the results, and researchers need to consider if that might be the case in their experiment.

Figure 1.4: Preventing virus attachment and invasion of cells. Preventing activation of the immune system response. Other answers may be acceptable.

Figure 1.10: There may not have been enough participants at a particular stress level to allow effective analysis of the correlation. If individuals at stress level 3 have the same cold susceptibility as those at stress level 4, that should cause us to question the correlation because we'd expect those at stress level 4 to have greater susceptibility.

Figure 1.15: Hypothesis: true. Difference between control and experimental groups: large. Sample size: large. This result is very likely to be both statistically and practically significant.

Figure 1.16: On the figure, it would the follow the path through "hypothesis true", "difference between groups large", and "sample size large," meaning that the result is highly likely to be both statistically and practically significant.

SAVVY READER

1. Red flags for questions 1, 2, 9, 10, and 11.
2. Most other questions from the checklist do not pertain to this particular excerpt.

LEARNING THE BASICS

1. A placebo allows members of the control group to be treated exactly like the members of the experimental group *except* for their exposure to the independent variable.
2. b; **3.** b; **4.** d; **5.** a; **6.** d; **7.** e; **8.** c; **9.** b; **10.** c; **11.** b; **12.** e

ANALYZING AND APPLYING THE BASICS

1. No, another factor that causes type 2 diabetes could lead to obesity as well, or it is possible that type 2 diabetes causes obesity, not vice versa.
2. Neither the participants nor the researchers were blind, meaning that both subject expectation and observer bias might influence the results. If participants felt that vitamin C was likely to be effective, they may underreport their cold symptoms. If the clinic workers thought one or the other treatment was less effective, they might let that bias show when they were talking to the participants, influencing the results.
3. See the answer to number 1; BDNF may increase as a result of exercise, or excess BDNF might cause an increased interest in exercise.

Chapter 2

STOP & STRETCH

p. 37: The carbons should each be bonded to two hydrogens to complete their valence shells.

p. 48: Yes, because ribosomes are not bounded by membranes and because prokaryotic cells need to synthesize proteins

VISUALIZE THIS

Figure 2.4: Oxygen

Figure 2.7: 100×

Figure 2.15: Red spheres are sugars; purple are phospahtes. Purines (blue) are double ring structures and thus longer in the diagram, while pyrimidines (yellow) are shorter.

Figure 2.18: Central vacuole, cell wall, and chloroplast

Figure 2.20: Yes

1. Not really; how would a negative charge produce hydroxyl ions? While it is true that hydroxyl ions are negatively charged, if you remove the hydroxyl ion from water you would be left with the positively charged hydrogen ion. Hydroxyl ions and the hydrogen ions would bind to each other to produce water again.
2. Red flags—this is an untested assertion put our by a company that stands to gain financially.
3. Amazing, once and for all, hugely improved etc.
4. No
5. Error bars

LEARNING THE BASICS
1. Carbohydrates, lipids, proteins, nucleic acids
2. See Table 2.1.
3. Phospholipid bilayer with embedded proteins
4. b; **5.** a; **6.** a; **7.** d; **8.** d; **9.** d; **10.** d **11.** B; **12.** C

ANALYZING AND APPLYING THE BASICS
1. No. A virus is not capable of reproduction without help from the host cell.
2. Oxygen partial negative charge; carbon partial positive.
3. Silicon, like carbon, has four spaces in its valence shell and would also be tetravalent.

Chapter 3

STOP & STRETCH
p. 58: These food pairings together provide all the essential amino acids.
p. 67: The concentration of dissolved particles outside of cells will increase after drinking this beverage and water can leave the cells.
p. 67: Active transport.

VISUALIZE THIS
Figure 3.4: The top tail of part b.
Figure 3.10: Fat

SAVVY READER
1. Published by a University, analyzes claims made by a manufacturer.
2. You should be more skeptical of a marketing claim. The University has nothing to lose if you don't use a supplement while a company loses money.
3. You should be more skeptical of a study published by scientists hired by a manufacturer of a product. Studies published by government and university scientists gave undergone peer review. A claim made on a web site has not necessarily undergone peer review.

LEARNING THE BASICS
1. Nutrients serve as energy stores and as building blocks for cellular structures.

2. Carbon dioxide, oxygen, and water
3. d; **4.** c; **5.** b; **6.** b; **7.** d; **8.** d; **9.** e; **10.** b

ANALYZING AND APPLYING THE BASICS
1. Eat lots of protein rich foods like beans.
2. Fatigue is one of the ways your body responds to toxic levels of alcohol. Energy drinks may override fatigue signals, allowing consumption of alcohol past a level at which one would normally pass out.
3. We would likely produce fewer toxins if we knew we could not simply wash other vitamins, minerals, and fiber away.

Chapter 4

STOP & STRETCH
p. 76: 200 calories per day.
p. 83: More oxygen is available for the conversion of lactic acid to pyruvic acid.
p. 87: The amount of the total cholesterol that is LDL.

VISUALIZE THIS
Figure 4.2: No
Figure 4.3: 278.4 Calories
Figure 4.12: *Glyco* means sugar, and *lysis* means cutting.
Figure 4.13: A double bond is broken to allow for the bonding of hydrogen.
Figure 4.17: Oxygen exists in cells as molecular oxygen (O_2). Because only one oxygen atom is required to make water (H_2O), the 1/2 is placed before O_2.

SAVVY READER
1. No evidence is presented for any claims.
2. No, pathogenic bacteria, plant toxins, snake venom, hurricanes and ice storms are all natural but can be very bad for humans.
3. No, there needs to be evidence presented from the results of controlled studies.
4. No. No.

LEARNING THE BASICS
1. All of the building up and breaking down chemical reactions that occur in a cell
2. When a substrate binds to the active site of an enzyme, the enzyme clamps down on the substrate, causing increased stress on the bonds holding the substrate together.
3. Reactants of photosynthesis are carbon dioxide and water. Reactants of respiration are oxygen and glucose. Products of photosynthesis are oxygen and glucose. Products of respiration are carbon dioxide and water.
4. c; **5.** b; **6.** c; **7.** e; **8.** b; **9.** c; **10.** c

ANALYZING AND APPLYING THE BASICS
1. No. Fats and proteins also feed into cellular respiration at various points in the cycle and fats actually store more energy than carbohydrates.

2. An enzyme will be digested in the stomach, like any other protein.
3. Being very muscular can lead to an inaccurate BMI because muscle weighs more than fat.

Chapter 5

STOP & STRETCH

p. 95: Because the moon has no or very little greenhouse gas, heat from the sun is not retained to be reradiated overnight, which is why the nighttime temperatures are cold. The warm temperatures during the day occur because water is not present to absorb some of the heat energy without changing temperature.

p. 100: The chlorophyll is reabsorbed because it is so valuable to the plant—its components can be stored and the chlorophyll can be reproduced in the spring.

p. 101: The products of respiration are used during photosynthesis. The products of photosynthesis are reactants in respiration.

p. 102: ATP is a short-lived molecule that cannot "store" energy. Therefore it cannot be used to transport energy to parts of a plant where photosynthesis cannot occur (e.g. roots) or provide energy that can be used when the sun is not shining.

p. 103: Nitrogen, sulfur, and phosphorous

VISUALIZE THIS

Figure 5.1: Increased carbon dioxide will absorb more heat and keep it in the atmosphere, allowing more heat to radiate back to Earth.

Figure 5.2: Water levels would rise and shorelines would erode.

Figure 5.4: Human activity

Figure 5.6: The increase in concentration is steeper during the period 2000-2005 when compared to 1960-1970.

Figure 5.8: The left axis is for the blue line and the right for the red line. They are on the same graph to better illustrate the relationship between carbon dioxide level and temperature over time.

Figure 5.10: The absorption of light would be small in the lower wavelengths (red) and thus the line would be closer to the x axis on the left of the graph. The absorption of light should be higher in the middle and perhaps highest wavelengths and thus the line would be far from the x axis in the middle and right side of the graph.

Figure 5.12: The reactions are similar in that energized electrons are used to create a gradient of hydrogen ions, which is then used to generate ATP. However, the energized electrons in photosynthesis come from chlorophyll that has been energized by the sun; in respiration the electrons come from NADP, which was produced by the breakdown of sugar. At the end of the chain, electrons in photosynthesis are used to produce sugar; in respiration, the electrons are picked up by oxygen, which reacts with hydrogen to produce water.

Figure 5.13: Without sunlight, ATP and NADPH inputs from the light reactions would be absent and the cycle could not function.

SAVVY READER

1. Rising seas from global warming are causing flooding on low-lying Panamanian islands.
2. Yes, it may be that the flooding is caused by coral mining.
3. No, because other alternative hypotheses have not been ruled out as causes for the flooding.

LEARNING THE BASICS

1. Reactants of photosynthesis are carbon dioxide and water. Products of photosynthesis are oxygen and glucose.
2. c; **3.** d; **4.** e; **5.** d; **6.** a; **7.** c; **8.** a; **9.** d; **10.** d

ANALYZING AND APPLYING THE BASICS

1. Because colder water has the capacity to absorb more heat before water molecules can evaporate into the atmosphere, colder temperatures mean that less water circulates in the water cycle.
2. The water cycle would remain basically unchanged. However, the carbon cycle would be quite different. The only way carbon dioxide would move from the atmosphere to Earth's surface would be by absorption into water and perhaps incorporation into rocks. It would return to the atmosphere via diffusion from water and volcanic activity.
3. High rates of photosynthesis are continually pumping oxygen into the air. So even though the molecule is being "used up" through reactions with other compounds, its continual production ensures high levels in the atmosphere.

Chapter 6

STOP & STRETCH

p. 114: If a cell has many mutations, it is better that it is not allowed to continue to replicate.

p. 120: DNA is semi-conservatively replicated; that is, each strand serves as a template for the synthesis of a new strand. Photocopying is completely conservative. One copy is conserved and the other is new.

p. 122: Descent with modification explains the existence of similar processes in distantly related organisms.

p. 128: Tissues and organs will cease functioning as cells reach their cell division limit.

p. 131: Because it damages DNA.

p. 135: Both meiosis I and mitosis produce two daughter cells. Meiosis I differs from mitosis in that meiosis I separates homologous pairs of chromosomes from each other. Mitosis does not do this.

VISUALIZE THIS

Figure 6.5: Four DNA molecules would be produced. Two of those would be all purple; two would be half purple, half red.

Figure 6.10: The cell should be prevented from passing the G2 checkpoint.

Figure 6.18: The X chromosome is the larger of the two and carries more genetic information.

Figure 6.22: No

Figure 6.25: 4

SAVVY READER

1. No
2. Check some highly regarded sources. Ideally peer reviewed literature. Otherwise check reputable websites and see if there is a consensus.
3. Subtract
4. In addition to controlling for type of cancer, age of patient, etc., it is imperative to know what percentage of people not taking the supplement were alive one year after taking the supplement.
5. more skeptical; more skeptical
6. An advanced degree is not a guarantee that a person will act in your best interest.
7. Desperation is one reason.

LEARNING THE BASICS

1. Mitosis occurs in somatic cells, meiosis in cells that give rise to gametes. Mitosis produces daughter cells that are genetically identical to parent cells, while meiosis produces daughter cells with novel combinations of chromosomes and half as many chromosomes as compared to parent cells.
2. Rapid cell division
3. d; **4.** a; **5.** c; **6.** d; **7.** b; **8.** c; **9.** d
   10. a, diploid; b, diploid; c, haploid; d, haploid
   11. Crossing over and random alignment of the homologues
   12. Oncogenes are mutated versions of proto-oncogenes.
   13. Cytokinesis in plant cells requires the building of a cell wall.

ANALYZING AND APPLYING THE BASICS

1. No. Only mutations in cells that produce gametes are passed to offspring.
2. No, but more cell divisions means more opportunities for mutations to occur.
3. Cancers that are thought to be localized (not yet spread) are treated with radiation.

Chapter 7

STOP & STRETCH

p. 151: A mutation in a promoter portion of the DNA may affect whether the gene is copied or translated. A mutation in a structural portion of a chromosome may weaken the chromosome or change its ability to be transcribed. A mutation in "nonsense" DNA is likely to have no effect.

p. 152: Eight different gametes. The relationship is 2^n, where n is the number of chromosome pairs.

p. 157: Its effects are not seen until later adulthood, after an individual is likely to have had children.

p. 159: Eight possible gamete types (e.g. T Y R; T y R; t Y R, etc.) and 64 possible cells inside the Punnett square.

p. 161: Immigration from countries where average height is shorter may be changing the genetics of height in the United States such that there are a larger number of "small" alleles today than in the past. Dietary changes may also be leading to reduced growth or early puberty, which stunts growth in height.

p. 167: Even a trait that is highly heritable can be strongly influenced by the environment. If the current environment results in some individuals having very low IQ scores relative to others, it is certainly possible that a changed environment would lead to smaller differences among individuals. See the case of the "maze-dull" and "maze-bright" rats.

VISUALIZE THIS

Figure 7.3: Many possibilities

Figure 7.6: Blood group gene from his dad and eye color gene from his dad; Blood group gene from his mom and eye color gene from his mom.

Figure 7.14: 50%

Figure 7.15: 9/16 or 56%

Figure 7.16: No, the majority of men were not 5 feet 10 inches. There is a large range of heights in this population.

Figure 7.19: Points would be more widely scattered, and the correlation line would be flatter.

SAVVY READER

1. There is some evidence that we will select friends that are similar to us in some genetically-influenced behaviors and dissimilar to us in other genetically-influenced behaviors. Whether the friendship itself is genetically determined is difficult to tease out from the fact that the personality trait is partially genetically-determined.
2. If people with a genetic predisposition to alcoholism tend to have a genetic predisposition to become friends, part of understanding how to prevent and treat alcoholism may include understanding the social dynamics of these friendships.
3. Clearly, your choice of friends is influenced by the environment. This study supports the idea that there is a genetic basis favoring particular friendships as well.

LEARNING THE BASICS

1. Genotype refers to the alleles you possess for a particular gene or set of genes. Phenotype is the physical trait itself, which may be influenced by genotype and environmental factors.
2. Multiple genes or multiple alleles influencing a trait, environmental effects that affect the expression of a trait, or a mixture of both factors
3. b; **4.** a; **5.** b; **6.** a; **7.** b; **8.** c; **9.** e; **10.** d

GENETICS PROBLEMS

1. All have genotype Tt and are tall.
2. One-quarter are TT and are tall; one-half are Tt and are tall; and one-quarter are tt and are short.

3. Both are alleles for the same trait; P is dominant, and p is recessive.
4. Both parents must be heterozygous, Aa.
5. 50%
6. a. yellow; b.YY and yy
7. Yellow, round parent: Yy Rr; green, wrinkled parent: yy rr.
8. a. 50%; b. 50%; c. 25%; d. yes
9. a. dominant; b. The phenotype does not inevitably surface when the allele is present.

ANALYZING AND APPLYING THE BASICS
1. The trait could be recessive, and both parents could be heterozygotes. If the trait is coded for by a single gene with two alleles, one would expect 25% of the children (1 of 4) to have blue eyes. That 50% (2 of 4) have blue eyes does not refute Mendel because the probability of any genotype is independent for each child and is not dependent on the genotype of his or her siblings. (Consider that some families may have 4 boys and no girls; this does not refute the idea that the Y chromosome is carried by only 50% of a man's sperm cells.)
2. No, most quantitative traits are influenced by the environment. Even if differences in genes account for most of the differences among individuals in a trait, if the environment changed, the trait would very likely change as well.
3. No, heritability doesn't explain why any two individuals differ. For instance, John and Jerry could be identical twins, but Jerry might suffered a minor accidental head injury that reduced his IQ.

Chapter 8

STOP & STRETCH
p. 177: $I^C I^C$ type C; $I^A I^C$ type AC; $I^B I^C$ type BC; $I^C I$ type C; along with previously described AB types
p. 181: Her father must have the disease, and her mother is a carrier or has the disease.
p. 183: Crossing over; they may be infertile.
p. 187: They may or may not carry different alleles of gene. The VNTRs are adjacent to the allele, they are not part of the allele.
p. 188: Because each parent only contributes half of their genetic information to a child

VISUALIZE THIS
Figure 8.4: 50% (1/2); 50% (1/2)
Figure 8.6: A cross involving a hemophilic female
Figure 8.9: This pedigree would have a square connected to a circle connected to a square in the first generation (the parents). The second generation (their kids) would have a square from one set of parents and a circle from the second set of parents.

SAVVY READER
1. 50%; No.
2. It is difficult to imagine how this could happen.
3. No. Your gametes do not choose which traits to select. Your environment does.

1. More than two alleles are present in the population. Each individual can have (at most) two different alleles.
2. Females inherit one X chromosome from their mom in the egg cell and one from their dad in the sperm cell. Males inherit an X chromosome from their mom in the egg cell and a Y chromosome from their dad in the sperm cell.
3. DNA is isolated, then the amount of DNA can be amplified using PCR. DNA is cut with restriction enzymes and the fragments are loaded on a gel, to which an electrical current is applied. The separated fragments of DNA produce a pattern unique to an individual.
4. $I^a i$ (dad), $I^B i$ (mom)
5. ¼A1, ¼A2, ⅛AB2, ⅛AB1, ⅛B1, ⅛B2
6. c; **7.** b; **8.** c; **9.** e; **10.** d; **11.** b; **12.** d

ANALYZING AND APPLYING THE BASICS
1. Alleles that are co-dominant are equally expressed. An allele that is incompletely dominant does not completely mask the presence of the recessive allele.
2. Answers will vary. Just be certain that each band found in a daughter is present in one of the parents. About 50% of the bands in a given daughter will be present in each parent. The sisters should have around 50% of their bands in common.
3. Pedigrees will vary. However, the first cousins should have one set of common grandparents. If a grandparent is a carrier of a rare recessive allele, he or she could pass the same rare recessive allele to each of the cousins. If these cousins (both carriers of a rare recessive allele) mate, their child has a 1 in 4 chance of having the recessive disease.

Chapter 9

STOP & STRETCH
p. 202: Proline-Asparagine
p. 203: New alleles, or mutations, are a source of variation for natural selection to act upon.
p. 204: Gene expression would no longer be negatively regulated. A gene may be on all the time.
p. 207: There could be more growth hormone in the milk of treated cows.
p. 211: Preembryonic cells are capable of becoming any cell type.
p. 216: Dolly's chromosomes were that of an older sheep and therefore shortened. Compare the lengths of chromosomes from Dolly's cells with those of a newborn lamb.

VISUALIZE THIS
Figure 9.1: The second carbon of the sugar, ribose, is bonded to -OH while the same carbon in deoxyribose is attached to hydrogen. There is a CH3 group in thymine that is not present in uracil.
Figure 9.3: Sample answer starting after the fork in the DNA: AGCTGGGCAGGTAC. From this particular DNA, the mRNA would be UCGACCCGUCCAUG.

Figure 9.5: Adenine paired with Uracil and Cytosine paired with Guanine

Figure 9.9: Bottom

SAVVY READER
1. No. Cartoons are not peer reviewed
2. No. This is not an issue that can be resolved by hypothesis testing.
3. Answers will vary.

LEARNING THE BASICS
1. GCUAAUGAAU
2. gln, arg, ile, leu
3. mRNA, ribosome, amino acids, tRNAs
4. b; **5.** c; **6.** b; **7.** c; **8.** c; **9.** b; **10.** d ; **11.** T; **12.** F; **13.** F

ANALYZING AND APPLYING THE BASICS
1. The same amino acid is often coded for by codons that differ in the third position. For instance, UUU and UUC both code for phenylalanine.
2. Transcription would be slowed or stopped in that cell.
3. GAATTC; CTTAAG

Chapter 10

STOP & STRETCH

p. 227: Because statements of faith are based on nonrepeatable, mostly unobserved phenomenon, they cannot be falsified. Thus, they are not subject to scientific hypothesis testing.

p. 232: It may be that the human diet was harder on teeth in the past, so that teeth were lost due to breakage, leaving "room" for the wisdom teeth to erupt. These teeth may have been beneficial, in fact, replacing the role of missing molars.

p. 238: They are very close. The oldest human ancestor is slightly younger than the estimated date of divergence between chimpanzees and humans, as is expected.

VISUALIZE THIS

Figure 10.1: All lice would have been eliminated by the pesticide, and the population would not have evolved.

Figure 10.2: These branches represent organisms that are extinct.

Figure 10.6: Perhaps the presence of plant-eating tortoises on the islands made those cacti with longer stems (thus out of reach of the tortoises) more likely to survive, leading to the evolution of longer stems. There also may be a lack of other trees on the islands, meaning that taller cacti were able to capture more sunlight and thus reproduce more than smaller plants.

Figure 10.9: Hair-covered bodies, birth to live young, other characteristics of all mammals (also forward-facing eyes and opposable thumbs)

Figure 10.10: Squirrel would be the farthest left branch, originating above the mammal ancestor. The shared ancestor is the junction point.

Figure 10.20: 3 million years

SAVVY READER
1. The argument is essentially that the theory of evolution is suspect because several hundred scientists are skeptical of the theory.
2. Science is not a democratic process, but is based on evidence. The theory of evolution has support from many different subfields of biology, as well as chemistry and geology. A scientific theory that fits the evidence more effectively has not been proposed by these skeptics. In addition, the article provides little evidence that the skeptics are experts in evolutionary theory. For example, the scientist quoted is an expert in human medicine.
3. Look at the "About" page of an Internet site to learn about the mission of the host organization. An organization with a particular agenda is more likely to cherry pick data that supports their agenda than to present a balanced view of an issue.

LEARNING THE BASICS
1. The theory of common descent describes that the similarities among living species can be explained as a result of their descent from a common ancestor.
2. b; **3.** d; **4.** d; **5.** b; **6.** e; **7.** b; **8.** a; **9.** d; **10.** d

ANALYZING AND APPLYING THE BASICS
1. Parentheses surround groups that share a common ancestor ((((pasture rose and multiflora rose) spring avens) live forever) spring vetch)
2. (((((((Humans and chimpanzee) mouse) donkey) carp) yeast) corn) green algae)
3. The comparison is only between humans and these organisms in genetic sequence. There is no comparison on the degree of relatedness among the organisms. The differences between humans and yeast may be in totally different segments of DNA than the differences between humans and algae.

Chapter 11

STOP & STRETCH

p. 254: The TB bacteria can continue to multiply, doing damage to the lungs and resulting in a larger population. More cells mean more likelihood of resistant variants, as well as a more difficult time clearing out the infection.

p. 258: Humans have invented technologies and disease cures that decrease the differences in fitness among individuals with different genetics (for example, corrective lenses have increased the survival of individuals with poor eyesight). We are still subject to natural selection for traits that are less amenable to technological fixes and even in our ability to respond to technology.

p. 261: The cystic fibrosis allele may provide protection from diseases that are common in more urban environments. Individuals carrying the allele in these environments were more likely to survive, thus causing the allele to become more common in those populations.

p. 267: Mutations appear by chance and may not have much affect on fitness unless the environment changes. The resistance allele appeared randomly.

VISUALIZE THIS

Figure 11.8: Because the elephants are using most of their resources in Generation 3, a percentage of Generation 4 will die from starvation.

Figure 11.9: It decreased by a large amount.

Figure 11.11: The population would change in the direction of longer legs and coats.

Figure 11.13: The individual with the fast metabolism of alcohol would not have proportionally more offspring than those without the allele, so the ratio of fast to slow metabolizers would stay the same on each "shelf."

SAVVY READER

1. Airline air is polluted and leads to virus spreading. Andrew Speaker knowingly put his bride and others at risk of contracting a dangerous form of TB. There is no evidence for either statement.
2. Other postings by this author; evidence that the author linked to credible sites to back up statements of fact
3. Students may provide various acceptable answers.

LEARNING THE BASICS

1. Antibiotics have helped reduce deaths due to tuberculosis, but they have become less effective as strains of antibiotic-resistant *M. tuberculosis* have appeared in human populations.
2. Artificial selection is a process of selection of plants and animals by humans who control the survival and reproduction of members of a population to increase the frequency of human-preferred traits. Artificial selection is like natural selection in that it causes evolution; however, it differs because humans are directly choosing which organisms reproduce. In natural selection, environmental conditions cause one variant to have higher fitness than other variants.
3. Within a single patient, the *Mycobacterium tuberculosis* consists of a large number of genetic variants. These variants differ in a number of traits, including their susceptibility to a particular antibiotic. When the patient takes an antibiotic, the individual bacteria that are more resistant to the drug have higher fitness, and natural selection leads to the evolution of a bacteria population that is resistant to the drug.
4. c; **5.** d; **6.** e; **7.** e; **8.** b; **9.** c; **10.** d

ANALYZING AND APPLYING THE BASICS

1. Farmers who desired larger strawberries must have saved seeds from the wild plants that produced the largest berries. Over the course of many generations of the same type of selection in domesticated strawberries, average berry size increased dramatically. Tradeoffs include: domestic strawberries have less energy to put into roots and leaves than do wild strawberries; domesticated strawberries require a larger supply of water and nutrients, and perhaps domesticated

strawberries have fewer chemical defenses against insects and other predators.

2. Individual zebras with some sort of striping must have had greater survival than those who were not striped. Over the course of many generations of this type of selection, individuals in the population evolved dramatic stripes.
3. No, not all features are adaptations; some might arise by chance. A trait is an adaptation if it increases fitness relative to others without the trait. One can look for ways that the trait increases fitness or examine individuals who lack or have a less-dramatic version of the trait to measure its fitness effects.

Chapter 12

STOP & STRETCH

p. 277: Individuals with fewer or lighter spots had higher fitness, perhaps because they blended better into the savannah background, allowing them to sneak up on prey.

p. 279: No, because the varieties are the same species, so the number of chromosomes should be the same in both parents.

p. 286: Morphological concept, because they would be identifying birds by sight (and sound) without information about mating habits or genetic relatedness.

p. 293: For example: average height varies but is not typically used to identify similar groups. Adding this category might lead to identifying a larger number of races.

p. 297: "Ma" or "mmmm" is one of the first sounds an infant can make. Since the mother is so associated with infant care, this sound became associated with her. ("Da" or "Pa" is a very early sound as well.)

p. 300: Females put more investment (energy) in offspring compared to males, so have a higher incentive for making sure that the offspring has the greatest chance of survival and later reproduction.

VISUALIZE THIS

Figure 12.2: No, because the chromosomes from the two different parents are not homologous and thus still could not pair properly.

Figure 12.10: Heterozygotes can form via (p)males × (q)females AND by (p)females × (q)males.

Figure 12.17: Various possibilities as visualized on the graph (for the first question, dots of the same color that are distant from each other on the y-axis; for the second question, dots of different color that are in close proximity on the y-axis).

Figure 12.18: An individual with dark skin in a low-UV environment could take vitamin D to ensure proper bone growth, while an individual with light skin in a high-UV environment could avoid sun exposure, use sunscreen, and/or take folate supplements.

SAVVY READER

1. The drug BiDil is ineffective in whites, but effective in African-Americans. Asthma drugs are less well absorbed by African-Americans compared to whites.

In studies of hypertension, some populations identified as black have very low rates of the disease, despite the fact that high rates of hypertension is assumed to be common to all blacks.

2. Race is associated with lower quality health care, even taking into account poverty, age, and condition severity, indicating that racism is at play. Race is often a proxy for socioeconomic status, which may be causing differences in health.

3. The existence of studies that identify certain alleles or groups of alleles that are more common in one race; studies that divide people by genetic markers rather than self-identified racial groupings and show a correlation between group and disease outcome

LEARNING THE BASICS

1. The gene pools of populations must become isolated from each other, they must diverge (as a result of natural selection or genetic drift) while they are isolated, and reproductive incompatibility must evolve.

2. Because it takes time for mutations to accumulate in a population and/or species, the relative lack of mutations is evidence of recent origin.

3. Genetic drift can occur as a result of founder's effect, the bottleneck effect, or by the chance loss of alleles from small, isolated gene pools. By changing allele frequencies, all of these events result in the evolution of a population.

4. d; **5.** b; **6.** a; **7.** c; **8.** c; **9.** d; **10.** c

ANALYZING AND APPLYING THE BASICS

1. It depends on the species definition one uses. The two populations of wolves are not different biological species. They are likely different genealogical species—the definition that corresponds to "race." To determine if they are different races, one would have to look for the standard evidence: unique alleles or unique allele frequencies.

2. Ireland, 0.01; Britain 0.007; Scandinavia 0.005. The allele frequency could be different as a result of genetic drift, particularly the founder effect, in these different populations. It also could be different because the allele has some yet unknown fitness advantage in the environment of Ireland relative to the environment of Britain or Scandinavia.

3. If the early mutation becomes widespread, the timing difference between the early and normal populations should cause reproductive isolation. The same is possible with the expanded mutation if certain pollinators become associated with the individuals that bloom earlier than expected.

Chapter 13

STOP & STRETCH

p. 314: Superficial similarities allow observers to group things quickly, by obviously visible characteristics, and without knowing much biology.

p. 317: Mitochondria in plant and animal cells are identical, indicating that they evolved before plants and animals diverged.

p. 323: It is highly visible and easy to catch. Predators would quickly pick these animals off if they were not poisonous, and the poisons they have might have medicinal properties.

p. 328: Early human societies were probably more likely to test plants as possible food and drug sources; the drug use of other organisms would not be as obvious (and microorganisms would be unknown).

p. 332: A beak that can tear meat, long wings, large size, and the ability to soar. This classification can help beginner birders learn to quickly identify all birds that share these easily visible characteristics.

VISUALIZE THIS

Figure 13.5: They use bases that are common to all groups as a framework and then infer that older species are more likely to possess the ancestral bases than younger species.

Figure 13.10: To advertise its poisonous nature, "scaring" predators away and preventing injury to the frog caused by a predator "testing" it

Figure 13.13: To allow dispersal of spores away from the food source that is already in use

Figure 13.21: It probably had little head striping, like the junco, but a sparrow colored body (brown striped) like all of the sparrows.

SAVVY READER

1. He has authored books and somehow has insider knowledge about pharmaceutical companies and government regulators. It is not clear what his training or other experience is. He argues that he must be credible because he has sold millions of books. There is no evidence presented for his statement that the medical business is a fraud, but his statement plays on the frustrations people have when doctors cannot offer an easy cure or solution to their condition and that doctors are withholding effective treatments because it is more profitable to continue to treat sick people.

2. He appeals to concerns regarding the medical business, believing that they withhold information that can make life better inexpensively; he appeals to the preconceived idea that someone who risks criticism from the medical community to help others deserves to be trusted (and supported by purchase of their book).

3. Mr. Trudeau says that a doctor's statement that a particular drug is typically only prescribed for people with pancreatic problems is an opinion, and the interviewer agrees. This statement is either fact or not true—it is not an opinion. The same is the case for the statement that purging could lead to dehydration, electrolyte problems, and disturbance of heart rhythms. However, Mr. Trudeau says this statement is false (he is treating it as a potential fact), but then labels it as an opinion. It is important to distinguish facts from opinions because a proposed fact, like a hypothesis, can be "tested" and found to be true or untrue. An opinion should be based on facts and what is known or not known about a subject.

1. Scientists have identified over 1.5 million different species, but at least 10 million, and possibly over 100 million, different species might exist on Earth today.
2. After flowers evolved, the number of different kinds of plants with flowers increased rapidly. One hypothesis for why this occurred is that flowering plants were better adapted to dry conditions, another is that the relationship between plants and animals (as pollinators and dispersers) contributed to the diversity of forms through co-evolution.
3. By comparing the DNA sequences of organisms to see if the pattern of similarity and difference matches what is predicted and by examination of the fossil record, looking for the record of evolutionary change in the group of organisms
4. b; **5.** e; **6.** e; **7.** d; **8.** c; **9.** d; **10.** b

ANALYZING AND APPLYING THE BASICS

1. An oxygen-sensitive cell that kept an oxygen-consuming prokaryote alive inside of it would be protected from the damaging effects of oxygen, as any nearby oxygen would quickly be "used up" by the prokaryote. This would be an advantage over cells that did not have this prokaryote.
2. Although it is not very motile, this is probably an animal since it consumes other organisms (is not photosynthetic) by ingesting.
3. Vertebrates are unlikely to compete with bacteria for food resources, so they are unlikely to need to "kill them off." Vertebrates also have an immune system for dealing with infections, so typically do not need to invest in manufacturing antibiotics.

Chapter 14

STOP & STRETCH

p. 341: 10,000 (solve for x: 1/100 = 100/x)

p. 344: Average life span equals the sum of the ages at death divided by the number of deaths. Deaths of infants thus greatly bring down the average life span, and reducing these deaths will raise life expectancy dramatically, even if the overall population health has not changed.

p. 346: Infectious diseases require a population of susceptible individuals to survive and spread. If the population is small, the infectious organism cannot maintain a population.

p. 350: A large population with a high birth rate produces a large number of offspring, potentially overshooting the carrying capacity. A large population with a low birth rate produces many fewer young, allowing a gradual approach to the carrying capacity.

VISUALIZE THIS

Figure 14.1: Trap-happiness would cause the population estimate to be too small because a greater proportion of returnees will be marked. Trap-shyness causes the opposite effect because fewer returnees will be marked.

Figure 14.5: Where birth rates and death rates are closest to equal; that is, before the transition and after

Figure 14.6: Where the line is most vertical; that is, at the point where the "S" turns from rising to flattening out

Figure 14.13: The United States has about one quarter the population of India.

Figure 14.14: Several reasons: larger immigration rate, younger population on average, religious traditions

SAVVY READER

1. The Christmas Bird Count and Breeding Bird Surveys, "citizen science" programs (i.e., birdwatchers participate in data collection)
2. Many answers possible, such as, evidence that the causes Audubon identifies are truly responsible, the types of actions that are needed, how effective these actions are in preserving bird populations, etc.

LEARNING THE BASICS

1. A decrease in death rate, especially a decrease in infant mortality, has led to increasing growth rates. In addition, the process of exponential growth lends itself to these population explosions as the number of people added in any year is a function of the population of the previous year.
2. In most populations, growth rate declines because death rate increases (or birth rate decreases) as resources are "used up" by the population.
3. b; **4.** c; **5.** b; **6.** a; **7.** b; **8.** d; **9.** d; **10.** d

ANALYZING AND APPLYING THE BASICS

1. 2500. Solve for x: 1/50 = 50/x.
2. −0.002 (r = Birth rate − Death rate = 25/1000 − 27/1000). This population is likely near carrying capacity because it is not growing, but in fact declining slightly, as would be expected if resources are limited.
3. The carrying capacity is expected to rise since more resources are available. However, because the flies are in a restricted environment (a culture bottle), the accumulation of waste may become a problem, and a large population of flies may not be able to survive this accumulation. Depending on fly behavior, there may also be an increase in aggressive interactions or a decrease in mating in more dense situations.

Chapter 15

STOP & STRETCH

p. 365: Habitat destruction, pollution, introduction of exotics, even over-exploitation of colorful or dramatic species. It would help to know the environmental situation of many of the species at risk to determine if they are threatened by a common cause. It may be just that these animals are especially vulnerable to all threats because of their unique biology.

p. 371: All species pairs are competitors; the organisms they compete for are a) fish b) krill c) penguins d) squid e) herbivorous zooplankton. The most straightforward way to test if competition is occurring is to remove one of the competing species and see if the remaining species increases in population size.

p. 374: The cycle is changed because more nitrogen is made available for cycling, although because the nitrogen does not stay in he fields where it is spread, it may be

p. 380: Moving animals around so that close relatives are less likely to mate; importing animals from other areas for the same reason

p. 382: If homozygous individuals are more common in small populations, individuals who express the deleterious allele (homozygous recessives) are also more common. If these individuals do not survive, the allele is less likely to be passed on to the next generation, so change in allele frequency is more rapid.

VISUALIZE THIS

Figure 15.2: 10 to 50 million years

Figure 15.7: If the island lost 67% of its habitat (10,000 of 15,000 square kilometers), you'd expect about 25% of the species to become extinct.

Figure 15.12: If baleen whales disappeared, krill populations would probably rise, leading to higher populations of fish and other carnivorous zooplankton that compete with whales for krill. However, it might lead to a decrease in smaller toothed whales, which prey on the Baleen whales.

Figure 15.16: The total number of bacteria reaches the carrying capacity of the environment (that is, the intestines)

Figure 15.18: In man-made systems, nutrients typically flow in one direction (from crops to humans to waste water and cemeteries), so we can no longer refer to the process as a nutrient cycle.

Figure 15.24: Because the normal allele does not cause disease, homozygotes for the normal allele are healthy. This is not the case for individuals who are homozygous for the mutant allele, who are ill.

SAVVY READER

1. The representative cites unnamed "scientists and wolf recovery advocates" as experts who agree that the wolf is "back." Without listing who these individuals are and their qualifications for making these statements, the politician's statement is not convincing.

2. It does not seem likely that the return of the wolf will wipe out elk and moose. There is always greater biomass of prey than of predator species, and the experience in Yellowstone indicates that wolves do not cause the loss of their prey, but may change the prey's behavior.

LEARNING THE BASICS

1. Habitat fragmentation endangers species because it interferes with their ability to disperse from unsuitable to suitable habitat and because it increases their exposure to humans and human-modified environments. Species that have very specialized habitat requirements, those that cannot move rapidly, and those that are very susceptible to human disturbance are most negatively affected by fragmentation.

2. Mutualism is a relationship among species in which all partners benefit. Predation is a relationship among species in which one benefits and others are consumed. Competition is a relationship among species in which all partners are harmed by the presence of the others.

3. d; **4.** d; **5.** a; **6.** a; **7.** b; **8.** c; **9.** d; **10.** a **11.** d; **12.** d

ANALYZING AND APPLYING THE BASICS

1. Kelp → sea urchins → sea otters. Kelp also provides habitat for fishes, clams, and abalones. When sea otters are removed, sea urchins devastate the kelp forest, leading to the loss of these associated species. Sea otter thus serves as a keystone.

2. Even though there are more individuals in situation A, with so few fathers contributing genes to the next generation, the likelihood of large changes in allele frequency due to drift (i.e., the founder effect) is much higher in situation A than in situation B.

3. How uncertain are the environmental conditions—that is, do density-independent factors cause the death of large numbers of individuals when they occur? What is the mating system— are a few males mating with the majority of the females, or is the species more monogamous? What is the likely carrying capacity for piping plovers in this environment?

Chapter 16

STOP & STRETCH

p. 392: A reduction in tilt would reduce seasonality because areas would be receiving the same solar irradiance throughout the year. An increase in tilt would increase the yearly range of temperatures in seasonal areas.

p. 397: The ocean is dark and thus absorbs light and heats up. Ice reflects the light and keeps areas around it cold.

p. 401: Chlorophyll may be "expensive" for the trees to make in some way. The most likely expense is nitrogen, which is limited in the soil and may not be easy to replace if simply dropped to the soil.

p. 408: Wetlands slow water flow and thus can absorb extra nutrients before they accumulate in the open water body.

p. 411: Fossil fuels were easy to access just by digging. The technology to convert the fuels to electricity (burning and generating steam) or to power an engine (controlled explosions) were a lot simpler than the technology needed to convert sunlight into electricity. Wind and solar power is not as reliable as the power generated from fossil fuels.

VISUALIZE THIS

Figure 16.2: Solar irradiance is higher in Florida compared to Vermont because it is closer to the equator. Thus Florida is warmer than Vermont.

Figure 16.3: Because when the Northern Hemisphere is tilted away from the sun, the Southern Hemisphere (where Australia is located) is tilted toward it.

Figure 16.5: Near 30° north and south and near the equator, where air is more likely to be moving vertically rather than horizontally, thus not producing a directional breeze.

Figure 16.9: This period is the hottest part of the summer, and areas near the sea will be cooler than areas inland.

Figure 16.11: The east side is windward, and the west side is leeward.

Figure 16.17: Cattle are not nearly as large or strong as elephants and cannot uproot these large trees.

SAVVY READER

1. None of the statements include objective support, but some imply that the company has evidence to support their claim.
2. On the website where the claims are made and obviously and easily located, like one hyperlink click away or on a clearly labeled drop down menu item.
3. Answers will vary, but should focus on objective evidence.

LEARNING THE BASICS

1. The tilt of Earth's axis means that as the planet revolves around the sun, the Northern Hemisphere is tilted toward the sun during part of the year and away during part of the year. The temperature at a given point on Earth's surface is determined in large part by the solar irradiance (i.e., the strength of sunlight) striking the surface. Because the Northern Hemisphere is tilted away in the winter, solar irradiance is low, as are temperatures. The opposite occurs in the summer.
2. Water is slow to gain and lose heat, so when surrounding land areas cool as a result of declining solar irradiance, warmer air from over the water keeps the temperature high. The opposite is true as the surrounding land warms as a result of increasing solar irradiance.
3. b; **4.** c; **5.** e; **6.** e; **7.** c; **8.** a; **9.** b; **10.** d

ANALYZING AND APPLYING THE BASICS

1. Semitropical with seasonal wet and dry periods as a result of latitude. Relatively rainy due to location on windward side of mountain range.
2. The temperature will become less moderate because warm water will not be carried to the coasts of Europe, warming nearby landmasses. Seasonality will be more pronounced and winters colder and longer.
3. Bend, Yakima, and perhaps Walla Walla are on the leeward side of the Cascades, while the other cities are on the windward side.

Chapter 17

STOP & STRETCH

p. 422: The lack of blood vessels in cartilage makes it difficult for transplanted tissues to receive the nutrients they need to survive.

p. 433: If gastrin release is inhibited, digestion slows.

VISUALIZE THIS

Figure 17.3: Actin and myosin deposits are visible.

Figure 17.5: Cardiovascular and urinary are two examples.

Figure 17.11: Damage to the liver or kidney

SAVVY READER

1. Both scientists are affiliated with universities—Goyal with Geisinger Health System in State College, Pa. and Sehgal with the CWRU School of Medicine.
2. They reported the data even though it was not what they expected.
3. Less credible than the JAMA article because the organ-trafficking agency stands to benefit financially from organ trade

LEARNING THE BASICS

1. Epithelia are tightly packed sheets of cells that cover and line organs, vessels, and body cavities. Connective tissues are loosely organized tissues that bind organs and tissues to each other. Muscle tissue consists of bundles of long, thin, cylindrical muscle fibers that are able to contract. Nervous tissue is composed mainly of neurons that can conduct and transmit electrical impulses.
2. Loose connective tissue, adipose tissue, blood, fibrous connective tissue, cartilage, and bone
3. Muscle tissues differ based on the presence or absence of striations and whether the muscle is voluntary or involuntary. Cardiac muscle is striated and involuntary; skeletal is striated and voluntary; smooth muscle is not striated and involuntary.
4. c; **5.** e; **6.** a; **7.** a; **8.** d; **9.** b; **10.** c; **11.** d; **12.** False

ANALYZING AND APPLYING THE BASICS

1. See Figure 17.8.
2. Liver, pancreas, and gallbladder
3. Answers will vary. An example of negative feedback is regulating home temperature. An example of positive feedback is getting a good grade makes a student work harder.

Chapter 18

STOP & STRETCH

p. 444: By forcing the diaphragm upward violently, the volume of the lungs is quickly reduced, resulting in a large exhalation of air.

p. 447: Less surfactant means that alveolar walls are sticking together, requiring greater force to allow air to gain access to the respiratory surface.

p. 449: The only "cure" is lung transplant. Increasing respiratory capacity (i.e., strength of the respiratory muscles) can also help by drawing more air into the lungs.

p. 452: The brain

p. 454: The valves that prevent backflow

p. 460: It would contain larger materials such as whole proteins and perhaps some blood cells.

VISUALIZE THIS

Figure 18.2: It causes a rapid intake of air.

Figure 18.5: Less oxygen crosses the respiratory membrane, meaning that individuals must breathe more rapidly to obtain adequate oxygen.

Figure 18.8: The disadvantages would be greater likelihood that the alveoli walls will stick together, more friction impeding the exchange of gases with the outside air, and by devoting more energy to build cells in the lung there is less energy for other organs.

Figure 18.11: An abundance of one or more blood cell type, typically ones that are not fully differentiated

Figure 18.13: More force is required to move blood to the body extremities than to the lungs, and too much force sending blood to the lungs can cause fluid to leak from capillaries into alveoli, interfering with breathing.

Figure 18.18: It crushes the plaque while increasing the diameter of the artery

SAVVY READER

1. The majority (52%) oppose the ban.
2. Only those individuals who feel strongly about the ban are likely to participate in the survey. This means that smokers may be overrepresented in the sample.
3. Answers will vary, but could focus on developing a survey that takes a randomized sample of the student population and focuses on younger students who would experience the effects of the ban rather than students who would graduate before the ban is in place.

LEARNING THE BASICS

1. Contraction of the diaphragm flattens out this dome-shaped muscle at the base of the chest cavity. This has the effect of increasing the size of the lung cavity, lowering air pressure and allowing outside, oxygen-rich air to flow through the upper respiratory system and into the lungs.
2. The first step, filtration, occurs when blood plasma is forced out of tiny holes in the capillaries of the glomerulus and into the nephron. The second stage is reabsorption, when water and other important constituents are actively recovered from the filtrate and returned to the bloodstream. Secretion occurs near the end of the nephron, when toxins and other wastes in low concentration in the bloodstream are actively removed.
3. e; **4.** d; **5.** a; **6.** d; **7.** b; **8.** d; **9.** d; **10.** e

ANALYZING AND APPLYING THE BASICS

1. The blood after doping has greater oxygen-carrying capacity and can supply muscles with oxygen more effectively. The danger is the risk of clots forming from excess blood cells.
2. Blood that is low in oxygen is allowed to mix in with the blood pumped to the body tissues, reducing the rate of oxygen delivery.
3. By crossing into the bloodstream at the respiratory membrane and then being carried throughout the body by the circulatory system

Chapter 19

STOP & STRETCH

p. 468: Infectious but not contagious

p. 474: All the unvaccinated would be at risk.

p. 481: It's possible that an individual that can fight off the infection has slightly different set of genes than those that can't.

VISUALIZE THIS

Figure 19.1: Yes, because ribosomes are not bounded by membranes.

Figure 19.4: Surface proteins

Figure 19.6: The odds of an epidemic increase as the numbers of vaccinated people decreases.

SAVVY READER

1. Sometimes there are subtle differences in interpretation of data. For example, in the case of AIDS, there might be some debate about the rate at which mutations occur in HIV. However, there is no debate about the larger issue—that HIV causes AIDS.
2. Answers will vary
3. Answers will vary

LEARNING THE BASICS

1. Bacteria are single-celled organisms that have no nucleus or membrane-bounded organelles. They house their DNA in the nucleoid region and are surrounded by a cell wall. Viruses are composed of genetic material surrounded by a protein coat called the capsid. Some viruses are also surrounded by a membranous envelope.
2. They can remove the infectious organism from the tissue it infects
3. B cells produce receptors that are found on cell surfaces or secreted as antibodies and destroy invaders. T cells also produce receptors but not ones that are secreted.
4. d; **5.** b; **6.** c; **7.** d; **8.** b; **9.** c; **10.** c ; **11.** e; **12.** F; **13.** F

ANALYZING AND APPLYING THE BASICS

1. They cannot replicate themselves without the aid of a host cell.
2. Because the flu virus is a rapidly mutating virus.
3. Because the presence of memory cells allows a stronger and more rapid response

Chapter 20

STOP & STRETCH

p. 497: Androgens provide negative feedback on hypothalamus, resulting in decreased FSH and LH production

p. 503: One example would be to compare elite male and female athletes in the same sport. For control, compare rates of knee injury in de-conditioned males and females.

p. 503: Marathoners may have more slow twitch, and weight lifters more fast twitch.

VISUALIZE THIS

Figure 20.2: Males have less estrogen overall and fewer estrogen receptors in susceptible tissues such as breast tissue.

Figure 20.8: Positive

Figure 20.14: No

1. Random selection of participants; collecting data from same set of colleges over different years
2. Symptoms of steroid abuse are hard to accept. May lead to escapism via drug and alcohol use, or students who are likely to abuse alcohol and drugs may be more likely to use steroids also
3. Answers will vary. It may be competition for scholarships increased or steroid availability increased.

LEARNING THE BASICS

1. The hypothalamus regulates body temperature and affects behaviors such as hunger, thirst, and reproduction. The pituitary gland secretes two hormones involved with producing sex differences: follicle-stimulating hormones (FSH) for sperm production and egg cell development, and luteinizing hormones (LH) for testosterone production and the release of an egg cell during ovulation. Adrenal glands secrete the hormone adrenaline (epinephrine) in response to stress or excitement and androgens, including testosterone and estrogen. Testes secrete testosterone, the hormone that aids in sperm production, hair thickness and distribution, increased muscle mass, and voice deepening. Ovaries produce and secrete estrogen, which regulates menstruation, the maturation of egg cells, breast development, pregnancy, and menopause.
2. Bone tissue can be tightly or loosely packed. Compact bone forms the hard outer shell of bones, and spongy bone is the porous honeycomb-like inner bone. Blood vessels run throughout bone. Bone marrow inside bones helps to produce blood cells.
3. A muscle contains bundles of parallel muscle fibers. Within each muscle fiber is a parallel array of thread-like filaments called myofibrils. Each myofibril is a linear arrangement of sarcomeres. A sarcomere is composed of thin filaments of the protein actin and thick filaments of the protein myosin.
4. c; **5.** d; **6.** a; **7.** b; **8.** b; **9.** b; **10.** b

ANALYZING AND APPLYING THE BASICS

1. Differences between members of the same sex prevent generalizations from having much meaning.
2. Muscles would not increase in size.
3. Because puberty starts later in males and testosterone has a larger skeleton to start working on

Chapter 21

STOP & STRETCH

p. 516: Using the same amount of estrogen found in a typical waterway, comparing numbers of feminized fish before and after treatment in the same lake, and comparing numbers against fish in untreated lakes

p. 518: Warmer temperature in abdomen is not optimal for sperm production.

p. 528: The experiment could simply track start dates of menstrual cycles for the women over a year or so to see whether their cycles converge.

VISUALIZE THIS

Figure 21.3: See the figure.

Figure 21.7: No, because each egg cell would differ genetically as would each sperm cell. These fraternal twins are no more alike genetically than any two siblings.

Figure 21.11: Binding of the sperm head to the zona pellucida is the species-specific step.

Figure 21.15: The drug causes increased uterine contractions.

SAVVY READER

1. No data outlining the rate of precocious puberty in other years is presented. Based on the data from one year presented in this article, it is not possible to determine whether rates are increasing, decreasing, or staying the same.
2. No.
3. Check reputable medical and government websites.
4. No. It just means that more research is needed before any real conclusions can be drawn.

LEARNING THE BASICS

1. Asexual reproduction occurs when one parent produces offspring that are genetically identical to the parent and to each other. Sexual reproduction requires genetic input from two parents.
2. See Figures 21.3 and 21.4.
3. See Figures 21.5 and 21.6.
4. a; **5.** a; **6.** c; **7.** e; **8.** c; **9.** a; **10.** e; **11.** c

ANALYZING AND APPLYING THE BASICS

1. The level of estrogen in the pill is not enough to stimulate FSH and LH release.
2. Damage to the fetus can occur low doses. Developing organs are more susceptible than already formed organs.
3. Positive feedback. Oxytocin stimulates contractions and contractions stimulate oxytocin release.

Chapter 22

STOP & STRETCH

p. 547: The distance to the brain is longer than the distance to the spinal cord. Therefore, by the time the pain message reaches your brain, your spinal cord has already sent a message to your hand to remove it from the hot surface.

p. 551: Yes, because taking medicines could alter a person's brain structure.

p. 556: Blocking sodium channels would prevent action potentials from being generated.

VISUALIZE THIS

Figure 22.6: See figure.

Figure 22.7: Cerebral cortex and caudate nuclei because altered functioning in these structures could lead to some of the symptoms of ADD

Figure 22.11: A defective sodium potassium pump could lead to an absence of nerve cell firing because no gradient in ions would form.

1. Using all four-year colleges instead of a mixture of two- and four-year colleges; randomly selecting students
2. Yes.
3. No.

LEARNING THE BASICS
1. The CNS is composed of the brain and spinal cord and is responsible for integrating, processing, and coordinating information taken in by the senses. It is also the seat of functions such as intelligence, learning, memory, and emotion. The PNS includes the network of nerves outside the brain and spinal cord and functions to link the CNS with senses.
2. See Figure 22.2 and Figure 22.3.
3. General senses do not have special organs, but arise from all over the body. These include sense of touch, pain and temperature.
4. a; **5.** d; **6.** e; **7.** d; **8.** d; **9.** d; **10.** e; **11.** True

ANALYZING AND APPLYING THE BASICS
1. A more strongly affected individual may have less dopamine or fewer dopamine receptors
2. Blocking neurotransmitter receptors slows down transmission.
3. Reflexes are prewired and automatic responses, whereas other nervous system responses, require sensory activity followed by integration and motor output.

Chapter 23

STOP & STRETCH
p. 569: Plant cells can transform from one type (e.g., root) into others (e.g., stem).
p. 573: The successful adult is probably using all of the resources available to it quite well. The young plants would generally have a difficult time establishing under these conditions.
p. 576: The requirement of an extended freezing spell ensures that the seeds will not germinate in fall, which would doom the seedling to death during winter conditions.
p. 578: Transfer of nutrients from the leaves to the roots of the plant. If only a patch is removed, phloem still remains intact between the leaves and roots.
p. 583: They do get larger and flower and thus use up nutrients. They also do not get additional nutrients from a natural cycle.
p. 588: If some individuals in a population are resistant to the pesticide, they survive and produce the next generation, passing the "resistance" trait on to the next generation.

VISUALIZE THIS
Figure 23.1: By saving seeds from the plants that produced the largest ears and those that kept their seeds, they drove the evolution of the crop toward larger and larger, tightly bound fruits.

Figure 23.5: By increasing the surface area for absorption many-fold
Figure 23.9: Advantages are that a pollinator can both pick up pollen for dispersal and facilitate fertilization of eggs in a single visit and that a flower that is attractive because it contains pollen is also able to be fertilized. Disadvantages include increased risk of self-pollination (and thus inbreeding).
Figure 23.14: Needs to be indicated on the photo.
Figure 23.15: So that the vascular tissue is continuous from the branch root to the stem
Figure 23.20: The plant would wilt because water would be drawn osmotically from the cells.
Figure 23.24: Nutrients would return to the soil after consumption by organisms present on the site.

SAVVY READER
1. Organic food is measurably better for consumers than conventionally produced food.
2. As asserted by a scientist at the University of Minnesota, organic food does contain less pesticide residue than conventionally produced food, but there is no evidence that organic food is more nutritious.
3. What are the health costs associated with conventional agriculture that are less direct than what you eat (e.g., air, soil, water pollution)? How much more expensive is organic produce than conventionally grown produce?

LEARNING THE BASICS
1. See Figures 23.2, 23.3, and 23.9
2. There are two stages in the life cycles of land plants. The sporophyte is a diploid plant that produces haploid spores via meiosis. Spores undergo mitosis to produce a multicellular gametophyte that produces gametes (sperm and eggs) by mitosis. Sperm and egg fuse to produce a diploid zygote that goes on to develop into the sporophyte. Since land plants first evolved, the gametophyte has become less and less prominent in the life cycle. In the earliest land plants, the gametophyte is much more prominent than the sporophyte. The immediate descendents of these plants had a much more prominent sporophyte than gametophyte, but the gametophyte is still free living. In the most recently evolved land plants, the gametophyte is only a small number of cells and is completely dependent on the sporophyte for survival.
3. e; **4.** d; **5.** d; **6.** d; **7.** b; **8.** b; **9.** a; **10.** e

ANALYZING AND APPLYING THE BASICS
1. Most plants will only produce fruit if the ovules are fertilized. It is difficult to trick a plant into thinking it has been fertilized when it has not.
2. By releasing the axillary buds from suppression, allowing side branches to emerge
3. The soil may be depleted of nutrients that are especially helpful to tomatoes, and pests specific to tomatoes may

have established populations in that area of his garden so they are ready to attack the plants as soon as the plants appear. The easiest solution is to move the tomatoes to a new area of the yard or garden. He could also add organic fertilizer to the soil to replenish it.

Chapter 24

STOP & STRETCH

p. 600: Because they are all transpiring, humidity is higher in the area of their leaves, reducing the evaporation rate from all plants.

p. 602: If many of the roots have died as a result of oxygen deprivation, the rate of transpiration may be too high for them to adequately replace water lost from the plant. Removing leaves reduces this rate.

p. 606: Before the tap, the sugar source is roots and the sink is tree leaf buds; after the tap, the sugar source is still the roots, but some of the sink is the taps.

p. 609: So that when the tree "wakes up" and begins sending sugars to the buds, the only buds that are supplied are those that remain and that the gardener wants to grow

p. 614: Ripening fruit produce ethylene, which promotes petal abscission.

VISUALIZE THIS

Figure 24.1: Evaporation is much reduced and the stomata are most likely closed (since the light reactions of photosynthesis cannot proceed without sunlight). Transpiration is unlikely.

Figure 24.6: The grain is "tighter" in trees where xylem cells are mainly tracheids, like in pine.

Figure 24.13: Because the plant will be triggered to flower, taking away energy from leaf production and reducing crop yield.).

SAVVY READER

1. Use a library database to search for articles by Bill Wolverton.
2. Houseplants may reduce air quality if the wet soil or pots promotes mold growth, leading to an increase in airborne mold spores.
3. No. Interested individuals would need to perform a thorough literature review.

LEARNING THE BASICS

1. Hydrogen bonding among water molecules (cohesion) and between water molecules and xylem cell walls (adhesion) provides a force that counters the strong negative pressure and keeps the water column together.
2. The C_4 adaptation pumps carbon dioxide toward actively photosynthetic cells, allowing stomata to remain only partially open. The CAM adaptation allows plants to accumulate carbon dioxide during the cooler night when less water will evaporate so that they may keep their stomata closed during of side the hottest parts of the day.
3. Auxin is destroyed on the illuminated part of the stem, preventing cell elongation. As the cells on the darker side continue to elongate, the stem tip is pushed toward the light source.
4. d; **5.** c; **6.** b; **7.** d; **8.** a; **9.** a; **10.** c

ANALYZING AND APPLYING THE BASICS

1. Make sure that the soil in the site is well watered and mulched. The lily should have some of its leaves removed to reduce transpiration.
2. Keep it in a dark closet for a good portion of the day.
3. Prune competing apical meristems so that only one remains to promote a single tip. Prune apical meristems of side branches to encourage these side branches to branch more to improve business.

Glossary

ABO blood system A system for categorizing human blood based on the presence or absence of carbohydrates on the surface of red blood cells. (Chapter 8)

abscisic acid A hormone in plants associated with dormancy in seeds and buds. (Chapter 23)

abscission The dropping off of leaves, flowers, fruits, or other plant organs. (Chapter 23)

accessory organs Organs including the pancreas, liver, and gallbladder, which aid the digestive system. (Chapter 16)

acetylcholine A neurotransmitter with many functions, including facilitating muscle movements, and thought to be involved in development of Alzheimer's disease. (Chapter 21)

acetylcholinesterase An enzyme in nerve cells that breaks down acetylcholine. (Chapter 21)

acid A substance that increases the concentration of hydrogen ions in a solution. (Chapter 2)

acid rain Rain (or other precipitation) that is unusually acidic; caused by air pollution in the form of sulfur and nitrogen dioxides. (Chapter 16)

acquired immune deficiency syndrome (AIDS) Syndrome characterized by severely reduced immune system function and numerous opportunistic infections. Results from infection with HIV. (Chapter 18)

acrosome An organelle at the tip of the sperm cell containing enzymes that help the sperm penetrate the egg cell. (Chapter 20)

actin A protein found in muscle tissue that, together with myosin, facilitates contraction. (Chapters 16, 19)

action potential Wave of depolarization in a neuron propagated to the end of the axon—also called a nerve impulse. (Chapter 21)

activation energy The amount of energy that reactants in a chemical reaction must absorb before the reaction can start. (Chapter 4)

activator A protein that serves to enhance the transcription of a gene. (Chapter 8)

active immunity Immunity that results from the production of antibodies, as differentiated from passive immunity. (Chapter 19)

active site Substrate-binding region of an enzyme. (Chapter 4)

active smoker An individual who smokes tobacco. (Chapter 18)

active transport The ATP-requiring movement of substances across a membrane against their concentration gradient. (Chapter 3)

adaptation Trait that is favored by natural selection and increases an individual's fitness in a particular environment. (Chapters 10, 11)

adaptive radiation Diversification of one or a few species into large and very diverse groups of descendant species. (Chapter 13)

ADD See attention deficit disorder. (Chapter 22)

adenine Nitrogenous base in DNA, a purine. (Chapters 2, 4, 5, 9)

adenosine diphosphate (ADP) A nucleotide composed of adenine, a sugar, and two phosphate groups. Produced by the hydrolysis of the terminal phosphate bond of ATP. (Chapter 4)

adenosine triphosphate (ATP) A nucleotide composed of adenine, the sugar ribose, and three phosphate groups that can be hydrolyzed to release energy. Form of energy that cells can use. (Chapters 2, 3, 4)

adhesion The sticking together of unlike materials—often water and a particular surface. (Chapter 24)

adipose tissue Fat-storing connective tissue. (Chapter 17)

adrenal gland Either of two endocrine glands, one located atop each kidney, that secrete adrenaline in response to stress or excitement, help maintain water and salt balance, and secrete small amounts of sex hormones. (Chapter 20)

adrenaline A hormone secreted by the adrenal glands in response to stress or excitement. (Chapter 20)

aerobic An organism, environment, or cellular process that requires oxygen. (Chapter 4)

aerobic respiration Cellular respiration that uses oxygen as the electron acceptor. (Chapter 4)

agarose gel A jelly-like slab used to separate molecules on the basis of molecular weight. (Chapter 8)

agriculture Cultivation of crops, raising of livestock; farming. (Chapter 23)

algae Photosynthetic protists. (Chapter 13)

alimentary canal Part of the digestive system that forms a tube extending from the mouth to the anus. Also called the digestive tract. (Chapter 17)

allele Alternate versions of the same gene, produced by mutations. (Chapters 6, 7, 9, 11, 15)

allele frequency The percentage of the gene copies in a population that are of a particular form, or allele. (Chapters 12, 15)

allergy An abnormally high sensitivity to allergens such as pollen or microorganisms. Can cause sneezing, itching, runny nose, and watery eyes. (Chapter 19)

allopatric Geographic separation of a population of organisms from others of the same species. Usually in reference to speciation. (Chapter 12)

alternation of generations A reproductive cycle in which a haploid phase, called a gametophyte, alternates with a diploid phase, called a sporophyte. (Chapter 23)

alternative hypothesis Factor other than the tested hypothesis that may explain observations. (Chapter 1)

alveoli (singular: alveolus) Sacs inside lungs, making up the respiratory surface in land vertebrates and some fish. (Chapter 18)

Alzheimer's disease Progressive mental deterioration in which there is memory loss along with the loss of control of bodily functions, ultimately resulting in death. (Chapter 22)

amenorrhea Abnormal cessation of menstrual cycle. (Chapters 4, 20)

amino acid Monomer subunit of a protein. Contains an amino, a carboxyl, and a unique side group. (Chapters 2, 3, 9)

amnion The fluid-filled sac in which a developing embryo is suspended. (Chapter 21)

anabolic steroid A derivative, usually synthetic, of the steroid hormone testosterone. (Chapter 20)

anaerobic Said of an organism, environment, or cellular process that does not require oxygen. (Chapter 4)

anaerobic respiration A process of energy generation that uses molecules other than oxygen as electron acceptors. (Chapter 4)

anaphase Stage of mitosis during which microtubules contract and separate sister chromatids. (Chapter 6)

anchorage dependence Phenomenon that holds normal cells in place. Cancer cells can lose anchorage dependence and migrate into other tissues or metastasize. (Chapter 6)

androgen A masculinizing hormone, such as testosterone, secreted by the adrenal glands. (Chapters 20, 21)

anecdotal evidence Information based on one person's personal experience. (Chapter 1)

angiogenesis Formation of new blood vessels. (Chapter 6)

angiosperm Plant in the phyla Angiospermae, which produce seeds borne within fruit. (Chapters 13, 23, 24)

animal an organism that obtains energy and carbon by ingesting other organisms and is typically motile for part of its life cycle (Chapter 13)

Animalia Kingdom of Eukarya containing organisms that ingest others and are typically motile for at least part of their life cycle. (Chapters 10, 13)

annual plant Plant that completes its life cycle in a single growing season. (Chapters 16, 23, 24)

annual growth rate Proportional change in population size over a single year. Growth rate is a function of the birth rate minus the death rate of the population. (Chapter 14)

anorexia Self-starvation. (Chapter 4)

antagonistic muscle pair A set of muscles whose actions oppose each other. (Chapter 20)

anther The pollen-containing structure on the stamen of a flower. (Chapter 23)

antibiotic A chemical that kills or disables bacteria. (Chapter 11, 13, 19)

antibiotic resistant Characteristic of certain bacteria; a physiological characteristic that permits them to survive in the presence of particular antibiotics. (Chapter 11)

antibody Protein made by the immune system in response to the presence of foreign substances or antigens. Can serve as a receptor on a B cell or be secreted by plasma cells. (Chapter 19)

antibody-mediated immunity Immunity that occurs via secreted antibodies, versus immunity mediated by T cells. (Chapter 19)

anticodon Region of tRNA that binds to an mRNA codon. (Chapter 9)

antigen Short for antibody-generating substances, an antigen is a molecule that is foreign to the host and stimulates the immune system to react. (Chapter 19)

antigen receptor Protein in B- and T-cell membrane that bind to specific antigens. (Chapter 19)

antioxidant Certain vitamins and other substances that protect the body from the damaging effects of free radicals. (Chapter 3)

antiparallel Feature of DNA double helix in which nucleotides face "up" on one side of the helix and "down" on the other. (Chapter 2)

apical dominance The influence exerted by the terminal bud in suppressing the growth of axillary or lateral buds. Mediated by the hormone, auxin. (Chapter 24)

apical meristem The actively dividing cells at the tip of stems and roots in plants. (Chapter 23)

appendicular skeleton The part of the skeleton composed of the bones of the hip, shoulder, and limbs. (Chapter 20)

aquaporin A transport protein in the membrane of a plant or animal cell that facilitates the diffusion of water across the membrane (osmosis). (Chapter 3)

aquatic Of, or relating to, water. (Chapter 16)

Archaea Domain of prokaryotic organisms made up of species known from extreme environments. (Chapter 13)

artery Blood vessel that carries oxygenated blood from the heart to body tissues. (Chapter 18)

artificial selection Selective breeding of domesticated animals and plants to increase the frequency of desirable traits. (Chapter 11)

asexual reproduction A type of reproduction in which one parent gives rise to genetically identical offspring. (Chapters 6, 12, 21)

assortative mating Tendency for individuals to mate with other individuals who are like themselves. (Chapter 12)

asthma A respiratory disease characterized by spasmodic inflammation of the air passages in the lungs and overproduction of mucus. Often triggered by air contaminants. (Chapter 18)

asymptomatic Stage in an infection that is characterized by relatively unnoticeable or absent symptoms of illness. (Chapter 19)

atherosclerosis Accumulation of fatty deposits within blood vessels; hardening of the arteries. (Chapter 18)

atom The smallest unit of matter that retains the properties of an element. (Chapter 2)

atomic number The number of protons in the nucleus of an atom. Unique to each element, this number is designated by a subscript to the left of the symbol for the element. (Chapter 2)

ATP synthase Enzyme found in the mitochondrial membrane that helps synthesize ATP. (Chapter 4)

atrium An upper chamber of the heart that receives blood from the body or lungs and pumps it to a ventricle. (Chapter 18)

attention deficit disorder (ADD) Syndrome characterized by forgetfulness; distractibility, fidgeting; restlessness; impatience; difficulty sustaining attention in work, play, or conversation; or difficulty following instructions and completing tasks in more than one setting. (Chapter 22)

autoimmune disease Any of the diseases that result from an attack by the immune system on normal body cells. (Chapter 19)

autosome Non-sex chromosome, of which there are 22 pairs in humans. (Chapters 6, 8)

auxin A class of plant hormones that control cell elongation, among other effects. (Chapter 24)

AV valves Heart valves between the atria and the ventricles. (Chapter 18)

axial skeleton Part of the skeleton that supports the trunk of the body and consists largely of the bones making up the vertebral column or spine and much of the skull. (Chapter 20)

axillary bud A bud located at the junction between a plant stem and leaf. (Chapters 23, 24)

axon Long, wire-like portion of the neuron that ends in a terminal bouton. (Chapter 22)

B lymphocyte (B cell) The type of white blood cell responsible for antibody-mediated immunity. (Chapter 19)

background extinction rate The rate of extinction resulting from the normal process of species turnover. (Chapter 15)

bacteria Domain of prokaryotic organisms. (Chapters 2, 4, 9, 13, 15, 19)

ball and socket joint Type of joint in the hips and shoulders that enables arms and legs to move in three dimensions. (Chapter 20)

bark The outer layer of a woody stem, consisting of cork, cork cambium, and phloem. (Chapter 23)

basal metabolic rate Resting energy use of an awake, alert person. (Chapter 3)

base A substance that reduces the concentration of hydrogen ions in a solution. (Chapter 2)

base-pairing rule A rule governing the pairing of nitrogenous bases. In DNA, adenine pairs with thymine and cytosine with guanine. (Chapter 2)

behavioral isolation Prevention of mating between individuals in two different populations based on differences in behavior. (Chapter 11)

benign Tumor that stays in one place and does not affect surrounding tissues. (Chapter 6)

bias Influence of research participants' opinions on experimental results. (Chapter 1)

biennial Plant that completes its life cycle over the course of two growing seasons. (Chapter 23)

bile Mixture of substances produced in the liver that aids in digestion by emulsifying fats. (Chapter 16)

binary fission An asexual form of bacterial reproduction. (Chapters 18, 20)

bioaccumulation A phenomenon that results in the concentration of persistent (i.e., slow to degrade) pollutants in the bodies of animals at high levels of a food chain. (Chapter 16)

biodiversity Variety within and among living organisms. (Chapters 13, 15)

biogeography The study of the geographic distribution of organisms. (Chapter 10)

biological classification Field of science attempting to organize biodiversity into discrete, logical categories. (Chapters 10, 13)

biological diversity Entire variety of living organisms. (Chapter 13)

biological evolution See evolution. (Chapter 10)

biological population Individuals of the same species that live and breed in the same geographic area. (Chapters 10, 12)

biological race Populations of a single species that have diverged from each other. Biologists do not agree on a definition of "race." See also subspecies. (Chapter 12)

biological species concept Definition of a species as a group of individuals that can interbreed and produce fertile offspring but typically cannot breed with members of another species. (Chapter 12)

biology The study of living organisms. (Chapter 1)

biomagnification Concentration of toxic chemicals in the higher levels of a food web. (Chapter 23)

biomass The mass of all individuals of a species, or of all individuals on a level of a food web, within an ecosystem. (Chapter 15)

biome A broad ecological community defined by a particular vegetation type (e.g. temperate forest, prairie), which is typically determined by climate factors. (Chapter 16)

biophilia Humans' innate desire to be surrounded by natural landscapes and objects. (Chapter 15)

biopiracy Using the knowledge of the native people in developing countries to discover compounds for use in developed countries. (Chapter 13)

bioprospecting Hunting for new organisms and new uses of old organisms. (Chapter 13)

biopsy Surgical removal of some cells, tissue, or fluid to determine whether cells are cancerous. (Chapter 6)

bipedal Walking upright on two limbs. (Chapter 10)

bladder An organ of the excretory system that stores urine after it is excreted from the kidneys. (Chapter 18)

blade The broad part of a leaf. (Chapter 23)

blastocyst An embryonic stage consisting of a hollow ball of cells. (Chapter 21)

blind experiment Test in which subjects are not aware of exactly what they are predicted to experience. (Chapter 1)

blood The combination of cells and liquid that flows through blood vessels in the cardiovascular system; made up of red blood cells, plasma, white blood cells, and platelets. (Chapters 17, 18, 19)

blood clotting The process by which a blood clot, a mass of the protein fibrin and dead blood cells, forms in the region of blood vessel damage. (Chapter 18)

blood pressure The force of the blood as it travels through the arteries; partially determined by artery diameter and elasticity. (Chapter 18)

blood vessel The structure that carries blood via the circulatory system throughout the body; arteries, capillaries, and veins. (Chapter 18)

body mass index (BMI) Calculation using height and weight to determine a number that correlates to an estimate of a person's amount of body fat with health risks. (Chapter 3)

bolus In digestion, a soft ball of chewed food. (Chapter 17)

bone A type of connective tissue consisting of living cells in a matrix rich in collagen and calcium. (Chapter 17)

bone marrow Network of soft connective tissue that fills bones and is involved in the production of red blood cells. (Chapters 18, 19, 20)

boreal forest A biome type found in regions with long, cold winters and short, cool summers. Characterized by coniferous trees. (Chapter 16)

botanist Plant biologist. (Chapter 13)

brain stem Region of the brain that lies below the thalamus and hypothalamus; it governs reflexes and some involuntary functions such as breathing and swallowing. (Chapter 22)

bronchi The large air passageways from the trachea into the lungs. (Chapter 18)

bronchiole The branching air passageway inside the lungs. (Chapter 18)

bronchitis Inflammation of the bronchi and bronchioles in the lungs. (Chapter 18)

budding An asexual mechanism of propagation in which outgrowths of the parent form and pinch off, producing new individuals. (Chapter 21)

bulb Plant organ consisting of short stems and multiple leaves modified for food storage. (Chapter 23).

bulbourethral gland Either of the two glands at the base of the penis that secrete acid-neutralizing fluids into semen. (Chapter 21)

bulimia Binge eating followed by purging. (Chapter 3)

C$_3$ plant Plant that uses the Calvin cycle of photosynthesis to incorporate carbon dioxide into a 3-carbon compound. (Chapter 5)

C$_4$ plant Plant that performs reactions incorporating carbon dioxide into a 4-carbon compound that ultimately provides carbon dioxide for the Calvin cycle. (Chapter 5)

calcitonin A hormone produced by the thyroid that helps lower blood calcium levels. (Chapter 20)

calcium Nutrient required in plant cells for the production of cell walls and for bone strength and blood clotting in humans. (Chapter 3)

Calorie A kilocalorie or 1000 calories. (Chapter 3)

calorie Amount of energy required to raise the temperature of one gram of water by 1 °C. (Chapter 3)

Calvin cycle A series of reactions that occur in the stroma of plants during photosynthesis that utilize NADPH and ATP to reduce carbon dioxide and produce sugars. (Chapter 5)

CAM plant A plant that uses Crassulacean acid metabolism, a variant of photosynthesis during which carbon dioxide is stored in sugars at night (when stomata are open) and released during the day (when stomata are closed) to prevent water loss. (Chapter 5)

Cambrian explosion Relatively rapid evolution of the modern forms of multicellular life that occurred approximately 550 million years ago. (Chapter 13)

cancer A disease that occurs when cell division escapes regulatory controls. (Chapter 6)

capillary The smallest blood vessel of the cardiovascular system, connecting arteries to veins and allowing material exchange across their thin walls. (Chapter 18)

capillary bed A branching network of capillaries supplying a particular organ or region of the body. (Chapter 18)

capsid Protein coat that surrounds a virus. (Chapter 19)

capsule Gelatinous outer covering of bacterial cells that aids in attachment to host cells during an infection. (Chapter 19)

carbohydrate Energy-rich molecule that is the major source of energy for the cell. Consists of carbon, hydrogen, and oxygen in the ratio CH_2O. (Chapters 2, 3)

carcinogen Substance that causes cancer or increases the rate of its development. (Chapters 6, 18)

cardiac cycle The cycle of contraction and relaxation that a normally functioning heart undergoes over the course of a single heart beat. (Chapter 18)

cardiac muscle Muscle that forms the contractile wall of the heart. (Chapter 17)

cardiovascular disease Malfunction of the cardiovascular system, including, but not limited to, heart attack, stroke, and hypertension. (Chapter 18)

cardiovascular system The organ system made up of the heart and circulatory system, including arteries, capillaries, and veins. (Chapter 18)

carpel The female structure of a flower, containing the ovary, style, and stigma. (Chapter 23)

carrier Individual who is heterozygous for a recessive allele. (Chapter 8)

carrying capacity Maximum population that the environment can support. (Chapter 14)

cartilage Connective tissue found in the skeletal system that is rich in collagen fibers. (Chapter 17)

catalyst A substance that lowers the activation energy of a chemical reaction, thereby speeding up the reaction. (Chapter 4)

catalyze To speed up the rate of a chemical reaction. Enzymes are biological catalysts. (Chapter 4)

caudate nucleus Structure within each cerebral hemisphere that functions as part of the pathway that coordinates movement patterns, learning, and memory. (Chapter 22)

cell Basic unit of life, an organism's fundamental building-block units. (Chapters 2, 3)

cell body Portion of the neuron that houses the nucleus and organelles. (Chapter 22)

cell cycle An ordered sequence of events in the life cycle of a eukaryotic cell from its origin until its division to produce daughter cells. Consists of M, G$_1$, S, and G$_2$ phases. (Chapter 6)

cell division Process a cell undergoes when it makes copies of itself. Production of daughter cells from an original parent cell. (Chapter 6)

cell plate A double layer of new cell membrane that appears in the middle of a dividing plant cell and divides the cytoplasm of the dividing cell. (Chapter 6)

cell wall Tough but elastic structure surrounding plant and bacterial cell membranes. (Chapters 2, 6, 19, 23, 24)

cell-mediated immunity Type of specific immune response carried out by T cells. (Chapter 19)

cellular respiration Metabolic reactions occurring in cells that result in the oxidation of macromolecules to produce ATP. (Chapter 4)

cellulose A structural polysaccharide found in cell walls and composed of glucose molecules. (Chapters 2, 6)

central nervous system (CNS) Includes brain and spinal cord and is responsible for integrating, processing, and coordinating information taken in by the senses. It is the seat of functions such as intelligence, learning, memory, and emotion. (Chapter 22)

central vacuole A membrane-enclosed sac in a plant cell that functions to store many different substances. (Chapters 2, 23, 24)

centriole A structure in animal cells that helps anchor for microtubules during cell division. (Chapters 2, 6)

centromere Region of a chromosome where sister chromatids are attached and to which microtubules bind. (Chapter 6)

cereal grains Staples of the human diet that are from the grass family, namely wheat, corn, oats, rye, and barley. (Chapter 23)

cerebellum Region of the brain that controls balance, muscle movement, and coordination. (Chapter 22)

cerebral cortex Deeply wrinkled outer surface of the cerebrum where conscious activity and higher thought originate. (Chapter 22)

cerebrospinal fluid Protective liquid bath that surrounds the brain within the skull. (Chapter 22)

cerebrum Portion of the brain in which language, memory, sensations, and decision making are controlled. The cerebrum has two hemispheres, each of which has four lobes. (Chapter 22)

cervix The lower narrow portion of the uterus at the top end of the vagina. (Chapter 21)

chaparral A biome characteristic of climates with hot, dry summers and mild, wet winters and a dominant vegetation of aromatic shrubs. (Chapter 16)

checkpoint Stoppage during cell division that occurs to verify that division is proceeding correctly. (Chapter 6)

chemical reaction A process by which one or more chemical substances is transformed into one or more different chemical substances. (Chapter 2)

chemotherapy Using chemicals to try to kill rapidly dividing (cancerous) cells. (Chapter 6)

chlorophyll Green pigment found in the chloroplast of plant cells. (Chapter 5)

chloroplast An organelle found in plant cells that absorbs sunlight and uses the energy derived to produce sugars. (Chapters 2, 5, 13, 24)

cholesterol A steroid found in animal cell membranes that affects membrane fluidity. Serves as the precursor to estrogen and testosterone. (Chapters 2, 3, 20)

chondrocyte A type of cartilage cell that produces collagen and proteoglycans. (Chapter 17)

chromosome Subcellular structure composed of a long single molecule of DNA and associated proteins, housed inside the nucleus. (Chapters 6, 7, 8, 9, 12)

chyme Partially digested food and enzymes that are passed from the stomach to the intestine. (Chapter 17)

circulatory system The vessels that transport blood, nutrients, and waste around the body. (Chapters 6, 17, 18, 19)

citric acid cycle A chemical cycle occurring in the matrix of the mitochondria that breaks the remains of sugars down to produce carbon dioxide. (Chapter 4)

cladistic analysis A technique for determining the evolutionary relationships among organisms that relies on identification and comparison of newly evolved traits. (Chapter 13)

classification system Method for organizing biological diversity. (Chapters 10, 13)

cleavage In embryology, the period of rapid cell division that occurs during animal development. (Chapter 21)

climate The average temperature and precipitation as well as seasonality. (Chapter 16)

climax community The group of species that is stable over time in a particular set of environmental conditions. (Chapter 16)

clinical trial Controlled scientific experiment to determine the effectiveness of novel treatments. (Chapter 6)

clitoris Sensitive erectile tissue found in the external genitalia of females that functions in sexual arousal. (Chapter 21)

clonal population Population of identical cells copied from the immune cell that first encounters an antigen. The entire clonal population has the same DNA arrangement, and all cells in a clonal population carry the same receptor on their membrane. (Chapter 19)

cloning Producing copies of a gene or an organism that are genetically identical. (Chapter 9)

closed circulatory system A circulatory system in which vessels are continuous and retain the circulating fluid. (Chapter 18)

clumped distribution A spatial arrangement of individuals in a population where large numbers are concentrated in patches with intervening, sparsely populated areas separating them. (Chapter 14)

codominant Two different alleles of a gene that are equally expressed in the heterozygote. (Chapters 7, 8)

codon A triplet of mRNA nucleotides. Transfer RNA molecules bind to codons during protein synthesis. (Chapter 9)

coenzyme (or cofactor) Substances such as vitamins that help enzymes catalyze chemical reactions. (Chapter 4)

coevolution Occurs when change in one biological species triggers a change in an associated species. Coevolution commonly occurs between predator and prey or parasite and host. (Chapter 13)

cohesion The tendency for molecules of the same material to stick together. (Chapters 2, 24)

collecting duct A structure in the kidney that accepts filtrate from multiple nephrons and transmits it to the renal pelvis. Each kidney contains hundreds of collecting ducts. The amount of water retained by the kidneys is largely controlled at the collecting ducts. (Chapter 18)

colon Tube between the small intestine and anus of the digestive system. (Chapter 17)

combination drug therapy The use of more than one drug simultaneously to treat a disease. Often used for disease organisms that mutate quickly or are difficult to control to combat the problem of drug resistance. (Chapter 11)

commensalism In ecology, a relationship between two species in which one is benefitted and the other is neither harmed nor benefitted. (Chapter 15)

common descent The theory that all living organisms on Earth descended from a single common ancestor that appeared in the distant past. (Chapter 10)

community A group of interacting species in the same geographic area. (Chapter 15)

compact bone The hard, outer shell of bones. (Chapter 20)

competition Interaction that occurs when two species of organisms both require the same resources within a habitat; competition tends to limit the size of populations. (Chapter 15)

competitive exclusion Reduction or elimination of one species in an environment resulting from the presence of another species that requires the same or similar resources. (Chapter 15)

complementary Complementary bases pair with each other by hydrogen bonding across the DNA helix. Adenine is complementary to thymine and cytosine to guanine. (Chapter 2)

complement protein Type of blood protein with which an antibody-antigen complex can combine in order to kill bacterial cells. Enhances the immune response on many levels. (Chapter 19)

complementary base pair Nitrogenous bases that hydrogen bond to each other. In DNA, adenine is complementary to thymine, and cytosine is complementary to guanine. In RNA, adenine is complementary to uracil and guanine to cytosine. (Chapters 2, 8, 9)

complete digestive tract A digestive tube running from the mouth to anus. (Chapter 16)

complete protein Dietary protein that contains all the essential amino acids. (Chapter 3)

complex carbohydrate Carbohydrate consisting of two or more monosaccharides. (Chapter 3)

compound A substance consisting of two or more elements in a fixed ratio. (Chapter 2)

confidence interval In statistics, a range of values calculated to have a given probability (usually 95%) of containing the true population mean. (Chapter 1)

connective tissue Animal tissue that functions to bind and support other tissues. Composed of a small number of cells embedded in a matrix. (Chapter 17)

consilience The unity of knowledge. Used to describe a scientific theory that has multiple lines of evidence to support it. (Chapter 10)

contact inhibition Property of cells that prevents them from invading surrounding tissues. Cancer cells may lose this property. (Chapter 6)

contagious Spreading from one organism to another. (Chapter 18)

continuous variation A range of slightly different values for a trait in a population. (Chapter 7)

control Subject for an experiment who is similar to experimental subject except is not exposed to the experimental treatment. Used as baseline values for comparison. (Chapter 1)

convergent evolution Evolution of the same trait or set of traits in different populations as a result of shared environmental conditions rather than shared ancestry. (Chapters 10, 12)

copulation Sexual intercourse. (Chapter 21)

coral reef Highly diverse biome found in warm, shallow salt water, dominated by the limestone structures created by coral animals. (Chapters 13, 16)

cork A tissue produced by the cork cambium that provides protection to a woody stem by forming outer bark. (Chapter 23)

cork cambium A meristematic tissue in bark that arises from phloem cells and creates cork cells before being shed. (Chapter 23)

corpus callosum Bundle of nerve fibers at the base of the cerebral fissure that provides a communication link between the cerebral hemispheres. (Chapter 22)

corpus luteum Hormone-producing tissue (the ovarian follicle after ovulation) that makes progesterone and estrogen and degenerates about 12 days after ovulation if fertilization does not occur. (Chapter 21)

correlation Describes a relationship between two factors. (Chapters 1, 7, 12)

cotyledon The first leaves produced by an embryonic seed plant. (Chapter 23)

countercurrent exchange The process by which streams of fluid move in parallel but opposite directions, maximizing material, gas, or heat exchange between them. (Chapter 18)

covalent bond A type of strong chemical bond in which two atoms share electrons. (Chapter 2)

critical night length The night length that a plant must experience in order to trigger a photoperiodic response. (Chapter 24)

crop rotation The practice of growing different crops at different times on the same field in order to maintain soil fertility and decrease pest damage. (Chapter 23)

cross In genetics, the mating of two organisms. (Chapter 7)

crossing over Gene for gene exchange of genetic information between members of a homologous pair of chromosomes. (Chapter 6)

cryptorchidism A developmental defect in which the testes fail to descend to the scrotum. (Chapter 21)

cultural control Control of agricultural pests through environmental techniques, such as crop rotation and maintenance of native vegetation. (Chapter 23)

cuticle The waxy layer on the outer surface of plant epidermal cells. (Chapter 23)

cyst Noncancerous, fluid-filled growth. (Chapter 6)

cytokinesis Part of the cell cycle during which two daughter cells are formed by the cytoplasm splitting. (Chapter 6)

cytoplasm The entire contents of the cell (except the nucleus) surrounded by the plasma membrane. (Chapters 2, 9)

cytosine Nitrogenous base, a pyrimidine. (Chapters 2, 6, 8)

cytoskeleton A network of tubules and fibers that branch throughout the cytoplasm. (Chapter 2)

cytosol The semifluid portion of the cytoplasm. (Chapters 2, 4)

cytotoxic T cell Immune-system cell that attacks and kills body cells that have become infected with a virus before the virus has had time to replicate. (Chapter 19)

data Information collected by scientists during hypothesis testing. (Chapter 1)

daughter cells The offspring cells that are produced by the process of cell division. (Chapter 6)

death rate Number of deaths averaged over the population as a whole. (Chapter 14)

deciduous Pertaining to woody plants that drop their leaves at the end of a growing season. (Chapters 16, 24)

decomposer An organism, typically bacteria and fungi in the soil, whose action breaks down complex molecules into simpler ones. (Chapter 15)

decomposition The breakdown of organic material into smaller molecules. (Chapter 16)

deductive reasoning Making a prediction about the outcome of a test; "if/then" statements. (Chapter 1)

defensive protein Nonspecific protein of the immune system including interferons and complement proteins. (Chapter 19)

deforestation The removal of forest lands, often to enable the development of agriculture. (Chapters 5, 15)

degenerative disease Disease characterized by progressive deterioration. (Chapter 9)

dehydration Loss of water. (Chapters 3, 24)

deleterious In genetics, said of a mutation that reduces an individual's fitness. (Chapter 15)

demographic momentum Lag between the time that humans reduce birth rates and the time that population numbers respond. (Chapter 14)

demographic transition The period of time between when death rates in a human population fall (as a result of improved technology) and when birth rates fall (as a result of voluntary limitation of pregnancy). (Chapter 14)

denature (1) In proteins, the process where proteins unravel and change their native shape, thus losing their biological activity. (2) For DNA, the breaking of hydrogen bonds between the two strands of the double-stranded DNA helix, resulting in single-stranded DNA. (Chapter 8)

dendrite Short extension of the neuron that receives signals from other cells. (Chapter 22)

density-dependent factor Any of the factors related to a population's size that influence the current growth rate of a population—for example, communicable disease or starvation. (Chapter 14)

density-independent factor Any of the factors unrelated to a population's size that influence the current growth rate of a population—for example, natural disasters or poor weather conditions. (Chapter 14)

deoxyribonucleic acid (DNA) Molecule of heredity that stores the information required for making all of the proteins required by the cell. (Chapters 2, 6, 7, 8, 9, 11, 13)

deoxyribose The five-carbon sugar in DNA. (Chapters 2, 6, 7, 9)

dependent variable The variable in a study that is expected to change in response to changes in the independent variable. (Chapter 1)

depolarization Reduction in the charge difference across the neuronal membrane. (Chapter 22)

depression Disease that involves feelings of helplessness and despair, and sometimes thoughts of suicide. (Chapter 22)

desert Biome found in areas of minimal rainfall. Characterized by sparse vegetation. (Chapter 16)

desertification The process by which formerly productive land is converted to unproductive land, typically by overgrazing of cattle or unsustainable agricultural practices, but also increasingly by the effects of climate change. (Chapter 16)

determinate growth Growth of limited duration. (Chapter 23)

developed country A term coined by the United Nations to describe a country with high per-capita income and significant industrial development. (Chapter 14)

development All of the progressive changes that produce an organism's body. (Chapters 13, 21)

diabetes Disorder of carbohydrate metabolism characterized by impaired ability to produce or respond to the hormone insulin. (Chapter 3)

diaphragm (1) Dome-shaped muscle at the base of the chest cavity. Contraction of this muscle helps draw air into the lungs. (Chapter 18) (2) A birth control device consisting of a flexible contraceptive disk that covers the cervix to prevent the entry of sperm. (Chapter 21)

diastole The stage of the cardiac cycle when the heart relaxes and fills with blood. (Chapter 18)

diastolic blood pressure The lowest blood pressure in the arteries, occurring during diastole of the cardiac cycle. (Chapters 4, 18)

dicot The class of angiosperms characterized by having two cotyledons. (Chapter 23)

differentiation Structural and functional divergence of cells as they become specialized. (Chapter 21)

diffusion The spontaneous movement of substances from a region of their own high concentration to a region of their own low concentration. (Chapters 3, 18)

digestive system or tract See alimentary canal. (Chapter 17)

dihybrid cross A genetic cross involving the alleles of two different genes. For example AaBb x AaBb. (Chapter 8)

diploid cell A cell containing homologous pairs of chromosomes (2n). (Chapters 6, 21)

directional selection Natural selection for individuals at one end of a range of phenotypes. (Chapter 11)

disaccharide A double sugar consisting of two monosaccharides joined together by a glycosidic linkage. (Chapter 2)

diverge See divergence. (Chapter 12)

divergence Occurs when gene flow is eliminated between two populations. Over time, traits found in one population begin to differ from traits found in the other population. (Chapters 10, 12, 13)

diversifying selection Natural selection for individuals at both ends of a range of phenotypes but against the "average" phenotype. (Chapter 11)

dizygotic twins Fraternal twins (non-identical) that develop when two different sperm fertilize two different egg cells. (Chapter 7)

DNA See deoxyribonucleic acid. (Chapters 2, 6, 7)

DNA fingerprinting Powerful genetic identification technique that takes advantage of differences in DNA sequences between all people other than identical twins. (Chapter 8)

DNA polymerase Enzyme that facilitates base pairing during DNA synthesis. (Chapter 6)

DNA replication The synthesis of two daughter DNA molecules from one original parent molecule. Takes place during the S phase of interphase. (Chapters 6,7)

domain Most inclusive biological category. Biologists group life into three major domains. (Chapters 10, 13)

domesticated Referring to animals or plants that are now largely dependent on humans for reproduction and survival as a result of artificial selection. (Chapter 23)

dominant Applies to an allele with an effect that is visible in a heterozygote. (Chapter 7)

dopamine Neurotransmitter active in pathways that control emotions and complex movements. (Chapter 22)

dormancy A condition of arrested growth and development in plants. (Chapter 24)

dormant In a state of dormancy; i.e. alive, but not growing or developing. (Chapter 23)

double blind Experimental design protocol when both research subjects and scientists performing the measurements are unaware of either the experimental hypothesis or who is in the control or experimental group. (Chapter 1)

double circulation The flow of blood through two circuits in an animal body, from heart to lungs, back to heart, and to the body. (Chapter 18)

double fertilization The fusion of egg and sperm and the simultaneous fusion of a sperm with nuclei in the ovule of angiosperms. (Chapter 23)

ecological footprint A measure of the natural resources used by a human population or society. (Chapter 16)

ecological niche The functional role of a species within a community or ecosystem, including its resource use and interactions with other species. (Chapter 15)

ecology Field of biology that focuses on the interactions between organisms and their environment. (Chapters 13, 14)

ecosystem All of the organisms and natural features in a given area. (Chapter 15)

ecotourism The visitation of specific geographical sites by tourists interested in natural attractions, especially animals and plants. (Chapter 15)

ectoderm The outermost of the three germ layers that arise during animal development. (Chapter 21)

ectotherm An animal that must use energy from the sun and other environmental sources to regulate its body temperature. (Chapter 17)

effector Muscle, gland, or organ stimulated by a nerve. (Chapter 22)

egg cell Gamete produced by a female organism. (Chapters 6, 7, 8, 12, 13, 20, 21)

electron A negatively charged subatomic particle. (Chapters 2, 4)

electron shell An energy level representing the distance of an electron from the nucleus of an atom. (Chapter 2)

electron transport chain A series of proteins in the mitochondrial and chloroplast membranes that move electrons during the redox reactions that release energy to produce ATP. (Chapter 4)

electronegative The tendency to attract electrons to form a chemical bond. (Chapter 2)

element A substance that cannot be broken down into any other substance. (Chapter 2)

embryo The developmental stage commencing after the first mitotic divisions of the zygote and ending when body structures begin to appear; from about the second week after fertilization to about the ninth week. (Chapters 6, 7, 9, 10, 13, 21)

emphysema A lung disease caused by the breakdown of alveoli walls; characterized by shortness of breath and an expanded chest cavity. (Chapter 18)

encephalitis Pathology, or disease, of the brain. (Chapter 19)

Endangered Species Act (ESA) U.S. law intended to protect and encourage the population growth of threatened and endangered species enacted in 1973. (Chapter 15)

endocrine disrupter Chemical that disrupts the functions of the hormone-producing endocrine system. (Chapter 21)

endocrine gland Any of the glands that secrete hormones into the bloodstream. (Chapter 20)

endocrine system The internal system of chemical signals involving hormones, the organs and glands that produce and secrete them, and the target cells that respond to them. (Chapter 20)

endocytosis The uptake of substances into cells by a pinching inward of the plasma membrane. (Chapter 3)

endoderm The innermost of the three germ layers that arise during animal development. (Chapter 21)

endometriosis The abnormal occurrence of functional endometrial tissue outside the uterus, resulting in painful menstrual cycles. (Chapter 21)

endometrium Lining of the uterus, shed during menstruation. (Chapter 21)

endoplasmic reticulum (ER) A network of membranes in eukaryotic cells. When rough, or studded with ribosomes, functions as a workbench for protein synthesis. When devoid of ribosomes, or smooth, it functions in phospholipid and steroid synthesis and detoxification. (Chapter 2)

endoskeleton A hard skeleton buried within the soft tissues of an animal. (Chapter 20)

endosperm The triploid structure formed by the fusion of a sperm and two nuclei in the ovule of an angiosperm plant. (Chapter 23)

endosymbiotic theory Theory that organelles such as mitochondria and chloroplasts in eukaryotic cells evolved from prokaryotic cells that took up residence inside ancestral eukaryotes. (Chapter 13)

endotherm An animal that uses its own metabolic energy to maintain a constant body temperature. (Chapter 17)

enveloped viruses A virus that is surrounded by a membrane. (Chapter 19)

environmental tobacco smoke (ETS) The tobacco smoke in the air that results from smoldering tobacco on the lit ends of cigarettes and pipes as well as the smoke exhaled by active smokers. (Chapter 18)

enzyme Protein that catalyzes and regulates the rate of metabolic reactions. (Chapters 2, 3, 4, 11)

epidemic Contagious disease that spreads rapidly and extensively among many individuals. (Chapter 19)

epidermal tissue The outermost layer of cells of the leaf and of young stems and roots. (Chapter 23)

epididymis A coiled tube located adjacent to the testes where sperm are stored. (Chapter 21)

epiglottis Flap that blocks the windpipe so food goes down the pharynx, not into the lungs. (Chapter 17)

epithelial tissue (epithelia) Tightly packed sheets of cells that line organs and body cavities. (Chapter 17)

equator The circle around Earth that is equidistant to both poles. (Chapter 16)

esophagus Tube that conducts food from the pharynx to the stomach. (Chapter 17)

essential amino acid Any of the amino acids that humans cannot synthesize and thus must be obtained from the diet. (Chapter 3)

essential fatty acid Any of the fatty acids that animals cannot synthesize and must be obtained from the diet. (Chapter 3)

estrogen Any of the feminizing hormones secreted by the ovary in females and adrenal glands in both sexes. (Chapters 20, 21)

estuary An aquatic biome that forms at the outlet of a river into a larger body of water such as a lake or ocean. (Chapter 16)

ethylene A plant hormone involved in fruit ripening and abscission. (Chapter 24)

eukaryote Cell that has a nucleus and membrane-bounded organelles. (Chapters 2, 9, 13)

eutrophication Process resulting in periods of dangerously low oxygen levels in water, sometimes caused by high levels of nitrogen and phosphorus from fertilizer runoff, that result in increased growth of algae in waterways. (Chapters 15, 23)

evolution Changes in the features (traits) of individuals in a biological population that occur over the course of generations. See also theory of evolution. (Chapters 1,2, 10, 11)

evolutionary classification System of organizing biodiversity according to the evolutionary relationships among living organisms. (Chapter 13)

excretion The release of urine from the kidneys into the storage organ of the bladder. (Chapter 18)

exocytosis The secretion of molecules from a cell via fusion of membrane-bounded vesicles with the plasma membrane. (Chapter 3)

exoskeleton A hard outer encasement on the surface of an animal. (Chapter 20)

experiment Contrived situation designed to test specific hypotheses. (Chapter 1)

exponential growth Growth that occurs in proportion to the current total. (Chapter 14)

external fertilization Fertilization that does not require that the parents ever have physical contact with each other. Seen in many aquatic invertebrates and most fish and amphibians. (Chapter 21)

extinction Complete loss of a species. (Chapter 15)

extinction vortex A process by which an endangered population is driven toward extinction via loss of genetic diversity, increased vulnerability to the effects of density-independent factors, and inbreeding depression. (Chapter 15)

facilitated diffusion The spontaneous passage of molecules, through membrane proteins, down their concentration gradient. (Chapter 3)

falsifiable Able to be proved false. (Chapter 1)

fat Energy-rich, hydrophobic lipid molecule composed of a three-carbon glycerol skeleton bonded to three fatty acids. (Chapters 2, 3, 4)

fatty acid A long acidic chain of hydrocarbons bonded to glycerol. Fatty acids vary on the basis of their length and on the number and placement of double bonds. (Chapters 2, 3)

femur Bone extending from the pelvis to the knee; also called thighbone. (Chapter 20)

fermentation A process that makes a small amount of ATP from glucose without using an electron transport chain. Ethyl alcohol and lactic acid are produced by this process. (Chapter 4)

fertility Ability to produce viable gametes. (Chapter 21)

fertilization The fusion of haploid gametes (in humans, egg and sperm) to produce a diploid zygote. (Chapters 6, 7, 8, 21)

fertilizer Any of a variety of growth medium amendments that increase the level of plant nutrients. (Chapter 23)

fetus The term used to describe a developing human from the ninth week of development until birth. (Chapter 21)

fever Abnormally high body temperature. (Chapter 19)

fibrin A protein produced during the process of blood clotting that makes up a net to trap and block blood flow from a damaged blood vessel. (Chapter 18)

fibroblast A protein-secreting cell found in loose connective tissue. (Chapter 17)

fibrous connective tissue A dense tissue found in tendons and ligaments composed largely of collagen fibers. (Chapter 17)

filament The stalk of a stamen. (Chapter 23)

filtration In the kidneys, the removal of plasma from the bloodstream through capillaries surrounding a nephron. (Chapter 18)

fitness Relative survival and reproduction of one variant compared to others in the same population. (Chapters 11, 15)

flagellum (*plural*: flagella) A long cellular projection that aids in motility. (Chapter 19)

flower Reproductive structure of a flowering plant. (Chapters 23, 24)

flowering plant Member of the kingdom Plantae, which produce flowers and fruit. See also angiosperm. (Chapter 13)

fluid mosaic model The accepted model for how membranes are structured with proteins bobbing in a sea of phospholipids. (Chapter 2)

follicle Structure in the ovary that contains the developing ovum and secretes estrogen. (Chapter 21)

follicle cell A flattened cell within the single layer that surrounds each primary oocyte. (Chapter 21)

follicle-stimulating hormone (FSH) Hormone secreted by the pituitary gland involved in sperm production, regulation of ovulation, and regulation of menstruation. (Chapters 20, 21)

food chain The linear relationship between trophic levels from producers to primary consumers, and so on. (Chapter 15)

food web The feeding connections between and among organisms in an environment. (Chapter 15)

foramen magnum Hole in the skull that allows passage of the spinal cord. (Chapter 10)

forest Terrestrial community characterized by the presence of trees. (Chapter 16)

fossil Remains of plants or animals that once existed, left in soil or rock. (Chapters 10, 13, 15)

fossil fuel Nonrenewable resource consisting of the buried remains of ancient plants that have been transformed by heat and pressure into coal and oil. (Chapters 5, 15)

fossil record Physical evidence left by organisms that existed in the past. (Chapter 10)

founder effect Type of sampling error that occurs when a small subset of individuals emigrates from the main population to found a new population, leading to differences in the gene pools of both. (Chapter 12)

founder hypothesis The hypothesis that the diversity of unique species in isolated habitats results from divergence from a single founding population. (Chapter 12)

frameshift mutation A mutation that occurs when the number of nucleotides inserted or deleted from a DNA sequence is not a multiple of three. (Chapter 9)

free radical A substance containing an unpaired electron that is therefore unstable and highly reactive, causing damage to cells. (Chapter 3)

frontal bone Upper front portion of the cranium, or the forehead. (Chapter 20)

frontal lobe The largest and most anterior portion of each cerebral hemisphere. (Chapter 22)

fruit A mature ovary of an angiosperm, containing the seeds. (Chapters 23, 24)

Fungi Kingdom of eukaryotes made up of members that are immobile, rely on other organisms as their food source, and are made up of hyphae that secrete digestive enzymes into the environment and that absorb the digested materials. (Chapters 10, 13, 15)

gall bladder Organ that stores bile and empties into the small intestine. (Chapter 17)

galls Tumor growths on a plant. (Chapter 9)

gamete Specialized sex cell (sperm and egg in humans) that contain half as many chromosomes as other body cells and is therefore haploid. (Chapters 6, 7, 21)

gamete incompatibility An isolating mechanism between species in which sperm from one cannot fertilize eggs from another. (Chapter 12)

gametogenesis The production of gametes. (Chapter 21)

gametophyte In terrestrial plants, the haploid, gamete-producing generation or phase. (Chapter 23)

gas exchange The passage of gases, as a result of diffusion, from one compartment to another. (Chapter 18)

gastrula The 2-layered, cup-shaped stage of embryonic development. (Chapter 21)

gel electrophoresis The separation of biological molecules on the basis of their size and charge by measuring their rate of movement through an electric field. (Chapter 8)

gene Discrete unit of heritable information about genetic traits. Consists of a sequence of DNA that codes for a specific polypeptide—a protein or part of a protein. (Chapters 6, 7)

gene expression Turning a gene on or off. A gene is expressed when the protein it encodes is synthesized. (Chapter 9)

gene flow Spread of an allele throughout a species' gene pool. (Chapter 12)

gene gun Device used to shoot DNA-coated pellets into plant cells. (Chapter 9)

gene pool All of the alleles found in all of the individuals of a species. (Chapter 12)

gene therapy Replacing defective genes (or their protein products) with functional ones. (Chapter 9)

genealogical species concept A scheme that identifies as separate species all populations with a unique lineage. (Chapter 12)

general sense Also called proprioception. Any of the senses including temperature, pain, touch, pressure, and body position, with sensory receptors scattered throughout the body. (Chapter 22)

genetic code Table showing which mRNA codons code for which amino acids. (Chapters 9, 10)

genetic drift Change in allele frequency that occurs as a result of chance. (Chapters 12, 15)

genetic variability All of the forms of genes, and the distribution of these forms, found within a species. (Chapter 15)

genetic variation Differences in alleles that exist among individuals in a population. (Chapter 7)

genetically modified organisms (GMOs) Organisms whose genome incorporates genes from another organism: also called transgenic or genetically engineered organisms. (Chapter 9)

genome Entire suite of genes present in an organism. (Chapters 9, 19)

genotype Genetic composition of an individual. (Chapters 6, 7, 8)

genus Broader biological category to which several similar species may belong. (Chapters 10, 12)

geologic period A unit of time defined according to the rocks and fossils characteristic of that period. (Chapter 13)

germ line therapy Gene therapy that changes genes in a zygote or early embryo, so that the embryo will pass on the engineered genes to their offspring. (Chapter 9)

germ theory The scientific theory that all infectious diseases are caused by microorganisms. (Chapter 1)

germination The beginning of growth by a seed. (Chapter 23)

gibberellin A plant hormone that causes the elongation of stem cells. (Chapter 24)

gill The respiratory exchange surface in many aquatic animals, consisting of thin sheets of tissue containing many capillaries. (Chapter 18)

glans penis The head of the penis. (Chapter 21)

glial cell A type of cells within the brain that does not carry messages but rather supports neurons by supplying nutrients, repairing the brain after injury, and attacking invading bacteria. (Chapter 22)

global warming Increase in average temperatures as a result of the release of increased amounts of carbon dioxide and other greenhouse gases into the atmosphere. (Chapters 5, 15)

glycolysis The splitting of glucose into pyruvate, which helps drive the synthesis of a small amount of ATP. (Chapter 4)

Golgi apparatus An organelle in eukaryotic cells consisting of flattened membranous sacs that modify and sort proteins and other substances. (Chapter 2)

gonads The male and female sex organs; testicles in human males or ovaries in human females. (Chapters 5, 20, 21)

gonadotropin-releasing hormone (GnRH) Hormone produced by the hypothalamus that stimulates the pituitary gland to release FSH and LH, thereby stimulating the activities of the gonads. (Chapters 20, 21)

gradualism The hypothesis that evolutionary change occurs in tiny increments over long periods of time. (Chapter 12)

grana Stacks of thylakoids in the chloroplast. (Chapters 2, 5)

grassland Biome characterized by the dominance of grasses, usually found in regions of lower precipitation. (Chapter 16)

gravitropism Directional plant growth in response to gravity. (Chapter 24)

gray matter Unmyelinated axons, combined with dendrites and cell bodies of other neurons that appear gray in cross section. (Chapter 22)

greenhouse effect The retention of heat in the atmosphere by carbon dioxide and other greenhouse gases. (Chapter 5)

ground tissue The plant tissues other than the vascular tissue, epidermis, and meristem. (Chapter 23)

growing season The length of time from last freeze in spring to first freeze in autumn. (Chapter 24)

growth factor Protein that stimulates cell division. (Chapter 6)

growth rate Annual death rate in a population subtracted from the annual birth rate. (Chapter 14)

guanine Nitrogenous base in DNA, a purine. (Chapters 2, 6, 8, 9)

guard cell Either of the paired cells encircling stomata that serve to regulate the size of the stomatal pore, helping to minimize water loss under dry conditions and maximize carbon dioxide uptake under wet conditions. (Chapters 5, 23)

gymnosperm Member of a group of plants made up of several phyla that produce seeds, but not true flowers. (Chapter 23, 24)

habitat Place where an organism lives. (Chapter 15)

habitat destruction Modification and degradation of natural forests, grasslands, wetlands, and waterways by people; primary cause of species loss. (Chapter 15)

habitat fragmentation Threat to biodiversity caused by humans that occurs when large areas of intact natural habitat are subdivided by human activities. (Chapter 15)

half-life Amount of time required for one-half the amount of a radioactive element that is originally present to decay into the daughter product. (Chapter 10)

haploid Describes cells containing only one member of each homologous pair of chromosomes ("); in humans, these cells are eggs and sperm. (Chapters 6, 21)

hardened Relating to plants that have undergone the appropriate physiological changes that allow them to survive freezing temperatures. (Chapter 24)

hardiness In plants, the ability to survive freezing temperatures. (Chapter 24)

Hardy-Weinberg theorem Theorem that holds that allele frequencies remain stable in populations that are large in size, randomly mating, and experiencing no migration or natural selection. Used as a baseline to predict how allele frequencies would change if any of its assumptions were violated. (Chapter 12)

heart The muscular organ that pumps blood via the circulatory system to the lungs and body. (Chapters 17, 18)

heart attack An acute condition, during which blood flow is blocked to a portion of the heart muscle, causing part of the muscle to be damaged or die. (Chapters 4, 18)

heat The total amount of energy associated with the movement of atoms and molecules in a substance. (Chapter 4)

helper T cell A type of immune-system cell that enhances cell-mediated immunity and humoral immunity by secreting a substance that increases the strength of the immune response. See also T4 cell. (Chapter 19)

hemoglobin An iron-containing protein that carries oxygen in red blood cells. (Chapter 18)

hemophilia Rare genetic disorder caused by a sex-linked recessive allele that prevents normal blood clotting. (Chapter 8)

hepatocyte Liver cell. (Chapter 17)

herbicide A chemical that kills plants. (Chapter 23)

heritability The amount of variation for a trait in a population that can be explained by differences in genes among individuals. (Chapter 7)

hermaphrodite An individual with both male and female reproductive organs. (Chapter 21)

heterozygote Individual carrying two different alleles of a particular gene. (Chapters 7, 8, 12)

heterozygous Said of a genotype containing two different alleles of a gene. (Chapter 7)

high-density lipoprotein (HDL) A cholesterol-carrying particle in the blood that is high in protein and low in cholesterol. (Chapters 4, 18)

hinge joint A joint that allows back and forth movement. (Chapter 20)

histamine Chemcial released from mast cells during allergic reactions. Causes blood vessels to dilate and become more permeable and lowers blood pressure. (Chapter 19)

HIV See human immunodeficiency virus. (Chapter 19)

homeostasis The steady state condition an organism works to maintain. (Chapters 2, 17)

hominin Referring to humans and human ancestors. (Chapters 10, 12)

homologous pair Set of two chromosomes of the same size and shape with centromeres in the same position. Homologous pairs of chromosomes carry the same genes in the same locations but may carry different alleles. (Chapters 6, 7)

homology Similarity in characteristics as a result of common ancestry. (Chapter 10)

homozygous Having two copies of the same allele of a gene. (Chapters 7, 15)

hormones A protein or steroid produced in one tissue that travels through the circulatory system to act on another tissue to produce some physiological effect. (Chapter 20)

Human Genome Project Effort to determine the nucleotide base sequences and chromosomal locations of all human genes. (Chapter 9)

human immunodeficiency virus (HIV) Agent identified as causing the transmission and symptoms of AIDS. (Chapter 19)

humoral immunity B-cell-mediated immunity that occurs when a B-cell receptor binds to an antigen. The B cell divides to produce a clonal population of memory cells; the cell also produces plasma cells. (Chapter 19)

hybrid Offspring of two different strains of an agricultural crop. See also interspecies hybrid. (Chapters 12, 23)

hydrocarbon A compound consisting of carbons and hydrogens. (Chapter 2)

hydrogen atom One negatively charged electron and one positively charged proton. (Chapters 2, 5)

hydrogen bond A type of weak chemical bond in which a hydrogen atom of one molecule is attracted to an electronegative atom of another molecule. (Chapters 2, 5)

hydrogen ion The positively charged ion of hydrogen (H+) formed by removal of the electron from a hydrogen atom. (Chapters 2, 5)

hydrogenation Adding hydrogen gas under pressure to make liquid oils more solid. (Chapter 3)

hydrophilic Readily dissolving in water. (Chapter 2)

hydrophobic Not able to dissolve in water. (Chapter 2)

hydrostatic skeleton A skeletal system composed of fluid in a closed body compartment. (Chapter 20)

hypertension High blood pressure. (Chapters 4, 18)

hyphae Thin, stringy fungal material that grows over and within a food source. (Chapter 13)

hypothalamus Gland that helps regulate body temperature; influences behaviors such as hunger, thirst, and reproduction; and secretes a hormone (GnRH) that stimulates the activities of the gonads. (Chapters 20, 21, 22)

hypothesis Tentative explanation for an observation that requires testing to validate. (Chapters 1,10, 12)

immortal Property of cancer cells that allows them to divide more times than normal cells. (Chapter 6)

immune response Ability of the immune system to respond to an infection, resulting from increased production of B cells and T cells. (Chapter 19)

immune system The organ system that produces cells and cell products, such as antibodies, that help remove pathogenic organisms. (Chapter 19)

in vitro fertilization Fertilization that takes place when sperm and egg are combined in glass or a test tube. (Chapter 9)

inbreeding Mating between related individuals. (Chapter 15)

inbreeding depression Negative effect of homozygosity on the fitness of members of a population. (Chapter 15)

incomplete digestive system A digestive system in which there is only one opening that serves as mouth and anus. (Chapter 17)

incomplete dominance A type of inheritance where the heterozygote has a phenotype intermediate to both homozygotes. (Chapter 8)

independent assortment The separation of homologous pairs of chromosomes into gametes independently of one another during meiosis. (Chapters 7, 8)

independent variable A factor whose value influences the value of the dependent variable, but is not influenced by it. In experiments, the variable that is manipulated. (Chapter 1)

indeterminate growth Unrestricted or unlimited growth. (Chapter 23)

indirect effect In ecology, a condition where one species affects another indirectly through it intervening species. (Chapter 15)

induced fit A change in shape of the active site of an enzyme so that it binds tightly to a substrate. (Chapter 4)

inductive reasoning A logical process that argues from specific instances to a general conclusion. (Chapter 1)

infant mortality Death rate of infants and children under the age of 5. (Chapter 14)

infectious Applies to a pathogen that finds a tissue inside the body that will support its growth. (Chapter 19)

infertility The inability to conceive after one year of unprotected intercourse. (Chapter 21)

inflammatory response A line of defense triggered by a pathogen penetrating the skin or mucus membranes. (Chapter 19)

inner cell mass A cluster of cells in the blastocyst that eventually develops into the embryo. (Chapter 21)

inorganic Said of chemical compounds that do not contain carbon. (Chapter 23)

insulin A hormone secreted by the pancreas that lowers blood glucose levels by promoting the uptake of glucose by cells and the storage of glucose as glycogen in the liver. (Chapter 4)

integrated pest management (IPM) A strategy for pest control that relies on a mix of pesticides, cultural control, and better knowledge of pest populations. (Chapter 23)

interferon A chemical messenger produced by virus-infected cells that helps other cells resist infection. (Chapter 19)

intermembrane space The space between two membranes, e.g., the space between the inner and outer mitochondrial membrane. (Chapter 4)

internal fertilization The union of gametes inside the body of the female. (Chapter 21)

interneuron A neuron located between sensory and motor neurons that functions to integrate sensory input and motor output. (Chapter 22)

internode The region of a plant stem between nodes. (Chapter 23)

interphase Part of the cell cycle when a cell is preparing for division and the DNA is duplicated. Consists of G_1, S, and G_2. (Chapter 6)

interspecies hybrid Organism with parents from two different species. (Chapter 12)

intertidal zone The biome that forms on ocean shorelines between the high tide elevation and the low tide elevation. (Chapter 16)

introduced species A nonnative species that was intentionally or unintentionally brought to a new environment by humans. (Chapter 15)

invertebrate Animal without backbones. (Chapter 13)

involuntary muscle Muscle tissue whose action requires no conscious thought. (Chapter 17)

ion electrically charged atom. (Chapter 2)

ionic bond A chemical bond resulting from the attraction of oppositely charged ions. (Chapter 2)

irrigation The technique of supplying additional water to crop plants, typically via flooding or spraying. (Chapter 23)

karyotype The chromosomes of a cell, displayed with chromosomes arranged in homologous pairs and according to size. (Chapter 8)

keystone species A species that has an unusually strong effect on the structure of the community it inhabits. (Chapter 15)

kidney Major organ of the excretory system, responsible for filtering liquid waste from the blood. (Chapters 17, 18)

kingdom In some classifications, the most inclusive group of organisms; usually five or six. In other classification systems, the level below domain on the hierarchy. (Chapters 10, 13)

labia majora Paired thick folds of skin that enclose and protect the labia minor of the vulva. (Chapter 21)

labia minora Paired thin folds of the vulva; enclose the urinary and vaginal openings and clitoris. (Chapter 21)

labor Strong rhythmic contractions that force a developing baby from the uterus through the vagina during childbirth. (Chapter 21)

lactation Production of milk to nurse offspring. (Chapter 21)

lake An aquatic biome that is completely landlocked. (Chapter 16)

laparoscope A thin tubular instrument inserted through an abdominal incision and used to view organs in the pelvic cavity and abdomen. (Chapter 6)

large intestine Colon, portion of the digestive system located between the small intestine and the anus that absorbs water and forms feces. (Chapter 17)

larynx A portion of the upper respiratory tract made up primarily of stiff cartilage. Also known as the voice box. (Chapter 18)

latent Said of the inactive or dormant phase of a disease. (Chapter 19)

latent virus An inactive virus that is integrated into the host genome. (Chapter 19)

leaf The primary photosynthetic organ in plants. (Chapters 23, 24)

legumes A family of plants that produces seeds in pods, like peas and beans. (Chapter 23)

leptin A hormone by fat cells that may be involved in the regulation of appetite. (Chapter 4)

less developed country A term coined by the United Nations to describe a country with low per-capita income and often poor population health and economic prospects. (Chapter 14)

Leydig cell Cells scattered between the seminiferous tubules of the testicles that produce testosterone and other androgens. (Chapter 21)

life cycle Description of the growth and reproduction of an individual. (Chapter 7)

light reaction A series of reactions that occur on thylakoid membranes during photosynthesis and serve to convert energy from the sun into the energy stored in the bonds of ATP and evolve oxygen. (Chapter 5)

linked gene Genes located on the same chromosome. (Chapter 6)

lipid Hydrophobic molecule, including fats, phospholipids, and steroids. (Chapters 2, 3)

liver Organ with many functions, including the production of bile to aid in the absorption of fats. (Chapter 17)

lobule Subdivision of the lobes of the liver. (Chapter 17)

logistic growth Pattern of growth seen in populations that are limited by resources available in the environment. A graph of logistic growth over time typically takes the form of an S-shaped curve. (Chapter 14)

loose connective tissue Connective tissue that serves to bind epithelia to underlying tissues and to hold organs in place. (Chapter 17)

low-density lipoprotein (LDL) Cholesterol-carrying substance in the blood that is high in cholesterol and low in protein. (Chapters 4, 18)

lung The primary organ of the respiratory system; the site where gas exchange occurs. (Chapter 18)

luteinizing hormone (LH) Hormone involved in sperm production, regulation of ovulation, and regulation of menstruation. (Chapters 20, 21)

lymph node Organ located along lymph vessels that filter lymph and help defend against bacteria and viruses. (Chapters 6, 19)

lymphatic system A system of vessels and nodes that return fluid and protein to the blood. (Chapters 6, 19)

lymphocyte White blood cells that make up part of the immune system. (Chapter 19)

lysosome A membrane-bounded sac of hydrolytic enzymes found in the cytoplasm of many cells. (Chapter 2)

macroevolution Large-scale evolutionary change, usually referring to the origin of new species. (Chapter 10)

macromolecule Any of the large molecules including polysaccharides, proteins, and nucleic acids, composed of subunits joined by dehydration synthesis. (Chapters 2, 4)

macronutrient Nutrient required in large quantities. (Chapters 3, 23)

macrophage Phagocytic white blood cell that swells and releases toxins to kill bacteria. (Chapter 18)

malignant Describes a tumor that is cancerous, whether it is invasive or metastatic. (Chapter 6)

mandible Bone of the lower jaw. (Chapter 20)

marine Of, or pertaining to, salt water. (Chapter 16)

mark-recapture method A technique for estimating population size, consisting of capturing and marking a number of individuals, releasing them, and recapturing more individuals to determine what proportion are marked. (Chapter 14)

mass extinction Loss of species that is rapid, global in scale, and affects a wide variety of organisms. (Chapter 15)

matrix (1) In a mitochondrion, the semifluid substance inside the inner mitochondrial membrane, which houses the enzymes of the citric acid cycle. (Chapters 2, 4). (2) In connective tissue, a nonliving substance between cells, ranging from fluid blood plasma to fibrous matrix in tendons to solid bone matrix. (Chapter 17)

mean Average value of a group of measurements. (Chapter 1)

mechanical isolation A form of reproductive isolation between species that depends on the incompatibility of the genitalia of individuals of different species. (Chapter 12)

medulla The center of an organ or gland, such as the kidney or adrenal gland. (Chapter 17)

medulla oblongata Region of the brain stem that is a continuation of the spinal cord and conveys information between the spinal cord and other parts of the brain. (Chapter 22)

meiosis Process that diploid sex cells undergo in order to produce haploid daughter cells. Occurs during gametogenesis. (Chapters 6, 7, 8, 11)

memory cell Cell that is part of a clonal population, programmed to respond to a specific antigen, that helps the body respond quickly if the infectious agent is encountered again. (Chapter 19)

menopause Cessation of menstruation. (Chapters 20, 21)

menstrual cycle Changes that occur in the uterus and depend on intricate interrelationships among the brain, ovaries, and lining of the uterus. (Chapter 21)

menstruation The shedding of the lining of the uterus during the menstrual cycle. (Chapter 21)

meristematic tissue Undifferentiated plant tissue from which new plant cells arise. (Chapter 23)

mesoderm The middle of three germ layers that arise during animal development. (Chapter 21)

messenger RNA (mRNA) Complementary RNA copy of a DNA gene, produced during transcription. The mRNA undergoes translation to synthesize a protein. (Chapters 9, 11)

metabolic rate Measure of an individual's energy use. (Chapter 4)

metabolism All of the physical and chemical reactions that produce and use energy. (Chapters 2, 4, 5)

metaphase Stage of mitosis during which duplicated chromosomes align across the middle of the cell. (Chapter 6)

metastasis When cells from a tumor break away and start new cancers at distant locations. (Chapter 6)

microbe Microscopic organism, especially Bacteria and Archaea. (Chapters 13, 19)

microbiologists Scientists who study microscopic organisms, especially referring to those who study prokaryotes. (Chapter 13)

microevolution Changes that occur in the characteristics of a population. (Chapter 10)

micronutrient Nutrient needed in small quantities. (Chapters 3, 23)

microorganism See microbe. (Chapter 13)

microtubule Protein structure that moves chromosomes around during mitosis and meiosis. (Chapters 2, 6)

microvillus Fine fingerlike projection composed of epithelial cells that function in absorption. (Chapter 17)

micturation Release of urine from the bladder. Also known as urination. (Chapter 18)

midbrain Uppermost region of the brain stem, which adjusts the sensitivity of the eyes to light and of the ears to sound. (Chapter 22)

mineral Inorganic nutrient essential to many cell functions. (Chapter 3)

mitochondria Organelles in which products of the digestive system are converted to ATP. (Chapters 2, 4)

mitosis The division of the nucleus that produces daughter cells that are genetically identical to the parent cell. Also, portion of the cell cycle in which DNA is apportioned into two daughter cells. (Chapter 6)

model organism Any nonhuman organism used in genetic studies to help scientists understand human genes because they share genes with humans. (Chapters 1, 9)

mold A fungal form characterized by rapid, asexual reproduction. (Chapters 10, 13)

molecular clock Principle that DNA mutations accumulate in the genome of a species at a constant rate, permitting estimates of when the common ancestor of two species existed. (Chapter 10)

molecule Two or more atoms held together by covalent bonds. (Chapter 2)

monocot One of the two classes of flowering plants, also called narrow-leaved plants. Monocot seeds contain one leaf (cotyledon). (Chapter 23)

monoculture Practice of planting a single crop over a wide acreage. (Chapter 23)

monosaccharide Simple sugar. (Chapter 2)

monosomy A chromosomal condition in which only one member of a homologous pair is present. (Chapter 6)

monozygotic twins Identical twins that developed from one zygote. (Chapter 7)

morphological species concept Definition of species that relies on differences in physical characteristics among them. (Chapter 12)

morphology Appearance or outward physical characteristics. (Chapter 12)

motor neuron Neuron that carries information away from the brain or spinal cord to muscles or glands. (Chapter 22)

mulching Covering the ground around favored plants with light-blocking material to prevent weed growth. (Chapter 23)

multicellular The condition of being composed of many coordinated cells. (Chapter 13)

multiple allelism A gene for which there are more than 2 alleles segregating in the population. (Chapter 8)

multiple hit model The notion that many different genetic mutations are required for a cancer to develop. (Chapter 6)

multiple sclerosis Chronic, degenerative nervous system disease caused by breakdown of myelin. (Chapter 19)

muscle fiber Single cell that aligns with others in parallel bundles to form muscles. (Chapter 20)

muscle tissue Specialized contractile tissue. (Chapter 17)

mutagen Substance that increases the likelihood of mutation occurring; increases the likelihood of cancer. (Chapter 6)

mutation Change to a DNA sequence that may result in the production of altered proteins. (Chapters 6, 7, 9, 11)

mutualism Interaction between two species that provides benefits to both species. (Chapter 15)

mycologist Scientist who specializes in the study of fungi. (Chapter 13)

myelin sheath Protective layer that coats many axons, formed by supporting cells such as Schwann cells. The myelin sheath increases the speed at which the electrochemical impulse travels down the axon. (Chapter 22)

myofibril Structure found in muscle cells, composed of thin filaments of actin and thick filaments of myosin. (Chapter 20)

myosin A type of protein filament that, along with actin, causes muscle cells to contract. (Chapters 17, 20)

natural experiment Situation where unique circumstances allow a hypothesis test without prior intervention by researchers. (Chapter 7)

natural killer cell A cell that attacks virus-infected cells or tumor cells without being activated by an immune system cell or antibody. (Chapter 19)

natural selection Process by which individuals with certain traits have greater survival and reproduction than individuals who lack these traits, resulting in an increase in the frequency of successful alleles and a decrease in the frequency of unsuccessful ones. (Chapters 10, 11, 12, 21)

negative feedback A mechanism of maintaining homeostasis in which the product of the process inhibits the process. (Chapter 17)

nephron The functional structure within a kidney where waste filtration and urine concentration occurs. (Chapter 18)

nerve Bundle of neurons; nerves branch out from the brain and spinal cord to eyes, ears, internal organs, skin, and bones. (Chapter 22)

nerve impulse Electrochemical signal that controls the activities of muscles, glands, organs, and organ systems. (Chapter 22)

nervous system Brain, spinal cord, sense organs, and nerves that connect organs and link this system with other organ systems. (Chapter 22)

nervous tissue Tissue composed of neurons and associated cells. (Chapters 17, 22)

net primary production (NPP) Amount of solar energy converted to chemical energy by plants, minus the amount of this chemical energy plants need to support themselves. A measure of plant growth, typically over the course of a single year. (Chapter 14)

neurobiologist A scientist that studies the nervous system. (Chapter 22)

neuron Specialized message-carrying cell of the nervous system. (Chapters 17, 22)

neurotransmitter One of many chemicals released by the presynaptic neuron into the synapse, which then diffuse across the synapse and bind to receptors on the membrane of the postsynaptic neuron. (Chapter 22)

neutral mutation A genetic mutation that confers no selective advantage or disadvantage. (Chapters 9, 22)

neutron An electrically neutral particle found in the nucleus of an atom. (Chapter 2)

nutrients Atoms other than carbon, hydrogen, and oxygen that must be obtained from a plant's environment for photosynthesis to occur. (Chapters 3, 23, 24)

nicotinamide adenine dinucleotide Intracellular electron carrier. Oxidized form is NAD+; reduced form is NADH. (Chapter 4)

nicotine The active drug in tobacco that stimulates dopamine receptors in the brain. (Chapter 18)

nitrogen-fixing bacteria Organisms that convert nitrogen gas from the atmosphere into a form that can be taken up by plant roots; some species live in the root nodules of legumes. (Chapters 15, 23)

nitrogenous base Nitrogen-containing base found in DNA: A, C, G, and T and in RNA : U. (Chapters 2, 4, 6)

node Point on a plant stem where a leaf and axillary bud arise. (Chapter 23)

node of Ranvier Small indentation separating segments of the myelin sheath. Nerve impulses "jump" successively from one node of Ranvier to the next. (Chapter 22)

nodule Compartment housing nitrogen-fixing bacteria, produced on the roots of legume plants such as beans and alfalfa. (Chapter 23)

nondisjunction The failure of members of a homologous pair of chromosomes to separate from each other during meiosis. (Chapter 6)

nonpolar Won't dissolve in water. Hydrophobic. (Chapter 2)

nonrenewable resource Resource that is a one-time supply and cannot be easily replaced. (Chapter 14)

nonspecific defenses Defense system against infection that does not distinguish one pathogen from another. Includes the skin, secretions, and mucous membranes. (Chapter 19)

normal distribution Bell-shaped curve, as for the distribution of quantitative traits in a population. (Chapter 7)

notochord A long, flexible rod that runs through the axis of the vertebral body in the future position of the spinal cord. (Chapter 22)

nuclear envelope The double membrane enclosing the nucleus in eukaryotes. (Chapters 2, 6)

nuclear transfer Transfer of a nucleus from one cell to another cell that has had its nucleus removed. (Chapter)

nucleic acids Polymers of nucleotides that comprise DNA and RNA. (Chapters 2, 6)

nucleoid region The region of a prokaryotic cell where the DNA is located. (Chapter 19)

nucleotides Building blocks of nucleic acids that include a sugar, a phosphate, and a nitrogenous base. (Chapters 2, 4, 6, 8, 9)

nucleus Cell structure that houses DNA; found in eukaryotes. (Chapters 2, 6, 9, 11, 23, 24)

nutrient cycling Process by which nutrients become available to plants. Nutrient cycling in a natural environment relies on a healthy community of decomposers within the soil. (Chapter 15)

nutrients Substances that provide nourishment. (Chapter 3)

obesity Condition of having a BMI of 30 or greater. (Chapter 4)

objective Without bias. (Chapter 1)

observation Measurement of nature. (Chapters 1, 11)

occipital lobe The posterior lobe of each cerebral hemisphere, containing the visual center of the brain. (Chapter 22)

ocean A biome consisting of open stretches of salt water. (Chapter 16)

oncogene Mutant version of a cell cycle controlling proto-oncogene. (Chapter 6)

oogenesis Formation and development of female gametes, which occurs in the ovaries and results in the production of egg cells. (Chapter 21)

open circulatory system A circulatory system in which the circulating liquid is released into the body cavity and returned to the system via pores in the vessels. (Chapter 18)

opportunistic infections Disease caused by a pathogen that does not normally cause illness in a healthy host, only when the host is compromised. (Chapter 19)

organ A specialized structure composed of several different types of tissues. (Chapter 17)

organ systems Suites of organs working together to perform a function or functions. (Chapter 17)

organelle Subcellular structure found in the cytoplasm of eukaryotic cells that performs a specific job. (Chapters 2, 4)

organic When pertaining to agriculture, refers to products grown without the use of manufactured pesticides and inorganic fertilizer. (Chapter 23)

organic chemistry The chemistry of carbon-containing substances. (Chapter 2)

organic farming Agricultural technique that requires no manufactured chemical inputs (such as inorganic fertilizer or chemical pesticides), and relies on developing and maintaining the health of soil. (Chapter 22)

osmosis The diffusion of water across a selectively permeable membrane. (Chapter 3)

ossa coxae The paired bones that form the bony pelvis. (Chapter 20)

osteoblast Bone-forming cell responsible for the deposition of collagen. (Chapter 20)

osteoclast Bone-reabsorbing cell that liberates calcium. (Chapter 20)

osteocyte Highly branched cell found in bone. (Chapter 17)

osteoporosis A condition resulting in an elevated risk of bone breakage from weakened bones. (Chapter 3)

ovary (1) In animals, the paired abdominal structures that produce egg cells and secrete female hormones. (Chapters 6, 20, 21) (2) In plants, a chamber of the carpel containing the ovules. (Chapter 23)

overexploitation Threat to biodiversity caused by humans that encompasses overhunting and overharvesting. (Chapter 15)

oviduct Egg-carrying duct that brings egg cells from ovaries to uterus. (Chapter 21)

ovulation Release of an egg cell from the ovary. (Chapter 21)

ovule Structure in flowering plants consisting of eggs and accessory tissues. After fertilization, will develop into a seed. (Chapter 23)

ovum An egg; the female gamete. Cell produced during oogenesis that receives the majority of the cytoplasmic nutrients and organelles. (Chapter 21)

oxytocin Pituitary hormone that stimulates the contraction of smooth muscle of the uterus during labor and facilitates secretion of milk from the breast during nursing. (Chapter 21)

pacemaker A patch of heart tissue or an implanted device that produces a regular electrical signal, setting the rhythm for the cardiac cycle. (Chapter 18)

paleontologist Scientist who searches for, describes, and studies ancient organisms. (Chapter 9)

pancreas Gland that secretes digestive enzymes and insulin. (Chapters 17, 20)

parasite An organism that benefits from an association with another organism that is harmed by the association. (Chapters 15, 19)

parathyroid gland One of four endocrine glands located on the thyroid that secrete parathyroid hormone in order to regulate blood calcium levels. (Chapter 20)

parietal lobe Part of the brain that processes information about touch and is involved in self-awareness.(Chapter 22)

Parkinson's disease Disease that results in tremors, rigidity, and slowed movements. May be due to faulty dopamine production. (Chapter 22)

particulate Tiny airborne particle found in smoke and other pollutants. (Chapter 18)

passive immunity Immunity acquired when antibodies are passed from one individual to another, as from mother to child during breast-feeding. (Chapter 19)

passive smoker An individual who inhales environmental tobacco smoke but who does not actively consume tobacco. (Chapter 18)

passive transport The diffusion of substances across a membrane with their concentration gradient and not requiring an input of ATP. (Chapter 3)

pathogen Disease-causing organism. (Chapter 19)

pedigree Family tree that follows the inheritance of a genetic trait for many generations. (Chapter 8)

peer review The process by which reports of scientific research are examined and critiqued by other researchers before they are published in scholarly journals. (Chapter 1)

penis The copulatory structure in males. (Chapter 21)

peptide bond Covalent bond that joins the amino group and carboxyl group of adjacent amino acids. (Chapter 2)

perennial plant Plant that lives for many years. (Chapters 16, 23, 24)

peripheral nervous system (PNS) Network of nerves outside the brain and spinal cord that links the CNS with sense organs. (Chapter 22)

peristalsis Rhythmic muscle contractions that move food through the digestive system. (Chapter 17)

permafrost Permanently frozen soil. (Chapter 16)

pest Any organism that competes with humans for agricultural production or other resources. (Chapter 23)

pesticide Chemical that kills or disables agricultural pests. (Chapter 23)

pesticide treadmill The tendency for pesticide toxicity and total applications to increase once a farmer begins to employ pesticides. (Chapter 23)

petal A flower part that typically functions to attract pollinators. (Chapter 23)

petiole The stalk of a leaf. (Chapter 23)

pH A logarithmic measure of the hydrogen ion concentration ranging from 0–14. Lower numbers indicate higher hydrogen ion concentrations. (Chapter 2)

phagocytosis Ingestion of food or pathogens by cells. (Chapter 19)

pharynx Tube and muscles connecting the mouth to the esophagus; throat. (Chapter 17)

phenotype Physical and physiological traits of an individual. (Chapters 6, 7, 8)

phenotypic ratio Proportion of individuals produced by a genetic cross who possess each of the various phenotypes that cross can generate. (Chapter 7)

phloem The portion of a plant's vascular tissue modified for multidirectional transport of nutrients throughout the body of the plant. (Chapter 23)

phloem sap The liquid, often containing dissolved carbohydrates, carried in the phloem tubes in a plant. (Chapter 24)

phospholipid One of three types of lipids, phospholipids are components of cell membranes. (Chapter 2)

phospholipid bilayer The membrane that surrounds cells and organelles and is composed of two layers of phospholipids. (Chapter 2)

phosphorylation To introduce a phosphoryl group into an organic compound. (Chapter 4)

photoperiodism Response to the duration and timing of day and night length in plants. (Chapter 24)

photorespiration A series of reactions triggered by the closing of stomatal openings to prevent water loss. (Chapter 5)

photosynthesis Process by which plants, along with algae and some bacteria, transform light energy to chemical energy. (Chapter 5)

phototropism Growth in response to directional light. (Chapter 24)

phyla (singular: phylum) The taxonomic category below kingdom and above class. (Chapters 10, 13)

phylogeny Evolutionary history of a group of organisms. (Chapter 13)

phytochrome The light-sensitive chemical in plants responsible for photoperiodic responses. (Chapter 24)

pituitary gland Small gland attached by a stalk to the base of the brain that secretes growth hormone, reproductive hormones, and other hormones. (Chapters 9, 20)

pivot joint A type of joint that allows freedom of movement. (Chapter 20)

placebo Sham treatments in experiments. (Chapter 1)

placenta Membrane produced by a developing fetus that releases a hormone to extend the life of the corpus luteum. (Chapter 21)

plant hormone Substance in plants that helps tune the organisms' response to the environment. (Chapter 24)

Plantae Multicellular photosynthetic eukaryotes, excluding algae. (Chapter 13)

plasma The liquid portion of blood. (Chapter 18)

plasma cell Cell produced by a clonal population that secretes antibodies specific to an antigen. (Chapter 19)

plasma membrane Structure that encloses a cell, defining the cell's outer boundary. (Chapters 2, 3, 19)

plasmid Circular piece of bacterial DNA that normally exists separate from the bacterial chromosome and can make copies of itself. (Chapter 9)

platelet Cell in the blood that carries constituents required for the clotting response. (Chapter 18)

pleiotropy The ability of one gene to affect many different functions. (Chapter 8)

polar Describes a molecule with regions having different charges; capable of ionizing. (Chapter 2)

polar body Smaller of the two cells produced during oogenesis; it therefore does not have enough nutrients to undergo further development. (Chapter 21)

polarization The difference in charge between the inside and outside of the resting neuron cell. (Chapter 22)

poles Opposite ends of a sphere, such as of a cell (Chapter 6) or of a planet such as Earth (Chapter 16).

pollen The male gametophyte of seed plants. (Chapters 12, 13, 15, 23)

pollination The process by which pollen is transferred to the female structures of a plant. (Chapter 23)

pollinator An organism that transfers sperm (pollen grains) from one flower to the female reproductive structures of another flower. (Chapter 15)

pollution Human-caused threat to biodiversity involving the release of poisons, excess nutrients, and other wastes into the environment. (Chapter 15)

polyculture Practice of planting many different crop plants over a single farm's acreage. (Chapter 23)

polygenic trait A trait influenced by many genes.(Chapters 7, 8)

polymer General term for a macromolecule composed of many chemically bonded monomers. (Chapter 2)

polymerase An enzyme that catalyzes phosphodiester bond formation between nucleotides. (Chapter 8)

polymerase chain reaction (PCR) A laboratory technique that allows the production of many identical DNA molecules. (Chapter 8)

polyploidy A chromosomal condition involving more than two sets of chromosomes. (Chapter 12)

polysaccharide A carbohydrate composed of three or more monosaccharides. (Chapter 2)

polysaturated Containing several double bonds between carbon atoms. (Chapter 3)

polyunsaturated Relating to fats consisting of carbon chains with many double bonds unsaturated by hydrogen atoms. (Chapter 3)

pond An aquatic biome that is completely landlocked. (Chapter 16)

pons A structure located on the brain stem, between the brain and spinal cord. (Chapter 22)

population Subgroup of a species that is somewhat independent from other groups. (Chapters 12, 14)

population bottleneck Dramatic but short-lived reduction in population size followed by an increase in population. (Chapter 12)

population crash Steep decline in number that may occur when a population grows larger than the carrying capacity of its environment. (Chapter 14)

population cycle In some populations, the tendency to increase in number above the environment's carrying capacity, resulting in a crash, following by an overshoot of the carrying capacity and another crash, continuing indefinitely. (Chapter 14)

population genetics Study of the factors in a population that determine allele frequencies and their change over time. (Chapter 12)

population pyramid A visual representation of the number of individuals in different age categories in a population. (Chapter 14)

positive feedback A relatively uncommon homeostatic mechanism in which the product of a process intensifies the process, thereby promoting change. (Chapter 17)

postsynaptic neuron The neuron that responds to neurotransmitter released from the presynaptic neuron. (Chapter 22)

prairie A grassland biome. (Chapter 16)

precipitation When water vapor in the atmosphere turns to liquid or solid form and falls to Earth's surface. (Chapter 16)

predation Act of capturing and consuming an individual of another species. (Chapter 15)

predator Organism that eats other organisms. (Chapter 15)

prediction Result expected from a particular test of a hypothesis if the hypothesis were true. (Chapter 1)

pressure flow mechanism The process by which phloem sap moves from a sugar source to a sugar sink within a plant. (Chapter 24)

presynaptic The neuron that secretes neurotransmitter into a synapse, transmitting a signal. (Chapter 22)

primary consumer Organism that eats plants. (Chapter 15)

primary growth Growth occurring at the tips of a plant, originating in the apical meristems. (Chapter 23)

primary source Article reporting research results, written by researchers, and reviewed by the scientific community. (Chapter 1)

primary spermatocyte Diploid cell that begins meiosis in the production of sperm. (Chapter 21)

primers Short nucleotide sequences used to help initiate replication of nucleic acids. (Chapter 8)

probability Likelihood that something is the case or will happen. (Chapter 1)

probe Single-stranded nucleic acid that has been radioactively labeled. (Chapter 8)

processed food Food that has been modified from its original form to increase shelf life, transportability, or the like. (Chapter 3)

producer Organism that produces carbohydrates from inorganic carbon, typically via photosynthesis. (Chapter 15)

product The modified chemical that results from a chemical or enzymatic reaction. (Chapter 2)

progesterone Ovarian hormone. High levels have a negative feedback effect on the hypothalamus, causing GnRH secretion to decrease. (Chapter 21)

prokaryote Type of cell that does not have a nucleus or membrane-bounded organelles. (Chapters 2, 9, 13)

prolactin A hormone produced by the pituitary gland that stimulates the development of mammary glands.(Chapter 21)

promoter Sequence of nucleotides to which the polymerase binds to start transcription. (Chapter 9)

prophase Stage of mitosis during which duplicated chromosomes condense. (Chapter 6)

prostate A gland in human males that secretes an acid-neutralizing fluid into semen. (Chapter 21)

protein Cellular constituent made of amino acids coded for by genes. Proteins can have structural, transport, or enzymatic roles. (Chapters 2, 3, 4, 9, 11, 15)

protein synthesis Joining amino acids together, in an order dictated by a gene, to produce a protein. (Chapter 9)

Protista Kingdom in the domain Eukarya containing a diversity of eukaryotic organisms, most of which are unicellular. (Chapter 13)

proton A positively charged subatomic particle. (Chapters 2, 5)

proto-oncogenes Genes that encode proteins that regulate the cell cycle. Mutated proto-oncogenes (oncogenes) can lead to cancer. (Chapter 6)

prune To trim back the growing tips of a plant. (Chapter 24)

pseudopodia A cellular extension of amoebas used for feeding and in motility. (Chapter 19)

puberty The point in human development when male and female hormones are triggered in the body. Males produce sperm at this time, and females begin the menstrual cycle. (Chapter 20)

pulmonary circuit The path of blood through vessels from the heart, through the lungs, and back to the heart; one half of the double circulation. (Chapter 18)

pulse The volume of blood that passes into the arteries as a result of the heart's contraction. (Chapter 18)

punctuated equilibrium The hypothesis that evolutionary changes occur rapidly and in short bursts, followed by long periods of little change. (Chapter 12)

Punnett square Table that lists the different kinds of sperm or eggs parents can produce relative to the gene or genes in question and predicts the possible outcomes of a cross between these parents. (Chapter 7)

purine Nitrogenous base (A or G) with a two-ring structure. (Chapter 2)

pyrimidine Nitrogenous base (C, T, or U) with a single-ring structure. (Chapter 2)

pyruvic acid The 3-carbon molecule produced by glycolysis. (Chapter 4)

Q angle Angle of the femur in relation to a horizontal line drawn through the kneecap. (Chapter 20)

qualitative trait Trait that produces phenotypes in distinct categories. (Chapter 7)

quantitative trait Trait that has many possible values. (Chapter 7)

race See biological race. (Chapter 12)

racism Idea that some groups of people are naturally superior to others. (Chapter 12)

radiation therapy Focusing beams of reactive particles at a tumor to kill the dividing cells. (Chapter 6)

radioactive decay Natural, spontaneous breakdown of radioactive elements into different elements, or "daughter products." (Chapter 10)

radioimmunotherapy Experimental cancer treatment with the goal of delivering radioactive substances directly to tumors without affecting other tissues. (Chapter 6)

radiometric dating Technique that relies on radioactive decay to estimate a fossil's age. (Chapters 10, 15)

random alignment When members of a homologous pair line up randomly with respect to maternal or paternal origin during metaphase I of meiosis, thus increasing the genetic diversity of offspring. (Chapters 6, 7)

random assignment Placing individuals into experimental and control groups randomly to eliminate systematic differences between the groups. (Chapter 1)

random distribution The dispersion of individuals in a population without pattern. (Chapter 14)

random fertilization The unpredictability of exactly which gametes will fuse during the process of sexual reproduction. (Chapter 7)

reabsorption Reuptake of water and other essential substances by a nephron from the filtrate initially squeezed into the kidney. (Chapter 18)

reactant Any starting material in a chemical reaction. (Chapter 2)

reading frame The grouping of mRNAs into 3 base codons for translation. (Chapter 9)

receptor (1) Protein on the surface of a cell that recognizes and binds to a specific chemical signal. (Chapters 6, 10) (2) See sensory receptor. (Chapter 22)

recessive Applies to an allele with an effect that is not visible in a heterozygote. (Chapter 7)

recombinant Produced by manipulating a DNA sequence. (Chapter 9)

recombinant bovine growth hormone (rBGH) Growth hormone produced in a laboratory and injected into cows to increase their size and ability to produce milk. (Chapter 9)

red blood cell Primary cellular component of blood, responsible for ferrying oxygen throughout the body. (Chapters 6, 18)

reflex Automatic response to a stimulus. (Chapter 22)

reflex arc Nerve pathway followed during a reflex consisting of a sensory receptor, a sensory neuron, an interneuron, a motor neuron, and an effector. (Chapter 22)

remission The period during which the symptoms of a disease subside. (Chapter 6)

renal artery The vessel that carries blood to the kidney for filtering. (Chapter 18)

renal vein The vessel that carries filtered blood from the kidney. (Chapter 18)

repolarization The restoration of a charge difference across a membrane. (Chapter 22)

repressor A protein that suppresses the expression of a gene. (Chapter 9)

reproductive cloning Transferring the nucleus from a donor adult cell to an egg cell without a nucleus in order to clone the adult. (Chapter 9)

reproductive isolation Prevention of gene flow between different biological species due to failure to produce fertile offspring; can include premating barriers and postmating barriers. (Chapter 12)

reproductive organ Internal and external genitalia involved in production and delivery of gametes. (Chapters 21, 23)

respiratory surface Body surface across which gas exchange occurs. (Chapter 18)

respiratory system The organ system involved in gas exchange between an animal and its environment. In humans, the lungs and air passages. (Chapter 18)

restriction enzyme An enzyme that cleaves DNA at specific nucleotide sequences. (Chapters 7, 9)

reticular formation Extensive network of neurons that runs through the medulla of the brain and projects toward the cerebral cortex. It functions as a filter for sensory input and activates the cerebral cortex. (Chapter 22)

reuptake In neurons, the process by which neurotransmitters are reabsorbed by the neuron that secreted them. (Chapter 22)

reverse transcriptase Enzyme in RNA viruses that produces DNA by transcription of viral RNA. (Chapter 19)

Rh factor Surface molecule found on some red blood cells. (Chapter 8)

rhizomes Underground stems in plants. (Chapter 23)

ribose The five-carbon sugar in RNA. (Chapter 8)

ribosomal RNA (rRNA) RNA that makes up part of the structure of ribosomes. (Chapter 2)

ribosome Subcellular structure that helps translate genetic material into proteins by anchoring and exposing small sequences of mRNA. (Chapters 2, 9, 13)

risk factor Any exposure or behavior that increases the likelihood of disease. (Chapter 6)

Ritalin Stimulant used to treat ADD. (Chapter 22)

river Aquatic biome characterized by flowing water. (Chapter 16)

RNA (ribonucleic acid) Information-carrying molecule composed of nucleotides. (Chapters 2, 9)

RNA polymerase Enzyme that synthesizes mRNA from a DNA template during transcription. (Chapter 9)

root cap Loosely organized cells that cover the growing tip of a root and are continually shed as the root pushes through the soil. (Chapter 23)

root hair Extension of an epidermal cell on young roots, which maximize the surface area for water and nutrient uptake. (Chapter 23)

root system The plant organ system responsible for water and nutrient uptake and anchorage. (Chapter 23)

rough endoplasmic reticulum Ribosome-studded subcellular membranes found in the c ytoplasm and responsible for some protein synthesis. (Chapter 2)

RuBisCo Abbreviation for ribulose bisphosphate carboxylase oxygenase, the enzyme that catalyzes the first step in the Calvin cycle of photosynthesis. (Chapter 5)

runoff Water moving across a land surface, picking up contaminants and eventually flowing into lakes, streams, or groundwater. (Chapter 23)

salinization Degradation of soil by mineral salts deposited as a result of irrigation. (Chapter 23)

salivary amylase An enzyme secreted by salivary glands of the mouth to break down starch. (Chapter 17)

salt A charged substance that ionizes in solution. (Chapter 2)

sample Small subgroup of a population used in an experimental test. (Chapter 1)

sample size Number of individuals in both the experimental and control groups. (Chapter 1)

sampling error Effect of chance on experimental results. (Chapter 1)

sarcomere Any of the repeating units in a myofibril of striated muscle, bounded by Z discs. (Chapter 20)

saturated fat Type of lipid rich in single bonds. Found in butter and other fats that are solids at room temperature. This type of fat is associated with higher blood cholesterol levels. (Chapter 3)

savanna Grassland biome containing scattered trees. (Chapter 16)

Schwann cell Cells that form the myelin sheath along the axons of nerve cells in the peripheral nervous system. (Chapter 22)

scientific method A systematic method of research consisting of putting a hypothesis to a test designed to disprove it, if it is in fact false. (Chapter 1)

scientific theory Body of scientifically accepted general principles that explain natural phenomena. (Chapters 1, 10)

scrotum The pouch of skin that houses the testes. (Chapter 21)

secondary compounds Chemicals produced by plants and some other organisms as side reactions to normal metabolic pathways and that typically have an antipredator or antibiotic function. (Chapter 13)

secondary consumers Animals that eat primary consumers; predators. (Chapter 15)

secondary growth Growth in girth of plants, as a result of cell division in lateral meristems. (Chapter 23)

secondary sources Books, news media, and advertisements as sources of scientific information. (Chapter 1)

secondary spermatocyte Haploid cell produced by the first meiotic division in the production of sperm. (Chapter 21)

secretion A step in waste removal in the kidney, in which contaminants in low concentration in the blood are actively absorbed into the urine. (Chapter 18)

seed A plant embryo packaged with a food source and surrounded by a seed coat. (Chapters 13, 23)

seed coat Layer of tissue surrounding a plant embryo and endosperm, arising from accessory tissue in the ovule. (Chapter 23)

seedling A young plant that develops from a germinating seed. (Chapter 23)

segregation Separation of pairs of alleles during the production of gametes. Results in a 50% probability that a given gamete contains one allele rather than the other. (Chapter 7)

semen Sperm and energy-rich associated fluids. (Chapter 21)

semilunar valve Heart valve controlling blood flow from the ventricles into blood vessels leading away from the heart. (Chapter 18)

seminal vesicle Either of two pouchlike glands located on both sides of the bladder that add a fructose-rich fluid to semen prior to ejaculation. (Chapter 21)

seminiferous tubule Highly coiled tube in the testicles where sperm are formed. (Chapter 21)

semipermeable In biological membranes, a membrane that allows some substances to pass but prohibits the passage of others. (Chapter 2)

sensory neuron A neuron that conducts impulses from a sense organ to the central nervous system. (Chapter 22)

sensory receptor Any of the cellular systems that collect information about the environment inside or outside the body and transmit that information to the brain. (Chapter 22)

sepal The outermost floral structure, usually enclosing the other flower parts in a bud. (Chapter 23)

Sertoli cell Testicular cell that secretes substances that aid in the development of mature sperm. (Chapter 21)

severe combined immunodeficiency (SCID) Illness caused by a genetic mutation that results in the absence of an enzyme and a severely weakened immune system. (Chapter 9)

sex chromosome Any of the sex-determining chromosomes (X and Y in humans). (Chapters 6, 7)

sex determination Determining the biological sex of an offspring. Humans have a chromosomal mechanism of sex determination in which two X chromosomes produce a female and an X and a Y chromosome produce a male. (Chapter 8)

sex hormone Any of the steroid hormones that affect development and functions of reproductive structures and secondary sex characteristics. (Chapter 20)

sex-linked gene Any of the genes found on the X or Y sex chromosomes. (Chapter 8)

sexual reproduction Reproduction involving two parents that gives rise to offspring that have unique combinations of genes. (Chapters 6, 21)

sexual selection Form of natural selection that occurs when a trait influences the likelihood of mating. (Chapter 12)

shoot system The organs of a plant that are modified for photosynthesis: stem and leaf. (Chapter 23)

signal transduction When a change in a cell or its environment is relayed through various molecules and results in a cellular response. (Chapter 20)

single nucleotide polymorphism A DNA sequence variation that occurs when members of species differ from each other at a single nucleotide (A, T, C, or G) locus. (Chapter 12)

sink A plant organ that is using carbohydrate, either for growth and metabolism or for conversion into starch. (Chapter 24)

sinoatrial (SA) node Region of the heart muscle that generates an electrical signal that controls heart rate. See pacemaker. (Chapter 18)

sister chromatid Either of the two duplicated, identical copies of a chromosome formed after DNA synthesis. (Chapter 6)

skeletal muscle Striated muscle involved with voluntary movements. (Chapter 17)

sliding filament model The theory that muscles contract when actin filaments slide across myosin filaments, shortening the sarcomere. (Chapter 20)

small intestine The narrow, twisting, upper part of the intestine where nutrients are absorbed into the blood. (Chapter 17)

smog Products of fossil fuel combustion in combination with sunlight, producing a brownish haze in still air. (Chapter 16)

smooth endoplasmic reticulum The subcellular, cytoplasmic membrane system responsible for lipid and steroid biosynthesis. (Chapter 2)

smooth muscle Nonstriated, spindle-shaped muscle cells that line organs and blood vessels. (Chapter 17)

sodium potassium pump A protein pump in a cell membrane that moves sodium out of the cell and potassium into the cell, both against their concentration gradients. (Chapter 22)

soil erosion Loss of topsoil. (Chapter 23)

soil Medium for plant growth made up of mineral particles, partially decayed organic matter, and living organisms. (Chapter 23)

solar irradiance The amount of solar energy hitting Earth's surface at any given point. (Chapter 16)

solid waste Garbage. (Chapter 16)

solstice When the sun reaches its maximum and minimum elevation in the sky. (Chapter 16)

solute The substance that is dissolved in a solution. (Chapter 2)

solution A mixture of two or more substances. (Chapter 2)

solvent A substance, such as water, that a solute is dissolved in to make a solution. (Chapter 2)

somatic cell Any of the body cells in an organism. Any cell that is not a gamete. (Chapter 6)

somatic cell gene therapy Changes to malfunctioning genes in somatic or body cells. These changes will not be passed to offspring. (Chapter 9)

somatomammotropin A hormone produced by the placenta that plays a role in stimulating milk production. (Chapter 21)

source Plant organs that are actively generating sugar. (Chapter 24)

spatial isolation A mechanism for reproductive isolation that depends on the geographic separation of populations. (Chapter 12)

special creation The hypothesis that all organisms on Earth arose as a result of the actions of a supernatural creator. (Chapter 10)

special senses Senses that have specialized organs. These include sight, hearing, equilibrium, taste, touch, and smell. (Chapter 22)

speciation Evolution of one or more species from an ancestral form: macroevolution. (Chapter 12)

species A group of individuals that regularly breed together and are generally distinct from other species in appearance or behavior. (Chapters 2, 10, 12, 13, 15)

species-area curve Graph describing the relationship between the size of a natural landscape and the relative number of species it contains. (Chapter 15)

specific defense Defense against pathogens that utilizes white blood cells of the immune system. (Chapter 19)

specificity Phenomenon of enzyme shape determining the reaction the enzyme catalyzes. (Chapters 4, 19)

sperm Gametes produced by males. (Chapters 6, 7, 8, 13, 20, 21)

spermatid Cell produced when secondary spermatocytes undergo meiosis II to produce haploid cells that no longer have duplicated chromosomes. (Chapter 21)

spermatogenesis Production of sperm. (Chapter 21)

spermatozoa Mature sperm composed of a small head containing DNA, a midpiece that contains mitochondria, and a tail (flagellum). (Chapter 21)

spinal cord Thick cord of nervous tissue that extends from the base of the brain through the spinal column. (Chapters 17, 22)

spongy bone The porous, honeycomblike material of the inner bone. (Chapter 20)

spore Reproductive cell in plants and fungi that is capable of developing into an adult without fusing with another cell. (Chapters 13, 23)

sporophyte Diploid stage in the alternation of generations in plants, which produces spores via meiosis. (Chapter 23)

stabilizing selection Natural selection that favors the average phenotype and selects against the extremes in the population. (Chapter 11)

stamen Male flower part, producing the pollen. (Chapter 23)

standard error A measure of the variance of a sample; essentially the average distance a single data point is from the mean value for the sample. (Chapter 1)

staple crop One of a number of agricultural crops that make up the majority of calories in human societies. (Chapter 23)

statistical test Mathematical formulation that helps scientists evaluate whether the results of a single experiment demonstrate the effect of treatment. (Chapter 1)

statistically significant Said of results for which there is a low probability that experimental groups differ simply by chance. (Chapter 1)

statistics Specialized branch of mathematics used in the evaluation of experimental data. (Chapter 1)

stem The part of vascular plants that is above ground, as well as similar structures found underground. (Chapter 23)

stem cell Cells that can divide indefinitely and can differentiate into other cell types. (Chapters 9, 19)

steppe Biome characterized by short grasses, found in regions with relatively little annual precipitation. (Chapter 16)

steroid naturally occurring or synthetic organic fat-soluble substance that produces physiologic effects. (Chapters 2, 20)

stigma Sticky pad on the tip of a flower's carpel where pollen is deposited. (Chapter 23)

stomach a sac-like enlargement of the alimentary canal involved in digestion. (Chapter 17)

stomata Pores on the photosynthetic surfaces of plants that allow air into the internal structure of leaves and green stems. Stomata also provide portals through which water can escape. (Chapters 5, 23, 24)

stop codon An mRNA codon that does not code for an amino acid and causes the amino acid chain to be released into the cytoplasm. (Chapter 9)

stream Biome characterized by flowing water, sometimes seasonal. Typically smaller than rivers. (Chapter 16)

striated muscle A voluntary muscle made up of elongated, multinucleated fibers. Typically skeletal and cardiac muscle

that is distinguished from smooth muscle by the in-register banding patterns of actin and myosin filaments. (Chapter 17)

stroke Acute condition caused by a blood clot that blocks blood flow to an organ or other region of the body. (Chapters 4, 18)

stroma The semi-fluid matrix inside a chloroplast where the Calvin cycle of photosynthesis occurs. (Chapters 2, 5)

style Stalk connecting stigma to ovary in a flower's carpel. (Chapter 23)

subspecies Subdivision of a species that is not reproductively isolated but represents a population or set of populations with a unique evolutionary history. See also biological race, variety. (Chapter 12)

substrate The substance upon which an enzyme reacts. (Chapter 4)

succession Replacement of ecological communities over time since a disturbance, until finally reaching a stable state. (Chapter 16)

sugar-phosphate backbone Series of alternating sugars and phosphates along the length of the DNA helix. (Chapters 2, 5, 6)

supernatural Not constrained by the laws of nature. (Chapter 1)

sustainable Referring to human activities that are able to continue indefinitely with no degradation of the resources on which they depend. (Chapter 23)

symbiosis A relationship between two species. (Chapter 13)

sympatric In the same geographic region. (Chapter 12)

synapse Gap between neurons consisting of the terminal boutons of the presynaptic neuron, the space between the two adjacent neurons, and the membrane of the postsynaptic neuron. (Chapter 22)

systematist Biologist who specializes in describing and categorizing a particular group of organisms. (Chapter 13)

systemic circuit Flow of blood from the heart to body capillaries and back to the heart. (Chapter 18)

systole Portion of the cardiac cycle when the heart is contracting, forcing blood into arteries. (Chapter 18)

systolic blood pressure Force of blood on artery walls when heart is contracting. (Chapters 4, 18)

T lymphocyte (T cell) Immune-system cell that develops in the thymus gland it facilitates cell-mediated immunity. (Chapter 19)

Taq polymerase An enzyme that can withstand high temperatures, which is used during polymerase chain reactions. (Chapter 8)

target cell A cell that responds to regulatory signals such as hormones. (Chapter 20)

telomerase An enzyme that helps prevent the degradation of the tips of chromosomes, active during development and sometimes reactivated in cancer cells. (Chapter 6)

telophase Stage of mitosis during which the nuclear envelope forms around the newly produced daughter nucleus, and chromosomes decondense. (Chapter 6)

temperate forest Biome dominated by deciduous trees. (Chapter 16)

temperature A measure of the intensity of heat or kinetic energy. (Chapters 4, 16)

template Something that serves as a pattern or mold. (Chapter 8)

temporal bone Either of the paired bones forming the sides and base of the cranium. (Chapter 20)

temporal isolation Reproductive isolation between populations maintained by differences in the timing of mating or emergence. (Chapter 12)

temporal lobe Part of the cerebral hemisphere that processes auditory and visual information, memory, and emotion. Located in front of the occipital lobe. (Chapter 22)

tension In plants, negative water pressure. (Chapter 24)

terminal bouton Knoblike structure at the end of an axon. (Chapter 22)

terminal bud Apical meristem on the tip of a stem or branch. (Chapter 23)

testable Possible to evaluate through observations of the measurable universe. (Chapter 1)

testicle See testis. (Chapters 20, 21)

testis (*plural:* testes) Either of the paired male gonads, involved in gametogenesis and secretion of reproductive hormones. (Chapters 20, 21)

testosterone Masculinizing hormone secreted by the testes. (Chapters 20, 21)

thalamus Main relay center between the spinal cord and the cerebrum. (Chapter 22)

theory See scientific theory. (Chapters 1, 10)

theory of evolution Theory that all organisms on Earth today are descendants of a single ancestor that arose in the distant past. See also evolution. (Chapters 1, 2, 10)

therapeutic cloning Using early embryos as donors of stem cells for the replacement of damaged tissues and organs in another individual. (Chapter 9)

thermoregulation The ability to maintain a body temperatures within a narrow range. (Chapter 17)

thickening meristem Dividing tissue near the apical meristem in monocots that increases stem girth. (Chapter 23)

thigmotropism Growth in response to contact with a solid object. (Chapter 24)

thylakoid Flattened membranous sac located in the chloroplast stroma. Function in photosynthesis. (Chapters 2, 5)

thymine Nitrogenous base in DNA, a pyrimidine. (Chapters 2, 8)

thymus An endocrine gland located in the neck region that helps establish the immune system. (Chapters 19, 20)

thyroid gland An endocrine gland in the neck region that stimulates metabolism and regulates blood calcium levels. (Chapter 20)

Ti plasmid Tumor-inducing plasmid used to genetically modify crop plants. (Chapter 9)

tilling Turning over the soil to kill weed seedlings, with the goal of removing competitors before a crop is planted. (Chapter 23)

tissue A group of cells with a common function. (Chapter 17)

tongue A muscular structure in the mouth that has taste buds that help you taste food. It aids in breaking down food for the digestive process. (Chapter 17)

totipotent Describes a cell able to specialize into any cell type of its species, including embryonic membrane. Compare with multipotent, pluripotent. (Chapter 9)

trachea Air passage from upper respiratory system into lower respiratory system. Also called windpipe. (Chapters 17, 18)

tracheid Narrow xylem cell with pitted walls. (Chapter 24)

trans fat Contains unsaturated fatty acids that have been hydrogenated, which changes the fat from a liquid to a solid at room temperature. (Chapter 3)

transcription Production of an RNA copy of the protein coding DNA gene sequence. (Chapters 9, 11, 19)

transfer RNA (tRNA) Amino-acid-carrying RNA structure with an anti-codon that binds to an mRNA codon. (Chapter 9)

transgenic organism Organism whose genome incorporates genes from another organism; also called genetically modified organism (GMO). (Chapter 9)

translation Process by which an mRNA sequence is used to produce a protein. (Chapters 9, 11, 19)

translocation Movement of phloem sap around the body of a plant. (Chapter 24)

transpiration Movement of water from the roots to the leaves of a plant, powered by evaporation of water at the leaves and the cohesive and adhesive properties of water. (Chapters 5, 24)

trisomy A chromosomal condition in which three copies of a chromosome exist instead of the two copies of a chromosome normally present in a diploid organism. (Chapter 6)

trophic level Feeding level or position on a food chain; e.g. producers, primary consumers, etc. (Chapter 15)

trophic pyramid Relationship among the mass of populations at each level of a food web. (Chapter 15)

trophoblast The outer layer of a developing blastocyst that supplies nutrition to the embryo. (Chapter 21)

tropical forest Biome dominated by broad-leaved, evergreen trees; found in areas where temperatures never drop below the freezing point of water. (Chapter 16)

tropism In plants, directional growth. (Chapter 24)

tuber Underground storage structure in plants, made up of modified stem. (Chapter 23)

tuberculosis (TB) Degenerative lung disease caused by infection with the bacterium *Mycobacterium tuberculosis*. (Chapter 11)

tumor Mass of tissue that has no apparent function in the body. (Chapter 6)

tumor suppressor Cellular protein that stops tumor formation by suppressing cell division. When mutated leads to increased likelihood of cancer. (Chapter 6)

tundra Biome that forms under very low temperature conditions. Characterized by low-growing plants. (Chapter 16)

turgid In plants, cells that are filled with enough water so that the cell walls are deformed. (Chapter 23)

type I diabetes Diabetes that results from inability to produce inuslin. (Chapter 4)

understory The level of a forest below the canopy trees and shrub cover, typically consisting of herbaceous perennial plants and tree and shrub seedlings. (Chapter 16)

undifferentiated A cell that is not specialized. (Chapter 9)

unicellular Made up of a single cell. (Chapter 13)

uniform distribution Occurs when individuals in a population are disbursed in a uniform manner across a habitat. (Chapter 14)

unsaturated fat Type of lipid containing many carbon-to-carbon double bonds; liquid at room temperature. (Chapter 3)

unsustainable Relating to practices that may compromise the ability of future generations to support a large human population with an adequate quality of life. (Chapter 23)

uracil Nitrogenous base in RNA, a pyrimidine. (Chapters 2, 9)

urban sprawl The tendency for the boundaries of urban areas to grow over time as people build housing and commercial districts farther and farther from an urban core. (Chapter 16)

ureter Tube that delivers urine from a kidney to the bladder. (Chapter 18)

urethra Urine-carrying duct that also carries sperm in males. (Chapters 18, 21)

urinary system The organ system responsible for the filtering, collection, and excretion of liquid waste; consisting of the kidneys, ureters, bladder, and urethra. (Chapter 18)

urine Liquid expressed by the kidneys and expelled from the bladder in mammals, containing the soluble waste products of metabolism. (Chapter 18)

uterus Pear-shaped muscular organ in females that can support pregnancy and that undergoes menstruation when its lining is shed. (Chapter 21)

vaccination A preparation of a weakened or killed pathogen, or portion of a pathogen, that will stimulate the immune system of a recipient to prepare a long-term defense (memory cells) against that pathogen. (Chapter 19)

vagina Muscular canal in females leading from the cervix to the vulva. (Chapter 21)

valence shell The outermost energy shell of an atom containing the valence electrons which are most involved in the chemical reactions of the atom. (Chapter 2)

variable A factor that varies in a population or over time. (Chapter 1)

variable number tandem repeat (VNTR) A DNA sequence that varies in number between individuals. Used during the process of DNA fingerprinting. (Chapter 8)

variance Mathematical term for the amount of variation in a population. (Chapter 7)

variant An individual in a population that differs genetically from other individuals in the population. (Chapter 11)

vas deferens Either of the two ducts in males that carry sperm from the epididymis to the urethra. (Chapter 21)

vascular cambium Meristematic tissue that forms in vascular bundles of dicot roots and stems and permits secondary growth. (Chapter 23)

vascular system Plant tissue system responsible for the delivery of water and dissolved solutes throughout the plant body. (Chapter 18)

vascular tissue Cells that transport water and other materials within a plant. (Chapters 13, 23)

vasodilation Widening of blood vessels, caused by relaxation of smooth muscle lining vessel walls. (Chapter 19)

vector An organism that carries a pathogen from one host to another, such as a mosquito carrying West Nile virus. (Chapter 19)

vegetative organ Plant organ that is not involved in sexual reproduction. (Chapter 23)

vegetative reproduction Asexual cloning of plants. (Chapter 21)

vein Vessel that carries blood from the body tissues back to the heart. (Chapter 18)

ventricle Chamber of the heart that pumps blood from the heart to the lungs or systemic circulation. (Chapter 18)

vertebra (*plural:* vertebrae) Bone of the spinal column through which the spinal cord passes. (Chapter 22)

vertebral column Also called the spine, the series of vertebrae and cartilaginous disks extending from the brain to the pelvis. (Chapter 22)

vertebrate Animal with a backbone. (Chapter 13)

vesicle Membrane-bounded sac-like structure. In neurons, these structures are found in the terminal bouton and store neurotransmitters. (Chapter 22)

vessel element Wide xylem cell with perforated ends, found only in angiosperms. (Chapter 24)

vestigial trait Modified with no or relatively minor function compared to the function in other descendants of the same ancestor. (Chapter 10)

villus (*plural:* villi) Small fingerlike projections on the inside of the small intestine that function in nutrient absorption.

viral envelope Layer formed around some virus protein coats (capsids) that is derived from the cell membrane of the host cell and may also contain some proteins encoded by the viral genome. (Chapter 19)

virus Infectious intracellular parasite composed of a strand of genetic material and a protein or fatty coating that can only reproduce by forcing its host to make copies of it. (Chapters 1, 9, 19)

vitamin Organic nutrient needed in small amounts. Most vitamins function as coenzymes. (Chapter 3)

voluntary Muscle normally under conscious control. Mainly skeletal muscle. (Chapter 17)

vulva The outer portion of the female external genitalia including the labia majora, labia minora, clitoris, and vaginal and urethral openings. (Chapter 21)

wastewater Liquid waste produced by residential, commercial, or industrial activities. (Chapter 16)

water One molecule of water consists of one oxygen and two hydrogen atoms. (Chapters 2, 3, 4, 16, 23, 24)

weather Current temperature and precipitation conditions. (Chapter 16)

weed Common term for nonpreferred plant. (Chapter 23)

wetland Biome characterized by standing water, shallow enough to permit plant rooting. (Chapter 16)

white blood cell General term for cell of the immune system. (Chapter 18)

white matter Nervous system tissue, especially in the brain and spinal cord, made of myelinated cells. (Chapter 22)

whole food Any food that has not undergone processing. Includes grains, beans, nuts, seeds, fruits, and vegetables. (Chapter 3)

wood Xylem cells produced by secondary growth of stems and roots. (Chapter 23)

woody plant Plant that produces stiffened stems via secondary production of xylem. (Chapter 24)

X inactivation The inactivation of one of two chromosomes in the XX female. (Chapter 8)

x-axis The horizontal axis of a graph. Typically describes the independent variable. (Chapter 1)

X-linked gene Any of the genes located on the X chromosome. (Chapter 8)

xylem Plant vascular tissue, dead at maturity, that carries water and dissolved minerals in a one-way flow from the roots to the shoots of a plant. (Chapters 23, 24)

xylem sap Water and dissolved minerals flowing in the xylem vessels. (Chapter 24)

y-axis The vertical axis of a graph. Typically describes the dependent variable. (Chapter 1)

yeast Single-celled eukaryotic organisms found in bread dough. Often used as model organisms and in genetic engineering. (Chapters 4, 10, 13)

Y-linked gene Any of the genes located on the Y chromosome. (Chapter 8)

Z disc The border of a sarcomere in muscle. (Chapter 20)

zona pellucida The substance surrounding an egg cell that a sperm must penetrate for fertilization to occur. (Chapter 21)

zoologist Scientist who specializes in the study of animals. (Chapter 13)

zygote Single cell resulting from the fusion of gametes (egg and sperm). (Chapters 6, 8, 21)

Credits

Text and Art Credits

CHAPTER 1 **Savvy Reader.** Gupta, Terry, "Get back your health with vitamin C, tea" *The Telegraph* Nashua, NH, Oct. 26, 2007. Used with permission.

Figure 1.7. Data from: Lindenmuth, G. Frank, Ph.D. and Elise B. Lindenmuth, Ph.D., "The efficacy of echinacea compound herbal tea preparation on the severity and duration of upper respiratory and flu symptoms: A randomized, double-blind placebo-controlled study," *The Journal of Alternative and Complementary Medicine,* Vol. 6: 4, pp. 327–334, 2000, Mary Ann Liebert, Inc.

Figure 1.10. Cohen, Sheldon, Ph.D., et al., "Psychological Stress and Susceptibility to the Common Cold," *New England Journal of Medicine,* Vol. 325: 9, pp. 606–612, fig. 1, Aug. 29, 1991. Copyright © 1991 Massachusetts Medical Society. Used with permission.

Figure 1.13. Data from: Mossad, Sherif B., MD, et al., "Zinc Gluconate Lozenges for Treating the Common Cold, A Randomized, Double-Blind, Placebo-Controlled Study," *Annals of Internal Medicine,* Vol. 125: 2, July 15, 1996.

CHAPTER 2 **Opener (L)** joaquin croxatto/iStockphoto

CHAPTER 3 **Savvy Reader.** "Wellness Guide to Dietary Supplements: Açaí Juice," *UC Berkeley Wellness Letter,* June 2009, Copyright Clearance Center, http://www.wellnessletter.com/html/ds/dsAcai.php.Used with permission.

CHAPTER 5 **Savvy Reader.** Sean Mattson, "Rising sea drives Panama islanders to mainland" from reuters.com, July 11, 2010. Used by permission. All rights reserved. Republication or redistribution of Thomson Reuters content, including by framing or similar means, is expressly prohibited without the prior written consent of Thomson Reuters. Thomson Reuters and its logo are registered trademarks or trademarks of the Thomson Reuters group of companies around the world. © Thomson Reuters 2010. Thomson Reuters journalists are subject to an Editorial Handbook which requires fair presentation and disclosure of relevant interests.

Figure 5.6. Data from: "Trends in Atmospheric Carbon Dioxide—Mauna Loa," NOAA/ESRL Global Monitoring Division, www.esrl.noaa.gov/gmd/ccgg/trends/.

Figure 5.8. Reprinted by permission from Macmillan Publishers Ltd., from J.R. Petit, et al., "Climate and atmospheric history of the past 420,000 years from the Vostok ice core, Antarctica," Nature, 399 (3): 429–436, June 1999. Copyright © 1999 Nature Publishing, Inc.

CHAPTER 7 **Savvy Reader.** Jennifer Warner, "Genes May Link Friends, not Just Family" from WebMD Health News, Jan. 18, 2011. Reviewed by Laura J. Martin, MD. © 2011 WebMD, LLC. All rights reserved. Reproduced by permission.

Figure 7.19: Data from: Raberg, L., "Immune responsiveness in adult blue tits: Heritability and effects of nutritional status during ontogeny," *Oecologia,* 136: 360–4, Fig. 1, 2003.

CHAPTER 10 **Savvy Reader.** "Ranks of Scientists Doubting Darwin's Theory on the Rise," Discovery Institute, Feb. 8, 2007, Discovery Institute. Used with permission.

CHAPTER 11 **Savvy Reader.** Tandy, Florence, "What Gall!" http://fwtandy.wordpress.com/2007/06/08/what-gall/. Used with permission.

Figure 11.9. Data from: Grant, P. R. and B. R. Grant. "Evolution of character displacement in Darwin's finches," *Science,* 313: 224–226, 2006.

CHAPTER 12 **Savvy Reader.** "A place for race in medicine?" by Gregory M. Lamb. Reprinted with permission from the March 3, 2005 issue of The Christian Science Monitor. © 2005 The Christian Science Monitor (www.CSMonitor.com).

Figure 12.11. Data from: Weatherall, D. J. and J. B. Clegg. "Inherited haemoglobin disorders: an increasing global health problem," 79 (8): 704–12, Oct. 24, 2001, http://www.ncbi.nlm.nih.gov/pmc/articles/PMC2566499/pdf/11545326.pdf.

Figure 12.13. Data from: Beckman, Lars, "A Contribution to the Physical Anthropology and Population Genetics of Sweden: Variations of the ABO, Rh, MN and P Blood Groups," Lund: Berlingska Boktryck, 1959; http://aboblood.com/.

Figure 12.14. Data from: Weatherall, D. J. and J. B. Clegg. "Inherited haemoglobin disorders: an increasing global health problem," 79 (8): 704–12, Oct. 24, 2001, http://www.ncbi.nlm.nih.gov/pmc/articles/PMC2566499/pdf/11545326.pdf; "World Malaria Report 2010," World Health Organization, http://www.who.int/malaria/world_malaria_report_2010/en/index.html.

Figure 12.17. Data from: Jablonski, N. and G. Chaplin, "The evolution of human skin coloration," *Journal of Human Evolution,* 39: 57–106, 2000.

CHAPTER 13 **Savvy Reader.** From CNN, Oct. 13, 2005, © 2005 Turner Broadcasting System, Inc. All rights reserved. Used by permission and protected by the Copyright Laws of the United States. The printing, copying, redistribution, or retransmission of this Content without express written permission is prohibited.

CHAPTER 14 **Savvy Reader.** "Common Birds in Decline: What's happening to birds we know and love?" *State of the Birds,* National Audubon Society, Summer 2007 update, http://stateofthebirds.audubon.org/cbid/. Used with permission.

Figure 14.13. U.S. Census, http://www.census.gov/ipc/www/idb/informationGateway.php.

Figure 14.14. U.S. Census, http://www.census.gov/ipc/www/idb/informationGateway.php.

CHAPTER 15 **Figure 15.4.** Primack, R.B. *Conservation Biology,* 1st edition, fig. 4.5, p. 81, Sinauer Associates, 1993. Used with permission; Data from: Nilsson, G., *The Endangered Species Handbook. Animal Welfare Institute,* Washington DC, IUCN, 1988; "Red List of Threatened Animals," IUCN, Gland, Switzerland, 1988.

Figure 15.21. Reprinted by permission from Macmillan Publishers Ltd., from N. Myers et al., "Biodiversity hotspots for conservation priorities," Nature, Vol. 403, p. 853, 2/24/2000. Copyright © 2000 Nature Publishing, Inc.

CHAPTER 16 **Figure 16.24.** Harrison, Paul and Fred Pearce, "Earth in Peril," *American Association for the Advancement of Science Atlas of Population and Environment in 2000.* Used with permission.

CHAPTER 17 **Savvy Reader.** Stamatis, George, "Study reveals consequences for people who sell kidneys," Case Western Reserve University.

CHAPTER 18 **Savvy Reader.** Porter, Sabrina. "Students speak out in smoking ban survey," *The Tartan,* Feb. 26, 2007. Used with permission.

CHAPTER 21 **Savvy Reader.** Jim Ritter, "Much too soon—Disorder seems to be striking more preschoolers" from Chicago Sun-Times, Oct. 29, 2007 © 2007 Chicago Sun-Times. All rights reserved. Used by permission and protected by the Copyright Laws of the United States. The printing, copying, redistribution, or retransmission of this content without express written permission is prohibited.

CHAPTER 23 **Savvy Reader.** "Is Going Organic Worth the Price?" reported by Angela Davis, WCCO-TV, Oct. 8, 2007. Used with permission.

Figure 23.27. Fischer, Brian C. and Virginia L. McGuire, "Wyoming, 1980 to 1995," USGS, Report 99–197[1][2], Feb. 27, 2009.

CHAPTER 24 **Savvy Reader.** "Houseplants Offer Health Benefits" reported by Dennis Douda, WCCO-TV, May 21, 2006. Used with permission.

Photo Credits

CHAPTER 1 Opener (L) Paul Bradbury/OJO Images/Getty, Opener (R) Top Photolibrary, Opener (R) Middle Edyta Pawlowska/Shutterstock, Opener (R) Bottom iStockphoto, 1.2a Eye of Science/Photo Researchers, Inc., 1.2b Anders Wiklund/Reuters/Landov, 1.4a Custom Medical Stock Photo, Inc., 1.5 The Natural History Museum, London/Alamy Images, 1.6 Zina Seletskaya/Shutterstock, 1.9a Leonard Lessin/Photolibrary, 1.9b Joel Sartore/National Geographic/Getty, 1.9c Photo Researchers, Inc., 1.17 Graca Victoria/Shutterstock.

CHAPTER 2 Opener (R) Top NASA, Opener (R) Middle NASA, Opener (R) Bottom NASA, 2.1 Photos.com, 2.2a NASA, 2.2b DLR/FU Berlin (G. Neukum)/European Space Agency, 2.16 National Cancer Institute, 2.17a Juergen Berger/Photo Researchers, Inc., 2.19a Dr. Gary Gaugler/Photo Researchers, Inc., 2.19b Eye of Science/Photo Researchers, Inc., 2.19c Brittany Carter Courville/istockphoto, 2.19d Jan Krejci/Shutterstock, 2.19e webphotographeer/istockphoto, 2.19f M. I. Walker/Photo Researchers, Inc., 2.19g Dusan Zidar/istockphoto, 2.19h Vicki Beaver/istockphoto, 2.20 King County Deartment of Natural Resources and Parks, Water and Land Resources Division.

CHAPTER 3 Opener (L) Ina Fassbender/Reuters, Opener (R) Top Thinkstock, Opener (R) Middle Tony Gentile/CORBIS, Opener (R) Bottom Fuse/Getty Images, 3.3a fonats/Shutterstock, 3.3b Craig Veltri/istockphoto.

CHAPTER 4 Opener (L) John Schults/Reuters, Opener (R) Top Newscom, Opener (R) Middle Chistopher LaMarca/Redux Pictures, Opener (R) Bottom Guang Niu/Reuters, 4.10a Professors P. Motta & T. Naguro/SPL/Photo Researchers, Inc., 4.18a Suza Scalora/Photodisc/Getty Images, 4.18b SciMAT/Photo Researchers, Inc., 4.19a Mike Blake/Reuters, 4.19b Renn Sminkey, 4.19c Michael Ochs Archives/Getty Images Inc., 4.19d Everett Collection Inc/Alamy, SR4.1a Theo Wargo/WireImage/Getty Images, SR4.1b Steve Grantiz/WireImage/Getty Images.

CHAPTER 5 Opener (L) STR News/Reuters, Opener (R) Top Fotolia, Opener (R) Middle Milos Peric/istockphoto, Opener (R) Bottom Shutterstock, 5.5 STR News/Reuters, 5.7 British Antarctic Survey/Science Photo Library/Photo Researchers, Inc., 5.9a David Furness, Keele University/Photo Researchers, Inc., 5.14 Winton Patnode/Photo Researchers, Inc, T5.1a TongRo Image Stock/Jupiter Images, T5.1b iStockphoto, T5.1c James Forte/National Geographic Society.

CHAPTER 6 Opener (L) SPL/Photo Researchers, Inc., Opener (R) Top Photos by Jim and Mary Whitmer, Opener (R) Middle Photos by Jim and Mary Whitmer, Opener (R) Bottom Photos by Jim and Mary Whitmer, 6.2a Peter Arnold, Inc./Alamy, 6.2b rossco/Shutterstock, 6.3a Biophoto Associates/Photo Researchers, Inc., 6.3b Biophoto Associates/Photo Researchers, Inc., 6.8a Photo Researchers, Inc., 6.8b Credit: Kent Wood/Photo Researchers, Inc., 6.10a Fotolia, 6.10b From: Science, September 17, 1999 in the article entitled "Antiangiogenic Activity of the Cleaved Conformation of Serpin Antithrombin" written by Judah Folkman, Michael S. O'Reilly, Steven Pirie-Shepherd and William S. Lane, 6.15 Photos by Jim and Mary Whitmer, 6.16a CNRI/SPL/Photo Researchers, Inc., 6.16b CNRI/SPL/Photo Researchers, Inc., 6.16c CNRI/SPL/Photo Researchers, Inc.

CHAPTER 7 Opener (L) Tony Brain/Science Photo Library/Photo Researchers, Inc., Opener (R) Top Melissa Schalke/Shutterstock, Opener (R) Middle Digital Vision/Getty Images, Opener (R) Bottom Thinkstock/Superstock, 7.2a Photo courtesy of Gerald McCormack, 7.2b Ninan, C.A., 1958. Studies on the cytology and phylogeny of the pteridophytes VI. Observations on the Ophioglossaceae. Cytologia, 23: 291–316., 7.8 Alamy Images, 7.10 Photo by Linda Gordon, 7.11 Simon Fraser/Royal Victoria Infrimary/Photo Researchers, Inc., 7.12 Dr. Kathryn Lovell, Ph.D., 7.17a Photos courtesy of New York plastic surgeon, Dr. Darrick E. Antell. www.Antell-MD.com, 7.17b Photos courtesy of New York plastic surgeon, Dr. Darrick E. Antell. www.Antell-MD.com, 7.18a Getty Images, 7.18b Photonica/Getty Images, 7.21 Angel Franco/The New York Times/Redux Pictures, 7.23 Juice Images/Alamy.

CHAPTER 8 Opener (L) World History Archive/Alamy, Opener (R) Top V. Velengurin/R.P.G./Corbis, Opener (R) Middle Photo courtesy of Dr. Sergey Nikitin, Opener (R) Bottom Adam Gault/Alamy, 8.2a iStockphoto, 8.2b luri/Shutterstocki, 8.2c AGStockUSA/Alamy, 8.5 Andrew Syred/Photo Researchers, Inc., T8.3a Johan Pienaar/Shutterstock, T8.3b Shutterstock, T8.3c Vasilkin/Shutterstock, T8.3d Arnon Ayal/Shutterstock, T8.3e James King-Holmes/Photo Researchers, Inc., 8.8a Joellen L Armstrong/Shutterstock, 8.8b Oleg Kozlov & Sophy Kozlova/Shutterstock, 8.8c Justyna Furmanczyk/Shutterstock, 8.10a Biophoto Associates/Photo Researchers, Inc., 8.10b Fotolia, 8.10c Ellen B. Senisi/Photo Researchers, Inc., 8.15 David Parker/Photo Researchers, Inc., 8.17b Pictorial Press Ltd/Alamy.

CHAPTER 9 Opener (L) Sion Touhig/Getty Images, Inc., Opener (R) Top brue/istockphoto, Opener (R) Middle Newscom, Opener (R) Bottom H. Kuboto, R. Brinster and J.E. Hayden, RBP, School of Veterinary Medicine, University of Pennsylvania. Proc. Natl. Acad. Sci. USA 101:16489–16494, 2004, 9.11a Biology Media/Photo Researchers, Inc., 9.11b David McCarthy/Science Photo Library/Photo Researchers, Inc., 9.14 Pearson Education, 9.15 From Doebley, J. Plant Cell, 2005 Nov; 17 (11): 2859–72. Courtesy of John Doebley/University of Wisconsin, 9.16 Jon Christensen, 9.17a Leonard Lessin/Photo Researchers, Inc., 9.19 Corbis/Superstock, 9.20 Pascal Goetgheluck/SPL/Photo Researchers, Inc., 9.22b Ted Thai/Time Life Pictures/Getty Images, 9.23 Photograph courtesy of The Roslin Institute, SR9-1a (Best & Wittiest) Jimmy Margulies © North America Syndicate, SR9-1b Rob Rogers: © The Pittsburgh Post-Gazette/Dist. by United Features Syndicate, Inc.

CHAPTER 10 Opener (L) Balog, James/Getty Images Inc., Opener (R) Top Michelangelo Buonaroti (1475–1564). "Creation of Adam" Detail of Sistine Chapel ceiling. Sistine Chapel, Vatican Palace, Vatican State. © Scala/Art Resource, NY, Opener (R) Middle David Gifford/Photo Researchers, Inc., Opener (R) Bottom William Campbell/CORBIS, 10.3 Lebrecht Music and Arts Photo Library/Alamy, 10.5a Shutterstock, 10.5b Arterra Picture Library/Alamy, 10.6a Shutterstock, 10.6b (R) Worldwide Picture Library/Alamy, 10.8 Shutterstock, 10.12a Shutterstock, 10.12b Jerome Wexler/Photo Researchers, Inc., 10.12c Martin Shields/Alamy, 10.13b1 Thinkstock, 10.13b2 imagebroker/Alamy, 10.14 From: Richardson, M. K., et al., Science. Vol. 280: Pg 983c, Issue # 5366, May 15, 1998. Embryo from Professor R. O'Rahilly. National Museum of Health and Medicine/Armed Forces Institute of Pathology, 10.15a

CHAPTER 18 Opener (L) Christian Zachariasen/PhotoAlto/Getty Images, Opener (R) Top Newscom, Opener (R) Middle Simone van den Berg/Shutterstock, Opener (R) Bottom Shutterstock, 18-03a Vinatali/Shutterstock, 18.3a Sebastian Kaulitzki/Shutterstock, 18.3b Allsport Photography/Getty Images, Inc., 18.4b Eye of Science/Photo Researchers, Inc., 18.7a Biophoto Associates/Photo Researchers, Inc., 18.7b Biophoto Associates/Photo Researchers, Inc., 18.8 Science Source/Photo Researchers, Inc., 18.9a St. Bartholomew's Hospital/Science Photo Library/Photo Researchers, Inc., 18.9b Medical-on-Line/Alamy, 18.9c Science Photo Library/Alamy, 18.9d NIBSC/Photo Researchers, Inc., 18.11 CNRI/Photo Researchers, Inc., E18.1 Richard L. Carlton/Photo Researchers, Inc., E18.2 Vaughan Fleming/Photo Researchers, Inc., E18.3 Photos.com, E18.4 Jasenka Luksa/Shutterstock, E18.5 Rod Planck/Photo Researchers, Inc., E18.6 Shutterstock, 18.17a Alfred Pasieka/SPL/Photo Researchers, Inc., 18.17b R. Roseman/Custom Medical Stock Photo, Inc.

CHAPTER 19 Opener (L) Gene Young/Splash News/Newscom, Opener (R) Top Frank Polich/Reuters, Opener (R) Middle Alexander Raths/Shutterstock, Opener (R) Bottom Science Photo Library/Alamy, 19.1a Shutterstock, 19.1b Shutterstock, 19.1c Gary Gaugler/Getty/Photolibrary, T19.3a Eye of Science/Photo Researchers, Inc., T19.3b David Scharf/Photo Researchers, Inc., T19.3c Peter Arnold, Inc./Photolibrary, T19.3d Medical-on-Line/Alamy, T19.3e SPL/Photo Researchers, Inc., 19.8 Eye of Science/Photo Researchers, Inc., 19.9 Photo Lennart Nilsson/Scanpix.

CHAPTER 20 Opener (L) Steve Burmeister, Opener (R) Top Josh Meltzer/Duluth News Tribune, Opener (R) Middle Jerry Zitterman/Shutterstock.com, Opener (R) Bottom Courtesy IMS Photo by Ron McQueeney/Indy Racing League, E20.1 Duncan McEwan/Nature Picture Library, E20.2 Shutterstock, 20.10 Nin Zanetti/Pearson Science, 20.13 Pete Soloutos/ZEFA/age fotostock/SuperStock, Inc.

CHAPTER 21 Opener (L) Suzanne L. & Joseph T. Collins/Photo Researchers, Inc., Opener (R) Top Dennis MacDonald/age fotostock/SuperStock, Inc., Opener (R) Middle Stone Nature Photography/Alamy, Opener (R) Bottom Adam Gryko/iStockphoto, 21.1a Shutterstock, 21.1b Tom Branch/Photo Researchers, Inc., 21.2 Sea Studios Foundation, 21.5 Eye of Science/Photo Researchers, Inc., 21.9 C. Edelmann/La Villette/Photo Researchers, Inc., T21.2(2) Jacob Kearns/Shutterstock, T21.2(3) Andrew McClenaghan/Science Photo Library/Photo Researchers, Inc., T21.2(4) Don Farrall/Photodisc/Getty Images, Inc., T21.2(5) Alamy Images, T21.2(6) Custom Medical Stock Photo Custom Medical Stock Photo/Newscom, T21.2(7) Kumar Sriskandan/Alamy, T21.2(8) Pearson Science, T21.2(9) Conceptus Incorporated, T21.2(10) Frank LaBua, T21.2(11) Renn Sminkey, T21.2(12) Allendale Pharmaceuticals, Inc., T21.2(13) Renn Sminkey, T21.2(14) Renn Sminkey, T21.2(15) Renn Sminkey, T21.2(16) Ever/iStockphoto, T21.2(17) Duramed Pharmaceuticals, Inc., a subsidiary of Barr Pharmaceuticals, Inc., 21.12 Steve Allen/Science Photo Library/Photo Researchers, Inc., 21.13 G. Moscoso/Photo Researchers, Inc.

CHAPTER 22 Opener (L) Upper Cut Images/AGE Fotostock America, Inc., Opener (R) Top Shutterstock, Opener (R) Middle Monkey Business Images/Shutterstock, Opener (R) Bottom Arnd Wiegmann/Reuters, 22.5 Pascal Goetgheluck/Photo Researchers, Inc., E22.1 Tom Branch/Photo Researchers, Inc., E22.2 Gabriela Schaufelberger/iStockphoto, E22.3 Photos.com, 22.6b Martin M. Rotker/Science Source/Photo Researchers, Inc., 22.10a SPL/Photo Researchers, Inc., T22.2(1) Peter Raneri/iStockphoto, T22.2(2) Sarah Skiba/iStockphoto, T22.2(3) Sergei Didyk/Shutterstock, T22.2(4) U.S. Department of Justice/Drug Enforcement Agency, T22.2(5) Andrew Burns/Shutterstock, T22.2(6) George Post/Science Photo Library/Photo Researchers, Inc., T22.2(7) David Buffington/Photodisc/Getty Images, T22.2(8) Kiseleve Andrew Valerevich/Shutterstock, T22.2(9) MaxPhoto/Shutterstock.

CHAPTER 23 Opener (L) National Oceanic and Atmospheric Administration NOAA, Opener (R) Top © Bettmann/Corbis, Opener (R) Middle Fang xinwu/Color China/AP Images, Opener (R) Bottom Digital Vision/Getty Images, 23.1a Virginia Borden, 23.1b zonesix/Shutterstock, 23.4a 23.04a John Kaprielian/Photo Researchers, Inc., Shutterstock, 23.4b John Kaprielian/Photo Researchers, Inc., 23.5 Dr. Jeremy Burgess/Photo Researchers, Inc., 23.6 Photo by Dr. Steve O'Hair. Courtesy of Professor Stephen M. Olson, North Florida Research and Education Center, 23.7 Kathy Burns-Millyard/Shutterstock, 23.8a Ranald MacKechnie/Dorling Kindersley, 23.8b Dorling Kindersley, 23.8c Roger Phillips/Dorling Kindersley, 23.8d C. Squared Studios/Photodisc/Getty Images, 23.8e Denis Pepin/Shutterstock, 23.8f Siede Preis/Photodisc/Getty Images, Inc., 23.8g Corbis, 23.8h Dave King/Dorling Kindersley, 23.10a Jerome Wexler/Photo Researchers, Inc., 23.10b Shutterstock, 23.10c Shutterstock, 23.10d Merlin Tuttle/Photo Researchers, Inc., 23.10e Anthony Mercieca/Photo Researchers, Inc., 23.11a Steve Bloom Images/Alamy Images, 23.11b Kevin Schafer/Alamy, 23.11c Scott Camazine/Alamy Images, 23.11d Bruce Coleman, Inc./Alamy Images, E23.2a Shutterstock, E23.2b Niall Benvie/Nature Picture Library, E23.2c Garry DeLong/Photo Researchers, Inc., E23.2d Anna Galejeva/Shutterstock, E23.2e Shutterstock, E23.2f Peter Weber/Shutterstock, E23.2g David M. Phillips/Photo Researchers, Inc., E23.2h Vinicius Ramalh Tupinamba/iStockphoto, 23.13a Shane Kennedy/Shutterstock, 23.13b Dr. J. Burgess/Photo Researchers, Inc., 23.13c Shutterstock, 23.13d Adam Hart-Davis/Photo Researchers, Inc., 23.14 Peter Arnold, Inc./Photolibrary, 23.15 Patrick Lynch/Scence Photo Library/Photo Researchers, Inc., 23.18a Aqua Image/Alamy, 23.18b Tim Aylen/AP Wide World Photos, 23.20a Nigel Cattlin/Photo Researchers, Inc., 23.20b Nigel Cattlin/Photo Researchers, Inc., 23.21 GSFC/METI/ERSDAC/JAROS, and U.S./Japan ASTER Science Team/NASA, 23.22a Dr. Jeremy Burgess/SPL/Photo Researchers, Inc., 23.22b Dr. Jeremy Burgess/SPL/Photo Researchers, Inc., 23.25a Jens Ottoson/Shutterstock, 23.25b Science Photo Library/Photo Researchers, Inc., 23.26a Terry W. Eggers/CORBIS, 23.26b Herbert Kehrer/ZEFA/CORBIS, 23.28 NASA, 23.29 Erich Hartmann/Magnum Photos, Inc., 23-30a: Pearson Science, 23.30b Nigel Cattlin/Alamy, 23.31 Gary Kazanjian/AP Wide World Photos, 23.32 i love images/Alamy Images Royalty Free.

CHAPTER 24 Opener (L) Clay Perry/Corbis, Opener (R) Top PureStockX, Opener (R) Middle AGE Fotostock America, Inc., Opener (R) Bottom Simone van den Berg/iStockphoto, 24.2 Dennis Oblander/iStockphoto, 24.3 Elena Elisseeva/Shutterstock, 24.4 Radius Images/Alamy, 24.5 Dr. Keith Wheeler/Photo Researchers, Inc., 24.6a Dr. Keith Wheeler/Photo Researchers, Inc., 24.6b Science Photo Library/Photo Researchers, Inc., E24.1 Dr. Morley Read/Shutterstock, E24.2 Bjorn Svensson/Science Photo Library/Photo Researchers, Inc., E24.3 Martin Maun/Shutterstock, E24.4 Jupiterimages, E24.5 Rob Broek/iStockphoto, 24.9a Tracy Tucker/iStockphoto, 24.9b Richard Goerg/iStockphoto, 24.9c zorani/iStockphoto, 24.14 Dr. Keith Wheeler/Photo Researchers, Inc., 24.15a Martin Shields/Alamy, 24.15b John Kaprielian/Photo Researchers, Inc., 24.15c Charles D. Winters/Photo Researchers, Inc., 24.18 Shutterstock.

Index

Page references that refer to figures are in **bold**. Page references that refer to tables are in *italic*.